ISBN 978-1-5285-2376-9
PIBN 10900594

ANALYTICAL KE

TO

ELLIS & EVERHART'S

NORTH AMERIC

PYRENOMYCET

ANALYTICAL KEY

TO THE SUBORDERS, FAMILIES AND GENERA OF

THE

NORTH AMERICAN PYRENOMYCETES AND

HYSTERIACEAE.

The Nat. *ORDER PYRENOMYCETES* embraces those fungi in which the asci and sporidia are enclosed in a perithecium, and is divided into 4 *SUBORDERS* as follows:

Perithecia superficial, astomous, membranaceous, mostly on a subiculum. - - - *Perisporiaceae.*

Perithecia carnose, or membranaceo-carnose, ostiolate, bright colored (red, yellow, etc.,) not black. - - - - - *Hypocreaceae.*

Perithecia black or brown, mostly ostiolate, of various consistence, membranaceous, coriaceous, carbonaceous, woody or corky, but not carnose; stroma when present not carnose, dark colored outside, often white within. - - - - - *Sphaeriaceae.*

Stroma always present, of firm texture not carnose; black outside. Perithecia reduced to mere cells in the stroma. - - *Dothideaceae.*

KEY TO THE FAMILIES.

SUBORDER PERISPORIACEAE.

Mycelium of colorless, interwoven threads. Perithecia seated singly on the mycelium, subglobose, astomous, bearing variously formed, radiating, thread-like appendages. - - *Erysipheae.*

Mycelium persistent, brown, often bearing conidia, sometimes inconspicuous or evanescent. Perithecia globose or depressed, astomous, always superficial. - - - *Perisporieae.*

SUBORDER HYPOCREACEAE.

The genera included in this suborder are all closely related and are not separâted into families.

SUBORDER SPHAERIACEAE

A. **Perithecia scattered, stroma wanting** (*simple Sphaeriae.*)

* *Perithecia, superficial or erumpent-superficial.*

(*a*) *Perithecia hairy.*

Perithecia membranaceous, or coriaceo-membranaceous. - - - - - - *Trichosphaerieae.*

Perithecia thin-membranaceous, with an apical tuft of hairs, lignicolous. - - - *Chaetomieae.*

Perithecia carbonaceous or subcarbonaceous. - - - *Melanommeae* (partly.)

Perithecia glabrous or hairy, fimicolous. - *Sordarieae.*

(*b*) *Perithecia not hairy.*

Perithecia thin-membranaceous, with a subulate ostiolum, - - - *Ceratostomeae.*

Perithecia woody or corky. - *Melanommeae* (partly.)

Perithecia subcarbonaceous, ostiolum compressed. *Lophiostomeae.*

Perithecia subcarbonaceous, ostiolum not compressed. - - *Amphisphaerieae.*

Perithecia carbonaceous, ostiolum soon deciduous and perithecia then broadly perforated above. - - - *Trematosphaerieae.*

** *Perithecia buried or onlμ subemergent.*

Perithecia small, simply perforated above. *Sphaerelleae.*

Perithecia membranaceous or coraceo-membranaceous, ostiolum subulate. - - - *Gnomonieae.*

Perithecia membranaceous, or coracio membranaceous, ostiolum conical or papilliform. - *Pleosporeae.*

Perithecia coriaceous, thick walled. - *Massarieae.*

Perithecia covered with a stromatic shield. *Clypeosphaerieae.*

*** *Perithecia cespitose (or sometimes simply gregarious,) connected at base by a more or less distinct stromatic layer.* · *Cucurbitarieae.*

B. Perithecia included in a stroma (*Compound Sphaeriae.*)

Stroma effused, pulvinate or tuberculiform; perithecia buried in the stroma, their ostiola (typically) convergent and erumpent in a central fascicle. *Valseae.*

Stroma valsoid, pulvinate, conical or hemispherical, often inconspicuous or wanting, accompanied mostly by a conidial stroma, · *Melanconideae.*

Stroma various; pulvinate, tuberculiform, etc., erumpent or superficial. Spermatia and stylospores produced in cavities within the perithecial stroma. · · (1.) *Melogrammeae.*

Stroma effused, scutellate, pulvinate or tuberculiform. Perithecia monostichous or polystichous, with their ostiola separately erumpent. · (2.) *Diatrypeae.*

Stroma various; upright, dendroid, pulvinate, or globose; mostly carbonaceous and superficial, black, or becoming black. Sporidia, brown or black. · · · (3.) *Xylarieae.*

SUBORDER DOTHIDEACEAE.

Stroma pulvinate or effused, coriaceous, or subcarbonaceous; perithecia reduced to mere cells or cavities in the stroma. Sporidia hyaline, yellowish or brown · · *Dothideaceae.*

SUBORDER HYSTERIACEAE.

Stroma none. Perithecia simple, erumpent-superficial, oblong or linear, membranaceous, coriaceous or carbonaceous, becoming black, opening by a narrow, longitudinal cleft. · *Hysteriaceae.*†

---

1. Includes *Botryosphaeria*, *Myrmaecium*, *Endothia*, *Melogramma*, and *Valsaria*. (pp. 546 564.)

2. Includes *Diatrype*, *Anthostoma* and *Diatrypella* (pp. 565-595.) The diagnoses of these two families were omitted in their proper place in the N. Am. Pyrenomycetes.

3 This family should precede *Dothideaceae* in the systematic arrangement,

†This suborder does not properly belong to the Pyrenomycetes but is included here for the present as a matter of convenience.

KEY TO THE GENERA.

ERYSIPHEAE.

* *Appendages to the perithecia simple and similar to the threads of the mycelium.*

Only one ascus in a perithecium. - *Sphaerotheca.*

Several asci in each perithecium. - *Erysiphe.*

** *Appendages branched at their tips*

Only one ascus in a perithecium. - *Podosphaera.*

Several asci in each " - *Microsphaera.*

*** *Appendages not branched but different from the threads of the mycelium.*

Appendages swollen at the base, tips straight. *Phyllactinia.*

Appendages not swollen at the base, tips coiled. *Uncinula.*

PERISPORIEAE.

* *Mycelium present.*

Mycelium thin, arachnoid. - *Saccardia.*

" yellow. - - *Eurotium.*

" light colored. - - *Myriococcum.*

** *Mycelium scanty or wanting.*

Perithecia scattered, subglobose. - - *Perisporium.*

" " flattened. - - *Microthyrium.*

" stromatic. - - - *Lasiobotrys.*

*** *Mycelium black.*

Perithecia subglobose, appendiculate. - *Meliola*

" " not " - *Dimerosporium*

" flattened. - - *Asterina.*

" elongated, subulate etc. *Capnodium.*

**** *Mycelium, massive, coarse.* - - - *Scorias.*

HYPOCREACEAE.

A. Perithecia sunk in a stroma.

* *Stroma erect, sporidia filiform, hyaline.*

Stroma capitate, sclerotigenous. - - *Claviceps.*

" elavate, entomogenous. - - *Cordyceps.*

" " herbicolous. - - *Acrospermum.*

***Stroma tuberculiform.*

Fimicolous, sporidia elliptical. - *Selinia.*

Herbicolous, sporidia filiform. - *Oomyces.*

****Stroma effused, dark colored.*

Stroma byssoid. - - *Byssonectria.*
Stroma valsoid. - - - *Thyronectria.*
Parasitic on other fungi. *Hypomyces.*
Growing on wood or bark. - - *Hypocrea.*
Sporidia filiform. - - - - *Hypocrella.*
*****Stroma effused, bright colored.*
Sporidia ovoid, (foliicolous.) - - *Polystigma.*
Sporidia filiform, stroma, yellow, amplexicaul. - *Epichloe.*

B. Perithecia buried, without any distinct stroma. *Hyponectria.*

C. Perithecia superficial stroma, scanty or none.

* *Sporidia hyaline or subhyaline.*

Sporidia appendiculate, perithecia subulate. *Eleutheromyces.*
Sporidia uniseptate, conidial stroma tubercular. - *Nectria.*
Sporidia uniseptate, conidial stroma capitate. *Sphaerostilbe.*
Sporidia, 2-or more-septate. - - *Calonectria.*
Sporidia muriform. - - - *Pleonectria.*
Sporidia allantoid, minute, asci polysporous. *Chilonectria.*
Sporidia filiform, fimicolous. - - - *Barya.*
Sporidia filiform, xylogenous. - - *Ophionectria.*
Sporidia 3-pluriseptate, perithecia blue or violet. *Gibberella.*
Sporidia brown, perithecia beaked, mostly continuous. - - *Melanospora.*

CHAETOMIEAE.

Characters same as the family; includes the single genus. - - - - *Chaetomium.*

SORDARIEAE.

A. Stroma wanting.

* *Sporidia continuous brown.*

Asci 8-spored, sporidia without appendages - *Sordaria.*
Asci 8-spored, sporidia appendiculate - *Podospora.*
Asci polysporous - - - - - *Philocopra.*
** *Sporidia, 1-or more-septate.*
Sporidia uniseptate. - *Delitschia.*
Sporidia 4-20, septate. *Sporormia.*

B. Stroma present.

Sporidia continuous. *Hypocopra.*
Sporidia 4-20-septate. - *Sporormiella.*

TRICHOSPHAERIEAE.

Sporidia oblong or elliptical, uniseptate. - *Venturia.*

Sporidia cylindrical or vermiform, 1-or more-septate. - - - - *Lasiosphaeria.*

Sporidia oblong, ovate, or cylindrical, continuous. *Trichosphaeria.*

Sporidia 1-many-septate, subfusoid, hyaline or subhyaline. - - - *Acanthostigma.*

Sporidia fusoid or oblong, hyaline or brown, 1-7 septate. - - - - *Herpotrichia.*

Sporidia fusoid or oblong, brown, or with the terminal cells hyaline; perithecia seated on or surrounded by a felt-like or hairy subiculum. - - - *Chaetosphaeria.*

MELANOMMEAE.

Perithecia globose or subglobose, sporidia brown, continuous. - - - *Rosellinia.*

Perithecia fasciculate, elongated, sporidia, cylindrical, hyaline, with a brown head. - *Bombardia.*

Perithecia subdepressed, smooth, sporidia elliptical, or fusoid, uniseptate, pale brown or hyaline. *Melanopsamma.*

Perithecia tubercular-roughened, sporidia subcylindrical, uniseptate hyaline. - - - *Bertia.*

Perithecia globose or ovoid, sporidia oblong or fusoid, 1-or more-septate, brown. - - *Melanomma.*

Perithecia subsuperficial, ovate or oblong, pluriseptate, hyaline. - *Zignoella.*

CERATOSTOMEAE.

Sporidia continuous, brown. *Ceratostoma.*

Sporidia continuous, hyaline. - - *Ceratostomella.*

Sporidia elliptical, uniseptate, brownish. - *Rhyncostoma.*

Sporidia uniseptate, hyaline. - - *Lentomita.*

Sporidia elongated, pluriseptate, sub-hyaline or brownish. - - - *Ceratosphaeria.*

Sporidia filiform, 2-or more-septate, hyaline. *Ophioceras.*

Sporidia fusiform, hyaline, or subhyaline. *Ramphoria.*

Perithecia subulate, swollen below the tip, sporidia brownish, continuous. - - *Hypsotheca.*

AMPHISPHAERIEAE.

A Perithecia not collapsing.

Sporidia elliptical, or oblong, uniseptate, brown. - - - *Amphisphaeria.*

Sporidia oblong or fusoid, 3-or more-septate, brown. - - - *Trematosphaeria.*

Sporidia subelliptical, uniseptate. brown, with a snout-like (often 1-3-septate) prolongation at each end. - - - *Caryospora.*

Sporidia 3-septate, separating in the middle, brown. *Ohleria.*

B Perithecia often collapsing.

Sporidia 2-or more-septate, hyaline. *Winteria.*

Sporidia muriform, brown. - - *Teichospora.*

LOPHIOSTOMEAE.

Sporidia oblong or subfusoid, 1-multiseptate, hyaline or brown. - - *Lophiostoma.*

Sporidia ovate or oblong, brown, muriform. - *Lophidium.*

Sporidia vermicular or filiform, septate, subhyaline. *Lophionema.*

CUCURBITARIEAE.

A. Perithecia mostly cespitose.

Sporidia muriform, brown. - - *Cucurbitaria.*

Sporidia continuous, short cylindrical, hyaline. *Nitschkia.*

Sporidia continuous ovoid, elliptical, or oblong, hyaline. - - - *Wallrothiella.*

Sporidia continuous oblong or allantoid, subhyaline, more than 8 in an ascus. - *Fracchiaea.*

Sporidia oblong-elliptical, uniseptate, subhyaline. *Gibbera.*

B. Perithecia gregarious.

Sporidia short-cylindrical, continuous, hyaline. *Coelosphaeria.*

Sporidia elliptical or fusoid, uniseptate, brown or yellowish-brown, lignicolous. - *Otthia.*

Sporidia subelliptical, uniseptate, brown. *Parodiella.*

Sporidia subfusoid, 3-septate, brownish. - *Montagnella.*

Perithecia subelliptical, uniseptate, brown. *Parodiella.*

SPHAERELLEAE.

A. Paraphyses present.

Sporidia elliptical, continuous. hyaline. - *Physalospora.*

Sporidia subelliptical, yellowish-hyaline, unequally, uniseptate. - - *Stigmatea.*

Sporidia piriform, uniseptate near the lower (narrow) end, hyaline. - - *Apiospora.*

Sporidia ovoid, ellipsoid or suboblong, uniseptate in the middle. - - *Didymella.*

B. Paraphyses wanting.

Sporidia elliptical, continuous, hyaline. *Laestadia.*

Sporidia elliptical, continuous, 2-nucleate, hyaline. *Phomatospora.*

Sporidia elliptical, uniseptate, hyaline. *Sphaerella.*

Sporidia continuous, brown. - *Mulerella.*

Sporidia inequilateral, large, uniseptate, hyaline, becoming darker. - - *Lizonia.*

Sporidia 1-3-septate, hyaline. *Pharcidia.*

Sporidia 1-3-septate, brown. - *Tichothecium.*

Sporidia cylindrical or fusoid, 3-multiseptate, hyaline. - - *Sphaerulina.*

GNOMONIEAE.

Asci 8-spored, sporidia elliptical, etc., continuous or uniseptate, hyaline. - *Gnomonia.*

Asei polysporous, sporidia oblong or fusoid, continuous, subhyaline. - - *Ditopella.*

PLEOSPOREAE.

Sporidia ovoid-oblong, uniseptate, brown. *Didymosphaeria.*

Sporidia elliptical, muriform, brown. - *Pleospora.*

Sporidia as in *Pleospora*, but perithecia hairy or bristly. - - *Pyrenophora.*

Sporidia fusoid, 1-many-septate, yellow or brown. *Leptosphaeria.*

Sporidia as in *Leptosphaeria*, but hyaline. *Metasphaeria.*

Sporidia as in *Leptosphaeria* but appendiculate at each end. - - *Ceriospora.*

Sporidia elongated, appendiculate, multiseptate, hyaline. - - *Saccardoella.*

Sporidia large, muriform, only one or two in an ascus. *Julella.*

Sporidia filiform, guttulate or septate, subhyaline. *Ophiobolus.*

MASSARIEAE.

Sporidia large, with a hyaline envelope, 1-several-septate, mostly brown. - *Massaria.*

Sporidia constantly uniseptate, otherwise as in *Massaria.* - - *Massariella.*

Sporidia more or less distinctly muriform, brown. *Pleomassaria.*
Sporidia as in *Massariella*, but perithecia circinate. . . *Massariovalsa.*

CLYPEOSPHAERIEAE.

Sporidia ovate-oblong, continuous, subhyaline. *Trabutia.*
Sporidia ovoid or oblong continuous, brown. *Anthostomella.*
Sporidia oblong cylindrical, 1-3-septate, brown. *Clypeosphaeria.*
Sporidia fusoid, 1-3-septate, hyaline. *Hypospila.*
Sporidia filiform, hyaline. . *Linospora.*
Sporidia short-fusoid, muriform (foliicolous.) *Isothea.*
Sporidia subelliptical, muriform, hyaline or brown (lignicolous) . . *Thyridium.*

VALSEAE.

A. **Sporidia allantoid, continuous, hyaline or subhyaline, asci mostly 8-spored.**

* *Stroma convex or conical, sunk in the matrix with only the apex erumpent.*

Ostiola smooth, sporidia hyaline. *Valsa.*
Ostiola sulcate, sporidia yellowish. . *Eutypella.*
Perithecia circinate, asci polysporous. *Valsella.*

** *Stroma effused.*

Perithecia scattered in an effused or subpulvinate stroma, sporidia hyaline. . *Eutypa.*
Perithecia scattered or subcircinate in a (mostly) cortical stroma, sporidia yellowish. *Cryptovalsa.*

*** *Stroma wanting.*

Perithecia circinate on the surface of the inner bark. . . *Calosphaeria.*
Perithecia as in *Calosphaeria*, asci polysporous. *Coronophora.*
Perithecia and sporidia as in *Eutypa*, but stroma none. *Cryptosphaeria.*

B. **Sporidia 1-or more-septate.**

* *Stroma convex, conical or effused.*

Sporidia uniseptate, hyaline. . *Diaporthe.*

** *Stroma wanting.*

Perithecia buried, subcircinate or scattered, sporidia 1-3-septate, brown. . *Endoxyla.*

MELANCONIDEAE.

Sporidia uniseptate, hyaline or colored. . *Melanconis.*

Sporidia continuous, or 2 or more septate or pseudoseptate, hyaline.	*Cryptospora.*
Sporidia muriform, brown.	*Fenestella.*

MELOGRAMMEAE.

Sporidia continuous, hyaline.	*Botryosphaeria.*
Sporidia ovoid-oblong, uniseptate, hyaline, paraphyses present.	*Myrmaecium.*
Sporidia broad-fusoid, or elliptical, hyaline, paraphyses none.	*Endothia.*
Sporidia uniseptate, brown.	*Valsaria.*
Sporidia fusoid or cylindrical; with several septa.	*Melogramma.*

DIATRYPEAE.

Sporidia allantoid, yellowish, asei 8-spored.	*Diatrype.*
Sporidia allantoid, yellowish, asci polysporous.	*Diatrypella.*
Sporidia oblong or elliptical, continuous, brown.	*Anthostoma.*

SUBORDER DOTHIDEACEAE.

Sporidia continuous, hyaline or yellowish, stroma sclerotoid.	*Mazzantia.*
Sporidia continuous, hyaline or yellowish, stroma not sclerotoid.	*Phyllachora.*
Sporidia uniseptate, hyaline or yellowish, stroma thin, effused or discoïd.	*Dothidella.*
Sporidia uniseptate, hyaline or yellowish, stroma tubercular or pulvinate.	*Dothidea.*
Sporidia 3-septate, brown.	*Homostegia.*
Sporidia 3-5-septate, yellowish.	*Rhopographus.*
Sporidia filiform, continuous or nucleate.	*Ophiodothis.*
Sporidia muriform, brown.	*Curreya.*
Sporidia muriform, yellowish-hyaline.	*Myriangium.*

XYLARIEAE.

Stroma concave, or convex or subeffused.	*Nummularia.*
Stroma globose, pulvinate or subeffused.	*Hypoxylon.*
Stroma superficial, crustaceous, perithecia large.	*Ustulina.*
Stroma vertical, divided by a horizontal layer.	*Camillea.*
Stroma dendroid, mostly white within.	*Xylaria.*
Stroma globose, concentrically zoned within.	*Daldinia.*
Stroma clavate, becoming cup-shaped or discoid.	*Poronia.*

Stroma superficial, effused, perithecia immersed with elongated necks. - - *Bolinia.*

SUBORDER HYSTERIACEAE.

A. Sporidia hyaline.

* *Sporidia contiuuous.*

Perithecia subcarbonaceous, flattened or convex, minute. - - *Schizothyrium.*

** *Sporidia uniseptate.*

Perithecia membranaceous, minute, simple or branched, flattened. - *Aylographum.*

Perithecia carbonaceous, simple or obscurely branched. - - *Glonium.*

Perithecia stellate. *Actidium.*

Perithecia subcarnose. - *Angelina.*

*** *Sporidia 3-pluriseptate.*

Perithecia carbonaceous. *Gloniella.*

Perithecia subcoriaceous. *Dichaena.*

**** *Sporidia muriform.*

Perithecia carbonaceous - (*Gloniopsis.*)

***** *Sporidia filiform.*

(*a*) *Sporidia much shorter than the ascus.*

Perithecia membranaceous, flattened. - *Hypoderma.*

(*b*) *Sporidia nearly as long as the ascus.*

Perithecia elongated, flattened, membranaceous. *Lophodermium.*

Perithecia elongated, conchiform, subcarbonaceous. *Lophium.*

Perithecia elongated, coriaceo-subcarnose. *Clithris.*

Perithecia subsphæroid. - *Ostropa.*

B. Sporidia brown.

* *Sporidia uniseptate.*

Perithecia coriaceous, widely dehiscent. - *Tryblidium.*

Perithecia elongated, coriaceous. (*Lembosia.*)

** *Sporidia 3-pluriseptate.*

Perithecia carbonaceous. - *Hysterium.*

Perithecia coriaceous. (*Tryblidiella.*)

Perithecia conchiform. - *Mytilidion.*

Perithecia subcarbonaceous, striate. *Ostreion.*

*** *Sporidia muriform.*

Perithecia carbonaceous. *Hysterographium.*

THE NORTH AMERICAN PYRENOMYCETES.

A CONTRIBUTION TO

MYCOLOGIC BOTANY,

BY

J. B. ELLIS

AND

B. M. EVERHART.

WITH ORIGINAL ILLUSTRATIONS

BY

F. W. ANDERSON.

PUBLISHED BY ELLIS & EVERHART,
NEWFIELD, NEW JERSEY.
1892.

PRESS OF
WILLIAM H. CLOYD.
VINELAND, N. J.

PREFACE.

In the present state of mycological knowledge, the classification and description of the species of North American Pyrenomycetes is attended with many difficulties; chief among which is the fact that many of the published diagnoses are too imperfect to enable one to recognize the species, of which many of the types are either lost or practically inaccessible. An examination of the Schweinitzian Herbarium, at the Academy of Natural Sciences at Philadelphia, reveals the fact that about two hundred of the most valuable species described in Schweinitz' Synopsis of North American Fungi are not represented in the Herbarium at all, and that many of the specimens still remaining are too meager or too imperfect to be of much service. Coming to the species described by Berkeley & Curtis, the case is no better, but, as far as specimens are concerned, even worse, the types being entirely beyond the reach of the ordinary student. Our knowledge of these species is derived from the brief descriptions in Grevillea and the supplementary notes by Dr. M. C. Cooke, who has also examined and determined, by comparison with the original types, many of the species included in the present work. We are also indebted to Drs. Rehm, Winter, and Saccardo for the identification of many doubtful species; but above all to the many collectors who have furnished abundant material from all parts of the country, thus making it possible to give an approximately complete synopsis of the North American species of this extensive Order. The names of these contributors appear in connection with the species they have furnished. The name of Ravenel is attached to most of the species collected by him and described by Berkeley & Curtis in Grevillea; and Schweinitz or "Schw." to his species published in the Synopsis of North American Fungi.

The system of classification adopted (as stated elsewhere) is mainly that of Dr. Winter in *Rabenhorst's Kryptogamen Flora* (quoted mostly in this vol. as "Die Pilze"). The system differs from that of Saccardo in the Sylloge, mainly in the different grouping of the genera, in this respect standing intermediate between the Saccardian system and that adopted by Cooke in his "Synopsis Pyrenomycetum."

As a guide to the correct pronunciation, the long sound of the accented vowel in generic and specific names, is indicated by the grave accent (`) and the short sound by the acute (´).

The name of the author first publishing any species has been retained, placed in parenthesis in case the species has been removed from the genus in which it was first placed. The name after the parenthesis has been omitted as too cumbersome and unnecessary. The piratical practice of omitting the first name and substituting the second in its place can not be too strongly condemned.

In the case of old and well known species we have adopted the published diagnoses supplemented with notes, in cases where our observations do not entirely agree with the published characters.

The measurements (in centimeters, cm., millimeters, mm., and micromillimeters, μ,) have all been made with the same optical combination and with the same micrometrical scale. They agree generally with those given by Saccardo, Karsten, and Winter, but are mostly less than those given by Cooke, so that it seems probable that the scale we have used differs somewhat from his. In the measurements of asci, the length of the spore-bearing part (p. sp.) has, in most cases, been given.

The illustrations by our much lamented friend, F. W. Anderson, do credit to his artistic talent, but cause us to regret that a career of such brilliant promise should end so soon. The figures were made from actual specimens carefully examined and accurately drawn, to show the more salient characters of the genera.

The fungi of Greenland, enumerated by Professor Rostrup (see page 32) have been included, as it is probable that most of the species found in that great continental island, will also be met with in the northern part of British America and Alaska, and in the elevated regions along the Rocky Mountain chain further South.

The Erysipheæ in this Vol. (pp. 2–30) have been elaborated and prepared by Professor T. J. Burrill, of the University of Illinois, who wishes here to express his thanks for contributions of much service mostly specimens, to the following named gentlemen, viz.: F. W. Anderson, J. C. Arthur, Geo. F. Atkinson, W. J. Beal, H. W. Harkness, W. A. Kellerman, T. H. McBride, L. H. Pammel, C. H. Peck, S. M. Tracy, Wm. Trelease, A. B. Seymour, and Geo. P. Clinton.

We regard it as extremely fortunate that Prof. B. was willing to undertake the editing of the *Erysipheæ*, as his previous studies in this difficult Family have made him specially competent to do the work thoroughly.

The index of genera and species at the end of the volume is the work of our friend W. C. Stevenson, Jr., to whom we are already indebted for the indices of species and their habitats in the "North American Fungi."

PYRENOMYCÈTES.

This order includes those fungi in which the hymenium is inclosed in a subglobose envelope or shell (perithecium), which either remains closed or, more generally, is pierced above with a small, round opening (ostiolum) which is often more or less prolonged, so as to form a short tube or beak.

From the lower part, and often from the sides of the inner surface of the perithecium, spring numerous transparent, membranaceous sacs (asci) containing minute, globose, oblong, cylindrical or thread-like bodies (sporidia).

Between the asci arise usually, slender, filiform bodies (paraphyses) whose offiee is not well understood. They have been supposed to be abortive asci.

The perithecia are either developed directly from the mycelium, separate and distinct from each other (simple), or are more or less imbedded in a carnose, coriaceous or carbonaceous substance called the stroma (compound). The stroma assumes various forms, effused, pulvinate, tubercular or vertically elongated, cylindrical or dendroid.

Often the stroma consists of the slightly altered substance of the wood or bark.

The surface of the stroma before the development of the perithecia is often clothed with a hyphomycetous growth, producing aerial spores (conidia). Also, accompanying or preceding the ascigerous perithecia, are others (spermogonia or pycnidia) producing in their hymenial cavities spores (sporules) borne on pedicels and not in asci.

The conidia, spermogonia and pycnidia are supposed to be generically connected with the ascigerous perithecia, but this matter is not yet well understood.

Dr. Winter, whose systematic arrangement we have mostly adopted, divides the Pyrenomycetes into the following suborders.

1. **PERISPORIÀCEÆ.** Perithecia with the ostiolum obscure or wanting, coriaceous or brittle-carbonaceous, opening irregularly, generally without any stroma, but mostly seated on a well developed, superficial mycelium.

2. **HYPOCREÀCEÆ.** Perithecia with an ostiolum, and with the stroma (when present) carnose or membranaceo-carnose, and bright colored (yellow, red, &c.).

3. **SPHÆRIÀCEÆ.** Perithecia mostly with a distinct ostiolum, of various consistence, but not carnose or membranaceo-carnose, brown or black. Stroma, when present, not carnose, dark colored outside, and often white within.

4. **DOTHIDEÀCEÆ.** Stroma always present, not fleshy, black, or dark-colored. Perithecia mostly reduced to mere cells in the stroma, and not separable from it. Ostiolum always present.

PERISPORIÀCEÆ.

This suborder is divided into two families. The *Erysipheæ* and *Perisporieæ*.

FAMILY. ERYSÌPHEÆ, Lév.

Ann. Sci. Nat. Ser. III, Tome XV.

On living plants. Mycelium superficial, consisting of numerous branching, septate, usually white, much interwoven threads, which extend widely over the epidermis of the host, adhering to it by means of haustoria. Conidia simple, colorless, cylindrical, oval or ovate, borne one above the other, on erect, simple, septate, colorless hyphæ. Perithecia seated singly on the mycelium, membranaceous, indehiscent, globose or sometimes depressed, at first colorless, then yellow, becoming dark brown or black when mature, bearing various thread-like, radiating appendages. Asci arising from the base of the perithecium, delicate, thin-walled, colorless, oblong, oval, ovate or suborbicular, usually pedicellate, containing 2–8 sporidia, which are simple, colorless, granular, oblong or oval.

The *Erysipheæ*, commonly known as "white mildews" or "blights," may be easily recognized by the white, dusty or web-like coating they form on the leaves, or other succulent parts of many common plants. They frequently grow throughout the summer, but usually only reach their full development in the fall, when the perithecia, or little fruit-balls, may be seen by good unaided eyes, scattered over the whitened surface of the leaves.

The very abundant mycelium consists of numerous slender, white or colorless, septate threads that branch widely, and extend over the leaf in every direction, frequently crossing and interlacing. These threads are usually pressed close to the host, but they do not themselves enter it. They send out at intervals, however, short, special branches called haustoria, that penetrate the epidermal cells, serving for the secure attachment of the fungus, and probably also for its nourishment. These haustoria present several forms, and they are of some importance in the classification of the species. In some cases the haustorium simply consists of a slender tube which penetrates the epidermal cell of the host, within which it swells to an oval or club-shaped sac, filled with granular protoplasm. More often there is an external appendage or sucker, that is pressed close to the surface of the epidermal cell, and from this, or from near it on the mycelial thread, the haustorium proper takes its rise and penetrates the epidermis. This external appendage may be smooth and entire, merely constituting a hemispherical swelling on the mycelial thread, or, it may take the form of a flattened disk with an indented margin. In the latter case it is said to be "lobed," in the former, "not lobed."

The conidia, or asexual reproductive bodies, are cylindrical, oval or nearly orbicular, simple, colorless cells filled with protoplasm. They are formed by constriction at the ends of short, simple, erect, rather stout, septate, colorless branches of the mycelium, called fertile hyphæ or conidiophores. A septum forms near the end of the young hypha, and the walls at this point become constricted. The cell thus cut off usually swells a little, and at length falls away as a mature conidium. Before this happens, however, other constrictions have taken place below, thus forming a chain of nearly mature conidia adhering end to end. Under favorable conditions they germinate quickly, sending out a slender tube, which, on the proper host, soon develops into a new mycelium. They are produced in immense numbers throughout the growing season, and, as they are very light and easily carried by the wind, they serve for the rapid increase and wide distribution of the parasite.

The perithecium remains on the fallen leaves over winter. It is not provided with a mouth or ostiolum of any kind. The contained asci and sporidia only escape on its decay in the spring.

Delicate membranaceous conceptacles, other than the perithecia, are sometimes found in connection with the mycelium of the *Erysipheæ*. They are thin-walled, and, on slight pressure, rupture irregularly, emitting immense numbers of minute, oblong, nucleated spores, immersed in a gelatinous fluid. They were noticed by Cesati, in connection with the grape mildew. Supposing them to be independent organisms, he named them *Ampelomyces quisqualis*, and specimens were published under that name as No. 1669 in Rabenhorst's Herbarium Mycologicum. Later they were called *Cicinobolus florentinus* by Ehrenberg, and *Byssocystis textilis* by Riess. Tulasne, Mohl, and others, finding that these conceptacles were borne on the same mycelium as the conidia and perithecia, naturally concluded that they were organs of the same plant, and, from their analogy to certain asexual reproductive bodies in allied groups of the *Ascomycetes*, called them pycnidia, and the minute bodies they contain, stylospores or pycnidiospores. This is still the accepted belief of many botanists. De Bary (Morph. und Phys. der Pilze, III, pp. 53–75, Tafeln VI, VII), shows that the pycnidia instead of being reproductive organs of the *Erysiphe*, are, in reality, the fructification of a fungus that is parasitic on the *Erysiphe*. He calls it *Cicinobolus Cesatii*, and gives numerous figures showing its delicate, septate mycelium, developing within the mycelial threads of the *Erysiphe*, and sending up branches which, by repeated division, form the cellular wall of the pycnidium.

There is much confusion in regard to the nomenclature of this group. The earlier authors, with poor magnifiers or none at all, made meager descriptions of their supposed species, for which reason it is often impossible to determine from their writings what they had in hand. Host plants are not so much of a guide here as in most cases for parasites, though many specific names have been proposed, it seems solely upon the observed habitat of the fungus. In what follows, an earnest endeavor has been made to identify and describe true species wherever they grow. The Bulletin of the Illinois State Laboratory of Natural History, Volume II, Article VI, *Parasitic Fungi of Illinois*, part II, by T. J. Burrill and F. S. Earle, has been made the basis of this work. This bulletin was founded upon collections made in Illinois, mostly by A. B. Seymour who also made studies upon the species.

KEY TO THE GENERA.

1. Appendages to the perithecia simple, and similar to the threads of the mycelium. - - - - - 2.
   Appendages various, readily distinguished from the mycelium. - - - - - - - - - - 3.
2. Only one ascus in a perithecium. - - *Sphærotheca.*
   Several asci in each perithecium. - - - *Erysiphe.*
3. Appendages branched at their tips. - - - - 4.
   Appendages not branched. - - - - - - 5.
4. Only one ascus in a perithecium. - - *Podosphæra.*
   Several asci in each perithecium. - - *Microsphæra.*
5. Appendages swollen at base, tips straight. *Phyllactinia.*

SPHÆROTHÈCA, Lév.

Ann. Sci. Nat. Series III, Tome XV, p. 138.

Perithecia containing only one ascus. Appendages simple threads not unlike the mycelium with which they are frequently interwoven. Ascus suborbicular, usually containing eight sporidia. Very rarely two asci have been observed.

S. pruinòsa, C. & P. Erysiphei of the U. S. in Journ. Bot. 1872.

Hypogenous. Mycelium thin, effuse, persistent. Perithecia scattered, 80–100 μ; cell-reticulations small, appendages few, simple, rigid, even, hyaline, 3–4 times the diameter of the perithecium. Ascus ovate. Sporidia 8, 20–25 μ long.

On *Rhus typhina*, *R. glabra* and *R. copallina*. From New York to Missouri, and no doubt more widely distributed; not very common. (The perithecia appear to develop earlier and better on the galls of a Phytoptus affecting these host-plants).

S. Hùmuli, (DC.)

Erysiphe Humuli, DC. Flore Franc. VI, p. 106.
Sphærotheca Castagnei, Lev. in part, Ann. Sci. Nat. XV, p. 139
Sphærotheca Humuli, Burrill, Parasitic Fungi of Illinois, II, p. 400.

Mostly hypophyllous. Mycelium inconspicuous or evanescent. Perithecia scattered, abundant, mostly rather small, 75–95 μ, wall-texture firm and compact, though thin, surface smooth, reticulations small, often obscure, usually less than 15 μ; appendages slender, three or more times as long as the diameter of the perithecium, usually colored throughout when mature, mostly free from the mycelium. Ascus broadly elliptical or suborbicular. Sporidia usually 8, large, averaging 20 μ long.

On *Viola canina* var. *sylvestris*, *Geranium maculatum*, *G. Richardsoni*, *G. incisum*, *Spiræa*, *Physocarpus opulifolia*, *Rubus odoratus*, *R. triflorus*, *R. strigosus*, *R. hispidus*, *Geum album*, *G. Virginianum*, *G. macrophyllum*, *Fragaria*, *Potentilla palustris*, *P. anserina*, *Poterium*, *Agrimonia Eupatoria*, *Gilia gracilis*, *G. linearis*, *Humulus lupulus*.

This species probably occurs on many other hosts, especially upon other Rosaceæ, upon which the fungus has usually been identified as *S. Castagnei*.

Common throughout North America east of the Rocky Mountains. In California Dr. Harkness reports *S. Castagnei* on hop leaves. This is no doubt what is here called *S. Humuli*.

It is exceedingly difficult to determine from literature upon what hosts *S. Humuli* occurs, because it is very rarely separated from

S. Castagnei in published lists. The first named hosts are those upon which the parasite has been considered distinct by Earle and the writer; the others are given upon the authority of the authors quoted. The distinction between the two species in question is, however, apparently real in the fact that the characteristically large, irregular cells of the wall of the perithecium of *S. Castagnei* are not found in those called *S. Humuli*, and the sporidia of the latter constantly average larger than those of the former. The difference also in the appendages seems to indicate specific distinction. Tulasne's figure (Select Fung. Carp. I, tab. IV, fig. 9) clearly shows these peculiarities of *S. Humuli*, the name then used for the fungus. The much more abundant mycelium of *S. pannosa*, and the smaller appendages sufficiently separate that from *S. Humuli*, though the affinities appear to be quite as close here as between the latter and *S. Castagnei*, with which only our plant has been confounded.

This is a very destructive parasite, especially on cultivated hops, in the Old as well as in the New World. On raspberries it also does considerable damage, though in many cases only the conidia are produced.

On *Geranium maculatum* the cells of the wall of the perithecium vary much, even in the same mount under the microscope, but the sporidia and other characteristics are those of *S. Humuli.*

S. pannòsa, (Wallr.)

Alphitomorpha pannosa, Wallr. Verhand. d. Naturf. Freunde, I, p. 43.
Erysibe pannosa, Lk. Species Plant. VI, I, p. 104.
Eurotium Rosarum, Grev. Scott. Crypt. Fl. III, p. 164, fig. 2.
Sphærotheca pannosa, Lev. Ann. Sci. Nat. XV, p. 138.

Mycelium abundant on the leaves, stems, etc., often sterile. Perithecia more often occurring on the branches, scattered, delicate, 90–100 μ, reticulations evident, small, 10–15 μ; appendages short and delicate, much interwoven with the mycelium, sometimes colored. Ascus large, delicate, ovate, expanding, when free from the perithecium, to a length greater than its diameter. Sporidia 8, large, 29 μ long.

On *Rosa blanda, R. Arkansana, R. lucida, R. humilis, R. parviflora.* Rather common from New England to California (Harkness), and Texas (Jennings).

Winter (Die Pilze, II, p. 26) and Saccardo (Syll. Fung. I, p. 2) describe this species with hyaline appendages; but Tulasne (Select Fung. Carp. I, p. 208) describes them as colored. They frequently are colored in our specimens. De Bary (Morph. und Phys. der Pilze, II, p. 48) says "colorless or brown at base."

S. Màli, (Duby).

Erysiphe Mali, Duby Bot. Gall. p. 869.
Sphærotheca leucotricha, E. & E. Jour. Myc. IV, p. 58.
Sphærotheca Mali, Burrill, (this publication).

Amphigenous. Mycelium white, submembranaceous, persistent.

Perithecia few or numerous, immersed in the mycelium, small 75–85 μ, globose or subpyriform; appendages of two kinds, in part rudimentary, floccose, deeply colored, attached in a broad tuft to the smaller end of the pyriform perithecium, the others strongly developed, rigid, straight or curved, continuous or septate, simple or rarely forked at the extremity, deeply colored at base, becoming pale outwardly, attached in a loose cluster opposite the rudimentary ones, 2–5 times the diameter of the perithecium. Ascus oval to subglobose. Sporidia 8, 20–30 μ long.

On upper parts of the twigs of *Pyrus Malus*, especially in nurseries of young trees, and upon the suckers from old ones. Not apparently very frequent but exceedingly abundant at times. Mississippi Valley, and probably eastward.

This exceedingly interesting species has not been well separated from *Podosphæra Oxyacanthæ* which occurs on the same host and to casual observation has much the same appearance. In our species the tips of the large appendages are occasionally forked (once or even slightly twice), which again may have been confusing. But these vague, stiff branches are totally unlike the dichotomous divisions of *Podosphæra*, and otherwise the species are very distinct. The tuft of short, interwoven, rudimentary appendages, like a dense cluster of short roots, is a very characteristic mark.

There is still some doubt as to the name. Evidently *Erysiphe Mali*, Moug. usually given as a variety of *Erysiphe adunca*, (Fries Syst. Myc. III, p. 245, Wallr. Flora Germ. IV, p. 755, &c.) is a different thing, but Duby's description (Botanicon Gallicum I, p. 869) so far as it goes, is sufficiently correct for our species. *Erysiphe Mali*, Duby, in Roumeguere's *Fungi Gallici Exsiccati* is a *Sphærotheca*, and seems to be the same as our plant, though the specimen examined was insufficient for satisfactory comparison. It is moreover scarcely possible that this host should have a solely American parasite of this kind upon it, hence the preference given in the nomenclature.

S. mors-ùvæ, (Schw.)

Erysiphe mors-uvæ, Schw. Syn. N. Am. 2494.
Sphærotheca mors-uvæ, B. & C. Grevillea IV, p. 158.

Mycelium abundant, at first white, becoming dark brown, densely covering the leaves, stems and fruit. Perithecia most abundant on the stems and fruit, densely aggregated, imbedded in the thick, felted mycelium, variable in size, 19–120 μ, dark brown, reticulations obscure; appendages short, delicate, hyaline or slightly colored, interwoven with and overrun by the dense mycelium. Ascus broadly elliptical, 8-spored, both ascus and sporidia smaller than in *S. pannosa*.

On *Ribes Cynosbati, R. gracile, R. rotundifolium, R. divaricatum* var. *irriguum, R. floridum, R. uva-crispa, R. cereum.*

From the eastern seaboard to Nebraska (Webber), and Montana (Anderson). In Berkeley's Notiees of N. A. Fungi (Grevillea IV, p. 158) it is said to occur "on grapes." From what we now know of the limitation of the species, this is probably incorrect.

This is the common "gooseberry mildew." It has been referred to *S. pannosa*, but it is sufficiently distinguished by its dense, dark-colored mycelium, which is strikingly unlike that of most of the *Erysipheæ.*

S. Epilòbii, (Link.)

Erysibe Epilobii, Link. Species Plantarum VI, p. 102.
Podosphæra (Sphærotheca) Epilobii, De Bary Beitrage Zur Morph. und Phys. d. Pilze. III, p. 48.

Amphigenous. Mycelium dense or arachnoid. Perithecia densely aggregated, small 70–80 μ, cell-walls thin with evident reticulations; appendages not numerous, very distinct, deeply colored, septate, simple, slender, 6–11 times the diameter of the perithecium. Ascus oval, wall medium. Sporidia usually 8, 15–21 μ long.

On *Epilobium alpinum* and *E. coloratum*, White Mts. (Farlow); Wisconsin (Davis). Conidiiferous specimens, probably of this species, have been collected in Illinois. Apparently rare.

S. Castágnei, Lev.

Synon.: Compare De Candolle Flore Franc. VI, pp. 106–108, Wallr. Flora Germanica pp. 753–76. Rabenhorst Deutschl. Krypt. Flora I, p. 230 et seq. It seems impossible to separate this species from others in the numerous descriptions by these and other authors. The name, as given above, is found in Ann. Sci. Nat. Ser. III, Tome XV, p. 139. *Erysiphe Gerardiæ* Schw. N. Am. Fungi p. 269, and *Erysiphe fuscata*, B. & C. Grev. IV, p. 159 belong to this species.

Mycelium abundant and persistent or sometimes inconspicuous, occurring on either or both sides of the leaves. Perithecia abundant, scattered or somewhat aggregated, small, usually about 75 μ, but varying from 60–100 μ. Texture soft, surface uneven, reticulations very large and irregular, 20–30 μ; appendages long, stout, usually colored throughout, but sometimes colorless, flexuous, somewhat uneven in width, more or less interwoven with the mycelium. Ascus rather small, elliptical or suborbicular. Sporidia usually 8, small, about 15 μ long.

On *Vernonia Noveboracensis, Erigeron Canadensis, Coreopsis aurea, C. aristosa, Bidens frondosa, B. connata, B. cernua,*

B. chrysanthemoides, B. bipinnata, Cacalia reniformis, C. tuberosa, Erechtites hieracifolia, Hieracium, Prenanthes altissima, Lactuca, Taraxacum officinale, Phlox divaricata, Hydrophyllum Virginicum, Veronica Virginica, Gerardia grandiflora, Brunella vulgaris, Pedicularis lanceolata, Shepherdia argentea.

Very abundant and widely distributed over the continent, on many widely different hosts. It is, however, variable, and it is quite possible that some forms should be separated as specifically distinct. This is very difficult to do on account of the intermediate forms in perhaps every particular. For the differences between this and *S. Humuli*, see note appended to the latter. Most specimens on *Rosaceæ* are easily determined as *S. Humuli*, while in some others this distinction is not so evident. The size of the perithecia varies much on different hosts,—larger than the average on *Erechtites* and some other *Compositæ*, smaller on *Veronica*. The appendages vary much in color; sometimes they are nearly or quite hyaline, even when certainly mature, but more often are tinged, sometimes deeply, throughout their entire length, with brown.

[In figure 3, plate 1, the germinal tube of a conidium is incorrectly represented as penetrating a stoma, instead of creeping over the surface of the leaf.]

S. phytoptóphila, Kell. & Swingle, Journ. Mycol. IV, p. 93.

Amphigenous. Mycelium sparse. Conidia about 15 x 27 μ. Perithecia globular, dark colored, with obscure reticulations, and with wall rather fragile, 60–85 μ diam; appendages few, hyaline or more often fuliginous, irregular, interwoven, sometimes septate, mostly longer than the diameter of the perithecium. Ascus large, broadly oval. Sporidia 8, 18–35 μ long.

On *Celtis occidentalis* affected by a Phytoptus forming bushy-branched tufts of twigs. The fungus grows upon these distorted branchlets and apparently not elsewhere on the tree, Kansas (Kellerman and Swingle), and Illinois. Doubtless widely distributed, but not yet reported elsewhere.

S. lanéstris, Hark. Trans. California Acad. of Sci. 1884, p. 20.

Hypophyllous. Mycelium abundant, firm, felt-like, becoming dull chocolate-brown. Conidia very firm, barrel-shaped, bulging much in the middle. Perithecia variable, about 100 μ, wall lined with a distinct, separable layer of hyaline cells; appendages none. Ascus elliptical, thick-walled except at summit. Sporidia 8, oval or subglobose, about 20–24 μ long.

On *Quercus agrifolia*, California (Harkness).

A very distinct and characteristic species. The filaments of the mycelium are exceedingly robust, forming a dense, coherent stratum. The perithecia are imbedded in this wooly mass, but when carefully separated appear to be absolutely without appendages. The thick threads of the mycelium do not adhere to the perithecia when thus transferred to a microscopic slide.

ERYSÌPHE, (Hedw.)

Lev. Ann. Sci. Nat. Ser. III, Tome XV.

Perithecium containing several asci. Appendages simple, threads similar to and frequently interwoven with the mycelium.

E. Liriodéndri, Schw. Syn. N. Am. p. 209.

On leaves and succulent stems. Mycelium abundant, dense, white, persistent. Perithecia developing late, mostly after the leaves have fallen, rather large, 100 μ or more, delicate, thin-walled, imbedded in and partially covered by the dense mycelium, reticulations small and indistinct; appendages several, hyaline, rather long, much interwoven with the mycelium. Asci several, eight or more. Sporidia 6–8, small.

On *Liriodendron tulipifera*, New York (Peck) to Illinois and probably westward. The species is not uncommon on the host named, though it appears to be slow in fruiting. May be identified by its abundant, white mycelium, especially on the young stems.

E. commùnis, (Wallr.)

Alphitomorpha communis, horridula, Wallr. in part, Verhandl. Naturf. Freund. L
Erysibe communis, nitida, Rabh. Deutschl. Krypt. Flora.
Erysibe communis, Lk. in part.
Erysiphe Aquilegiæ, DC. Flore Franc. VI, p. 105.
Erysiphe Pisi, DC. l. c. II, p. 274.
Erysiphe Convolvuli, DC. l. c. II, p. 274.
Erysiphe Polygoni, DC. l. c. II, p. 273.
Erysiphe communis, Fr Summa Veg. Scand. p. 406.
Erysiphe communis, Martii, Lev. in part, Ann. Sci. Nat. Ser. III, Tome XV.

Amphigenous. Mycelium abundant, persistent or sometimes evanescent. Perithecia variable in size and reticulations; appendages variable in length, often long, lying on the mycelium or more or less interwoven with it, usually colored in part or throughout, but occasionally all hyaline, sometimes branched. Asci 4–8, or more. Sporidia mostly 4–8, variable in size.

Very common from ocean to ocean on a very large number of host plants, among which are the following; *Clematis Virginiana, C. ligusticifolia, Anemone Virginiana, Anemonella thalictroides,*

Thalictrum polygamum, T. purpurascens, Ranunculus Cymbalaria, R. abortivus, R. sceleratus, R. septentrionalis, R. macranthus, R. acris, Aquilegia Canadensis, Delphinium elatum, Geranium maculatum, G. Richardsoni, Baptisia tinctoria. Thermopsis montana. Lupinus parviflorus, L. perennis, Trifolium longipes, T. involucratum, T. monanthum, Psoralea floribunda, Astragalus caryocarpus, A. Canadensis. A. multiflorus, A. decumbens, A. junceus, A. triphyllus, Oxytropis Lamberti, Lathyrus polymorphus, L. venosus, Pisum sativum, Phaseolus perennis, P. helvolus, Amphicarpæa monoica, Desmanthus brachylobus, Amelanchier alnifolia, Œnothera biennis, Œ. sinuata, Œ. albicaulis.

The form on *Clematis* is referred by authors to *E. tortilis*, (Wallr.), or as often written, *E. tortilis*, Link. It seems a mistake to separate it from the other forms occurring on *Ranunculaceæ*, some of which have equally long appendages; especially as on *Clematis*, these are radiant and more or less interwoven with the mycelium, as is usual in *E. communis*, while in European specimens of *E. tortilis* on *Cornus* (Rabh. Fungi Europ. No. 2033, J. Kunze, Fungi Selecti Exsicc. No. 577, etc.), the appendages are fasciculate and assurgent. (See also Tulasne, Sel. Fung. Carp. I, pp. 213–216).

The forms on *Leguminosæ*, etc., are often referred to *E. Martii*, Lév. De Bary (Morph. und Phys. der Pilze, III, p. 40) and Tulasne (l. c. p. 215) agree in considering this a synonym of *E. communis*. Winter, however, (Die Pilze, II, p. 31) retains *E. Martii* and refers to it all forms having hyaline appendages; but he says that he cannot decide whether this character is always constant and sufficient for their separation. Careful examination and comparison of the herbarium specimens specially mentioned by Winter, show that this character is not constant, for some of those given by him under *E. Martii* have distinctly colored appendages, while in some of those given under *E. communis* they are very slightly, if at all, colored. In fact the coloring of the appendages seems to depend to a considerable extent on the age and vigor of the specimen, being light colored or hyaline in the young, and often quite dark in fully matured. vigorous specimens. A portion, at least, of the appendages often remains hyaline of those on *Leguminosæ*, while in the case of those on *Ranunculaceæ*, they are usually all quite dark.

E. tórtilis, (Wallr.)

Alphitomorpha tortilis, Wallr. Verhandl Nat Freunde, I, p. 31.
Erysibe tortilis, Link, Sp. Plant. VI, part I, p. 3.
Erysiphe Corni, Duby Botan. Gall. II, p. 870.
Erysiphe tortilis, Fries Syst. Myc. III, p. 243.

Hypophyllous. Mycelium arachnoid, effuse, evanescent. Perithecia scattered, 80–100 μ; appendages not numerous, about 8–15, dark colored, flexuous, very long, 10–14 times the diameter of the perithecium, usually fasciculate by irregularly bending near the base, and clustered towards one side, very distinct from the mycelium. Asci 3–5. pedicellate. Sporidia 4–6, 22–30 μ long.

On *Cornus sanguinea*, Missouri (Tracy and Galloway).

This is here included solely upon the authority given, and apparently upon one collection. In Europe the fungus is common on the same host, and as this is abundant in cultivation with us, it is altogether probable that this parasite may be frequently found. It is very distinct from those forms of *E. communis* with long appendages, as on *Clematis*. These latter have indeed been referred to *E. tortilis*, evidently by mistake.

E. Cichoracearum. DC. Flore Franc. II, p. 274.

Alphitomorpha communis var. *depressa*, *horridula*, Wallr. Verhandl. Naturf. Freunde, IV.
Alphitomorpha lamprocarpa, Schl. Verhandl. Naturf. Freunde, VII, p. 49.
Erysibe communis, *lamprocarpa*, *depressa*, *horridula*, Lk. and Rabh.
Erysiphe horridula, *Montagnei*, *lamprocarpa*, Lev., in part
Erysiphe Ambrosiæ, *Verbenæ*, *Phlogis*, *Asterum*, Schweinitz, Syn. N. Am. p. 270.
Erysiphe Linkii, Lev. Ann. Sci. Nat. XV, p. 161.
Erysiphe spadicea, B. & C. Grev. IV, p. 159.

Amphigenous. Mycelium abundant, persistent, haustoria rounded, not lobed. Perithecia variable; appendages numerous, mostly short, 1–2 times the diameter of the perithecium, hyaline or mostly colored, much bent and interwoven with the mycelium. Asci variable, mostly numerous, 4 or 5 to 20. Sporidia large, quite uniformly 2, but occasionally varying to 3, or even 4 (on *Bigelovia* 5 or 6).

On *Napæa dioica*, *Vernonia Noveboracensis*, *V. fasciculata*, *V. Baldwinii*, *Stevia*, *Eupatorium purpureum*, *E. perfoliatum*, *Gutierrezia Euthamiæ*, *Grindelia squarrosa*, *Chrysopsis villosa*, *Bigelovia Douglasii*, *B. graveolens* var. *albicaulis*, *Solidago Missouriensis*, *S. serotina*, *S. Canadensis*, *S. nana*, *S. rigida*, *S. occidentalis*, *Aster conspicuus*, *A. corymbosus*, *A. macrophyllus*, *A. oblongifolius*, *A. cordifolius*, *A. sagittifolius*, *A. Drummondii*, *A. lævis*, *A. ericoides*, *A. multiflorus*, *A. viminosus* var. *foliolosus*, *A. salicifolius*, *A. diffusus*, *A. commutatus*, *A. junceus*, *A. longifolius*, *A. foliaceus* var. *Eatoni*, *A. umbellatus*, *A. adscendens*, *A. canescens*, *Erigeron Canadensis*, *E. divaricatus*, *E. strigosus*, *E. macranthus*, *E. glabellus*, *E. corymbosus*, *E. armeriæfolius*, *Inula Helenium*, *Silphium terebinthinaceum*, *Iva frutescens*, *I. xanthiifolia*, *Ambrosia trifida*, *A. artemisiæfolia*, *A. psilostachya*, *Xanthium strumarium*, *X. Canadense*, *Rudbeckia occidentalis*, *Helianthus annuus*, *H. rigidus*, *H. grosse-serratus*, *H. Californicus*,

H. doronicoides, *H. strumosus*, *H. decapetalus*, *H. tuberosus*, *Helianthella Parryi*, *Verbesina encelioides*, *Actinomeris squarrosa*, *Helenium autumnale*, *Gaillardia aristata*, *Dysodia chrysanthemoides*, *Artemisia dracunculoides*, *A. biennis*, *A. Ludoviciana*, *A. discolor*, *Cnicus undulatus*, *C. undulatus* var. *canescens*, *C. altissimus*, *C. altissimus* var. *discolor*, *Hieracium Canadense*, *Prenanthes alba*, *Lactuca pulchella*. *Asclepias variegata*, *Hydrophyllum Virginicum*, *H. Canadense*, *Phacelia circinata*, *P. Menziesii*, *Phlox paniculata*, *P. Drummondii*, *Cynoglossum*, *Echinospermum Virginicum*, *E. Redowskii*, *Mertensia Sibirica*, *Lithospermum arvense*. *Mimulus luteus*, *Tecoma radicans*, *Verbena officinalis*, *V. urticæfolia*, *V. angustifolia*, *V. hastata*, *V. stricta*, *V. bracteosa*, *Plantago major*, *Galium Aparine*, *Humulus Lupulus*, *Pilea pumila*, *Parietaria debilis*, *P. Pennsylvanica*.

This exceedingly common species is abundant in all sections of our country, and is found upon an extraordinary number of widely distinct host-species, as the preceding list shows, though this is undoubtedly incomplete. Various names have been given to somewhat different forms included herein, but after a careful examination of a large amount of material from widely separated regions, it seems impossible to admit specific distinctions among even the most divergent forms. Unusual variations seem to occur in the Rocky Mountain regions, as Anderson and Kelsey have noticed in Montana. For instance, a form on *Bigelovia graveolens* has 20–30 asci, many of which have 3–5, and perhaps more, sporidia, and the appendages are short and almost hyaline. Taken by itself, it could hardly be admitted as belonging to the present species. Ellis & Everhart (Botanical *Gazette*, XIV, p. 286), provisionally propose the name *E. sepulta* for it. But on *Bigelovia Douglasii*, growing with the preceding, the fungus is in all characteristics the same, except that the sporidia are uniformly 2, in the specimens examined, and so reported by others. The asci are often as many as 30 in both cases,—a number much greater than commonly given for typical *E. Cichoracearum*. On other host-species the number of asci is exceedingly variable, mostly only 4–8, but in some collections east of the Mississippi river, reaching 20, with apparently no way of distinguishing different species among the variable forms. Those on *Bigelovia* are indeed further aberrant, but it does not seem wise to separate one or both as specifically distinct, either from each other or from those with which they are undoubtedly allied on the host-plants enumerated above.

E. Galeópsidis, DC. Flore Franc. VI, p. 108.

Erysiphe lamprocarpa, Lev. in part.
Erysiphe Labiatarum, Chev. Flora Paris, III, p 380.
Erysiphe Chelones, Schw. Syn. N. Am p. 270

Amphigenous. Mycelium abundant, persistent, haustoria of the mycelial threads lobed. Perithecia somewhat aggregated, appendages numerous, short, flexuous, colored, interwoven with the mycelium. Asci numerous, often 12 or more. Sporidia 2, mostly formed late.

On *Teucrium Canadense*, *Mentha*, *Scutellaria lateriflora*, *S. aspera*, *Galeopsis Tetrahit*, *Chelone glabra*.

Widely distributed east of the Rocky Mountains and in Montana (Anderson), but not often distinguished from *E. Cichoracearum* in published lists. It can scarcely be separated from the latter by the characters of the perithecia, but the difference in the haustoria, first pointed out by De Bary (Morph. und Phys. der Pilze, III, p. 49), can be observed by first soaking a portion of the leaf in caustic potash and then removing a little of the mycelium to the slide. The perithecia and appendages are often rather lighter colored than is usual in *E. Cichoracearum*.

E. aggregàta, (Peck).

Erysiphella aggregata, Peck 28th Rep. N. Y. State Mus. Nat. Hist. p. 63.
Erysiphe aggregata, Farlow, Bull Bussey Inst. II, (1878), p. 227.

Mycelium dense, felt-like, white, but becoming yellowish. Perithecia very numerous. closely crowded, opake, thick-walled, with rather small cells, 140–180 μ; appendages very numerous, interwoven, hyaline or nearly so, rather slender. Asci numerous, 10–50, oblong-ovate or sometimes narrower, thick-walled. Sporidia crowded in the ascus, oval, mostly 15–20 μ long.

On the fertile aments of *Alnus serrulata* and of *A. incana*, in autumn and spring.

Reported from Massachusetts (Farlow), New York (Peck), New Jersey and Pennsylvania (Ellis); said to be common in these localities. The genus *Erysiphella* was proposed for this when it was supposed there were no appendages.

E. trìna, Hark. Trans. Cal. Acad. of Sci. 1884, p. 41.

Epiphyllous. Mycelium covering orbicular spots, pruinose, fugacious. Perithecia clustered, minute, yellowish-brown, 56–70 μ; appendages none. Asci 3, nearly globular, 31–38 μ. Sporidia 2, oblong-elliptical, or somewhat boat-shaped, very large, sometimes filling the ascus, 18–20 x 28–32 μ.

On *Quercus agrifolia*, California (Harkness). Inserted from paper by Dr. Harkness, read before the California Academy of Sciences, February 4, 1884.

No specimens examined. In the original, *Erysiphella* is given as an alternate genus.

E. gráminis, DC. Flore Franc. VI, p. 106.

Amphigenous, often epiphyllous. Mycelium dense, felt-like, persistent, white or gray, sometimes tinted brown. Perithecia immersed in the mycelium, few and scattered or many and crowded, depressed, large, about 225 μ (150–255 μ); appendages numerous, rather short, simple or occasionally branched, rigid, but variously curved and interwoven with the mycelium, hyaline or tinted. Asci 16–25, oblong to oval, pedicellate. Sporidia 8, or rarely 4, maturing late, often only on old plants in the spring.

On *Beckmannia erucæformis*, *Panicum sanguinale*, *Agrostis exarata*, *Poa tenuifolia*, *P. pratensis*, *Glyceria nervata*, *G. aquatica*. *Bromus unioloides*, *Hordeum jubatum*, *Elymus condensatus*, *Triticum vulgare*.

Common, Massachusetts to California, mostly observed in the conidial state (*Oidium monilioides*, Link) in shady places. Perithecia usually forming only late in summer or autumn, and sporidia much later. In Montana, however, the former have been collected in July, and ripe sporidia found in November, or even earlier (Anderson). In California it has been destructive to wheat (Harkness). Anderson says there are sometimes as many as 20 sporidia in an ascus,—a variation not reported elsewhere.

UNCÍNULA. Lév.

Ann. Sci. Nat. Ser. III, Tome XV.

Perithecium containing several asci. Appendages free from the mycelium, recurved or coiled at the tip.

U. Clintònii, Peck, Trans. Albany Inst. VII, p. 216, 25th Rep. N. Y. State Mus. p. 106.

Amphigenous. Mycelium thin, rather persistent. Perithecia scattered, 90–120 μ, firm, nearly black; appendages 15–30, 1–1½ times diameter of perithecium in length, hyaline or nearly so, uncinate-coiled at the tip. Asci 4–6, oval, very short-beaked. Sporidia 4–8, mostly 18–21 μ long, completely filling the ascus.

On *Tilia Americana*, not very common. Apparently often escaping observation. Reported from New York (Peck), Wisconsin (Davis), Iowa (Pammel), and Illinois.

U. necàtor, (Schw.)

Erysiphe necator, Schw. Syn. N. Am. 2495.
Uncinula spiralis, Berk. Crypt. Bot. p. 268, fig. 64.
Uncinula Ampelopsidis, Peck, Trans. Albany Inst. VII, p. 216.
Uncinula Americana, Howe, Erysiph. U. S. Journ. Bot. 1872.
Uncinula subfusca, B. & C. Grev. IV, p. 160.
Uncinula necator, Burrill (this publication).

Amphigenous or frequently epiphyllous. Perithecia 85–120 μ, dark brown, opake, reticulations small, rather obscure; appendages from 10 or 12 to 20 or more, varying in length from once and a half to four or more times the diameter of the perithecium, colored for more than half their length, frequently septate, occasionally forked, tips loosely and somewhat spirally coiled. Asci ovate, pedicellate. Sporidia 4–6.

Very rare on vines of cultivated varieties of *Vitis labrusca* in native [American] vineyards. Mycelium very thin, whitish, with very slender flocci, orbicular, not close pressed. Perithecia very minute, scattered, brownish-black, globose. When abundant, this species also destroys the fruit.—Schw. l. c.

On *Vitis aestivalis*, *V. cinerea*, *V. labrusca*, *V. riparia* and *V. vinifera*, *Ampelopsis cuspidata* and *A. quinquefolia*.

This is a very common species, widely distributed throughout the country, from the Atlantic to the Pacific coast. It is the powdery mildew of cultivated grape vines, and there is good reason to suppose that the so-called *Oidium Tuckeri* of European vineyards is the same thing, without, however, the development of the perithecia. The conidia found with the perithecia in America are indistinguishable from those on European vines in their native regions, and these same European vines grown in this country have both conidia and perithecia altogether similar to those on American grapes. If it is true that the European conidia-bearing parasite is really the present species, the non-development of the perithecia in the Old World is a curious biological phenomenon, though similar peculiarities exist in regard to other species of fungi in contrasted regions in our country.

The plant varies considerably, but nearly or quite as much on leaves of *Vitis* as upon this on the one hand, and *Ampelopsis* species on the other. The name most commonly adopted is *U. spiralis*, Berk., from the named figure, without description, in Berkeley's Introduction to Cryptogamic Botany, but subsequently described in Grevillea. Prior to the latter, Peck described *U. Ampelopsidis*, hence this name has been used in some cases for the specimens on both *Vitis* and *Ampelopsis* since these fungi have been recognized as the same species. But there is no reasonable doubt that Schweinitz had before him specimens of this same species, and if so, his neglected name must be accepted. It should be remembered that he worked before the era of the compound microscope, and his description is necessarily meager, but it is correct as far as it goes.

U. flexuòsa, Peck, Trans. Albany Inst. VII, p. 215.

Hypophyllous. Perithecia large (110–125 μ), dark, opake, reticulations obscure; appendages numerous, 40 or more, about equaling the diameter of the perithecium, hyaline, minutely roughened, thickened and irregularly flexuous toward the tip. Asci about 10, ovate or pyriform, strongly pedicellate. Sporidia 3, small, 15–20 μ long.

On *Æsculus glabra*, *Æ. flava*, and *Æ. Hippocastanum*. Apparently not common, but reported from New York (Peck), Illinois (Earle), Missouri (Demetrio).

The species is easily determined by the abrupt, wavy crooks in the outer half of the appendages.

U. circinàta, C. & P. Erysiphei of the U. S. in Journ. Bot. 1872.

Hypophyllous or sometimes amphigenous. Perithecia very large, depressed, 150–225 μ in greatest diameter, texture soft, reticulations very small and irregular; appendages very numerous, slender, simple, about equal to the diameter of the perithecium, hyaline, smooth, tips not swollen, ascending from the upper half of the perithecium. Asci numerous, 14 or more, long and slender, oblong or narrowly ovate, pedicellate, about 30 x 75 μ. Sporidia 8, small, about 10 x 15 μ.

On *Acer Pennsylvanicum*, *A. spicatum*, *A. saccharinum*, *A. dàsycarpum*, *A. rubrum*.

Not rare throughout the country east of the Rocky Mountains, though not reported in the extreme northwest.

This is readily distinguished from *U. Aceris*, (DC.) by its simple appendages and more numerous, very narrow asci. In some specimens the mycelium is inconspicuous, but in specimens from Massachusetts (Seymour) it is more abundant. The leaves affected by it can often be distinguished at a distance, as the areas covered by it remain green after the rest of the leaf has assumed its autumnal tint.

U. Àceris, (DC.)

Erysiphe Aceris, DC. Flore Franc. VI, p. 104.
Alphitomorpha bicornis, Wallr. Verhandl. I, p. 38.
Uncinula bicornis, Lev. Ann. Sci. Nat. XV, p. 153.
Uncinula Aceris, Sacc. Syll. I, p. 8.

Amphigenous. Mycelium variable, sometimes in conspicuous, dense patches, sometimes spreading over one or both surfaces of the entire leaf. Perithecia scattered, large, 150–200 μ; appendages very numerous, short, less than diameter of perithecium, once or more rarely twice forked mostly beyond the middle, the elongated tips somewhat tapering and strongly recurved or spirally wound. Asci 8–12, somewhat pear-shaped. Sporidia 8, about 14 x 30 μ.

On maple leaves, California (Harkness). This exceedingly interesting and characteristic species is inserted as American upon the authority of Harkness and Moore (Pacific Coast Fungi, p. 32) where it is named *U. bicornis*, Lév.

The appendages are so unique and the forks so conspicuous that it seems impossible a mistake should be made in identification. In some instances, when the division is near the end, an approach to *Microsphæra* is apparent, but the characteristic curve or coil of the tips, leaves no room to doubt the proper generic position.

U. macróspora, Peck, Trans. Albany Inst. VII, p 215.—25th Rep. N. Y. State Mus. p. 96.

Amphigenous. Mycelium conspicuous, abundant. Perithecia large, 110–165 μ, wall tissue soft, reticulations very small, usually 5–10 μ, and rather obscure; appendages very numerous, 50 or more, hyaline, slender, smooth, usually shorter than the diameter of the perithecium, tips closely coiled, not enlarged. Asci several, 8–10. Sporidia 2, large, 20 x 30–35 μ.

On *Ulmus fulva*, *U. Americana*, *U. alata*, *Ostrya Virginica*.

Rather common east of the Rocky Mountains. On the last named host (collected by Tracy in Wisconsin) the fungus is undoubtedly the same as that much more commonly found on elms.

It differs sufficiently from European specimens of *U. Bivonæ*, Lév. on *Ulmus campestris* (Thüm. Mycoth. Univer. No. 755). In these latter the perithecia are smaller (80–90 μ), and the reticulations are much larger (10–15 μ), and more distinct. The fewer (less than 20) appendages are stouter, somewhat roughened and conspicuously swollen at their tips. The usually four asci each contain two sporidia about 30 μ long but narrower than in *U. macrospora*.

U. intermedia, B. & C. appears from description to be the same as *U. macrospora*.

U. párvula, C. & P. Erysiphei of the U. S. Journ. Bot. 1872.

Amphigenous. Perithecia usually small, 90–100 μ, rarely 135 μ, delicate, reticulations distinct, small and regular, averaging about 10 μ; appendages 60–100, delicate, slender, hyaline, commonly shorter than the diameter of the perithecium. Asci 5–7, broadly elliptical. Sporidia 5–8, mostly 6, about 20–25 μ long.

On *Celtis occidentalis*. From the Atlantic coast to Washington (Seymour). Not apparently abundant, but often collected from widely separated localities.

Distinct from *U. polychæta*, B. & C., on same host. Washington (State) specimens have perithecia of larger size than usual, attaining rather more than 125 μ, while 100 μ seems to be about the extreme for eastern specimens.

U. polychæ̀ta, (B. & C.)

Erysiphe polychæta, B. & C. Grev. IV, p. 159.
Uncinula Lynchii, Speg. Fung. Arg. II, p. 17.
Pleochæta Curtisii, Sacc. & Speg. Fung. Arg. II, p. 44.
Uncinula polychæta, Rav. F. Car. IV, No. 68.

Hypophyllous. Mycelium usually dense, forming irregular, whitish spots or patches. Perithecia scattered, flattened or depressed above, large, 225–280 μ; appendages very numerous, 250–300 or more, about half as long as the diameter of the perithecium, at first clavate, then attenuate and once coiled at the apex, hyaline. Asci numerous, 50 or more, about 27 x 80 μ. Sporidia 2, or rarely more, large, about 17 x 27 μ, nearly filling the ascus.

On *Celtis occidentalis*, Carolina (Ravenel), Mississippi (Tracy). South America (Spegazzini). Apparently not abundant.

U. confùsa, Massee, Grev. XVII, p. 78.

Uncinula polychæta, B. & C. Grev. IV, p. 159 (No. 993).
Pleochæta Curtisii, Sacc. and Speg. l. c.

Hypophyllous. Mycelium very scanty, not forming spots. Perithecia scattered, usually not more than 2–3 on a leaf, 150–200 μ; appendages 25–28, simple, colorless, very slender, about 300 x 2–3 μ; apices strongly involute, not at all thickened. Asci about 25, cylindric-clavate, 4-spored. Sporidia colorless, simple, elliptical, oblong, 20 x 10 μ.

On leaves of *Celtis occidentalis*, Carolina.

U. geniculàta, Gerard, Bull. Torr. Bot. Club, IV, p. 48.

Epiphyllous. Mycelium thin, in definite spots or overspreading the leaf. Perithecia scattered, usually few, 90–120 μ; appendages 15–30, hyaline, somewhat roughened, often geniculate, about once to twice the diameter of the perithecium in length, about 1½ times spirally coiled at the apex. Asci 6–8. Sporidia 4–6, about 14–20 μ long.

On *Morus rubra*, rare or missed by collectors. New York, Illinois (Pammel).

U. Sàlicis, (DC.) **(Plate 2)**

Erysiphe Salicis, DC. Flore Franc. II, p. 273.
Erysiphe Populi, DC. Flore Franc. VI, p. 104.
Alphitomorpha adunca, *guttata*, Wallr. Verh. Naturf. Freunde, I, pp 37, 42.
Erysibe adunca, *obtusata*, Lk. Spec. Plant. VI, I, p. 117.
Erysiphe adunca, Grev. Scott. Crypt. Flora, V, tab. 296.
Uncinula adunca, Lev. Ann. Sci. Nat. Ser III, Tome XV.
Uncinula leuculenta, Howe, Trans. Albany Inst. VII, quoted in Amer Nat VII, p. 58.
Uncinula heliciformis, Howe, Torr. Bull. V, p. 4.
Uncinula Salicis, DC. Winter, Die Pilze, II, p 40

Amphigenous. Mycelium abundant, persistent. Perithecia usually large, 100–160 μ, wall tissue soft, elastic, reticulations rather small and indistinct; appendages variable in number, usually very numerous, hyaline, not much swollen at the tip, once to twice as long as the diameter of the perithecium. Asci from 4 or 5 to 12 or more, ovate. Sporidia usually 4 or 5, sometimes 6–8.

On *Salix nigra*, *S. nigra* var. *falcata*, *S. amygdaloides*, *S. longifolia*, *S. rostrata*, *S. flavescens*, *S. discolor*, *S. humilis*, *S. petiolaris*, *S. corduta*, *S. glauca*, *Populus tremuloides*, *P. angulata*, *P. grandidentata*, *P. heterophylla*; *P. balsamifera* var. *candicans*, *P. monilifera*.

PHYLLACTÍNIA, Lév. (Plate 3)

Ann. Sci. Nat. Ser. III, Tome XV, p. 144.

Perithecium containing several asci. Appendages free from the mycelium, acicular, acute at the tip, abruptly swollen at base.

P. suffúlta, (Reb.)

Sclerotium suffultum, Reb. Flor. Neom. p. 360.
Erysiphe Coryli, *Fraxini*, DC. Flore Franc. II, p. 273.
Erysiphe vagans, Bivon. Stirp. rar. Sicil, III, p. 197.
Alphitomorpha guttata, Wallr. Verh. Naturf. Freunde, I, p. 42.
Erysibe guttata, Lk. Spec. Plant. VI, I, p. 116.
Erysibe guttata, Fr. Syst. Mycol. III, p. 245.
Phyllactinia Candollei, Lev. Grev. IV, p. 158.
Phyllactinia guttata, Lev. Ann. Sci. Nat. Ser. III, Tome XV.
Phyllactinia suffulta, Sacc. Michelia II, p. 50.

Mostly hypophyllous. Mycelium abundant, persistent, or scant and evanescent. Perithecia very large, 150–275 μ, wall tissue soft, cellular structure, and reticulations obscure; appendages few, usually 8–12, easily detached, hyaline, varying in length from less than, to three or four times the diameter of the perithecium. Asci 4 or 5 to 20 or more, ovate, pedicellate. Sporidia normally 2, occasionally 3 or 4, variable in size, mostly large.

On *Magnolia acuminata*, *Liriodendron tulipifera*, *Berberis*, *Xanthoxylum Americanum*, *Ilex decidua*, *Celastrus scandens*, *Acer saccharinum*, *Desmodium Canadense*, *Cratægus coccinea*, *C. tomentosa*, *C. punctata*, *C. crus-galli*, *Heuchera parvifolia*, *Ribes Cynosbati*, *Hamamelis Virginiana*, *Fraxinus Americana*, *F. pubescens*, *F. viridis*, *F. sambucifolia*, *Asclepias Cornuti*, *Catalpa speciosa*, *C. bignonioides*, *Cornus florida*, *C. circinata*, *C. stolonifera*, *C. sericea*, *C. paniculata*, *Ulmus Americana*, *U. alata*, *Betula papyrifera*, *B. nigra*, *B. occidentalis*, *Alnus serrulata*, *A. incana*, *Corylus Americana*, *Ostrya Virginica*, *Carpinus Caroliniana*, *Quercus macrocarpa*, *Q. rubra*, *Q. coccinea*, *Q. coccinea* var. *tinctoria*, *Q. falcata*, *Castanea sativa* var. *Americana*, *Fagus ferruginea*, *Typha latifolia*.

This everywhere common species presents many variations in the size of the perithecia, the length of the appendages, the number and size of the asci, and the size of the sporidia; but none of these forms seem constant enough to justify their separation. On *Liriodendron* the mycelium is usually inconspicuous, the appendages but little longer than the diameter of the perithecium, and the few (8–10) asci are large and broadly ovate. On *Ulmus* the mycelium is abundant and persistent, the perithecia and appendages medium, and the very numerous (20–30) asei are small and narrow. On *Quercus* the perithecia are very large, and the 10–15 asei and the sporidia are much larger than on *Ulmus*. On *Corylus* the perithecia are small, but the appendages are very long. It is remarkable for the exceedingly great diversity of the host species which it affects. Scarcely a deciduous leafed tree seems to be proof against it. The most peculiar thing in this connection is its appearance on *Typha latifolia* (Anderson, Journ. Mycol. V, p. 193).

In a large number of instances the perithecia have a dense layer of short, branched, fine, hyaline, radiating hyphæ, totally distinct from the conspicuously bulbous appendages. There are projections from the wall of the perithecium, issuing from the lower side and forming a cushion-like mass.

PODOSPHÆRA, Kunze.

Mycol. Hefte II, p. 111.

Perithecium containing a single ascus. Appendages free from the mycelium, dichotomously branched at the end.

P. Oxyacánthæ, (DC.) **(Plate 4)**

Erysiphe Oxyacanthæ, DC. Flore Franc. VI, p. 106.
Alphitomorpha clandestina, Wallr. Flora Crypt. Germ. III, p. 753.
Erysibe clandestina, Lk. Spec. Plant. VI, I, p. 103.
Podosphæra Kunzei, clandestina, Lev. Ann. Sci. Nat. Ser. III, Tome XV, p. 19.
Podosphæra minor, Howe, Torr. Bull. V, p. 3.
Microsphæra fulvo-fulcra, Cke. Grev. VI, p. 110.
Podosphæra Oxyacanthæ, DBy. Morph. und Phys. der Pilze. III, p. 480.

Amphigenous. Mycelium variable, often abundant, persistent. Perithecia 65–110 μ, dark, opake, reticulations regular, about 10–15 μ. evident when young, scarcely observable when old, except by the uneven surface; appendages 8–20, dark brown for more than half their length, frequently septate, 1–4 times as long as the diameter of the perithecium, 3–5 times dichotomously forked, branches short, often swollen, tips recurved. Ascus broadly elliptical or orbicular, about 50 x 60 μ, thick walled. Sporidia usually 8.

On *Prunus domestica, P. Americana, P. Cerasus, P. pumila, P. Virginiana, P. demissa, Spiræa salicifolia, S. tomentosa, S. Douglasii* var. *dumosa, Pyrus Malus, Cratægus Oxyacantha, C. coccinea, C. tomentosa, C. punctata, C. crus-galli, Amelanchier Canadensis, Diospyros Virginiana.*

In Europe three species are described as follows;

P. Oxyacanthæ, (DC.) DBy.—Appendages 8 or more, about equal to the diameter of the perithecium, standing erect on its upper surface. On *Cratægus*, *Sorbus* and *Mespilus*.

P. tridactyla, (Wallr.) DBy.—Appendages 3–7, standing erect in a parallel bundle on the summit of the perithecium. On *Prunus* sps.

P. myrtillina, (Schubert) Kunze.—Appendages 6–10, arising from the upper surface of the perithecium, but radiating divergently or reflexed. On *Vaccinium*.

European specimens on the above hosts show these distinguishing characters sufficiently well, but American specimens on *Prunus* cannot be separated from those on *Cratægus*, etc. There appear to be none on *Vaccinium*. Whatever may be done with European forms, the American ones must be considered one species.

P. biuncinàta, C. & P. Erysiphei of the U. S. Journ. Bot. 1872. Pk. 25th Rep. p. 94.

Amphigenous. Mycelium thin, arachnoid, rather persistent. Perithecia small, 70–90 μ, scattered; appendages 6–12, 3–5 times as long as the diameter of the perithecium, hyaline, with a conspicuous, widely spreading fork at the apex, each branch of which is sometimes divided. Ascus globose. Sporidia mostly 18–21 μ long.

On *Hamamelis Virginiana*, Massachusetts (Seymour), New York (Peek), Illinois (Waite).

An easily recognized species, by the unique character of the tips of the appendages, which approach, though they are easily different from those of *Uncinula Aceris*. The mycelium shows much more on the upper surface of the leaves, but is also common below.

MICROSPHÆRA, Lév.

Ann. Sci. Nat. Ser. III, Tome XV, p. 381.

Perithecium containing several asci. Appendages free from the mycelium, more or less dichotomously branched at the end.

M. Menispérmi, Howe, Bull. Torr. Bot. Club, V, p. 3.

Epiphyllous. Mycelium rather abundant, thin and widely effused, but thickened in certain circular spots, of which there are usually not more than three or four on a leaf. Perithecia aggregated upon the special denser spots of the mycelium, otherwise remotely scattered, black, very variable in size, 60–115 μ diam; appendages not numerous, 8–15, rather rigid, tinted at base, exceedingly variable in length and amount of branching, 1–7 times diameter of perithecium in length and 1–7 times dichotomously branched, the branches of the first order short or often much elongated, the others usually short and compact, tips strongly recurved, very ornate. Asci 1–6 or more, variable in shape and size. Sporidia 4–6.

On *Menispermum Canadense.* Not frequent. New York to Iowa (Holway).

Remarkable for the variation in structure and size. In some perithecia only one ascus is found, while in others in the same microscopical preparation at least seven have been seen. The appendages on a single perithecium are somewhat equal in length but are often exceedingly variable in the division of the tips. Sometimes there is only a single fork with two equal, straight, obtuse branches and again the exceedingly ornamental tip fills the field of the microscope with its complex scroli-work.

M. Russéllii, Clinton, 26th Rep. N. Y. State Mus. p. 80.

Amphigenous. Mycelium inconspicuous. Perithecia small, 75–100 μ, delicate, reticulations regular, distinct, about 10 μ; appendages 8–18, many times longer than the diameter of the perithecium, colored for half or two thirds of their length, occasionally septate, simple, bifid, or two or three times irregularly branched, branches long, often distorted, tips not swollen or recurved. Asci 4–8. Sporidia usually 4, small.

On *Oxalis violacea*, and *O. corniculata* var. *stricta.* Not uncommon east of the Mississippi; not reported westward. A well characterized species.

M. Ravenélii, Berk. Grev. IV, p. 160.

Amphigenous. Mycelium usually abundant, persistent. Perithecia abundant, usually large, 100–130 μ, reticulations small and irregular, about 10 μ; appendages 10–20, somewhat roughened, usually hyaline, occasionally colored for a distance, the color ending at an abrupt line like a septum, once or twice as long as the diameter of the perithecium, 5–7 times dichotomous, branches short, forming a more or less compact head, tips usually acute and recurved. Asci 6–10, frequently 8, ovate, pedicellate, about 45 x 60 μ. Sporidia 4–6 (Saccardo says 8).

On *Astragalus adsurgens*, *Lathyrus palustris*, *Gleditschia triacanthos*, *Vicia Americana*, *V. Americana* var. *linearis.*

Rather common and very widely distributed. It is reported more often from the valley of the Mississippi and westward to Montana and Texas.

M. diffùsa, C. & P. Erysiphei of U. S. in Journ. of Bot. 1872, 25th Rep. N. Y. State Mus. p. 95.

Usually epiphyllous. Perithecia scattered, 90–120 μ, dark, opake, reticulations rather obscure, 10–15 μ, appendages 15–25, hyaline, or slightly tinted at the base, 2–4 or more times as long as the diameter of the perithecium, 1 to 4 or 5 times irregularly or dichotomously branched, branches long and diffusely spreading, not at all swollen or recurved. Asei 4–7, ovate, pedicellate, rather small, 30–35 x 60–65 μ. Sporidia 4–8, mostly 4–5.

On *Desmodium canescens*, *D. cuspidatum*, *D. paniculatum*, *D. Canadense*, *D. sessilifolium*, *Lespedeza*, *violacea*, *L. hirta*, *L. capitata*, *Lathyrus ochroleucus*, *Vicia*, and *Phaseolus perennis*.

Reported from the eastern seacoast to Minnesota (Seymour) and Missouri (Tracy and Galloway). It is abundant and frequent, and though variable is well characterized by the elongated branches of the appendages.

M. Grossulàriæ, (Wallr.)

Alphitomorpha penicillata var. *Grossulariæ*, Wallr. Verh. Naturf. Freunde I, p. 40,
Microsphæra Grossulariæ, Lev. Ann. Sci. Nat. Ser. III, T. XV, p. 160.
Microsphæra Van-Bruntiana, Ger. Bull. Torr. Bot. Club. VI, p. 31.

Amphigenous. Mycelium thin or rather dense, white, persistent. Perithecia scattered, 75–120 μ, reticulations about 15 μ; appendages 10–20, once to twice as long as the diameter of the perithecium, hyaline or tinted at the base, very conspicuously 4–6 times dichotomously forked, with straight, obtuse digitate branchlets. Asci 3–5. Sporidia 4–6, variable, 15–30 μ long.

On *Ribes rotundifolium*, *R. floridum* and *R. nigrum*, *Sambucus Canadensis* and *S. racemosa*.

Not very uncommon from the Atlantic to Montana (Anderson). The branched tips of the appendages are very characteristic, though the branchlets vary considerably in length.

M. Symphoricàrpi, Howe, Bull. Torr. Bot. Club. V, p. 3.

Amphigenous. Mycelium abundant, persistent. Perithecia small, 80–100 μ, delicate, reticulations large, regular, 15–20 μ; appendages 8–16, hyaline or slightly colored at base, 2–4 times as long as the diameter of the perithecium, 4–5 times dichotomous, branches short, compact, tips truncate, somewhat swollen, not recurved. Asci 4–10, small, 50 μ long. Sporidia 4–6, small and narrow, 10–18 μ.

On *Symphoricarpus vulgaris*, *S. occidentalis* and *S. racemosus*. Common across the continent.

Much like some forms of *M. Vaccinii*, but the mycelium is more abundant and the reticulations are larger and more evident.

M. semitósa, B. & C. Grev. IV, p. 160.

Epiphyllous. Mycelium persistent. Perithecia few, somewhat aggregated, 90–100 μ, delicate, reticulations regular and distinct, about 10 μ; appendages 12 or more, about equal to the diameter of the perithecium, colored throughout, paler toward the tip, or the color stopping at a distinct line like a septum, 3 or 4 times dichotomously branched, primary branches long, others short, tips obtuse, not recurved. Asci several. Sporidia small, 10 x 15 μ.

On *Cephalanthus occidentalis*, rare, Carolina (Curtis)? Illinois (Waite, Pammel).

In the specimens examined, the number of asci varied from four to six, and the sporidia in each seemed to be about six, but were poorly developed This fact is evidently characteristic of the species, though in some other species the sporidia mature quite as tardily.

M. Vaccínii, (Schw.)

Erysiphe Vaccinii, Schw. Syn. N. Am. No. 2491.
Microsphæra Vaccinii, C. & P. Erysiphei of the U. S. in Journ. of Bot. 1872.

Amphigenous. Mycelium thin and delicate, often evanescent, or sometimes abundant, persistent. Perithecia variable, often small, 80–90 μ, or large, 110–120 μ, fragile; appendages 6–20, hyaline, smooth, slightly colored at base, 2 or 3 to as many as 6 times the diameter of the perithecium, branching various, usually 3 or 4 times forked, with the tips truncate or bifid, not recurved, occasionally more ornate, with tip distinctly recurved. Asci 4–8, small and broad, about 40 x 55 μ. Sporidia 4–6, small.

On *Gaylussacia resinosa*, *Vaccinium Pennsylvanicum*, *V. Canadense*, *V. vacillans*, *V. corymbosum*, *Andromeda*, and *Epigæa repens*. Evidently not often collected. Massachusetts (Seymour) to Illinois: New Jersey (Arthur).

This is a variable species, not only in the character of the mycelium, but in the length and branching of the appendages. In most cases the tips are swollen and not at all recurved. It has sometimes been referred to *Erysiphe* because the appendages were not found to be forked, but careful search has revealed some forked tips in all the specimens at hand, including those named *Erysiphe* by others. Moreover, in all cases the appendages are stiff and somewhat straight, not having the floccose character of typical *Erysiphe* species. Peek reports that he has never seen what has been called *Erysiphe Vaccinii*, Schw., on *Epigæa repens* in fruit in New York, but a specimen collected in that State showed us perithecia with dichotomously forked appendages. As so many specimens on *Vaccinium* all prove to be true *Microsphæra*, that collected by Schweinitz in Pennsylvania evidently was the same thing, hence the synonymy as given above. *Microsphæra Vaccinii* is described as a new species in the XXIII Rep. N. Y. State Mus. p. 65, but though the MS. was completed in 1870, it was not printed until after the publication by Cooke & Peek in Journal of Botany.

M. elevàta, Burrill, Bull. Ill. St. Lab. Nat. Hist. Vol. I, No. 1, p. 58.

Mostly epiphyllous. Mycelium abundant, persistent, frequently covering the leaves for some time before the appearance of perithecia, which are usually few, though occasionally abundant, 100–120 μ, reticulations large, evident when young; apppendages 6–12, sometimes more, 3–4 times as long as the diameter of the perithecium, hyaline, slightly colored at base, smooth, 2–4 times dichotomous, branches short, not swollen, tips at first truncate, divergent, becoming acute and recurved. Asei 4–8, ovate, about 33 x 60 μ. Sporidia 4–6, mostly 4.

On *Catalpa speciosa* and *C. bignonioides*, not uncommon in Illinois, also collected in Missouri (Demetrio), and New Jersey (Ellis). Probably widely distributed.

This sometimes involves the foliage of an entire tree, giving it a gray color noticeable at some distance, and causing the leaves to fall prematurely.

M. Euphórbiæ, (Peck).

Erysiphe Euphorbiæ, Peck, 26th Rep. N. Y. State Mus. p. 80.
Microsphæra Euphorbiæ, B. & C. Grev. IV, p. 160,

Amphigenous. Mycelium abundant, persistent. Perithecia scattered, abundant, usually small, 80–100 μ, but often larger (120 μ), texture soft, elastic, reticulations 10–15 μ, frequently obscure; appendages 15–20, very long, 5–6 or more times the diameter of the perithecium, hyaline, often slightly tinted at base, irregularly flexuous and often nodularly swollen, at first simple, then part of them bifid or 3 or 4 times dichotomous, branches long, lax, tips sometimes bifid, but not swollen or recurved. Asei 4–8, frequently 6, pedicellate, 35–40 x 65 μ. Sporidia 4–6.

On *Euphorbia Preslii*, *E. marginata*, *E. corollata*.

This is a common species throughout the country east of the Rocky Mountains. Easily recognized by its very long, nearly colorless appendages. *Erysiphe Euphorbiæ*, Peek, is evidently this same thing. The name seems to have been founded upon specimens in which the appendages were not branched.

M. Alni, (DC.)

Erysiphe Alni, Betulæ, DC. Flore Franc. VI, p 104.
Alphitomorpha penicillata, Wallr Verhandl Naturf Freunde I, p. 40
Erysibe penicillata, Lk. Spec. Plant VI, I, p 113
Erysiphe Viburni, Duby, Bot. Gall. II, p. 872.
Erysiphe Ceanothi, Viburni, Syringæ, Schw. N A. Fungi, pp 269, 270.
Microsphæra Hedwigii, penicillata, Friesii, Lev. Ann. Sci. Nat. Ser. III, Tome XV.
Microsphæra Platani, Howe, Torr. Bull. V, p. 4.
Microsphæra Viburni, Howe, Torr. Bull. V, p. 43.
Microsphæra pulchra, C. & P. Erysiphei of U. S. in Journ of Bot 1872.
Microsphæra Nemopanthis, Peck, 38th Rep N. Y. State Mus. p. 102.
Microsphæra Alni, Winter, Die Pilze II. p. 38.

Amphigenous. Mycelium often delicate and evanescent, sometimes abundant and persistent. Perithecia usually small, 75–100 μ, sometimes larger, 100–130 μ, wall tissue compact, rather fragile, reticulations not large, 10–15 μ; appendages 6 or 8 to 15 or 20, hyaline, usually tinted at base, often somewhat roughened, usually about equaling, but varying from less than, to more than twice the diameter of the perithecium, 4–6 times dichotomous, branches varying in length and angle of divergence, but always regular and symmetrical, tips acute, distinctly, often strongly recurved. Asci varying with the size of the perithecium from 2 or 3 to 8 or more, usually 4 or 5, ovate when numerous, suborbicular when few. Sporidia 4–8, variable, mostly small, averaging about 20 μ long.

On *Ilex decidua, Nemopanthes fascicularis, Euonymus atropurpureus, Celastrus scandens, Ceanothus Americanus, Syringa vulgaris, Cornus stolonifera, C. sericea, C. alternifolia, Viburnum acerifolium, V. pubescens, V. dentatum, V. lentago, V. prunifolium. Lonicera sempervirens, L. Sullivantii, L. hirsuta, L. glauca, Andromeda ligustrina, Rhododendron nudiflorum, Forestiera acuminata, Ulmus Americana, Platanus occidentalis, Juglans cinerea, J. nigra, Carya alba, Betula lenta, B. lutea, B. pumila, Alnus incana, A. serrulata, Corylus Americana, C. rostrata, Ostrya Virginica, Carpinus Caroliniana, Castanea sativa* var. *Americana, Fagus ferruginea.*

The forms here included under *M. Alni* have been assigned by different authors to various species, distinguished, for the most part, by the number of the asci and sporidia. In all of these forms, the size of the perithecia, even when standing side by side on the same leaf, is quite variable, and, as a consequence, the number and shape of the asci they contain vary equally widely. Very small perithecia contain only a few (2–4) suborbicular asci, while larger ones contain a greater number, which, owing to lateral crowding, are narrower and longer. The sporidia are by no means constant in number, even in asci from the same perithecium. It is manifestly impossible to maintain specific distinctions based on such variable characteristics, and it becomes necessary, as in other genera of the family, to combine these rather widely varying forms. Aside from the number of asci and sporidia, the forms included here do not, however, present any very wide variations. In fact, the branching of the appendages, and the cellular structure of the wall of the perithecium, are strikingly alike in all of them. Specimens on *Juglans cinerea* and *J. nigra*, are sometimes very different from the type, having appendages less than the diameter of the perithecium. But on these same hosts other forms imperceptibly grade into the characteristic ones, leaving no room for specific distinction.

The form on *Syringa* is usually known as *M. Friesii*, Lév., that on *Viburnum* as *M. Viburni*, Howe, that on *Platanus* as *M. Platani*, Howe, that on *Nemopanthes* as *M. Nemopanthis*, Peck, and that on *Euonymus* as *M. Euonymi* (DC.) or *M. comata*, Lév. Others are mostly referred to *M. penicillata*, Lév.

In several American lists and collections we find *M. Lonicerœ*, (DC.) or its synonym *M. Dubyi*, Lév. In Europe the fungus on species of *Lonicera* is evidently quite distinct and is appropriately named; but American specimens on our species of this genus of host plants are certainly different. Whatever else we do with them, they must not be confounded with European species on allied hosts. Unfortunately, no opportunity has been available to examine collections on foreign cultivated species of *Lonicera*. The parasite on American honeysuckles, is *M. Alni*, as above described.

M. quércina, (Schw.)

Erysiphe quercinum, Schw. Syn. N. Am. No. 2492.
Microsphæra extensa, C. & P. Erysiphei of U. S. in Journ. of Bot. 1872.
Microsphæra abbreviata, Peck, 28th Rep. N. Y. State Mus. p. 64.
Microsphæra quercina, Burrill, Bull. Ill. State Lab. Nat. Hist. II, p. 324.

Epiphyllous, hypophyllous, or amphigenous. Mycelium abundant, rather thin and pruinose, forming orbicular patches or spreading over the whole surface of the leaf. Perithecia abundant, scattered, varying from 80–140 μ, reticulations evident, small and irregular; appendages less than 20, varying in length from less than, to 4 or 5 times the diameter of the perithecium, tinted at base, smooth or sometimes roughened, usually regularly 5–6 times dichotomous, branches usually short and tips strongly recurved, but presenting many curious and ornate variations caused by the more extended or unequal growth of some of the branches. Asci 3–8, often rupturing by slight pressure. Sporidia 4–8, variable, usually large, 20–30 μ long.

On *Quercus alba*, *Q. stellata*, *Q. macrocarpa*, *Q. lyrata*, *Q. bicolor*, *Q. Prinus*, *Q. rubra*, *Q. ilicifolia*, *Q. coccinea*, *Q. coccinea* var. *tinctoria*, *Q. palustris*, *Q. falcata*, *Q. Catesbœi*, *Q. aquatica*, *Q. nigra*, *Q. imbricaria*, *Q. robur*.

Common on oaks across the continent. A very variable species as here recognized. If it seemed possible it would be much more satisfactory to distribute it under several specific names, some of which are already in use. Often the form on a certain host species is sufficiently distinct when considered by itself to merit this treatment. For example the type on *Quercus rubra* known as *M. extensa*, C. & P., is everywhere on this host sufficiently alike to be easily recognized; but, upon examining specimens even from the same localities on allied oaks, it soon becomes impossible to find dividing bounds between this form and others which at first sight are very different, like that called *M. abbreviata*, Peck on *Quercus imbricaria*, etc. It must also be acknowledged that it is well nigh impossible to distinguish some forms referred to *M. Alni* from certain specimens placed under *M. quercina*, except by reference to the host plants. Usually however the latter differs from the former by longer, commonly more numerous appendages and larger sporidia. The perithecia also average larger.

M. calocladóphora, Atkinson, in Journ. Elisha Mitchell Soc. VII, p. 13.

Microsphæra densissima, (Schw.) Journ. Mycol. I, p. 101.

Hypophyllous. Mycelium thin and diffuse or in dense, orbicular spots. Perithecia black, at length depressed in the center, walls thick, reticulations rather distinct, 100–140 μ; appendages not numerous, about 7–12, subhyaline, 1–2 times diameter of the perithecium, 2–8 times dichotomously branched, or the axis continuous, bearing two or more sets of opposite branches and the lowermost of these sometimes showing the same axial elongation; tips strongly incurved. Asci 4–6. Sporidia granular and nucleate, 6–8, 20–25 μ in length.

On *Quercus aquatica*, *Q. laurifolia*, Florida (Martin), South Carolina and Alabama (Atkinson).

This species is especially distinguished by the peculiar branching of some of the appendages first described by Ellis and Martin (Journal of Mycology I, p. 101), but referred to *M. densissima*, (Schw.) It cannot now be positively ascertained to what Schweinitz applied this name, but his plant was collected at Bethlehem, Pa., upon oak leaves, species not given. This is a strong evidence against his specimen belonging to the present species. In Cooke and Peck's Erysiphei of the United States (1872), *M. densissima*, (Schw.) was identified with a specimen on *Quercus tinctoria* in New York. This was presumably correct but if so our present species is certainly distinct. as it differs conspicuously from the New York specimens. There can however be no doubt that the description in the Journal of Mycology was drawn from specimens belonging to this newly named species. (The specimen in Ell. N. A. F. 1238 is Atkinson's new species.)

M. erineóphila, Peck, Bull. Torr. Bot. Club. X, p. 75.

Mycelium thin. Perithecia 90–100 μ, fragile, dark, opake, reticulations obscure; appendages few, 8–12, dark colored, except the branches, scarcely equal to the diameter of the perithecium, 4-6 times regularly dichotomous, branches short and rather thick, tips recurved. Asci 5–8, oval or ovate, pedicellate, rather small, 35 x 55 μ. Sporidia uniformly 8, small.

On the "erineum" caused by a Phytoptus (mite), on the lower sides of leaves of *Fagus ferruginea.*

The erineum is usually very common wherever the tree grows, but the fungus seems to be rare. It has also been collected by Earle in Illinois and Indiana. See *Fungi Europæi,* No. 3245.

M. Astrágali, (DC.)

Peck reports *M. holosericea,* (Wallr.) Lév.—a synonym for *M. Astragali*—(25th Report N. Y. State Mus., p. 95), on *Astragalus Cooperi,* but this species does not seem to have been elsewhere collected in America. Is it possible that a poorly marked specimen of *M. diffusa* was thus identified?

M. Lýcii, (Lasch.)

Microsphæra Mougeotii, Lev.

This is inserted on the sole authority of the publication in Grevillea IV, p. 160, where the fungus is said to have been found on *Desmodium Dillenii.* Undoubtedly an incorrect determination.

FAMILY. PERISPORIÈÆ.

Mostly without any stroma but with a strongly developed, brown, persistent, conidia-bearing mycelium, which, however is sometimes inconspicuous or evanescent. Perithecia spherical or depressed, membranaceous or coriaceous, generally astomous, always superficial.

This Family is made up of rather heterogeneous material, the different members not being closely allied so as to form a well characterized natural group, as in the preceding family.

KEY TO THE GENERA.

	Mycelium present. - - - - - - -	1.
	Mycelium scanty or wanting. - - - - - -	2.
	Mycelium black, dense. - - - - -	3.
1.	Mycelium thin, arachnoid. - - - -	*Saccardia.*
	Mycelium yellow. - - - - -	*Eurotium.*
	Mycelium light colored. - - - -	*Myriococcum.*
2.	Perithecia scattered, subglobose. - -	*Perisporium.*

*Winter, in Die Pilze, includes here also *Aspergillus* and *Penicillium,* of which till recently, only the conidial stage was known. In these genera perithecia are wanting, the asci being inclosed in a tuber-like sclerotium which is developed from the mycelium.

Perithecia scattered, flattened. - - *Microthyrium.*
Perithecia stromatic. - - - - - *Lasiobotrys.*
3. Perithecia subglobose, appendiculate. - - *Meliola.*
Perithecia subglobose, not appendiculate. *Dimerosporium.*
Perithecia flattened. - - - - - - *Asterina.*
Perithecia elongated, subulate, etc. - - *Capnodium.*
Mycelium massive, coarse. - - - - - *Scorias.*

DIMEROSPÒRIUM, Fckl.

Symb. Myc. p. 89.

Perithecia superficial, subglobose, subastomous, membranaceo-carbonaceous. Mycelium copious, black, bearing conidia. Asci short, 8-spored. Sporidia didymous, hyaline or brown.

D. púlchrum, Sacc. F. Ven. II, p. 299.

Apiosporium pulchrum, Sacc. in Thum. M. U. No 52.
Exsicc. Rab. F. E 2149, 2684 —Thum. l. c.

Mycelium mostly epiphyllous, dark brown, often nearly covering the surface of the leaf. Conidia lateral on the branches of the mycelium, at first spherical with two septa at right angles to each other, and then 4-celled, at length dark brown, many-celled and more irregular in shape, 20–25 μ diam. With these are also subfalcate, 3-septate, hyaline conidia 35–40 x 8–9 μ, not constricted at the septa. Perithecia scattered among the mycelium, clear yellow-brown, 80–100 μ diam. Asci elliptical, 46 x 30 μ, 8-spored. Sporidia elongated-ovate, uniseptate, constricted in the middle, hyaline, 22 x 11 μ.

Only the conidial form has yet been found in this country, but this is quite common west and southwest, from Iowa to Texas, on leaves of *Cornus*, *Fraxinus* and other trees.

D. Collínsii, (Schw.)

Sphæria Collinsii, Schw. Syn. N. Am. 1512.
Dimerosporium Collinsii, Sacc. Syll. I, p. 54.
Sphæria papilionacea, B. & C. Grev. IV, p. 106.
Sphæria Russellii, B. & C. Grev. l. c. (sec. Cke. Grev. XV, p. 82).
Exsicc. Thum. M. U. 849.—N. A. F. 488.—Rab. F. E. 3541.

Mycelium brown-black, septate, crustaceous, hypophyllous, covering the whole surface. Perithecia black, globose, closely aggregated, 150–160 μ. Asci cylindric-clavate, 8-spored, 45–60 x 10 μ. Sporidia hyaline, oval, uniseptate, 2-seriate, 12–15 x 3–4 μ.

On leaves of *Amelanchier Canadensis*, and *A. alnifolia*, New York, Massachusetts, Prince Edward's Island, and Sierra Nevada Mts., California.

D. oreóphilum, Speg.* Mich. II, p. 160.

Perithecia gregarious or scattered, superficial, very minute, 110–120 μ, globose, clothed on all sides with short, dark hyphæ, scarcely equaling the diameter of the perithecia. Ostiola scarcely discernible. Texture thin, membranaceous, dark brown. Asci cylindric-clavate, rounded above, narrowed below into a very short, nodulose stipe, paraphysate, 8-spored, 60–65 x 20 μ. Sporidia ovoid, uniseptate and constricted in the middle, 15–16 x 7–8 μ, the upper cell larger, hyaline.

On living leaves of *Rhododendron Lapponicum*, Godhavn, etc., Greenland.

D. Ellísii, Sacc. Syll. 241.

Meliola maculosa, Ell. Bull. Torr. Bot. Club, VIII, p. 91.
Venturia maculosa, Ell. N. A. F. 200.

Spots black, suborbicular, 2–3 mm. diam. Perithecia black, globose, 90–115 μ, borne upon a brown, flexuous, remotely septate, mycelium, with a circle of straight, black setæ at the base; structure cellular, setæ 100 x 5 μ, apices entire. Asci cylindrical, 8-spored, 50–60 μ long. Sporidia ellipsoid, didymous, hyaline, mostly uniseriate, 10–12 x 4 μ.

On fallen leaves of *Andromeda racemosa*, Newfield, N. J.

D. melioloídes, (B. & C.)

Asterina melioloides, B. & C. Grev. IV, p. 10.
(*Meliola Baccharidis*, B. & Rav. Grev. IV, p. 158)?

Perithecia brown-black, globose, rugulose, astomous, epiphyllous, clustered, 80–95 μ, borne on brown, radiating hyphæ. Asci cylindric-clavate, 8-spored, 33–40 x 10–13 μ. Sporidia hyaline, subcymbiform, uniseptate, 1–2-seriate, 10 x 3 μ.

On leaves of *Baccharis halimifolia*, Florida (Martin).

D. orbiculàre, (B. & C.)

Asterina orbicularis, B. & C. Grev. IV, p. 9, Cuban Fungi, 784.
Exsicc. Ell. N. A. F. 1362.

Mycelium black, branching, remotely septate, adnate, forming orbicular, crustaceous patches, ½–1 cm. diam. Perithecia black, subglobose, 80–100 μ diam., amphigenous, but those on the upper surface are of a dull black color, and often sterile. Spreading from the apex to the circumference, are numerous moniliform threads of subglobose,

*This and all other Greenland species hereafter noted in this work were collected along the western coast of Greenland from 1812–1886, mostly by the various expeditions sent out to explore that country; viz. The Nares Arct. Exp., 2d Dutch North Polar Exp., Hammer's Exp., Danish Exp. 1876–79, Fylla Exp. 1884–86, etc., etc., and described by Prof. E. Rostrup in his "Fungi Groenlandiæ," published at Copenhagen in 1888.

dusky cells 10–12 μ diam. Asci ovate or obovate, 8-spored. Sporidia ovate-oblong, 33–35 x 15–18 μ, hyaline at first, then light brown.

On leaves of *Ilex coriacea* and *I. opaca*, Carolina to Florida.

D. erysipheoìdes, E. & E. Journ. Mycol. IV, p. 121. (Plate 8)

Exsicc. Ell. & Evrht N. A. F. 2d Ser. 2341.

Amphigenous. Perithecia scattered, astomous, globose, 100–130 μ diam., sparingly clothed with short (30–40 x 3 μ), erect, spreading hairs, and seated on a rather scanty mycelium of slender, brown, branching hyphæ. Asci oblong, subsessile, 35–40 x 12–14 μ, with filiform paraphyses. Sporidia biseriate, acutely elliptical, uniseptate and constricted, each cell with a large nucleus, 15–18 x 6–7 μ.

On leaves of *Cynodon dactylon*, Louisiana (Langlois).

With the ascigerous perithecia are many smaller ones (spermogonia) containing a few globose, brownish sporules 4–5 μ diam.

D. Langloìsii, E. & E. Journ. Mycol. II, p. 129.

Exsicc. Ell. & Evrht. N. A. F 2d Ser. 1786.

Perithecia gregarious, depressed-spherical, rough, black, subastomous, 112–120 μ diam., seated on a thin mycelium of brown, branching threads, forming small, dark colored patches thickly scattered over the upper surface of the leaf, and giving it a mottled appearance. Asci subsessile, oblong, often inequilateral or bulging on one side, 25–30 x 7–9 μ, without paraphyses. Sporidia biseriate, clavate-oblong, yellowish-brown, 4-nucleate, uniseptate and slightly constricted at the septum, 9–10½ x 4 μ. Some of the perithecia contain oblong-cylindrical, 2-nucleate, subhyaline, 7–8 x 2 μ stylospores. The perithecia have a radiate-cellular structure.

On living leaves of *Dianthera humilis*, Louisiana (Langlois).

D. nimbòsum, E & M. Journ. Mycol. II, p. 129.

Mycelium composed of prostrate, brown, branching, septate threads, with short, erect branches bearing oblong-clavate, 3–4-septate, brown conidia 35–40 x 6–8 μ, and longer (70–80 x 5–6 μ). erect, straight, septate, opake, sterile branches, the whole forming orbicular, velutinous, black patches ½–1 cm. across, mostly soon confluent, extending along and enveloping the stem for 5 cm. or more. The mycelium finally disappears, leaving a black, smooth, shining surface. Perithecia collected mostly in the center of the spots, erumpent, conical, black, carbonaceo-membranaceous, rough, about ⅓ mm. broad and high, sometimes imperfectly sulcate-striate around the prominent, mammose ostiolum. Asci subcylindrical, 70–80 x 10–14 μ, nearly sessile, surrounded with abundant filiform paraphyses and containing 8 biseriate, oblong-cylindrical, 16–20 x 5–6 μ sporidia yellowish and 2-nucleate at first, finally brown and uniseptate, and more or less constricted at the septum.

On living stems of *Smilax*, near Jacksonville, Florida (Calkins).

Mystrosporium aterrimum, B. & C., appears to be the conidial stage.

D. Spartinæ, E. & E. Journ. Mycol. II, p. 102.

The mycelium forms small (2–4 mm. long), oblong or elliptical, black patches consisting of a thick growth of erect, simple, septate, subnodulose, sterile hyphæ, nearly hyaline at first, but soon opake, 130–175 x 6–9 μ, and pale yellowish, prostrate hyphæ, producing fusoid-cylindrical or subfalcate, nucleate conidia 40–60 x 3 μ. Nestling among the sterile hyphæ are black, membranaceous, subovate perithecia ¼–⅓ mm. diam., with a very large opening above. Asci clavate-cylindrical, 75 x 15 μ, with imperfectly developed paraphyses. Sporidia 8 in an ascus, oblong-cylindrical, yellowish (nearly hyaline), slightly curved, 3–4-nucleate, becoming uniseptate, 18–20 x 4–5 μ. Some sporidia were seen imperfectly 3-septate, but one septum seems to be the normal state. This is remarkable for the large apical opening more like a half-grown *Cenangium* than like an ostiolum.

On dead sheaths of *Spartina polystachya*, Louisiana (Langlois).

D. capnoides, (Ell.)

Asterina capnoides, Ell. Am. Nat. 1883, p. 318.

Mycelium brown-black, branching, septate, epiphyllous. Conidia brown, ovate, uniseptate, 10–12 x 6 μ. Macroconidia brown, pedicellate, submuriform, 3-septate, 35 x 15 μ. Perithecia brown-black, subglobose, 50–100 μ diam. Asci oblong-ovate, sessile, 8-spored, 35–40 x 11–16 μ. Sporidia hyaline, subcymbiform, uniseptate, 2-seriate, 10–17 x 4–6 μ.

On living leaves of *Asclepias Cornuti*, Kansas.

D. conglobàtum, (B. & C.)

Asterina conglobata, B. & C. Grev. IV, p. 9.

Perithecia hypophyllous, globose, about 50 μ, astomous, coarsely cellular, membranaceous, in compact, subconfluent groups of 4–6 on a subcrustaceous mycelium. Asci obovate, 20–25 x 10 μ, without paraphyses. Sporidia oblong, uniseptate, hyaline, 6–8 x 2–2½ μ.

Description made out from specimens sent from Maine by Rev. Jos. Blake, on leaves of *Arbutus Uva-Ursi*.

D. xylógenum, E. & E. Journ. Mycol. II, p. 102.

Perithecia superficial, scattered, depressed-hemispherical, rough, 165–250 μ diam., with an obscure, papilliform ostiolum. Asci obovate, contracted below into a short stipe, 35–40 x 20–24 μ, without paraphyses, and containing 8 oblong-elliptical, uniseptate, granular, subhyaline, 15–16 x 8 μ sporidia.

On decaying wood of *Salix*, Louisiana (Langlois).

Differs from the usual type of *Asterina* and *Dimerosporium* in its habitat, and the absence of any definite mycelium.

D. anómalum, (Cke. & Hark.)

Asterina anomala, Cke. & Hark. Grev. IX, p. 87

"Effused, black, velvety. Perithecia hemispherical or globose-depressed. Mycelium intricate, brown, with erect, rigid, scattered setæ. Asci clavate. Sporidia biseriate, lanceolate, 1–5-septate, hyaline, 20–22 x 4 μ. Perithecia 80 μ diam.; setæ about twice as long."

On living laurel leaves, California.

Dimerospòrium Pópuli, E. & E., mentioned in Journ. Mycol. V, p. 81, was not published on account of the specc. proving unsatisfactory.

ASTÉRINA, Lév.

Ann Sci. Nat. 1845, III, p 59.

Perithecia globose-depressed or lenticular, membranaceous, subastomous, seated on spots of black, radiating, subsuperficial (rarely subinnate) mycelium. Asci typically short and thick, mostly 8-spored. Sporidia two-celled, pluriseptate or continuous, hyaline or brown.

The perithecia vary from an entire membranaceous sac to a mere covering of coalesced, radiating hyphæ, and the mycelium, which typically forms black spots, is often light colored, scant, evanescent or entirely wanting. The genus therefore contains some species nearly approaching *Sphærella* and *Microthyrium* on the one hand, and *Ascomycetella* on the other.

A. *Perithecia seated on a distinctly developed mycelium.*

A. rubícola, E. & E. (in Herb.)

Perithecia epiphyllous, erumpent-superficial, single or 2–3 together, conic-hemispherical, membranaceous, 75–100 μ diam., with a distinct ostiolum. Asci oblong, sessile, 35–45 x 12–15 μ, with obscure paraphyses. Sporidia biseriate, ovate-elliptical, yellowish-brown, uniseptate, constricted at the septum, rounded at the ends, 12–15 x 6–8 μ.

On leaves of *Rubus strigosus* and *R. occidentalis*, Canada and Wisconsin.

A. alièna, Ell. & Galw. (in Herb.)

Mycelium crustaceous, forming subelliptical patches, $\frac{1}{2}$–1 x $\frac{1}{4}$–$\frac{1}{2}$ cm., on dead areas of the leaf which are separated from the living part by a narrow, reddish border. Perithecia amphigenous, lenticular, sub-confluent, 150–200 μ diam., of cellular-fibrous structure, with a reticulate-fimbriate margin. Asci at first ovate, finally subelongated, 50–60 x 20 μ, short-stipitate, 8-spored. Sporidia inordinate, obovate, granular, hyaline, (becoming uniseptate)? 13–15 x 4–5 μ.

On leaves of pineapple (cult.), Washington, D. C.

A. (Asterélla) Chamænérii, Rostr. Fungi Grönl. p. 545 (No. 105).

Spots black, amphigenous or caulicolous, 3–4 mm. across. Mycelium subsuperficial, pseudoparenchymatic, margin radiate. Perithecia globose-depressed, subastomous, lying in the central part of the mycelium. Asci ovate-cylindrical, 50–60 x 12–14 μ. Sporidia oblong-clavate, hyaline, unequally uniseptate, the upper cell three times larger than the lower, nucleate, 16–20 x 6–7 μ.

On partly living stems and leaves of *Chamœnerium latifolium*, Sukkertoppen, etc., Greenland.

A. pìcea, B. & C. Linn. Journ. X, p. 374.

Perithecia epiphyllous, hemispherical, collapsing, about 150 μ diam., seated 4–8 together on small (1–1$\frac{1}{2}$ mm.), orbicular patches of black, crustose mycelium. Asci oblong, 40–50 x 10–12 μ, without paraphyses. Sporidia biseriate, brown, clavate-oblong, uniseptate and constricted, 10–12 x 3–4 μ.

On dead leaves of *Magnolia*, Louisiana (Langlois).

The patches of mycelium are at first covered by the cuticle of the leaf, which gives them a shining appearance, but when the cuticle is thrown off the mycelium is of a dead black color.

A. nùda, Pk. 38th Rep. N. Y. State Mus. p. 102.

Mycelium brown, branching, scanty. Perithecia black, at first subglobose, afterwards depressed, thickly clustered near the midrib, mostly hypophyllous, 75–100 μ diam. Asci oblong or subcylindrical, 8-spored, 35–40 x 12 μ. Sporidia oblong, hyaline, uniseptate, biseriate, 10–12 x 3–4 μ.

On dead leaves of *Abies balsamea*, Adirondack Mts., N. Y.

A. delitéscens, E. & M. Am. Nat. 17, p. 1284. **(Plate 6)**

Exsicc. Ell. N. A. F. No. 1291.

Mycelium thin, black, epiphyllous, forming small (2–4 mm.), orbicular patches composed of much branched, closely appressed hyphæ, on which are seated the flattened, crowded perithecia 75–100 μ diam, of radiate-cellular structure. Asci obovoid or subglobose, 8-spored, 30–35 x 18–22 μ. Sporidia subhyaline, ovate-oblong, uniseptate, 15–18 x 6–7 μ.

On living leaves of *Persea palustris*, Florida (Martin).

A. pelliculòsa, Berk. Ant. Voyage, Crypt. p. 137.

Exsicc. Rav. F. Am. 75.

Mycelium epiphyllous, radiate and branching in a dendroid manner, forming black, suborbicular patches ½–1 cm. diam., often more or less confluent. Perithecia seated on thickened portions of the main branches of the mycelium, depressed-hemispherical, imperfectly collapsed above. Asci subglobose, 50–60 μ diam., 8-spored. Sporidia oblong-elliptical, uniseptate and constricted, hyaline, 25 x 19 μ (16–20 μ long, Sacc.) becoming dark brown, about the same as in *Dimerosporium orbiculare*.

On leaves of *Ilex coriacea*, Florida (Ravenel).

A. paupércula, E. & E. Journ. Mycol. IV, p. 121.

Epiphyllous. Perithecia scutellate, brownish-black, 90–120 μ diam., seated on and surrounded by a thin network of brown, branching mycelium. Asci subelliptical, 22–25 x 12–15 μ, contracted below into a short, stipe-like base, 8-spored. Sporidia crowded, fusoid, hyaline, uniseptate, 12–15 x 2 μ.

On living leaves of *Jacquinia armillaris*, Florida (Curtis, com. A. Commons).

B. *Mycelium a marginal fringe around the base of the perithecia.*

A. discoidea, E. & M. Am. Nat. 18, p. 1148.

Perithecia hypophyllous, orbicular, slightly depressed in the center, olivaceous, thin, 500–800 μ diam., with an indistinct, reticulated margin. Asci obovate or globose, 30–40 x 30–35 μ. Sporidia crowded, clavate-oblong, uniseptate, 12–16 x 4–5 μ.

On living leaves of *Quercus laurifolia* and of *Olea Americana*, Florida (Martin).

This may not be distinct from *A. oleina*, Cke., but as that species was described and distributed in an immature state, it is now impossible to decide.

A. cupréssina, Cke. Grev. VI, p. 17.

Venturia cupressina, Rehm Asc. 494.

Exsicc. Ell. N. A. F. 160, 500.—Thum. M. U. 1543.

Mycelium nearly obsolete. Perithecia dark brown, hemispherical or lenticular, adnate, perforated above, 185–200 μ diam., with a short, scanty fringe of mycelium around the base. Asci cylindric-clavate, 60–75 x 10–12 μ, short-stipitate, 8-spored. Sporidia biseriate, uniseptate, upper cell broadest, often with a small nucleus in each cell, sometimes oblique or uniseriate, pale yellowish-brown, 12–15 x 6–7 μ.

On dead foliage of *Cupressus thyoides*, Newfield, N. J.

Sometimes there are a few scattered bristles on the perithecia, and some of them are filled with a mass of minute, oblong, hyaline spermatia, while others contain brown, elliptical stylospores (*Sphæropsis*) 20 x 10 μ.

A. Ílicis, Ell. Am. Nat. 17, p. 319.

Exsicc. Ell. N. A. F. No. 1357.

Perithecia brown-black, hypophyllous, scattered, adnate, at first hemispherical, then flattened and depressed, 100–120 μ diam., with a circular opening, forming a disc of brown, interlacing hyphæ, covering the nucleus and forming a narrow margin beyond. Asci obovate, 8-spored, 22–30 x 9–15 μ. Sporidia biseriate, subhyaline, oblong, uniseptate, 11 x 4 μ.

On living leaves of *Ilex glabra*, Newfield, N. J.

A. intricàta, E. & M. Journ. Mycol. I, p. 136.

Mycelium pale, scanty, evanescent. Perithecia brown, flat, orbicular, soft, very thin, hypophyllous, 500 μ diam. Asci globose, stipitate, 18 x 15–18 μ. Sporidia hyaline, obovate or ovate, uniseptate, 7–12 x 2–3 μ.

On living leaves of *Quercus arenaria*, Florida (Martin).

A. patelloìdes, E. & M. Journ. Mycol. I, p. 136.

Asterina erysipheoides, E. & M. Ell. N. A. F. No. 1358.

Perithecia dark brown, soft, orbicular, flattened, depressed in the center, hypophyllous, 275–300 μ diam., with a narrow border of scanty, radiating, pale mycelium. Asci ovate or ovate-oblong, 8-spored, 36 x 15 μ. Sporidia obovate, 2-seriate, uniseptate, hyaline, 12–15 x 4½–5 μ. Like the preceding species, closely allied to *Ascomycetella*.

On living leaves of *Quercus laurifolia*, Florida (Martin).

A. cárnea, E. & M. Am. Nat. 17, p. 1284.

Exsicc Ell. N. A. F. No. 1290.

Mycelium thin, brown, hypophyllous, adnate, mostly near the margin of the leaf or in orbicular spots about 5 mm. diam. Perithecia flesh-colored, flattened, soft, crowded, 60–100 μ diam. Asci obovate, sessile, 8-spored, 30–40 x 22–35 μ. Sporidia subhyaline, ovate, 2-celled, uniseptate, 16–17 x 7–8 μ.

On *Persea palustris*, Florida (Martin).

A. Celástri, E. & K. Journ. Mycol. I, p. 3.

Perithecia hypophyllous, in groups or scattered, convex, orbicular, black, 250 μ diam., with brown, radiating threads of mycelium around the base. Asci oblong-ovate, 12–15 x 6–7 μ, filled with granular matter. The parts of the leaf occupied by the groups of perithecia are a little darker than the surrounding portions.

On living leaves of *Celastrus scandens*, Kansas.

The specc. examined were immature.

A. comàta, B. & Rav. Grev. IV, p. 10.

Exsicc. Rav. F. Am. 73—Ell. & Evrht. N. A. F. 2d Ser. 2339.—Roum. F. G 5036.

Perithecia epiphyllous, superficial, scattered, large ($\frac{1}{2}$–$\frac{3}{4}$ mm. diam.), depressed-hemispherical, clothed with black, spreading, subfasciculate. bristle-like hairs which are deflexed above, leaving the variously ruptured ostiolum bare. Asci oblong, 70–75 x 20 μ. Sporidia biseriate. clavate-oblong, hyaline at first, then brown, about 20 x 5–7 μ, uniseptate.

On leaves of various species of *Magnolia* in the Southern States. Often sterile.

A. pinástri, Sacc. & Ell. Mich. II, p. 567.

Parodiella rigida, E & E Journ. Mycol. IV, p 62.
Exsicc. Ell. N. A. F. 789.

Perithecia gregarious, superficial, depressed-spherical, roughish, without any prominent ostiolum, 100–120 μ diam., with a sparing, brown, creeping mycelium around the base. Asci oblong, sessile, 60–70 x 12–15 μ, very evanescent. Paraphyses (?). Sporidia subbiseriate, ovate-oblong, brown, uniseptate and deeply constricted (the two cells sometimes separating), 15–20 x 7–9 μ. Spermogonia similar, with hyaline, oblong sporules 15–18 x 7–8 μ, with a large central nucleus.

On dead leaves of *Pinus rigida*, still attached to limbs cut off about eighteen months ago, Newfield, N. J., April 26, 1888.

A. tenèlla, Cke. Grev. XIII, p. 67.

"Epiphyllous, effused, thin, black. Perithecia minute, 220–300 μ diam., applanate, mingled with brown, creeping mycelium. Asci saccate, 4–8-spored. Sporidia 28–30 x 12–14 μ, in the 8-spored asci, 40 x 22 μ in the 4-spored, light brown."

On *Persea Caroliniensis*, Carolina.

A. Xerophýlli, Ell. Am. Nat. 17, p. 319.

Mycelium scanty. Perithecia entirely superficial, orbicular or subelongated, slightly depressed, 167 μ diam. Asci obovate, 35 x 15 μ, contracted into a thick, stipe-like base. Sporidia hyaline, fusiform or clavate-fusiform, faintly 3-septate, 18–20 x 3–3½ μ.

On fading leaves of *Xerophyllum asphodeloides*, New Jersey.

A. lepidigenoìdes, E. & E. Journ. Mycol. IV, p. 121.

Mycelium obsolete. Perithecia hypophyllous, scattered, attached to the scales on the leaf, small, pierced above, scutellate. Asci oblong, sessile, 60 x 12 μ, mostly broader below. Sporidia biseriate, fusoid, 3-septate, hyaline, 12–14 x 2–2½ μ.

On living leaves of *Capparis Jamaicensis*, Key West, Florida (Coll. A. H. Curtis, Com. A. Commons).

Closely allied to *A. lepidigena*, E. & E., but differs in its smaller perithecia, longer asci, and 3-septate sporidia.

A. decólorans, B. & C. Grev. IV, p. 9.

"Spots orbicular, red, undulate, bullate. Mycelium scanty, consisting of a few moniliform threads and others entire. Perithecia punctiform. Asci short, oblong. Sporidia uniseptate, 10 μ long."

On an unknown leaf, New Jersey.

A. cuticulòsa, Cke. Grev. VII, p. 49.

Exsicc. Rav. F. Am, No. 328.

Perithecia brown, orbicular, applanate, adnate, hypophyllous, clustered near the margin of the leaf, 500–800 μ diam., structure cellular-membranaceous. Mycelium scanty. Asci globose, 25 μ diam. Sporidia elliptical, ends obtuse, uniseptate, subconstricted, hyaline, 10 x 5 μ. A somewhat abnormal species.

On leaves of *Ilex opaca*, Georgia (Ravenel).

A. plantáginis, Ell. Bull. Torr. Bot. Club, 9, p. 74.

Exsicc. Ell. N. A. F. No. 790.

Spots brownish, immarginate. Perithecia brown-black, subglobose, membranaceous, innate, clustered in the spots, mostly epiphyllous, 70–80 μ. Asci ovate, 26–33 x 13–16 μ. Sporidia hyaline, oblong, obtuse, uniseptate, slightly constricted in the middle, or 2-nucleate, 9–10 x 3–5 μ.

On living leaves of *Plantago major*, Philadelphia, Pa., and Newfield, N. J. Approaches *Sphærella*.

A. ramulàris, Ell. Bull. Torr. Bot. Club, 9, p. 20.

Exsicc. Ell. N. A. F. No. 720.

Mycelium pale, subhyaline, very scanty. Perithecia flattened, orbicular, clustered, frequently coalescing, subinnate, 250–300 μ, consisting of dark brown, moniliform hyphæ covering the nucleus, obsolete beneath, mostly sterile. Asci oblong, spore-bearing portion 50 x 25 μ.

Sporidia pale, crowded, elliptical, coarsely granular, about 15 x 10 μ, with 1–2 large vacuoles at first.

On dead twigs of *Lindera Benzoin*, West Chester, Pa. (Everhart).

A. minor, E. & E. Journ. Mycol. II, p. 42.

Perithecia applanate, superficial, orbicular, ¼ mm. diam., black, mycelium obsolete. Asci obovate, 25–30 x 12–15 μ. Sporidia crowded, oblong-obovate, uniseptate, yellowish, 10–13 x 3–3½ μ (becoming brown?). Differs from *A. ramularis*, Ell., in its smaller perithecia and sporidia.

On dead twigs, Texas (Ravenel).

A. inquinans, E. & E. Journ. Mycol. III, p. 41.

Exsicc. Ell. & Evrht. N. A. F. 2d Ser. 1785.

Perithecia scutelliform, black, umbonate, of radiate-cellular structure, the marginal cells subelongated and slightly enlarged at their extremities. Asci ovate or obovate, 35–40 x 18–22 μ, contracted at the base into a short stipe. Sporidia irregularly crowded, ovate-elliptical or oblong-elliptical, yellowish and faintly uniseptate (?), 10–12 x 5–7 μ. The perithecia are thickly scattered over both surfaces of the leaf, and look much like the masses of exuded spores of some *Pestalozzia* or *Melanconium*.

On dead leaves of *Sabal Palmetto*, Louisiana (Langlois).

A. lepidígena, E. & M. Am. Nat. 18, p. 1148.

Exsicc. Ell. N. A. F. No. 1361.

Mycelium hyaline, scanty, hypophyllous. Perithecia black, subglobose, at length flat, very thin and fragile, 200–300 μ diam. Asci ovate, 8-spored, 30 x 15 or 42 x 12 μ. Sporidia obovate, uniseptate, 12 x 4 μ.

Attached to the epidermal scales on old, living leaves of *Andromeda ferruginea*, Florida (Martin).

A. pustulàta, E. & M. Am. Nat. 18, p. 1148.

Exsicc. Ell. & Evrht. N. A. F. 2d Ser. 1543.

Perithecia brown, soft, flattened, hypophyllous, adnate, 200–500 μ diam., consisting of a membranaceous disk of brown, branching, coalesced hyphæ, covering the nucleus and forming a narrow border beyond. Asci subglobose, 8-spored, 50–60 μ diam. Sporidia hyaline, obovate, uniseptate, 30–40 x 10–12 μ.

On living leaves of *Quercus laurifolia*, Florida.

A. stomatóphora, E. & M. Journ. Mycol. I, p. 98.

Perithecia lenticular, scattered, small, 170–185 μ diam., with a thin, reticulated margin, and indistinctly perforated in the center; texture cellular. Asci 30–35 x 6–8 μ, oblong, rather broader below and abruptly contracted into a short, stipitate base. Paraphyses none. Sporidia biseriate, oblong, uniseptate, rather narrower and more acute at the lower end, 7–12 x 2½–3 μ, hyaline. When the perithecium is removed from the leaf, a piece of the epidermis often adheres to its lower surface, so that under the microscope the stomata are visible through the thin edge of the perithecium, appearing as if they actually formed a part of it. It is to be noted that in this and most of the other species with flattened perithecia, the wall of the perithecium is nearly obsolete below, so that the perithecium is in fact hardly more than a shield-like disk covering the asci.

On living leaves of *Quercus laurifolia*, Florida (Martin).

A. subcyànea, E. & M. Am. Nat. 18, p. 1148.

Exsicc. Ell. N. A. F. No. 1360.

Perithecia hypophyllous, convex, depressed, ostiolate, obsolete beneath, 250–300 μ diam., subglobose, composed of dark greenish-blue cells 5–7 μ diam., which cover the nucleus and extend beyond in a thin, membranaceous border closely adnate to the leaf. Ostiolum papilliform, collapsing, with a broad, circular opening when dry. Asci slightly narrower at each end, sessile, 8-spored, 75 x 15 μ. Sporidia hyaline, oblong-clavate, uniseptate, 2-seriate, 20 x 4–7 μ.

On living leaves of *Quercus laurifolia*, Florida (Martin).

A. Gaulthèriæ, Curtis, (in Herb. Curtis).

Exsicc. Ell. N. A. F. No. 1358.

Perithecia brown-black, flattened, slightly elevated in the center, hypophyllous, scattered, 170–250 μ diam., surrounded by a narrow border of brown, branching mycelium. Asci ovate, 22–25 x 13–16 μ. Sporidia hyaline, obovate, uniseptate, the upper cell larger, 9 x 3 μ.

On living leaves of *Gaultheria procumbens*, common.

A. Pearsòni, E. & M. Journ. Mycol. I, p. 92.

Perithecia minute (100 μ), flat, superficial, obscurely perforated above, of close, cellular structure, with a scanty, subradiating mycelium around the margin. Asci oblong, obtuse, sessile, 40 x 15 μ, without paraphyses. Sporidia biseriate, clavate-oblong, granular, becoming uniseptate and slightly constricted at the septa, 15–20 x 3½–4½ μ, acute below, obtuse above, hyaline.

On blackberry canes, Vineland, N. J. (Pearson).

A. Bignòniæ, E. & E. Proc. Acad. Nat. Sci. Phil. July, 1890.

Perithecia hypophyllous, thin membranaceous, 115 μ diam., sub-hemispherical, becoming slightly depressed above, thickly scattered over the surface of the leaf. Ostiolum papilliform. Asci obovate, 20 x 15 μ, or elongated to 30 x 15 μ, without paraphyses. Sporidia 8, clavate-oblong, uniseptate, hyaline, scarcely constricted, 10–12 x 3–4 μ.

On leaves of *Bignonia capreolata*, Louisiana (Langlois).

A. purpùrea, E. & M. Journ. Mycol. II, p. 128.

Perithecia hypophyllous, convex-scutellate, scattered or gregarious, often collected along the midrib towards the base of the leaf, subastomous, of radiate-cellular structure, 130–150 μ diam., margined with a narrow fringe of pale purplish-black hyphæ closely appressed to the surface of the leaf, which is of a reddish-purple tint for a little distance around. Asci obovate, 30–35 x 18–22 μ, 8-spored. Sporidia crowded, ovate-oblong or oblong-elliptical, 12–16 x 5–6 μ, hyaline, with the endochrome three times divided, and often one of the cells with an imperfect longitudinal division, thus varying from the usual type.

On leaves of *Olea Americana*, Florida (Calkins).

A. clavulígera, Cke. Grev. VI, p. 142.

Exsicc. Rav. F. Am. 76.

Mycelium dark brown. Conidia elliptical, brown, white-banded. ("albo-fasciatis"), septate, constricted in the middle, pedicels elongated, hyaline. Perithecia scutellate. Asci subglobose, apiculate at the base. Sporidia elliptical (immature). Dr. Martin, in Journ. Mycol. I, p. 145, states that the sporidia are oval or obovate, uniseptate, hyaline, 21 x 6 μ. We have seen no specimen except that in Rav. F. Am. and that is without perithecia.

C. *Species not well-known.*

A. olèina, Cke. Grev. XI, p. 38.

Exsicc. Rav. F. Am. No. 757.

Perithecia hypophyllous, scattered, flattened, discoid with a narrow margin of brown, radiating hyphæ. Asci clavate, 24–30 x 9–10 μ. "Sporidia hyaline, small, uniseptate, (immature). Pycnidia similar but smaller, stylospores minute, oval, hyaline, 5 μ long." (Cke.)

On leaves of *Olea Americana*, Georgia.

A. nigérrima, Ell. Bull. Torr. Bot. Club, VIII, p. 91.

The specimens of this species are poor and unsatisfactory, and its habitat (old decaying stems of *Erigeron*) indicates that it can hardly belong here.

A. diplodioìdes, B. & C. Grev. IV, p. 9.

"Spots orbicular, mycelium interrupted. Perithecia minute. Sporidia oblong, obtuse, uniseptate, light brown, 8 μ long."

On leaves of *Andromeda acuminata*, Alabama.

A. spúrca, B. & C. Grev. IV, p. 9.

"Perithecia scattered, dot-like, surrounded by short, articulated, submoniliform, radiating threads, which are joined together laterally in twos, sometimes forked at the apex."

On leaves and stems of *Hyptis radiata*, Carolina and Alabama.

A. Wrìghtii, B. & C. Grev. IV, p. 10.

"Mycelium very thin. Perithecia brown, granular, crowded, like little grains of gunpowder, surrounded by cirrhate threads. Asci clavate, short," Texas (C. Wright.)

"Apparently on some smooth Cucurbit."

A. congregàta, B. & C. N. Pac. Expl. Exp. p. 129, No. 169.

Perithecia very small, shining, gregarious. Mycelium (subiculum) very scanty, mouth round. Asci yellowish ("helvolus"). Sporidia narrow.

On leaves, Nicaragua.

With the habit of *Sphærella maculiformis.*

A. bullàta, B. & C. N. Pac. Expl. Exp. p. 129, No. 170.

Spots orbicular, on projecting portions of the matrix, hyphæ interwoven into a compact but thin, and here and there cellulose stratum. Perithecia prominent, scutellate, margin fimbriate.

On leaves, Nicaragua.

With the habit of *Strigula.*

A. ostiolàta, B. & C. N. Pac. Expl. Exp. p. 129, No. 166.

Punctiform, black. Ostiolum distinct, papilliform. Mycelium reduced to a mere border around the perithecia ("subiculo nullo nisi peritheciali.")

On the upper surface of leaves with *Cephaleurus virescens*, Nicaragua.

MICROTHÝRIUM, Desm.

Ann. Sci. Nat. XV, p. 137.

Perithecia superficial, flat or plano-convex, membranaceous, perforated in the center, margin subfimbriate. Asci mostly obovate, 8-spored, without paraphyses. Sporidia oblong or fusoid, continuous or uniseptate, hyaline.

M. Smilacis, De Not. Mier. Dec. IV, p. 22.

Myiocopron Smilacis, Sacc. Syll. II, p. 660.

Exsicc. Rabh. Herb. Myc. 654.—Rehm. Ascom. 447.—Thum. M. U. 1448.—Ell. N. A. F. 600.

Perithecia superficial, flat, orbicular, ½–¾ mm. diam., often confluent, forming a continuous, rough, black crust. Asci oblong or ovate, 50–60 x 18–20 μ (10 x 6 μ, Sacc.) Sporidia biseriate, ovate-elliptical, 1-celled, hyaline, 12–18 x 8–10 μ.

On dead stems of *Smilax*, common.

M. microscópicum, Desm. Ann. Sci. Nat. Ser. II, XV, p. 138.

Sacc. Fungi Italici, tab. 562.

Exsicc. Rab. F. E. 967, 1963, 2943

Perithecia flattened, brown, about 150 μ diam., pierced in the center, margin subfimbriate. Asci clavate-oblong, sessile, 8-spored, 25–30 x 7–9 μ. Sporidia biseriate or irregularly crowded, elongated-ovate, subinequilateral, hyaline, and when mature, with a septum near the lower end, 8–10 x 3–3½ μ.

On dead dry leaves of *Azalea* and *Myrica*, New Jersey. Probably common in other sections and on other leaves.

M. Pinastri, Fckl. and *M. litigiosum*, Sacc. should also be found in this country as their spermogonial stage is often met with.

MELÌOLA, Fr.

Elench Fung. II, p. 109.

Perithecia globose, astomous, membranaceous, surrounded by stout, bristle-like, simple or branched appendages, and seated on orbicular patches of radiately-branched, or subcrustaceous mycelium. Asci mostly short and thick, 2–8-spored; paraphyses none. Sporidia oblong, 2–5-septate, dark, exceptionally hyaline, continuous or fenestrate.

A. *Sporidia 4-septate.*

M. amphitricha, Fr. Elench. Fung. II, p. 109.

See also Grev. XI, p. 37.—Mont. Cuba, p. 326.—Revue Mycol. 1888, p. 134.—Journ. Mycol. I, p. 146.

Syn. *Sphæria amphitricha*, Fr. S. M. II, p. 513.

Meliola tenuis, B. & C. (sec. Cooke, Grev. VII, p. 49).

Rav. F. Am. 83.—Ell. N. A. F. 1296.

Spots black, crustaceous, orbicular, often confluent, amphigenous, but mostly hypophyllous, mycelium brown-black, remotely septate, radiating, with short, pyriform, uniseptate, alternate branches (hyphopodia). Perithecia black, globose, then depressed and at length collapsing, rugulose, 200–300 μ diam., surrounded by black, opake, rigid, erect, entire setæ 300–500 x 9–12 μ. Asci oval, 2-spored, evanescent. Sporidia oblong, dusky, 3-septate, constricted at the septa, 50–60 x 16–24 μ.

On *Persea Olea Americana*, *Magnolia*, &c., Carolina to Louisiana.

M. furcàta, Lev. Ann. Sci. Nat. 1846, p. 266. **(Plate 5)**

Bornet Org. des Mel. tab. 22, fig. 14.

Perithecia black, wrinkled, globose, 200–250 μ diam., surrounded by numerous, erect, dark brown, continuous or obscurely septate appendages 10–12 μ thick, divided at the tip into two recurved spreading branches 30–40 μ long, and each with 2–3 divergent lobes 10–15 μ long. Mycelium forming round, black patches, 2–6 mm. in diam., scattered or confluent, composed of brown, radiating, septate hyphæ 8–10 μ thick, with short, alternate, two-celled branches (hyphopodia), the lower cell short and cylindrical, the upper globose or ovoid and longer. Asci 2–4-spored. Sporidia brown, cylindrical, 4-septate, slightly constricted at the septa, rounded at the ends, 35–45 x 14–16 μ.

The foregoing description is from specimens in the Herb. of the U. S. Dept. of Ag. Washington, D. C., collected by Wright in Nicaragua, on some coriaceous leaf. N. A. F. 1297 (*a* & *b*) are *M. bidentata*, Cke., and 1297 (*c*) is *M. palmicola*, Winter.

M. Cookeàna, Speg. F. Arg. Pug. IV, p. 42.

Meliola amphitricha, Fr. Ravenel F. Am. No. 84, and Ellis N. A. F. No. 1295.

Mycelium amphigenous, broadly and irregularly effused, subcrnstaceous, black, easily separating when mature, hyphæ thick, branching, intricate, remotely septate. Hyphopodia short-pyriform, uniseptate; hyphæ few, rigid, often only in a circle around the perithecia, 120–250 x 8–10 μ, apices entire. Perithecia scattered or aggregated, globose, black, carbonaceous, bald, not collapsing, scaly, granular, 150–200 μ diam. Asci elliptical, short and thick, stipitate, 2-spored, often immature. Sporidia at first hyaline, then brown, oblong, 4-septate, constricted at the septa, 30–40 x 10–12 μ.

On living leaves of *Callicarpa Americana*, Florida.

M. bidentàta, Cke. Grev. XI, p. 37. **(Plate 5)**

Exsicc. Rav. F. Am. 330.—Ell. N. A. F. 1297 *a* & *b*.—Rabh. Winter Fungi, 3546.

Perithecia globose, (150 μ). Appendages few, erect, 2–4-cleft at the tips. Asci saccate. Sporidia oblong-cylindrical, 3-septate, slightly constricted at the septa, rounded at the ends, brown, 45–50 x 16–18 μ, usually two in an ascus.

On leaves of *Bignonia capreolata*, Carolina to Texas.

M. palmícola, Winter, Rabh-Winter, Fungi Eur. 3547. **(Plate 5)**

Exsicc. Ell. N. A. F. 1297 *c*.—Rav. F. Am. 81.—Rabh-Winter F. E. 2846, 3547

Mycelium forming irregularly rounded spots 1–15 mm. broad,

amphigenous, black, at length broadly confluent, formed of loosely branched, interwoven, creeping hyphæ 10–11 μ thick. Hyphopodia scattered, capitate, generally entire or a little crenate, often curved, with a short, cylindrical stipe. Bristles numerous, erect, straight, tips di-trichotomously divided, the branches again sometimes bifid, or more rarely with 3–4 short branches near the apex.

Perithecia gregarious, globose, tardily collapsing, rough, black, 175–240 μ diam. Asci evanescent. Sporidia cylindrical, rounded at the ends, 4-septate, constricted at the septa, dark, the middle cell mostly swollen, 52–62 x 19–23 μ.

On leaves of *Sabal serrulata*, Florida to Louisiana and Texas.

B. *Sporidia 3–5-septate.*

M. cryptocárpa, E. & M. Am. Nat. 17, p. 1284.

Exsicc Ell N. A. F. No. 1293.

Spots tomentose, mostly epiphyllous, suborbicular, 2–4 mm. diam., numerous and often confluent. Mycelium pale brown, creeping, septate, irregularly branched, bearing numerous, oblong-fusiform conidia, which are pale brown, 3–4-septate, 30–40 x 5–9 μ, obtuse or acute above, and contracted below into a short, hyaline stipe. Erect bristles abundant, simple, multiseptate, black, tips entire and paler. Perithecia black, subglobose, not abundant and often sterile, 180–200 μ diam., collapsing, surrounded at base with a few diverging, brown, septate appendages, which, like the bristles, are more or less crisped or undulate above. Asci oblong-ovate, 8-spored. Sporidia brown, oblong or oblong-clavate, 2-seriate, 3–5-septate, 30–50 x 10–12 μ.

On leaves of *Gordonia lasianthus*, Florida (Martin).

M. mánca, E. & M. Am. Nat. 17, p. 1284.

Meliola sanguinea, E. & E. Journ. Myc. II, p. 42

Exsicc. Ell. N. A. F. 1292.

Perithecia membranaceous, $\frac{1}{2}$–$\frac{1}{3}$ mm. diam., subastomous, smooth, or at least without any bristle-like appendages, mostly epiphyllous, and either solitary or several in a small, rather compact group, on orbicular, subindeterminate patches of black, branching mycelium with alternate, obovate hyphopodia, much the same as those of *M. amphitricha*, Fr. Asci oblong-obovate, 2-spored. Sporidia oblong-cylindrical, slightly curved, 3-septate, 38–44 x 12–15 μ, obtuse, hyaline, becoming brown.

On living leaves of *Myrica cerifera*, Florida, and on leaves of *Rubus trivialis* (*M. sanguinea*), in Louisiana.

M. Mitchéllæ, Cke. Grev. VI, p. 143.

Exsicc Rav F. Am. 88.—Ell. N. A. F. 1294.

Spots black, thin, mostly epiphyllous, often covering the entire surface, mycelium dark brown, branching, intricate. Hyphopodia short, ovate, alternate, uniseptate. Erect hyphæ, simple, dark brown, setaceous, apices entire, 250 x 6 μ. Conidia light brown, obovate or clavate, 3-septate, 27–30 x 4 μ, borne on erect, light brown, subhyaline hyphæ. Perithecia black, globose, smooth, 100–125 μ diam. Asci cylindric-clavate, 39 x 9 μ. Sporidia oblong-elliptical, brown, 4-septate, 35 x 15 μ.

On leaves of *Mitchella repens*, Florida.

M. nidulans, (Schw.)

Sphæria nidulans, Schw. Syn. Car. 185, Fr. S. M. II, p. 443.
Chætosphæria nidulans, Rehm Ascom. 287.
Meliola nidulans, Cke. Grev. XI, p. 37.

Exsicc Ell N. A. F. 192.—Rabh-Winter F. Eur. 3544.—Rav. Fung. Car. I, 50.

Perithecia gregarious, globose, rough, collapsing, black, seated on a subiculum of black, erect, simple, rigid, bristle-like hairs which form at first small, orbicular patches, soon confluent and often surrounding the limb and extending continuously for 2–4 inches. Asci elliptic-oblong, 2–4-spored, 70–80 x 25–30 μ. Paraphyses stout, branching. Sporidia oblong, obtuse, 3-septate, (4-septate, Cke.), slightly constricted, 35–50 x 14–16 μ, brown.

On living branches of *Vaccinium corymbosum*, New Jersey to Florida.

M. Heterómeles, (Cke. & Hark.)

Meliolopsis Heteromeles, Cke. & Hark. Grev. XIII, p. 21.
Meliola Heteromeles, Berl. & Vogl. Sacc Syll 6243.

Effused, black. Mycelium subcrustaceous, moniliform, branched, interwoven, mixed with *Capnodium*. Perithecia globose, 200 μ diam., membranaceous, free. Asci clavate, 8-spored. Sporidia lanceolate, 3–5-septate, hyaline, 40 x 8 μ.

On leaves of *Heteromeles*, mixed with *Capnodium Heteromeles*, California.

C. *Sporidia muriform, yellowish or brown; mycelium scanty or obsolete.*

M. fenestràta, C. & E. Grev. V, p. 95, Journ. Mycol. I, p. 147.

Perithecia subgregarious, subglobose, black, smooth or clothed with a few spreading hairs, about 150 μ diam., with a fringe of brown, spreading, septate hyphæ and a circle of short, spreading, mostly incurved, black, bristle-like, 40–60 x 3 μ appendages (entire at the tip) around the base. Asci evanescent and not seen. Sporidia rather acutely elliptical, brown, 7–8-septate and muriform, 20–27 x 12–14 μ (30–40 x 12 μ, Martin).

On scales of pine cones, New Jersey.

CAPNÒDIUM, Mont.

Ann. Sci. Nat. Ser. III, XI, p. 233.

Mycelium effused, persistent, forming a black, felt-like coating on living leaves and limbs. Perithecia vertically elongated, sometimes branched, generally fimbriate-lacerate at the vertex. Asci obovate or elongated, 8-spored. Sporidia 4- or more-celled, often with both transverse and longitudinal septa, yellow or yellow-brown.

The members of this genus are but impefectly known; many of the so-called species being only mycelium without any ascigerous perithecia, and often without even pycnidial perithecia—sometimes even without conidia.

A. *Asci present; sporidia muriform.*

C. salícinum, (A. & S.) **(Plate 10)**

Dematium salicinum, Alb. & Schw. Consp. p. 368.
Fumago vagans, Pers Myc. Eur. I, p 9
Cladosporium fumago, Lk. in Linn. Spec. Plant.
Torula fumago, Chevall. Flor. Paris, I, p. 34.
Capnodium salicinum, Mont. l c.
Capnodium Sphæroideum, De Lacr. in Rab. F. E. 352
C. Citri, B & Desm. Molds. p. 11.

Exsicc Rab. F. E. 352.—Thuemen F. Austr. 169, 435 —Thum. M. U. 1146.—Cke. F. Brit. 2d Ser. 291.—Linh. Fungi Hung 358.—Myc. March. 2231.—Vize Micr. Fungi, 100.

Mycelium widely effused, covering the whole surface of the leaves and young twigs and even the trunk and branches to some extent. Perithecia fleshy, greenish-black, simple or branched, sessile or contracted below, thickened above and opening with a subfimbriate or lobed orifice (ostiolum). Asci obovate, sessile, 40–60 x 20–25 μ, 6–8-spored. Sporidia crowded, obovate, obtuse at the ends, 3–4-septate and often with a longitudinal septum, subconstricted at the septa, dark brown, 22–26 x 9–13 μ.

In the specimens figured on pl. 10, the sporidia were elliptical, hyaline, 3-septate, one of the cells sometimes divided by a longitudinal septum, 12–15 x 4–5 μ, crowded in asci about 60 x 10 μ, with abundant paraphyses. Perithecia subovate, with a short, conical ostiolum.

On *Populus tremuloides* (limbs) and on *Negundo aceroides*, Montana (Kelsey).

Winter, in Die Pilze, gives as probable forms of this species, *C. elongatum*, Berk. & Desm., *C. expansum*, Berk. & Desm., and *C. Persoonii*, Berk. & Desm., at least as they are represented in

Rab. F. Eur. 663. 665 and 677.—Thüm. F. Austr. 486, 992, 1076, 1166, and Thüm. M. U. 664.

C. grandísporum, E. & M. (in Herb.) **(Plate 7)**

Mycelium smoky-black, thin, composed of creeping, branching, multiseptate, nucleate threads. Pycnidial perithecia brown-black, setaceous, apices enlarged and subentire, 900 x 35 μ, containing hyaline, uniseptate, oval or elliptical sporules 10–12 x 2 μ. Ascigerous perithecia brown-black, globose, at length depressed-sublentiform, (120 μ), with 10–12 straight, brown appendages, 100–120 x 4 μ. Asci subclavate, 8-spored, 75–120 x 15–20 μ. Sporidia hyaline, clavate-cylindrical or fusoid-oblong, 4–5-nucleate and more or less constricted between the nuclei, at length distinctly 4–5-septate, 30–45 x 6–7 μ, lying in 3 or 4 series in the asci.

On leaves of *Gelsemium sempervirens* and *Myrica*, Florida (Martin). Whether the subulate, pycnidial perithecia belong to this is uncertain.

B. *Asci unknown; sporidia muriform.*

C. elongàtum, Berk. & Desm. Molds referred to *Fumago*, p. 251, fig. 5 (1849).

Exsicc. Rav. F. Am. 80.—Ell. N. A. F. 1544.

Setose. Perithecia elongated, acuminate, generally simple, fimbriate at the apex. Sporidia 2–4-septate, finally constricted at the septa and longitudinally divided.

On leaves of *Persica, Smilax, Populus, Liriodendron, Pyrus, Bignonia, Tussilago*, etc.

C. Rhámni, Cke. & Hark. Grev. XIII, p. 21.

Maculiform, black. Hyphæ creeping, generally moniliform, branching, forming a thin stratum. Perithecia erect, cylindrical, attenuated above, 300 x 40 μ, simple, mouth fimbriate. Sporidia elliptical, triseptate, brown, 18–20 x 8 μ, one cell divided by a longitudinal septum. Conidia free, dark, uniseptate, 12 x 6 μ.

On leaves of *Rhamnus*, California. (Not *C. rhamnicolum*, Rab.)

C. Heterómeles, Cke. & Hark. Grev. XIII, p. 21.

Effused, black, subvelutinous. Hyphæ densely interwoven, branching, septate and moniliform. Perithecia ventricose-cylindrical, erect, rather slender, simple, 200 x 20 μ. Sporidia uncertain, apparently 3-septate, muriform, brown, 18 x 9 μ.

On leaves of *Heteromeles*, California.

C. pelliculòsum, B. & Rav. Grev. IV, p. 156, Journ. Myc. I, p. 98.
Exsicc. Rav. F. Am 79 —Thum. M. U. 2059.—Ell. N. A. F. 1544.

Mycelium epiphyllous, forming a thin, sooty-colored layer on the surface of the leaves and consisting of closely septate, brown, subrectangularly branched and interwoven threads, 5–8 μ thick, with each cell or joint nucleate, and bearing when well developed, stellately 3–4 parted conidia, much like those of *Triposporium*, nearly hyaline at first, becoming brown, each arm 4–5-septate and nucleate, 7–9 μ thick at the base and 50–75 μ long, tapering to an obtuse point at the apex. Pycnidial perithecia growing like thick branches from the sides of the prostrate threads, membranaceous, of rather coarse, cellular, structure, oblong or flask-shaped, 75–208 x 30–50 μ, apex subobtuse and subfimbriate, discharging countless, minute, hyaline, oblong sporules 3–4 x 1 μ. Sometimes these perithecia are quite globose and formed by the enlargement of one of the component cells of a thread or hypha. There are also produced from the mycelium, cylindrical, brown, multiseptate conidia 70–80 x 6–7 μ, like the conidia of *Helminthosporium*. Ascigerous, perithecia 100–150 μ diam., with brown, septate appendages like those of an *Erysiphe*, 15–25 in number, 75–100 μ long. Asci at first oblong, becoming ellipsoidal, and about 40 x 25 μ. Sporidia crowded, broad fusiform, hyaline, uniseptate at first, becoming 3-septate at maturity and 15–22 x 4–7 μ.

On leaves of *Prunus Chicasa*, South Carolina (Ravenel), and on leaves of *Magnolia glauca*, Florida (Martin). The above description is from the Florida specimens.

The description in Grevillea is very brief, viz. "Threads of the mycelium erect, trifid at the apex after the fashion of *Triposporium*, shorter than the oblong, constricted perithecia."

Whether the Carolina and Florida specimens are the same we can not positively state.

C. Schweinítzii, Berk. & Desm. l. c.

Velutinous. Flocci (hyphæ) subcylindrical. Perithecia subsimple, elongated, smooth. Sporidia obovate, cellulose-muriform, scarcely constricted at the septa.

On leaves of herbs, Pennsylvania.

C. *Asci unknown; sporidia 2–5-septate.*

C. Caroliniénse, Berk. & Desm. l. c p. 12, fig. 7.

Scattered, setose, mycelium scanty. Perithecia with elongated, lateral, flask-shaped branches. Sporidia oblong, 2–3-septate.

On the lower surface of leaves of *Quercus obtusiloba*, South Carolina (Curtis).

C. expánsum, Berk. & Desm. l. c. p. 12, fig. 8.

Exsicc. Thum. M. U 664.—Roum. Fungi Gall 3660.

Widely effused, velutinous. Perithecia conical, short, connate, often beset with short, moniliform filaments. Sporidia triseptate, oblong, curved.

On bark of *Acer saccharinum* var. *nigrum*, Ohio.

D. *Asci unknown; sporidia uniseptate.*

C. Fùligo, Berk. & Desm. l. c. p. 9, fig. 2.

Mycelium tolerably thick, compact, black, separable from the matrix. Perithecia conic-cylindrical, roughened with black, floccose tufts and ostioliform projections. Sporidia rather small, ovate-oblong, often spuriously uniseptate, subhyaline.

On leaves of *Uvaria triloba*, generally on the upper side, Ohio, and on leaves of various herbs, Pennsylvania.

C. Nèrii, Rabh. F. Eur. 662, Hedw. 1864, p. 73.

Mycelium moniliform, ramose-reticulate, much like that of *C. Citri*. Perithecia elongated, subconical. Sporidia oblong, uniseptate.

On leaves and branches of *Nerium Oleander*, Florida (Martin).

E. *Asci unknown; sporidia (spermatia)? continuous or unknown.*

C. tùba, Cke. & Hark. Grev. XIII, p. 21.

Effused, crustaceous, black. Hyphæ interwoven, branching, creeping, some septate, others moniliform, forming a thick, deciduous crust. Perithecia erect, numerous, cylindrical, 120 x 14 μ, divided, ciliate above and expanded in a subinfundibuliform manner, filled with ovate, continuous, hyaline sporules.

On leaves of *Umbellularia*, California.

C. axillàtum, Cke. Hedw. 1878, p. 40.

Exsicc. Rav. F. Am. 77.—Ell. N. A. F. 2111.

Black, velutinous, situated in the axils of the nerves. Perithecia elongated, bristle-like, attenuated above, filled with minute, oblong sporules. Mycelium moniliform.

On leaves of *Catalpa cordifolia*, California, New Jersey and Pennsylvania.

C. Cítri, Berk. & Desm. l. c. p. 11.

Exsicc. Rav. F. Am. 329.—Thum. M. U. 2243.—Fungi Gall. 1097

Scattered, setose, scarcely adhering. Perithecia elongated, rarely bifurcate. Mycelium branching, moniliform, beautifully reticulated. Spermatia minute, oblong, hyaline.

On orange leaves, Southern States and California.

Farlow, Bull. Bussey Inst. March, 1876, considers this as only imperfectly developed *Capnodium salicinum.* See also Tul. Sel. Carp. II, p. 283.

C. pomòrum, B. & C. Grev. IV, p. 157.

Mycelium obsolete. Perithecia smooth, variable, ovate-lanceolate, obovate or forked, pedicellate. Stipe cylindrical, black.

On decaying apples, South Carolina (Ravenel).

C. quèrcinum, Berk. & Desm. l. c. p. 11.

Exsicc. Thum M. U. 1451?—Rav. F. Am 78.—M. March. 698.—F. G. 5147

Compact and thick. Perithecia fasciculate, branched, the outer stratum separating transversely. Mycelium scanty, scarcely constricted.

On leaves of *Quercus obtusiloba*, South Carolina (Ravenel).

C. puccinioìdes, E. & E. Journ. Mycol. IV, p. 65.

Perithecia amphigenous, cylindrical, obtuse, 75–80 x 20 μ, hyaline, becoming opake, erumpent in minute black tufts, which are either scattered or collected in groups or patches 1–3 mm. across. In the central portion of these groups the leaf becomes dead, and a whitish, bare spot is formed, the margin of which is fringed with a black border of perithecia, the whole presenting the general appearance of an effused *Puccinia.* The perithecia on the specimens examined were entirely sterile.

On living leaves of *Frasera speciosa*, Pike's Peak, Colo., August, 1887 (Tracy).

Apiospòrium and Antennària, which include mostly only undeveloped forms, have been omitted.

SACCÁRDIA, Cke.

Grev. VII, p. 49.

Mycelium arachnoid, very delicate and evanescent. Perithecia globose. Asci globose or ovate, 8-spored. Sporidia oblong or elliptical, muriform, hyaline.

S. quércina, Cke. l. c.

Hypophyllous. Mycelium evanescent. Perithecia scattered or gregarious, globose, thin-walled, 80–100 μ diam., loosely attached by a few delicate, spreading, thread-like hyphæ at the base. Asci globose or ovate, about 30 μ diam. Sporidia oblong-elliptical, hyaline, muriform, 18–20 x 8–10 μ.

On leaves of *Quercus virens*, Georgia (Ravenel).

S. Martini, Ell. & Sacc. Mich. II, p. 574. **(Plate 9)**

Exsicc. Ell. N. A. F. 1289.

Perithecia seated on a mucedinous, evanescent subiculum of interwoven hyphæ, gregarious, superficial, globose-depressed, and finally umbilicate-collapsing, 100–150 μ, texture minutely angular-parenchymatic. Asci fasciculate, oblong-clavate, briefly stipitate, rounded above, 65–70 x 20–28 μ, without paraphyses. Sporidia 2–3 seriate, oblong-cylindrical, slightly curved, hyaline, 3–6-septate, and when fully developed, with two or more cells divided by a longitudinal septum, sometimes slightly constricted at the septa, 16–25 x 3–6 μ. Var. *major*, on leaves of *Magnolia glauca*, has the perithecia and sporidia larger, the latter 30–40 x 7–8 μ, and 6–9-septate.

On leaves of *Magnolia*, *Gelsemium sempervirens*, and *Quercus laurifolia*, Florida (Martin), and on leaves of *Ilex opaca*, Texas (Ravenel).

LASIOBÓTRYS. (Plate 8)

Kunze Mycol. Hefte, II, p 88.

Perithecia attached to the border of a small, tough-coriaceous, plano-convex or finally concave, black stroma, superficial, small, brownish, surrounded and nearly hidden by tufts of erect, brown hairs. Asci cylindrical, 8-spored. Sporidia oblong, hyaline, becoming uniseptate.

L. Lonicèræ, Kunze, l. c.

Xyloma Xylostei, DC. Syn. p. 99.
Xyloma Loniceræ, Fr. Obs. I, p. 198.
Sphæria ruboidea, Fr. Vet. Akad. Handl. 1817, p. 269
Lasiobotrys Xylostei, Lk. Handbuch, III, p. 389.
Dothidea Loniceræ, Fr. S. M. II, p. 557.
Lasiobotrys affinis, Hark. New Cal. Fungi, p. 22.

Exsicc. Fckl. F. Rhen. 1749.—Kunze F. Sel. 573.—Rab. Herb Mycol. 668.—Rab. F. E. 1434.—Rehm Asc. 132.—Thum. F. Austr. 1045.—id. M. U. 957.

Stromata flat, convex, becoming concave, small (250–300 μ diam.), black, shining, usually collected in compact, circular groups scattered irregularly over the surface of the leaf, or ranged in a circle, with a vacant space in the center. The margin of each stroma is fringed with copious, bristle-like, straight, septate, olive-brown hairs 60–130 x 3–4 μ, at the base of which nestle the globose (50 μ) perithecia astomous and of tough, cellular texture. Asci fasciculate, cylindric-clavate, 40–50 x 10–12 μ, without paraphyses. Sporidia ovate-oblong, 8–14 x 4–5 μ, hyaline, continuous at first, becoming yellow-brown and uniseptate. The small stromata, after the perithecia have fallen, collapse, and then much resemble a single cupulate perithecium.

On living leaves of *Lonicera hispidula*, Tamalpais, California (Harkness).

As the California fungus differs from the European only in the yellow-brown, uniseptate sporidia, it seems better to regard it as the mature state of *L. Loniceræ* instead of making it specifically distinct. In all the European specc. the sporidia are continuous and hyaline as represented in Pl. 8, figs. 8 and 9.

SCÒRIAS, Fr. (Plate 10)

Syst. Orb. Veg. p. 171, Syst. Mycol. III, p. 290.

Mycelium of greenish-black, much branched, rigid, septate hyphæ, enveloped and glued together by an abundant mucilaginous substance, and forming a loosely compacted, black, rough, spongy mass of considerable extent, more compact on the surface, and bearing an abundance of ovate or pyriform, coriaceous perithecia.

S. spongiòsa, (Schw.)

Botrytis (?) *spongiosa*, Schw. Syn. Car. No. 1311.
Scorias spongiosa, Fr. S. M. III, p 290.
Exsicc. Rav. Fung Car. I. 81.--Rav. F. Am. 334.—Ell N. A. F. 1363.—Rab. F. E 3052. Thum. M. U. 697.

Perithecia abundant, attached to the fibers of the mycelium, near their extremities, ovate or pyriform, obtuse or acuminate, 100–120 x 70–80 μ, mostly contracted below into a short, thick stipe, coriaceous, greenish-black, texture subradiate-cellular. Asci narrow obovate-clavate, thick-walled, 40–45 x 7 μ, 8-spored. Sporidia fusoid, yellowish-hyaline, 2-seriate, 3-septate, 12–15 x $2\frac{1}{2}$–3 μ.

On living beech limbs associated with some species of wooly aphis, or on the ground and on various other things, weeds, grass, &c., where the aphides or their droppings have fallen.

PERISPÒRIUM, Fr. (Plate 7)

Syst. Orb. Veg. I, p. 161, Syst. Myc. III, p. 248.

Perithecia scattered or gregarious, superficial, carbonaceo-coriaceous, opening at length irregularly or with a circular mouth, globose or depressed-globose, bare and black. Asci clavate, with a long, slender base. Paraphyses none. Sporidia oblong-cylindrical, 3-septate, dark brown.

P. vulgàre, Cda. Icones, II, fig. 97.

Exsicc. Thum. M. U. 1941.—Cke. F. Brit. 2d Ser. 289.—Vize. Micr. Fungi 83.

Gregarious, superficial, black and shining. Asci broad-clavate, with a short, nodose stipe, 35–40 x 17–19 μ. Sporidia 8, 3-septate, brown, the two middle cells oblong-cubical, the end cells subconical. 28 x 5 μ.

On wood, decaying paper and cloth, and various other decaying materials.

Saccardo in Sylloge I, p. 55, gives the following measurements from specimens on different hosts, viz. On bark of dead limbs, asci 35–40 x 17–19 μ, stipe 7–8 μ long, sporidia 28 x 5 μ; on decaying paper, asci 80–100 x 18 μ, stipe 20–30 μ long, spor. 26 x 7–8 μ; on sheep dung, asci either short-stipitate, 45 x 15–17 μ, stipe 10 μ long, spor. 25 x 5½–6 μ, or asci 170–190 x 20 μ, stipe 70–100 μ long, spor. 35 x 6½–7 μ.

The species seems to be common in Europe but, as far as we know, has not yet been found in this country.

The species figured on plate 7 (*P. funiculatum*, Preuss.) is from specimens in Kriegers Saxon Fungi 426, on straw of an old thatched bee-hive.

P. fimèti, B. & C. Grev. IV, p. 157.

Minute, scattered. Asci elliptical. Sporidia oblong or short-elliptical, numerous, brown.

On rabbits' dung, South Carolina.

"Sporidia brown, 4-celled, rod-shaped, pointed at the ends, not constricted at the septa, 12½–15 x 2½–3 μ." (Farlow in literis.)

P. Zèæ, B. & C. Grev. IV, p. 157.

Perithecia seriate, forming short, black lines, and surrounded by short "villose" hairs. Asci lanceolate. Sporidia short-fusiform, triseptate, slightly constricted at the septa.

On *Zea Mays*, South Carolina.

Dr. Farlow gives the following notes from spece. in Herb. Curtis:

"Asci slightly clavate, averaging 60 x 8–9 μ. Sporidia distichous, 4-celled, brown, acute at the ends, somewhat swollen in the middle, 14–16 x 5–6 μ." He adds that the specimens of "*Perisporium Zeæ*, Desm.," in Rav. Fungi Car. III, No. 65, show nothing like *Perisporium*, and do not resemble *P. Zeæ*, B. & C. at all.

P. Wrìghtii, B. & C. Grev. IV, p. 157.

Perithecia carbonacco-membranaceous, thickly gregarious, in orbicular patches 1 cm, or more in diam., rough, seated on a scanty, inconspicuous, colorless mycelium. Ostiola papilliform, smooth. Asci ovate or subglobose, 35 x 40 μ, or 40–50 x 30 μ, 8-spored and without paraphyses. Sporidia crowded, yellowish, oblong, obtuse, 3-septate, with a longitudinal septum running through one or more cells, 18–22 x 8–10 μ, slightly curved.

On stems of *Opuntia Macrorrhiza*, Texas, (Wright).

Described from specc. in the Curtis Collection (com. Farlow).

P. Caládii, (Schw.) Syn. Car. 316.

This is *Sclerotium Caladii*, Fr. S. M. II, p. 261.

EUROTIUM, Lk.

Spec Plant. VI, I, p. 79.

Perithecia superficial, globose, astomous, membranaceo-coriaceous. Asci subspherical or pear-shaped, scattered. Sporidia lens-shaped or spherical, subhyaline.

E. herbariòrum, (Wigg.) (**Plate 8**)

Mucor herbariorum, Wigg. Prim Flor. Hols. p. 111.
Eurotium epixylon, Schm & Kze. Deutschl Schwamme 83
Eurotium Aspergillus glaucus, de Bary Beitr. III, p. 19.
Exsicc. Rab. Herb Mycol 488 —Thum F. Austr 656, 848 —Ell N A F 383.

Mycelium superficial, loose, at first white, then red or reddish-yellow. Perithecia globose, sulphur yellow, 75–100 μ diam. Asci spherical or pear-shaped, 12–15 μ diam., 8-spored. Sporidia lens-shaped, with a groove around the edge and the margin on each side briefly radiate-striate, 8–10 μ diam.

On poorly preserved herbarium specimens &c., in damp places.

MYRIOCÓCCUM, Fr.

Systema Mycologicum II, p 304

Perithecia globose, membranaceous, subastomous, firm, gregarious. seated on an interwoven, mucedinous subiculum.

M. Everhártii, Sacc. & Ell. Syll. I, p. 760. (**Plate 9**)

Exsicc Ell N. A. F 766.

Perithecia densely gregarious, subglobose, obtuse, astomous. 150–200 μ diam. densely clothed with a very thin, pale cinereous. cottony coat. Texture of the perithecia areolate, thin, membranaceous, subrufous. Asci? Spores? conglutinate, globose or angular. and of various shapes, 8–12 μ diam., granular-nucleate, hyaline. Differs from *M. præcox*, Fr. principally in the cottony layer enveloping the perithecia.

On rotten wood, New Jersey—inside a hollow willow log, Pennsylvania (Everhart).

M. consimile, E. & E. Proc. of the Acad. of Arts and Sci. Phila.. July, 1890.

Perithecia gregarious, globose, 80–100 μ diam., carbonaceo-membranaceous, nearly black, collapsing, pierced with a small round opening above, the upper half finely radiate-striate, texture close. finely radiate-cellular, filled with olivaceous, oblong, 4–4½ x 1½ μ spores, without any evident asci or basidia. The perithecia are enveloped in a loose, glauco-cinereous mycelium of the same color and character as in the preceding species from which it differs in its smaller, subolivaceous spores and smaller, striate perithecia with an apical opening.

On the basswood bottom and elm hoops of a barrel standing in a cellar, Newfield, N. J., July, 1889.

It is uncertain whether the perithecia in this genus ever become ascigerous.

M. spársum, Hark. Bull. Cal. Acad. Sci. Feb., 1884, p. 42.

Perithecia scattered, yellowish-brown, 170–180 μ diam., surrounded by a scanty, white subiculum. Spores numerous, unequally elliptical, hyaline, apiculate at each end, with a large vacuole, 7–9 x 6 μ.

On dead trunks of *Acer macrophyllum*, between the bark and wood, Sunol, Cala.

SUBORDER. HYPOCREÀCEÆ.

Simple, or compound. Perithecia subcarnose, or coriaceo-membranaceous, never carbonaceous, bright-colored (red, yellow, blue etc.), opening by a subcentral ostiolum. Stroma, when present, soft, waxy-carnose, or occasionally cottony. Sacc. Syll. II, p. 447.

The *Hypocreaceæ* form a very extensive and natural group, all the different members composing it showing an unmistakable relationship easily recognized.

KEY TO THE GENERA.

A. *Perithecia sunk in the stroma.*

*	Stroma erect, sporidia filiform, hyaline. - - -	1.
**	Stroma effused, sporidia various, hyaline. - - -	3.
1.	Stroma capitate. sclerotigenous. - - -	*Claviceps.*
	" elavate, entomogenous. - - -	*Cordμceps.*
	" " herbicolous. - - -	*Acrospermum.*
2.	Stroma tubercular, sporidia filiform, fimicolous. -	*Selinia.*
	" " " " herbicolous.	*Oomyces.*
3.	Stroma suborbicular, flat. - - - - -	*Polystigma.*
	" byssoid. - - - - - -	*Byssonectria.*
	Parasitic on other fungi. - - - - -	*Hypomyces.*
	Growing on wood, bark etc. - - -	*Hμpocrea.*
	Sporidia filiform. - - - - - -	*Hypocrella.*
4.	Stroma amplexicaul, sporidia filiform. - -	*Epichloe.*
5.	Stroma valsoid. - - - - - -	*Thyronectria.*

B. *Perithecia superficial.*

***	Stroma none or scanty, sporidia hyaline, subelliptical. -	6.
6.	Perithecia subulate, sporidia appendiculate.	*Eleutheromyces.*
	Conidial stroma tubercular, sporidia uniseptate.	*Nectria.*
	" " capitate " "	*Sphærostilbe.*
	Sporidia 2-or more-septate. - - -	*Calonectria.*
	" muriform. - - - - -	*Pleonectria.*
	Asci polysporous. - - - - -	*Chilonectria.*
	Sporidia filiform, fimicolous. - - - -	*Barya.*
	" " xylogenous. . . .	*Ophionectria.*
7.	Perithecia blue or violet. - - - -	*Gibberella.*
8.	Perithecia beaked, sporidia brown. - - -	*Melanospora.*

CLÁVICEPS, Tul.

Ann Sci Nat II, 1853, Ser XX, p 43

Stroma erect, consisting of a sterile stem and subglobose, fertile head, growing from a subcylindrical, black, hard sclerotium. Perithecia sunk in the stroma, flask-shaped. Asci clavate-cylindrical, 8-spored. Sporidia filiform, colorless, continuous.

The development of this genus begins in early summer in the flowers of wheat, rye and various grasses, the mycelium penetrating the ovary of the affected flower and transforming it by degrees into a soft, smutty-white substance filled with numerous cavities whose walls are lined with sterigmata bearing numerous ovate, colorless stylospores. This soft amorphous body is gradually changed into the cylindrical, horn-shaped sclerotium (ergot) which falling from the head of grain and lying on or in the ground through the winter, puts forth in the following spring, the perfect ascigerous fungus.

C. purpùrea (Fr.)

Sphaeria purpurea, Fr S M II, p 325 (partly)
Claviceps purpurea, Tul. l. c., Winter Die Pilze, II. p. 146
Exsicc Rab. Herb. Myc. 431 —Thum. F Austr 555, 875, 975.—Schweiz. Krypt. 630, 631. Krgr. Fung. Sàx 73, 369, 489, 490.

Sclerotium 1½–2½ cm. long, cylindrical-horn-shaped, mostly a little curved, wrinkled, purplish-black outside, white within. Stromata usually several (seldom only one) from the same sclerotium. Heads sphæroid, tuberculose from the prominent perithecia, borne on short flexuous stems. Asci narrow, linear, 8 spored. Sporidia filiform, continuous, attenuated toward each end, hyaline, 50–76 μ long. *Fusarium heterosporum*, Nees. and *Oidium abortifaciens*, B. and Br. are considered to be the conidia of this species.

In heads of rye and various other species of the order Gramineæ.

C. microcéphala, (Wallr.) **(Plate 15)**

Kentrosporium microcephalum, Wallr Beitrage II, tab 3
Claviceps microcephala, Winter Die Pilze II, p 147.
Exsicc Rab. Herb Myc 430 —id F. E 2667.—Thum F Austr 1087 —id. M U 699, 970, 998, 1798 —Linh Fung Hung 69.—Erikss F Scand 93 —Krgr F. Sax 126, 127, 163, 164, 368 —Sydow M March. 996, 1548, 2599.

Differs from *C. purpurea*, principally in its smaller size (stem filiform, 10–16 mm. long), head globose, rufous, (½ mm.), otherwise like the preceding species. Grows from the ergot of *Phragmites communis*, which, however, also produces the first-mentioned species. Both these may be raised by cultivation of their sclerotia, which may be lightly covered with earth, kept properly moistened in a flower pot.

CÓRDYCEPS, Fries.

Syst. Mycol. II, p 324.

Stroma erect, clavate (mostly), simple or branched, sterile below, bearing the perithecia (and often conidia) above. Perithecia buried or more or less nearly superficial. Asci cylindrical, with 8 filiform, septate sporidia which finally separate at the septa.

The species of Cordyceps grow for the most part from dead insects, but some also from fungi (species of *Elaphomyces*). They abound in the tropical regions, but are found also in the temperate zones. Their conidial stage is represented by the various species of *Isaria*.

C. entomorrhìza (Dicks.)

Sphaeria entomorrhiza, Dicks. Pl. Cr Brit I, p. 22.
Kentrosporium granulatum, Wallr. Beitr II, p. 166.
Cordiceps entomorrhizus, Lk. Hndbk III, p. 347.
Torrubia entomorrhiza, Tul. Sel. Carp I, p. 61.
Cordyceps Menesteridis, Mull. & Berk. Gard. Chron. 1878.
Exsicc Rab. F. E 1218?

Carnose. Head subglobose, golden yellow, becoming darker. Stipe subcompressed, 2 inches long, and over. Asci cylindrical. Sporidia filiform, hyaline, breaking up into cylindrical joints, or sections, 7–8 μ long.

Growing from larvæ of insects. Carolina (Ravenel).

C. armeniàca. B. & C. Journ Linn. Soc. I, p. 159, tab. 1, fig. 1.

Apricot-colored, stipe flexuous, rather short, 8 mm. long. Head subglobose, rather pale, roughened by the perithecia. Asci elongated, subinflated at the apex. Sporida linear, immature.

On dung of birds, probably from the remains of insects eaten, Carolina (Ravenel).

C. palústris, Berk. & Br. Linn. Journ. l. c. fig. 5.

Carnose-suberose, dark, dirty flesh-colored, stipe cylindrical, bifid, or trifid above, 25–50 mm. long, including the clavate, subcylindrical head which is roughened by the projecting ostiola. Sporidia filiform, separating into small ($1\frac{1}{2}$ μ,) globose joints.

On dead larvæ in damp ground, Carolina (Ravenel).

C. stylóphora, Berk. & Br. Linn. Journ. l. c. fig. 3.

Yellow. Stipe slender, 12–18 mm. long, $\frac{1}{2}$ mm. thick. Head much elongated, with the surface nearly smooth. Perithecia immersed.

On dead larvæ, Carolina (Ravenel).

The specimen in Ravenel's Fungi Car. Exsicc. V, No. 49, has the slender stem a little over 2 cm. long, the ascigerous part occupying a medial position, cylindrical, and slightly enlarged, about 8 mm. long by 1 mm. thick, with a sterile, slender beak, about $\frac{1}{2}$ cm. long, being a prolongation of the stipe, but the specimen is apparently immature, being without asci or sporidia.

C. clavulàta, (Schw.) **(Plate 15)**

Sphæria clavulata, Schw Syn N Am 1155
Torrubia pistillariæformis, B & Br. Brit Fung, 969?
The Syn. "*Torrubia cinerea*, Ell" in Sacc. Syll. rests on some error
Exsicc Thum M. U 1258

From specimens collected by Prof. Peck, and distributed in de Thuemen's Mycotheca Universalis, No. 1258, we have drawn the following description. Stroma simple, clavate, about $3 \times \frac{1}{4}$ mm., consisting of a light-cinereous stipe, surmounted by a black, ovate, or elliptical head, about 1 mm. high and $\frac{1}{2}$ mm. thick, roughened by the rounded, prominent perithecia, which are of coarse cellular structure, and only imperfectly perforated above. Asci subsessile, broadest in the middle, contracted above, and rounded at the apex, 80–95 x 8–10 μ. Sporidia filiform, multiseptate, 40–70 x $1\frac{1}{2}$–2 μ, joints 3–5 μ long.

On dead scale insects (Lecanium), on living branches of *Fraxinus* and *Prinos*, N. Y. (Peck). On branches of *Clethra*, Newfield, N. J., on *Carpinus*, Canada (Dearness).

In Sacc. Sylloge II, p. 568, the species represented by the above specimens is made a synonym of *C. pistillariæformis*, B. & Br., but if the two species are the same, the name of Schweinitz has priority, and it is quite certain that the specimens in M. U. 1258, are the genuine *C. clavulata* Schw.

C. Langloisii, E. & E. (in Herb.)

Stromata solitary, simple, capitate, about 3 mm. high. Stipe about 2 mm. high and 1 mm. thick, round or subcompressed, head depressed-globose, soft and spongy, about 2 mm. across, white at first, soon becoming reddish-purple. The perithecia which occupy the upper convex surface of the head are of a tough membranaceous texture, ovate-conical, 250–300 μ high and 100–150 μ broad at the widest part, nearly half the upper part projecting and more deeply colored. Asci fusoid-linear, 180–200 x 2–2½ μ, attenuated towards each end. Sporidia filiform, interwoven, nearly as long as the asci. less than ½ μ thick.

On dead larvæ of the "mason wasp," near St. Martinsville, La. (Langlois 2295.)

C. militàris (Linn.)

Clavaria militaris, Linn Sp Pl. Ed. III, tom II, p. 1652.
Clavaria granulosa, Bull. Champ 1, p. 199
Sphæria militaris, Ehr. Beitr. z. Naturkde. tom. III, p 86.
Cordiceps militaris, Lk Hndbk III, p 347.
Kentrosporium militare & *K clavatum*, Wallr. Beitr pp 166 & 167.
Torrubia militaris, Tul. Sel Carp III, p 6.
Cordyceps militaris, Sacc Syll 5031
Exsicc Plowr. Sph. Brit. 501 —Sydow M. March 654.—Rab. F. E 3548
Roum. F. Gall. 3157.

Stromata solitary, or sometimes several, issuing usually from the head, but sometimes from the articulations of the pupa, orange-colored, 4–5 cm. high, including the elongated-clavate head, which is 1–1½ cm. long, and minutely tuberculose from the subconical, emergent, orange-red perithecia. Asci slender, 115–150 x 4–5 μ, containing eight slender, filiform, closely-jointed sporidia nearly as long as the asci, and breaking up into minute, hyaline, subelliptical segments 2–3 μ long. The conidial stage (*Isaria farinosa*, Fr.) is often met with and resembles a small white plume of about the same height as the ascigerous stroma and more or less branched above.

Growing from dead pupæ of moths buried just below the surface of the ground. Massachusetts (Farlow), Carolina (Ravenel), Pennsylvania (Everhart), New York (Peck), New Jersey (Ellis), California (Harkness), Wisconsin (Trelease) conidial stage.

C. Ravenélii, B. & C. Journ. Linn. Soc. 1, p. 159, tab. 1, fig. 4.

Exsicc Rav Fungi Car IV, no. 28

Stroma (stipe) elongated, flexuous, compressed and sulcate when dry, at first minutely tomentose, finally nearly glabrous, 5 inches or more high, (see Riley, in American Entomologist, 1880), including the elongated-cylindrical head, which is roughened by the superficial, black, subhemispherical, large (175–200 μ) perithecia. Asci linear-cylindrical, 150–200 x 7–9 μ, slightly narrowed above and rounded at the apex, containing 8 filiform sporidia nearly as long as the asci, about 2 μ thick, and breaking up into joints 3–5 μ long. The specimens in Rav. Fungi Car. Exsicc. IV, No. 28, are 8–10 cm. high, the yellowish-brown stem about 2 mm. thick, enlarged above, in that part occupied by the perithecia, to about 3 mm. thick; but the specimens are no doubt considerably smaller than when fresh.

Growing from dead larvæ of the "June beetle" (*Lachnosterna fusca*) and other larvæ (?) buried in the ground, Carolina (Ravenel), Iowa (Bessey), Pennsylvania (Everhart).

Descriptions and good drawings of this and the two preceding species are given in Journ. N. Y. Microscop. Soc. Vol. I, p. 91, et seq., by Rev. J. L. Zabriskie.

C. insignis, Cke. & Rav. Grev. XII, p. 38.

Livid-purple. Stipe straight, 3–4 cm. high, pale, sulcate, equal. Head subglobose or ovate, slightly roughened by the perithecia, which are minute, crowded, ovate; the punctiform ostiola a little darker. Asci cylindrical, very long (600 μ), erumpent. Somewhat resembles *C. Entomorrhiza*, but is larger and more robust. Stem about 4–5 mm. thick and longitudinally sulcate. Head 1½ cm. long and 1 cm. broad. Sporidia filiform (450 μ long), breaking up into segments 12 μ long.

On dead larvæ buried in the ground, Carolina (Ravenel).

C. herculea, (Schw.)

Sphæria herculea, Schw. Syn N. Am. 1153.

Head large (12 mm. thick), ovate-clavate, obtuse, decurrent on the attenuate-elongated stem, alutaceous (leather-color), yellow within, stipe also yellow. Perithecia rather small, concolorous. Height of the whole fungus, about 1½ inches (36 mm.)

On the ground, among fragments of decaying wood, Salem, N. C. (Schweinitz).

A fine specimen of this species has been sent from Ohio by Prof. A. P. Morgan. When fresh it was about three inches high and half an inch thick, growing from some dead larva of considerable size. The fertile head, which occupies about an inch of the upper part of the stem, leaving a short, rather obtuse, sterile tip, is of a light yellow color and roughened by the somewhat prominent, closely-packed perithecia, which are about 150 μ in diameter, with slightly prominent ostiola, of a pale, radiate-fibrous structure. Asci 200–225 x 6-7 μ, gradually attenuated to the base and containing eight filiform sporidia which separate into joints 6–8 x $\frac{3}{4}$–1 μ, with the ends slightly swollen.

The Ohio specimens are to all outward appearances identical with the specimen in Herb. Schw.

C. aciculàris, Rav. Linn. Journ. l. c. fig. 2.

Cordyceps Carolinensis, B & Rav. in Rav Fungi Car. Exsicc. IV, No. 29

Stipe slender, elongated, brown. Head cylindrical, with a long, acuminate, sterile apex. Perithecia superficial, free. Asci very long, flexuous. Sporidia linear, breaking up into truncate segments about 5 μ long.

On larvæ buried a little distance below the surface of the ground, Carolina (Ravenel).

We have copied the above description from Saccardo's Sylloge II, p. 574.

The specimens in Rav. Exsicc. have a filiform, flexuous stem, yellowish-brown below, cinereous and attenuated above, 8–10 cm. high, and (in our copy) entirely sterile.

C. Sphíngum, (Tul.) (Plate 15)

Torrubia Sphingum, Tul. Sel. Carp. III, p. 12.
Isaria Sphingum, Schw. Syn. Car. 1298 (conidia).
Cordyceps Sphingum, Sacc. Mich. I, p. 321, Cke. Syn. 46.

Stromata numerous, about thirty in the single specimen found, thread-like, about 5 cm. high and rather less than 1 mm. thick, cinereous, nearly smooth and glabrous, or slightly white farinose-tomentose, bulbous at the base and more or less undulate and bent, especially below and within the cocoon, which they seem to have penetrated with some difficulty. Perithecia superficial, cylindric-conical, 200–225 μ high, 125–150 μ thick, rounded above, chestnut color. Ostiolum not prominent. Asci linear-lanceolate, 150–200 x 6–7 μ, when young, with a depressed, conical tip about 4 μ wide. Sporidia filiform, nucleate, about as long as the asci and about 2 μ wide, probably finally separating into joints or segments. The larva from which the fungus grows is about 3 cm. long and $\frac{1}{2}$ cm. thick, and the stipes or stromata arise from all the segments of the body. Some of the stromata were sparingly branched above.

On a dead larva in its cocoon, attached to a rotten limb lying on the ground in the swamp, Newfield, N. J.

In Tulasne's figure the fungus is represented as growing from the perfect insect, and the perithecia are said to be of a pale red color ("pallide rubentia"), but the Newfield specimen does not seem to us specifically distinct. This species is also reported from Massachusetts by Dr. Farlow.

C. superficialis, (Pk.)

Torrubia superficialis, Pk. 28th Rep. N. Y. State Mus. p. 70.
Cordyceps superficialis, Sacc. Syll, 5036, Cke. Syn. 36.

"Slender, about one inch high, smooth, brown, the sterile apex gradually tapering to a point. Perithecia crowded, superficial, subglobose, blackish-brown, sometimes collapsed, with a small, papilliform ostiolum. Asci cylindrical. Sporidia long, slender, filiform. Related to and intermediate between *C. Ravenelii* and *C. acicularis*. The stem of the plant is about equal in length to the club, or perithecia-bearing part. The perithecia are more loosely placed at the extremities of the club; thereby giving it a subfusiform shape. The sporidia are more slender than those of *C. acicularis*, but the plant itself is less elongated and slender." We have seen no specimens, and copy the above from the report cited.

Under hemlock trees, on buried larvæ, Northville, N. Y. (Peck).

C. sobolifera, (Berk.)

Sphæria sobolifera, Berk. Hook. Lond. Journ. Bot. II, p 207.
Cordyceps sobolifera, Sacc Syll. 5021, Cke. Syn. 23.

Head 5–8 mm. long, ovate-oblong or tongue-shaped, dotted with the minute ostiola of the buried perithecia, and a little thicker than the stipe, which is stout, round, rigid, erect, simple or with rudimentary branches, 15–20 mm. or more in height. Asci cylindrical. Sporidia separating into linear joints about eight times as long as broad.

On the larvæ of some lamellicorn insects, in the West Indies, and in Mexico.

Some of the Mexican specc. were much branched; instead of the fertile head, presenting a contracted panicle, or brush-like tuft of deformed branches. The anterior portion of the affected larvæ is enveloped in a white mold.

B. *Mycogenous.*

C. ophioglossoides, (Ehr.)

Sphæria ophioglossoides, Ehr. Beitrag, III, p. 88.
Clavaria parasitica, Willd. Fl. Ber. Prod. p. 405.
Clavaria radicosa, Bull. Champ. I, p. 195.
Sphæria radicosa, DC. Fl. Franc. II, p. 283.
Cordiceps ophioglossoides, Lk. Hndbk. III, p 347.
Torrubia ophioglossoides, Tul. Sel. Carp. III, p. 20.
Cordyceps ophioglossoides, Sacc. Syll. 5038, Cke. Syn. 48.

Exsicc. Rab. Herb. Mycol. 427.—id. F. Eur. 442.—Rehm. Asc. 471.—Thum. M. U. 569. Sydow M. March, 280.

Stromata solitary, rarely cespitose, simple or very rarely branched, flexuous, subcompressed, carnose, yellow within, 8–12 cm. high, 5–8 mm.

thick. Head oblong, obtuse or attenuated above, often hollow, roughened by the slightly projecting, densely crowded perithecia, dark rufous, about 2 cm. long and 6–8 mm. thick. Stem olivaceous, becoming black, sending out from its base, yellow, fibrous rootlets which embrace the matrix and penetrate the soil for 2 or 3 inches around. Asci cylindrical, 250–300 x 7–9 μ, 8-spored. Sporidia crowded, filiform, 150–180 x 7–9 μ, multiseptate, at length breaking up into subellipsoid, yellowish-hyaline joints 3–4 x 2–3 μ; paraphyses very slender.

Parasitic on *Elaphomyces granulatus* and *E. muricatus*, Massachuetts (Farlow), New Jersey (Ellis), Pennsylvania (Everhart).

C. capitàta, (Holmsk.)

Clavaria capitata, Holmsk. Otia, t. I, p. 38.
Sphæria agariciformis, Bolt. Fungi of Halifax, III, p. 61.
Sphæria capitata, Pers. Comment. p. 145, Pers. Myc. Eur. tab. 10, figs. 1-4.
Cordiceps capitatus, Lk. Hndbk. III, p. 347.
Torrubia capitata, Tul. Sel. Carp. III, p. 22.
Cordyceps capitata, Sacc. Syll. 5039, Cke. Syn. 49.
Exsicc. Rav. Fung. Car. V, 48.—Sydow M. March. 279.

Stromata cespitose or solitary, simple, 3–8 cm. high. Club, or head ovoid-sphæroid, roughened by the slightly prominent, ovoid, densely crowded perithecia, liver-color or reddish-yellow, about 1 cm. thick. Stipe equal, glabrous, citron-color, or yellow, at length fibrose-strigose and yellowish-black, 3–4 mm. thick. Asci cylindrical, very long, 15 μ thick. Sporidia filiform, very long, at length breaking up into fusoid-elongated, or subbacillary joints, greenish-yellow and 25–40 x 5–6 μ.

Parasitic on *Scleroderma*, Carolina (Ravenel), and on some tuberaceous fungus, Florida (Calkins).

C. *Species imperfectly known.*

In Curtis' Catalogue, pp. 138 and 139, two other species are mentioned, but not described:

Cordyceps gryllotalpæ, M. A. C., on buried sand moles.
Cordyceps isarioides, M. A. C., on dead moths.

C. Melolònthæ, Tul.

Torrubia Melolonthæ, Tul. Sel. Carp. III, p. 12.
Cordyceps Melolonthæ, Sacc. Syll. 5044, Cke. Syn. 50.

A figure is given in Silliman's Journ. VIII (1824), tab. IV, representing the general appearance of the fungus found in Pennsylvania, growing from the cervical portion of buried larvæ of the "May bug" (Melolontha), but there are no notes of the asci and sporidia.

ACROSPÉRMUM, Todè.

Fung. Meckl. I, p. 8 (partly).

Perithecia vertical, elongated-clavate, sessile or stipitate, carnose but firm, and of a horn-like texture when dry. Asci filiform, 8-spored. Sporidia parallel, filiform, continuous (or jointed in one species).

A. comprèssum, Tode, Fungi Meekl. I, p. 8, tab. II, fig. 3.

Clavaria herbarum, Pers. Comment. Clav. p. 68, tab. III, fig. 4.
Scleroglossum lanceolatum, Pers. (in Moug. Exs.)
Exsicc. Moug. & Nestl. Stirp. Vosg. 671.—Rab. Herb. Mycol. 35.—id. F. E. 2847.—Kriegr. F. Sax. 438.—Ell. N. A. F. 1318.—Vize. Micr. Fungi, 107.

Perithecia solitary or subcespitose, sessile, club-shaped, attenuated above and generally compressed, pale at first, finally dark or olive-black, shining, smooth at first, becoming longitudinally subsulcate, 1–3 mm. high. Asci filiform, very long (200–400 x 3–6 μ), 8-spored. Sporidia packed side by side, filiform, pale yellowish-hyaline, 100–300 x $\frac{3}{4}$–1 μ; paraphyses slender.

On dead herbaceous stems, and culms of grasses, common.

A. virídulum, B. & C. Grev. IV, p. 161.

Exsicc. Ell. N. A. F. 857.

Perithecia scattered, ovate-conical, $\frac{1}{3}$–$\frac{1}{2}$ mm. high, abruptly contracted below into a short, stipe-like base, obtuse above, greenish-cinereous, subfurfuraceous. Asci linear, 150–200 x 5–6 μ. Sporidia filiform, hyaline or slightly yellowish, about as long as the asci. The specimens on decaying hickory limbs have the asci narrower ($3\frac{1}{2}$–4 μ), but do not appear to differ otherwise.

On decayed herbaceous stems, So. Carolina (Ravenel), on fallen hickory limbs, and on fallen pear leaves, New Jersey (Ellis), on white oak leaves, Texas (Ravenel).

A. Gráminum, Lib. Exsicc. Ard. No. 33, Corda Icones, III, p. 27.

Exsicc. Rab. Herb. Mycol. 776.

Perithecia at first linear, then clavate, often thicker and somewhat bent in the middle, more or less compressed, often blackish with the apex lighter, mostly about 1 mm. high by 200–300 μ thick. Asci cylindrical, with indistinct paraphyses. Sporidia filiform, very slender, 150 x $\frac{3}{4}$ μ, hyaline, continuous.

On decaying leaves of grasses, N. America (Sacc. in Syll.)

Dr. Rehm in Die Pilze considers this only a var. of *A. compressum*.

A. foliícolum, Berk. Grev. IV, p. 161.

Exsicc. Rav. Car. II, 65.—Rav. F. Am. 734.—Ell. & Evrht. N. A. F. 2d Ser. 2140.

Perithecia erect, clavate, 1–$1\frac{1}{2}$ mm. high, yellowish-chestnut, of coriaceo-membranaceous texture. Asci cylindrical, 300–400 x 4–5 μ, with thread-like paraphyses longer than the asci. Sporidia 8 in an ascus, filiform, continuous, lying parallel.

On dead leaves of *Ulmus*, *Celtis* and *Vitis*, South Carolina, and on leaves of *Smilax*, Texas (Ravenel), on leaves of *Vitis*, New Jersey.

The specimens on leaves of *Vitis* and *Celtis* are darker and smaller than those on *Smilax* and *Ulmus*, but do not appear to differ otherwise.

A. Ravenélii, B. & C. Grev. l. c.

Perithecia short, nearly cylindrical at first, then clavate, 300–450 μ high, and 70–80 μ thick (while retaining the cylindrical form), becoming as much as 150 μ thick when clavate, color quite dark. Asci 180–210 x 3–4 μ, with filiform paraphyses a little longer than the asci. Sporidia filiform, nearly as long as the asci, separating into joints 12–15 μ long.

This is a smaller species than *A. foliicolum*, Berk. from which it is safely distinguished by its *jointed sporidia.*

In Rav. F. Am. 735, on leaves of *Fraxinus*, the specimens in the copies we have seen are too poor to enable one to decide with certainty, but as far as can be ascertained, the sporidia are continuous and if so this No. is referable to *A. foliicolum*, Berk.

SELÍNIA, Karst.

Symb. ad Mycol. Fenn. III, (1876).

Stromata carnose, at first small, elliptical or tubercular, becoming confluent, irregular, red, covered with a rusty-red conidial layer. Perithecia one or at least only a few in each stroma, buried, globose, pale, carnose, with a thick, conical, prominent ostiolum. Asci elongated-ventricose, containing 4–8 continuous, elliptical, hyaline, sporidia. Paraphyses filiform, septate.

S. púlchra, (Winter).

Hypocreopsis pulchra, Winter in Hedwigia 1875, p. 26.
Winteria pulchra, Sacc. Mich. I, p. 281.
Selinia pulchra, Sacc. Syll. 4586.

Stromata at first small, subconical in the center, mostly becoming confluent and forming a rusty-red crust, the surface of which is overspread with a rusty-brown, conidial layer bearing short, cylindrical conidia. Perithecia buried, globose, about ½ mm. diam., with thick, conical ostiola. Asci ventricose below, narrower above, sessile, 270 x

52 μ. Sporidia 4–8, inordinate, elliptical, narrowed at each end, continuous, smooth, hyaline, 55 x 30 μ.

On dung of cows and sheep.

We have included this species, though not aware that it has been actually found, as yet, in this country.

OOMỲCES, B. & Br.

Brit. Fungi No. 590 (1851).

Perithecia included in a common, carnose-membranaceous stroma. Asci linear. Sporidia filiform, continuous, hyaline.

O. Langloìsii, (E. & E.) **(Plate 17)**

Coscinaria Langloisii, E. & E. Journ. Mycol. II, p. 88.

Stromata tuberculiform, erumpent, soft, $\frac{1}{3}$–$\frac{1}{2}$ mm. diam., pale flesh-color or horn-color when fresh, becoming nearly black when dry, of rather close, membranaceous texture on the surface, softer and looser within, surrounded by the ruptured epidermis, convex above. Perithecia ovate, minute, with thin, transparent walls, 250–300 μ high, and 150–200 μ broad, 20-30 in a stroma. Ostiola punctiform, not prominent. Asci linear, 150–200 x 5 μ, with filiform paraphyses branched above. Sporidia filiform, nearly as long as the asci, hyaline, continuous, 1 μ thick.

On dead stems of *Vigna luteola*, Louisiana (Langlois).

A more careful examination of this curious production shows no essential character to separate it from *Oomyces* to which we now refer it. By some oversight this was placed on plate 17 with *Sordarieæ*. and the following species on plate 7 with the *Perisporieæ*.

POLYSTÍGMA, De Cand.

Comment Mus. Hist. Nat. t. III, p. 330 (et seq.)

Stroma subcarnose, effused, colored (ochraceous, red or yellow), phyllogenous. Perithecia immersed. Asci 8-spored. Sporidia ovoid, continuous, hyaline.

P. rùbrum, (Pers.) **(Plate 7)**

Xyloma rubrum, Pers. Syn. p. 105.
Dothidea rubra, Pers. Fr. S. M. II, p. 553.
Sphæria rubra, Fr. Obs. I, p. 172.
Séptoria rubra, Desm. Ann. Sci. Nat. Ser. II, t. XIX, p. 342.
Polystigma rubrum, DC. l. c. p. 337.

Exsicc. Kunze F. Sel. 271.—Rab. Herb. Mycol. 562, 580.—Thum. Fungi Austr. 179, 180. id. M. U. 973.—M. March. 258.—Briosi & Cav. F. Paras. 12.—Linhart Fungi Hung. 272, &c.

Hypophyllous, suborbicular, slightly convex, 2–4 mm. across, with an even margin, red, at length darker. Perithecia (cells) small, immersed, reddish. Ostiola impressed, punctiform, becoming more prominent. Asci clavate, 8-spored. Sporidia ovoid, subobtuse, straight, subhyaline, 10 x 6 μ. The spermogonial stage is *Libertella rubra*, Bon. Spermatia filiform, uncinate, 30 μ long.

On living leaves of several species of *Prunus* (*P. domestica*, *P. spinosa*). Credited to N. America by Sacc. in Syll.

P.(?) Bumèliæ, (Schw.)

Dothidea Bumeliæ, Schw. Syn. N. Am. No. 1884.
Polystigma Bumeliæ, Sacc. Syll. 4589, Cke. Syn. 177.

Stroma orbicular, large, $\frac{3}{4}$ cm. and over, ovate or of somewhat irregular shape, conspicuous on both sides of the leaf; on the upper side brick-red, and somewhat shining as if varnished, paler and without any varnished appearance below. Perithecia (cells) few, scattered, minutely pseudo-ostiolate and subprominent.

Epiphyllous, on living leaves of *Bumelia oblongifolia;* found in the Arkansas region by Nuttall.

BYSSONÉCTRIA, Karsten.

Symb. Mycol. Fenn. VII, p. 6 (1879).

Stroma more or less byssoid. Perithecia subsuperficial, crowded. Sporidia (in the American species) hyaline. Asci paraphysate.

A. *Sporidia continuous.*

B. fimèti, (Cke.)

Nectria fimeti, Cke. Grev. XI, p. 108.
Byssonectria fimeti, Sacc. Syll. 4584, Cke. Syn. 349.
Exsicc. Rav. F. Am. 646.

Perithecia gregarious or scattered, golden-yellow, subglobose, seated on a byssoid, golden-yellow stroma, bare and glabrous above, tomentose below, finally collapsing. Asci cylindrical, 160–190 (p. sp. 112–115) x 10–12 μ. Sporidia uniseriate, elliptical, continuous, hyaline, 15 x 8 μ.

On cow dung, Aiken, S. C. (Ravenel).

B. *Sporidia uniseptate.*

B. chrysócoma, Cke. & Hark. Grev. XII, p. 101.

Stroma fibrose-byssoid, golden-yellow, effused. Perithecia gregarious, minute (100–150 μ), obovate, thin, dark yellow, semiimmersed in the stroma. Asci clavate, 8-spored. Sporidia biseriate, narrowly elliptical, uniseptate, hyaline, 10 x 2–2½ μ.

On wood of *Eucalyptus*, California (Harkness).

B. rosélla, Cke. & Hark. Grev. l. c.

Delicate, effused, with a rose-colored tint. Hyphæ creeping, interwoven, with the minute, obscure perithecia scattered on it. Conidia lunate, like those of *Fusarium*, acute at each end, 5-septate, mostly nodulose and hyaline, 40 x 5–6 μ. Unfortunately in an immature condition.

On dead grass, California (Harkness).

HYPONÉCTRIA, Sacc.

Syll. II, p. 455.

Perithecia covered, otherwise as in *Nectria*.

H. Gossýpii, (Schw.)

Sphæria Gossypii, Schw. Syn. Car. 207.
Hyponectria (?) *Gossypii*, Sacc. Syll. 4582 and Cke. Syn. 542.

Scattered, rather soft, immersed. Perithecia globose, purplish flesh-color. Ostiolum elongated to the surface and discharging gelatinous matter. The minute perithecia are deeply sunk in the substance of the immature capsules so as not to be visible outwardly, but, through the elongated ostiola, a gelatinous substance is discharged, which hardens on the surface of the matrix and gives it a purplish color. In the mature specimens the surface of the capsules is granulose or papillose from the subjacent perithecia.

On dead capsules of the cotton plant, So. Carolina (Schweinitz).

We have seen no specimens of this species, but have received from Prof. F. L. Scribner a *Fusarium* on capsules of cotton, from South Carolina, which may be the conidial stage.

HYPOMỲCES, Fries.

Summa Veg. Scand. p. 382.

Perithecia gregarious, with a cottony stroma in which they are more or less immersed, mostly parasitic on various *Hymenomycetes* or *Discomycetes*, bright colored, with papilliform or slightly elongated ostiola. Asci mostly cylindrical, 8-spored, without paraphyses. Sporidia oblong or fusoid, uniseptate, hyaline. Conidial stage represented by *Asterophora*, *Sepedonium*, *Dactylium*, *Verticillium* &c.

A. *Sporidia continuous.*

H. Van Bruntiànus, Ger. Bull. Torr. Bot. Club, IV, p. 64.

Perithecia globose, densely crowded, pallid, hygrophanous, immersed, with a short, thick, exserted, obtuse ostiolum; subiculum white. Sporidia oblong, hyaline, shortly apiculate at the broad end, subobtuse at the other, 14–16 x $3\frac{1}{2}$–$4\frac{1}{2}$ μ. Asci cylindrical.

On the pileus, stipe and gills of an unknown *Agaric*, Poughkeepsie, N. Y. (Gerard), Iowa (Holway).

H. víridis, (A. & S.)

Sphæria viridis, Albt. & Schw. Consp. p. 8.
Hypomyces viridis, Sacc. Syll. 4633.
Sphæria luteo-virens, b. Fries. S. M. II, p. 339.
Hypomyces viridis, Berk. & Br. Brit. Fungi No. 1101.
Hypomyces luteo-virens, Plowr. Grev. XI, p. 46.
Exsicc. Rehm Asc. 586.

Stroma broadly effused, with a dirty yellowish-green tomentum and sterile margin. Perithecia closely packed, ovoid or sphæroid, pale, with their conical apices projecting and becoming dark brown or black. Asci cylindrical, 170–180 x 5 μ, yellowish-hyaline, continuous or with the endochrome sometimes 2-parted.

On *Agaricus alutaceus*, Carolina (Ravenel), Pennsylvania (Michener & Everhart), New England (Sprague).

H. polypòrinus, Pk. 26th Rep. N. Y. State Mus. p. 84.

Exsicc. Ell. & Evrht. N. A. F. 2d Ser. 1946.

Perithecia minute, ovate or subconical, seated on a pallid subienlum, smooth, yellowish or pale amber. Asci narrow, linear. Sporidia fusiform, acuminate at each end, nucleate, 15–18 μ long. The outward appearance is almost exactly the same as that of *Hypocrea pallida*, E. & E.

On *Polyporus versicolor*, New York (Peek), New Jersey (Ellis).

H. apiósporus, Cke. Grev. XII, p. 80.

Effused, pale, thin. Perithecia semiimmersed, slightly papillate, (honey-color when dry). Ostiola indistinct. Asci cylindrical, 8-spored. Sporidia lanceolate, apiculate above, rounded below, minutely roughened, continuous, yellowish (except the apiculus), 18 x 6½ μ.

On *Clavaria pistillaris*(?), New York (Gerard).

H. Banníngii, Pk. Bot. Gaz. IV, p. 139.

"Subiculum white, then sordid. Perithecia crowded, ovate, with a papilliform ostiolum, pale amber or honey-color. Asci slender, cylindrical. Sporidia uniseriate, oblong-fusiform, yellowish in the mass, 30–38 x 4–5 μ. On some decaying Agarie, apparently a *Lactarius*, Baltimore, Md. (Miss Banning). The sporidia in our specimens are simple, but they may possibly become uniseptate when old."

On *Lactarius* sp., Pennsylvania (Everhart), sporidia 27–40 x 4–5 μ, slightly curved, nucleate.

H. Geoglóssi, E. & E. Journ. Mycol, II, p. 73.

Perithecia immersed, subglobose (75–80 μ), subangular from mutual pressure. Ostiola papilliform, black. Asci clavate-cylindrical, subsessile, yellowish, 75–80 x 6–7 μ, containing eight crowded, sub-biseriate, clavate-oblong, continuous, hyaline, 6–8 x $2\frac{1}{2}$–3 μ sporidia. The subiculum appears to be obsolete, the perithecia being immersed in the hymenium of the host. As no notes were taken from the fresh specimens, the color of the perithecia can not be certainly stated, but in the dry specimens they are nearly or quite black.

On *Geoglossum hirsutum*, Newfield, N. J. Occupying the greater part of the hymenium of the host, the proper fruit of which is mostly suppressed.

H. xylóphilus, Pk. Bull. Tor. Bot. Club, XI, p. 28.

Subiculum effused, whitish. Perithecia numerous, crowded, small, yellowish, with a blunt ostiolum. Asci cylindrical, 80–112 x 6–$7\frac{1}{2}$ μ. Sporidia continuous, uniseriate, subfusiform, 15–17 x 5–6 μ.

On decaying wood, Ohio (Morgan).

H. hyálinus, (Schw.)

Sphæria hyalina, Schw Syn. Car. No. 35.

Effused, thin, glabrous, hyaline-fuscous, roughened with the crowded, dark-colored ostiola. Asci cylindrical, 8-spored. Sporidia elliptic-fusoid, 15–20 x $4\frac{1}{2}$–5 μ, uniseriate, continuous, greenish-hyaline, granular-roughened, and minutely apiculate at each end.

On *Russula fœtens*, Carolina (Schw.), on some Agarie, Pennsylvania (Everhart), Connecticut (Thaxter).

B. *Sporidia uniseptate.*

H. lateritius, (Fr.)

Sphæria lateritia, Fr S M. II. p. 338.
Hypocrea lateritia, Fr Summa Veg Sc p. 383
Hypomyces lateritius, Tul Ann. Sci. Nat Ser IV, tom XII, p. 11, Plow Grev. XI, p 41, tab. 148.
Exsicc. Rab. F E. 317—Thum. M. U 2164.

The mycelium forms a dense, white, felt-like stratum, which finally becomes more compact and of a pale brick color. Perithecia spherical or subovate, abundant, sunk in the stroma, except their slightly prominent, smooth, brownish ostiola. Asci cylindrical, 200–250 μ long, spore-bearing part 112 x 6–7 μ. Sporidia uniseriate, elongated-fusiform, uniseptate, acuminate at each end, yellowish-hyaline, 18–20 x 4–$4\frac{1}{2}$ μ.

On the hymenium of several species of *Lactarius*, New England (Murray & Sprague), on *Lactarius Indigo*, Carolina (Ravenel), on *Lactarius sp.* Potsdam, N. Y. (Ellis).

H. aurántius, (Pers.)

Sphæria aurantia, Pers. Syn. p. 68.
Sphæria aurea, Grev Scot. Cr Fl tab. 47
Nectria aurantia, Fr. Summa Veg. Sc. p. 388.
Hypomyces aurantius, Tul. Sel. Carp III, p 43, Plowr Grev. XI, p. 44, tab 150.
Exsicc Rab F. E. 138 —Thum. M. U. 1747.

Perithecia gregarious, subsphæroid, 300 μ diam., their apices projecting from the effused, orange-colored subiculum. Asci cylindrical, 8-spored, 110–140 x 6 μ. Sporidia uniseriate, uniseptate, fusoid, with the ends subapiculate, slightly curved, nearly smooth, 15–24 x 5–6 μ, hyaline. The white mold, *Diplocladium minus*, Bon., is said to be its conidial stage.

On *Stereum* and on the under side of logs. Carolina (Ravenel).

H. Lactifluòrum, (Schw.) **(Plate 11)**

Sphæria Lactifluorum, Schw Syn Car No. 34.
Hypomyces Lactifluorum, Tul. Sel. Carp. III, p 63.
Exsicc Rav. Car. 1 54 —id V 64 —Ell N A. F 467.

In the affected specimens of *Lactarius*, the gills are entirely obliterated, so that the hymenium of the Agaric presents an even, orange-colored surface in which the subglobose perithecia are thickly bedded, with only their slightly prominent reddish ostiola visible. Asci long and slender. Sporidia uniseriate, fusiform, straight or slightly curved, rough, hyaline, uniseptate, cuspidate-pointed at the ends, 30–38 x 6–8 μ, The general appearance is much the same as that of the preceding species, but the sporidia are larger, rough and warted and the felt-like mycelium is wanting. In decay, the color changes to a purplish red.

On *Lactarius*, especially *L. piperatus*. Carolina, (Ravenel), Pennsylvania (Everhart), New Jersey (Ellis).

H. roséllus, (A. & S.)

Sphæria rosella, A & S Consp. p 35.
Nectria Albertini, Berk. Brit. Fungi. 971.
Nectria rosella, Fr Summa Veg. Sc p 388
Hypomyces rosellus, Tul. Sel. Carp. III, p. 45
Exsicc Thum. M. U. 1953.—Fckl F. Rh 987 —Plowr. Sph Brit. 204.—Roum. F. Gall. 1273.—Krgr. F. Sax. 228.

Mycelium lax, effuse, consisting of loosely-woven threads, at first white and bearing conidia (*Tricothecium agaricinum*, Bon.), then deep rose-color or nearly blood-red. Perithecia deep rose-red, subspherical or ovoid, of variable size, buried in the stroma, except the projecting, rather obtuse, papilliform ostiola. Asci linear, 150 x 6–7 μ, with eight uniseriate, narrowly lanceolate, apiculate, straight or slightly curved, hyaline, 22–37 x 5–7 μ sporidia nucleate or spuriously 1–3-septate and often subinequilateral.

On various decaying fungi and on leaves and rubbish near where fungi have decayed. Pennsylvania (Michener & Everhart), Plainfield, N. J. (Meschutt).

Dr. C. B. Plowright, in his valuable monograph of this genus in Grevillea, Vol. XI, says: "There are two varieties of this species, one with larger, pointed perithecia, as figured by Greville and by Albertini & Schweinitz; the other, with smaller and more obtuse perithecia. The sporidia vary a good deal in size, as does the color of the subiculum, which is sometimes nearly absent. Sometimes it is almost white, but mostly rose colored, with a whitish margin." The specimens from Mr Meschutt were on decaying leaves, forming little patches ½–1 cm. across.

H. ochràceus, (Pérs.)

Sphæria ochracea, Pers. Syn. p. 18.
Cryptosphæria aurantia Grev. Scot. Cr. Fl. tab 78.
Hypomyces armeniacus, Tul. Ann Sc Nat. IV, Ser. tom. XIII, p 12.
Hypomyces ochraceus, Tul. Sel. Carp. III, p. 41, Plowr Grev XI, p 45.

Perithecia crowded, subglobose, yellowish, immersed, with a short, thick, obtuse, exserted ostiolum. Subiculum (*Verticillium agaricinum*, Cda.) at first white, then straw-colored, ochraceous and yellow. Asci cylindrical, 25–30 x 6½ μ, containing eight oblong-lanceolate, uniseptate, constricted, 35 x 6½ μ sporidia, which are mucronate at each end.

On decaying *Agaricus*, Pennsylvania (Michener).

H. asterópliorus, Tul. Sel. Carp. III, p. 55 (partly), Plowr. Grev. XI, p. 6, tab. 147.

Perithecia originating in an effused, byssoid stroma, in which they are thickly strewn, ovoid or sphæroid, narrowed above into a more or less elongated neck, with an acute, pervious, ciliate ostiolum, pale yellowish-brown, subhyaline, 150 μ high by 70–90 μ broad. Asci broadly ovate, abruptly attenuated below, 40–50 x 18–20 μ, containing 4–6 narrowly lanceolate, slightly curved, mucronate at each end, uniseptate, subhyaline, then dirty-yellow, 25–35 x 5 μ sporidia.

Parasitic on *Nyctalis*, Carolina (Ravenel),

Dr. Plowright, in his monograph already cited, says: "The perithecia of this species of *Hypomyces* differ considerably from those of the other members of the genus. They are formed of very large polygonal cells and originate from the intertwining of the dilated and convoluted bases of the conidia-bearing hyphæ that compose the stroma. These (the perithecia) are most frequently found upon the inside of the stem of the *Nyctalis*, but they are by no means of common occurrence."

H. insignis, B. & C. Fungi of Mexico, No. 6, p. 424.

Hypomyces transformans, Pk. 39th Rep, N. Y State Mus. p 57

Mycelium red, effused. Perithecia oblong, more deeply colored. Sporidia fusiform, apiculate at each end, 37 μ long, spuriously 1-septate, hyaline. On the hymenium and pileus of *Cantharellus*, which it covers with a red stratum and obliterates the gills. In Mexico, near Orizaba (Botteri).

The above is from Saccardo's Sylloge, Vol. II, p. 472.

The following is Peck's description of his *H. transformans.*

Subiculum effused, variable in color, pallid, golden-yellow, ochraceous or brick-red. Perithecia ovate or subglobose, papillate, sunk in the subiculum. Ostiola prominent, obtuse, amber or orange. Asci cylindrical. Sporidia fusiform, apiculate at each end, somewhat rough, continuous or rarely with the endochrome obscurely divided, colorless, 33–38 μ long.

On *Cantherellus cibarius* which it transforms into an irregular mass.

We have what appears to be the same, on *Cantharellus cibarius*, from Massachusetts (S. J. Harkness) and from Pennsylvania (Everhart).

C. *species not well known.*

H. tegíllum, B. & C. Grev. IV, p. 15.

"Perithecia ovate, rufous-brown, scattered over a continuous, white mycelium, like thin parchment." On pine, Carolina (Ravenel).

H. flavéscens, (Schw.), Grev. XII, p. 80.

Perithecia gregarious, distinct, globose-ovate, papillate, whitish, covered with a villose coat that finally disappears, seated on a milk-white, broadly effused, pubescent subiculum; asci cylindrical. Sporidia narrowly elliptical, hyaline (uniseptate)? On hymenium of some resupinate *Polyporus*. Pennsylvania (Schweinitz).

H. pannòsus, (Schw.), Grev. XII, p 80.

Stroma effused, whitish, shaggy, thin, margin fimbriate-cottony. Perithecia semi-immersed, pale, with dark-colored, punctiform ostiola. Asci cylindrical; sporidia (?) On rotten wood, Ex. Herb. Schw.

The descriptions of this and the preceding species are taken from Grevillea l. c. and were, apparently, made from specimens in Herb. Berkeley. They are not in Schw. Synopsis N. Am. The *H. pannosus* here described is said to be a different thing from the *Sphæria pannosa*, Fr.

In the next two species, the fructification is unknown.

H. tubericola, (Schw.)

Sphæria tubericola, Schw. Syn. N. Am. No 1192.
Hypomyces tubericola, Cke. Syn. 335.

Effused, thin, brown-black, margin scarcely determinate. Perithecia at first subimmersed, finally erumpent ("omnino prominulis"), brown-black, rugose-tuberculate, minute, subconical, densely crowded, at length collapsing. The interior of the perithecia is said to be entirely similar to that of the preceding species, *i. e. H. Lactifluorum*, etc.

On an old "White Tuber" infesting the outer bark. Found at Pocono, Pa. (Schw.)

H. boleticola, (Schw.)

Sphæria boleticola Schw. Syn N. Am No 1494
Hypomyces boleticola, Cke, Syn. 336.

Allied to *H. aurantius* (Pers.), but distinguished by its paler color, by the character of the subiculum (stroma) and shape of the perithecia. Subiculum effused, interrupted, of loose texture, with irregular spaces ("plagis irregularibus"), pale orange with the margin whitening out Perithecia only partially immersed, conic-ovate, more or less scattered, at first colored like the stroma, at length orange-red, rather soft, papillate, easily separating from the stroma, leaving small cavities; contents of perithecia very white, at length oozing out and remaining attached to the ostiola, like a white villous pubescence.

Found but seldom on very much decayed *Polyporus citrinus*, at Bethlehem, Pa.

H. tomentósus, Fr. Berkeley, Notices of N. Am. Fungi, in Grevillea IV, p. 15.

Of this species, but little appears to be known. It is stated by Cooke, in Grevillea XII, p. 80, that the asci are cylindrical and the sporidia lanceolate, mucronate at each end, uniseptate, hyaline and 42–60 x 6–7 μ. Found on some Agaric.

Mycogone cervina Ditm., which is found on *Helvella* and on various decaying *Pezizas* and *Asterophora Pezizæ*, Cda., on *P. hemispherica*, both of which are found here, are, with several similar molds, regarded as the conidial stage of different species of *Hypomyces*, of which the ascigerous stage has not yet been found.

HYPÓCREA, Fr.

Summ. Veg. Scand p 383 (in part.)

Stroma fleshy, pulvinate, hemispherical or effused, sometimes reduced to a mere membrane or to a loose felt-like stratum. Perithecia ovate-globose, mostly only partly sunk in the stroma. Asci cylindrical, 8-spored. Sporidia (typically) two-celled, the two cells which are generally spherical, soon separating so that the ascus appears to contain 16 sporidia.

A. *Sporidia hyaline, stroma subpulvinate.*

H. rùfa (Pers.)

Sphæria rufa, Pers. Obs. I. p. 20
Hypocrea rufa, Fr. Summ. Veg. Sc p, 383.
Cyttaria rufa, Bon. Abhandl. der Mycol 1864, p 166.
Exsicc Ell. N. A. F. 157 and Ell. & Evrht. N. A. F. 2d Ser 1552, (issued as *H. gelatinosa*, Tode.

Stromata gregarious, superficial, subhemispherical, elliptical, or irregular in shape, mostly ½–1 cm. diam., convex when fresh, contracted and rugose when dry, nearly brick-colored, punctulate from the slightly projecting ostiola, whitish within. Asci cylindrical, nearly sessile, 65–75 x 4–5 μ, with 8 uniseriate, didymous sporidia, composed of two subequal, hyaline, subglobose, or subcubical cells, each 3–4 μ in diam.

On dead limbs of *Andromeda ligustrina*, Newfield, N. J., on decaying wood and on old *Polyporus*, Carolina (Ravenel), also on decaying wood and bark from various localities. Probably found throughout the U. S.

H. lénta (Tode).

Sphæria lenta, Tode Fungi Meckl. II, p 30.
Hypocrea lenta, B. & Br Fungi of Ceylon, no 992.

"Stromata gregarious, 2–3 lin., broad, thick, margin repand, disk nearly plain, partially free from the matrix. Perithecia minute, globose, immersed, ostiola punctate, minute."

The only specimen of this species in our possession was sent from California by Dr. Harkness and is on wood of fir (Herb. Hark. 3496). In this specimen the stromata are ¼–½ cm. in diam., nearly round, central portion adnate, leaving a narrow, free margin closely applied to the surface of the wood. The pale, globose, carnose-membranaceous, peripheric perithecia (200–220 μ in diam.) lie in a single layer, their ostiola very prominent and distinctly roughening the surface of the dirty-black stroma. Asci cylindrical, 80–100 x 5–6 μ, without paraphyses, containing 8 two-celled sporidia, each cell subcubical, or nearly globose and 4–4½ μ in diam. or slightly ovoid, 4–5 x 3½ μ.

H. Schweinítzii, (Fr.)

Sphæria Sahweinitzii, Fr. Elench. II, p. 60.
Sphæria contorta, Schw. Syn. N. Am. 1224.
Hypocrea contorta, B. & C. Grev. III, p. 14
Sphæria rigens, Fr. l. c.
Sphæria lenta, Schw. Car. no. 28. (non Tode).
Hypocrea rufa, forma umbrina, Sacc. F. Ven. IV, p. 24.
Hypocrea Schweinitzii, Sacc. Syll. 4840.
Exsicc. Ell. N. A. F. 156.—Rav. F Am. 642.—Sacc.M. V. 124.

Stroma as in the preceding species, except that the central adnate portion is smaller, and, in well-grown specimens, the margin is distinctly undulate and sublobate. Perithecia immersed, globose, or subovate (150 μ), scarcely roughening the surface, which is merely punctate from the minute, slightly prominent ostiola. Asci cylindrical, 60–65 x $3\frac{1}{2}$ μ, without paraphyses, containing 8 two-celled sporidia composed of two globose, hyaline cells, 3–$3\frac{1}{2}$ μ in diam., and readily separable. On bark and wood, common.

In the above synonyms we have included *Sphæria rigens* Fr. (Sacc. Syll. II, p. 523) which is said to differ from *H. Schweinitzii* in its smaller and more regularly-shaped stroma, with the perithecia confined mostly to the central portion of the disc and by its habitat on bare wood, and not on bark. The stroma is also said to be of a darker color, without any olive or greenish shade. We are inclined to think that these points of distinction are due to imperfect development and are not of specific value, and we have examined many specimens, as the species (as represented in N. A F. and Rav. F. Am.) is very common, both around Newfield and West Chester. We have also repeatedly received from Carolina, Florida and Louisiana, specimens of what is evidently the same as the "*Hypocrea contorta*" N. A. F. 156, but never anything that we could refer to the "*Hypocrea rigens*, (Fr.)" as distinct from *H. Schweinitzii.*

H. chiònea. E. & E. (in Herb.)

Stromata gregarious, pulvinate, subhemispherical, 1–2 mm. diam., snow-white, dotted with the minute, punctate, horn-colored ostiola. Perithecia ovate, minute, horn-colored. Asci about 60 x 3 μ. Sporidia 8 in an ascus, hyaline, two-celled, each cell $2\frac{1}{4}$–3 μ diam.

On decaying, decorticated wood. London, Canada (Dearness).

Differs from *H. lactea*, Fr. in its pulvinate stroma.

H. lati-zonàta, Pk.

Subiculum dirty-white, forming a broad ($\frac{1}{2}$–$\frac{3}{4}$ cm.) band around the outside of the cups of the *Cyathus*, thickly punctate with the dark-brown, slightly prominent ostiola. Asci cylindrical, 75–80 x $3\frac{1}{2}$–4 μ, containing eight didymous sporidia, the cells separable, subglobose, hyaline and 3–3½ μ in diam. A very curious species sent from Ohio, under the above name, by Prof. A. P. Morgan.

Parasitic on *Cyathus striatus*, Hoff.

H. armeniàca, B. & C. Grev. IV, p. 15.

"Forming a thin, apricot-colored stratum which, when barren, looks like *Corticium ochroleucum*, at length fertile. Perithecia superficial, scattered, of a deeper tint." The sporidia (in Grev. XII, p. 78) are said to be 4 μ diam.

Specimens found at Newfield, N. J., on bark of dead *Magnolia glauca* have been doubtfully referred to this species. The stroma is of a rather loose, membranaceous-byssoid nature, thin and dull or dirty-white, extending more or less continuously for several inches and soon covered with a more or less compact layer of apricot-colored perithecia which in drying shrink away from each other and leave the hymenium more or less cracked. Asci cylindrical, about 75 x 4 μ, containing 8 didymous, hyaline sporidia which separate into two subglobose or subovate cells 3–3½ μ in their longer diameter.

H. ochroleùca, B. & Rav. Grev. l. c.

"Effused, thin, ochroleucous, seated on a pale mycelium, with a barren border, often cracked when old." The cells of the sporidia are said by Cooke, in Grev. l. c., to be 6 μ diam., hyaline.

On *Myrica cerifera*, Carolina (Ravenel).

H. scutellæfórmis, B. & Rav. in Rav. Fungi Car., IV, No. 31.

Stromata scutellate, centrally attached, margin free and, in the larger specimens, undulate and sublobate, 1–2 mm. across, convex, nearly smooth, only slightly punctate from the scarcely prominent ostiola, color dull red. Cells of the sporidia subglobose, 3–4 μ diam. Our knowledge of this species is derived from the specimen cited, in which the asci had disappeared, but the globose cells of the sporidia were abundant.

On bark of *Acer rubrum*, Carolina (Ravenel).

H. patélla, C. & P. 29th Rep. N. Y. State Mus. p. 57.

"Fleshy, patellate, discoid, 1–2 lines broad, pale ochraceous. Asci cylindrical. Sporidia globose, sixteen, hyaline, 3–4 μ diam. Resembles, externally, some species of Helotium."

On decaying wood, New York (Peck).

H. minima, Sacc. & Ell. Mich. II, p. 570.

Stromata scattered, superficial, pulvinate, discoid, olivaceous, becoming nearly black when dry, hardly 1 mm. in diam., minutely punctulate from the slightly prominent ostiola, texture finely cellular, dark-olivaceous. Asci cylindrical, without paraphyses, 75 x 3½–4 μ, sub-sessile, containing 8 didymous, hyaline sporidia composed of two globose-cuboidal cells about 3½–4 μ and readily separating.

On bark of dead *Magnolia glauca*, Newfield, N. J.

H. olivàcea, C. & E. Grev. II, p. 92.

Stromata scattered, consisting at first, of patches of thin white tomentum ½–1 cm. diam., becoming carnose and subpulvinate and of an olive-yellow shade, at length dark olive or nearly black, and punctate from the slightly prominent ostiola. Asci cylindrical, 65–75 x 3½–4 μ, contracted below into a substipitate base, and containing 8 two-celled, hyaline sporidia, the cells nearly globose, about 3 μ diam. and readily separating.

On decaying pine wood, Newfield, N. J. What appears to be the same was found on decaying bark of *Sassafras* lying on the ground.

H. Stereòrum, Schw.

Sphæria Stereorum, Schw. Syn N. Am 1183
Hypocrea Stereorum, (Schw) Cke Syn No 92

"Undulate-confluent, applanate, margin sublobate, surface plicate, subpulvinate, flesh-color, becoming brown; when young, covered with a white tomentum and then more distinctly pulvinate, sometimes solitary, but generally confluent in elongated strips in the folds of the matrix; substance quite soft, but not gelatinous; surface granular from the prominent perithecia, which are distinctly ostiolate and *not* immersed in the whitish subjacent stroma; seminal dust (sporidia) copious. Often confluent for an inch in length, the separate, cushion-like stromata 3–4 lin. broad; margin partially free."

On *Stereum fasciatum*, Bethlehem, Pa. (Schw.), on *Polyporus Curtisii* South Carolina (Ravenel).

We have seen no specimens, and copy the above description from Schw. Cooke, in Grev. XII. p. 78, says the cells of the didymous sporidia are subglobose and hyaline.

H. fungícola, Karst. Myc. Fenn. II, p. 204.

Hypocrea Karsteniana, Niessl in Rehm's Asc 678

Stroma irregularly effused, incrusting, even, waxy, tolerably thick, 3–12 cm. in extent, but sometimes interrupted and shorter. At first white farinose with a floccose margin, finally citron- or lemon-yellow, pale inside, thickly punctate from the rather prominent ostiola. Perithecia sunk in the stroma, crowded, ovate, pale. Asci cylindrical.

briefly stipitate, 60–75 μ long (p. sp.), 4–5 μ thick. Sporidia 8 in an ascus, uniseriate, composed of two similar, ovate, hyaline cells about 4½ x 3½ μ.

On old *Polyporus*, Ohio (Morgan).

H. polyporoìdea, B. & C. Grev. IV, p. 15.

"Fawn-colored. Perithecia free, tomentose, with a naked ostiolum, seated on a pale crust, here and there elevated and thinner towards the margin. A very curious species." Cooke, in Grev. l. c., gives the hyaline, subglobose cells of the sporidia as 5 μ in diam.

H. sulphùrea (Schw.)

Sphæria sulphurea, Schw. Syn. N Am. 1221.

"Rather thin, subcarnose, at length horn-like in texture (when dried), the thin, partially free margin variously lobed, sulphur-color, white within. Perithecia crowded, globose-depressed, immersed, dirty-yellow. Ostiola concolorous, papillate, situated in little pit-like depressions of the otherwise smooth surface. On bark; rare; separable when fresh, subrotund, 1½ inches across." In Grev. l. c., the globose cells of the hyaline sporidia are said to be 5 μ in diam.

H. pállida, E. & E. Journ. Mycol. II, p. 65, and Proc. Acad. Nat. Sci. Phila. July, 1890, p. 245.

Perithecia pale horn-color, subglobose (250 μ), immersed in a rather scanty, loose, white tomentose mycelium (stroma), which overspreads the surface of the pores and covers the sides of the perithecia themselves, leaving their apices and papilliform ostiola bare. Asci cylindrical, 65–75 x 4–4½ μ, containing eight oblong-elliptical, 2-celled, hyaline sporidia, the cells subcubical or nearly globose, 3 μ diam. and readily separating. The upper part of the perithecium collapses when dry, and, in old or weather-beaten specimens, the tomentose stroma disappears, leaving the perithecia sessile on the pores. We have seen no specimens of *H. polyporoidea*, B. & C., but our species will be distinct from that, in the absence of any crust-like stroma and in its smaller sporidia. It was first found in October, 1880, and again in October, 1886.

Parasitic on decaying *Polyporus cæsius*, Newfield, N. J.

Specimens found by Dr. Macoun in Prince Edward's Island, on *Pol. chioneus*, Fr., agree perfectly with the Newfield specimens, only there is an orange-colored mycelium which stains the *Polyporus* within of a fine, light yellow.

H. melaleùca, E. & E. Proc. Phila. Acad. July, 1890.

Subiculum membranaceous, thin, white, covered except the margin with a single layer of minute (110–130 μ), slate-colored perithecia filled with globose sporidia? (or perhaps stylospores, as no asci were seen).

This seems to be a distinct species, but requires further observation with more perfectly developed specimens. Seems different from *H. olivacea*, C. & E.

H. corticiícola, E. & E. Journ. Mycol. I, p. 140. **(Plate 11)**

Perithecia globose, pale, 75–100 μ diam., buried in the stroma and visible under the lens as horn-colored specks. Asci clavate-cylindrical, 20–22 x 3½ μ, sessile, without paraphyses. Sporidia biseriate, eight in an ascus, each consisting of two globose, hyaline cells, easily separating and 1–1½ μ diam. (mostly 1 μ or a little over). This is closely allied to *H. hypomycella*, Sacc. Mich. I, p. 302, Syll. II, p. 529, but differs in its asci and sporidia being only about half as large as in that species.

Parasitic on *Corticium scutellare*, B. & C., or some closely allied species, on dead limbs of *Magnolia* and oak, Newfield, N. J.

H. subcárnea, E. & E. Journ. Mycol. III, p. 114.

Perithecia carnose, pale, minute (80 μ), buried and barely visible under the lens as minute specks, giving the surface of the *Corticium* a punctate appearance. Asci subcylindrical, sessile, without paraphyses, 30–35 x 5–7 μ. Sporidia uniseriate or partly biseriate above, subhyaline (with a yellowish tint), oblong-elliptical, 1–2-nucleate, 3½–4½ x 2–2½ μ. Outwardly this scarcely differs from *H. corticiicola*, E. & E., but the sporidia are very different, much like those of *H. consimilis*, Ell., from which, however, it is quite distinct.

Parasitic on some thin *Corticium*, on dead limbs of *Lonicera* lying on the ground, Newfield, N. J.

H. láctea, Fr. Summa. Veg. Scand. p. 383.

"Carnose, broadly effused, bare, milk white, ostiola punctiform. Asci cylindrical, 56 x 4 μ, subequal cells of the didymous, hyaline sporidia, globose, 3 μ diam." In Sacc. Syll. II, p. 529, the species is credited to North America. We have seen no specimens. The habitat is given as on rotten wood, on *Polyporus medulla-panis* and on the ground.

H. viridirùfa, B. & Rav. Grev. IV, p. 14.

"Subglobose, congested, or confluent, greenish-rufous. Ostiola impressed. Sporidia oblong, with two nuclei."

On dead alders, South Carolina (Ravenel).

In Grevillea XI, p. 129, this is referred to Hypoxylon, but if the specimen of *H. rufo-viridis*, B. & Rav., in Rav. Car. fasc. V. No. 53, is the same as *H. viridi-rufa*, B. & Rav., in Grev. l. c., the stroma is not carbonaceous (as it should be in *Hypoxylon*), but carnose. The specimen referred to is, in our copy, without fruit—apparently immature.

H. solenóstoma, B. & Rav. Grev. IV, p. 14.

"Subglobose, pale rufous, rather irregular. Ostiola cylindrical, elongated. Sporidia globose, 4 μ diam."

On decaying *Pachyma cocos*, Schw., Carolina.

B. Stroma pulvinate, sporidia colored.

H. gelatinòsa (Tode).

Sphæria gelatinosa, Tode. Fungi Meckl. II, pp 48 & 49.
Hypocrea gelatinosa, Fr. Summ. Veg. Scand. p. 382
Exsicc. Thum. M. U. 2163

Stromata gregarious, superficial, pulvinate, or subhemispherical, carnose, soft, punctate from the slightly prominent ostiola, 1½–3 mm. in diam. at first with a thin, light-colored tomentum at the base, pale, becoming yellowish or at length greenish, whitish within, subrugose, and partially collapsing when dry. Asci cylindrical, contracted into a short pedicel at the base, 80–90 x 3½–4½ μ, 8-spored. Sporidia composed of two unequal cells, the upper nearly spherical (4 μ), the lower ellipsoid, or ovoid, (3 μ), yellowish. Probably common throughout. Var. *viridis* (Tode) is reported by Peck on maple chips, New York State.

On rotten wood (*Carya &c.*), South Carolina (Ravenel), Pennsylvania (Everhart), Connecticut (Thaxter). The stroma is at first of a yellowish-horn-color, becoming dirty-yellowish and dusted with the greenish sporidia (brownish-yellow under the microscope).

H. chloróspora, B. & C. Grev. IV, p. 14.

Stromata small, greenish-black, nearly round, sessile, convex, (1–1½ μ), roughened by the rather prominent ostiola. Asci narrow-cylindrical, about 75 x 4 μ, with eight two-celled sporidia, each cell subcubical or nearly globose, of an olivaceous color and 3–3½ μ in diam.

On decaying bark, Newfield, N. J. also reported from New York.

H. chromospérma, C. & P. 29th. Rep. N. Y. State Mus. p. 57.

"Fleshy, soft, convex, orbicular, 1–2 lines broad, flattened and patellate when dry, whitish or watery tan-color. Ostiola slightly prominent. Asci cylindrical. Sporidia quadrate-globose, brownish when mature, 4–5 μ."

On decaying wood, Buffalo and Greenbush, N. Y.

H. cerámica, E, & E. (in Herb.)

Stroma convex 1–2 mm. diam. round, pale brick-red, (white inside), wrinkled when dry and finally dusted with the greenish sporidia as in the preceding species. The first appearance is a speck of white tomentum which soon shows the brick-red color in the center and remains for some time around the base of the stromata and more or less effused on the matrix. Asci cylindrical, p. sp. 55–60 x 4 μ. Sporidia two-celled, the upper cell subquadrate-globose, about 4 μ diam., the lower mostly ovate 4–4½ x 3–3½μ, brownish.

On bark of decaying limbs of *Juniperus*, Connecticut (Thaxter).

H. rufa has the stroma of the same color but has hyaline sporidia. *H. gelatinosa* has about the same sporidia but the color of the stroma is different as it also is in *H. chlorospora* (black or at least dark.) *H, scutellæformis* has the stroma of a deeper red and smooth, with a thin, free margin.

C. *Stroma effused, sporidia hyaline.*

H. cítrina, (Pers.)

Sphæria citrina, Pers. Syn p. 18
Hypocrea citrina, Fr. Summ Veg. Sc p. 185
Exsicc, Rab F E. 629.—Rehm Asc 677.

Stroma effused, thin, carnose, lemon-yellow, punctate from the rather prominent ostiola, forming a thin crust overspreading decaying wood and bark, or sometimes decaying leaves and mosses for several inches or even a foot in extent. Perithecia entirely sunk in the stroma, crowded, spherical, yellowish. Asci cylindrical, yellowish, 80–90 x 5–6 μ. Sporidia 8 in an ascus, two-celled, the cells soon separating, the upper one spherical 4½ μ diam., the lower one ovate, about 5 x 4½ μ.

Common throughout the U. S. and Canada.

H. tremellícola, E. & E. (in Herb.)

Perithecia ovate, minute, 100–115 μ diam., sunk in the thin, cinereous-white, byssoid-crustaceous stroma which forms a continuous layer on the matrix, the margin at first byssoid and loose but soon smooth and subcrustaceous or nearly vanishing. Asci cylindrical, p. sp. about 75 x 5 μ. Sporidia uniseriate, elliptical, hyaline, 1–2-nucleate (becoming uniseptate), 7–8 x 3 μ.

Parasitic on *Tremella albida*, Ohio Morgan, 894

The perforated apices of the perithecia just visible give the surface of the stroma a minutely punctate appearance.

D. *Stroma discoid. Sporidia continuous, hyaline.*

H. consímilis, Ell. Grev. XII, p. 79. **(Plate 11)**

Exsicc. Ell, N. A. F., 158.

Stroma orbicular or elliptical, convex, 2–3 mm. across, brick-red, wrinkled, carnose. Asci clavate-cylindrical, 60–70 x 4 μ. Sporidia uniseriate. hyaline, 10–12 x 3½–4 μ.

On dead *Azalea viscosa*, Newfield, N. J.

H. Richardsòni, Berk. & Mont. Grev. IV, p. 14.

Exsicc Ell. N A. F., 1329.

Discoid-tubercular, scattered or gregarious, dull purplish-red, centrally attached, ½ cm. across, deeply wrinkled, margin sublobate and free, whitish within. In Grevillea l. c., Berkeley states that the asci are clavate and the sporidia elliptical, and that it was first found in one of the Arctic expeditions by Sir J. Richardson. All the specimens we have seen are entirely sterile, like those in N. A. F., 1329. *Tubercularia pezizoidea*, Schw., is said to be the same. Its range appears to be northward from Maine to Wisconsin and west to Colorado and Utah.

On bark of dead poplar.

E. *Stroma pulvinate or effused, sporidia fusoid, hyaline, uniseptate.*

H. apiculàta, C. & P. 29th Rep. N. Y. State Mus. p. 57.

"Fleshy, soft, growing in irregular patches, smooth, ochraceous, inclining to orange, the extreme margin barren. Asci cylindrical. Sporidia fusiform with an apiculus at each end, uniseptate, colorless, 27–37 x 7½–10 μ".

On the ground and on rocks, Catskill mountains, New York (Peck).

H. papyràcea, Ell. & Hol. Journ. Mycol. II, p. 66.

Stroma membranaceous, thin, separable, 2–3 cm. across, white with a yellow, substerile margin. Perithecia superficial, fawn-colored, small (150 μ), thickly scattered on the stroma. Asci slender, about 75 x 3 μ, (spore-bearing part about 60 μ), without paraphyses. Sporidia fusoid, hyaline, uniseptate, 8–10 x $2\frac{1}{2}$–3 μ, readily separating at the septum. The yellow margin may be only accidental. This differs from *H. corticioides*, B. & Br., in its larger sporidia and different color.

Under side of an old log, Decorah, Iowa (Holway), Ohio (Morgan).

H. citrinélla, Ell. Bull. Torr. Bot. Club, VI, p. 108. **(Plate 11)**

Stromata scattered or subconflnent, minute (1–2 mm.), thin-pulvinate, bright lemon-color, atro-punetate from the minute ostiola. Asci slender, 100–120 x 5–6 μ. Sporidia fusoid, hyaline, uniseptate, uniseriate, 12–14 x $3\frac{1}{2}$ μ.

On dead twigs and limbs of *Vaccinium corymbosum* not yet fallen and not much decayed.

In the original description, the true character of the sporidia was overlooked, the specimens first found being rather old and the cells of the sporidia separated.

H. lichenoìdes, (Tode.) **(Plate 11)**

Acrospermum lichenoides, Tode Fungi Meckl I, p. 9.
Sphæria riccioidea, Bolt. Fungi Halifax IV, p. 174.
Hypocrea parmelioides, Mont Syll 210
Hypocreopsis riccioidea, Karst Myc Fenn II, p 221, 251.
Hypocrea digitata, E. & E Journ Mycol I, p. 42.

Stromata carnose, yellowish, digitate, radiating from a central point and dividing into numerous (2 mm. diam.) semicylindrical, finger-like lobes, closely appressed to and surrounding the matrix and extending longitudinally for about 5 cm. The rounded ends and the sides of the lobes are sterile, the perithecia being found only on the upper or outer surface; perithecia immersed, their position being indicated only by prominent but minute black ostiola. Asci cylindrical, 80–90 μ long. Sporidia uniseriate, ends mostly overlapping, hyaline, uniseptate, oblong or narrow-elliptical, often subinequilateral, 20–26 x 6–8 μ. The stroma is like the fingers of a hand clasping the limb.

On a dead limb, White mountains, N. H. (Miss Minns).

F. *Sporidia continuous, brown.*

H. bìcolor, E. & E. Journ. Mycol. IV, p. 58.

Stromata gregarious or crowded, convex, suborbicular, dull cinereous, becoming dull black, attached as in *H. Schweinitzii*, Fr., which when mature it much resembles. Perithecia peripherical, globose, about $\frac{1}{2}$ mm. diameter, buried in the stroma, which is of a dull white color within and has the surface minutely roughened by the punctiform ostiola. Asci cylindrical, 70 x 5 μ. Sporidia uniseriate or crowded above, elliptical, continuous, smoky-brown, about 5 x 2½ μ.

On a decaying elm log, Manhattan, Kansas (Kellerman & Swingle).

G. sporidia uniseptate, brown.

H. cubíspora, Ell. & Hol. Journ. Mycol. I, p. 4.

Stroma obconic-tuberculiform, subplicate below, about 1 cm. high, lemon-yellow within and without, surface minutely punctate with the black ostiola. Perithecia peripherical, globose, about 250 μ diam , contents black. Asci cylindrical, containing eight cubical or oblong-truncate, dark-olive or brownish-black, 2-nucleate, 4–7 x 3–4 μ sporidia, some of which are obscurely uniseptate.

On a decaying log, Iowa.

H. *sporidia 3-septate, yellowish or brown.*

H. chlòrina, Cke. Grev. VII, p. 49.

Exsicc. Rav. F. Am 342

Flattened, discoid, elliptical or elongated, 1–2 mm. diam., clay-colored (bright-yellow within). Perithecia immersed, brown, ostiola blackish, punctiform. Asci clavate, spore-bearing part about 75 x 15 μ, surrounded with abundant filiform paraphyses. Sporidia biseriate, narrow-elliptical, endochrome three times divided, yellowish (becoming brown ?), 20–25 x 8–9 μ. On bark of hickory, Darien, Ga. The stroma is of about the same color as the bark, flatter than in the preceeding species, but of about the same color.

I. *Stroma erect.*

H. Petérsii, B. & C. Grev. IV, p. 13.

"At first sight this looks like an Agaric invested with some *Hypomyces*, but the fructification is exactly that of a *Hypocrea*. Stem irregular, dilated upwards, about an inch high. Head orbicular, irregular, rufous. Perithecia both on the under and upper sides. Sporidia globose in linear asci."

Found in Alabama, by Hon. J. M. Peters.

There can be but little doubt that the stipe and agariciform head are a real Agaric on which the *Hypocrea* is parasitic.

H. alutàcea (Pers.)

Sphæria alutacea, Pers Comment. Clavar p 12.
Sphæria clavata, Sow. Brit. Fungi tab II, fig. 159.
Cordiceps alutaceus, Lk. Hndbk de Gewachse III, p. 374.
Hypocrea alutacea, Tul Sel. Carp. I, p. 62 Winter Die Pilze II, p 142.
Exsicc Rab F. E. 132, 246

"Perithecia immersed, obtusely papillate and at length subprominent, 200–225 μ diam. Asci cylindrical, 56 x 4 μ. Sporidia didymous, upper cell globose (4 μ), lower cell subovate, 4 x 3 μ, hyaline."

Specc. on bark of decaying (maple?) limb on the ground, Newfield, N. J., have (stroma?) 2 cm. high, clavate, leather-color. Only two specimens were found, and those were immature, so that the fructification could not be made out, though the surface of the club was finely punctate from the ostiola of the immersed perithecia. According to Tulasne & Broome, the club-shaped body, on the upper part of which the perithecia are borne, is not a true stroma, but either *Clavaria ligula* or *Spathularia flavida* bearing the *Hypocrea* as a parasite.

The following species are imperfectly described:

H. atramentòsa, B. & C. Journ. Linn. Soc. X, p. 377.

"Forming a thin, black stratum on the under side of the leaves of grasses in Cuba, and of *Andropogon* in Alabama. Perithecia globose and, with the ostiola, immersed. Sporidia filiform. (See p. 91).

H. parásitans, B. & C. Grev. IV, p 15. "Minute, pallid, subelliptical, sometimes winding around the teeth. Sporidia globose, rather large." On *Hydnum erinaceum*, South Carolina.

H. subvíridis, B. & C. l. c. "Effused. Perithecia pale dull green, tomentose, crowded, seated on a white mycelium. A curious species."

On dead grass leaves, South Carolina.

H. sterílior, (Schw.)

Sphæria sterilior, Schw Syn N. Am 1188. *Hypocrea sterilior*, Cke. Syn. 168.

Substance at first very soft; broadly effused, applanate, surface longitudinally striate, flesh-color, becoming light yellow when dry. The margin is very delicate, cottony, with interwoven fibers with which the whole surface appears lightly covered. Texture carnose, becoming horn-like when dry. Perithecia few, scattered.

On leaves, &c., Carolina (Schw.). Stroma about an inch in circumference, and two lines thick.

H. sublobàta, (Schw.)

Sphæria sublobata, Schw. S N. A 1225. *Hypocrea sublobata*, Cke. Syn. 169

Scutellate, small, slightly attached, margin obtuse, lobate-repand, black, then subolivaceous: surface flat, rugulose. Perithecia subperipherical, in a single layer, becoming yellow, immersed in the light yellow stroma, which is about 3 lines in diameter. Ostiola impressed.

On bark of *Platanus*, Bethlehem, Pa. (Schw.).

H. molliúscula, (Schw.)

Sphæria molliuscula, Schw. Fr. El. II, p. 66. *Hypocrea molliuscula*, Cke. Syn 172.

"Minute (1 line across), round, plano-convex. Perithecia small, entirely hidden, connate, surface of the stroma roughened by the ostiola, pruinose, sooty black. On rotten wood, Pennsylvania."

Hypocrea Ravenelii, Berk. Grev. IV, p. 14, Rav. Car. V, 51, *Diatrype lateritia*, Ell. Bull. Torr. Bot, Club, IX, p. 19, "*Hypoxylon myriangioides*, B. & C.," in Ell. N. A. F. 471, are all *Melogramma Bulliardi*, Tul. Roumeguere, in his Fungi Gallici, 1174, has issued the same thing as *Hypoxylon fuscum* (Pers.).

HYPOCRÉLLA, Sacc.

Syll. II, p. 579.

Stroma subcarnose, pulvinate, disciform or effused, bright colored or dark. Perithecia more or less perfectly buried in the stroma. Asci cylindrical, 8-spored. Sporidia filiform, nearly as long as the asci, frequently separating into short sections or joints.

H. tuberifórmis. (B. & Rav.)

Hypocrea tuberiformis, B. & Rav. Grev. IV, p. 13.
Hypocrella tuberiformis, Atkinson, Bot. Gaz. XVI, p. 282, pl. XXV, figs. 1-6.

Stroma subglobose, 1 cm. or more diam., entire, lobed or divided, fastened to the leaf by a mycelium of whitish, radiating threads. At first thickly covered with erect, fertile hyphæ 35–40 x 2–3 μ, bearing oval or broadly fusoid, inequilateral, hyaline, continuous conidia 7–10 x 3½–4 μ. When mature, a section shows three different colored strata; the inner white, the intermediate one light ochre, and the outer one cinnamon. Perithecia sessile, or only their rounded bases immersed. subcylindrical, a little broader in the middle, cinnamon-color, 3–20 or more together, 1 mm. long, ⅓ mm. broad, frequently branched, the bases of 2–3 joined, and the cavities confluent below. Asci very large, 450–750 x 14 μ, tapering to a slender point below, more gradually towards the truncated apex. Sporidia linear, hyaline, pluriguttulate and pluriseptate, rounded at the ends and separating at the septa.

On *Arundinaria*, Carolina (Ravenel), Alabama (Atkinson).

H. phyllógena, Mont. Syll. 711.

Stroma pulvinate, hemispherical, base constricted and orange-colored. Perithecia peripherical, erect, ovate and, with the punctiform ostiola, bright purple, sunk in the upper part of the stroma, which is of the same color. Asci linear with the apex cap-shaped or obtusely conical. Sporidia linear, curved, finally breaking up into segments 16–18 x 2 μ.

On living leaves of *Centaurea speciosa*, Cayenne, South America.

We have included this species, which will, not improbably, yet be found in Southern Florida or Mexico.

H. Hypóxylon, (Pk.)

Dothidea vorax, D. atramentaria and *D. pilulæformis*, B. & C. Grev IV, p. 105.
Epichloe Hypoxylon, Pk. 27th Rep. N. Y. State Mus p. 108
Hypocrea atramentosa, B. & C. Journ. Linn. Soc. X, p. 377?
Hypocrella Hypoxylon, Sacc. Syll. 5069.
Exsicc. Ell. N. A. F. 683.

Stroma effused, thin, grayish at first, becoming black, white inside, continuous or interrupted, extending along the upper surface of the leaf or enveloping the culm for 2 or more cm. in length. Perithecia small, crowded, semiemergent, with papilliform ostiola. Asci narrow, linear, 100–120 x 5–6 μ. Sporidia filiform, multinucleate, hyaline, nearly as long as the asci and about 1 μ thick.

On culms and leaves of living grasses, from Canada to Carolina and Louisiana.

Var. *pilulæformis*, B. & C., has the stroma tubercular or otherwise irregular. *Ephelis borealis*, E. & E., Journ. Mycol. I, p. 86, is only the stylosporous stage of this species.

EPÍCHLOE, Fr.

Summ. Veg. Scand. p 381.

Stroma effused, at first pale and conidiophorous, at length lemon-yellow, subcarnose, surrounding the culms like a sheath. Perithecia immersed, with the ostiola scarcely prominent. Asci with 8 filiform sporidia.

E. týphina, (Pers.) **(Plate 15)**

Sphæria typhina, Pers. Icon. et Descr. I, p 21.
Polystigma typhinum, De Cand. l. c. p. 338.
Dothidea typhina, Fr S. M. II, p. 553.
Epichloe typhina, Tul. in Ann. Sci. Nat. Ser IV, t. XIII, p 18
Stromatosphæria, Grev , *Typhodium*, Lk. in Fr. S. M. III, p. 362.
Exsicc Kz F. Sel. 344.—Rab. Herb. Myc. 578.—id. F E. 541, 2237.—Rehm Asc. 428 Thum. F. Austr. 254 —id. M. U. 1065 —Schweiz. Cr. 111 —Sydow M M 65 —Ell & Evrht. N. A. F. 2d Ser. 1553.

Stroma pale, thin, surrounding the sheaths and included culms of living grasses, *Phleum pratense*, *Dactylis glomerata* (and *Carex*, fide Peck), extending longitudinally for 2–5 cm., and bearing in the early stage of growth, small (4–5 x 3 μ), ovoid, hyaline conidia (*Sphacelia typhina*, Sacc.), finally covered with a layer of semiimmersed, soft, carnose-membranaceous, yellow perithecia, with somewhat prominent ostiola. Asci cylindrical, 150–230 x 7–9 μ, slightly narrowed above, with the apex truncate and capped with a subhemispherical, hyaline crest. Sporidia 8 in an ascus, filiform, yellowish, multinucleate and nearly as long as the asci.

The species is common in Europe, and appears also to be widely diffused in this country. It is reported from Carolina to Pennsylvania and Iowa; and from Northern New York and Canada.

THYRONÉCTRIA, Sacc.

Grev. IV, p. 21, and Mich. I, p. 325.

Stroma valsiform or linear, covered by the bark and only partially erumpent. Perithecia monostichous, subcarnose, reddish, but covered with a yellow-furfuraceous coat, except the short ostiola. Asci 8-spored, paraphysate. Sporidia oblong, muriform, hyaline or subhyaline, or at least in one American species, brown.

Th. Xanthóxyli, (Pk.)

Valsa Xanthoxyli, Pk. 31st Rep. p. 49.
Fenestella Xanthoxyli, Sacc. Syll. II, p. 332.

Pustules slightly prominent, with a yellowish-furfuraceous, lanceolate disk, which is dotted by the black ostiola. Perithecia 2–15 in a stroma, rarely single, fragile, pale outside, reddish inside, small (about ¼ mm.), subcircinate, enclosed in the slightly altered substance of the bark, with a tawny-yellowish, floccose-tomentose substance surrounding them and filling the spaces between them, attenuated above into slender necks, with the short, black, obtuse, papilliform, then narrowly perforated ostiola erumpent in the light, sulphur-yellow, elongated disk which bursts through longitudinal cracks in the bark, but does not rise above it. Asci oblong-clavate, p. sp. 65–75 x 12–15 μ, subsessile, paraphysate, 8-spored. Sporidia biseriate, yellowish, oblong-cylindrical, obtuse, slightly curved, faintly 3–5-septate and muriform, 18–22 x 6–9 μ. The septa are very faint, and in many cases scarcely visible. The ostiola are at first covered with a greenish-yellow powder.

On dead limbs of *Xanthoxylum Americanum*, Troy, N. Y. (Peck).

This is very near *Th. Patavina*, Sacc, but seems to differ in its 5-septate sporidia and ostiola united in a disk.

Th. virens, Hark. (in Herb.)

Perithecia cespitose, seated on the surface of the inner bark, bursting through and surrounded by the upturned epidermis, in compact clusters of 5–12, globose, ⅓ mm. diam., clothed with a dense, greenish-yellow, tomentose-furfuraceous coat. Ostiolum papilliform, black, tardily appearing. Asci clavate-cylindrical, 100–110 (p. sp. 70–75) x 12–15 μ; paraphyses evanescent. Sporidia biseriate, oblong-cylindrical, obtuse, curved, hyaline, 6–7-septate and muriform, 18–22 x 6–8 μ.

On dead limbs of *Rhus*, California (Harkness), Connecticut (Thaxter), on dead ash, Canada (Dearness).

Var. *chrysogramma* (*Thyronectria chrysogramma*, E. & E. in Proc. Phil. Acad. l. c. p. 245), on bark of dead elm limbs, Potsdam, N. Y. (Ellis), Missouri (Demetrio), and Kansas (Kellerman), differs from the type in its larger, brown sporidia, larger asci and more or less scattered perithecia.

ELEUTHEROMỲCES, Fckl.

Symb p. 183.

Perithecia superficial, gregarious, subulate, attenuated above from an ovate-cylindrical base. Asci cylindrical, 8-spored. Sporidia elliptical, with a bristle-like appendage at each end, hyaline.

E. subulàtus, (Tode.) **(Plate 14)**

Sphæria subulata, Tode. Fung Meckl. II, p. 44
Sphæronema subulatum, Fr. S. M. II, p. 536.
Eleutheromyces subulatus, Fckl. Symb. p. 183
Exsicc. Rab Herb Myc 662 —id. F E 1334.

Perithecia mostly thickly gregarious, but sometimes scattered, superficial, gradually attenuated upwards from the base, ½–1½ mm. high, yellowish, translucent, soft, becoming hard and brownish. Asci cylindrical, with a stipe-like base, 8-spored, 48–52 x 2½–3 μ. Sporidia uniseriate, elliptical, acute at the ends and prolonged into a bristle-like appendage, hyaline, 4–6 x 1½ μ.

On dried up and decaying Agarics, Massachusetts and New York.

NÉCTRIA, Fr.

Summa Veg. Scand. p. 387.

Perithecia free, cespitose, on a tuberculiform, carnose, conidial stroma (*Tubercularia*), or scattered without any definite stroma, carnose-membranaceous, mostly bright colored, (red &c.), smooth, subvillose, squamulose, &c. Asci cylindric-clavate, 8-spored, mostly without paraphyses. Sporidia oblong or elliptical, hyaline, uniseptate.

A. *Perithecia cespitose.*

N. cinnabárina, (Tode.)

Sphæria cinnabarina, Tode. Fung. Meckl. II, p. 9.
Sphæria decolorans, Pers. Syn. p. 49.
Sphæria pezizoidea, *a. rubrofusca*, DC. Flor. Franc. tom. VI, p. 125
Cucurbitaria cinnabarina, Grev. Scot. Crypt. Flor. tom. III, tab. CXXXV
Nectria cinnabarina, Fr. Summ. Veg. Scand. p. 388
Exsicc. Rab. Herb. Myc 633.—Rab. F. E. 324, 1631.—Rehm Asc. 184, 282, 635.—Thum F. Austr. 1050, 1052.—M. March. 347, 348, 349.

Perithecia densely cespitose, bright red, becoming darker, rough, with a papilliform ostiolum, seated on a pulvinate, tuberculiform, fleshy stroma (*Tubercularia vulgaris* Tode). Asci clavate-cylindrical, subattenuated above, 70–90 x 8–11 μ. Sporidia subbiseriate, oblong, obtuse at the ends, straight or slightly curved, uniseptate, hyaline, mostly 12–15 x 5–7 μ.

On dead limbs of almost every kind of deciduous trees, common and variable.

N. Sàmbuci, E. & E. Proc. Acad. Nat. Sci. Phil. July, 1890, p. 246.

Cespitose on a tubercular base (*Tubercularia Sambuci*, Cda.) Perithecia 4–12 on a stroma, ovate-globose, pruinose, pale red, about ⅓ mm. diam., strongly collapsed above when dry. Ostiolum papilliform, finely fimbriate. Asci oblong-clavate, 50–60 x 6–7 μ (p. sp.), without paraphyses. Sporidia biseriate, oblong, uniseptate, straight or slightly curved, hyaline, 12–20 x 3–4½ μ. The *Tubercularia* has allantoid, hyaline conidia 6–8 x 1½–2 μ, on basidia 35–40 μ long, branched above. This is, according to the specimens in De Thümen's Mycotheca and in Roumeguere's Fungi Gallici, the *T. Sambuci*, Cda.

On *Sambucus Canadensis*, Lincoln, Nebraska (Webber)

Possibly a var. of *N. cinnabarina.*

N. pithoìdes, E. & E. Proc. Phil. Acad. July, 1890, p. 247.

Densely cespitose, forming suborbicular tufts 1½–2½ mm. diam. Perithecia ovate, dark red, about 200 μ diam., muriculate-roughened, collapsing above so as to appear slightly truncate and slightly concave, appearing in profile like small jars. Ostiolum papilliform, only slightly prominent. Asci cylindrical, 75–80 x 5 μ. Paraphyses not seen. Sporidia uniseriate, oblong-elliptical, 2-nucleate, becoming uniseptate, 6–10 x 3–3½ μ, smoky -hyaline. The perithecia are seated on a convex, yellow stroma, 50–100 together, and when young are clothed with a few short, white, glandular hairs. Nearly allied to *N. microspora*, C. & E. which has less numerous, paler red, smoother, more irregularly collapsed perithecia. The specific name from Greek *pithos* a barrel.

On dead alders, British Columbia, May, 1889 (Macoun, 122).

N. Russéllii, B. & C. Grev. IV, p. 45.

"Cespitose, red, inclining to brown. Ostiolum papilliform, at length sunk from collapsing. Sporidia cymbiform, uniseptate, 15–20 μ long. On elm, New England, Russell)." Var. *Magnoliæ*, Sacc.

differs somewhat from the type in its shorter (10–11 x 5–6 μ) sporidia slightly constricted, with the lower cell a little narrower, and the perithecia at length collapsing.

On bark of *Magnolia*, South Carolina (Ravenel).

N. offuscàta, B. & C. Grev. IV, p. 45.

"Cespitose, dingy, dark brown-red, minutely granulated, ostiolum depressed. Asci clavate. Sporidia biseriate, oblong, about one-fourth as broad as long; externally resembling *N. Russellii*. On *Hibiscus Syriacus*, South Carolina."

N. coccinea, (Pers.)

Sphæria coccinea, Pers. Syn. p. 49.
Nectria coccinea, Fr. Summ Veg. Scand p. 368
Exsicc. Rab. F. E. 924, 1630.—Thum M. U. 1063, 1850 —Ell N. A. F. 161 —Plowr. F. Br. 8 Sacc M. V. 1482 —Rav. F. Am. 737.—Roum. F. G. 272, &c.

Perithecia cespitose, ovoid, smooth, bright red, papilliform, about 200 μ diam., usually not collapsing, seated on a yellowish, slightly erumpent stroma, which is often nearly obsolete. Asci subcylindrical, 80–95 x 6–7 μ. Sporidia uniseriate, uniseptate, hyaline or nearly so, scarcely constricted, rather acutely elliptical, 12–15 x 4–5 μ (12–16 x 5–7 μ, Sacc.)

On bark of various deciduous trees, common.

N. muscívora, Berk. in Rav. Fung. Car. I, p. 57.

Nectria subcoccinea, Sacc. & Ell. Mich. II, p. 570.
Nectria muscivora, B. & Br. Brit. Fungi No. 608?
Exsicc. Rav. Fung Car I, 57.—Ell. N A. F. 1333.

Perithecia mostly cespitose, bright red, subovate, more or less collapsing, small (200 μ diam.), seated mostly around the margin of the small, pale, tuberculiform stroma, which, together with its group of perithecia, is mostly less than 1 mm. diam. Asci cylindrical, 75 x 8–10 μ. Sporidia uniseriate, narrow-elliptical, 14–16 x 6–7 μ, yellowish-subhyaline, becoming uniseptate.

On bark of living alder, West Chester, Pa. (Everhart & Haines).

Distinguished from *N. coccinea* by its broader, more obtuse, yellowish sporidia and its more distinctly superficial stroma, which, in some cases at least, seems to arise from the remains of dead scale insects which are abundant on the bark.

The above description is from the Pennsylvania specimens of *N. subcoccinea*, S. & E., which are the same as the specimens of *N. muscivora*, Berk. cited, which is presumably the same as *N. muscivora*, B. & Br., in Cooke's Handbk. No. 2364, though neither the

Pennsylvania specimens nor those in Rav. Car. show anything of the "white lanose patches 2 in. or more in diameter."

N. díploa, B. & C. Journ. Linn. Soc. X, p. 378.

Perithecia cespitose, minute, ovate, subfurfuraceous, at length collapsing, light red, parasitic on some erumpent *Valsa?* Asci subsessile, oblong-cylindrical, about 65 x 10 μ. Sporidia obliquely uniseriate, oblong-elliptical, endochrome finally divided in the middle, 20–25 x 9–11 μ, hyaline or nearly so. In some of the asci, the sporidia are partially biseriate and somewhat smaller.

On bark of alder, South Carolina (Ravenel).

The description here given is from an examination of the specimens in Rav. Fungi Caroliniani, III, 55. In these specimens the nuclei have disappeared and the underlying (*Valsa?*) is so completely covered by the perithecia of the *Nectria* as to be easily overlooked. Differs from the preceding species in its more compactly clustered and less prominent perithecia, and its larger sporidia.

Var. *diminuta*, Grev. IV, p. 46, is (sec. Cooke) a *Calonectria* with 3-septate sporidia. See p. 114 of this work.

N. verrucòsa, (Schw.) **(Plate 12)**

Sphæria verrucosa, Schw. Syn. N. Am. 1401.
Sphæria dematiosa, Schw. l. c. 1424 (?)
Nectria verrucosa, Sacc. Syll. 479.
Exsicc. Rav. F. Car. I, 52.—Ell. & Evrht. N. A. F. 2d Ser. 2371.

Perithecia cespitose, globose, about ⅓ mm. diam., verrucose-roughened, light brick-red, becoming much darker and finally collapsing above, more or less distinctly clothed with weak, short, rudimentary, hyaline, glandular hairs, seated on an orange-red, concave or depressed stroma (*Tubercularia*), forming groups 1–2 mm. diam., at length more or less deciduous. Asci oblong-cylindrical, with a short, substipitate base, 60–70 x 10–12 μ. Sporidia biseriate, oblong, uniseptate, 12–18 x 4–5 μ. The stroma, as in most other *Nectrias*, is finally hidden and partially obliterated by the perithecia.

Common on *Morus* and *Sassafras*, Pennsylvania (Schweinitz), on *Morus*, *Melia*, etc., South Carolina (Ravenel), and on *Morus*, New Jersey (Ellis).

Differs from *N. cinnabarina* in its concave, scarcely prominent stroma, the peculiar roughening of the perithecia, and in its shorter asci, and mostly narrower sporidia. *N. coccinea* has the perithecia nearly smooth, or when dry, slightly furfuraceous.

N. ochroleùca, (Schw.)

Sphæria ochroleuca, Schw Syn N Am No. 1418
Nectria ochroleuca, Berk. in Grev. IV, p 16
Exsicc Ell. N. A. F. 773.

Stroma granulose-byssoid, subpulvinate and nearly white at first, becoming firmer and yellowish. Perithecia 3–15 on a stroma, ovate-globose, dull yellowish-white, mostly less than $\frac{1}{3}$ mm. diam., surface densely furfuraceo-squamulose, except at the apex, around the rather darker, slightly depressed, papilliform ostiolum, where the edge of the squamulose coat forms a miniature crown or wreath, giving the apex of the perithecium the appearance of the blossom end of an apple. Asci clavate-cylindrical, 45–55 x 10–12 μ. Sporidia biseriate, oblong-fusoid, uniseptate, 12–16 x 4–5 μ.

On limbs of various deciduous trees, Carolina and Pennsylvania (Schweinitz), on dead limbs of *Laurus Benzoin*, West Chester, Pa. (Everhart).

N. rubicàrpa, Cke. Grev. VII, p. 50., Journ. Myc. II, p. 79.

Exsicc. Rehm Asc. 337.—Ell N A F 80.—Rav F. Am. 341.

"Cespitose, red, scarcely papillate, obtusely verrucose-roughened, collapsing. Asci cylindrical, 65–75 x 6–7 μ. Sporidia uniseriate, elliptical, uniseptate, 10–12 x 4–4$\frac{1}{2}$ μ, mostly not much constricted. Looks like a miniature red raspberry, both in the clusters and individual perithecia, the latter becoming eventually nearly even."

On dead limbs of *Gelsemium*, South Carolina (Ravenel), on stems of *Ilex glabra*, New Jersey (Ellis).

The specimens in Rehm's Asc. and Ell. N. A. F. were erroneously issued as *N. punicea*, Kze. & Schm. In our copy of Rav. F. Am. the specimen (No. 341) has the perithecia distinctly collapsed. In his diagnosis of his *Ascomycetes*, under No. 337, Dr. Rehm refers the specimens of *Nectria punicea*, Kz. & Schm., in N. A. F. No. 80, to *N. rubicarpa*, Cke. We have carefully compared the N. A. F. specimens with *N. rubicarpa* in Rav. F. Am., and they seem to us to be the same thing. Referring to our exsiccati, we find in Plowright's Sph. Brit. No. 206, a specimen labeled *N. punicea*, Kz. & Schm., in which the perithecia are not collapsed and the sporidia 15–19 x 4–5 μ, which are about the measurements given in Sylloge. The specimen of *N. punicea* in Roumeguere's Fungi Gallici, No. 1465, we can not distinguish from *N. cinnabarina*, Fr. If Plowright's specimen is authentic, the N. A. F. specimens can hardly be that species, having most of the sporidia less than 12 μ long and the perithecia collapsed. N. A. F. 772 can not be *N. rubicarpa*, but is more like a pale var. of *N. ditissima*, Tul.

N. ditissima, Tul. Sel. Fung. Carp. III, p. 73.

Exsicc. Thum. M. U. 1156.—Rav. F. Am. 766.—Myc. March. 950, 1546.—Ell. & Evrht. N. A. F. 2d Ser. 1548.—Ell. N. A. F. 772?

Perithecia densely gregarious, small, subglobose, bright red, vertically collapsed when prematurely dried. Ostiolum papilliform, minute. Asci clavate, about 80 μ long and 8–10 μ broad above, contracted below into a slender base. Sporidia crowded, biseriate, fusoid-oblong, uniseptate, slightly curved, 14-16 x 4–4½ μ.

On dead *Acacia*, So. Ca. (Ravenel), on *Melia*, Louisiana (Langlois), on *Ilex*, Connecticut (Thaxter).

N. Celástri, (Schw.)

Sphæria Celastri, Schw. Syn. N. Am. 1421.
Nectria Celastri, Sacc. Syll. 4669.

Stroma tuberculiform, pale yellowish within, becoming black outside, subpulvinate. Perithecia densely cespitose on the stroma, globose-ovate, rugose, about the size of those of *N. cinnabarina*, easily falling from the stroma. Ostiola obsolete, but showing as dark-brown specks. The perithecia at length become indurated, but do not collapse.

Specimens on *Celastrus scandens*, collected in New York State by W. R. Gerard, agree well with the above-quoted characters, except in having the perithecia only 150–200 μ diameter and collapsed. We do not, however, consider this latter character in every case reliable, as it depends, in some measure, on the stage of growth at which the specimens are collected. In Gerard's specimens, the stroma is not well shown, as it is already covered with and partially obliterated by the superimposed perithecia, which are of a bright red color at first, but at length dark red and collapsed, and have the surface subverrucose-roughened. The asci are cylindrical, 55–60 x 6 μ, or sometimes enlarged above to 8 or 10 μ thick. Sporidia uniseriate, elliptical, uniseptate, not constricted, 7–8 x 3½–4 μ. The groups of perithecia are about 1 mm. across. Notwithstanding the slight discrepancies, we are inclined to regard this as the species meant by Schweinitz, who found it rather rare on *Celastrus*, about Bethlehem, Pa. Peck, in 26th Report, also mentions it as found by him at Greenbush, N. Y., but he gives no description of his specimens and we have not seen them.

N. vulgàris, Speg. Fung. Arg. Pug. IV, No. 198.

Perithecia erumpent-superficial, subsolitary or cespitose (10–30 together); at first covered with a conidial layer (*Verticillium tubercularioides* Speg.), then bare, when dry, contracted into various shapes, subconic-lenticular when moist, 250–300 μ diam., amber-yellow or yellow-orange, very smooth, ostiolum inconspicuous, texture thick-membranaceous, minutely and indistinctly parenchymatic, yellowish, inclining to reddish. Asci cylindric-clavate, obtusely rounded above, briefly stipitate below, 40–45, or more rarely 60–70 x 5–7 μ, without paraphyses, 8-spored. Sporidia distichous or obliquely monostichous, elliptic-cylindrical, uniseptate, 10–13 x 3–4 μ, cells equal, 2-nucleate, hyaline.

On stumps of orange trees, Louisiana (Langlois).

N. micróspora, C. & E. Grev. V, p. 53.

Cespitose, bursting out through cracks in the bark in groups of 3–15. Perithecia subglobose, 150–200 μ diameter, orange-red, seated on a scanty, white, web-like mycelium, collapsing above, with a papilliform ostiolum. Asci cylindrical, 50–60 x 5–6 μ (p. sp.). Sporidia elliptical, uniseptate, hyaline, 6–7 x 3½–4 μ. Groups of perithecia ½–1 mm. across. Stroma not conspicuous.

On bark of dead *Magnolia*, Newfield, N. J., and on bark of dead beech, Lake Nipigon, Canada.

In the Newfield specimens the perithecia are only partially collapsed above so as to appear truncate.

N. aureofúlva, C. & E. Grev. VII, p. 8.

Exsicc. Ell. N. A. F. 574.

With the same habit as the preceding species, but perithecia larger (¼mm.), subglobose, smooth, pale golden-yellow, with the minute ostiolum darker. Asci cylindrical, 50 x 6–7 μ. Sporidia biseriate, oblong-fusoid, 8–12 x 3 μ, 2-nucleate, becoming tardily uniseptate, hyaline. Stroma pale, flattish, scarcely rising above the epidermis.

On bark of *Magnolia glauca*, Newfield, N. J.

N. infusària, Cke. & Hark. Grev. XII, p. 101.

Cespitose, erumpent, pale red. Perithecia few, oval, soft-waxy, subconfluent, smooth, glabrous, 5–10 on a stroma. Asci cylindrical, 8-spored. Sporidia uniseriate, elliptical, obtuse, uniseptate, not constricted, hyaline, 10 x 4½ μ. Conidia (*Fusarium Acaciæ*, Cke. & Hark.), either accompanying the perithecia on the same stroma or produced in separate pustules of earlier growth, on a pale red, pulvinate stroma, curved, hyaline, acute at each end, 3-septate, 30–40 x 2½ μ.

On *Acacia* twigs, California (Harkness).

N. atrofúsca (Schw.)

Sphæria atrofusca, Schw. Syn. N. Am. 1429
Exsicc. Ell. & Evrht. N. A. F. 2d Ser. 1547.

Stroma pulvinate, subcarnose, wood-color inside, darker outside, small (½ mm. or about that), erumpent through the epidermis, by the ruptured margin of which it is closely embraced. Perithecia cespitose, nearly black, smooth and glabrous, small, mostly less than 165 μ, conical, becoming ovate or obovate, and finally collapsing above. Ostiolum obtuse-conical, black and shining, rather large. Asci subcylindrical, 45-55 x 7 μ, with abundant, imperfectly developed paraphyses. Sporidia subbiseriate, oblong-elliptical, lower cell sometimes a little narrower, uniseptate, hyaline, 10–12 x 4½ μ.

On dead stems of *Staphylea trifolia*, Bethlehem, Pa. (Schweinitz), West Chester, Pa. (Everhart).

The foregoing description is from specimens distributed in N. A. F. 1547, which agree with specimens in Herb. Schw. The groups of perithecia are often arranged in a subseriate manner, about 1 mm. diam. and, with the stroma to which they are attached, are finally deciduous.

N. nigréscens, Cke. Grev. VII, p. 50.

"Cespitose, red, at length turning black, glabrous. Ostiolum papilliform. Asci cylindrical. Sporidia elongated-elliptical, uniseptate, 18 x 6 μ. Stylospores on the same stroma, some ovate, brown, 5 x 3 μ, others linear, 6 x 2 μ, hyaline."

On *Gleditschia*, Aiken, So. Carolina (Ravenel).

N. curcurbítula, (Tode).

Sphæria curcurbitula, Tode Fung Meckl. II, p. 38 (partly).
Nectria curcurbitula, Fr. Summ. Veg. Sc. p. 338.
Exsicc. Kunze F Sel 105 —Sydow. M. March. 472.—Rehm Asc. 781.

Perithecia ovate-globose, smooth, dark red, with a distinct, papilliform ostiolum, not collapsing, scattered or oftener cespitose on a depressed tuberculiform stroma. Asci cylindrical, 90–100 x 6–7 μ, without paraphyses, 8-spored. Sporidia monostichous, biconic-elliptical, uniseptate, hyaline, 14–16 x 5–6½ μ.

The only American specimen we have seen was on bark of *Abies balsamea*, collected by Mr. Peek at North Elba, N. Y., and sent to us as *N. balsamea*, C. & P. It agrees with the characters above given and with the specimens in the Exsiccati above quoted. We have not seen the spermogonia mentioned by Saccardo in Syll. II, p. 484.

N. rhizógena, Cke. Grev. XI, p. 108.

"Cespitose, erumpent, stromatic, orange-red, at length scarlet. Perithecia subglobose, glabrous, scarcely papillate, breaking out in small groups of 10–12 together. Asci cylindrical. Sporidia narrowly elliptical, uniseptate, hyaline, 8–9 x 3 μ. The conidia are those of a *Tubercularia*, with a rose-colored tint and 5 x 2 μ.

On exposed roots of *Ulmus*, seaboard of South Carolina (Ravenel). Clusters 1 mm. in diam. Perithecia 165 μ." The specimens in Rav. F. Am. 645, show only the depressed-tuberculiform, flesh-colored stroma and conidia.

N. diplocárpa, E. & E. Proc. Phil. Acad. July, 1890, p. 244.

Perithecia gregarious or subcespitose (2–3 connate), superficial, ovate, about ¼ mm. diam., clothed with white, septate, sparingly branched, substrigose hairs, collapsing more or less distinctly above. deep flesh-color, ostiolum papilliform, large and distinct, smooth. Asci clavate, 40–50 x 8–12 μ, filled with reddish granular matter at first. then containing 4 oblong-elliptical, hyaline sporidia, 8–12 x 4–5 μ, uniseptate and more or less constricted at the septum, ends rounded and obtuse, lying irregularly in the asci. Paraphyses apparently present but obscure as are also the asci which are soon dissolved. Together with the sporidia already described are others much larger, 30–45 x 18–25 μ, granular, hyaline, uniseptate and strongly constricted at the septum, oblong-elliptical in shape with the ends obtuse and rounded.

On thallus of some foliaceous lichen (*Parmelia*) ? on trunk of a tree, Farmington, N. Y. (Edgar Brown).

In examining our Exsiccati we find that specimens collected in Missouri by Demetrio on thallus of *Parmelia* and issued by Dr. Winter in his Rabenhorst-Winter Fungi, No. 3252 as *Nectria lecanodes*. Rabh., are the same as this. The description, however, of *N. lecanodes* does not apply to this, that species having sporidia only 9–11 x 3–4 μ. and in fact the specimens of *N. lecanodes* in De Thumen's Mycotheca, 1746 and Fungi Gallici 665 (both collected by Madame Libert) as well as those in Rehm's Ascomycetes No. 38 and Plowright's Sphær. Britannici 212, have the sporidia 8–12 x 3–4 μ. The New York and Missouri specimens also differ from those just cited in their brighter red color and distinctly hairy perithecia and come nearer to *N. erythrinella*, Nyl., which again has the perithecia only partially emergent and sporidia 18–25 x 6–8 μ, much larger than in *N. lecanodes* it is true, but still far too small. Possibly this variability in the size of the sporidia is only accidental, but from its occurrence in specimens from such widely separated localities there is reason to consider it normal and if so, characteristic of a species not heretofore described.

B. *Perithecia scattered or gregarious.*

N. tremelloìdes, E. & E. Journ. Mycol. II, p. 121.

Perithecia gregarious, ovate, 300 μ diam., coarsely furfuraceous and subtuberculose-roughened, pale orange, with a distinctly papillose-conical ostiolum. Asci about 50 x 7–8 μ, cylindric-clavate, sessile, imperfectly paraphysate. Sporidia biseriate, oblong-elliptical, 2-nucleate, becoming uniseptate, 8–12 x 3–4 μ, hyaline. The asci are contracted at the apex into a short truncate apiculus.

On bark of dead willow, Plaquemines county, La. (Langlois).

N. perforàta, Ell. & Holw. in Geol. and Nat. Hist. Survey of Minn., Bull. No. 3, p. 33.

Perithecia gregarious and subconfluent, 150–200 μ diam., rough pruinose-furfuraceous, pale at first, becoming orange-red, depressed-globose, ostiolum papilliform and collapsing when dry, so as to appear broadly perforated above. Asci clavate-cylindrical, 75 x 7–8 μ, without paraphyses. Sporidia obliquely uniseriate, elliptical or subovate, uniseptate, hyaline or with a faint tinge of rose-color, 8–12 x 5–6 μ. This comes very near *N. vulpina*, Cke., and possibly may not be specifically distinct.

On a decaying *Agaricus*, Minnesota (Holway).

N. sanguínea (Sibth.)

Sphæria sanguinea, Sibth. Ox. p. 404, Fr. S. M. II, p. 453.
Nectria sanguinea, Fr. Summ. Veg. Scand. p. 388.
Exsicc. Th. M. U. 566.—Plowr. Sph. Brit. 208.—Vize Micr. Fung. 270.—Ell. N. A. F. 573.

Perithecia scattered, adnate, ovoid, rarely subsphæroid, smooth, blood-red, rarely flesh-color, soft, about 180 μ diam. Ostiolum papilliform. Asci cylindrical, 50–60 x 5–6 μ, 8-spored. Sporidia obliquely uniseriate, elliptical or subelliptical, unequally uniseptate, slightly constricted, hyaline or with a slight tinge of rose-color, 7–10 x 4–5 μ.

Common on moist, decaying wood and bark of various deciduous trees. Saccardo properly observes that this scarcely differs from *N. episphæria*, Fr., except in its ovoid, scarcely collapsing perithecia and its less distinctly septate sporidia.

N. truncàta Ell. Am. Nat. February, 1883, p. 194.

Exsicc. Ell. N. A. F. 1332.

Perithecia gregarious, minute (125–150 μ), flesh-colored, subglobose or ovate, the apex flattened into a circular, granular-roughened disc, with the edges slightly projecting. Ostiolum in the center of the disc, minute, papilliform, brown. Asci sublanceolate, 35 x 5 μ. Sporidia biseriate, oblong-fusiform, subhyaline, uniseptate and slightly constricted around the middle, 11–13 x 2½–3 μ.

On the inside of white cedar bark, stripped from the living tree and left lying on the ground, Newfield, N. J., Sept., 1882.

N. Umbellulàriæ Plow. & Hark. Trans. Cal. Acad. Sci. 1884, p. 26.

Perithecia superficial, scattered, 200–230 μ diam., globose, subhyaline, with a tinge of flesh-color, beset externally with a few hyaline. mycelial threads. Ostiola obtuse. Asci clavate, 50 x 10–15 μ. Sporidia hyaline, ovate, uniseptate, 10–12 x 5–8 μ.

On *Umbellularia*, California (Harkness).

N. vulpina, Cke. Grev. XII, p. 83.

Exsicc. Ell. N. A. F. 774.

Perithecia gregarious or scattered, light yellow, about ½ mm. diam., globose at first, subfurfuraceous, and thinly clothed with short, erect, subglandular hairs, finally collapsing so as to be easily mistaken for a *Peziza*. Asci clavate-cylindrical, 35–45 x 6–7 μ, 8-spored. Sporidia either biseriate, 8–11 x 3–3½ μ or obliquely uniseriate, 8–12 x 4–4½ μ, elliptical or oblong-elliptical, 2–4-nucleate, with some sporidia in old specimens, becoming uniseptate.

On rotten wood of various deciduous trees, not uncommon.

N. dispérsa, C. & E. Grev. V, p. 33.

Perithecia widely scattered, ovate, papillate, light red, about ½ mm. high, sparingly clothed with pale, weak, glandular hairs. Asci cylindrical, 70–80 x 10–12 μ. Sporidia subbiseriate or obliquely uniseriate. oblong-elliptical or almond-shaped, often more prominent on one side, 18–22 x 7–10 μ, ends subacute while lying in the asci, rounded when free, 2–4-nucleate, becoming uniseptate. Cooke finds triseptate, curved conidia 50 μ long. We have not seen them.

On (pine)? bark, Maine (Rev. Joseph Blake).

N. vitícola, B. & C. Grev. IV, p. 45.

"Scattered, bright crimson, soft, collapsing laterally, seated on a thin, white mycelium. Sporidia uniseriate, elliptical, uniseptate." In Grev. XII, p. 82, Cooke gives the measurements of the sporidia as 10 x 4 μ.

On branches of vine, Alabama (Peters).

N. Eucalýpti, Cke. & Hark. Grev. XII, p. 82.

Scattered, superficial, pale. Perithecia globose, at length subdepressed (2 mm.), at first beset with papillose, hyaline hairs, finally bare. Asci clavate, 8-spored. Sporidia lanceolate, uniseptate, not constricted, hyaline, 16–18 x 4 μ.

On bark of *Eucalyptus* branches, California (Harkness).

N. squamulòsa, Ell. Bull. Torr. Bot. Club IX, p. 20.

Gregarious, minute (75–100 μ), pale, ovate-globose, covered, except the brownish, obtuse, slightly prominent ostiolum, with a light colored, squamulose coat. Asci lanceolate, narrowed and subtruncate above, 25–30 x 5–6 μ, containing eight clavate or cylindric-oblong, biseriate sporidia, 5–6 x $1\frac{1}{4}$–$1\frac{1}{2}$ μ, 2-nucleate at first and probably becoming uniseptate.

On decaying wood of a fallen limb, Newfield, N. J.

N. Gálii, Plow. & Hark. Trans. Cal. Acad. Sci. 1884, p. 26.

Perithecia scattered, immersed, then erumpent, obtuse, pale red. Asci cylindrical, very delicate, 60 x 5–8 μ. Sporidia eight, uniseriate, uniseptate, pale straw-color, oblong-oval, with bluntly pointed ends, 10 x 5 μ.

On *Galium trifolium*, California (Harkness).

N. Pepònum, B. & C. Grev. IV, p. 16.

Nectria perpusilla, B. & C. Rav Fung. Car. IV, No. 51.

Very small, scattered, scarlet. Sporidia oblong, uniseptate. It looks at first sight as if seated on a smooth, white mycelium, but this is only the external coat of the gourd. Var. *aurelia* (l. c.), having sporidia continuous, is probably only the immature state of the same thing.

On dead gourds and on tomato, So. Carolina (Ravenel).

We add the following from an examination of the specimen in Ravenel's Fung. Car., above quoted. Perithecia depressed-globose, 100–120 μ diam., ostiolum broad, papillate. Asci clavate-cylindrical, 35–40 x 5–6 μ. Sporidia not well matured, but apparently about 10 x $3\frac{1}{2}$ μ.

N. conígena, E. & E. Bull. Torr. Bot. Club X, p. 77.

Minute, membranaceous, smooth, orange-yellow, lighter and collapsing when dry. Asci about 50 x 7 μ. Sporidia uniseriate or partially biseriate above, acutely elliptical, 2-nucleate, becoming uniseptate, 7–8 x 3–$3\frac{1}{2}$ μ. Ostiolum papilliform, minute. Perithecia with a few weak, white, radiating hairs at the base. Differs from *N. vulpina*, Cke., in its habitat, smaller and paler perithecia and rather narrower and more acute sporidia.

On an old decaying cone of *Magnolia glauca*, Newfield, N. J.

N. filicina, Cke. & Hark. Grev. XII, p. 101.

"Scattered or gregarious, orange-colored. Perithecia obovate, smooth, glabrous, subshining, scarcely papillate. Asci cylindrical,

8-spored. Sporidia elliptic-cylindrical, obtuse at each end, hyaline, 8 x 2½ μ." Our specimens from Dr. Harkness have the asci about 40 x 5 μ. Sporidia mostly biseriate, 3-4-nucleate, becoming uniseptate, 7-10 x 2½-3 μ.

On stipes of tree fern, California (Harkness).

N. Pezizæ, (Tode).

Sphæria Pezizæ, Tode Fung. Meckl. II, p. 46
Sphæria miniata, Hoff. Flor. Germ. III, tab. 12, fig. 1.
Peziza hydrophora, Bull. Champ. p. 243.
Nectria Pezizæ, Fr. Summ. Veg. Sc. p. 388.
Exsicc. Thum. F. Austr. 1262.—id. M. U. 654.

Perithecia gregarious, superficial, spherical, becoming concave by collapsing, subpapillate, reddish-orange, fading at length, subpilose at base, soft, about ⅓ mm. diam. Asci very shortly pedicellate, cylindrical or clavate-cylindrical, 80-90 x 8-10 μ, when young, subcristate at the apex, 8-spored. Sporidia subbiseriate, elliptical, obtuse at each end, uniseptate, but not constricted, each cell nucleate, hyaline, 10-14 x 5-6 μ.

On decaying wood and bark, So. Carolina (Ravenel), Massachusetts (Murray). See Grev. IV, p. 16.

N. sulphùrea, Ell. & Calkins. Journ. Mycol. IV, p. 57.

Scattered on a thin farinose-tomentose, yellow subiculum extending for a centimeter or more. Perithecia ovate-conical, pruinose, yellow (nearly sulphur yellow), with a papillose ostiolum, 125-165 μ diam. In the specimens thus far seen, the asci had disappeared, but there was an abundance of oblong or clavate-oblong, hyaline, uniseptate, 7-12 (mostly 8-9) x 2½-3½ μ sporidia, distinctly constricted at the septum, ends rounded or obtusely pointed.

Parasitic on old *Stereum rugosum*, near Jacksonville, Fla. (Calkins).

N. sulphuràta, E. & E. Proc. Phil. Acad. July, 1890, p. 248.

Perithecia gregarious, minute (200 μ), subglobose with a slightly contracted base, covered with a sulphur-yellow, granulose-pruinose coat which finally disappears and leaves the perithecia black; collapsed above when dry and more or less distinctly radiate-sulcate. Asci subcylindrical, 65-70 x 6-7 μ. Sporidia subbiseriate, allantoid, hyaline, with a small nucleus near each end, slightly curved, 7-12x2 μ. Differs from *N. aurea*, S. & S., in its smaller sporidia.

On dead wood of *Populus tremuloides*, Sand Coulee, Montana (Anderson).

N. áthroa, E. & E. Proc. Phil. Acad. July, 1890, p. 247.

Densely gregarious. Perithecia ovate, 150–200 x 110–120 μ, dark red, smooth or nearly so, not collapsing. Asci (p. sp.) about 35 μ long, cylindrical, evanescent. Sporidia uniseriate, oblong-elliptical, hyaline, obtuse at the ends, uniseptate and slightly constricted, 5–6 x 2½ μ. This has the general appearance of *N. ditissima*, Tul., but besides the more regularly shaped perithecia the sporidia are much smaller. The asci are with difficulty seen so that this might be taken for a stylosporous fungus only from the fact that here and there series of eight sporidia lying end to end indicate the presence of asci.

On a decaying sycamore log, Manhattan, Kansas (Kellerman and Swingle).

N. mammoidea, Plowr. Grev. III, p. 126, tab. 42, fig. 5.

Cespitose. Perithecia of medium size, ½ mm. diam., globose, minutely furfuraceous, of a brick-red color, sometimes collapsing, with a darker colored papilliform ostiolum. Asci subcylindrical, 100–115 x 12 μ. Sporidia 15 x 5–6 μ, oblong-elliptical, uniseptate.

This species was found at Newfield some years ago on outer bark of living quince trees. The Newfield specimens agree with authentic specimens from Plowright in all respects, only the sporidia are a little smaller (15 x 5–6 μ). In Plowright's specimens they are 18–22 x 6–7 μ and subinequilateral. The Newfield specimens were immature, the asci being mostly filled with granular matter. What appears to be the same was also sent by Dr. Macoun from British Columbia, on bark of maple.

N. thujàna, Rehm, Mich. I, p. 295.

Exsicc Thum. M. U 972.—Rehm Asc. 338.—Ell. N. A F. 160.

Perithecia very minute, scarcely visible to the naked eye, solitary or 2–3 together, conic-globose, slightly depressed at the apex, purplish-red. Asci oblong-elliptical, 60–80 x 10–14 μ, with eight biseriate, oblong, two-celled, hyaline sporidia, slightly constricted in the middle and about 11 x 7 μ, becoming at length subfuscous.

On dead foliage of white cedar not yet fallen from the branches of a tree cut some time previously, Newfield, N. J.

N. depauperàta, Cke. Grev. VII, p. 50.

Exsicc. Ell N. A. F. 677.

Perithecia globose, scarlet, scarcely papillate, 1–3 on a stroma. Asci clavate. Sporidia elliptical, uniseptate, 10 x 3½ μ. We have never seen the original specimens on *Yucca* and have taken the foregoing from Grevillea. The conidial stage is *Fusarium Yuccæ.*

On *Yucca aloifolia*, Aiken, So. Carolina.

From an examination of the specimens in N. A. F. (det. by Cke.) we add the following notes.—Perithecia ovoid-globose, small (160–190 μ), pale and furfuraceous at first, becoming bare and pale red. Ostiolum papillate and slightly darker. Asci clavate-cylindrical, 35–40 x 4 μ, sessile. Sporidia oblong-elliptical, subbiseriate, 9–11 x 3–3½ μ. There are often six and even ten perithecia on each erumpent, white byssoid-grumose stroma.

The N. A. F. specc. were found on dead *Clethra*, at Newfield, N. J.

N. aglæothèle, B. & C. Grev. IV, p. 45.

Pale; ostiolum distinct, papilliform, darker, then deeply sunk by collapsion. Asci linear. Sporidia elliptical, uniseriate, slightly attenuated at each end, 12–14 μlong.

On alder, apparently growing from the remains of some *Coccus*, New England (Sprague).

A stylosporous form (*N. crustulina*, B. & Rav., Santee Canal, So. Carolina), has cespitose, neat tan-colored perithecia with obovate, uniseptate sporules, 10–15 μ long.

N. rimíncola, Cke. Grev. XI, p. 108.

Exsicc. Rav. F. Am. 644.

Gregarious or scattered, superficial, scarlet, growing in cracks of the bark. Perithecia subglobose, finally depressed, glabrous, subshining, ⅓ mm. diam. Asci cylindrical. Sporidia uniseriate, uniseptate, hyaline, not constricted, 12 x 4 μ.

In cracks of the bark of *Liquidambar*, seaboard of Carolina (Ravenel).

N. Brássicae, Ell. & Sacc. Mich. II, p. 374.

Exsicc. Ell. N. A. F. 572.

Perithecia densely gregarious, globose-conoid, not collapsing, very small (125 μ), blood-red, ostiolum rather obtuse-conical, texture loosely cellular, rose-tinted, paler around the ostiolum. Asci clavate-cylindrical, 60 x 7–8, without paraphyses, briefly-stipitate, obtuse at the apex, containing eight oblong, subclavate, hyaline, uniseptate, 10–11 x 4–4½ μ sporidia.

This was first found on old cabbage stalks and at the same time the next year on old potato stalks lying scattered over the same ground previously occupied by the cabbages, at Newfield, N. J., also on sweet potato stems, at Vineland, N. J., and on old cabbage stalks and stems of *Sechium edule*, Louisiana (Langlois).

N. Apócyni, Pk. 26th Rep. N. Y. State Mus. p. 84.

"Cespitose or scattered, dull red. Perithecia minute, pale ochraceous and subglobose when moist, dull red, collapsed or latterly compressed and rough, with minute, whitish scales when dry; ostiola minute. Sporidia biseriate, uniseptate, filiform, usually constricted in the middle, nucleate, 16–20 μ long." Specimens found at Newfield, N. J., July, 1883, on dead stems of *Asclepias tuberosa*, with *Volutella flexuosa*, C. & E., had asci about 35 x 7 μ, sporidia 12–18 x 3½–4 μ, constricted at the septum and nucleate.

N. depállens, Cke. & Hark. Grev. XII, p. 82.

Scattered or gregarious, superficial, brick-red, fading out. Perithecia subglobose, smooth, bare, opake, 200–250 μ. Asci clavate, 8-spored. Sporidia lanceolate, subacute at each end, uniseptate, not constricted, hyaline, 22–24 x 4–4½ μ.

On stems of *Lupinus*, California (Harkness).

N. Curtísii, Berk. Grev. IV, p. 46.

"Minute, erumpent, scattered. Asci lanceolate. Sporidia oblong, curved, with four nuclei, 12 x 2 μ."

On *Zea*, So. Carolina (Ravenel).

N. episphæ̀ria (Tode).

Sphæria episphæria, Tode Fung. Meckl. II, p. 21.
Sphæria erythrococcus, Ehr. Sylv. Ber. p. 9.
Nectria episphæria. Fr. Summ. Veg. Scand. p. 388.
Exsicc. Rab. F. E. 262, 642.—Rehm Asc. 585.—Thum. M. U. 766.—Sydow M March. 352. Rav. F. Am. 340 —Roum F. Gall. 4655.—Ell. N. A. F. 469.

Perithecia gregarious or scattered, superficial, subsphæroid, collapsing and frequently subcompressed, soft, smooth, blood-red, about 180 μ diam., with a papilliform ostiolum. Asci cylindrical, 50–60 x 5–6 μ, 8-spored. Sporidia obliquely uniseriate, subellipsoid, rather unequally uniseptate, hyaline, slightly constricted, 7–10 x 4–5 μ.

On various sphæriaceous fungi—*Diatrype*, *Hypoxylon*, *Valsa*, etc., common.

N. rubefàciens, E. & E. Journ. Mycol. III, p. 116.

Perithecia globose, 80 μ diam., smooth, or roughened with scattered, rudimentary, gland-like hairs, subastomous, of fine cellular texture, pallid at first, becoming orange-red. Asci broad-clavate,

35–40 x 10–12 μ, without paraphyses. Sporidia irregularly crowded, oblong-cylindrical, hyaline, uniseptate and constricted at the septum, distinctly curved, 14–18 x $2\frac{1}{2}$–3 μ. The thallus of the lichen (*Parmelia tiliacea*)? turns dull red (bright red inside). The perithecia are scattered and superficial. This seems to be quite distinct from any of the other lichenicolous species.

Parasitic on thallus of some lichen, on various dead limbs lying on the ground, Newfield, N. J.

Sporidia continuous, (Nectriella, Sacc.)

N. mycetóphila, Pk. 26th Rep. N. Y. State Mus: p. 85.

"Perithecia crowded or scattered, minute, smooth, subglobose, pale yellow when young, then pinkish-ochre. Ostiola minute, papillate, distinct, darker colored. Asci subclavate. Sporidia oblong, simple, 12–13 x 4 μ."

On decaying fungi, New York (Peck).

Of the thirty species enumerated by Saccardo in Syll. & Additamenta, *N. mycetophila*, Pk., is the only American species with continuous sporidia, if in fact they are continuous. *N. microspora*, C. & E., and *N. vulpina* Cke., have the sporidia distinctly uniseptate.

Perithecia hairy, (Lasionectria).

N. poliòsa, E. & E. Journ. Mycol. II, p. 39.—id. III, p. 1.

Perithecia scattered, membranaceous, orange-red, ovate-globose, 165 μ diam., sparsely clothed, except the papilliform ostiolum, with straight, spreading, hyaline, septate, glandular hairs, about equal in length to half the diameter of the perithecium. Asci sessile, oblong-cylindrical, about 75 x 12 μ. Sporidia biseriate, oblong or subfusiform-oblong and subinequilateral, hyaline, uniseptate and slightly constricted at the septum, containing several nuclei irregularly placed, 18–22 x 7–8 μ, ends rounded or subacute. The hairs which clothe the perithecia are at first about 7 μ thick, with the ends obtuse and a little swollen, but at length they become elongated and attenuated above. This must be nearly allied to *N. tephrothele*, Berk., but in the description of that species the perithecia are not described as hairy.

Parasitic on *Diatrype platystoma*, (Schw.), Florida (Calkins).

N. lasiodérma, (Ell.) Am. Nat. February, 1883, p. 194.

Exsicc. Ell. N. A. F, 1177.

Perithecia mostly single, subamorphous, obtuse-conical, broadly perforated above, about $\frac{1}{4}$ mm. high, shaggy with short, septate, obtuse, imperfectly developed hairs, dull red when dry, pale orange when moist. Asci cylindrical, 75–80 x 7–8 μ. Sporidia uniseriate, elliptical, hyaline, uniseptate, scarcely constricted, 11–12 x 4–5 μ.

Parasitic on old *Valsa lutescens*, Ell., on dead limbs of *Quercus coccinea*, lying on the ground, Newfield, N. J.

N. Rexiàna, Ell. l. c. and Journ. Mycol. III, p. 2.

Perithecia minute, less than ¼ mm. diam., flesh-color, becoming darker, slightly compressed laterally, enveloped in white down, which forms little tufts, appearing under the lens like some minute, tufted mucedinous growth. Asci linear, 35–40 μ long, evanescent. Sporidia uniseriate, oblong, hyaline, 1–2-nucleate, becoming uniseptate?

Parasitic on *Myxogasters* (*Chondrioderma spumarioides*), Adirondack Mts., N. Y., August, 1882 (Dr. Geo. A. Rex).

N. láctea, Ell. & Morgan (in Herb.)

Densely gregarious. Perithecia globose, about 250 μ diam., yellowish horn-color, densely clothed, except the bare, papilliform ostiolum, with a dense white coat of glandular-pruinose pubescence, which finally disappears in part. There is also a sparing, web-like, white mycelium overrunning the matrix. Asci cylindrical, sessile, without paraphyses, 55–65 x 6 μ. Sporidia uniseriate, elliptical, hyaline, not constricted, 6–8 x 3½–4½ μ.

On pores of some old *Polyporus*, Ohio (Morgan), on old *Stereum subpileatum*, Florida (Calkins), and on decaying wood of *Melia*, Louisiana (Langlois).

Species not well known.

N. fibríseda, Schw. Syn. N. Am. 1542.

Very minute, scattered, blood-red, pellucid, globose-ovate, papillate, adhering in dense clusters to the bark of dead chestnut limbs from which the epidermis has peeled off, entirely glabrous, finally collapsing, scarcely visible to the naked eye. Of this species nothing is known to us except the description above quoted.

Among the loosened fibers of chestnut bark, Bethlehem, Pa. (Schweinitz). Allied to *Nectria sanguinea.*

Cooke, in his synopsis, mentions a *Nectria Smilacis*, Schw. We find no such species in Schw. Synopsis.

N. móbilis, (Tode).

Sphæria mobilis, Tode Fungi Meckl. II, tab. 9, fig. 71, Fr. S. M. II, p 461.

Perithecia very delicate, superficial, free, so as to be moveable ("*ut quaquaversus facile se moveri sinant.*") Ostiolum papilliform, finally deciduous. Color brown, becoming black, (reddish when young)?

On decaying limbs, Carolina and Pennsylvania (Schweinitz).

Nothing definite is known of this species.

SPHÆROSTÍLBE, Tul.

Sel Carp. III, p. 99.

Perithecia as in *Nectria*, globose, soft, bright colored, growing at the base of or in company with their conidia (*Stilbum*, *Atractium*, *Microcera*). Asci elongated, 8-spored. Sporidia oblong or ovoid, hyaline, uniseptate. Paraphyses none or spurious.

Sph. flámmea, Tul. l. c. p. 104.

Perithecia globose, bright red, nearly smooth, crowded on or near the conidiophorus stroma (*Atractium flammeum*, B. & Rav.) Asci obovate-oblong, 8-spored. Sporidia ovate, obtuse, uniseptate, hyaline, slightly constricted, 12–16 x 5–6 μ. The conidial fungus, as represented in Rav. Fungi Car. V. 86, has the stromata at first narrow-conical, becoming finally flattened above and subspezizoid; the conidia linear-lanceolate, very large (80–100 x 6½ μ), a little curved, 6-9-septate, hyaline with a tinge of rose color. On maple bark, Carolina (Ravenel), on *Salix*, Louisiana, (Langlois).

Berkeley, in Grev. IV, p. 47, adds.—"There is a very distinct species on *Magnolia glauca*, Car. Inf., No. 5005 (*Attractium pallidum*, B. & C.), with short, fusiform spores 13 μ long, with the endochrome retracted to either end."

Sph. coccóphila, Tul. l. c. p. 105.

Perithecia numerous, on and near the conidial stromata, very small, globose, obtuse, minutely papillate, very smooth, bright red, often 4–5 together, collapsing when old. Asci linear, 60–80 x 6½ μ. Sporidia oblique, uniseriate, ovate, 10 x 4 μ, uniseptate, subhyaline, slightly constricted.

On *Alnus serrulata*, Pennsylvania (Michener).

The conidial stage (*Microcera coccophila*, Desm.), which has been sent from Florida by Dr. Martin and collected in Carolina by Ravenel (F. Am. 286), has the stroma arising from various species of dead bark lice. It is red, obtuse and about 2 mm. high. The conidia are linear-lanceolate, 5–7-septate and 56–65 x 5–6 μ, nearly hyaline.

Sph. gracilipes, Tul. l. c. p. 105. **(Plate 12)**

Exsicc. Rav. F. Am. 285.—Ell. & Evrht. N. A. F. 2131.

Perithecia collected at the base of the conidial stroma, small, (⅓ mm.), red, light pruinose, fading out. Asci cylindric-clavate.

Sporidia uniseriate, ovoid, 12–16 x $4\frac{1}{2}$–6 μ. The conidiophorous fungus (*Stilbum corynoides*, E. & E.) has the stem slender, gray, becoming nearly black, $\frac{1}{4}$–$\frac{1}{2}$ cm. long, head globose, $\frac{1}{4}$–$\frac{1}{2}$ mm., orange-yellow, becoming fuscous. Conidia oblong-elliptical, hyaline, 5–6 x $1\frac{1}{2}$ μ.

On *Hibiscus*, *Carya*, *Melia* and *Platanus*, Carolina (Curtis and Ravenel), on *Melia*, Louisiana (Langlois).

Sph. cinnabárina, Tul. l. c. p. 103.

Perithecia growing at the base of the conidial stroma (*Stilbum cinnabarinum*, Mont.), few, sessile, globose, scarcely papillate, very smooth, orange-red, finally collapsing partially. Asci clavate-oblong, 80 x 13–16 μ. Sporidia biseriate, ovate-oblong, 22–26 x 7 μ, plurinucleate. Conidial stroma bacillary, clavate above, red, conidia ovoid, small, $3\frac{1}{2}$ x $1\frac{1}{2}$ μ.

On trunks of *Carya*, *Morus* and *Rhus*, Carolina, Louisiana and Mexico.

CALONÉCTRIA, De Not.

Reel. Pir. Ital. in Comm. Soc. Ital. Critt. II, p. 477.

Perithecia scattered or cespitose, superficial or erumpent, soft (carnose-membranaceous), bright colored. Sporidia oblong or fusoid, 2- or more-septate, hyaline.

C. erubéscens, (Desm.)

Sphæria erubescens, Desm. XIII, Not. 72.
Calonectria erubescens, Sacc. Mich. I, p. 309.

Scattered or gregarious, superficial, with white, woolly, radiating hairs at base. Perithecia minute, pale red, globose, finally collapsing, soft, glabrous, with a papilliform ostiolum. Asci clavate, 35–40 x 7–8 μ. Sporidia oblong-fusoid, straight or slightly curved, hyaline, 3–4-nucleate, becoming 1–3-septate, 10–16 x $2\frac{1}{2}$–3 μ (10 x 3 μ, Sacc.), ends subacute.

On the under side of living leaves of *Quercus laurifolia*, *Myrica cerifera* and *Olea Americana*, Florida (Martin and Calkins). Mostly on mycelium of *Meliola*.

Calonectria leucorrhodina (Mont.), Sacc. Syll. II, p. 548, according to So. American specimens from Spegazzini, scarcely differs from this, except in its epiphyllous growth.

C. Canadénsis, (E. & E.) Bull. Torr. Bot. Club XI, p. 74, Journ. Mycol. II, p. 122. **(Plate 13)**

Nectria Canadensis, E. & E. l. c.
Exsicc. Ell. & Evrht. N. A. F. 2d Ser. 2547.

Perithecia cespitose, ovate-globose, $\frac{1}{4}$–$\frac{1}{3}$ mm. diam., dull red, densely tuberculose-granular. Ostiola not prominent, more or less distinctly stellate-cleft and finally collapsing slightly. Asci sublanceolate, 75–80 x 10–12 μ, without any distinct paraphyses. Sporidia subbiseriate, hyaline, oblong-elliptical, 3-septate, slightly curved, ends obtuse, 18–22 x 7–9 μ. The conidial stage is a *Tubercularia* about 2 mm. high, bursting out in a seriate manner through cracks in the outer bark, having an orange-red head and brick-red, stipitate base. The perithecia originate from the lower or medial part of the stipitate base, and finally entirely surround and overtop the orange-colored heads of conidia, which are either entirely hidden or remain partially visible in the midst of the dense clusters of perithecia. Conidia oblong-cylindrical, hyaline, continuous, 5–7 x $2\frac{1}{2}$ μ.

On bark of dead elm, Ottawa, Canada (Macoun), London, Canada (Dearness).

C. polythálama, (Berk.)

Nectria polythalama, Berk Fl. New Zealand II, p. 203, Grev. IV, p. 46.
Nectria aurigera, B & Rav. Grev. l. c.
Exsicc. Rav. Car. III, 54.—id. V, 60 (under the name of *N. aurigera*, B. & Rav)—Ell N. A. F. 79.

Erumpent, cespitose, stroma pale brick-red, mostly subelliptical. Perithecia globose, about $\frac{1}{3}$ mm. diam., covered with a yellow powder. Ostiolum papilliform, collapsing. Asci sublanceolate, p. sp., 55–60 x 12–15 μ. Sporidia subbiseriate, oblong-cylindrical, straight or slightly curved, yellowish, 6–8-septate, 20–24 x 5–6 μ. Var. *flavitecta*, B. & C. (Grev. l. c.), on *Kerria Japonica*, has the "sporidia larger and quadriseptate, sometimes with a gelatinous coat."

On *Fraxinus* and *Chionanthus*, Carolina and New Jersey.

There is no shadow of difference in the asci and sporidia, at least as the two are represented in Rav. Fungi Car., between *Nectria aurigera*, B. & Rav., and *N. polythalama*, Berk., and if, as there is no reason to doubt, the specimens there distributed are authentic, the two species are the same, and, in that case, the specific name, *polythalama*, would have precedence.

C. chlorinélla, (Cke.)

Nectria chlorinella, Cke. Grev. XI, p. 108.
Exsicc. Rav. F. Am. 736.—Ell. & Evrht. N. A F. 2d Ser. 2546

Gregarious, ovate-globose, lemon-yellow, about $\frac{3}{4}$ or 1 mm. high, $\frac{1}{2}$ mm. broad, slightly contracted below, woolly-tomentose except the bare, black, strongly papillose ostiolum. Asci clavate-cylindrical,

100–125 x 12–15 μ, with abundant, filiform paraphyses. Sporidia biseriate, broad-fusiform, slightly curved, subhyaline. 1–3-septate and constricted at the middle septum, 25–35 x 7–8 μ.

Cooke makes the sporidia 18–20 x 5 μ. They vary considerably, but mature and well developed specimens have the sporidia as above.

On bark of *Ulmus*, seaboard of South Carolina (Ravenel), on various kinds of decaying wood and limbs, (oak, *Rhus*, &c.), Newfield, N. J., and on decaying elm, Canada (Macoun & Dearness). The ostiolum is sometimes more or less compressed, so as to resemble *Lophiostoma*.

C. Dearnéssii, E. & E. Proc. Phil. Acad. July, 1890, p. 245.

Perithecia 3–12 in a cluster, ovate, narrowed above and below, light orange-yellow, about 165 μ diam. and a little more than that in height, seated on a white, radiate-fibrous, silky mycelium which at first partially envelops and clothes the perithecia but finally disappears. Ostiola broad papilliform, not distinctly prominent, at length slightly collapsing. Asci 75–80 x 10–12 μ, with paraphyses. Sporidia oblong-cylindrical, obtuse, yellowish-hyaline, biseriate, 3–5-septate, more or less constricted at the septa, 25–35 x 6–7 μ, ends obtuse, slightly curved.

Cespitose on the ostiola of some *Massaria*, on *Fraxinus*? and *Ulmus*, Canada (Dearness).

C. balsámea, (C. & P.)

Nectria balsamea, C. & P. 26th. Rep. N. Y. State Mus. p. 84, Grev. XII, p. 81.

Erumpent, cespitose, suborbicular. Perithecia rather small, smooth, scarlet, papillate, at length collapsed, crowded on a pale stroma. Asci cylindric-clavate, 8-spored. Sporidia fusiform, hyaline, 5-septate, 30 x $7\frac{3}{4}$ μ.

On bark of *Abies balsamea*, New York (Peck).

C. diminúta, (Cke.)

Dialonectria diminuta, (B. & C.) Cke. Grev. XII, p. 83.
Nectria diploa var. *diminuta*, B. & C. Grev. IV, p. 46.

Very minute, scarlet. Asci lanceolate but obtuse. Sporidia sometimes larger, binucleate, at length triseptate (uniseptate, Berk.), uniseriate, or smaller, 4-nucleate, biseriate, 25–30 x $8\frac{1}{2}$ μ.

On some *Sphæria*, on alder, South Carolina (Ravenel).

PLEONÉCTRIA, Sacc.

F Ven. Nov. Ser. V. 178, Mich. I p. 324.

Perithecia cespitose, subglobose, bright colored, carnose-membranaceous, papillate. Asci 8-spored. Sporidia, when mature, pluriseptate-muriform, hyaline.

P. Berolinénsis, Sacc. Mich. I, p. 123. **(Plate 12)**

Nectria Ribis, Niessl, (non Tode)
Pleonectria Ribis, Karst, Symb. Mycol. Penn. VI, p. 42.
Exsicc. Ell. N. A. F. 470.

Perithecia erumpent-superficial, cespitose, globose-depressed, $\frac{1}{3}$–$\frac{1}{2}$ mm. diam., finally collapsing to cup-shaped. of a brick-red color and loosely-cellular structure. Ostiolum not prominent. Asci cylindrical, subsessile, 90–115 x 10–14 μ. subtruncate above. Sporidia uniseriate, oblong-elliptical, obtuse at each end, 5–7-septate and muriform, not constricted, minutely guttulate, hyaline, 16–22 x 7–8 μ.

On dead stems of *Ribes*, Canada and the northern United States west to Montana.

Sec. Dr. Farlow No. 6140, collected in Canada by Poe and marked in the Curtis collection *Nectria fenestrata*, B. & C., but in Grev. IV, p. 46, referred to *Sphærostilbe pseudotrichia*, (Schw.) and evidently the same as Ell. N. A. F. 470, is different from authentic specimens of *Sphæria pseudotrichia*, Schw. (from Surinam), which, besides its peculiar shaped conidia, has the ascospores larger, about 38 μ, with a crenulated outline.

P. denigràta, Winter, Bull. Torr. Bot. Club X, p. 48.

Exsicc. Rabh. Winter Fungi 2948.—Ell. and Evrht. N. A. F. 1334, 2372.

Perithecia erumpent-superficial, forming dense, pulvinate-hemispherical tufts 3–5 mm. diam., seated on a pulvinate stroma which is of a dirty yellow within, nearly globose, gray outside, becoming black. 360–470 μ diam., at length depressed and umbilicate at the vertex. Ostiolum black, papilliform. Asci cylindrical, briefly stipitate, 8-spored, about 70 x 8–10 μ (elongated finally in water to 110 μ). Sporidia generally monostichous, variable in shape, rounded-subangular, 3–5-septate and muriform, more or less constricted at the septa, pale yellowish, 10–16 x 7–10 μ. Paraphyses filiform, very long.

On dead branches of *Gleditschia triacanthus*, Kentucky (Kellerman), Ohio (Morgan), Delaware (Commons), Missouri (Webber).

P. Missouriénsis, (E. & E.)

Nectria Missouriensis, E, & E. Journ. Mycol IV, p 57.

Perithecia cespitose, erumpent-superficial (6–20), globose, $\frac{1}{4}$–$\frac{1}{3}$ mm. diam., seated on an inconspicuous stromatic base, dark red, sprinkled with small yellowish granules. Ostiolum papilliform and distinct. Asci clavate-oblong, 100–120 x 15–20 μ (p. sp. 75–85 μ).

with abundant filiform but evanescent paraphyses. Sporidia irregularly crowded, clavate-oblong or elliptic-oblong, subinequilateral, yellowish-hyaline, muriform, 20–30 x 10–12 μ.

On bark of *Carya alba*, Missouri (Demetrio) and on *Carya*, Delaware (Commons).

CHILONÉCTRIA, Sacc.

Mich. I, p. 270.

Perithecia cespitose, globose, papillate, red, yellowish or brown. Asci clavate-cylindrical, many-spored. Sporidia minute, allantoid, hyaline, accompanied by other larger sporidia? in the same ascus.

Ch. cucurbítula, Curr. **(Plate 12)**

Sphæria cucurbitula, Curr. Comp. Sphær. tab. 49, fig. 178 (partly).
Exsicc. Ell. & Evrht. N. A. F. 2d Ser. 1551.

Perithecia cespitose, orange-red, nearly smooth, at length collapsing. Ostiolum papilliform, minute. Asci clavate-cylindrical. 75–100 x 10–12 μ, filled with countless minute sporidia, 2–3 (mostly 2½ μ) x 1–1¼ μ. In the early stages of growth, the asci contain 2–4 cylindrical bodies about 3 μ thick and varying in length from 15 μ to nearly the entire length of the asci. These elongated bodies are subundulate and often appear distinctly multiseptate with the septa either running square across or with their ends a little curved so as to give the appearance of a series of hyaline, subglobose sporidia, but finally these cylindrical bodies are filled with the minute sporidia, which at length fill the entire cavity of the asci. Sometimes, instead of the elongated bodies, some of the asci will be seen to contain one or two series of faintly-outlined, subglobose cells, which, also, are at length filled with the minute sporidia and disappear. The peculiarity here noted has been observed in all the specimens on coniferous trees thus far examined and we fancy we see the same structure in the *N. cucurbitula*, issued by Fries in his Scler. Succ., which we have had the opportunity of examining.

On *Pinus rigida* and *P. Strobus*, Newfield, N. J., and on *Abies balsamea*, West Chester, Pa.

Mycologists are not agreed as to the nature of the minute allantoid sporidia in this and the following species. Fückel considered them spermatia. Winter (in "Die Pilze") regards them as secondary spores resulting from the germination of the true sporidia. In that case, the asci should contain in the earlier stages of growth only the larger "true sporidia"; but specimens on the same limb of *Pinus*

Strobus, observed from May to November, showed the two kinds of sporidia mixed from the first.

Ch. Córyli, (Fckl.)

Chilonectria cucurbitula Sacc. Syll. II, p. 453 (partly).
Nectria Coryli, Fckl. Symb. p. 180.
Exsicc. Ell. N. A. F. 159.—Sacc. M. V. 1446.

Perithecia cespitose, erumpent, smooth, subastomous, dark red, collapsing when dry, not differing in appearance from those of the preceding species, unless in being of a deeper shade of red. Asci clavate, 85–100 x 10–12 μ, mostly filled with minute sporidia, exactly as in the preceding species, but some contain cylindric-fusoid (sporidia)? 1-septate, 10–15 x 2½–3 μ, with a short, curved apiculus at each end. These fusoid sporidia, lying in two or three series in the asci, are so arranged with their contiguous ends in contact as to resemble closely the cylindrical bodies mentioned in the preceding species, and, like those, are often seen filled, more or less completely, with the minute, oblong sporidia. This species is found exclusively on bark and limbs of deciduous trees.

The specimens of *N. inaurata*, in Saccardo's Mycotheca Veneta, No. 1446, do not differ, as far as we can see, either in the perithecia or the fructification, from the specimens in N. A. F. 159. In both, the perithecia are distinctly collapsed.

Ch. crinígera, E. & E. Proc. Phil. Acad. July, 1890.

Nectria sphærospora, E. & E. in Bessey & Webber's Cat. Flora Nebr. 1890, p 53.

Perithecia cespitose on a tubercular stroma, in compact clusters of 3–12, the single perithecia subglobose and about ½ mm. diam., covered at first with a brownish, farinaceous coat, becoming nearly black, rounded and obtuse above, with a papilliform ostiolum which is slightly collapsed when dry. Asci clavate-cylindrical, 70–80 x 12–15 μ, attenuated above when young, but becoming rounded and obtuse, overtopped by the abundant but evanescent paraphyses and filled with innumerable spermatoid, hyaline sporidia? (2½–3 x ½–¾ μ), among which lie in a single series eight subglobose or subelliptical hyaline sporidia, 5–8 μ diam., granular at first, but at length faintly about 3-septate and muriform, or sometimes marked in a sarcinuliform manner by two septa at right angles to each other. When these sporidia have escaped from the asci it is seen that they are sparingly clothed with spreading, hyaline filaments, 8–10 μ long, 2–12 or more in number and standing out on all sides from the body of the sporidium like rays from a star.

On bark of dead *Fraxinus viridis*, Lincoln, Neb., Nov. 1888, (H. J. Webber, No. 18).

BÁRYA, Fckl.

Symb. p. 93

Perithecia carnose, subdiaphanous, at length of the consistence of horn, superficial, conic-acute, nestling in a loose, cottony, conidia-bearing mycelium. Conidia oblong, obscurely uniseptate, obtuse at each end. Asci elongated-lanceolate, acuminate above, obtuse below, 8-spored, stipe globular. Sporidia filiform, continuous, about as long as the asci, hyaline.

B. parasítica, Fckl. l. c.

Perithecia gregarious, free, conical, acute, subventricose in the middle, finally compressed, about the size of the perithecia of *Nectria coccinea*, glabrous, yellowish-green, diaphanous, becoming opake, dark brown, seated in a conidiophorous mycelium, first white, then yellowish-green, finally brown, crowned with a terminal white globule. Conidia oblong, 14 x 5 μ, hyaline. Asci lanceolate, acuminate, globose-stipitate at base. Sporidia nearly as long as the asci, filiform, continuous, hyaline. Parasitic on *Bertia moriformis*, (Germany); Var. *cespitosa*, Pk. 43d Rep. p. 33, has the perithecia crowded in dense tufts and sometimes tapering above into a rather long neck, asci and sporidia slenderer and longer.

On *Bertia moriformis*, Catskill Mountains, N. Y.

We have seen no specimens and quote from the authors cited.

OPHIONÉCTRIA, Sacc.

Mich. I. p. 323.

Perithecia globose-conical, superficial, papillate, subcarnose, bright-colored (red, yellow &c). Asci 8-spored. Sporidia filiform, multiseptate or multinucleate.

O. cèrea, (B. & C.)

Sphæria cerea, B. & C. Grev. IV, p. 108.
Calonectria cerea, Sacc. Syll. 4967.
Dialonectria cerea, Cke. Syn. 476.
Dialonectria fulvida, E. & E. Journ. Mycol. II, p. 136.
Ophionectria Everhartii, Ell. & Galw. l. c. VI, p. 32.

Perithecia gregarious, ovate-globose, 160–175 μ diam., dull, dirty yellow and granular-pruinose, except the rather acutely papilliform ostiolum. Asci oblong-cylindrical, 75–80 x 12–14 μ. with rather indistinct paraphyses. Sporidia fasciculate, fusoid-cylindrical, yellowish-hyaline, multiseptate, multinucleate at first, straight while lying in the asci, curved when free, 35–50 x 3–3½ μ, slightly tapering from the middle to each end.

On old *Diatrype stigma* and on the bark of decaying oak limbs, New Jersey and Carolina.

We can not say positively that *Sphæria cerea*, B. & C. and *Ophionectria Everhartii*, E. & G. are the same, but the probability of their identity is so strong that we have assumed it to be a fact. There may also be some question of generic relationship; but if *Nectria melina*, Mont. can be referred to *Ophionectria*, as is done in Sacc. Syll. II, p. 563, our species may properly go there too, having sporidia still narrower than in that species.

O. coccícola, (E. & E.)

Nectria coccicola, E. & E. Journ Mycol II, p. 39.
Dialonectria coccicola, E. & E. l. c. II, p. 137.

Perithecia cespitose, membranaceous, about ⅓ mm. diam. and ½ mm. high, flesh-color, becoming dirty buff when mature, obovate, astomous, surface roughish, with a few scattered, white, rudimentary hairs, or at length bald. Asci clavate-cylindrical, 150–190 x 20 μ, with abundant, rather stout paraphyses. Sporidia eight in an ascus, clavate-cylindrical, multinucleate, hyaline, 110–140 x 6–7 μ at the upper end, attenuated below.

On dead scale insects on bark of living orange trees, Florida (Scribner).

The groups of perithecia are seated either on the shells of dead insects or on the bark itself, with a subiculum more or less distinct, composed of white, decumbent or prostrate hairs of the same character as those found on the perithecia themselves. The species seem to be quite distinct.

GIBBERÉLLA, Sacc.

Mich. I, p. 43.

Perithecia either cespitose and stromatically connected or separate. Asci 8-spored. Sporidia subfusoid, 3-pluri-septate, subhyaline. The substance of the perithecia is of a blue or violet tint.

G. acervàlis, (Moug.)

Sphæria acervalis, Moug. in Fr. Elench. II, p. 83.
Gibberella acervalis, Sacc. Syll. 4979.

Emergent, bare and black, collected in small heaps or clusters. Perithecia subconnate, globose, rugulose, at length umbilicate. Asci oblong, 70 x 10 μ. Sporidia ovate-oblong, 3-septate, hyaline, 18 x 6 μ.

Spermogonia—perithecia cespitose, about one third as large as the ascigerous perithecia; spermatia minute, cylindrical, oscillatory.

Found in Europe on willow limbs; var. *Juniperi Virginianæ*, reported by Peek as found in New York State.

G. pulicàris, (Fr.) **(Plate 13)**

Sphæria pulicaris, Fr. in Kze. & Sehm. Mycol. Hefte II, p. 37.
Gibbera pulicaris, Fr. Summ. Veg. Sc. p. 402.
Botryosphæria pulicaris, Ces. & de Not. Schema p. 212.
Gibberella pulicaris, Sacc. Mich. I, p. 43.
Exsicc. Rab. Herb. Mycol. 735.—Rehm Asc. 230, 489.—Ell. N. A. F. 81.—Sydow Mycoth. March. 977. 1544. &c.

Cespitose, stroma cortical. Perithecia crowded, superficial, purplish, at length collapsing or laterally compressed. Asci oblong, sessile. 8-spored, 60–75 x 12–16 μ. Sporidia biseriate, ovate, elliptical or subpyriform, subobtuse, somewhat curved, 3-septate, pale yellowish, 18–20 x 6–8 μ. The conidial stage is considered to be *Fusarium sambucinum*, Fckl., or *F. roseum*, Lk. This is quite a common species and is found on bark of dead limbs of various deciduous trees and on dead herbaceous stems—especially on dead stalks of *Zea Mays* and also on the grain.

Tulasne, Sel Carp. III, p. 68. enumerates as conidia of this species, *Fusarium roseum*, Lk., *F. incarnatum*, (Rob,), *F. aurantiacum*, Lk., *Selenosporium sarcochroum* and *S. lateritium*, Mazerio.

G. Saubinètii, (Mont.)

Botryosphæria Saubinetii, Niessl. Beitrag. p. 45.
Gibbera Saubinetii, Mont. Syll. p. 252.
Botryosphæria dispersa, de Not. Sfer. Ital. p. 85. tab.92.
Gibberella Saubinetii, Sacc. Syll. 4977.

Perithecia gregarious, confluent-cespitose and concrescent, coriaceo-membranaceous, verrucose, at length flaccid, plicate, ovoid, subcontracted at the base, bluish, 200–300 x 170–220 μ, papillate. Asci oblong-lanceolate, acuminate above, contracted at the base into a short, thick stipe, 8-spored, 60–76 x 10–12 μ. Sporidia uniseriate or subbiseriate, fusiform, curved or straight, subacute, 3-septate, but scarcely constricted, nearly hyaline, 18–24 x 4–5 μ. Conidial stage, *Fusarium roseum*, Lk. The characters of this species, as here given, are taken from Sacc. Sylloge. From the specimens at our command, we have always found it difficult to separate this from the preceding species.

G. fícini, (Cke. & Hark.)

Gibbera ficini, Cke & Hark. Grev. IX, p. 87.

Perithecia crowded, cespitose, dark violet, smooth, scarcely papillate. Stylospores lanceolate, obtuse, slightly curved, 3-septate, hyaline, 30 x 8 μ.

On bark of *Ficus*, California (Harkness).

Asci and sporidia not seen. Stylospores evidently different from those of *G. pulicaris*.

MELANÓSPORA, Corda.

Icones I, p. 24.

Perithecia superficial but with the base more or less sunk in a thick layer of interwoven hyphæ, without any true stroma, of soft, fleshy substance, translucent, globose-pyriform with a long beak. Sporidia elliptical, mostly continuous, dark.

A. *Sporidia continuous.*

M. chiònea, (Fr.) **(Plate 14)**

Sphæria chionea, Fr. S M II, p 446.
Ceratostoma chioneum, Fr Obs II, p 340.
Melanospora chionea, Corda Icon. I, p 25
Exsicc. Fckl F. Rh 2015 —Thum. M. U 1844 —Krgr. Fungi. Sax 273

Perithecia gregarious or scattered, superficial, globose, white-woolly, 300–400 μ diam., with a straight, reddish-yellow, stout beak about 1 mm. long and having a few spreading hairs at its apex. Asci obovate-clavate, stipitate, 8-spored, 35-45 x 13–16 (p. sp.) Sporidia crowded or imperfectly biseriate, globose-elliptical, brown, 10–12 x 9–10 μ.

On decaying pine leaves and on leaves of deciduous trees, London, Canada (Dearness). The hairs at the end of the beak are not as distinct as in the European specimens.

M. lagenària, (Pers.)

Sphæria lagenaria, Pers Syn. p 58
Ceratostoma lagenarium, Fr. Summ. Veg Sc. p. 396
Auerswaldia lagenaria, Rab. Hedw. I, p 116. tab XV, fig 2
Melanospora lagenaria, Fckl. Symb p. 126
Exsicc. Fckl. F. Rh 807 —Rehm Asc 699

Perithecia scattered or gregarious, emergent-superficial, sphæroid or subovoid, rugulose, at first yellowish, then liver color and finally black, 400–420 μ diam., with a very long (2–2½ μ), straight or flexuous beak tipped with white hairs. Asci broad-clavate, stipitate, 35–40 x 12–15 μ. Sporidia irregularly crowded, elliptical with ends attenuated, 11–14 x 6 μ, becoming brown.

Given on the authority of Peek (27th Rep. p. 110), who found it in the Adirondack Mts., N. Y., on the hymenium of some old *Polyporus*, He says: "Asci very broad, spores crowded, simple, elliptical, colored 12½ x 7½ μ."

B. *Sporidia muriform.*

M. chrysomálla, B. & Br. Fung. Ceylon. No. 1125.

Perithecia globose, about ⅓ nnn. diam., subcespitose, densely yellow-tomentose, with a short-cylindrical, nearly black ostiolum which projects but little above the yellow tomentum. Asci cylindrical, 75x7 μ, sessile or nearly so, with abundant, evanescent paraphyses. Sporidia obliquely uniseriate, oblong-elliptical, olivaceous-brown, 3-septate, with occasionally one or two of the cells divided by a longitudinal septum, 12–14 x 5–6 μ.

On base of culms of *Andropogon muricatus*, Louisiana (Langlois).

SUBORDER. SPHAERIÀCEÆ.

Perithecia mostly with a distinct, papilliform, elongated or compressed ostiolum, not carnose or membranaceo-carnose, but eoriaceous, woody, carbonaceous, often brittle, dark colored outside, mostly light colored within, never carnose. Perithecia distinct from the stroma though seated on or buried in it.

FAMILY. CHAETOMÌEÆ.

Perithecia superficial, free, seated on a superficial mycelium, generally with an ostiolum and an apical tuft of hairs or bristles, brown, fragile. Asci clavate or cylindrical, 8-spored, very evanescent. Sporidia continuous, brown.

CHAETÓMIUM, Kunz. & Schmidt.

Mycologische Hefte I, p. 15.

Perithecia superficial, free or adnate, generally seated on a superficial mycelium and with a distinct ostiolum and an apical tuft of hair or bristles, brown, thin-membranaceous. Asci club-shaped or cylindrical, very evanescent, without paraphyses. Sporidia 1-celled and brown, more or less compressed, or flattened.

Ch. elàtum, Kunze Deutschl. Schwamme, No. 184.

Conoplea atra, Spreng. Syst IV, p. 554.
Chætomium lageniforme, Cda. Icones I, p. 24
Chætomium, Fieberi, Fckl. Symb. p. 90.
Exsicc. Thum M. U. 758.—id F. Austr. 989 —Rehm. Asc. 247 —Cke. F. Br 1st. Ser. 290. Kunze F. Sel. 66.—Sydow. M. March 343.—Roum F. G. 1428.—Rab. F E. 529 —Ell. N. A. F. 560.—Rav. Fung. Car. III, 66,

Perithecia elliptical, about 400 μ high and 320 μ broad, with a distinct ostiolum; terminal tuft of hairs squarrosely branched, brown, rough. Asci club-shaped, short-stipitate. Sporidia flattened, the broad side elliptical, apiculate, 10–12 x 9–10 μ, olive brown; when viewed edgewise, fusiform, 7–8 μ broad.

On decaying herbaceous stems common.

Ch. muròrum. Cda. Icon. I. p. 24. tab. VII, fig. 293. B.

Exsicc. Rab. F E. 234.

Perithecia globose, ovate or elliptical, varying in size from 160–280 μ high by 160–230 μ broad. Ostiolum short, papilliform, distinct, sometimes collapsing; apical hairs simple, 2–3 times as long as the height of the perithecium, granular-roughened, dark brown, broadly arched with tips incurved; sides of the perithecia more sparingly clothed with finer, subulate hairs brown below and subhyaline above. Asci broad-clavate, short-stemmed, 8-spored. Sporidia narrow-elliptical, scarcely apiculate, olive or chocolate-brown, 12–16 x 7–8 μ, slightly compressed.

On bark of dead poplar, Montana (Anderson No. 651).

Ch. sphærospérmum, C. & E. Grev. VIII, p. 16.

Perithecia superficial, black, strigose, crowded in elongated groups, and connected by a loose mycelium, ovate globose. Hairs rigid, erect, elongated, slender, smooth, brownish-black, divaricately branched above, the main axis prolonged at the point of branching or bifurcation exactly as in *Ch. pusillum.* Sporidia viewed in front nearly round, 7–8 μ diam. (10 μ Cke.), viewed edgewise elliptical, 7–8 x 3 μ.

On the basswood bottom of a barrel standing in a cellar at Newfield, N. J., and found in the same situation by Rev. J. L. Zabriskie, at Flatbush, Long Island. Differs from *Ch. pusillum* only in its larger, darker, nearly round sporidia.

Ch. pusíllum, E. & E. Proc. Phil. Acad., July, 1890. **(Plate 16)**

Perithecia gregarious, black, membranaceous, ovate, about 200 μ high by 150 μ thick, the lower part clothed with fine, loosely entangled, pale, branching, slate colored hairs, the upper part more sparingly with longer, darker, nearly smooth and simple, partially transparent, continuous or faintly septate hairs, about 4 μ thick at the base and tapering above. Asci narrow-cylindrical, 30 x 3½ μ (p. sp.), without paraphyses. Sporidia uniseriate, elliptical, brown, compressed, 3½–5 x 2½–3½ μ, and 1½–2 μ thick. The asci soon disappear and the sporidia are expelled in a cylindrical mass ½–¾ mm. long and carrying along with it the upper part of the perithecium clothed with its spreading hairs. *Ch. sphærospermum*, C. & E. has the same habit and otherwise much resembles this, but has the apical hairs more branching and the sporidia larger (7–8 μ) and globose.

On the basswood bottom and hickory hoops of a barrel standing in a cellar at Newfield, N. J. and on an old wooden churn in a cellar, at Manhattan, Kansas (Kellerman), on decaying paper, Columbia, Mo. (Galloway), and Alabama (Atkinson).

The mode of branching is similar to that in *Ch. chartarum*, Ehr., the main axis being prolonged 10–15 μ beyond the point of separation or bifurcation, the prolongation being somewhat enlarged and hyaline. In the Missouri specimens, the apical bristles were swollen at intervals. This peculiarity was not observed in the other specimens.

Ch. pannósum, Wallr. Fl. Cr. Ger. p. 267., Rab. Deutschl. Kr. Flora I, p. 227. Zopf. Entw. der Asc. (Chaetonium), p. 276.

Ch. velutinum, E. & E. Journ. Mycol. I, p. 90.

Exsicc Kl. Herb. Mycol. II, 748.—M. March 1548.—Roum. F. G. 51.—Rab. F. E. 2025.

Perithecia closely gregarious or also scattered and single, subglobose, large (⅓–½ mm.), very fragile, dirty black; ostiolum large, short and thick, papilliform and colorless; rhizoids (basal root-like hairs) more numerous than in the other species. The perithecia are covered with a dense coat of dark, rough, crisped and interwoven hairs, causing them to appear confluent, forming continuous or partially interrupted, dense, felt-like, black patches several centimeters in extent, much like the subiculum of *Rosellinia aquila*. The lateral hairs are mostly simple while the apical ones are longer, stouter and straighter, 7–8 μ thick at base and sparingly branched. Asci (sec. Zopf) club-shaped, with a stipe-like base, reaching 100 μ long by 15–20 μ broad. Sporidia viewed in front, broad-elliptical, plainly but briefly apiculate, viewed edgewise, fusoid, 11–13 x 7–8 μ.

On a rotten maple log, Oregon (Carpenter).

Ch. olivàceum, C. & E. Grev. VI, p. 96.

Exsicc Ell. N. A. F. 56.

Perithecia gregarious, ovate-globose, about 250 μ high by 200 μ broad, thickly and evenly clothed with soft, granular-roughened, flexuous, sparingly septate, simple, greenish-yellow hairs 3–4 μ thick, without any very distinct apical tuft of coarser hairs, as is usual in most of the species. Asci oblong-clavate, p. sp. 35–40 x 12 μ. Sporidia crowded, globose-elliptical, apiculate at each end, yellowish-brown, 9–12 x 8–9 μ, (mostly 9–11 x 7–8 μ), with one or more nuclei.

On decaying stems of *Erigeron*, New Jersey and Louisiana.

Var. *chartarum*, (*Ch. chartarum*, Ehr. Sylva. p. 27)? issued in Ell. & Evrht. N. A. F. 2d Ser. 1541 as *Ch. chartarum*, Ehr., in Thum. M. U. 1942 and Roum. F. G. 4930, as *Ch. olivaceum*, C. & E., differs only in its slightly smaller sporidia and color varying from nearly black to mouse-color, but sometimes greenish-yellow as in the type.

Judging from the description of that species and the figure in Corda's Icones, this var. on paper can hardly be the genuine *Ch. chartarum*, Ehr. which with *Ch. Fieberi*, Cda. and *Ch. globosum*, Kze., Dr. Zopf. includes in his *Ch. Kunzeanum*.

Ch. bostrychòdes, Zopf, Entw. d. Chaet. p. 81. tab. 7, figs. 14–28.

Ch. caninum, E. & E. Journ. Mycol IV, p. 79.
Exsicc Sydow M. March 43 ?—Rab. F. E. 3340.

Perithecia small, elliptical, fusoid, or cylindrical, reaching 340 μ high and 220 μ thick, brown, with a short, papillose, hyaline ostiolum. Root-like hairs from the base (rhizoids) not very numerous. Apical tuft consisting of regularly spirally bent, roughish hairs. Asci club-shaped, p. sp. 20–22 x 7–10 μ with a stipe-like base 20–30 μ long. Sporidia ovate-globose or elliptical, somewhat flattened, 5–6 x 4–5 μ, olive-brown. On dog's dung and on an old cast-off leather shoe. Louisiana (Langlois), on the first named host. Long Island, N. Y. (Zabriskie).

Ch. lanòsum, Pk., 28th Rep. N. Y. St. Mus. p. 64.

Exsicc. Roum F. G. 4437 ?

Perithecia minute, subglobose, scattered or gregarious, densely covered with long woolly hairs, which are either dingy olivaceous or mouse-colored. Asci short, broad, fugacious. Sporidia subglobose, at first pale, then slightly colored, $7\frac{1}{2}$–$8\frac{1}{2}$ μ diam. with a single large nucleus.

On herbarium specimens of grasses, Albany, N. Y. (Peek).

The specimens in Fungi Gallici above quoted have the asci (p. sp.) 35–40 x 12–15 μ with a stipe-like base 15–20 μ long. Sporidia closely packed, globose-elliptical, yellowish-hyaline, very short-apiculate-pointed at each end, 10–13 x 8–10 μ; apical hairs very long, spreading, smooth, very sparsely septate, brown, 5–6 μ diam., not abundant. Perithecia $\frac{1}{4}$–$\frac{1}{3}$ mm. diam. with apex bare, black and shining.

Ch. funícolum, Cke. Grev. I, p. 176.

Perithecia scattered, subovate, black; hairs of the vertex very long, dichotomous or simple, erect, slender, acute, black. Sporidia lemon-shaped, dingy brown.

On an old broom, Albany, N. Y. (Peck).

This species is closely allied to *Ch. elatum*, but much smaller and neater. It is wholly black and without the fibrous base of *Ch. elatum*. The hairs are more delicate and not half the diameter, and the sporidia are scarcely more than half as long and broad.

Ch. melioloìdes, C. & P. 27th Rep. N. Y. State Mus. p. 106.

Perithecia scattered, minute, brown, globose, arising from a branching, septate mycelium. Bristles black, rough, two or three times dichotomously branched above, the branches spreading. Sporidia ovoid or subglobose, pale brown, 4–5 x 3–3½ μ.

On culms of Zea Mays, New York, New Jersey and Louisiana.

Differs from *Ch. funicolum* principally in its more branching hairs which cover the perithecia about uniformly above and below.

FAMILY. SORDARÌEÆ.

Perithecia either superficial or sunk in the matrix, with their upper part more or less projecting or sometimes erumpent, without (or more seldom) with a stroma, in which, when present, they are immersed; of soft, watery substance, brown or black, often translucent, with a decided round ostiolum. Asci generally with the apical membrane strongly thickened; paraphyses present. Sporidia dark colored, one or more-celled.

Mostly coprophilous; short lived in comparison with the *Rosellinias* and other similar forms from which they are also distinguished by their membranaceous or coriaceo-membranaceous perithecia.

SORDÀRIA, Ces. & De Not.

Schema di Classif. d. Sferiacei, p. 51.

Without stroma. Perithecia immersed, often finally erumpent and free, or superficial from the first, membranaceous, soft, black. Sporidia one-celled, brown or black, without appendages but surrounded with a gelatinous envelope.

S. humàna, (Fckl.) **(Plate 17)**

Hypocopra humana, Fckl Symb. p. 241.
Sordaria humana, Winter Die Deutschen Sordarien p. 21.
Exsicc. Fckl. F. Rhen. 1801.

Perithecia thickly gregarious, buried, globose, with a short, cylindrical, top-shaped, bare, shining neck, black, glabrous, about ⅓ mm. diam. Asci cylindrical, short-stipitate, with an obtuse or flattened apex which is strongly thickened, about 130 x 15 μ; paraphyses ventricose. Sporidia 8, uniseriate, elliptical or subglobose, dark brown, with a gelatinous envelope, 15–21 x 10–15 μ.

On decaying human ordure, in damp woods, Newfield, N. J.

S. híppica, Sacc. Syll. 900.

Hypoxylon equinum, B. & Rav. Grev. IV, p. 93.

"Perithecia springing from a thin, effused, white mycelium. Ostiola black, papilliform. Sporidia short-cymbiform, 15 μ long."

On horse dung, So. Carolina (Ravenel).

S. leucopláca, (Berk. & Rav.)

Sphæria leucoplaca, B. & Rav. Grev. IV, p. 143.
Syn? *S. minima*, Sacc. & Speg Mich. I, 373, F. Ital. tab. 617.
Hypocopra leucoplaca, Sacc. Syll. 887.
Exsicc. Rav. Fungi Car. Exsicc. IV, 61.

"Minute, black, seated on a broad, white-floccose spot; asci linear; sporidia uniseriate, elliptical."

On cow dung, So. Carolina, (Ravenel).

The specimen in Rav. Car. affords the following characters.—Perithecia scattered, erumpent, membranaceous, minute (90–120 μ), at first clothed with short, spreading, black hairs, which soon disappear. Ostiolum obtusely conical. Asci cylindrical, rounded at the apex, p. sp. 40–45 x 7 μ; paraphyses not seen. Sporidia uniseriate, short-elliptical, becoming opake, 8–10 x 5–6 μ.

In Ravenel's specimen the manure is not much decayed, of a light color and thickly sprinkled with the minute perithecia, which much resemble those of *Sporormia minima*. *Sordaria minima*, S. & S., differs from this only in being glabrous, but it may have been, like this, hairy at first.

S. fimícola, (Rob.)

Sphæria fimicola, Rob in Ann. Sci. Nat. 3. XI, p. 353.
Sordaria fimicola, Ces & De Not. Schema. p. 52.
Hypocopra fimeti and *stercoraria*, Fckl. Symb. I, p 240 & 241 (sec. Winter.)
Hypocopra fimicola, Sacc. Syll. 869.
Sordaria Iowana, Ell & Holw. Journ. Mycol. IV, p. 65
Sordaria fimicola, Winter Pilze p. 166.
Exsicc. Fckl. F. Rh 903, 1001, 1802.—Kze. F. Sel 101.—Rehm. Asc. 137, 300.—Sacc. M. V. 1181.—Sydow. M. March. 539.—Rab. F.F. 3254 —Krieg. F. Sax. 275.—Desm. Plantes Crypt. de Fr. 2061.

Perithecia mostly densely gregarious, ovate, ⅓ mm. diam. the base sunk in the matrix (without any stroma), the upper half or three-fourths projecting, black, tuberculose-roughened, of coarse cellular structure and membranaceous. Asci cylindrical, 107–114 x 12–15 μ (p. sp.); paraphyses ventricose. Sporidia uniseriate, elliptical, becoming opake,

18–20 x 8–12 μ, without any appendages but surrounded by a gelatinous zone, ends subacute.

On horse dung, Iowa and New Jersey, and on rabbit's dung, Canada (Dearness), on decaying potato, Lyndonville, N. Y. (Fairman).

Winter in Die Pilze makes the spore-bearing part of the asci 120–140 x 17–19. The form on potato has asci (p. sp.) 120–134 x 12–15 μ, but in the fimicolous specimens, the asci are shorter. The breadth of the sporidia as given in Journ. Mycol. is too small.

S. bombardioides, (Awd.) in literis.

Hypocopra bombardioides, Sacc. Syll. 847.
Exsicc. Rab F. E. 1527 —Rehm. Asc. 233.

Perithecia superficial, scattered or collected in small, crowded clusters or heaps, very large (¾ to 1⅓ mm. high), ovate-oblong or ventricose-pyriform, mostly somewhat contracted below, broadly rounded above and subumbilicate, with a small papilla, of a tolerably firm, waxy-carnose texture, chestnut-brown. Asci clavate-cylindrical, 140–170 x 16–18 μ (p. sp.); paraphyses slender, filiform, jointed. Sporidia 8, obliquely, uniseriate, or lying irregularly, elliptical, generally subinequilateral, dark brown, with a gelatinous envelope, 22–26 x 12–14 μ.

On horse dung, Mt. Shasta, California (Harkness).

S. sphærόspora, E. & E. (in Herb.)

Perithecia densely gregarious, ovate-globose, 250–300 μ, covered, but raising the surface of the matrix into little hemispherical prominences, from the apices of which project the tubercular-papilliform, obtuse, rough, black, bare ostiola. Asci cylindrical, subsessile, 150–160 x 18–20 μ, with very obscure paraphyses. Sporidia uniseriate, obovate-globose, hyaline at first with a broad, gelatinous margin, finally opake and almost black, 20–22 x 18–20 μ.

On dung of dogs, Newfield, N. J.

Outwardly resembles *Podospora coprophila*, (Fr.), but sporidia not appendiculate and more nearly globose.

PODÓSPORA, Césati.

Rabh. Herb Mycol No. 259 and in Hedwigia I, p 103.

Without stroma. Perithecia at first more or less sunk in the matrix, becoming finally more or less free, membranaceous, soft, black or dark brown. Sporidia 1-celled, black or brown, with one or more appendages, mostly without any gelatinous envelope.

Pod. coprόphila, (Fries.)

Sphæria coprophila, Fr. S. M. II, p. 342.
Hypoxylon coprophilum, Fr. Summa Veg. Sc. p. 348
Sordaria coprophila, Ces & De Not. Schema, p. 52.
Podospora coprophila, Winter, Pilze, p. 172.
Cercophora mirabilis, Fckl. Symb. p. 245.
Exsicc Fckl. F Rh 2271.—Rab. Herb. Myc. 257.—id. F. E. 830.—Rehm Asc. 198, 234. Ell. N. A. F 409.—Sydow Myc. March. 864.—Sacc. M. Ven. 1492.

Perithecia crowded, subconfinent, at first covered, except the black ostiolum, with a thin, white tomentum and immersed, but soon more or less emergent and bare, ovate or pyriform, dirty black, ostiolum obtuse, subconical. Asci pedicellate, cylindric-clavate, p. sp. 100–150 x 16–20 μ, 8-spored. Sporidia subbiseriate, ellipsoid, opake (when mature), 18–25 x 8–10 μ, with a hyaline appendage below about twice as long as the sporidium and terminated by two short, secondary appendages, of which there is also one at the upper end of the sporidium.

On cow dung, common.

Sec. Woronin this species has *pycnidia* and *conidia*. The first consist of globose receptacles (perithecia), 50–70 μ diam., pierced above, with 10–12 bristles around the opening, discharging the minute. hyaline, globose or ovate stylospores in the form of cirrhi. The conidial apparatus consists of abundant, small, flask-shaped bodies arising from a mycelium, and filled with small, globose bodies which, when discharged from the apical opening, surround themselves with a membrane in which they are visible as a bright central nucleus. The further development of these so-called conidia is still unknown. A similar growth is also found with some other species of *Sordarieæ*.

Pod. cúrvula, (de Bary), var. aloides, Winter, Deutsch Sord. p. 39.

Sphæria valsoides, Pk. 28th Rep. p. 78.
Sordaria aloides, Fckl. Symb. Nachtr. II, p. 43.
Exsicc. Fckl. F. Rhen. 2549.

Perithecia scattered, semiimmersed, black, globose-conical, about 1 mm. high, nearly smooth below, surrounded just below the apex by 4–8 compact tufts of hairs closely glued together and forming little horn-like processes, which soon curve downward so as to bear some resemblance to the leaves of an Aloe (whence the name), but finally the hairs composing the tufts separate and stand erect. Asci long-stipitate (p. sp. 146 x 26–28 μ Fckl.) Sporidia biseriate, ovate-oblong, becoming opake,with a hyaline, straight appendage at the lower end, about as long as the sporidium. Sporidia 34 x 16 μ (Fckl.), 27–33 μ long, (Peck).

On cow dung, New York State (Peck), Iowa (Holway).

In the Iowa specimens, the perithecia are ovate-conical, about 1 mm. high and ⅔ mm. broad, the horn-like tufts of hair around the vertex numerous, short and erect. Asci (p. sp.) 115–140 x 15–20 μ. Sporidia elliptical, 22–27 x 12–14 μ, mostly biseriate above.

Pod. amphicórnis, (Ell.)

Sphæria amphicornis, Ell Bull. Torr. Bot. Vol. VI, p. 109.
Sphæria eximia, Pk. 28th Rep. N. Y. State Mus. p. 78.*

Perithecia superficial, coriaceo-membranaceous, ovate or ovate-conical, ½–¾ mm. high and about ⅔ as much broad, sparingly clothed with straight, spreading, brown, septate hairs 50–100 μ long. Ostiolum broad-papilliform, bare and black. Asci ventricose-clavate, attenuated above, paraphysate, 140–160 x 25–35 μ. Sporidia crowded-biseriate, cylindrical, hyaline and nucleate at first, with a slender appendage at each end about as long as the sporidium, finally acutely elliptical and opake, with the terminal appendage about 20 x 5 μ, rounded and obtuse at the end, and the lower one of about the same length or a trifle narrower, with a transverse septum about 12 μ from the end of the sporidium. Each of these appendages, at a certain stage of growth, has a slender, hyaline, thread-like prolongation several times longer than the thicker, basal appendage, but this slender prolongation soon disappears.

On rabbit's dung, New Jersey, and New York.

Differs from *S. curvula*, de Bary, in its sparsely scattered, not fasciculate hairs, and the different character of the appendages of the sporidia, which resemble more those of *P. fimiseda*, Ces. & De Not, from which our species may be distinguished by its smaller, and less densely hairy perithecia.

Pod, fimíseda, (Ces. & De Not.)

Sordaria fimiseda, Ces. & De Not. Schema, p. 52.
Podospora fimicola, Ces. in Rab. Herb. Mycol. 259, and in Hedw. l. c.
Podospora fimiseda, Winter Pilze, p. 170.
Cercophora fimiseda, Fckl. Symb. p. 244.
Exsicc. Fckl. F. Rh. 2037.—Rab. Herb. Mycol. 259.—Rehm Asc. 235.

Perithecia scattered or collected in small groups, erumpent-superficial, ovate-globose, attenuated above into a more or less distinct neck, dark brown, thickly clothed all over with dark brown, short, septate hairs, 1–1½ mm. high. Asci cylindrical, with a long, stipe-like base, somewhat attenuated above, about ½ mm. long, p. sp. 350–400 x 45–

*The 28th Report in which Peck's Sphæria eximia was published is dated Jan. 1875, but was not given to the public until after *Sphæria amphicornis* was published in the Torr Bull. in Sept. 1876.

70 μ, mostly 8-spored. Sporidia mostly biseriate, elliptical, 50–60 x 28–30 μ, dark brown, with a long, cylindrical, often curved appendage of cellulose membranaceous structure below, and at the end of this appendage, and also at the apex of the sporidium, with a gelatinous appendage of various form. Paraphyses filiform, septate, longer than the asci.

On cow dung, New York State (Peck).

Pod. striàta, (E. & E.)

Sordaria striata, E & E. Journ. Mycol IV, p 79.

Gregarious. Perithecia ovate-conical, $\frac{2}{3}$ mm. high and $\frac{1}{2}$ mm. broad, black, tubercular-roughened, the tubercles seriate above so as to cause the conical ostiola to appear striate. The tubercles are at first capped with a few light colored granules, like grains of white sugar, but these at length disappear. Asci linear-lanceolate, contracted towards each end and perforated above, 200 μ long and over, (including the filiform base) and 12–15 μ wide, with abundant paraphyses. Sporidia biseriate, elliptical, brown, 14–16 x 8–10 μ, the upper end acute or with a short hyaline appendage 8–12 μ long, the lower end prolonged into a yellowish-hyaline, cylindrical, curved appendage 35–40 x 5 μ.

On decaying stems of some large weed, St. Martinsville, La. (Langlois).

Pod. penicillàta, (E. & E.)

Sordaria penicillata, E. & E Journ Mycol IV, p. 78

Perithecia gregarious, ovate, $\frac{1}{3}$–$\frac{1}{2}$ mm. diam., at first entirely buried except the protruding ostiolum, at length with the upper half emergent; ostiolum short-cylindrical or obtusely conical and surrounded by a tuft of straight, erect, closely crowded, pale brown, continuous hairs $\frac{1}{3}$–$\frac{1}{2}$ mm. long. Asci oblong-cylindrical, about 150 x 20–25 μ, 4-spored, with indistinct paraphyses. Sporidia subhyaline and clavate-cylindrical at first, then olivaceous with a single large nucleus, finally elliptical and opake. 25–30 x 18–20 μ, with a cylindrical, nearly straight, hyaline appendage 12–15 x 4–5 μ at the lower end, and the upper end obtusely pointed or subtruncate.

On an old decaying Chinese mat., St. Martinsville, La. (Langlois).

The measurements of the perithecia given in Journ. Mycol. for this and *Philocopra lutea* were too small.

Pod. Brássicae. (Klotzsch).

Sphæria Brassicae, Klotzsch in Smith's Engl Flora V, p. 216.
Sphæria lanuginosa, Preuss in Linn. 1853, XXVI, p. 714.
Arnium lanuginosum, Nits in Fckl Symb Nachtr. I, p. 38.
Sordaria Curreyi, Awd. in Niessl Beitrag. p 42.
Sordaria lanuginosa, Sacc Syll. 859.
Podospora Brassicae, Winter Pilze 2985.

Perithecia globose-conical, often curved, subtruncate above, dirty gray-tomentose, about 1 mm. high and ¾ mm. thick, superficial. Asci cylindrical, p. sp. 150–160 x 25–26 μ, stipitate. Sporidia obliquely uniseriate, ovate-elliptical, 40–50 x 24 μ, dark brown, with a deciduous, hyaline appendage at each end, or sometimes only below.

Dr. Harkness reports this as found in California on dead branches of *Lupinus arboreus*. We have not seen the specimens and take the diagnosis from Saccardo's Sylloge. The synonymy is from Winter's Pilze, where the asci are given as 200–300 x 34–40 μ for the spore-bearing part, instead of 150–160 x 25–26 μ as stated in Sylloge.

Pod. Califórnica, (Plowr.)

Sordaria Californica. Plowr. Grev. VII, p. 72, tab. 120, fig. 2.

Perithecia ovate, clothed with a dusky brown, felt-like coat, about half sunk in the matrix, 1 mm. high, by about ¾ mm. broad, the projecting apex more or less distinctly ridged, with the intervening furrows more distinct and coarser than in *Pod. striata*, E. & E. Asci subcylindrical, 300–350 x 15–20 μ, with filiform paraphyses. Sporidia biseriate, oblong-elliptical, 30–35 x 15–18 μ, subtruncate at base, with a hyaline, curved appendage about as long as the sporidium.

On cow dung, California (Harkness).

Our specimen from Dr. Harkness has the perithecia immature, the sporidia being as yet hyaline, and consisting of a narrow-elliptical head 25 x 12 μ with a long cylindrical tail below.

PHILOCÓPRA, Speg.

Nova Syst. Carp Class. in F. Arg. Pug. I, Sacc. Syll. I, p. 249.

Perithecia semi-immersed, globose-conical, glabrous or hairy, membranaceous, turning black. Asci very large, polysporous. Sporidia polystichous, ovoid, brown, mostly with a hyaline appendage at one or both ends.

Ph. lútea, (E. & E.) **(Plate 18)***

Sordaria lutea, E. & E. Journ. Mycol. III, p. 118

Perithecia gregarious, ¾–1 mm. diam., membranaceous, conic-

*In explanation of Pl. 18, instead of *Podospora lutea* read *Philocopra lutea*.

globose, covered, except the papillose-conical, black ostiolum, with a dense, light yellow tomentum composed of branching, slightly roughened hairs. Asci lanceolate, rounded and perforated at the apex, 190–130 x 15 μ. Sporidia 12–16 in an ascus, at first vermiform and greenish-yellow, finally almond-shaped and opake, 14–16 x 7–8 μ, with a cylindrical, curved appendage, 30–35 x 4 μ, attached to its base and a short, slender appendage, about 15 x 2 μ, at the apex. The asci are very evanescent. The yellow coat also turns black at maturity. A closely allied species, with sporidia 22–25 x 12–15 μ, has been met with on dead herbaceous stems, but we have not sufficient material to give a full description.

On rotten wood (*Acer* and *Kalmia*), in swampy woods, Newfield, N. J., November, 1879, and August, 1887.

The perithecia are represented in the drawing (fig. 2) as attenuated above into a distinct, beak-like neck. This is exceptional. Usually there is only the globose-conical, black ostiolum without any distinct neck. The apical appendage is often only rudimentary or wanting.

Ph. canina, (Pk.)

Sphæria canina, Pk. 28th Rep. N. Y. State Mus. p. 78.
Philocopra canina, Sacc. Syll. 911.

"Perithecia minute, scattered or crowded, free, subglobose, reddish-brown or dark amber color, then blackish. Asci broad-oblong or oblanceolate. Sporidia numerous, elliptical, slightly colored, 6½–8 μ long."

On dung of dogs, Bethlehem, N. Y. (Peck).

Ph. multifera, (B. & Rav.)

Sphæria multifera, B. & Rav. Grev. IV, p. 143
Philocopra multifera, Sacc. Syll. 913.

"Minute, black, subglobose. Asci short, oblong, blunt, filled with numerous, subglobose, brown sporidia 6–7 μ diam."

On cow dung, Pennsylvania.

HYPOCÓPRA, Fr.

Summa Veg. Scand. p. 397, Winter Die Pilze, p. 177. (non Fuckel).

Stroma present. Perithecia more or less sunk in the stroma. Sporidia one-celled, brown or black, without appendages, but surrounded by a gelatinous zone. Paraphyses present, filiform. Distinguished from the species of *Sordaria* with gelatinous-zoned sporidia, by the presence of a stroma.

H. fimèti, (Pers.)

Sphæria fimeti, Pers. Syn. p. 64
Hypocopra fimeti, Fr. l. c.
Sordaria fimeti, Winter Deutsch. Sord. p. 14
Coprolepa fimeti, Sacc. Syll. 903.

Perithecia crowded or confluent, globose or ovate, 300–350 μ broad, ending above in a thick, conical, mostly oblique neck, and about half sunk in the thin, crustaceous, effused, black, slightly wrinkled, bare stroma. Asci cylindrical, short-stipitate, 140–150 x 17 μ (p. sp.), 8-spored, with filiform paraphyses. Sporidia elliptic-oblong, dark brown, 18–20 x 12 μ, with a gelatinous envelope.

On horse dung, New York (Peck).

We have never met with this species, and give it on the authority quoted. Diagnosis from Winter.

H. equòrum, (Fckl.) **(Plate 17)**

Hypoxylon equorum, Fckl F. Rh 1058.
Coprolepa equorum, Fckl. Symb. p. 240.
Sordaria equorum, Winter Deutsch Sordar p 13
Hypocopra equorum, Winter Die Pilze II. p. 178
Exsicc. Fckl F. Rh. 1058.—Cke. F. Brit. 2d Ser. 241 and 242 —Rehm Asc. 992.

Perithecia scattered, about ½ mm. diam., globose, pale brown, their short, black, shining, papillose ostiola erumpent through the thin, crustose, black, villose-tomentose, broadly effused stroma. Asci cylindrical, short-stipitate, thickened at the apex, 8-spored, 175–200 x 15–18 μ, shorter than the simple, filiform, septate paraphyses. Sporidia obliquely uniseriate, elliptical, or elongated-elliptical, subinequilateral, dark, 19–21 x 9–10 μ.

On horse dung in open woods, Newfield, N. J.

Differs from *H. fimeti* in its villose stroma, larger, buried perithecia and shorter ostiola.

SPORÓRMIA, De Not.

Mier. Ital. decas V, No. 6, (1849).

Perithecia innate-emergent, membranaceous, glabrous or subglabrous, black, with papilliform ostiolum. Asci elongated, 8-spored. Sporidia 4–20-celled, the cells soon separating, dark colored, mostly fimicolous.

Sp. minima, Awd. Hedwigia VIII, p. 66. **(Plate 18)**

Exsicc Rab. F. E. 1339.—Rehm Asc. 34 —Thum. M. U. 2167 —Ell. N. A. F. 198.

Perithecia scattered, about half sunk in the matrix, 100–114 μ diam., membranaceous, black, bare, with a very small, papilliform ostiolum. Asci oblong-cylindrical, with a short stipe, 8-spored, 80–90 x 12–15 μ. Sporidia overlapping-fasciculate, parallel, nearly straight, 4-celled, rounded at both ends, 25–30 x 4–6 μ, the two middle cells 6½–7½ μ long, the end cells mostly a little longer. The four cells of the sporidia readily separate from each other.

Common on cow dung, also on rabbits' and goats' dung, around Newfield, N. J., and probably throughout the country.

Sp. intermèdia, Awd. Hedw. VII, p. 67.

Exsicc. Kunze F. Sel 67.—Rab. F. E. 644 —Rehm Asc. 134.

Perithecia scattered or gregarious, buried, except the short, papilliform or subglobose ostiolum, small (150–200 μ), black, bare, carnose-membranaceous. Asci oblong, ovate, or nearly cylindrical, 112–125 x 20–24 μ (118–175 x 24–30 μ, Winter), briefly stipitate and paraphysate. Sporidia 2–3-seriate, partly overlapping each other, cylindrical, straight or slightly curved, rounded at the ends, 4-celled, 43–54 μ long, the end cells ovate-cylindrical, 12–15 μ long, the two intermediate cells barrel-shaped, about 12 x 8–10 μ, the joints readily separating.

On rabbits' dung, and cow dung, Newfield, N. J.

Sp. hercùlea, E. & E. (in Herb.)

Perithecia scattered, semiimmersed, ovate-globose. ½–¾ mm. diam., with a short, cylindrical or tuberculiform, rough ostiolum. Asci ventricose-oblong, 250–342 x 50–60 μ, with a short, nodulose stipe, and abundant paraphyses. Sporidia fusoid-cylindrical, 10–13-jointed, 112–152 x 14–16 μ, terminal joints ovate-conical, 15–18 x 12 μ, the others slightly flattened-globose, about 15 x 12 μ, the longer diameter being transverse. The 4th or 5th cell from the upper end of the sporidium is usually larger, 22 x 20 μ. The joints readily separate, and the entire sporidium is at first enclosed in a thin, gelatinous envelope.

This is easily distinct from all the allied species.

On cow dung, Newfield, N. J.

Sp. fimetària, De Not. Micr. Ital. dec. V, p. 10.

Sphæria fimetaria, Rab. Herb. Mycol. Ed. I, 1733.

Perithecia depressed-globose, membranaceous, with a simple, perforated ostiolum. Asci cylindrical, attenuated below into a short stipe, 80–100 x 14–16 μ. Sporidia 16–20-celled, rod-shaped, lying parallel, without any visible gelatinous envelope, 55–60 x 3–4 μ, medial cells about 2½ μ, and terminal ones 4 μ long

On cow dung, South Carolina (Ravenel).

SPORORMIÉLLA, E. & E. nov. gen.

Perithecia soft-carnose, embedded in a flattish, semiimmersed, subcarnose stroma. Asci and sporidia as in *Sporormia*.

S. nigropurpùrea, E. & E. (in Herb.)

Perithecia globose or slightly ovate-globose, about ⅓ mm diam., thick-walled, soft, and buried in the soft, flattish stroma, which is ½–1 cm. across, or by confluence more, slightly raised above the surface of the matrix, dark gray outside, and, like the perithecia, purplish-black within. Ostiolum subtubercular, erumpent, perforated. Asci clavate-cylindrical, 100–125 x 10–12 μ. Sporidia mostly biseriate, at least above, cylindrical, nearly straight, 4-jointed, hyaline at first, soon dark brown, 16–20 x 4–5 μ, terminal joints subovate, intermediate ones subglobose, about 4 μ diam. Asci with a short, narrow stipe Paraphyses linear, nucleate.

On cow dung, Newfield, N. J.

Delítschia bispórula, (Crouan).

Hormospora bisporula, Crouan, Finist. p. 21.
Delitschia bisporula, Hansen's Fungi Danici. p. 107, tab. IX, figs. 7–11.

This species has been figured (Plate 17) to illustrate the genus, though, as far as we know, it has not yet been found in this country. *Delitschia* may be briefly characterized as a *Sordaria* with uniseptate sporidia.

FAMILY. TRICHOSPHÆRÌEÆ.

Perithecia from the first superficial or erumpent, without any stroma, coriaceous or membranaceous, or sometimes subcarbonaceous, generally clothed with bristles or hairs, often with a more or less strongly developed, felt-like layer of conidia-bearing hyphæ overspreading the matrix.

VENTÙRIA, Ces. & De Not.

Schema d. Classif. p. 51, Sacc. Syll. I, p. 586.

Perithecia superficial or erumpent, bristly, ostiolate, membranaceous. Asci sessile or briefly stipitate, 8-spored, mostly without paraphyses. Sporidia oblong or ovoid-elliptical, uniseriate, hyaline or yellowish, mostly foliicolous.

Coleroe, (Fr.), which comprises only species with strictly superficial perithecia, is here included in *Venturia*.

A. *On woody dicotyledonous plants.*

V. pulchélla, C. & P. 25th Rep. N. Y. State Mus. p. 106.

Perithecia minute, hypophyllous, crowded, 15–40 together in small (1½–2½ mm.), orbicular or subangular patches, and clothed with short, stout, spine-like bristles 20–40 x 4–5 μ. Asci cylindrical, or clavate-cylindrical, 60 x 7 μ, with imperfectly developed paraphyses. Sporidia uniseriate, uniseptate, yellowish, 7–9 x 3 μ.

On living leaves of *Cassandra calyculata*, New Jersey to Canada.

V. orbícula, (Schw.)

Sphæria orbicula, Schw. Syn. N. Am. No. 1789.
Venturia orbicula, C. & P 25th Rep. N. Y. State Mus. p. 105.
Exsicc. Thum M. U. 855.—Rab. Winter F. E. 3143 —Rav. Car. IV, 64.—Ell. N. A. F. 700. Ell. & Evrht. N. A. F. 1687.

Perithecia minute, globoso, superficial, clothed with black, persistent, spreading hairs, collected in orbicular groups 3–6 mm. diam. The perithecia are often more crowded around the margin of the groups, so as to form a narrow, black circumscribing line. Asci short, subclavate. Sporidia oblong, uniseptate, the cells often unequal, 10 x 2–2½ μ. Often sterile.

On leaves of various species of oak, New Jersey and New York, west to Kansas.

A Var. occurs on fallen leaves of *Quercus coccinea*, with the perithecia irregularly scattered (var. *sparsa* E. & E.)

V. dítricha, (Fr.)

Sphæria ditricha, Fr. S. M. II, p. 515.
Vermicularia ditricha, Fr. Summa. Veg Sc. p. 420.
Venturia maculans, Pk. 28th Rep p. 81, id. 30th Rep. p. 77.
Venturia ditricha, Karst. Myc. Fenn. II, p. 188.
Sphæria ditricha, Fckl Symb. p. 100.
Exsicc, Fckl. F. Rh. 568 —Rehm Asc. 597, 792.—Thum. F. Austr. 247.—id M. U 350. Krgr. F. Sax. 232.—Sydow. M. March. 982.

Perithecia very small, emergent, mostly hypophyllous, subgregarious, seated on gray, indeterminate, confluent spots, depressed-shpæroid, black, with a papilliform ostiolum and a few (4–10), spreading, continuous, black, bristle-like hairs. Asci with a very short stipe, oblong-cylindrical, 35–45 x 10–12 μ, 8-spored, with obscure paraphyses. Sporidia biseriate, ovate-oblong, uniseptate, greenish-hyaline, 12–16 x 5–6 μ, lower cell narrower.

On fallen birch leaves, New York (Peck).

V. chloróspora, (Ces.)

Sphæria chlorospora, Ces. Erb. critt. Ital n. 296.
Sphærella chlorospora, De Not. Sfer Ital. p. 85, tab. xcvii.
Venturia chlorospora, Karst. Myc Fenn. II, p. 189 and Sacc. F. Ital. tab. 349.
Sphærella inaequalis, Cke. Journ. Bot. 1866, tab. 50, fig. 26.
Sphærella cinerascens, Fleischak Rab. F. E. 845.
Sphærella ditricha, Awd. Rab. F. E. 933 (pr. p.)
Sphærella canescens, Karst. Myc. Fenn. II, p. 189.
Didymosphaeria inaequalis, Niessl in Rab. F. E. 2663.
Venturia inaequalis, Winter in Thum. M. U. 1544.
Exsicc. Fckl. F. Rh. 1776.—Rab. F. E. 48, 845, 943, 2054, 2663.—Rehm Asc. 292, 445. Thum. M. U. 650, 1544.

Perithecia subglobose, 50 μ diam., clothed with rigid, very black hairs thickened at the base and 40–50 x 4 μ. Asci cylindric-clavate, 8-spored, aparaphysate, 50–60 x 13–14 μ. Sporidia subbiseriate, oblong-ovate, 14–16 x 6–6½ μ, uniseptate, scarcely constricted, of a clear, light yellow color.

On leaves of *Salix herbacea* and *S. glauca*, Godthaab, Greenland.

V. asterinoides, E. & M. (in Herb.)

Perithecia superficial, scattered, conic-convex, 110–140 μ diam., ostiolum papilliform, clothed with a few long (100–200 x 3 μ), spreading, continuous or faintly-septate, brown hairs, surrounded by a scanty mycelium. Asci oblong, subsessile, 35–40 x 12–15 μ, 8-spored. Sporidia biseriate, clavate-oblong, hyaline, uniseptate, 12–15 x 3–4 μ. This is an *Asterina*, but for the hairy perithecia.

On leaves of *Quercus laurifolia*, Florida (Dr. Martin).

V. Clintònii, Pk. 28th Rep. p. 82.

Gregarious in indeterminate, suborbicular patches. Perithecia nearly free, globose, black, hispid, with a few straight, black bristles. Asci linear. Sporidia uniseriate, obovate, uniseptate, yellowish or yellowish-brown, 10 μ long, the septum usually nearer the small end.

On the under surface of fallen leaves of *Cornus*, New York State (Clinton).

V. Kálmiæ, Pk. 28th Rep. p. 82.

Perithecia minute, prominent, centrally aggregated on small, orbicular, brown spots or scattered along the midrib, black-bristly, with straight, rigid, divergent, black hairs. Asci subcylindrical, 35 μ long. Sporidia oblong or subfusiform, minutely nucleate, 8½–9 μ long.

On the upper surface of leaves of *Kalmia glauca*, Kasoag, N. Y. (Peck).

Specimens found at Newfield, N. J., June, 1880, on leaves of *Kalmia angustifolia*, have the perithecia 65–75 μ diam., bristles 40–70 x 4 μ. Asci oblong-ovate, 30–35 x 12 μ, without paraphyses. Sporidia biseriate, fusoid, 9–12 x 3–4 μ, 4-nucleate when first found, now (1890) distinctly uniseptate.

V. applanàta, Ell. & Martin, Am. Nat. 1884, p. 69.

Perithecia hypophyllous, lenticular, 100 μ diam., pierced in the center, texture subradiate-cellular, with 12–15 erect, sparingly septate bristles 70–100 μ long. Asci oblong, about 40 x 10–12 μ, sessile, without paraphyses. Sporidia in two or three series, ovate-oblong, uniseptate, yellowish, 10–12 x 3 μ.

On living leaves of *Magnolia glauca*, Florida.

T. formòsa, Ell. & Martin, (in Herb.)

Perithecia hypophyllous, scattered, depressed-spherical, 220–300 μ diam., with long (100–175 x 5–6 μ), spreading, septate, obtuse, pale brown hairs around the sides and base, and a few shorter ones above. Ostiolum indistinct or wanting. Texture of perithecia thin-membranaceous, fragile, pale. Asci about 50 x 20 μ, sessile and without paraphyses. Sporidia biseriate, oblong-fusoid, yellowish, uniseptate, 20–23 x 7–8 μ.

On living leaves of *Olea Americana*, Florida, 1885.

V. myrtílli, Cke. in Journ. Bot. (1866), p. 245, tab. L, fig. 4.

Sphæria Vaccinii, Fckl. Symb p 106 (1869).

Perithecia amphigenous, scattered or gregarious, superficial, black, very small, armed with long, spine-like, black bristles. Asci broad at the base, narrowed above, 50 x 12–14 μ, 8-spored. Sporidia biseriate or crowded in the lower part of the asci, ovoid-oblong, uniseptate, not constricted, straight, greenish-hyaline, 12–14 x 4–5 μ.

On fallen leaves of *Vaccinium?* New York (Peck).

V. curvisèta, Pk. 35th Rep. p. 46.

Exsicc. Ell. N. A. F. 1356.

Perithecia numerous, often crowded, minute, 75–100 μ diam. (mostly not over 75 μ), globose, crowned above, with 5–8 divergent, broadly curved, black bristles 75–125 x 4 μ. Asci oblong, attenuated above, often slightly curved, 40–50 μ long. Sporidia crowded or biseriate, oblong, 10½–12½ x 4–5 μ, hyaline, the upper cell broader.

On fading leaves of *Nemopanthes Canadensis*, Center, N. Y.

V. Arctostáphyli, Cke. & Hark. Grev. XIII, p. 20.

Amphigenous, scattered, black. Perithecia subglobose, superficial, strigose, 120–150 μ diam., clothed with rigid, acicular hairs 80 μ long. Asci obclavate, sessile, 8-spored. Sporidia elliptical, rounded at the ends, scarcely constricted, uniseptate, yellowish, 12–15 x 5 μ.

On dead leaves of *Arctostaphylus*, California (Harkness).

V. Cassándrae, Pk. 38th Rep. p. 104.

Exsicc. Ell. & Evrht. N. A. F. 2d ser. 2363.

Spots reddish-brown or brownish, sometimes with a grayish center. Perithecia amphigenous, minute (70–80 μ), black, with a few short, straight, diverging, black bristles 30–40 μ long around the apex. Asci oblong, gradually and slightly narrowed above, 40–45 x 8–10 μ. Sporidia biseriate, oblong, quadrinucleate, 12 x 5 μ.

On living leaves of *Cassandra calyculata*, Caroga, N. Y., and on leaves of *Andromeda polifolia*, London, Canada (Dearness).

The perithecia are found mostly on the lower side of the leaf. They are so small as to be scarcely visible to the naked eye. Sometimes they emerge from beneath the scales of the leaf, so as to appear erumpent, but they are really superficial.

In the Canada specimens the asci are 40–45 x 15–20 μ. Sporidia ovate-oblong, 12–18 x 4–5 μ, which are about the same as our measurements of the asci and sporidia in a specimen from Peck.

V. pezizoìdea, Sacc. & Ell. Mich. II, p. 567.

Exsicc. Ell. N. A. F. 1355.

Perithecia scattered or subgregarious, hypophyllous, globose, 80–115 μ diam., thickly clothed with dark, bristle-like, spreading hairs 60–100 x 5 μ. Texture of the perithecia (which finally collapse), very thin and pale. Asci oblong-cylindrical, sessile, 30–35 x 7–8 μ, with paraphyses. Sporidia biseriate, subcylindrical, hyaline, (8–10 x 2 μ, Sacc.)

On fallen leaves of *Andromeda racemosa*, Newfield, N. J

The bristles are abundant, and those arising from the lower part of the perithecia are slightly curved inwards. The specc. examined were immature and the sporidia continuous.

V. cupulàta, E. & M. Am. Nat. 1884, p. 69.

Exsicc. Ell. N A F. 1298.

Perithecia hypophyllous, scattered, superficial, astomous, about 300 μ diam., bristly below, bare and collapsing above. Bristles dark brown, 180–210 x 5–6 μ. Asci oblong, about 48 x 12 μ, 8-spored, subsessile. Sporidia biseriate, obovate, uniseptate, hyaline, 12–13 x 4–4½ μ.

On living leaves of *Quercus laurifolia*, Florida (Martin).

V. erysipheoìdes, E. & E. Journ. Mycol. III, p. 128.

Perithecia gregarious, black, globose, about 100 μ diam., broadly perforated above, beset with scattering, rigid, black, continuous bristles 40–70 x 5–6 μ. Asci oblong, sessile, without paraphyses, 40–45 x 7–8 μ. Sporidia crowded, fusiform, hyaline, slightly curved, 5–6-nucleate, about 20 x 2½ μ. This differs from *V. graminicola*, Winter, in its smaller perithecia (80–110 μ) with shorter, lateral bristles, and in its narrower sporidia (2–2½ μ). The number of bristles on a perithecium is generally not over ten or twelve and they stand out horizontally or nearly so, reminding one of some of the *Erysipheæ*.

On dead culms or sheaths of *Panicum Curtisii*, Louisiana (Langlois).

V. Bárbula, (B. & Br.)

Sphæria Barbula, B. & Br. Ann. N. Hist. No 870, tab. 10, fig. 20.
Venturia Barbula, B. & Br Cke Hndbk. No. 2784.
Exsicc Rehm Asc. 293, Ell. N. A. F 792 (var. *foliicola*).

Perithecia subcespitose, globose, collapsing, rather thickly clothed with short (25–35 x 4 μ), straight spines. Asci cylindrical, 60 x 6 μ, with scanty paraphyses. Sporidia uniseriate, hyaline, uniseptate, 8–10 x 4½ μ, often surrounded by a thin, gelatinous zone.

On bark of pine trees, England and Germany.

Var. *foliicola* (N. A. F. l. c.) on dead leaves of *Pinus rigida*, still attached to the branches, in a fallen tree top, at Newfield, N. J., differs from the type in its foliicolous growth, scattered perithecia and narrower sporidia (8–10 x 3–4 μ.

The typical form, on bark, has not yet (as far as wé know) been found in this country.

V. cincinnàta, (Fr.)

Sphæria cincinnata, Fr. S. M. II, p. 451.
Venturia cincinnata, Fr. Summa Veg. Sc. p. 405.

Perithecia amphigenous, superficial, solitary, black, very small, conic-cylindrical, smooth, ostiolum hardly visible, crowned with a tuft of erect black hairs about as long as the diameter of the perithecium. Sometimes the perithecium is also surrounded by hairs at the base. On the upper surface of the leaves, the perithecia are less perfect and nearly globose.

On decaying leaves of *Oxycoccus* (*Vaccinium*) *palustris*, Tasermiut, Greenland, on green leaves of *Oxycoccus macrocarpus*, Pennsylvania (Schweinitz).

In the Greenland specc. the asci are narrowed above from a broad base, 85–95 x 20–25 μ. Sporidia ovoid-oblong, uniseptate and constricted, at first hyaline, then olivaceous, 24–32 x 8–10 μ.

V. compácta, Pk. 25th Rep. p. 106.

Perithecia minute, (200–230 μ), subglobose with a papilliform ostiolum, collected in dense clusters or groups, 15–30 together, on the lower side of the leaves, thickly covered with short, spine-like, black bristles 25–40 x 3–4 μ. Asci fasciculate, clavate-cylindrical, about 40 x 6 μ, with abundant paraphyses more or less branched. Sporidia obliquely uniseriate or subbiseriate above, oblong-ovate, subolivaceous, 12–15 μ long.

On living leaves of *Vaccinium macrocarpon*, New York (Peck), northern New Jersey (Halsted).

Specimens from Maine have the perithecia less compactly grouped; mostly in small, orbicular patches margined by a circle of perithecia, as in *V. orbicula*.

V. Gaulthèriæ, E. & E. Journ. Mycol. I, p. 153.

Exsicc. Ell. & Evrht. N. A. F. 2d Ser. 1686.

On orbicular, dark brown, ⅓ mm. spots, which are mostly of a lighter color (gray) in the center. Perithecia scattered, orbicular (75 μ), membranaceous and rather coarsely cellular, with a few black, continuous, straight, spreading, 35 x 3 μ bristles above. Asci ovate-oblong, 30–35 x 8–11 μ, broader and slightly curved below, sessile, without paraphyses. Sporidia biseriate, subhyaline (with a greenish-yellow tint), ovate-oblong, 3–4-nucleate, uniseptate and slightly constricted at the septum, 11–14 x 3 μ.

On living leaves of *Gaultheria procumbens*, Newfield, N. J.

B. *Growing on herbaceous plants.*

V. Dickìei, (B. & Br.)

Sphæria Dickiei, B. & Br. Ann. Nat. Hist. No. 617., tab X, fig. 8.
Lasiobotrys Linnææ Dickie, Berk. Outl, p. 404
Venturia Dickiei, de Not. Schema, Sfer. p. 51.

Perithecia crowded in orbicular groups 1–1½ mm. across, epiphyllous, erumpent, becoming superficial, seated on a filamentose subiculum, subspherical, black, clothed above with spine-like bristles 40–80 μ long. Ostiolum papilliform. Asci sessile, elongated, obtuse, 35–40 x 8–10 μ.

Sporidia biseriate, oblong, 4-nucleate, at length often uniseptate, greenish-hyaline (finally darker), straight, 10–14 x 3–4 μ.

On dying leaves of *Linnæa borealis*, New Hampshire (Farlow), New York State (Peck),

A. Alchemíllae, (Grev.)

Asteroma Alchemillae, Grev. Fl. Edin. p. 369.
Dothidea ceramioides, Duby. Bot. Gall. II, p. 715.
Chætomium Alchemillae, Wallr. Fl. Crypt. Germ. II, p. 873.
Dothidea Alchemillae, Rab. Deutschl Krypt. Flora I, p. 165.
Stigmatea Alchemillae, Fr. Summa. Veg. Sc. p. 423.
Venturia Alchemillae, B. & Br. Not. Brit. Fungi 1493.
Exsicc. Fckl. F. Rh. 425.—Rab. F. E. 986, 2056.—Thum. M. U. 1835.

Epiphyllous, black, seated on pallid spots. Perithecia subprominent, globose-conical, seriate on fibrils radiating from a central point, subconnate or somewhat scattered, subsuperficial, sparingly clothed with acute, small bristles 20 x 2 μ. Asci subclavate, 35–40 x 8 μ, 8-spored. Sporidia biseriate, oblong, uniseptate, slightly constricted, 4-nucleate, 8–10 x 3½–4½ μ.

On leaves of *Alchemilla vulgaris*, Godthab & Kobbefiord. Greenland.

C. *On Monocotyledonous plants and on Cryptogams.*

V. sabalícola, E. & E. Proc. Acad. Nat. Sci. Philada., July, 1890.

Perithecia scattered or subgregarious, globose, 125–135 μ diam., pierced above, beset with stout, straight, black bristles 50–80 x 6–8 μ. Asci oblong-clavate, 50–60 x 7–8 μ, without paraphyses. Sporidia crowded-biseriate, fusoid-oblong, hyaline, 4-nucleate, (becoming uniseptate)? 10–13 x 2½–3 μ.

On dead leaves of *Sabal Palmetto*, Bayou Chene, La. (Langlois).

V. parasítica, E. & E. Proc. Acad. Nat. Sci. Philada. l. c.

Perithecia densely gregarious, globose, 90–100 μ diam., collapsing above, sparingly clothed with spreading, straight, rigid, continuous spines or bristles about 35 μ long and 5 μ thick at the base. Asci clavate-cylindrical, about 25 x 5 μ, without any paraphyses. Sporidia oblique or subbiseriate, fusoid, slightly curved, about 4-nucleate, hyaline, 6–8 x 2 μ.

Parasitic on old *Hypoxylon* (*perforatum*)? on bark of *Magnolia*, near St. Martinsville, La. (Langlois).

LASIOSPHÆRIA, Ces. & De Not.

Schema Sfer. p. 55.

Perithecia superficial, villose, tomentose or strigose. Asci elongated, with fugacious paraphyses. Sporidia hyaline or subhyaline, cylindrical or vermiform, mostly one-or more-septate.

A. *Sporidia 2-or-more-septate.*

L. hirsùta, (Fr.)

Sphæria hirsuta, Fr. S. M. II, p. 449.
Lasiosphæria hirsuta, Ces. & De Not. l. c.
Exsicc. Fckl. F. Rh. 950.—Ell. N. A. F. 893.—Roum. F. Gall. 793.

Perithecia mostly gregarious, on a thin subiculum of brown, creeping, broadly effused hyphæ, globose or subovate, about ½ mm. or a little more in diam., tubercular-roughened, dull black, clothed with scattered, brownish-black, spreading hairs. Texture coriaceo-membranaceous. Asci cylindrical, mostly narrowed above, p. sp. 150 x 12–15 μ, 8-spored. Sporidia biseriate, vermiform-cylindrical, narrower and curved below, hyaline, nucleate, becoming yellowish and more or less distinctly 5–7-septate, 55–60 x 5–6 μ.

On rotten wood, Canada, Carolina, Louisiana, New Jersey and west to Iowa and Montana.

The Louisiana specc. have the head of the sporida ovate-elliptical, subolivaceous, but not brown, 8–10 μ thick. Sporidia about 55 μ long.

L. Rhacòdium, (Pers.)

Sphæria Racodium, Pers. Syn. p. 74.
Lasiosphæria Rhacodium, Ces. & De Not. Schema. p. 55.
Exsicc. Fckl. F. Rh. 951.—Rab. Herb. Myc. 649.—id. F. Eur. 829.—Sydow. M. March 381

Perithecia gregarious, ½ mm. diam., coriaceo-membranaceous, seated on a felt-like subiculum of black, branching, interwoven hyphæ, clothed with spreading, black hairs. Asci stipitate, fusoid-clavate, about 180 μ long by 10–15 μ thick. Sporidia biseriate, cylindrical, curved below, hyaline or yellowish, faintly septate, 50–60 x 4–6 μ.

On rotten wood, Carolina to Michigan and Canada.

The orthography "*Rhacodium*" is doubtless etymologically correct, though Persoon wrote *Racodium*. This is scarcely more than a form of *L. hirsuta*, Fr., with the subiculum more highly developed.

L. emérgens, (Schw.)

Sphæria emergens, Schw. Syn N. Am. No. 1534.
Lasiosphæria emergens, Schw. Grev. XV. p. 82.

Scattered, emerging from a subiculum resembling a loose tomentum, which consists at first of threads erect or creeping loosely on the wood, finally more compact and interwoven. Perithecia ovate-conical, rugulose, black or brown, confluent with the ostiolum, and clothed with numerous long, divergent, black, loose (not rigid) hairs. At a more advanced stage the perithecia are often found broken, leaving little pits in the subiculum. Sporidia (sec. Cooke, Grev. XV. p. 82) 30–32 μ long.

On rotten wood, Bethlehem, Pa. (Schweinitz).

Allied to *L. Rhacodium.* Differing principally (sec. Schw.) in the loose surface of the subiculum, the shape of the perithecia, and their loose, hairy covering.

L. hispida. (Tode.)

Sphæria hispida, Tode. Fungi Meckl. II, p 17.
Lasiosphæria hispida, Fckl. Symb. p. 147.
Exsicc. Fckl. F. Rh. 949, 2039.—Thum. M. U 1745 —Sydow M. March. 1482.

Perithecia mostly closely gregarious, ovate-globose, ½ mm. diam., with a large, subconical, often compressed or deeply quadrisulcate ostiolum, clothed with spreading, brownish-black, scattering, bristle-like hairs, which are longer below, and pass gradually into a loosely interwoven subiculum of branched, septate, creeping hyphæ forming a more or less distinct layer on the matrix. Asci cylindrical, with a short stipe, 150–160 x 14–16 μ. Sporidia vermiform-cylindrical, curved below, becoming brown, and 6–7-septate, 55–62 x 5–7 μ.

On wood of *Juglans nigra*, North Carolina (Curtis).

Winter makes the sporidia 70–80 x 7–8 μ. We cannot find any in specimens issued by Thümen & Sydow over 62 or 63 μ long, (mostly between 55 and 60 μ).

This seems to differ from *L. hirsuta* principally in its bristly perithecia. *L. hirsuta*, *L. Rhacodium*, and *L. hispida* are very closely allied. Fries (in S. M. II, p. 450), expresses some doubt as to whether they are specifically distinct, or at least, says that it is difficult to separate them.

L. orthótricha, (B. & C.)

Sphæria orthotricha, B & C. Grev. IV, p 108.
Lasiosphæria orthotricha, Sacc. Syll. 3551

Perithecia gregarious but scarcely crowded, clothed with long, straight, acute hairs. Sporidia linear, with 6 septa, 50–62 μ long, dark brown, sometimes slightly curved.

On decayed *Nyssa*, South Carolina.

The characters given do not separate this accurately from *L. hispida*, (Tode).

L. Cæsariàta, (C. & P.)

Sphæria Cæsariata, C. & P. 29th Rep N. Y. State Mus. p. 60.
Lasiosphæria cæsariata, Sacc. Syll 3541.

Perithecia gregarious, about 300 μ diam. subglobose, papillate, black, shining, beset with scattered, erect, rigid, septate, black hairs. Asci cylindrical or clavate. Sporidia biseriate, narrowly fusiform, 5–7-septate, greenish, 37–42 μ long, each cell nucleate.

On decaying wood, Portville, N. Y. (Peck).

L. mutábilis, (Pers.)

Sphæria mutabilis, Pers. Syn p. 72.
Lasiosphæria mutabilis, Fckl. Symb Nachtr. I, p. 14.

Perithecia scattered or crowded, tolerably large, spherical, black, clothed except the large, black, projecting, papilliform ostiolum, with a fine yellow-green tomentum which finally turns brown and disappears. Asci clavate. Sporidia biseriate, curved, fusoid, 4-celled, hyaline, 20 μ long.

On rotten wood, Pennsylvania (Michener).

L. viridícoma, (C. & P.)

Sphæria viridicoma, C. & P. 29th. Rep.N. Y. State Mus. p. 64.
Lasiosphæria viridicoma, Sacc. Syll. 3546.

Perithecia erumpent, then superficial, two or three together, ovate, black, 75–100 μ diam., clothed with a dense, greenish tomentum. Ostiola thick, prominent, naked. Asci clavate or cylindrical. Sporidia biseriate, lanceolate, uniseptate and nucleate, becoming 3-septate and deeply constricted at each septum, hyaline, 35–45 x 7½–10 μ, the two middle cells nearly globose.

On decaying beech wood, N. Y. (Peck).

L. sublanòsa. (Cke.)

Sphæria sublanosa, Cke. Grev. VII, p. 41.
Metasphæria sublanosa, (Cke.) Sacc. Syll 3439.

Perithecia crowded, superficial or erumpent-superficial, ovate, black, ½–¾ mm. diam., seated on the bark or on the wood under the bark, which is then thrown off, clothed at first with coarse, sparingly septate, crisped, gray hairs, 80–150 x 5–7 μ, at length partially bald, surface coarsely tubercular-roughened, especially above, and the apex generally deeply and coarsely 3–4-sulcate-cleft. Asci lanceolate, short-stipitate, 80–100 x 10–12 μ, attenuated above, and at first with a knob-like enlargement at the tip. Paraphyses indistinct. Sporidia biseriate, cylindrical, nucleate, slightly rounded at the ends, yellowish-hyaline, 22–28 x 4–5 μ (35 x 5 μ Cke.)

On decaying *Andromeda*, Newfield, N. J.

Allied to *L. canescens*, (Pers.) The sporidia become finally 1–3-septate.

B. *Sporidia uniseptate.*

L. Coùlteri, (Pk.)

Sphæria Coulteri, Pk. in Hayden's U. S. Geolog. Survey, 1872, p 792.
Lasiosphæria acicola, Cke Grev. VIII, p. 87.
Enchnosphæria Coulteri, Sacc. Syll 3600.
Amphisphæria ? acicola, Sacc. Syll. 2753
Neopeckia Coulteri, Sacc Bull. Torr. Bot. Club. X, p. 127
Exsicc. Ell. N. A. F 1342

Perithecia subglobose, ¼–⅓ mm. diam., thin and fragile, brown-black, partly immersed in a compact, brown, copious subiculum of interwoven threads; which envelop the leaves and bind them together; ostiolum papilliform, black. Asci cylindrical, p. sp. 150 x 15 μ, with a short stipe and abundant paraphyses. Sporidia uniseriate, oblong-elliptical, slightly narrowed at the ends, uniseptate and constricted, brown, 20–30 x 8–10 μ.

On dead leaves of pine. Common in the Rocky Mountain region.

Fine specc. were found by Mr. Suksdorf on *Pinus albicaulis*, on Mt. Paddo, Washington, at an altitude of 6000 or 7000 ft.

Herpotrichia nigra Hartig. in Allesch. and Schnabl. F. Bavar. No. 70 has the same general appearance as this but differs in its shorter asci and 3-septate, subbiseriate sporidia In the specc. of *H. Coulteri* we have found only one septum in the sporidia, which are decidedly brown, while in the Bavarian fungus, they are (at least in the specimen quoted) hyaline.

L. luteóbasis, (Ell.)

Sphæria luteobasis, Ell. Bull. Torr Bot Club VI, p 134.
Byssosphæria luteobasis, Cke Syn. 2572.
Eutypa luteobasis, Sacc. Syll 639.
Exsicc. Ell. N. A. F 90.—Ell. & Evrht. N A F. 2d Ser. 1959.

Perithecia superficial or subsuperficial, ovate, of medium size, either seriately arranged or collected in compact groups of 6–10, at first enveloped, except the black, broad, even or faintly radiate-sulcate ostiola, in a densely matted, light yellow tomentum composed of smooth, branching, sterile hairs. Asci long and narrow (80–100 x 5 μ), the spore-bearing part only about 35 μ long. Sporidia biseriate, cylindrical, slightly curved, olive-brown, 8–10 x 1½–2 μ, 3–4-nucleate and at length uniseptate. The yellow mycelium which penetrates deeply into the rotten wood is very distinct and noticeable.

On decaying, decorticated limbs of *Quercus coccinea* or perhaps *Q. rubra*, lying on the ground, Newfield, N. J.

The spece. on pine wood, in N. A. F. 90, are *Kalmusia parallela*. (Fr.), which differs in its buried perithecia and absence of the yellow mycelium.

L. Vermiculària, (Nees.)

Sphæria Vermicularia, Nees. Syst. p 311, fig. 347.
Trichosphæria Vermicularia, Fckl Symb. p. 145.
Eriosphæria Vermicularia, Sacc. Syll. 2328.
Lasiosphæria Vermicularia, Cke. Syn 2713.

Perithecia crowded, ovate, subdepressed, astomous, black, very small, covered all over with erect, black bristles. Asci oblong, stipitate, 8-spored, 56 x 7 μ. Sporidia biseriate, ovate-oblong, uniseptate, hyaline, 8 x 3 μ.

On rotten pine wood, Carolina.

L. xestothèle, (B. & C.)

Sphæria xestothele, B. & C. Grev. IV, p 107.
Lasiosphæria xestothele, Sacc. Syll. 3548

Perithecia crowded, connected with brown, woolly matter. Ostiolum naked, perforated, tinged with rufous. Sporidia biseriate, fusiform, uniseptate, hyaline, 20 μ long.

On branches of *Cornus florida*, South Carolina (Ravenel), New York (Peck).

L. canéscens, (Pers.)

Sphæria canescens, Pers Syn. p. 72, Fr. S M. II. p 448.
Lasiosphæria canescens, Karst. Mycol. Fennica II, p. 162

Perithecia gregarious, sometimes connected by a stromatic crust, ovoid, about ½ mm. diam., densely clothed with rigid, spreading, coarse, light brown, continuous hairs 100–150 x 10–12 μ thick near the base. Asci clavate-cylindrical, stipitate 80–100 x 10–12 μ (p. sp.). Sporidia biseriate, oblong-cylindrical, nearly straight, ends obtuse, with two large nuclei at first, finally uniseptate, hyaline, becoming brownish. 20–30 x 4–5 μ.

On rotten wood, Carolina to Canada and California.

Distinguished from its allied species by its coarse, brown hairs and shorter sporidia.

C. *Sporidia continuous.*

L. spermoìdes, (Hoffm.)

Sphæria spermoides, Hoffm. Veg. Crypt II, p. 12.
Hypoxylon miliaceum, Bull. Champ tab 444. fig. 3.
Sphæria globularis, Batsch Elench. I, fig. 180.
Lasiosphæria spermoides, Ces & De Not Schema p. 55.
Leptospora spermoides, Fckl. Symb. p. 143.

Exsicc. Rab. Herb. Mycol. 47, 651 —id. F. Eur. 2430.—Rehm Asc. 587.—Thum. M. U. 1546 —Lin. Fungi Hung. 274.

Perithecia usually densely crowded, forming a uniform crustaceous layer, spherical or a little narrowed below, black, lusterless, rough, carbonaceous and brittle, with a small, papilliform ostiolum or sometimes slightly umbilicate, or without any distinct ostiolum, 500–700 mm. diam., often overrun with brown, creeping hyphæ. Asci fusoid-clavate, slightly attenuated above and narrowed below into a long stipe-like base, 8-spored, 130–160 x 9–10½ μ. Sporidia imperfectly biseriate, cylindrical, bent below, continuous or indistinctly uniseptate, hyaline, 19–22 x 4 μ. Paraphyses filiform, branching.

On rotten wood, London, Canada (Dearness), New York State (Clinton).

L. strigòsa, (A. & S.)

Sphæria strigosa, A. & S. Consp. p. 37.
Leptospora strigosa, Fckl Symb. Mycol. p. 144.
Lasiosphæria strigosa, Sacc. Syll. 3574.

Perithecia crowded, rather large, subglobose or subovate, of a dirty black color, substance tough, clothed with long, stiff, divergent, hoary, bristle-like hairs. Ostiolum papilliform-conical or obsolete. Sporidia elongated, somewhat curved, pale brown, continuous, 40 x 5 μ.

On damp pine wood, Carolina.

Fries (S. M. II, p. 448) says this species is not easily distinguished from *L. canescens*, but gives these distinctive characters: Perithecia a little larger, tough, covered all over with long, rigid, divergent bristles, much crowded, fibres of the subiculum not becoming crustaceous-concrescent.

L. intricàta, Pk. Bull. N. Y. State Mus. Nat. Hist. No. 2, p. 23.

Perithecia scattered or crowded, somewhat elongated, 650–900 μ long, 500–600 μ broad, generally narrowed towards the base, obtuse, subfragile, tomentose-hairy, brown or blackish-brown; subiculum very thin or none. Asci slender, elongated, 150–200 x 10–12 μ. Sporidia crowded, linear, curved or flexuous, greenish-yellow, 40–65 x 4½–5 μ.

On decaying wood and leaves, in damp places, Sandlake, N. Y. (Peck).

The perithecia, though small, resemble in shape those of *Bombardia fasciculata*. The minutely papillate ostiolum is often concealed by the tomentum of the perithecia; this is composed of intricate, matted, slender, septate, brown filaments, which, by their soft, tomentose character, readily distinguish this species from *L. strigosa*, *L. hispida*, *L. hirsuta*, &c.

L. ovina, (Pers.)

Sphæria ovina, Pers. Syn. p. 71.
Sphæria mucida, a. b. Tode Fungi Meckl. II, p. 16.
Sphæria lichenoides, Sow. Eng Fungi tab. 373, fig. 12
Lasiosphæria ovina, Ces. & De Not Schema.
Leptospora ovina, Fckl Symb p 143
Exsicc. Rab. Herb. Mycol. 730.—id. F. E 1444.—Thum. M. U. 968 —Sacc. Myc. Ven. 1155. Rehm Asc. 788.—Sydow M. March. 852, 2144 —Ell. N. A. F. 892.

Perithecia scattered or gregarious, superficial, spherical, clothed with a thin, white, closely adnate, felt-like coat, except the black, papilliform ostiolum, 400–500 μ diam. Asci, long clavate-fusoid, with a long stipe, 8-spored, 140–200 x 12–17 μ. Sporidia irregularly arranged, vermiform-cylindrical, mostly curved, continuous, 35–53 x 3–4 μ. Paraphyses slender, conglutinated.

Common on rotten wood.

L. stùpea, E. & E. Bull. Wash. Coll. I, p. 4. **(Plate 19)**

Perithecia superficial, gregarious, ovate, about 1 mm. high, densely clothed with dull brown or tow-colored, continuous hairs 200 μ long and 5–6 μ thick, often imperfectly toothed towards their extremities. Ostiola obtuse, strongly 4-ribbed. Asci 170–200 x 18–20 μ (p. sp. 120–150 μ long), broadest in the middle, attenuated above and contracted into a stipe-like base. Paraphyses abundant, stout, granular, evanescent. Sporidia overlapping and crowded subbiseriate, cylindrical, curved, obtuse, granular with a large nucleus in the center, pale brown, 34–38 x 8–11 μ, ends rounded and obtuse.

On a dead limb of *Tsuga Pattoniana*, Mt. Paddo, Washington (Suksdorf).

D. *Sporidia swollen above, brown, continuous.*

L. Newfieldiàna, E. & E. (in Herb.)

Perithecia gregarious, superficial, at first depressed-globose, becoming ovate-conical, ¾–1 mm. high and ½–¾ mm. broad, clothed (more densely below) with soft, spreading, brown, septate hairs; similar hairs, more or less branched and loosely interwoven, overspread the matrix between the perithecia forming a loosely-felted, pilose subiculum, which finally disappears more or less, as also do the hairs that clothe the perithecia. Asci subcylindrical, 150–200 x 8 μ, with abundant paraphyses. Sporidia uniseriate, not crowded, at first hyaline and vermiform-cylindrical, nucleate, 35 x 4 μ, with a short, horn-like appendage at each end, the upper end finally enlarging to a narrow-elliptical, brown head 12–14 x 5–7 μ, with a short (7 μ), curved appendage at the apex and a longer, curved, cylindrical appendage 40–60 x 3 μ at the base, almost exactly like the sporidia of *Bombardia fasciculata.*

On rotten wood, Newfield, N. J.

The general aspect is much like that of *L. hirsuta.*

Doubtful species.

L. bifórmis, (Pers.)

Sphæria biformis, Pers Syn. p. 59.
Lasiosphæria biformis, Sacc. Syll. 3585.

Perithecia scattered or crowded, globose or ovate, brown when young, becoming black, clothed with tolerably long, rigid, scattering hairs, with a more or less elongated, thick ostiolum. Asci and sporidia unknown.

Credited to America by Fries, and reported from California by Harkness.

TRICHOSPHÆRIA, Fckl.

Symbolae Mycologicae, p 144.

Perithecia small, superficial, membranaceous or sometimes of firmer, subcarbonaceous texture, spherical or ovate, hairy or bristly. Asci oblong or cylindrical, 8-spored, paraphysate. Sporidia oblong, ovate or cylindrical, continuous, hyaline.

T. pilòsa, (Pers.) **(Plate 19)**

Sphæria pilosa, Pers Syn. p. 73.
Trichosphæria pilosa, Fckl Symb. p. 145.
Exsicc. Fckl. F. Rh. 946.

Perithecia small (about 200 μ diam.), generally crowded in broadly effused patches, but sometimes scattered, superficial, of firm, almost woody texture, black, globose-ovate, hairy, with a short, thick, prominent ostiolum. Asci cylindrical or subventricose, contracted into a stipe-like base, 50–60 x 4–5 μ, with numerous filiform paraphyses. Sporidia uniseriate, or subbiseriate, elliptical, hyaline, one-celled, 5–8 x 3–4 μ.

Under the shingles of a roof, and on pine wood, South Carolina (Ravenel).

Tr. pulchrisèta, (Pk.)

Sphæria pulchriseta, Pk. 31st Rep. p. 50.
Acanthostigma pulchrisetum, Sacc. Syll. 3605.

Perithecia superficial, globose, membranaceous, black, 75–110 μ diam., gregarious, clothed with short (35–40 μ), rigid, straight, black bristles, at length collapsing. Asci sublanceolate, sessile 30–35 x 5–6 μ. Paraphyses? Sporidia biseriate, clavate-oblong or subfusoid, hyaline, 2–3-nucleate, 7–8 x 2–2½ μ.

On chips in woods, New York (Peck), on bark of a hickory barrel hoop in a cellar, Newfield, N. J.

The New Jersey specc. agree entirely with specc. from Peck. We can not see any paraphyses or any septum in the sporidia.

Tr. acanthostròma, (Mont.)

Sphæria acanthostroma, Mont. Syll. Crypt. No. 792 et Guy. No. 558.
Sphæria culcitella, B. & Rav. in Rav. Car. IV, 53 and *Sphæria culcitella*, B & C. in N. Pac. Expl. Expd. p. 128, No. 158.
Sphæria aculeata, B. & Br. (Sec. Cke. in Grev. XV, p. 122)
Trichosphæria acanthostroma, Sacc. Syll. 1754.
Scortechinia acanthostroma, in Additamenta to Sacc. Syll. p. 68.
Exsicc. Rav. Fungi Car. IV, 53.— Rav. F Am. 671, 749.—Ell. & Evrht. N. A. F. 2d Ser. 2356.

Perithecia very small, globose, gregariously crowded, smooth, black, without any prominent ostiolum, depressed and perforated in the center, seated on and surrounded by a subiculum composed of interwoven, septate hyphæ with short, spine-like branches. Sporidia ovate-oblong, hyaline, 2–3-nucleate, 5–6 x 2–2½ μ.

On bark, Carolina to Louisiana.

The subiculum often remains sterile.

Tr. solàris, (C. & E.)

Sphæria solaris, C. & E. Grev. V, p. 53, tab. 81.
Byssosphæria solaris, C. & E. Grev. XV, p. 122.
Lasiosphæria solaris, Sacc. Syll. 3578.

Perithecia gregarious, globose, pierced above, black, emerging from a brown, pulverulent, conidiophorous mycelium; ostiolum radiate-sulcate, not prominent. Asci cylindrical. Sporidia uniseriate, 30 x 6 μ, curved, subclavate, hyaline, continuous. Conidia in clusters on the apices of short, septate, simple threads, which constitute the subiculum, 12–14 x 8 μ, almond-shaped, opake, brown (*Acrotheca solaris* Sacc.)

On decaying maple wood (not pine), Newfield, N. J.

This appears to be rare, as it has only been met with once, and then only very sparingly.

Tr. flàvida, E. & E. (in Herb.)

Perithecia thickly gregarious, superficial, membranaceous, brownish-black, 125 μ diam., globose, clothed rather sparingly with straight, black, continuous spines about 35 x 3 μ, astomous, collapsing when dry. Asci fasciculate, sublanceolate, about 35 x 3 μ, sessile, aparaphysate. Sporidia biseriate, cylindrical, straight, nucleate, hyaline, 5–6 x 1½ μ.

On rotten wood which is stained yellowish, London, Canada (Dearness).

Tr. Arundinàriæ, E. & E. (in Herb.)

Perithecia scattered or gregarious, superficial, rather thick-membranaceous, black, ovate or ovate-conical, mostly prolonged above into a short neck, clothed all over with brownish-black, septate, straight, even, spreading or rather erect hairs (150–200 x 4–6 μ). Asci cylindrical, stipitate, 80–110 x 7–8 μ. Paraphyses obscure. Sporidia uniseriate with the ends overlapping, hyaline, continuous, broad-fusiform or elongated, almond-shaped, with acute ends, 20–22 x 5–7 μ.

On dead culms of *Arundinaria*, Louisiana (Langlois 1412).

Tr. corynèphora, (Cke.)

Byssosphæria corynephora, Cke. Grev. XI, p. 109
Trichosphæria corynephora, Sacc. Syll. 6022.

Perithecia subglobose, depressed, glabrous above, interwoven-tomentose at base. Asci cylindric-clavate. Sporidia lanceolate, continuous, hyaline, 20 x 4 μ. The perithecia are seated on a very black, effused mycelium composed of septate, creeping, branched hyphæ, the branches assurgent, and bearing clavate conidia 150 x 18 μ, with a long, attenuated base and truncate above, multi- (9–11-) septate.

On branches of *Ostrya Virginica*, Carolina.

Tr. subcorticàlis, (Pk.)

Sphæria subcorticalis, Pk. 28th. Rep. p 77.
Trichosphæria subcorticalis, Sacc. Syll. 1753.

Perithecia rather large, thin, sometimes collapsed, black, involved in a dense, blackish-brown tomentum, which is sometimes confluent, forming a subiculum. Sporidia oblong, colorless, 7–8 μ long.

Dead bark of *Carpinus Americana*, North Greenbush, N. Y. (Peck).

The perithecia are seated on the inner bark and are entirely concealed by the epidermis; when this is torn away, the perithecia usually come off with it.

Tr. tràmes, (B. & C.)

Sphæria trames, B. &. C. Grev. IV, p 142.
Lasiosphæria trames, Cke Syn. 2681.
Botryosphæria trames, Sacc Syll. 1790.

Perithecia arranged in lines three inches or more long, globose, minute, obscurely tomentose. Asci clavate. Sporidia hyaline, cymbiform, resembling those of *Botryosphæria Quercuum*.

On *Acer*, South Carolina (Ravenel).

Tr. fissuràrum, (B. & C.)

Sphæria fissurarum, B. & C. Grev. IV, p. 144.
Trichosphæria fissurarum, Sacc. Syll, 1747.

Perithecia globose, covered with short villosity; ostiolum rather thick, attenuated upwards. Asci lanceolate, short. Sporidia minute, oblong, hyaline. No measurements of asci or sporidia given.

On pine rails, South Carolina (Ravenel).

ACANTHOSTÍGMA, De Not.

Sferiacei Italici, p. 85.

Perithecia superficial, membranaceous, globose or ovate, mostly small, clothed with stiff hairs or bristles. Sporidia two- or more-celled, hyaline.

A. scópula, (C. & P.)

Lasiosphæria scopula, C. & P. Grev. XV, p. 82.
Acanthostigma scopula, C. & P. Bull. N. Y. State Mus. No. 2, p. 22.
Exsicc. Ell. N. A. F. 184.

Perithecia small (150–200 μ), subglobose, very black, with short, divergent, rigid, black hairs or bristles, which are 75–125 μ long by 7–8 μ thick. Asci lanceolate or subclavate, 8-spored. Sporidia crowded or biseriate, elongated, gradually narrowed toward each end, straight or slightly curved, multinucleate, at length multi- (10–12-) septate, yellowish- or greenish-hyaline, 60–70 x 3–3½ μ.

On pine and hemlock wood, New York and New Jersey.

A. Berenìce, (B. & C.)

Sphæria Berenice, B. & C. Grev. IV, p. 108.
Venturia saccardioides, E. & M. Am. Nat. Jan. 1884, p. 69.
Acanthostigma Berenice, Sacc. Syll. 3612.

"Perithecia minute, ovate, clothed all over with radiant flocci. Asci lanceolate. Sporidia hyaline, oblong, slightly attenuated at each end or subcymbiform."

On the under side of leaves of *Magnolia macrophylla* (Carolina)?

Cooke, in his synopsis, Grev. XVI. p. 18, places this in the *Acanthostigma* Sect. of *Venturia*, having sporidia 1–5-septate. *Venturia saccardioides*, E. & M., which is evidently the same as this, has perithecia hypophyllous, scattered, brownish-black, collapsing above, 150–200 μ diam., clothed with loose, spreading hairs, which are longer below and form a fringe around the base Asci clavate-cylindrical, about 50 x 8 μ, briefly stipitate, paraphysate. Sporidia fusoid, 3-septate, hyaline, 12–15 x 2½–3 μ. On leaves of *Magnolia glauca*, Florida. Fully matured specc, on leaves of *Magnolia grandiflora* from Louisiana, have the asci 70–80 x 10–11 μ, and sporidia 15–20 x 3 μ, 3–6-septate, and the hairs on the perithecia shorter.

A. decástylum, (Cke.) **(Plate 19)**

Sphæria decastyla, Cke. Grev. VII, p. 52.
Lasiosphæria subvelutina, E. & E. Journ Mycol. III, p. 117.
Sphæria cariosa, C. & E. Grev. VI, p. 94, tab. 100, fig. 28.
Sphæria atriella, C. & E. Grev. VI, p. 94.
Acanthostigma decastylum, Sacc. Syll. 3614.
Exsicc. Ell N. A. F. 783.

Perithecia superficial, black, conic-hemispherical, 150–200 μ diam., sparingly clothed with spreading, straight, remotely septate, rather obtuse, black, deciduous hairs subdiaphanous above, 100–150 x 4–5 μ. Asci clavate-cylindrical, about 150 x 12 μ, without paraphyses. Sporidia fusoid, hyaline, biseriate, slightly curved, ends rather obtuse, granular, becoming 3–5-septate, 22–30 x 4–4½ μ. Closely allied to *A. atrobarbum*, (C. & E.), but hairs of perithecia longer and of equal diameter throughout, and sporidia 3–5-septate. The surface of the wood itself, in both of these species. is thinly clothed with hairs similar to those growing on the perithecia.

On oak wood, Darien, Ga. (Ravenel), on rotten *Magnolia* wood and bark, Newfield, N. J., and Bethlehem, Pa. (Rau).

We have never been able to find the sporidia of *Sphæria atriella* constricted in the middle, as described and figured in Grevillea, and regard that species as only a more robust form of *S. cariosa*. The mature perithecia become bald. It is not improbable that *Sphæria longispora*, Ell. Bull. Torr. Bot. Club, VI, p. 135, is the same as this, but the specc. are too poor and scanty to determine with certainty.

A. Clintònii, (Pk.)

Sphæria Clintonii, Pk. 30th Rep. p. 65, pl II. figs 19–23.
Acanthostigma Clintonii, Sacc. Syll. 3610.
Venturia Clintonii, Cke. Syn. 2889.

"Perithecia very small, 120–160 μ diam., subglobose, gregarious, black, clothed with erect, black, bristly hairs. Sporidia fusiform, multinucleate, then 5–7-septate, colorless, 40–45 μ long.

On decaying wood, Alden, N. Y. (Clinton).

Related to *A. scopula* (C. & P.), from which it differs in its smaller perithecia and broader sporidia with fewer septa."

A. atrobárbum. (C. & E.)

Sphæria atrobarba, C. & E. Grev. VIII, p. 15.
Chætosphæria atrobarba, Sacc. Syll. 3215, Cke. Syn 2660.
Exsicc. Ell. N A. F. 590.

Perithecia gregarious, scattered, superficial, small (100–150 μ high), ovate-globose, clothed with scattering, long, black, bristle-like, septate hairs with subhyaline tips. Asci clavate-cylindrical, 70–75 x

6–7 μ, subsessile, with obscure paraphyses. Sporidia biseriate, oblong-fusoid, hyaline, 3–4-nucleate, becoming 3-septate, 12–15 x 3 μ.

This species is very common around Newfield on fallen, decorticated oak limbs. The surface of the wood is rather thinly clothed with hairs of the same character as those on the perithecia, forming black, pubescent patches several centimeters in extent. We have always found the sporidia hyaline.

A. occidentàle, (E. & E.)

Chætomella(?) perforata, E. & E Journ. Mycol. I, p. 153.
Venturia occidentalis, E & E. 1 c II, p. 43, Cke. Syn. 2897.
Exsicc. Ell. & Evrht. N. A. F. 2d Ser. 2141.

Perithecia superficial, $\frac{1}{4}$–$\frac{1}{3}$ mm. diam., membranaceous and of coarse, cellular structure, subhemispherical, with• a small, circular opening above, sparingly clothed with straight, black, continuous, bristle-like hairs about equal in length to the diameter of the perithecia, more thickly set around the orifice, paler and more or less substellate-tufted below. Asci subcylindrical, 75–100 x 7–10 μ, with imperfect paraphyses. Sporidia oblong-fusiform, 3-septate when mature, and constricted at the middle septum, or often at all the septa, nearly hyaline, 20–25 x 4–5 μ. Most of the perithecia contain only stylospores. Var. *minor*, on *Artemisia Ludoviciana*, has asci smaller (50–60 x 6–8 μ). and paraphyses more robust; sporidia about 15 x 3 μ, oval, and the middle cells slightly colored.

On *Cirsium discolor*, Iowa (Arthur), on *C. altissimum* and *Artemisia Ludoviciana*, Ames, Iowa (Bessey), and Illinois (Burrill).

A. Sequoìæ, (Plowr.)

Venturia Sequoiæ, Plowr. Grev. VII, p. 74.
Acanthostigma Sequoiæ, Sacc. Syll. 3604.

Perithecia unequal, scattered, setulose above. Asci cylindrical, 70–100 μ long, 8-spored. Sporidia somewhat unequal in length, faintly triseptate, hyaline.

On decaying foliage of *Sequoia*, California (Harkness).

Specimens from Dr. Harkness have perithecia depressed-hemispherical, 200–220 μ diam., clothed above with short spines 30–35 μ long. Ostiolum conic-papilliform. Asci cylindrical, sessile, 55–65 x 5–6 μ. Sporidia biseriate, fusoid, yellowish, 1–3-septate, nearly straight, 14–16 x 2½ μ.

A. spinòsum, (Hark.)

Rosellinia spinosa, Hark. Bull. Cal. Acad. Sci. Feb., 1884, p. 42.

Perithecia superficial, gregarious, globose, 300 μ diam., bearing numerous, yellowish-white, acuminate spines often 1–3 times branched, 200–300 x 18–22 μ, very thick walled and enclosing a central canal filled with oil-globules. Asci 8-spored, cylindrical, short-stipitate, 50 x 12 μ; paraphyses filiform. Sporidia uniseriate, ovate-elliptical, 2-celled, 9 x 4 μ, upper cell $\frac{2}{3}$–$\frac{3}{4}$ the length of the sporidium, olive-brown, with a large nucleus, lower cell conical, hyaline.

On decaying bark of *Eucalyptus*, San Francisco, Cal. (Harkness).

This species is anomalous in this genus on account of the "olive-brown" upper cell of the sporidia, but if these are really two-celled, it cannot belong in *Rosellinia*, which includes only species with one-celled sporidia. The branching spines are also unusual. We have seen no specimens, and take the diagnosis from the publication cited.

HERPOTRÍCHIA, Fckl.

Symbolæ Mycol. p. 146.

Perithecia superficial, spherical or subspherical, mostly of firm texture, coriaceous to subcarbonaceous, hairy or bald. Sporidia fusiform or oblong, 2- or more-celled, hyaline or brown.

Winter in Die Pilze makes fusiform sporidia (*i. e.* thicker in the middle and narrowed towards each end) the distinctive character between *Herpotrichia* and *Lasiosphæria*. Some of the species with uniseptate sporidia here included in *Herpotrichia*, have the sporidia oblong and obtuse.

A. *Sporidia uniseptate, brown.*

H. Rhenàna, Fckl. Symb. p. 146, tab. III, fig. 7. **(Plate 19)**

Perithecia scattered or gregarious, nearly free, rather large, globose, rugulose, densely clothed with long, brown, decumbent, creeping hairs, flattened above, with minute, truncate ostiola. Asci oblong, 150 x 16 μ, 8-spored, paraphysate. Sporidia ovate-oblong, sometimes curved, rounded at the ends, constricted in the middle, hyaline, at first 2-celled, later 4-celled, brownish, 21–28 x 8 μ.

On decaying culms of grasses. Not yet reported as found here, but given as a typical species to illustrate the genus.

H. parietàlis, (B. & C.)

Sphæria parietalis, B. & C. Grev. IV, p. 107.
Byssosphæria parietalis, Cke. Grev. XV, p. 123.
Enchnosphæria parietalis, Sacc. Syll. 3601.

Perithecia cup-shaped, seated on a scanty byssus (subiculum). Asci clavate. Sporidia biseriate, oblong, slightly attenuated at each end so as to be almost biconical, uniseptate, constricted at the septum.

On the inside of a hollow oak, South Carolina (Ravenel).

H. diffusa, (Schw.)

Sphæria diffusa, Schw. Syn. N. Am. 1502.
Sphæria rhodomphala, Berk. Hook. Lond. Journ. Bot. IV, p. 312.
Amphisphæria subiculosa, E. & E. Journ Mycol. II, p. 103, and Sacc. Syll. 7470.
Byssosphæria diffusa, Cke. Syn. 2625.
Exsicc. Ell. & Evrht. N. A. F. 2d Ser. 2130.

Perithecia superficial, gregarious, depressed-globose, $\frac{1}{3}$–$\frac{1}{2}$ mm. diam., brownish-black, rough, with the ostiolum subradiate sulcate, not prominent, soon pierced with a round opening, and sometimes slightly collapsed above. The lower part of the perithecia is clothed with a coat of brown, branching hairs, which also cover thinly the surface of the matrix around and between the perithecia, some of which touch each other, but are not confluent. Asci clavate-cylindrical, 80–90 x 10–12 μ (p. sp. 60–65 μ long), with abundant, filiform paraphyses. Sporidia biseriate, oblong or clavate-oblong, slightly curved, uniseptate and constricted at the septum, 15–20 x 5–6 μ, pale brown, ends obtuse, each cell with a large nucleus.

In a hollow, standing trunk of *Juglans cinerea*, effused for a whole yard in extent, Bethlehem, Pa. (Schw.), on decorticated poplar, Missouri (Demetrio), Louisiana (Langlois), Ohio (Morgan).

The Louisiana specc. have the upper part of the perithecia quite red, and in the other specimens the red is more or less discernible in and around the ostiolum. The subiculum finally disappears more or less completely. The length of the asci, as given in the Journ. of Mycol., is too great, and the sporidia too narrow.

H. barbicincta, (E. & E.)

Byssosphæria barbicincta, E. & E. Journ. Mycol. IV, p. 63.
Exsicc. Ell. & Evrht. N. A. F. 2d Ser. 1958.

Perithecia carbonaceo-membranaceous, gregarious or sometimes scattered, ovate-globose, 150–200 μ high, bare or with a few bristle-like hairs. Ostiolum conic-papilliform, black and shining. Asci subcylindrical, 65–70 x 5–6 μ, with a short, stipe-like base and imperfectly developed paraphyses. Sporidia biseriate, fusoid, yellowish-hyaline, 2–4-nucleate at first, finally uniseptate, not curved or constricted, 10–12 x 2–2$\frac{1}{2}$ μ, ends subobtuse.

Parasitic on old *Diatrype tremellophora* and on the bark adjacent, Newfield, N. J.

The perithecia, especially when growing on the old *Diatrype*, are surrounded by a dense growth of erect, black, septate, bristle-like hairs 250 μ, or more, long. The perithecia sometimes collapse.

H. lanuginòsa, (B. & C.)

Sphæria lanuginosa, B. & C. Grev. IV, p. 108.
Byssosphæria lanuginosa, Cke. Grev. XV, p. 123.
Melanopsamma lanuginosa, Sacc. Syll. 2254.

Perithecia globose, slightly flattened above, lanuginous at the base. Sporidia oblong, uniseptate, curved, sometimes pointed at the ends, so as to be biconical.

On *Robinia*, South Carolina.

B. *Sporidia 3- or more-septate, hyaline.*

H. rhodóspila, (B. & C.)

Sphæria rhodospila, B. & C. Grev. IV, p. 141.
Herpotrichia rhodospila, Sacc. Syll. 3622.

"Perithecia convex, seated on a black crust, brick-colored at the apex. Sporidia between cymbaeform and fusiform, hyaline, 3-septate, 20 μ long."

On *Cyrilla*, South Carolina.

Specimens on old oak stumps, Louisiana (Langlois 1692), and on rotten maple, Newfield, N. J., have perithecia convex, rough, ¼ mm. diam., with a conic-papilliform ostiolum. Asci clavate-cylindrical, about 65 x 8 μ, with paraphyses. Sporidia biseriate, fusoid, slightly curved, hyaline, 3-septate, 20–22 x 4–5 μ. The "brick-color" in these spece. is very faint.

H. leucóstoma, Pk. Bull. N. Y. State Mus. No. 2, p. 23.

Perithecia small, 300–450 μ broad, numerous, somewhat crowded, subglobose, seated on or involved in a blackish-brown tomentum, the ostiola naked, not prominent, whitish when moist, grayish or sordid when dry. Asci cylindrical or subclavate, 150–200 x 10–15 μ. Sporidia crowded or biseriate, oblong-fusoid, at first uniseptate, constricted at the septum, with 2–3 nuclei in each cell, then 3–5-septate, hyaline, 30–40 μ long, 7½–8 μ broad.

On dead branches of *Acer spicatum*, Catskill Mountains, N. Y.

The whitish ostiola are a marked feature. It is distinguished from *H. Scheidermeyeriana*, Fckl., by its much smaller perithecia and the more numerous septa of the sporidia; nor were any globose appendages observed at the ends of the sporidia.

H. pezìzula, (B. & C.)

Sphæria pezizula, B. & C. Grev. IV, p 106.
Lasiosphæria pezizula. Sacc. Syll. 3554
Lasiosphæria pezizula, Cke. Syn. 2737.
Exsicc. Ell. N. A. F. 696.—Rav. F. Am. 196.

Perithecia gregarious, membranaceous, depressed-spherical, soon collapsing to cup-shaped, 350–400 μ diam., smooth above, seated on a thin, black or olivaceous-black subiculum of interwoven, septate hyphæ. Asci oblong-clavate, sessile, 80–112 x 20–25 μ. Sporidia biseriate, clavate-fusoid, hyaline, 5–7-septate, nearly straight, 30–40 x 5–7 μ (èxceptionally 8–10 μ thick).

On decaying wood and bark, Carolina, New Jersey, Ohio and Missouri.

Cooke, in his Synopsis, places this among the brown-spored species, but we have always found the sporidia *hyaline*.

CHÆTOSPHÆ̀RIA, Tulasne.

Sel. Carp. II, p. 252.

Perithecia spherical or pear-shaped, carbonaceous, more or less hairy, seated on and surrounded by a more or less dense, felt-like subiculum of brown, septate hyphæ. Sporidia mostly fusoid or fusoid-oblong, 2- or more-septate, brown or with the terminal cells hyaline.

Ch. phæostròma, (Mont.)

Sphæria phæostroma, Mont. Flor. Alg. I, p 491, tab. XVI, fig. 2.
Sphæria tristis, Tode Fung. Meckl. II, p 9.
Sphæria tristis b. fusca, A. & S. Consp. p. 44.
Chætosphæria phæostroma, Fckl. Symb. p. 166.
Chætosphæria phæostroma, Sacc. Syll. 3200.
Exsicc. Rav. Fung. Car. V, 63.—Fckl. F. Rh. 2040.—Rab. Fungi Eur. 51.

Perithecia crowded, often in patches of considerable extent, spherical or pear-shaped, black, rough, with a broad, wrinkled ostiolum, bare above, but sending out around the base abundant, stiff, straight, dark brown hairs, which also form a tangled, bristly subiculum. Asci slightly clavate-thickened above and a little narrowed below, sessile. 8-spored, 95–108 μ long, 16–21 μ thick. Sporidia biseriate, short-cylindrical, rounded at the ends, 3-septate, the two middle cells brown and longer, and the two terminal cells hyaline, shorter and more or less curved, 28–38 x 8–9 μ; paraphyses filiform.

On rotten elm wood, Carolina and Missouri.

Var. *phæostromoides*, (Pk.), (*Sphæria phæostromoides*, Pk. 28th Rep. p. 77, t. II, fig. 30–35) is said to differ in its shorter (25 μ long) sporidia and uniseptate conidia. Specimens from Missouri (Demetrio No. 18) also have the uniseptate, brown, 12–15 x 8–10 μ conidia, and agree with specimens of *Ch. fusca*, Fckl., in Saccardo's Mycotheca Veneta. but all these can hardly be anything more than forms of *Ch. phæostroma*.

Ch. pannícola, (B. & C.) **(Plate 19)**

Sphæria pannicola, B. & C. Grev. IV, p 107.
Chætosphæria holophæa, B & C. Grev. XV, p. 82.
Chætosphæria pannicola, Sacc. Syll 3210.

"Perithecia globose, seated on a rather thick, brown, pilose stratum consisting of straight, acute threads. Sporidia oblong, obtuse at each end, arcuate, 3-septate, 20 μ long.

On roots of birch, New Jersey."

We have not seen the original specimens described by Berkeley, but specimens sent by Mr. Stevenson from Pennsylvania, and by Mr. Commons from Delaware, and which seem undoubtedly to belong here, have the perithecia densely gregarious, more or less pilose, with pale brown, spreading hairs, seated on a dense, brown, pilose subiculum. Asci clavate-cylindrical, p. sp. 70–75 x 12–15 μ, with abundant paraphyses. Sporidia oblong-cylindrical, hyaline at first, becoming brown and 3-septate, curved, 18–20 x 6–7 μ, ends obtuse.

Ch. innùmera, (B. & Br.)

Sphæria innumera, B & Br. Berk. Outl p. 395.
Chætosphæria innumera, Tul. Sel. Carp. II, p. 253, tab 33. figs. 7–9.

Perithecia small (100–200 μ), thickly scattered among the hairs of the subiculum, and some of them, at least when young, sparingly clothed with similar hairs, finally black and subshining, globose, scarcely papillate, the ostiolum being so small as to be easily overlooked. Asci clavate-cylindrical, 8-spored, briefly stipitate, paraphysate, 80 x 6½ μ. Sporidia subbiseriate, oblong-fusoid, rather obtuse at each end, nearly straight, 2–3-nucleolate, finally 2-septate, pale, 13 x 4 μ.

On decaying wood of *Quercus*, California (Harkness).

The conidia (sec. Tul.) are borne in heads or spikes, at the tips of the hairs of the subiculum, and are of three kinds; ovate, pale brown, 6½ x 3 μ, others linear-cylindrical, 12 μ long, and still others subfusoid, 25–30 x 3 μ, 6–8-septate, olive-brown.

Ch. leonìna, (C. & P.)

Sphæria leonina, C. & P 29th Rep. N. Y. State Mus. p 60
Chætosphæria leonina, Sacc. Syll. 3205.

Perithecia subconfluent or rarely scattered, dark brown, oval, covered with a short, thick, tawny tomentum, the papillate apex naked. Asci clavate or cylindrical. Sporidia biseriate, lanceolate, uniseptate and constricted, at length triseptate, brown, 35–38 μ long. Paraphyses slender, filiform.

On the cut surface of wood, Portville, N. Y. (Peck).

Ch. longípila, Pk. 42d Rep. p. 35.

Perithecia gregarious or scattered, ovate, minute (150 x 115 μ), clothed, especially below, with long, slender, pale hairs. Asci clavate-cylindrical, p. sp. about 40 x 8–10 μ, contracted below into a short, stipe-like base, and overtopped by filiform, undulate paraphyses. Sporidia biseriate, oblong-elliptical, 11–13 x 5 μ, hyaline and uniseptate at first, soon 3-septate, with the two central cells opake and the end cells hyaline, the central septum being now scarcely discernible, and the two dark, middle cells appearing like a broad, dark band across the middle of the sporidium.

On the basswood bottom of a barrel in a cellar, Long Island, N. Y. (Zabriskie).

Ch. parvicápsa, (Cke.)

Sphæria parvicapsa. Cke. Grev. VII, p. 52.
Chætosphæria parvicapsa, Sacc. Syll. 3207.

Perithecia very small, seated on a dark brown, byssoid subiculum. Asci clavate-cylindrical. Sporidia elliptical, 3-septate, brown, 12–14 x 6 μ.

On dead wood, Aiken, So. Carolina (Ravenel).

Our specc. from Cooke are sterile.

Ch. flàvo-cómpta, (B. & C.)

Sphæria flavo-compta, B. & C. Grev. IV, p. 108.
Chætosphæria flavo-compta, Sacc. Syll. 3209.

Perithecia ovate, black, clothed with rigid, yellow hairs. Sporidia oblong, elliptical, triseptate, 20–25 μ long.

On *Cyrilla*, So. Carolina.

Ch. ornàta, Hark. New Cal. Fungi p. 45. Bull. Cal. Acad. Sci. No. 1. Feb. 1884.

Perithecia superficial, globose, black, not rugose, 360 μ diam., setae hyphoid, ½ mm. or more long, black, not acuminate, occasionally septate. Asci 8-spored, mucoid, clavate, with a long pedicel, 100 x 8–10 μ; paraphyses inconspicuous. Sporidia fusiform, triseptate, middle cells olive brown, ultimate ones hyaline, 16 x 6 μ.

On decaying bark of *Eucalyptus*, San Francisco, Cal.

We have not seen the three species last mentioned and have only copied the published diagnoses.

The following Schweinitzian species, apparently referable to the *Trichosphærieæ*, are doubtful: *Sphæria setosa*, Schw. Syn. N. Am. 1533; *Sphæria cladosporiosa*, Schw. Syn, N. Am. 1530; *Sphæria inconstans*, Schw. Syn. N. Am. 1564; *Sphæria aggregata*, Schw. Syn. N. Am. 1561. These are not in the Herbarium at Philadelphia, but specc. in Herb. Berk., examined by Cooke, were without fruit.

FAMILY. MELANÓMMEÆ.

Stroma none. Perithecia superficial, generally of firm texture, woody, corky or carbonaceous, seldom coriaceous or exceptionally submembranaceous, mostly glabrous, but sometimes bristly or surrounded by a conidiophorous or sterile subiculum.

ROSELLÍNIA, Ces. & De Not.

Schema Sferiacei, p. 53.

Perithecia typically superficial, but also with their lower part more or less sunk in the matrix, coriaceous or oftener carbonaceous and brittle, spherical or ovate, black, bare or bristly; ostiolum distinct. Asci cylindrical, 8-spored. Sporidia elliptical, oblong or fusiform, one-celled, brown or black. Paraphyses filiform.

A. *Perithecia large, seated on a subiculum.*

R. áquila. (Fr.)

Sphæria aquila, Fr. S. M. II, p. 442.
Sphæria byssiseda, b. Tode. Meckl. Fungi II, p. 10.
Rosellinia aquila, de Not. Sferiacei Ital. p. 21.
Exsicc. Fckl. F. Rh. 964, 1061.—Rab. Herb. Mycol. 648.—id. F. E. 1616.—Rehm Asc. 538 Ell. N. A. F. 181.—Sydow M. March. 1248, 1550.—Speg. Hong. Sud. Amer. No. 59. Plowr. Sph. Brit. 61.

Perithecia gregarious or crowded, sometimes confluent, globose, with a distinct, black, conic-papilliform ostiolum, dark brown, with a thin, brown, tomentose coat at first, but finally bare. Seated on a loosely felted, thick, purplish-brown, cottony subiculum by which they are at first nearly enveloped, but which, finally, to a greater or less extent, disappears. The perithecia are 1–1½ mm. diam., with an outer, rather thick but brittle carbonaceous wall, and an inner coriaceous one. Asci cylindrical with a tolerably long stipe, p. sp. 100–130 x 8–10 μ, with obscure paraphyses. Sporidia uniseriate, oblong-cymbiform, continuous, dark brown, 16–27 x 6–9 μ, with or without a short, mostly obtuse, hyaline apiculus (2–2½ μ long) at each end.

Common on decaying fallen limbs.

Of the specimens quoted, those in Plowright's Sphæriacei have a short, obtuse, hyaline appendage (2–2½ μ long), at each end of the sporidia. Those issued by Rehm, Sydow, Spegazzini, and Ellis, are not appreciably appendiculate.

The *Sphæria Corticium*, Schw. can hardly be more than a var. of *R. aquila*, the perithecia being scattered, and each seated on a separate, orbicular patch of densely interwoven subiculum, resembling an orbicular brown *Corticium*.

R. purpùreo-fúsca, (Schw.)

Sphæria purpureo-fusca, Schw Syn. N. Am. 1499.
Byssosphæria purpureo-fusca, Schw. in Cooke's Synopsis, 2602.

Subiculum tomentose, rhacodium-like, purplish-brown, extensively effused, and at first covering the perithecia, which however are soon cespitose-erumpent or longitudinally-seriate, very large, (much larger than those of *R. aquila*), globose and clothed, except around the ostiolum, with a delicate brownish-purple tomentum. Ostiola black, conical, bare and sometimes sublateral.

On oak limbs, Bethlehem, Pa.

R. megalocárpa, (Plowr.)

Sphæria megalocarpa, Plowr. Grev. VII, p. 73.
Rosellinia megalocarpa, Sacc. Syll. 952.

Perithecia superficial, very large, 1½–3 mm. diam., spherical, dull black, slightly rough, becoming rugulose with age. Ostiola very minute, scarcely prominent, surrounded by a paler areola. Asci cylindrical or clavate, 70–130 x 10–15 μ, with numerous, flexuous paraphyses. Sporidia dark brown, oval, colorless when young, then binucleate and brown, 12–15 x 3–5 μ.

On bark of dead maple, California (Harkness).

In the specimen sent us by Dr. H., the perithecia are crowded, 1½–2½ mm. diam., clothed below with brown, strigose hairs, which also form a subiculum, mostly collapsed. Ostiola papilliform, paler. Asci (p. sp.) cylindrical, 75–80 x 4½–5 μ. Sporidia uniseriate, lying end to end and not overlapping, oblong, brown, 2-nucleate, 10–12 x 3½–4 μ.

R. impósita, (Schw.)

Sphæria imposita, Schw. Syn. N. Am. No. 1503.
Byssosphæria imposita, Cke. Synopsis, 2610.

Perithecia large, scarcely immersed, of a brown color, rugulose, globose, scattered or seriate and subaggregated, or even subconfluent, seated on a scanty, longitudinally effused, brown subiculum. Ostiolum subconic-papillate.

Sporidia (sec. Cke. Grev. XV, p. 81) lanceolate, continuous, brown, 25 x 6 μ.

R. rhodómela, (Schw.)

Sphæria rhodomela, Schw. Syn. N. Am. 1511 (not Sacc. Syll 3263).
Byssosphæria rhodomela, Cke. Grev. XV, p. 80.

"Perithecia globose, black, seated on a thin, rose-colored mycelium. Asci linear. Sporidia uniseriate, elliptical, continuous, brown, 10 x 6 μ.

On rotten wood, United States, Herb. Berk. 9604." (Cke. in Grev. l. c.) Found in Carolina and Pennsylvania, sec. Schw., and referred by him to *Sphæria rhodomela*, Fr. S. M. II, p. 445, but this (sec. Saccardo) is a different thing, having 3-septate sporidia 15–18 μ long.

R. Desmazièrii, (B. & Br.)

Sphæria Desmazierii, B. & Br. Ann. Nat. Hist. No. 618, tab. 9, fig 1.—Curr. Linn. Trans. XXII, tab. 57, fig. 2.—Cooke Hndbk. No. 2551.
Byssosphæria Desmazierii, Cooke's Synopsis 2589.
Rosellinia Desmazierii, Sacc Syll. 922 and Sacc F. Ital. tab. 393.

Spreading widely over the ground, fallen leaves, &c., covering them with a mouse-colored, tomentose subiculum, which consists of somewhat branched, anastomosing threads, with tips often subdivided, forming little racemes surmounted by oblong conidia (*Graphium Desmazierii*, Sacc. F. Ital. tab. 394). Perithecia large, half immersed in the subiculum, which in age acquires a somewhat darker hue, subscabrous, dull pitchy black or plumbaginous, globose with a large, central, papilliform ostiolum. Asci elongated-clavate, inner membrane furnished with an oblong process at the tip. Sporidia large, cymbiform, elongated, subacuminate, at first hyaline, with two or three variously sized globules, at length dark brown, with 6–7 globose nuclei. Sporidia 30 μ long (38–45 x 8 μ, Sacc.)

We have seen no specimens and copy the above from Cooke's Handbook. Sec. Berk. in Grev. IV, 106, found on birch in Maine, and on *Cornus florida* in Pennsylvania.

R. subiculàta, (Schw.)

Sphæria subiculata, Schw. Syn. N. Am. No. 1504.
Hypoxylon subiculosum, Berk. Grev. IV, p. 52.
Rosellinia subiculata, Sacc. Syll. 925.
Exsicc. Rav. Fungi Car. V. 72.—Ell. N. A. F. 182.

Perithecia gregarious or crowded, about 1 mm. diam., globose, black and shining, with a small, papilliform ostiolum, seated on a sulphur-yellow, waxy-pruinose subiculum which finally disappears. Asci cylindrical, p. sp. 80–90 x 6–7 μ, with abundant paraphyses. Sporidia uniseriate or partly biseriate above, subinequilaterally elliptical, brown, subacute, 10–12 x 5–5½ μ.

On rotten wood. Common from Canada to Florida and Louisiana, but apparently not so common in the Rocky Mt. region and on the Pacific coast.

After the subiculum has disappeared this may still be distinguished by its thin, shining perithecia.

R. thelèna, (Fr.) ?

(*Sphæria thelena*, Fr. in Kunze Mycol. Hefte II, p. 36.)?
Rosellina thelena, Rab. F. Eur. No. 757, Winter Die Pilze, 3082.
Exsicc. Kunze Fungi Sel. 343.—Rab. F. Eur. 757, 1536.—Thumen M. U. 1949.

Perithecia scattered or gregarious or sometimes crowded, spherical, with a strongly papilliform ostiolum, smooth, thin and brittle, brownish-black, about 1 mm. diam., seated on a superficial, loosely adhering, densely interwoven but thin, brownish-purple subiculum. Asci cylindrical, stipitate, 8-spored, 90–130 x 8–9 μ (p. sp.), with conglutinated paraphyses. Sporidia uniseriate, oblong, subinequilateral, brown, 18–23 x 6–7 μ, with a short (6–8 μ), hyaline, spine-shaped appendage at each end.

This differs from *R. aquila* in its thinner and more fragile perithecia and the longer, spine-like appendages of the sporidia.

The typical form of this species has not yet been reported as found in this country, but var. *pinea*, Sacc. (Thüm. M. U. 1949), has been sent from British Columbia by Dr. Macoun. In mature specimens, the spine-like appendages of the sporidia mostly disappear, but even in this stage, this species may be readily distinguished from *R. aquila*, by its much thinner and more fragile perithecia.

B. *Perithecia not on a distinct subiculum.*

R. mammifórmis. (Pers.)

Sphæria mammæformis, Pers. Syn. p. 64.
Hypoxylon mammæforme, Berk. Grev. IV, p. 52.
Hypoxylon globulare, (Bull) Fckl. F. Rh. 1060 (sec Winter).
Rosellinia mammiformis, Sacc. Syll. 938.
Exsicc. Plowr. Sph. Brit. No. 70 —Fckl. F. Rh. 1060.

Perithecia gregarious or crowded, and often confluent (2 or 3 in one), globose, double walled, the outer wall thinner than in *R. aquila* and more fragile, glabrous, nearly black, 1–1½ μ diam., bare but not polished; ostiolum strongly and abruptly papilliform, black and subshining. Asci (p. sp.) 100–115 x 8–10 μ, with abundant paraphyses. Sporidia uniseriate, oblong-elliptical, inequilateral or slightly curved, 19–25 x 7–9 μ, mostly without any distinct apiculus.

On rotten wood of *Cyrilla*, South Carolina (sec. Berk. in Grev. IV, p. 52).

The foregoing diagnosis was drawn from the specimen in Plowright's Sphæriacei Britannici.

The absence of any decided subiculum will separate this from *R. aquila*, and the glabrous perithecia from *R. medullaris*. The absence or presence of an appendage on the ends of the sporidia, can not be relied on as a distinguishing character, since in all three of these species, the sporidia at one stage of growth, are more or less distinctly appendiculate. If *R. mastoidea*, Sacc., is really glabrous in all stages of its growth, it must be, as Winter puts it, a synonym of *R. mammiformis*.

R. medullàris, (Wallr.)

Sphæria medullaris, Wallr. Fl. Crypt. Germ. II, p. 792.
Rosellinia medullaris, Ces. & De Not. Schema. p. 177.
Rosellinia Macouniana, E. & E. Bull. Torr. Bot. Club XI, p. 74.

Perithecia gregarious or crowded, superficial, globose or ovate-globose, ¾–1 mm. diam., covered at first, except the conic-papilliform, black ostiolum, with a pruinose-tomentose or pruinose-pubescent coat of a dull reddish-purple or brick-color, but finally becoming bare and black. Asci (p. sp.) 100–120 x 7–8 μ. Sporidia uniseriate, oblong, only slightly curved, brown, 19–22 x 6–7 μ, sometimes obscurely appendiculate, but mostly without any appendages.

On decaying wood, Canada, New York and New Jersey.

The American specimens here referred to *R. medullaris*, become glabrous when mature, though the purplish brick-colored, pruinose coat is at first very distinct. As in *R. aquila* and *R. thelena*, the wall of the perithecium is double and in thickness intermediate between these two species. There is only a very slight subiculum or none at all, even in the early stage of growth.

R. mùtans, (C. & P.)

Sphæria mutans, C. & P. 29th Rep. N. Y. State Mus. p. 64.
Rosellinia mutans, Sacc. Syll. 944.

Perithecia rather large (½–¾ mm.), gregarious or crowded, globose, papillate, black, at first clothed with a thin, tawny, evanescent tomentum, at length naked, smooth and shining. Asci subcylindrical. Sporidia uniseriate, elliptical, brown, 10–13 μ long.

On rotten wood, Tyre, N. Y. (Peck), Louisiana (Langlois).

The Louisiana specimens have the sporidia 10–15 x 5–6 μ, but do not show the asci.

C. *Perithecia with the base sunk in the matrix, mostly setose at first.*

R. prinícola, (B. & C.)

Sphæria prinicola, B. & C Grev. IV, p. 142.
Rosellinia prinicola. Sacc. Syll. 959.

Perithecia erumpent, rather large, subglobose, very minutely granulated, very obtuse, pierced at the apex. Asci linear. Sporidia uniseriate, oblong, subcymbiform, brown, 12–13 μ long.

On branches of *Prinos verticillata*, Pennsylvania (Michener).

R. parasítica, E. & E. Proc. Acad. Nat. Sci. Phila. Pa. July, 1890.

Perithecia gregarious, seated on the wood in transverse cracks of the bark or often, on or among the collapsed perithecia of a sterile *Valsa* on the same limbs, ovate-globose, covered with short, black, spreading bristles at first, but these soon disappear, leaving the perithecia rough, $\frac{1}{4}$–$\frac{1}{3}$ mm. diam., smoother above, with a broad-papilliform, obtuse ostiolum. Asci cylindrical, 60–70 x 6 μ (p. sp.), with abundant paraphyses. Sporidia uniseriate, oblong-elliptical, subobtuse, dark brown, 7–10 x 4–5 μ. This is certainly very near *R. detonsa* (Cke.), which Sacc. in Sylloge considers a var. of *R. ligniaria* (Grev.); but it differs in its perithecia more flattened above, and in its constantly smaller sporidia.

On dead limbs of *Symphoricarpus occidentalis*, Helena, Montana (Kelsey).

R. Kellermánni, E. & E. Proc. Acad. Nat. Sci. Phil., July, 1890.

Perithecia gregarious, superficial, subglobose, about 200 μ diam., clothed, except the papilliform ostiolum, with short (15–22 μ), straight, spreading bristles, but finally, nearly bare. Asci cylindrical, 35–40 x 5 μ (p. sp.), 8-spored. Sporidia uniseriate, brown, elliptical or subglobose, 4–6 x 3–4 μ. Distinguished by its small sporidia.

On rotten wood of *Negundo aceroides*, Manhattan, Kansas.

R. albolanàta, E. & E. Proc. Acad. Nat. Sci. Phil., Pa., July, 1890.

Perithecia subseriate, erumpent, the lower part remaining sunk in the wood, about 1 mm. diam., clothed, except the black, papilliform ostiolum, with a thin, white, farinose coating which finally disappears, bicorticate, outer wall carbonaceous, inner submembranaceous. Asci cylindrical, about 100 x 10 μ.

On old rails, Emma, Mo. (Demetrio). Found also bursting through the bark on dead *Salix* limbs, at Mill Creek, near Sheridan, Montana, by Mr. and Mrs. H. M. Fitch (com. F. W. Anderson).

R. obliquàta (Somm.) var. Americana, E. & E.

Sphæria obliquata, Somm. Fl Lap. p. 213.
Sordaria obliquata, Ces. & De Not, Schema p. 52.
Rosellinia obliquata, Sacc Syll. 949.

Perithecia gregarious, ⅓ mm. diam., erumpent, depressed-globose, glabrous or with a few rudimentary, gland-like hairs below; ostiolum acutely papilliform or obsolete. Asci cylindrical, 120–130 x 6–7 μ. Sporidia uniseriate, inequilaterally elliptical, brown, continuous, 8–10 x 5–6 μ. Paraphyses filiform.

On scales of dead cones of *Pinus ponderosa*, Belt Mts., Montana (Anderson 613.)

The perithecia in the Montana specimens are not obliquely attached and the sporidia are rather smaller than stated by Dr. Winter in Die Pilze (10–12 x 7 μ). The perithecia also are mostly evenly rounded above, without any distinct ostiolum.

R. obtusíssima, (B. & C.)

Sphæria obtusissima, B. & C. Grev IV, p 142.
Rosellinia obtusissima, Sacc. Syll 966

Perithecia ovate, with a papilliform orifice, half buried amongst the fibers of the wood. Asci linear. Sporidia uniseriate, elliptical, brown, very obtuse, 12–14 μ long.

On bleached rotten wood, Pennsylvania (Michener).

D. *Perithecia small, gregarious, not bristly or hairy.*

R. pulveràcea, (Ehr.)

Sphæria pulveracea, Ehr. in Pers. Syn. p. 83.
Sordaria Friesii, Niessl. Vorarb z. Crypt. Flora Von Mahren, p 112.
Rosellinia pulveracea, Fckl. Symb p 149
Rosellinia Friesii, Niessl. Beitr z. Kentniss d. Pilze, p. 34.
Sphæria millegrana, Schw. Syn. N Am 1559.
Sphæria transversalis, Schw. Syn. N Am. 1560.
Exsicc. Fckl. F. Rh 936 —Kunze F Sel 260.—Rab. F. E. 338, 1246, 2766 —Rehm Asc. 192, 695.—Rav. Car IV, 58 —id F. Am. 672 —Sacc. M. Ven. 1158.—Rab. F. E. 2766. Mycoth March 1335 —Ell. N. A. F. 193.

Perithecia densely gregarious, often forming a continuous crustaceous layer, but sometimes scattered, ovate-globose, minutely tubercular-roughened, about ⅓ mm. diam. Ostiolum papilliform, soon perforated. Asci cylindrical, p. sp. mostly 60–70 x 10–12 μ, with a stipitate base 20–30 μ long, and filiform paraphyses. Sporidia uniseriate, elliptical, brown, 8–15 x 6–9 μ (mostly 10–12 x 7–8 μ).

Common on dead wood (and bark) of deciduous trees from Maine to Louisiana, and west to the Pacific coast.

Var. *millegràna* (*Sphæria millegrana*, Schw.), on dead trees, Washington, and on decorticated oak limbs, New Jersey, has asci p. sp. 80 x 10–12 μ, and sporidia 12–15 x 8–10 μ.

Var. *transversàlis* (*Sphæria transversalis*, Schw. l. c.), on rotten limbs of *Liriodendron*, South Carolina (growing transversely and surrounding the limb like a ring) differs from var. *millegrana* in its annular mode of growth.

The following is a list of measurements of asci and sporidia in some of the exsiccati above quoted.—Sacc. M. V. 1158, asci (p. sp.) 65–70 x 8–10 μ, spor. 10–12 x 7–8 μ. Kunze. 260, spor. 8–10 x 6–8 μ. Rabh-Winter F. E. 2766, asci (p. sp.) 60 x 8–10 μ, spor. 8–10 x 6–8 μ. Rehm Asc. 695, asci (p. sp.) 65–70 x 8–10 μ, spor. 8–12 x 7–8 μ. Sydow M. March. 1835, spor. 10–12 x 7–9 μ. Measurements of *R. millegrana*, Schw. Rav. Car. IV, 58, spor. 8–11 x 6–7 μ. Rav. F. Am., spor. 10–12 x 6–8 μ. Specc. in our Herb. from Delaware, have spor. 7–8 x 5–7 μ (mostly 6–8 x 5–6 μ).

Excepting the var. *millegrana*, the American specimens have asci and sporidia about as in European specimens. There is no appreciable difference in the perithecia from the various localities, except a little variation in size and roughness. Were there no intermediate forms, the Washington and Delaware specimens might consistently be separated, on account of the great disparity in the size of the sporidia, but the occurrence of almost every gradation in size between them, makes any specific distinction a matter of doubtful expediency, and we have therefore, placed *R. millegrana* as a Syn. of *R. pulveracea*, (Ehr.)

R. ulmatícolor, (B. & C.)

Sphæria ulmaticolor, B. & C. Grev. IV, p 152.
Rosellinia ulmaticolor, Sacc Syll. 980.

"Perithecia very minute, on effused, umber-colored spots. Asci linear. Sporidia uniseriate, brown, elliptical, 7½–8 μ long.

On smooth, decorticated limbs," South Carolina.

R. glandifórmis, E. & E. Proc. Acad. Nat. Sci. l. c.

Perithecia scattered, ovate-globose, black and glandular-roughened, 1–1¼ mm. diam., the lower part (about ¼) sunk in the wood, and generally with a slight reinforcement around the lower half of the projecting part, like the cup of an acorn, but this is sometimes wanting or reduced to a mere thin, granular coat. Ostiolum papilliform, mostly small, sometimes obsolete, the apex of the perithecium being then evenly rounded. Asci cylindrical, 100–114 x 8–10 μ, with abundant paraphyses. Sporidia uniseriate, acutely elliptical, opake, (subhyaline at first), 14–15 x 7–8 μ. Allied to *R. subiculata*, (Schw.)

On a live oak stump, Louisiana (Langlois), on rotten wood, Long Island, N. Y. (Zabriskie).

R. ovàlis, (Ell.) **(Plate 20)**

Sphæria ovalis, Ell. Bull. Torr. Bot. Club VIII, p. 125.
Rosellinia ovalis, Sacc. Syll. 989.
Exsicc. Ell. N. A. F. 896.

Perithecia gregarious or subcespitose, ovate, rough, 250–300 μ diam., with an obtusely papilliform ostiolum. Asci narrow, cylindrical, 75–85 (p. sp. 60–65) x 6 μ. Sporidia uniseriate, continuous, brown, variable in size and shape, from short-elliptical to oblong, 8–12 x 5–7 μ (mostly 8–9 x 5–6 μ).

On "sage brush", (*Artemisia*,) Utah (S. J. Harkness).

Closely allied to *R. pulveracea*, differing principally in its perithecia.

R. mycóphila, (Fr.)

Sphæria mycophila, Fr. S. M. II, p. 462.
Rosellinia mycophila, Sacc. Syll. 958

Perithecia scattered, erumpent-superficial, black, fragile, subshining, smooth, hemispherical or irregular in shape, ostiolum papilliform, finally deciduous, leaving the perithecia simply perforated. Asci cylindrical. Sporidia uniseriate, elliptical, brown. Sometimes several perithecia are confluent.

On decaying *Polypori*, Pennsylvania and Carolina.

R. deeràta, C. & E.

Sphæria deerata, C. & E. Grev V, p. 93.
Rosellinia deerata, Sacc Syll. 961.

Perithecia few, small, scattered, erumpent, globose, black. Asci cylindrical. Sporidia uniseriate, elliptical, brown, 12 x 7 μ. Some of the perithecia contain stylospores.

On dead wood of *Juniperus Virginiana*, Newfield, N. J.

All the specimens of this species in our Herb. are stylosporous (*Shæropsis*), and we suspect the specimens from which it was described were *R. abietina*, Fckl. with denuded perithecia.

R. umbrinélla, (B. & C.)

Sphæria umbrinella, B. & C. Grev. IV, p. 152.
Rosellinia umbrinella, Sacc. Syll. 981.

"Perithecia umber colored, with a black, papilliform ostiolum. Asci linear. Sporidia uniseriate, brown, elliptical, binucleate."

On *Eupatorium coronopifolium*, South Carolina.

E. *Perithecia small, mostly gregarious, bristly.*

R. ligniària, (Grev.)

Sphæria ligniaria, Grev. Crypt Fl Scot I, p 82.
Rosellinia ligniaria, Sacc Syll. 991.
Exsicc. Fckl. F. Rh. 1810.

Perithecia gregarious, superficial, subglobose, or ovate-conical, black, about ⅓ mm. diam., or rather less, clothed with short (20–30 μ), black bristles. Asci cylindrical, p. sp. about 75 x 9–10 μ. Sporidia obliquely uniseriate, elliptical, brown, 12–14 x 5–6 μ.

On wood of various deciduous trees, Canada, New Jersey, Louisiana, Carolina and Virginia.

The Louisiana specc., (on poplar), have asci p. sp. 75–80 x 10–12 μ, spor. 12–15 x 7–8 μ, bristles 20–30 μ long. On *Ostrya Virginica*, (Canada), spor. 10–12 x 7–8 μ. On wood of *Sambucus* (Canada) spor. 10–12 x 7–9 μ, bristles 50–75 μ long. *Pinus* and *Cupressus* are also given in the Sylloge as habitats of this species.

R. trichòta, (C. & E.)

Sphæria abietina, Fckl var. *trichota* C. & E Grev. VI, p. 13.
Sphæria hirtissima, Pk. 28th. Rep N. Y. State Mus. p. 78.
Exsicc Ell. N. A. F. 895.

Perithecia mostly densely gregarious, ovate, 380 μ high and 300 μ wide, densely clothed with straight, black, erect bristles, 75–90 μ long. Asci (p. sp.) 80–90 x 10–12 μ, with a narrow, stipitate base 30–40 μ long. Sporidia uniseriate, lying end to end, subglobose or short elliptical, opake, 12–14 x 10–12 μ.

On decorticated fallen limbs of *Pinus rigida*, and on wood of white cedar, Newfield, N. J. Mostly in the excavations made by larvæ burrowing under the bark.

On account of its longer bristles and subglobose sporidia we consider this worthy of specific rank.

R. abiètina, Fckl., Symb. p. 150.

Exsicc. Fckl. F. Rh. 1811.

Perithecia gregarious or crowded, a little larger than those of *R. pulveracea*, ovate, attenuated above into an obtusely conical ostiolum, black, covered with very short, black bristles. Asci cylindrical, 95–100 x 10–12 μ, stipitate. Sporidia obliquely uniseriate, oblong, subelliptical, continuous, brown, 14–16 x 6–8 μ.

On exposed dead wood of *Juniperus Virginiana*, Newfield, N. J.

The Newfield specc. have the asci (p. sp.) 75–80 x 8–10 μ. Sporidia 12–14 x 6–7 μ. Perithecia ovate, about ⅓ mm. diam.

Differs from *R. ligniaria* in its rather larger asci and sporidia.

R. ambígua, Sacc. Fungi Ven. Ser. II, p. 328, F. Ital. tab. 594.

Exsicc. Rehm's Ascomycetes, 790.

Perithecia densely gregarious, superficial, with the habit and general appearance of *R. pulveracea*, (Ehr.), globose-conoid, 120–140 μ diam., very black, clothed all over, but especially around the small, papilliform ostiolum, with acute, divergent bristles, finally bald. Asci cylindrical, attenuated below, 80 x 8 μ, 8-spored, and with filiform paraphyses. Sporidia directly or obliquely uniseriate, ellipsoid, or subglobose, 10–12 x 7–8 μ, slightly compressed laterally, greenish at first, becoming dark, continuous.

On branches of *Sambucus pubens*, New York (Peck).

Evidently closely allied to *R. ligniaria.*

R. heríciun, (Schw.)

Sphæria hericium, Schw. in Herb. Berk.
Coniochæta hericium, Cke Syn. 2785 and Grev. XV, p. 82.
Sphæria hericium, Wall. Cr. Fl. Germ. No 3871?

Perithecia scattered, 300 μ diam., subglobose, dark brown, clothed with short, subulate, scattered hairs. Asci cylindrical, 8-spored. Sporidia oval, continuous, brown, 12–14 x 8 μ.

On rotten wood, Bethlehem, Pa. (Schweinitz).

R. Clavàriæ, (Tul.) **(Plate 20)**

Sphæria Clavariæ, Tul. Ann. Sci. Nat. Ser. IV, tom. V, p 113.
Sphæria Clavariæ, Awd. Rab. F. Eur. No. 252.
Sordaria Clavariæ, Ces. & De Not. Schema, p. 52.
Pleospora Clavariarum, Tul Sel. Carp. II, p. 271.
Helminthosphæria Clavariæ, Fckl. Symb. p. 166.
Rosellinia Clavariæ, Winter, Die Pilze, 3096.
Exsicc. Rab. F. Eur. 252, 1023, 2666.—Fckl. F. Rh. 2443.—Ell. N. A. F. 786.

Perithecia gregarious, superficial, globose, without any evident papilliform ostiolum, dark brown, clothed all over with simple, spreading, sharp-pointed, rigid, brown hairs, and seated on a widely effused, felt-like, black, subiculum, the branches of which, as well as the hairs on the perithecia bear conidia at their tips. Asci cylindrical, attenuated below into a stipe-like base, 80–90 x 6–7 μ. Sporidia uniseriate or partly biseriate, elliptical, pointed at the ends, dark brown, continuous, 10–14 x 6–8 μ. Paraphyses filiform, longer than the asci.

On living *Clavaria* Newfield, N. J.

The conidial stage, *Helminthosporium Clavariarum*, Desm., has conidia elliptical or elongated-elliptical, 1–2-celled, brown, 15–20 x 8 μ, borne singly or 2–3 together on the ends of the fertile hyphæ.

R. rhyncòspora, Hark. Bull. Cal. Acad. Sci. Feb. 1884, p. 42.

Perithecia densely crowded, globose-conical, superficial, covering large areas, studded all over with short, black spines. Asci 8-spored, cylindrical, 96 x 10 μ, with filiform paraphyses. Sporidia oval or elliptical, dark brown, continuous, 15–18 x 9 μ, often prolonged above into a curved beak ⅓–½ as long as the body of the sporidium.

On decorticated branches of *Sambucus glauca*, California.

R. foveolàta, (B. & C.)

Sphæria foveolata, B. & C. Cuban Fungi, No. 847.
Rosellinia foveolata, Sacc. Syll. 996.

Perithecia globose, at length collapsing, clothed with rigid hairs, seated on an effused, black stroma composed of hyphæ with acutely pointed branches. Sporidia hemispherical, foveolate, 5 μ diam.

Sporidia resembling the seeds of a *Veronica*.

R. arctéspora, (C. & E.)

Sphæria arctespora, C. & E. Grev. V, p. 93.
Sphæria detonsa, Cke and *Sphæria Xylariæspora*, C & E. Grev. VI, p. 94.
Rosellinia arctispora, Sacc. Syll. 984.
Exsicc. Ell. N. A. F. 594.

Perithecia gregarious or crowded, ovate, about ⅓ mm. diam., clothed with a black, farinose tomentum, and with short spines about like those of *R. ligniaria*. Asci (p. sp.) about 75 x 6 μ, with a narrow, stipe-like base. Sporidia overlapping-uniseriate, oblong-fusoid, brown, subinequilateral, 12–14 x 3½–4 μ, resembling the sporidia of *Xylaria*.

On wood of *Andromeda*, under the bark, Newfield, N. J.

R. Ráttus, (Schw.)

Sphæria Rattus, Schw. Syn. N. Am. No. 1535.
Sphæria ranella, B. & Rav. Grev. IV, p. 107. (sec. Cke. Grev. XVI, p. 17).
Coniochæta Rattus, Cke. Syn. 2789.
Rebentischia ranella, Sacc. Syll. 2894.

Perithecia gregarious or scattered, rather large, very fragile and thin, glabrous, globose-conical, with a subobtuse, black, at length bare ostiolum, which, at first, as well as the perithecium, is covered with a black, densely hairy pellicle, the hairs cinereous-brown and divergent, resembling a rat-skin. When young, the perithecia are immersed, with only the hairs erumpent, but finally they emerge and become partially free. Seen at a distance, the color is very black, but lighter (cinereous) when viewed with a lens. Berkeley (l. c.) says of *Sphæria ranella.*—" Perithecia globose, collapsing, rugged, seated on a brown stratum; sporidia of two joints, the upper elliptical or pointed, dark, the lower elongated and attenuated, with several nuclei, 18–25 μ long."

On bare wood and branches, Bethlehem, Pa. (Schw.), on *Platanus*, South Carolina (Ravenel).

Cooke places this among species with brown, continuous, appendiculate sporidia, and for this reason it is here referred to *Rosellinia*, but its true generic location is still uncertain.

R. didérma (Schw.)

Sphæria diderma, Schw. Syn. N. Am 1593.
Amphisphærella diderma, Cke. Grev. XVI, p. 91

Scattered, simple, rather large, bicorticate. The outer wall has the shape of a hemispherical or conical, denuded, rugose, dark brown *Sphæria* with the papilliform ostiolum at first prominent, but finally deciduous. Within this outer wall is enclosed another dark brown *Sphæria*, frequently collapsed, immersed in the wood, and communicating with the outer ostiolum by an elongated neck. Sporidia (sec. Cke. l. c.), almond-shaped, continuous, 14 x 8 μ.

On dry rotten wood, Bethlehem, Pa. (Schw.)

R. apérta, (Schw.)

Sphæria aperta, Schw, Syn. N. Am. 1588.
Amphisphærella aperta, Cke. Grev, XVI, p. 91.

Perithecia black, connected by a black crust, crowded, but sometimes also scattered, semiimmersed, rather large, subcylindrically elongated, truncate at the apex, collapsing, crowned with a large, papilliform ostiolum, which is finally deciduous, leaving the perithecium broadly perforated above, concentrically striate outside, but scarcely rugose. Sporidia (sec. Cke. l. c.) continuous, brown, oval, 7 x 5 μ.

Rather rare, on rotten wood, Bethlehem, Pa. (Schw.)

R. cuticulàris, (Schw.)

Sphæria cuticularis, Schw. Syn. 1505.
Byssosphæria cuticularis, Cke. Syn. 2650.

Perithecia somewhat scattered, hemispherical, shining-black, papillate, $\frac{1}{2}$–$\frac{3}{4}$ mm. diam., seated on a shining-white, longitudinally effused, narrow, membranaceous subiculum, which is not pulverulent or manifestly floccose.

On bark, Bethlehem, Pa. (Schw.) Allied to *R. subiculata.*

We find no measurements of the sporidia of this species.

BOMBÁRDIA, Fr.

Summa Veg Scand. p 389.

Perithecia superficial, more or less vertically elongated, of firm, coriaceous or membranaceous consistence, black. Asci cylindrical, long-stipitate, 8-spored. Sporidia at first cylindrical, curved, hyaline, finally enlarged above into an ovate or elliptical, brown head, the lower part remaining cylindrical and subhyaline.

The true character of the sporidia was first pointed out by Winter, in Hedwigia (1874). Ambiguous between *Rosellinia* and *Lasiosphæria*, having the perithecia of the former, and the sporidia after the type of the latter.

B. fasciculàta, Fr. Summa Veg. Sc. p. 389. **(Plate 21)**

Sphæria bombarda, Batsch Elenchus Cont. I, p 271.
Bertia bombarda, Ces. & De Not Schema Sferiacei p. 51.
Exsicc Fckl F. Rh. 940.—Rab. F. E. 949.

Perithecia fasciculate or crowded in continuous patches of 1 cm. (more or less) in extent, short, oblong-cylindrical, slightly attenuated below, grayish-pulverulent, becoming bare and black, 1–1½ mm. high, of firm, coriaceous consistence, with a small papilliform ostiolum. Asci elongated-cylindrical, very long-stipitate, somewhat attenuated above, 120–150 x 9–10 μ (p. sp.) Sporidia at first cylindric-vermicular, about 35 μ long, with a short, subulate, curved appendage at each end, but when mature, consisting of an ovate or elliptical, brown head 13–15 x 8 μ (the spore proper), and a cylindrical, curved, uniseptate, subhyaline appendage, 24–26 x 4 μ, below.

On rotten wood, (especially on the smooth, cut surface), New York State (Underwood), Iowa (Holway).

MELANOPSÁMMA, Niessl.

Notiz, ueber Pyren. p. 40.

Perithecia superficial, or more or less sunk in the matrix, sometimes seated on a felt-like subiculum, subglobose, carbonaceous, glabrous or at first clothed with conidia-bearing hairs, black. Sporidia elliptical or fusiform, uniseptate, typically hyaline, but sometimes becoming pale brown.

M. pomifórmis, (Pers.)

Sphæria pomiformis, Pers Syn p. 65.
Melanomma pomiformis, Fckl. Symb. p 159
Eriosphæria raripila, Sacc. F Ven Ser. II. p. 326.
Melanopsamma pomiformis, Sacc. Syll. 2248.
Exsicc Rab E. E. 738.—Rehm. Asc. 36.—Plowr. Sph. Brit. 68.—Sacc. M. Ven. 1479. Roum. F. G 181.

Perithecia superficial, crowded, or sometimes scattered, ovate-globose, smooth, 400 μ diam., black, firm but thin-walled, finally slightly depressed around the papilliform ostiolum, at first sparingly clothed with conidia-bearing hairs. Asci clavate-fusoid, sessile, 8-spored, paraphysate, 60–75 (p. sp. 50–55) x 8–10 μ. Sporidia biseriate, elliptical, hyaline, uniseptate, 12–15 x 5–7 μ.

On rotten wood, Carolina and Pennsylvania (Schw.), on dry, decorticated wood of *Platanus*, Canada (Dearness).

M. subfasciculàta, (Schw.)

Sphæria subfasciculata, Schw Syn. N. Am. 1565.
Psilosphæria subfasciculata, Cke Syn 2991.

Perithecia at first somewhat covered, but at length entirely denuded. Allied to *Bombardia fasciculata*, but more scattered and longitudinally subseriate. Perithecia subfasciculate, black, rugose, subindurated, sometimes shaped just like those of *B. fasciculata*, but often less elongated, subrotund and obtusely papillate, of smaller size and subconfluent. Asci? Sporidia (sec. Cke. Grev. XV, p. 83) ellipsoid, uniseptate, hyaline. 20 x 8 μ.

On decaying shoots of grape vines, Bethlehem, Pa. (Schweinitz).

M. confertíssima, (Plowr.)

Sphæria confertissima, Plowr Grev. VII, p 73
Melanopsamma confertissima, Sacc. Syll 2253

Perithecia minute, densely crowded in patches, more scattered circumferentially, mutually compressed, subspherical, 80–100 x 10–15 μ Sporidia biseriate, uniseptate, subequal, 15–20 x 5 μ, hyaline.

On scales of cones of *Sequoia sempervirens*, California.

M. abscóndita, E. & E. (in Herb.)

Perithecia hemispherical, obtuse, black, about ⅓ mm. diam., scattered or crowded and subconnate connected at base more or less distinctly by a thin, black crust, ostiolum papilliform, obscure. Asci clavate-cylindrical, p. sp. about 75 x 12 μ, with stout paraphyses. Sporidia biseriate, oblong-fusoid, uniseptate, and strongly constricted at the septum, slightly curved, nucleate, sometimes with two additional septa, 20–22 x 5–6 μ.

Inside of old hickory nuts decaying on the ground.

This has the outward appearance of *Trematosphæria nuclearia*, but the sporidia are very different.

B. *Perithecia more or less erumpent-superficial.*

M. graópsis, (Ell.

Sphæria graopsis, Ell. Bull. Torr. Bot Club IX, p 73
Melanopsamma graopsis, Sacc. Syll. 6086.
Exsicc. Ell. N. A. F. 780.

Perithecia scattered, erumpent, submembranaceous, globose or subelongated, partially collapsing above, 250–370 μ diam., rough and more or less covered with the bleached fibres of the wood. Ostiolum papilliform, slightly prominent, surrounded by a slight depression. Asci oblong, mostly bulging on one side, sessile, 50–58 x 20–25 μ, without paraphyses?, 8-spored. Sporidia irregularly crowded, oblong-elliptical, or oblong-navicular, hyaline at first, becoming paie yellowish-brown, uniseptate, 20–25 x 8 μ.

On decorticated pine poles, Newfield, N. J.

M. Eckféldtii, (Ell.)

Sphæria Eckfeldtii, Ell. Bull. Torr. Bot. Club VIII, p. 91.
Melanomma Eckfeldtii, Sacc. Syll. 3247.
Exsicc. Ell. N A. F. 593.

Perithecia scattered, erumpent, minute (200–250 μ diam.), carbonaceo-coriaceous, glabrous, subglobose, with a minute, papilliform ostiolum. Asci oblong or obovate, sessile (paraphysate)? 70–80 x 25–30 μ. Sporidia inordinate, oblong or oblong-fusiform, yellowish-hyaline, uniseptate, sometimes a little curved, 25–30 x 6–9 μ.

On bleached wood of chestnut, near Philadelphia, Pa. (Eckfeldt).

A more careful examination of the original specimens shows that the sporidia do not become 3-septate and brown, as was at first stated. Three-septate, brown, free sporidia were seen, and it was too hastily concluded that these were the mature sporidia, when, in fact, their presence was only accidental.

M. papílla, (Schw.)

Sphæria papilla, Schw. Syn. Car. 159, Fr. S. M. II, p. 462.
Melanopsamma papilla, Sacc. Syll. 2269.

Perithecia scattered, black, opake, symmetrically rounded, hemispheric-conical, innate, large (1 mm. diam.). Ostiolum papilliform. Asci cylindrical, p. sp. 200–230 x 15–20 μ, with a short, abrupt stipe about 25 μ long, and abundant filiform paraphyses. Sporidia uniseriate, oblong-elliptical, hyaline, 30–50 x 15–18 μ, slightly constricted, 2-nucleate.

On rotten maple wood and bark, Carolina, on outer bark of living white oak, Newfield, N. J. Also reported from New York and Pennsylvania.

In Grev. XVI, p. 89, this is placed by Cooke in *Amphisphæria* as having brown sporidia, but what we have taken for this species (on white oak bark. Newfield), has hyaline sporidia. Berkeley, in Grev. IV. p. 144, gives the sporidia as "shortly fusiform, oblique, uniseptate, 33 μ long."

M. latericóllis, (Fr.)

Sphæria latericollis. Fr. S. M. II, p 464
Melanopsamma latericollis, Sacc. Syll. 2259

Perithecia scattered or subgregarious, rarely 2–3-confluent, of medium size, at first immersed, then nearly free, obliquely ovate or irregular, very black. Ostiolum generally lateral, straight or curved, cylindrical, about as long as the perithecium. Asci oblong, stipitate, 8-spored, 80 x 14 μ. Sporidia biseriate, oblong-fusoid, obtuse at the ends, straight or curved, uniseptate, sharply constricted at the septum, cells 1–2-nucleate, slightly brownish, 24 x 7–8 μ.

On rotten wood, Bethlehem, Pa.

M. recéssa, (C. & P.

Sphæria, recessa, C & P. 29th. Rep. N. Y. State Mus. p 61.
Melanopsamma recessa, Sacc. Syll. 2265.

Perithecia gregarious, at first semiimmersed, smooth, flattened, dark brown or black, 300 μ diam. Asci subclavate. Sporidia 1–2-seriate, elliptical, uniseptate, deeply constricted at the septum, hyaline, 13–20 x 7–10 μ. The perithecia have something of a discoid appearance.

On rotten wood, New York State.

M. cupréssina, E. & E. Journ. Mycol. II, p. 103.

Perithecia gregarious or scattered, emergent-superficial, subhysteriiform-elongated, $\frac{1}{3}$–$\frac{1}{2}$ mm. diam., rough, carbonaceo-membranaceous, with a broad and rather depressed opening above, (probably with a prominent ostiolum at first). Asci clavate-cylindrical, 55–70 x 10–12 μ, with filiform paraphyses, and 8 biseriate, yellowish-hyaline, clavate-oblong, uniseptate, about 12 x 4 μ sporidia.

On bleached wood of cypress and red cedar, Louisiana (Langlois.)

M. pæcilóstoma, (B. & Br.)

Sphæria pæcilostoma, B. & Br. Brit Fungi, No 876, tab. 10, fig. 23.
Zignoella pæcilostoma, Sacc. Syll. 3652.
Conisphæria pæcilostoma, Cke. Syn. 3334

Perithecia partly free, subglobose, opake. Ostiola variable, obsolete, conic-cylindrical, frequently somewhat compressed. Asci short, clavate. Sporidia cymbiform, acute at each end, 4-nucleate, becoming uniseptate, 25 μ long, hyaline.

On decaying *Vaccinium*, Newfield, N. J.

M. Texénsis, (Cke.)

Sphæria Texensis, Cke Texas Fungi, 143.
Melanopsamma Texensis, Sacc Syll. 2266

Scattered over grayish or bleached spots. Perithecia subglobose, immersed in the blackened wood, scarcely prominent. Asci cylindrical. Sporidia lanceolate, uniseptate, hyaline, 15 x 6 μ.

On old oak rails, Texas (Ravenel).

Specimens of this species in our Herb. from Ravenel (part of the original Texas collection) have the asci oblong-cylindrical, about 60 x 12 μ, nearly sessile and surrounded by abundant, filiform, branching paraphyses. Sporidia biseriate, clavate-oblong, uniseptate, hyaline, 10–12 x 3–3½ μ, ends subacute. Only the vertex of the perithecia projects above the surface of the wood.

M. segregàta, (B. & C.)

Sphæria segregata, B. & C. Grev. IV, p 141.
Zignoella segregata, Sacc Syll 3647.
Psilosphæria segregata, Cke Syn. 2980.

"Perithecia scattered over the bleached surface, free, ovate, acute at the apex. Sporidia hyaline, narrow, uniseptate, elongated-biconical."

On decayed wood, North Carolina.

BÉRTIA, De Not.

Giorn. Botan. Ital. I, p. 335

Perithecia superficial or nearly so, more or less irregular in shape, of a leathery-carbonaceous texture, surface tubercular or coarsely wrinkled, glabrous, black. Asci clavate, 4–8-spored. Sporidia elongated, subcylindrical, uniseptate, hyaline.

B. morifórmis, (Tode).

Sphæria moriformis, Tode Fungi Meckl II, p. 22, fig. 90.
Sphæria claviformis, Sow. Eng. Fungi tab. 337.
Sphæria rugosa, Grev. Flor. Edin. p. 364.
Bertia moriformis, De Not. l. c.

Exsicc. Fckl F. Rh. 999—Rab Herb Mycol. 637—Rehm Asc. 442—Sydow Mycoth. March, 1818

Perithecia crowded, superficial, ovate or ovate-globose, often compressed vertically or otherwise irregular, black, surface coarsely tubercular-roughened, ½–¾ mm. diam. Asci elongated-clavate, pedicellate, 130–165 x 14–16 μ. Paraphyses faintly septate. Sporidia 8 in an ascus, crowded irregularly, hyaline, cylindrical or fusoid-cylindrical, curved, with two large nuclei, and soon uniseptate, 25–30 x 6–7 μ. (30–40 x 5–6 μ, Winter).

On decaying wood of maple, Lyndonville, N. Y. (Fairman), Catskill Mountains (Peck), Pennsylvania (Schweinitz), and on old *Daedalea unicolor*, Canandaigua, N. Y. (Edgar Brown).

B. submorifórmis, (Plowr.)

Sphæria submoriformis, Plowr. Grev. V, p. 74.
Bertia submoriformis, Sacc. Syll. 2272.

Perithecia irregular, rugose, bursting through the cuticle, variable in size. Sporidia biseriate, hyaline, simple, curved, 20–24 x 3 μ, uniseptate. This has much the appearance of *B. moriformis*, but differs in habit and in the sporidia being little over half as large as in that species.

On dead bark, California.

The foregoing is from Grevillea, l. c. Specimens on a dead (beech)? limb, from West Chester, Pa., determined by Cooke as this species (Grev. V, p. 74), have the perithecia subovate, more or less rough, ½ mm. diam., or rather less, erumpent through the cuticle in loose clusters of 2–6 together, or sometimes singly. Asci 75–80 x 12–15 μ, with abundant filiform paraphyses. Sporidia biseriate, broad-fusoid, hyaline, uniseptate, 15–18 x 3–4 μ, not curved.

MELANÓMMA. Fckl.

Symbolæ Mycol. p. 159.

Perithecia gregarious or scattered, typically superficial, but sometimes erumpent-superficial, carbonaceous, mostly bare and black, spherical or ovate. Asci cylindrical or clavate, 8-spored. Sporidia oblong or fusoid, 2-or more-celled, brown.

M. púlvis-pýrius. (Pers.)

Sphæria pulvis-pyrius, Pers. Syn. p. 86
Melanomma pulvis-pyrius, Fckl. Symb. p. 160.
Sphæria myriocarpa, Fr. (in Berk. N. Am. Fungi Grev. IV, p. 143 sec. Sacc. in Syll. II, p. 98.)
Exsicc. Fckl. F. Rh. 937.—Rab. Herb. Mycol. 650.

Perithecia generally crowded, often forming a continuous layer of considerable extent, but sometimes scattered, superficial, irregularly spherical or ovate, wrinkled or otherwise roughened, either sulcate or smooth above, hard, black, 400 μ diam., with a small papilliform ostiolum. Asci cylindrical or subclavate, stipitate, 8-spored, 80–100 x 6–9 μ, with filiform paraphyses. Sporidia uniseriate, oblong, narrowed towards each end, rounded above, straight or a little curved, 3-septate, and slightly constricted at the septa, brownish, 16–18 x 4–6 μ.

On dead limbs and wood, Carolina, Pennsylvania, Virginia, New York, California andCanada.

M. cinèreum, (Karst.)

Sphæria cinerea, Karst. Myc. Fenn. II, p 91.
Melanomma cinereum, Sacc. Syll. 3259

Perithecia gregarious or crowded, attached to the inner bark, and erumpent through the variously lacerated epidermis, at length bare and free, sphæroid, smooth and subshining (at least when young), black, 390–400 μ diam., with a subpapillate or umbilicate-impressed ostiolum. Asci cylindric-elavate, 120–140 μ long, 8-spored, with abundant, filiform paraphyses. Sporidia subuniseriate or biseriate, oblong, obtuse, 3-septate, the second cell generally somewhat enlarged, slightly constricted at the septa, greenish-hyaline, then pale yellow, 20–28 x 6–10 μ.

On dead limbs of *Salix groenlandica*, Sukkertoppen, Greenland.

M. parasíticum, E. & E. Proc. Acad. Nat. Sci. Phil. Pa., July, 1890, p. 240.

Perithecia scattered or gregarious, superficial, ovate-hemispherical, 110–130 μ diam., roughish, black. Ostiolum papilliform, soon perforated. Asci oblong-cylindrical, 40-50 x 8–10 μ, sessile, without paraphyses. Sporidia crowded-biseriate, oblong-fusoid, 3-septate, and finally slightly constricted at the septa, 10–12 x 3–4 μ, pale olivaceous. *Sphæria nigerrima*, Blox. (Cke. Hndbk. No. 2612), which is parasitic "on various species of *Diatrype*," has sporidia 12½–20 μ long and at length multiseptate, and perithecia "sprinkled with short, stiff bristles."

On old *Diatrype stigma*, Newfield, N. J.

M. Commónsii, E. & E. Proe. Acad. Nat. Sci. Phil., Pa., July, 1890, p. 239.

Perithecia gregarious, ovate-globose, rough, black, minutely tomentose-pubescent when young, 110–125 μ diam. Ostiolum papilliform. Asci clavate-cylindrical, 50–55 x 7–8 μ, with abundant, filiform paraphyses. Sporidia biseriate, fusoid-oblong, 3-septate, slightly constricted at the septa, olive-brown, 12–14 x 3–3½ μ.

Parasitic on *Hypoxylon Sassafras*, Wilmington, Del. (Commons).

M. occidentàle, (Ell,) (Plate 21)

Cucurbitaria occidentalis, Ell. Bull. Torr. Bot. Club. VIII, p. 125.
Melanomma occidentale, Sacc. Syll, 3272.

Perithecia crowded, depressed-hemispherical, flattened above, 300 –350 μ diam., with a papilliform ostiolum. Asci clavate-cylindrical, 75–80 x 12–14 μ, with abundant paraphyses. Sporidia biseriate, oblong, slightly curved, obtuse, yellowish-brown, 3–5 (exceptionally 6–7-) septate, often slightly constricted near the middle, 20–25 x 5–6 μ.

On sage brush, Utah (S. J. Harkness).

Var. *Tetonense*, E. & E. (*Melanomma Tetonense*, E. & E. in Proc. Acad. Nat. Sci. Phil. July, 1890, p. 240.), on dead stems of *Artemisia cana*, valley of the Teton, in Montana (Anderson 551), has the perithecia and sporidia a little larger.

M. inspérsum, (Schw.)

Sphæria inspersa, Schw. in Currey's Simple Sphær. 334, tab LIX, fig. 112.

Perithecia simple, corticolous, erumpent. Asci terete-clavate, briefly stipitate, 8-spored. Sporidia biseriate or crowded, with the endochrome 4-parted, constricted at the partitions, usually rather more pointed at one end than at the other, 25–28 μ long.

On bark, North America? (Schweinitz in Herb. Hooker).

M. helicóphilum, (Cke.)

Sphæria helicophila, Cke Grev. VI, p 145.
Byssosphæria helicophila, Cke. Grev. XV, p. 123
Melanomma helicophilum, (Cke.) Sacc. Syll 3274.

Perithecia scattered among the hyphæ of *Helicoma Berkeleyi*. subglobose. Asci elongated, cylindrical. Sporidia fusiform, multinucleate. becoming multiseptate, yellowish, 60–70 x 6 μ.

On decaying wood, with *Helicoma Berkeleyi*, Carolina (Ravenel).

M. séminis, (Cke. & Hark.)

Sphæria seminis, Cke. & Hark. Grev. XIII, p. 18.
Melanomma seminis, Sacc. Syll. 7007.

Perithecia superficial, $\frac{1}{3}$–$\frac{1}{2}$ mm. diam., black, opake, subglobose, sometimes subconfluent, ostiolum perforated. Asci clavate. Sporidia cylindrical, straight or curved, 5-septate, brown, 50 x 6 μ.

On twigs of *Baccharis*, California (Harkness).

M. sulcàtum, (Ell.)

Sphæria sulcata, Ell. Bull. Torr. Bot. Club, X, p 53.
Melanomma sulcatum, Sacc. Syll. 7009.
Exsicc. Ell. & Evrht. N. A. F. 2d ser. 1663.

Perithecia superficial, densely gregarious, ovate, $\frac{1}{3}$ mm. diam., black, not polished. Ostiolum tuberculiform, large, with a rather large and nearly circular opening. Asci clavate-cylindrical, 130 x 22 μ, with abundant paraphyses. Sporidia biseriate, broad navicular-fusiform, 3-septate, yellow, becoming brown, 35–45 x 11–15 μ.

On dead sage brush, Utah (S. J. Harkness).

M. Porothèlia, (B. & C.)

Sphæria porothelia, B & C Grev IV, p 142.
Leptosphæria stereicola, Ell. Am Nat March, 1883 p. 317.
Melanomma Porothelia, Sacc Syll. 3244, Cke. Syn. 3126.

Perithecia innate-erumpent, minute, scattered, each seated in a little depression. Asci subcylindrical, 50–60 x 7–8 μ. Sporidia 1–2-seriate, shortly oblong-fusiform, rather obtuse at either end, triseptate, brownish, 11–13 x 3–3½ μ.

On the hymenium of some *Stereum*, South Carolina, on *Stereum bicolor*, Pers. Delaware (Commons), Iowa (Holway).

M. spiníferum, E. & E. (in Herb.)

Exsicc. Ell & Evrht. N A. F 2d Ser 2610.

Perithecia scattered, minute (50–75 μ), at first immersed, except the papilliform ostiolum, finally partially erumpent, clothed above with short, black spines, 20–25 x 4 μ long. Asci oblong-clavate, 35–40 x 10–11 μ, subsessile. Paraphyses filiform. Sporidia 2–3-seriate, fusoid, 3-septate, not constricted, olive-brown, 12–14 x 3–3½ μ.

On some thick, effused, *Corticium*? growing on the base of trunks of *Morus rubra*, Delaware (Commons).

Differs from *M. Porothelia*, (B. & C.), in its spiny perithecia, which are also only partially erumpent, and in its shorter, broader asci. In *M. Porothelia* the perithecia are finally almost superficial, and 110–120 μ diam., and not spiny at any stage of growth.

M. exile, (Schw.)

Sphæria exilis, Schw. (in Herb Berk.) Grev. XV, p. 82.

A specimen from Schweinitz, in Herb. Berk., has small (⅛ mm.), pilose perithecia, with fusiform, brown, 3-septate sporidia, 16–18 x 5 μ.

On rotten wood, Nazareth, Pa.

This is a different thing from *Nitschkia exilis*, Fckl. Symb. p. 165, of which *Sphæria exilis*, A. & S. is quoted as a synonym.

M. squamulàtum, (Schw.)

Sphæria squamulata, Schw Syn. N Am 1538.
Coniochæta (Chætomastia) squamulata, Cke. Grev XV, p 83

Perithecia carbonaceous, ovate-globose or irregular, scattered or subconcrescent, above the medium size, seated on a black, even, broadly effused, or sometimes scanty and very rough, black crust overspreading the wood. Ostiolum distinct, thick, at length perforated. The perithecia, except the bare, black ostiolum, are clothed with a subshining, light yellow-brown, villose-squamulose, pulverulent coat, which finally becomes black. Sporidia (Grev. l. c.), lanceolate, 3–5-septate, brown, 20–40 μ long.

On decorticated wood, Bethlehem, Pa. (Schw.)

M. Aspergrènii, (Fr.)

Sphæria Aspergrenii, Fr. in Kze Mycol Hefte, II. p. 40.
Melanomma Aspergrenii, Fckl Symb p 159.

Perithecia gregarious or scattered, of firm texture, superficial, or with the base slightly sunk in the matrix, hemispherical, with a perforated, scarcely papilliform ostiolum, finally umbilicate-depressed, black and shining. Asci cylindrical, stipitate, 8-spored. Sporidia uniseriate, oblong, obtuse at the ends, 4-celled, scarcely constricted at the septa, olivaceous, 12 x 4 μ.

On decaying wood, Bethlehem, Pa. (Schw.)

The diagnosis is from Winter.

M. rubiginòsum (Cke.)

Byssosphæria (Melanomma) rubiginosa, Cke Grev XV, p. 80.

Perithecia gregarious, superficial, seated on a thin, filamentose, brown subiculum, globose, dark rust-color, subrugose. Ostiolum impressed. Asci clavate-cylindrical, 8-spored. Sporidia biseriate, lanceolate, triseptate, constricted in the middle, hyaline, becoming pale-brown. 20–24 x 4–6 μ.

On rotten wood, Poughkeepsie. New York (Gerard).

M. ramíncolum, (Schw.)

Sphæria ramincola, Schw. in Herb. Berk.
Melanomma ramincola, Cke. Grev. XV, p 83.

Perithecia subcespitose, subglobose, finally collapsing, black, opake. Asci clavate. Sporidia lanceolate, triseptate, brown, 25 x 6 μ.

On *Pinus pinea.*

M. inspíssum, (Schw.)

Sphæria inspissa, Schw. Syn. N. Am. 1566.
Melanomma inspissum, Cke. Grev. XV. p. 83.

Perithecia very thickly scattered among the loosened, weather-beaten fibres of the wood, and at first partly covered by them, but at length bare, black, globose, rugose, with a papilliform deciduous ostiolum, very minute and finally collapsing. Sometimes cespitose-crowded, but also scattered. Sporidia (sec. Cke. Grev. XV, p. 83), triseptate, brown, 12–14 μ long.

On rotten wood, Pennsylvania (Schw.)

M. sporádicum, E. & E. (in Herb.)

Perithecia subconic-globose, 300–350 μ diam., rough, black, with a rather broad, papilliform, at length perforated ostiolum, erumpent-superficial, standing singly or oftener 2–4 subseriate in cracks between the fibers of the wood. Asci clavate, subsessile, paraphysate, 8-spored, 100–110 x 15–20 μ. Sporidia subbiseriate, oblong or clavate-oblong, 3–5- (mostly 3-) septate, brown, obtuse, slightly constricted at the septa, especially at the middle one, 20–30 x 8–12 μ, very variable in size and shape.

On decorticated wood of *Platanus*, Canada (Dearness).

Differs from *M. cinereum*, Karst, in its rough perithecia and variable, 3-5-septate sporidia; from *M. squamulatum*, (Schw.), in its smaller perithecia and absence of any subiculum; from *M. inspissum*, (Schw.), in its much larger sporidia.

ZIGNOÉLLA, Sacc.

Sylloge II, p. 214.

Perithecia subsuperficial, mostly gregarious, carbonaceous, black, papillate, generally small, the lower part more or less sunk in the matrix. Asci 8-spored, generally paraphysate. Sporidia ovate or oblong, hyaline, pluriseptate. Differs from *Metasphæria* in its superficial or subsuperficial perithecia.

Z. pulvíscula, (Curr.)

Sphæria pulviscula, Curr. Linn. Trans. XXII. tab LVIII, fig. 52.
Melanomma pulviscula, Sacc. Mycol. Ven. Spec. p. 114.
Zignoella pulviscula, Sacc. Syll. 3627

Perithecia usually gregarious, often crowded, sometimes scattered superficial, hemispheric-conoid, black, smooth and bare, carbonaceous, $\frac{1}{4}$–$\frac{1}{3}$ mm. diam., with a small, papilliform ostiolum. Asci clavate-cylindrical, narrowed below, 8-spored, 75–80 x 9–12 μ. Sporidia biseriate, fusiform, often slightly curved, hyaline, finally 3-septate, 18–20 x 3$\frac{1}{2}$–4$\frac{1}{2}$. μ.

On rotten wood, London, Canada (Dearness).

Sec. Winter this should be included in *Z. ovoidea*, Fr.

Z. quercètis, (Cke. & Massee).

Conisphæria (Zignoella) quercetis, Cke. & Mass. Grev XVI, p. 91.

Perithecia scattered or gregarious, semiimmersed, laterally compressed, hysteriiform, black, ostiolum papilliform and finally deciduous. Asci clavate-cylindrical. Sporidia biseriate, lanceolate, 4-nucleate, at length triseptate, hyaline, 35 x 8 μ.

On naked, bleached wood, South Carolina (Curtis, 1915).

Z. exígua, (C. & P.)

Sphæria exigua, C. & P. 30th Rep N. Y State Mus. p 65.
Zignoella exigua, Sacc. Syll. 3629.

Perithecia subgregarious, small, 300 μ diam., globose, sometimes collapsed, smooth, shining, black, papillate. Asci clavate or cylindrical. Sporidia elliptical, binucleate, then 1–3-septate, hyaline, 15–18 x 7½ μ.

On decaying wood, New York (Clinton).

Z. funícola, Ell.

Sphæria funicola, Ell. Bull Torr. Bot Club, VIII, p. 90.
Zignoella funicola, Sacc. Syll. 3639

Perithecia scattered, minute (150–200 μ), ovate-globose, partly covered by the fibers of the cotton, pierced above, but without any prominent ostiolum, rough, black, carbonaceo-membranaceous. Asci oblong-cylindrical, contracted below into a short stipe, 65–75 x 15–18 μ, with abundant conglutinated paraphyses. Sporidia crowded-biseriate, oblong-elliptical, obtuse, 3-septate, hyaline 15–20 x 7–8 μ.

On old cotton cord on a grape trellis, New Jersey.

Too near the next species.

Z. diáphana, (C. & E.)

Sphæria diaphana, C. & E Grev. V, p. 53 tab 80 fig. 15.
Zignoella diaphana, Sacc. Syll. 3653.
Exsicc. Ell. N. A. F. 781.

Perithecia gregarious, mostly on bleached spots, suberumpent, membranaceous, finally collapsing above, subelongated or subhysteriiform, ½–¾ mm. in the longer diam., ostiolum not prominent. Asci subsessile, cylindrical, paraphysate, about 75 x 12 μ. Sporidia overlapping or biseriate, clavate-oblong or oblong-elliptical, hyaline, 3-septate, 15–22 x 6–7 μ.

Var. *albocincta* (*Sphæria albocincta*, C. & E. Grev. VII, p. 9), has the perithecia surrounded at base by a narrow zone of bleached wood, but does not differ otherwise, unless the sporidia may average a little larger (20–23 μ long).

Var. *soluta* (*Sphæria soluta*, C. & E. Grev. V, p. 54, tab. 80, fig. 16), has the perithecia subovate, with the sporidia averaging rather smaller (15–18 x 4–5 μ).

On exposed wood, both of coniferous and deciduous trees, Newfield, N. J.

Z. Sequoiæ, (Plowr.)

Sphæria Sequoiæ, Plowr. Grev. VII, p. 73.
Zignoella Sequoiæ, Sacc Syll. 3670.

Perithecia rather large, immersed, then suberumpent, seated on bleached spots. Asci cylindrical. Sporidia hyaline, obscurely 3-septate, slightly unequal, the upper half being the larger, 25–30 x 8–10 μ.

On dead bark of *Sequoia gigantea*, California.

Z. cyrillicola, (B. & C.)

Sphæria cyrillicola, B. & C. Grev. IV, p. 143.
Melanomma cyrillicolum, Sacc. Syll. 3267.
Conisphæria cyrillicola, Cke. Grev. XVI, p. 87.

Perithecia scattered, black, ovate, with an acute apex. Asci elongated, linear. Sporidia lanceolate, slightly curved, 4-septate, 20 μ long.

On *Cyrilla*, North Carolina.

Specimens from Pennsylvania (Michener) have the sporidia a little stouter, but of the same length.

Cooke, in Grev. XVI, p. 87, places this with hyaline-spored species.

Z. macrásca, Sacc. Mich. II, p. 138.

Perithecia scattered, semiimmersed, ½ mm. diam., globose-conical, carbonaceous, black. Ostiolum papilliform, at length deciduous, texture parenchymatic, dense, sooty-black. Asci cylindrical, very large, 200 x 10 μ, with a short stipe and surrounded by short (genuine)? paraphyses, 8-spored, rounded and entire above. Sporidia obliquely uniseriate, fusoid, slightly curved, 3-septate, 30 x 6 μ, not constricted, hyaline, 4-nucleate.

On rotten wood, Carolina (Ravenel).

Sec. Cooke (Syll. l. c.) this species approaches *Sphæria picastra*, Schw., in its sporidia.

Z. humùlina, Pk. Bull. N. Y. State Mus. No. 2, p. 24.

Perithecia small, 260–310 μ broad, depressed-hemispherical, slightly sunk in the matrix, subglabrous, black, with a minute, papillate ostiolum. Asci cylindrical, 60–70 x 8–10 μ. Sporidia uniseriate, elliptical, 4-celled, appearing obscurely 3-septate, hyaline, 12–15 x $6\frac{1}{4}$ –$7\frac{1}{2}$ μ.

On dead stems of *Humulus lupulus*, Carlise, N. Y. (Peck).

The perithecia have a dull, squalid, unpolished or subscabrous appearance.

Z. subvestìta, (E. & E.)

Sphæria subvestita, E. & E. Journ. Mycol. II, p. 100.
Zignoella subvestita, Sacc. Syll. 7495.

Perithecia seated on the bare wood or on the bark, with the base sunk in the matrix, ovate-conical, 165–250 μ diam., $\frac{1}{2}$ mm. or more high, black, rough, except the smooth, subshining, short-cylindrical or subconical, broadly perforated, subtruncate ostiolum. Asci clavate-cylindrical, 50 x 6–7 μ, with filiform paraphyses and a slender, stipitate base. Sporidia fusiform, yellowish-hyaline, slightly curved, 3-septate and constricted slightly at the middle septum, 12–15 x 2–3 μ, crowded-biseriate. The sporidia are much the same as in *Clypeosphæria Hendersoniæ*, Ell., except in having the ends slightly curved, but in that species, the perithecia are depressed, subcuticular and subastomous. *Zignoella conica*, (Fckl.), has much larger sporidia (28 x 4 μ).

On dry, bleached roots of *Vaccinium*, Newfield, N. J.

FAMILY. CERATOSTÒMEÆ.

Stroma none. Perithecia superficial, or at first immersed and finally erumpent, with an awl-shaped or at least more or less elongated ostiolum. Texture thin-membranaceous, or sometimes subcoriaceous. Asci mostly very evanescent, with the membrane thickened at the apex. Paraphyses generally present.

The most obvious character of this family is the elongated (subulate, cylindrical or conical) ostiolum. It is distinguished from the *Gnomonieæ* by the superficial or erumpent perithecia, and the presence of paraphyses.

Ceratóstoma, Fr. Obs. II, p. 340.

Perithecia more or less superficial, with a distinct, and generally, strongly developed beak; substance firm, coriaceous or carbonaceous. Sporidia ovate, oblong or elliptical, brown, continuous.

C. brevirόstre, (Fr.)

Sphæria brevirostris, Fr. S. M. II, p. 474, and Curr. Linn. Trans. XXII. p. 322. tab. 58, fig. 68.
Ceratostoma, brevirostre, Sacc. Syll. 775.

Perithecia scattered, semiimmersed, globose, smooth, glabrous, black. Ostiolum cylindrical, equal, about half as long as the perithecium. Asci cylindrical, briefly pedicellate, 100 x 10 μ. Sporidia uniseriate, elliptical, pale brown, 15–17 μ long.

On rotten wood, Bethlehem, Pa.

At first immersed, finally emergent and bare, very smooth, almost shining, fragile.

C. rubefàciens. (Pk.)

Sphæria rubefaciens, Pk 28th Rep p 79.
Ceratostoma rubefaciens, Sacc. Syll. 777.

Perithecia minute, scattered, subglobose, smooth, black, nearly free, abruptly tapering into the long, slender, subulate ostiola. Asci clavate, fugacious. Sporidia elliptical, brown, 4–5 x 4½ μ.

On decorticated wood of deciduous trees, New York State (Peek). The surface of the wood is variegated with red stains.

C. setígerum, E. & E. (in Herb.)

Perithecia densely gregarious, superficial, hemispherical, ⅓–½ mm. diam., cinereous-black, clothed with long, spreading, black bristles. Ostiola erect, stout, rough, slightly swollen at the tip, black, ¾–1 mm. long. Asci oblong-cylindrical, broadest in the middle, p. sp. 12–15 x 4 μ, aparaphysate. Sporidia conglomerate-biseriate, elliptical, olive-brown, continuous, about 3 x 1½ μ.

On decaying oak wood, Ohio (Morgan, 895).

C. Avocétta, (C. & E.)

Sphæria Avocetta, C. & E. Grev. VIII, p 15.
Ceratostoma Avocetta, Sacc. Syll. 774.
Sphæria caminata, C. & E. Grev. VI, p. 94.
Exsicc Ell. N. A. F. 779.—Rav. F. Am. 673.

Perithecia scattered, globose, ½ mm. diam., entirely buried in the wood, with the tips of the cylindrical (½–1½ mm. long) ostiola projecting or, by the decay or weathering away of the wood, becoming superficial; often in cracks of the wood, one side of the perithecium being buried and the other projecting into the open fissure; in that case bald and smooth, but when buried in the soft wood, more or less fibrous-strigose, with brownish hyphæ. Asci p. sp. cylindrical, 45–50 x 7–8 μ, with paraphyses. Sporidia uniseriate, often oblique, brown, oblong-elliptical, 1-2-nucleate, 8–10 x 3½ μ.

On rotten wood of deciduous trees, common around Newfield, and probably in other localities.

The specimens of *S. caminata*, C. & E., in my Herb., and of *S. caminata*, Cke. in Rav. F. Am., do not afford any characters to distinguish them from this.

C. carpóphilum. Ell. Bull. Torr. Bot. Club. IX, p. 73.

Perithecia subglobose, ½ mm. or over in diam., rough. Ostiolum capillary, three times or more as long as the perithecium, and more or less crooked or bent. Asci slender, about 50 μ long, with filiform paraphyses. Sporidia uniseriate or biseriate, fusiform, yellowish. 4½–6 μ long.

On old chestnut burs, with *Sphæria* (*Læstadia*) *echinophila*, Schw. which has the asci and sporidia hardly distinguishable from those of this species.

C. subulàtum, Ell. Am. Nat. March, 1883, p. 318.

Perithecia covered by the epidermis, globose, 150 μ diam. Ostiolum erumpent, subulate, ½ mm. long, apex subfimbriate. Asci 8-spored, evanescent, elliptical, 15 x 11 μ. Sporidia subcubical in the asci, becoming ovate or subglobose when free, brownish, 4½–5½ x 3–3½ μ, often oozing out in a globule at the tip of the ostiolum.

On dead stems and capsules of *Oenothera biennis*, and on the brush of an old broom-corn broom, Newfield, N. J.

In the description of *C. subulatum* in Am. Nat. the perithecia were erroneously described as "subulate." The ostiolum is subulate, but the perithecia are globose and immersed. *C. caulincolum*, Fckl. has larger perithecia and sporidia.

C. cónicum, E. & E. Proc. Acad. Nat. Sci. Phil., Pa., July, 1890.

Exsicc. Ell. & Evrht. N. A. F 2d Ser. 2348.

Perithecia immersed or superficial by the falling away of the surrounding wood, gregarious, subovate, about ½ mm. diam., rough, prolonged above into a conical, rough ostiolum projecting above the surface of the wood, and finally elongated to about 1 mm. in length. Asci 75–80 x 8 μ (p. sp.), with stout, lance-pointed paraphyses much longer than the asci. Sporidia biseriate, oblong-fusoid, pale brown, about 5-nucleate, straight, 18–20 x 3 μ.

On rotten pine logs, Newfield, N. J.

C. parasíticum, E. & E. l. c.

Perithecia membranaceous, subhemispherical, ⅓ mm. diam., reddish-brown, becoming slaty-black, prolonged above into a stout beak 2–2½ mm. long, 150 μ thick below, narrowing to about 75 μ at the paler, subfimbriate tip. Asci oblong-ovate, p sp. about 20–25 x 7–8 μ. Sporidia crowded, acutely elliptical, hyaline, becoming dark, 7–8 x 4½–5 μ. The asci and sporidia often ooze out at the tip of the long beak or ostiolum, and form a dark colored globule, which inclines to flatten out, and thus gives the appearance of an enlarged, truncate tip.

On old *Fomes applanatus*, West Chester, Pa.

Specimens of *Periconia sphærophila*, Pk., found by Mr. Meschutt in Northern New Jersey, and by Miss Minns in New Hampshire, are ascigerous and much resemble this.

C. junipérinum, E. & E. l. c.

Perithecia gregarious, awl-shaped, black, 700–800 μ high, slightly enlarged at the tip, swollen and about 150 μ thick below. Asci included in the ovate-swollen base, oblong-elliptical, about 12 x 6 μ, with a slender stipe 12–15 μ long. Sporidia crowded-biseriate, ovate-globose, brown, 4 x 3½ μ, or a little less. The sporidia exude and form a little brown head at the apex of the perithecium, thus giving the appearance of a *Calicium*.

On a wounded, dead place on a limb of *Juniperus Virginiana*, Flatbush, Long Island (Zabriskie).

C. subrùfum, E. & E. (in Herb.) **(Plate 22)**

Perithecia gregarious, minute, erumpent-superficial, ovate, 112 μ diam., with a cylindrical ostiolum about 300 μ long. Asci ovate, stipitate, p. sp. 10 x 6 μ, evanescent. Sporidia crowded-biseriate, 8 in an ascus, subcubical, brown, about 4 x 3 μ, with a single large nucleus.

On dead places in oak limbs partly killed by fire, in company with *Wallwrothiella melanostigma*, (C. & E.), Newfield, N. J.

C.? folitcolum, Fckl. 2d. Dutch North Polar Exp. p. 94, tab. I, fig. 7, Sacc. Syll. Addit. I, p. 37.

Perithecia with the habit of *Gnomonia*, scattered, at first immersed, finally suberumpent, globose, minute, black. Ostiolum very long and black. Asci not seen. Sporidia perfectly globose, nucleate, epispore smooth, 8 μ diam.

On the lower surface of dry leaves of *Salix arctica* Sabine Island, Arctic America.

Species imperfectly known.

C. fállax, Cke. & Sacc. Grev. VII, p. 8.

Exsicc. Ell. N. A. F. 788.

Perithecia loosely gregarious, superficial, globose-depressed, ¼ mm.

diam., prolonged above into a filiform, subflexuous ostiolum nearly 1 mm. long. The perithecia contain an abundance of minute, hyaline, oblong spermatia 4 x 1–1½ μ, often collected into a subglobose mass.

On pine lumber recently cut and closely packed, Newfield, N. J.

Doubtfully distinct from *C. piliferum*, (Fr.), which is also usually sterile.

C. piliferum, (Fr.)

Sphæria pilifera, Fr. S. M. II, p. 472.
Sphæria rostrata, Schum. Enum. Fl. Saell. p. 128.
Ceratostoma piliferum, Fckl. Symb p. 128.
Exsicc. Rab. F. E. 1525. 2327.

Perithecia gregarious, naked, black, very small, globose, smooth. Ostiolum capillary, very long, acuminate. Sporidia (sec. Berk. Grev. IV, p. 146) sausage-shaped, 5 μ long.

On pine rails, Carolina, on oak, Alabama. Found also in Pennsylvania by Dr. Michener.

C. mucronàtum, (Schw.)

Sphæria mucronata, Schw. Syn. Car. No. 4.—Fr. S. M. II, p. 475.
Ceratostoma mucronatum, Cke. Syn. 3814.

Perithecia gregarious, minute, black, bare, conical, glabrous, sometimes confluent but not crowded, innate-superficial, smooth, rigid, contracted above into a short-cylindrical, rather obtuse ostiolum.

On dead wood, Carolina. Rare.

C. drupìvorum, (Schw.)

Sphæria drupivora, Schw. Syn. N. Am. 1613.
Ceratostoma drupivorum, Cke. Syn. 3819.

Perithecia rather large, hemispherical, densely crowded, black, involved in a white, byssoid tomentum (which perhaps is only accidental). Ostiola very long, dark liver-color, straight.

On decaying nuts of *Juglans cinerea*, very rare, Bethlehem, Pa.

C. investìtum. (Schw.)

Sphæria investita, Schw. Syn. N. Am. 1621.

Perithecia globose, rather large, at length free, black, at first immersed in the wood and covered with tomentum. Ostiola large, long, rigid, sulcate-tuberculose, thickened above, black.

On rotten wood, emerging from a rather thick, cottony, brownish rust-colored, or light yellow tomentum, with which it is entirely covered. In other respects like *C. rostrata*.

C. hystricina, Cke.

Ceratostoma hystricina, Cke Grev XI, p 199
Ceratostomella hystricina, Sacc. Syll. 5995, Cke. Syn. 3766.

"Perithecia gregarious, semiimmersed, at length emergent and subsuperficial, dark brown, tomentose, beaked. Ostiola cylindrical, subflexuous. Asci cylindrical. Sporidia elliptic-lanceolate, hyaline, 16–18 x 6 μ.

On bark of *Ficus*, Aiken, So. Carolina.

The specimens distributed under this name in Rav. F. Am. 674, and Ell. & Evrht. N. A. F., 2d Ser., 2349, do not agree at all with the above quoted description, being a species of *Valsa*, with minute sporidia.

C. spina, (Schw.)

Sphæria spina, Schw. Syn. Car No. 154.
Ceratostoma spina, Schw. in Cooke's Synopsis, 3818
Sphæronema Fraxini, Pk in 29th Rep. p. 71, and in Ell. N. A. F. 737.
Sphærographium Fraxini. Sacc. Syll. III, p. 598.

According to specimens in Herb. Schw., these are all the same thing, *Sphærographium spina*, (Schw.)

CERATOSTOMÉLLA, Sacc.

Michelia I, p 370.

Characters the same as in *Ceratostoma*, only sporidia hyaline.

C. Màli, E. & E. Proc. Phil. Acad. l. c.

Perithecia scattered, membranaceous, globose, 400 μ diam., barely covered by the bark which is slightly raised above them and pierced by the short-cylindrical, obtuse, 150 x 75 μ ostiolum, with a rather large round opening at its apex. Asci clavate, subtruncate above and narrowed gradually to the acute base, about 40 x 5 μ, 8-spored. Paraphyses none. Sporidia biseriate, oblong-cylindrical, scarcely curved, faintly 2-nucleate, 6–8 x 1½ μ. Differs from *C. dispersa*, Karst. in its rather smaller, straight, bald ostiolum.

On the inner surface of loose, hanging bark of partly dead apple trees, Newfield, N. J.

C. nyssæcola, (B. & C.)

Sphæria nyssæcola, B. & C. Grev. IV, p. 143
Ceratostomella nyssæcola, Sacc. Syll. 1564.

Perithecia semiimmersed, subglobose, with an abrupt and at length elongated neck. Sporidia elliptical or shortly cymbiform, 7–8 μ long.

On *Nyssa* in Pennsylvania.

C. strícta, (Pers.)

Sphæria stricta, Pers Syn p. 59
Ceratostoma strictum, Fckl Symb p 127
Ceratostomella stricta, Sacc. Syll 1555

Perithecia crowded or seriate-conglomerate, scarcely emergent, sphæroid, smooth, rigid, fragile, glabrous, black. Ostiolum straight, rigid, erect, rather obtuse, twice longer than the perithecium. Asci attenuated into a slender pedicel, cylindric-clavate, 30–35 x 5 μ (p. sp.). 8-spored. Sporidia biseriate, elongated, slightly curved, continuous, 6–8 x 1½–2 μ, 2–4-nucleate, hyaline.

On rotten wood, Carolina and Pennsylvania, common.

C. echinélla, E. & E. (in Herb.)

Gregarious. Perithecia about ¼ mm. diam., globose, covered, but not deeply, membranaceo-carbonaceous. Ostiola exserted, 1 mm. long, rough and glandular-pubescent below, becoming glabrous, more or less bent or undulate, slightly thickened at the tips. Asci clavate, 25–30 x 4–5 μ. Sporidia crowded-biseriate, cylindrical, hyaline, curved, 3–4 x 1 μ.

On the inner surface of cast-off maple bark, New Jersey, and Ohio (Morgan, 928).

C. albocoronàta, (Ell.)

Ceratostoma albocoronatum, Ell, Am Nat 1883, p 318.
Ceratostomella albocoronata, Sacc Syll 5997

Perithecia conical, minute, vomiting forth the ascigerous nucleus in the form of a white globule. Asci cylindrical, 35 x 7 μ. Sporidia uniseriate or partly biseriate, elliptical, hyaline, 1–2-nucleate, 7½–9½ x 3–3½ μ. Accompanied by obovate, brown, 2–3-septate conidia borne singly on the apices of erect, brown, simple hyphæ thickly scattered over the matrix.

On rotten wood, Newfield, N. J.

C. rostràta, (Fr.)

Sphæria rostrata, Fr. S. M II, p 473
Ceratostoma rostratum, Fckl. Symb. p 127
Ceratostomella rostrata, Sacc Syll 1564.

Perithecia gregarious or scattered, immersed or free, globose, rugose, bald, black, variable in size from 300–700 μ diam., with a very long, unequal, rather obtuse, longitudinally striate, 4-cornered, mostly curved, rigid beak. Asci clavate-cylindrical, sessile, 8-spored, 40–45 x 8–9 μ. Sporidia oblong or elliptical, continuous or obscurely uniseptate, hyaline, 9–12 x 4 μ.

Found in New York State (sec. Peck in 22d Rep.) and on maple in South Carolina (Berk. in Grev. IV, p. 146).

C. cirrhòsa. (Pers.)

Sphæria cirrhosa, Pers. Syn p. 59
Ceratostoma cirrhosum, Fckl. Symb p 127.
Ceratostomella cirrhosa, Sacc. Syll. 1547.

Perithecia scattered or sometimes crowded, at first buried, finally semierumpent, globose, with brown, creeping, mycelioid hairs, which penetrate the surrounding wood, the projecting part becoming bald, black and smooth, about 400 μ or over, broad, with a long, mostly curved and tubercular, bald ostiolum. Asci subcylindrical, sessile, 8-spored, 65–75 x 7–9 μ. Sporidia uniseriate, oblong-elliptical, with two nuclei, hyaline, 9–12 x 3½ μ.

On rotten wood, Carolina.

C. capillàris, (Ell.)

Ceratostoma capillare, Ell. Bull. Torr. Bot. Club, IX, p. 20.
Ceratostomella capillaris, Sacc. Syll. 5996.

Perithecia capillary, nodulose, clothed below with short, spreading, hyaline hairs which are scarcely enlarged at the base. Asci elliptical, 30–35 x 9 μ. Sporidia 8 in an ascus, crowded, fusiform, nearly hyaline, indistinctly nucleate, straight or slightly curved, 11–14 x 3–3½ μ.

On decaying catkins of *Alnus serrulata*, Newfield, N. J.

RHYNCÓSTOMA, Karst.

Myc. Fenn II, p. 7.

Perithecia scattered or gregarious, immersed or erumpent, sphæroid, with a beak-like ostiolum equaling or exceeding in length the diameter of the carbonaceous, black, glabrous perithecium. Asci cylindrical or clavate, 8-spored, generally paraphysate. Sporidia ellipsoid, fuscous, uniseptate.

Rh. cornígerum, Karst. Myc. Fenn. II, p. 57.

Perithecia somewhat scattered, immersed, finally half emergent, ovate-globose, rather more than ½ mm. diam., black, surrounded below with long, concolorous filaments. Ostiolum stout, rough, mostly a little curved, conic-cylindrical, 3–6 mm. long, generally a little contracted just below the tip. Sporidia ellipsoid or oblong-ellipsoid, uniseptate and mostly a little constricted at the septum, brown, 8–10 x 3–4 μ.

Var. *Americana*, E. & E., on rotten wood, Ohio (Morgan, 260) and

Louisiana (Langlois, 2283), has the perithecia densely gregarious and emergent-superficial, the hairs around the base tow-colored and finally disappearing.

Rh. tínctum. (E. & E.)

Ceratostoma tinctum, E. & E. Bull Wash. Coll. vol I, p. 5.
Rhyncostoma tinctum, Sacc. Syll. 6620.

Perithecia erumpent-superficial, ovate-globose, 165–250 μ diam,, roughened with projecting points. Ostiolum filiform, ⅔–¾ mm. long, nearly smooth, sometimes a little swollen just below the apex, which is often crowned with a globule of ejected sporidia half as large as the perithecium itself. Asci subcylindrical, about 30 x 7 μ, subsessile, with filiform paraphyses. Sporidia biseriate, oblong-elliptical, subacute, uniseptate, 6–7 x 2½–3 μ, brown.

On dead wood of *Acer macrophyllum*, Klikitat Co., Washington (Suksdorf).

Closely allied to *Rh. minutum*, Karst.

Rh. altípetum. (Pk.)

Sphæria altipeta, Pk. Bot. Gaz. vol. V, p. 36.
Rhyncostoma altipetum, Sacc. Syll 2769.

Perithecia minute, subglobose, black, immersed. Ostiola emergent, subconical or cylindrical, nearly as long as the perithecia. Asci cylindrical. Sporidia uniseriate. oblong, colored, 16–17 x 5 μ, hyaline at one end, uniseptate near the other.

On decaying wood, Mt. Washington, N. Y.

The perithecia sometimes occur in long lines. When young, the sporidia are colorless, but they soon become colored, except at one end, and contain two nuclei. Finally a septum is formed near the colored end of the sporidium.

Rh. sphærincola, (Schw.)

Sphæria sphærincola, Schw. Syn. N Am. No 1611.

Perithecia densely gregarious, rather soft, erect. dark brown, very rough. Ostiolum of medium length, rigid, oblique, only a little longer than the ovate perithecia.

On decaying *Nummularia Bulliardi*, Carolina. Very rare quite different (sec. Schweinitz) from *Melanospora lagenaria*.

CERATOSPHÆRIA, Niessl.

Not. Pyr p. 43, (emended).

Perithecia semiimmersed or superficial, coriaceous or subcarbonaceous, more or less rostrate. Asci 8-spored. Sporidia elongated transversely pluriseptate, subhyaline or rarely brownish. Paraphyses distinct.

C. micródoma, E. & E. Journ. Mycol. IV, p. 78. **(Plate 22)**

Perithecia barely covered by the bark, densely gregarious, minute, not over 165 μ diam. Ostiola projecting, cylindrical, ⅓ mm. long, rough and more or less overrun with a brown tomentum (which, however, may be only accidental). Asci oblong-cylindrical or clavate-cylindrical, 50–55 x 7–8 μ, subsessile, with rather stout, filiform paraphyses. Sporidia biseriate or crowded, oblong or clavate-oblong, 3-septate and slightly constricted at the septa, olive-brown, 11–12 x 2½–3 μ.

On bark of decaying (*Sambucus*)? St. Martinsville, La., June 1888 (Langlois, No. 1310).

OPHIÓCERAS, Sacc.

Syll. II, p. 358

Perithecia immersed or emergent, subcarbonaceous, globose, with a conic-cylindrical ostiolum more or less elongated. Asci subcylindrical. Sporidia filiform or cylindrical, 2- or more-septate, hyaline.

This genus differs from *Ceratostoma* in its elongated, hyaline, septate sporidia.

O. Ohiénse, E. & E. (in Herb.)

Perithecia gregarious, at first immersed in the wood which is slightly pustulate-elevated over them and pierced by the stout, black, rough, beak-like ostiola, which are 1–2 mm. long, carbonaceous and brittle, their smooth, rounded tips at first pierced in the center with a small, round opening, at length broadly perforated. The perithecia are finally emergent and bare, only the base remaining sunk in the wood, about 1 mm. diam., with thick, carbonaceous walls, rough and brownish-black outside, the upper, projecting part finally breaking or falling away, leaving only the cup-shaped base sunk in the surface of the wood. Asci clavate-cylindrical, gradually attenuated below, 90–100 x 7–8 μ, pseudo-paraphysate, 8-spored. Sporidia subfasciculate, fusoid-cylindrical, hyaline, 8–12-septate, slightly and rather abruptly attenuated at the ends, which are mostly a little incurved and subobtuse, 40–50 x 3½–4 μ.

On rotten wood, Ohio (Morgan, No. 528).

This seems to be quite distinct from any other described species.

HYPSOTHÈCA, Ell. & Evrht.

Journ. Mycol. I, p 128.

Perithecia subulate, stylosporiferous at base. and with a medial or subapical enlargement above, containing the ascigerous nucleus. This latter character will distinguish this genus from *Ceratostoma*. *Caliciopsis*, Pk. is also closely allied, but is placed by its author among the *Discomycetes*.

H. subcorticàlis, (C. & E.) l. c. p. 128. **(Plate 22)**

Sphæronema suborticale, C. & E , Grev. VI, p 83
Hypsotheca subcorticalis, E & E Journ. Mycol. I, p. 129

Perithecia subcylindrical, about 1½ mm. high and 100 μ thick, only slightly enlarged at base, and containing subglobose, continuous, brown stylospores 3–5 μ diam., or oblong-ovate, 5–10 x 4–5 μ and 2–3-septate. Ascigerous cavity subapical, formed by a gradual, subovate enlargement 150–175 μ diam., in the middle, and closely packed with the oblong-ovate or subelliptical, 8-spored asci, 12–15 x 7–9 μ, on slender pedicels 15–20 μ long, and without paraphyses. The asci are hyaline and smaller at first. Sporidia conglomerated, subglobose, brownish, 3–3½ μ.

Growing from the inner surface of the loosened bark or from the exposed edges of loosened pieces of bark, on dry, decaying oak limbs lying on the ground, Newfield, N. J., Sept., 1877, and March, 1883.

This description was made from a reexamination of the original specimens.

H. calicioìdes, (Fr.)

Sporocybe calicioides, Fr. S M. III, p 342.
Hypsotheca calicioides, E. & E l. c.
Exsicc Rav. Fungi Car. I, p. 83

Gregarious, subulate, black and smooth, about 1½ mm. high, 115 μ thick, the swollen, subelliptical, ascigerous cavity, about midway between the base and the apex, 250 μ thick. Asci (spore bearing part) oblong or ovate, 15–20 x 8–11 μ, with a thread-like stipe of about the same length. Sporidia conglomerated, elliptical, hyaline, becoming brown, continuous, 6–7 x 3–3½ μ.

Ravenel's specimens are on poplar bark. Mr. C. J. Sprague has sent specimens collected by W. N. Suksdorf in Washington, and which agree in all respects with Ravenel's specimens. Suksdorf's specimens are also, apparently, on bark of poplar.

H. thùjina, E. & E. l. c. p. 128.

Perithecia subulate, 700–800 μ high and 55–60 μ thick below, with a gradually enlarged or swollen place near the top about 150 μ long and 90–100 μ thick, and above this, again contracted to about the same size as below, forming a truncate beak 75–80 μ long, the whole being, in fact, a hollow cylinder of fibrous-cellular structure, slightly enlarged at base and filled with minute, hyaline, oblong or cylindrical, straight or slightly curved, 2–3 x ½ μ spermatia and the swollen part above filled with an abundance of club-shaped, 8-spored asci, without paraphyses, and consisting of an obovate or subelliptical head or spore-bearing part 9–12 x 4½–5½ μ, and a filiform base or stipe 15–20 μ long.

On partly dead foliage of *Cupressus thyoides*, Newfield, N. J.

The membrane of the asci is very delicate, and scarcely discernible after the sporidia have matured, but easily seen in the young state. Sporidia globose, pale brown, 2½ μ diam., collected in a mass.

LENTOMÌTA, Niessl.

Notiz. uber Pyren. p. 44.

Perithecia sunk in the matrix, finally erumpent or free, soft-coriaceous or even membranaceous, prolonged above into a more or less evident beak. Asci delicate, with the membrane thickened at the apex. Sporidia uniseptate, hyaline. Paraphyses present.

RHAMPHÓRIA, Niessl. l. c.

Perithecia as in *Ceratosphæria*. Sporidia oblong, elliptical or ovate, muriform.

No representatives of either of these two genera have yet been reported in this country.

FAMILY. AMPHISPHÆRÌEÆ.

Stroma wanting. Perithecia either closely adnate or with the base sunk in the matrix, becoming erumpent-superficial, or sometimes superficial from the first; generally of hard, carbonaceous texture, with a papilliform ostiolum. Asci elongated, paraphysate. Distinguished from the next preceding and following families by the papilliform ostiolum.

AMPHISPHÆRIA, Ces. & De Not.

Schema. Sferiacei, p. 49.

Perithecia at first adnate or more or less sunk in the matrix, finally erumpent and mostly superficial, of firm, often carbonaceous texture, bald, with a small ostiolum. Asci paraphysate, 8-spored. Sporidia uniseptate, brown.

A. atrogràna, (C. & E.)

Sphæria (Psilosphæria) atrograna, C. & E. Grev. VIII, p. 15.
Amphisphæria atrograna, Sacc. Syll. 2734.

Perithecia thickly gregarious, depressed-globose, rough, black, subcarbonaceous, pierced above, 110--120 μ diam. Asci cylindrical, sessile, paraphysate, 40–45 x 6–7 μ. Sporidia uniseriate or subbiseriate, ovate-elliptical, brown, uniseptate, 9–12 x 4–5 μ (12–14 x 5 μ, Cke.)

On rotten wood of *Liquidambar*, Malaga, N. J.

The surface of the wood is blackened.

A. bisphérica, (C. & E.)

Sphæria bisphærica, C & E. Grev. VII, p. 41.
Amphisphæria bisphærica, Sacc Syll. 2724, Cke. Syn. 3432.

Perithecia gregarious, semiimmersed, grayish-black, somewhat flattened, 175–200 μ diam. 'Ostiolum indistinct. Asci cylindrical, paraphysate, subsessile, 100–120 x 5–6 μ. Sporidia uniseriate, pale brown, uniseptate and constricted, cells subglobose or conical, 10-12 x 4-5 μ (12 x 7 μ, Cke.)

On decorticated wood of apple tree limbs, Newfield, N. J. Not abundant.

A. incrústans, E. & E., in Webber's Cat. Fl. Nebraska, 1889, p. 53.

Amphisphæria saprogena, E. & K (in Herb.)

Perithecia gregarious or scattered, ovate-globose, roughish, 250–300 μ diam., with a papilliform or obtusely-conical ostiolum, texture subcarbonaceous, base slightly sunk in the wood, which is blackened and overrun with a species of *Dendryphium* (*D. crustaceum*, E. & E.) having simple, septate, erect hyphæ 90–115 x 7–8 μ, bearing terminal, subcatenulate, oblong or cylindrical, 2–4-septate, 25–45 x 10–12 μ conidia. Sometimes the hyphæ are branched above, the short, spreading branches also bearing conidia. Asci sessile, paraphysate, oblong-cylindrical, 70–75 x 12–15 μ. Sporidia biseriate, clavate-oblong, often slightly curved, uniseptate, brown, 22–27 x 7–8 μ.

On rotten wood, Nebraska (Webber), and Kansas (Kellerman).

The Kansas specimens (*A. saprogena*, E. & K.) are not accompanied by any conidial growth.

A. defórmis, Ell. & Lang. (in Herb.)

Perithecia superficial or at least, with the base simply adnate, ovate, obtuse, roughish, black, about ⅓ mm. diam., at length broadly perforated or irregularly ruptured above, carbonaceous. Asci clavate-cylindrical, paraphysate, 75 x 7–8 μ (p. sp. about 65 μ long). Sporidia uniseriate, ovate-elliptical, brown, uniseptate and slightly constricted, 8–10 x 3½–4 μ.

On the bare wood of an old cedar post, Louisiana (Langlois).

A. decorticàta, (Cke. & Hark.)

Sphæria decorticata, Cke. & Hark. Grev. XIII, p 19.
Amphisphæria decorticata, Sacc. Syll. 6613.

Perithecia seriate or irregularly gregarious, nestling among the loosened fibres of the wood, at length superficial or nearly so, subglobose, black, rather less than ½ mm. diam. Ostiolum inconspicuous. Asci, including the slender base, 150 x 12 μ, with abundant paraphyses. Sporidia uniseriate, oblong or ovate-oblong, uniseptate and slightly constricted at the septum, 15–20 x 8–9 μ, brown.

On decorticated limbs of oak, California.

A. dothideáspora, Cke. & Hark. Grev. XIV, p. 9.

Perithecia gregarious or scattered, emergent, covered by the blackened epidermis, conic-convex, black and shining, ½ mm. diam., with a papilliform ostiolum. Asci p. sp. 65–75 x 10–12 μ, with abundant paraphyses. Sporidia mostly biseriate, ovate-oblong, brown, uniseptate, the septum being below the middle, slightly constricted at the septum, 15–20 (mostly 15–17) x 7–8 μ (30 x 12 μ Cke.)

On dead stems of *Mimulus glutinosus*, California.

The epidermis is not uniformly blackened, but only directly over the perithecia.

A. salebròsa, (C. & P.)

Sphæria salebrosa. C. & P. 29th Rep N. Y. State Mus. p. 61.
Amphisphæria salebrosa, Sacc. Syll. 2747.

Perithecia gregarious or crowded, globose, rough, black, depressed and umbilicate, pierced at the apex and faintly radiately sulcate, ½–¾ mm. diam. Asci cylindrical or clavate. Sporidia lanceolate, acute, uniseptate, brown, constricted at the septum, 35–50 x 8–10 μ. Paraphyses numerous, filiform.

On dead shrubs, New York State (Peck).

A. melaspérma, (Cke.)

Psilosphæria melasperma, Cke. Grev. VIII, p. 118.
Amphisphæria melasperma, Sacc. Syll. 2745.

Erumpent, at length exposed. Perithecia globose, black, smooth and somewhat shining. Ostiolum thick and subprominent. Asci cylindrical. Sporidia elliptic-acuminate, dark, nucleate, at length uniseptate and nearly black and opake, so that the septum is hardly visible, 35 x 12 μ.

On naked wood, New York (Gerard).

A. Oronoénsis, (E. & E.)

Sphæria Oronoensis, E. & E. Journ. Mycol. III, p. 117.

Perithecia scattered, subcarbonaceous, black, roughish, subsuperficial, the base only slightly sunk in the wood, small (about one sixth mm.), globose or slightly depressed-globose. Ostiolum papilliform. Asci linear, 75 x 5 μ (spore-bearing part about 50 μ long), surrounded with abundant paraphyses. Sporidia uniseriate, oblong-elliptical, brown, uniseptate but not constricted, 6–8 x 2½–3 μ, cells equal or the lower one a little narrower.

On rotten wood, Orono, Maine (Harvey).

Apparently near *Sphæria sardoa*, De Not.

A. botulæspora, (Cke.)

Sphæria botulæspora, Cke Texas Fungi, No. 133.
Amphisphæria botulispora, Sacc Syll. 2719.

Perithecia gregarious, semiimmersed, slightly compressed laterally, about ⅓ mm. diam. Ostiolum papilliform, perforated. Asci saccate. Sporidia straight or slightly curved, obtuse at the ends, uniseptate, the upper cell being twice the length of the lower, constricted at the septum, brownish, 70–80 x 12 μ.

On old rails, Houston, Texas (Ravenel).

Specimens of this species in our Herb. from Ravenel, are sterile or at most afford only a few hyaline, oblong, 3-nucleate, stylospores 7–8 x 2½ μ.

A. phileùra, (C. & P.)

Sphæria phileura, C. & P. Grev. V, p 55, pl. 81, fig. 6.
Amphisphæria phileura, Sacc. Syll 2774, Cke. Syn. 3443.

Perithecia punctiform, small, scattered, orbicular, depressed, semi-immersed, black. Ostiolum simple, pierced. Asci clavate. Sporidia biseriate, elliptical, brown, uniseptate and slightly constricted at the septum, 22 x 10 μ.

On bark of *Tilia Americana*, New York State (Peck).

A. tumulàta, (Cke.)

Sphæria tumulata, Cke. Grev. VII, p. 4.
Amphisphæria tumulata, Sacc. Syll. 2751.

Perithecia immersed, subglobose, black. Ostiola erumpent. Asci cylindrical. Sporidia sublanceolate, obtuse, uniseptate, dark brown, 35–40 x 12 μ, with a hyaline apiculus at each end.

On wood of *Pinus contorta*, Sierra Nevada Mts., California, alt. 8000 feet.

A. thùjina, (Pk.)

Sphæria thujina, Pk. 27th Rep. p. 110.
Amphisphæria thujina, Sacc. Syll. 2750, Cke. Syn. 3446.

Perithecia scattered, nearly free, hemispherical or conical, slightly rugulose, thin and fragile. Ostiola at first slightly papilliform, then perforated. Sporidia large, uniseptate, oblong-elliptical, constricted at the septum, colored, 36–45 μ long.

On decaying wood of *Thuja occidentalis*, Adirondack Mts., New York.

As no mention is made of any asci, this, as far as the description goes, may be only a *Diplodia.*

A. æ̀thiops, (B. & C.)

Sphæria æthiops, B. & C. Grev. IV, p. 143.
Amphisphæria æthiops, Sacc. Syll. 2731, Cke. Syn. 3436.

Perithecia jet-black, subglobose, with a small, nipple-like ostiolum. Asci clavate or oblong. Sporidia clavate-oblong, uniseptate, sometimes with a gelatinous coat, 33 μ long.

On old logs, mountains of New York.

Nothing is said of the color of the sporidia, but Cooke (l. c.) places this in *Amphisphæria*, with species having brown, uniseptate sporidia.

A. albomáculans, (Schw.)

Sphæria albomaculans, Schw. Syn. N. Am. 1592.
Amphisphæria albomaculans, Cke. Syn. 3449.

Perithecia scattered on a well-defined white spot, very black and generally surrounded at the base by green, pulverulent matter, irregularly hemispherical, of medium size, semiimmersed, punctate, acutely papillate. Sporidia (sec. Cke. Grev. XVI, p. 91) uniseptate, brown, 12 x 3 μ.

Schweinitz also observes that often the ostiola of 2–3 perithecia are, as it were, connate, united into a single ostiolum, round or variously compressed and finally regularly perforated or subrimose.

On dead decorticated trunks of *Syringa*, Bethlehem, Pa. (Schw.)

A. aquática, (E. & E.)

Sphæria aquatica, E. & E. Bull. Torr. Bot. Club, XI, p. 42.
Amphisphæria aquatica, Sacc. Syll 6617, Cke. Syn. 3450.

Perithecia scattered, membranaceous, buried or suberumpent, 250 μ diam., subglobose or slightly elongated, with a papilliform ostiolum. Asci narrow-cylindrical, 150 x 8–10 μ, with abundant paraphyses. Sporidia uniseriate, oblong, uniseptate and slightly constricted at the septum, straight or slightly curved, pale at first with a single large nucleus in each cell, at length dark brown, 15–20 x 6–7 μ.

Inside a cedar water-pail in constant use, Newfield, N. J., 1883.

A. Langloisii, E. & E. in Herb.

Perithecia gregarious, hemispherical, 1 mm. diam., base sunk in the wood, black, roughish, thin and brittle. Ostiolum prominent, stout, tuberculo-conical, obtuse and at length perforated. Asci clavate-cylindrical, p. sp. 100–125 x 15–20 μ, with filiform paraphyses. Sporidia overlapping-uniseriate, or biseriate above, fusoid-oblong, uniseptate and subconstricted, brown, subinequilateral, 30–40 x 10–12 μ.

On a decaying log of *Carya* (?), Louisiana (Langlois 2171).

The perithecia resemble those of *A. umbrina*, but the sporidia are much larger.

A. Plátaní, E. & E. in Herb.)

Perithecia scattered, convex, hemispherical, about ½ mm. diam., covered by the slightly blackened epidermis which is pierced by the tuberculo-papilliform ostiolum. Asci stipitate, p. sp. 40–50 x 10–12 μ, with abundant paraphyses. Sporidia oblong-ovate or oblong-elliptical, brown, uniseptate, slightly constricted, sometimes slightly curved and slightly attenuated at the ends, 11–15 x 4–5 μ, crowded-biseriate, pale brown at first, then dark brown.

On loose bark of *Platanus*, Louisiana (Langlois 2213).

Differs from A. *fallax*, *A. umbrina* and *A. pseudo-umbrina*, in its smaller, biseriate sporidia, and from the last two, in its smaller, covered perithecia. Allied to *A. dothideaspora*, Cke. & Hark.

A. quercètis, Cke. & Mass. Grev. XVI, p. 92.

Perithecia gregarious, large (1 mm. diam.), at first immersed then semiemergent, conic-convex, for a long time covered by the epidermis, papillate, black, flattened at base. Asci cylindrical, 8-spored. Sporidia uniseriate, elliptical, rounded at the ends, slightly constricted and uniseptate in the middle, brown, 28 x 10 μ.

On bark of *Quercus alba*, North Carolina (Curtis).

A. conférta, (Schw.)

Sphæria confertula, Schw. Syn N. Am. 1508.
Sphæria conferta, Schw. Syn. Car. 187, Fr. S. M II, p. 444.
Byssosphæria (*Amphisphæria*) *conferta*, Cke. Grev. XV, p. 81.

Perithecia crowded, globose, granular, depressed at the apex and subpapillate, very black, seated on a radiate-fibrose, creeping mycelium. When the perithecia occur singly, seated in the center of the radiate subiculum, they present a very elegant appearance. Sporidia (sec. Cke. Grev. XV, p. 81) uniseptate, brown, obtuse at the ends, constricted in the middle, 12 x 4 μ.

On dead *Laurus Benzoin*, North Carolina (Schweinitz).

A, papillàta, (Schum.)

Sphæria papillata, Schum. Enum. Sæll. II. p. 161.
Amphisphæria papillata, De Not. Sfer. Ital. 68, fig. 71, Sacc. Myc Ven. Spec. p. 112, tab XI, fig. 24.

Perithecia scattered or sometimes 2 or 3 confluent, half sunk in the matrix, spherical, depressed around the papilliform ostiolum, about 1 mm. diam., hard, smooth, dark brown, clothed with a thin, floccose coat. Asci cylindrical, 8-spored, with filiform paraphyses. Sporidia elliptical, sometimes rather acute at the ends, uniseptate, brown, not constricted, 15 x 8 μ.

Common on wood of *Salix*, Bethlehem, Pa. (Schw.)

A. applanàta, (Fr,)

Sphæria applanata, Fr. S. M. II, p. 463.
Trematosphæria applanata, Fckl. Symb. p 162.
Amphisphæria applanata, Ces. & De Not. Schema. 224.
Exsicc. Fckl. F. Rh. 932.

Perithecia scattered, black, innate-superficial, flattened at base, smooth, convex, at length collapsing to plano-concave, rugose under the lens, sometimes confluent. (Asci)? Sporidia oblong, unequally uniseptate, upper cell smaller, lower larger and ventricose, acuminate, 24 x 8 μ.

Frequent on decorticated trunks of *Robinia*, Bethlehem, Pa. (Schw.)

A. Wellingtòniæ, (Cke. & Hark.)

Sphæria (*Amphisphæria*) *Wellingtoniæ*, Cke & Hark. Grev XIII, p 19.
Amphisphæria Wellingtoniæ, Sacc. Syll. 6615.

Perithecia gregarious, immersed, black, elongate-compressed, hysteriiform, striate, opake, 150–180 μ long, 80 μ wide, perforated above. Asci cylindrical, 8-spored. Sporidia elliptical, uniseptate, brown, scarcely constricted, each cell uninucleate, 12–14 x 8 μ.

On bleached wood of *Sequoia*, California.

TREMATOSPHÆRIA, Fckl.

Symb p. 161, emend Sacc. Syll. II, p 115

Perithecia simple, carbonaceous, hard, superficial or subsuperficial, at length broadly perforated above, conical or globose, generally large. Asci 8-spored. Sporidia oblong or fusoid, 3- or more-septate.

The genus as defined in Symb. Myc. included species with the sporidia 1–3-septate (*Amphisphæria*, in part).

Tr. pertùsa, (Pers.) **(Plate 22)**

Sphæria pertusa, Pers. Syn. p. 83
Trematosphæria pertusa. Fckl Symb. p 162.
Exsicc. Fckl. F. Rh. 931

Perithecia scattered, mostly only slightly sunk in the matrix or simply adnate, spherical-conoid, black, hard, subrugose, rather less than ¾ mm. diam. Ostiolum mostly conical and soon deciduous, leaving the perithecium perforated. Asci elongated, 110–140 x 15–20 μ. with filiform paraphyses. Sporidia biseriate or obliquely uniseriate, elongated-biconical, unequally uniseptate, finally 3-septate and more or less constricted at the septa, brown, 24–32 x 8–10 μ.

Schweinitz (Syn. N. Am. p. 214) says this species is common in Carolina and Pennsylvania. It is also mentioned by Peck, in 22d Rep., as found in New York, but we have never seen any American specimens which we could refer to this species.

Tr. nucleària, (De Not.)

Sphæria nuclearia, De Not. Micr. Ital. Dec. 9. p 462, fig. IV.
Hypoxylon nucitena, B & C Grev. IV, p. 52.
Sphæria caryophaga, Schw. Syn. N. Am. No. 1594.
Trematosphæria nuclearia. Sacc. Syll. 3308.
Sphæria Curtisii, Berk. in Curtis Cat. p 145.

Perithecia superficial, with their bases slightly sunk in the matrix, and connected by a thin, black, carbonaceous crust, hemispherical, rough, black, with a papilliform ostiolum. Asci clavate-cylindrical, 60–70 x 7–8 μ (p. sp.), with filiform paraphyses. Sporidia irregularly biseriate, oblong, slightly curved, narrowed and rounded at the ends, 12–16 x 4–5 μ, biseptate, end cells subhyaline, middle cell with a broad, black band across the center.

On decaying hickory nuts, Pennsylvania (Everhart).

Tr. penióphora, (Cke.)

Conisphæria peniophora, Cke. Grev. VIII, p. 119.
Trematosphæria peniophora, Sacc. Syll. 3292.

Perithecia scattered, black, conical, flattened at the base, smooth. Asci ample, clavate. Sporidia fusiform, constricted in the middle and then faintly 3–5-septate, acuminate at each end, brown, 100 x 14 μ.

On bark, New York (Gerard).

The sporidia remain for some time hyaline, with a granular endochrome and uniseptate, but at length become pale brown and faintly 3–5 septate.

Tr. subcollápsa, (E. & E.)

Lophiostoma subcollapsum, E. & E. Journ. Mycol. II, p. 100.

Perithecia cartilagino-membranaceous, black, globose, $\frac{3}{4}$–1 mm. in diam., buried in the substance of the bark, the epidermis slightly elevated and blackened over them, and pierced by the papilliform ostiolum, which finally collapses. Asci clavate-cylindrical, about 150 x 12 μ, with abundant paraphyses. Sporidia obliquely uniseriate or more or less distinctly biseriate above, oblong-elliptical, 20–26 x 6–10 μ, or regularly elliptical, 20–22 x 12–15 μ, hyaline at first, becoming brown and 3–7-septate, but not constricted at the septa. The sporidia are sometimes a little curved, or at least more prominent on one side. The ostiolum is quite inconspicuous and only slightly prominent. Sometimes one or two of the cells of the sporidia are divided by a longitudinal septum

On outer bark of living *Nyssa multiflora*, Newfield, N. J.

This evidently belongs more properly in *Trematosphæria*, on account of the inconspicuous, deciduous ostiolum.

Tr. mastoidea, (Fr.)

Sphæria mastoidea, Fr. S. M. II, p 463.
Sphæria revelata, B. & Br. Not. of Brit. Fungi, No. 634, tab. 11, fig. 18.
Sphæria Opuli, Fckl Symb. p. 115, tab. 111, fig. 24.
Melomastia, Friesii, Nits. in Fckl. Symb. Nachtr. I, p. 18.
Trematosphæria mastoidea, Winter Die Pilze p. 274.
Exsicc. Fckl. F. Rh. 2322.—Rab. F. E. 1937?

Perithecia scattered or gregarious, at first buried in the matrix, with only the ostiolum projecting, finally erumpent and nearly, but not entirely, superficial, spherical or subspherical, with a distinct, short-conical, perforated ostiolum, smooth and bare, $\frac{1}{2}$–1 mm. diam. Asci cylindrical, stipitate, 8-spored, 130–150 x 8–9 μ, with abundant, filiform paraphyses. Sporidia uniseriate, oblong, rounded at the ends, 2-septate, hyaline, 15–20 x 6–8 μ.

In Grev. IV, p. 144, Berkeley reports this from North Carolina, on *Fraxinus*, with sporidia shortly fusiform, 3-septate, 12½–15 μ long. The description and synonyms above are from Winter's Pilze. We have seen no specimens. This species appears to differ from the other members of the genus in its hyaline sporidia.

Tr. seminúda, (Pers.)

Sphæria seminuda, Pers Syn p. 70.
Trematosphæria seminuda, Fckl Symb p. 162.

Perithecia scattered or 2–3 connate, about half sunk in a white, felt-like subiculum, the upper, projecting part bald, with a sharp, conical ostiolum. Asci elongated-cylindrical, 8-spored. Sporidia uniseriate, brown, ovate or oblong, obtuse above and broader, the lower end attenuated, 3-septate, constricted at the septa, 14 x 6 μ.

Found (sec. Schw.) on bark, Carolina and Pennsylvania.

The above description is from Winter's Pilze. This must not be confounded with *Teichospora seminuda*, De Not. (Syll. II, p. 297).

CARYÓSPORA, De Not.

Micromycetes Ital. Dec. IX.

Perithecia as in *Trematosphæria*. Asci large, 2–8-spored. Sporidia large, biconical, with a snout-like (often 1–3-septate) prolongation at each end. This is hardly more than a subgenus of *Trematosphæria*.

C. putáminum, (Schw.) **(Plate 24)**

Sphæria putaminum, Schw Syn. Car No. 163.
Caryospora putaminum, De Not. Micr Ital. IX.
Exsicc. Rab-Winter F Eur. 3343 —Ell. N. A. F. 898.

Perithecia scattered, adnate-superficial, about 1 mm. diam., hemispherical, with a large tuberculiform, broadly perforated and at length deciduous ostiolum, carbonaceous, smooth or concentrically wrinkled, black. Asci broad-oblong, ventricose, stipitate, paraphysate, (280–340 x 70 μ, Winter). Sporidia overlapping-uniseriate, 2–8 in an ascus, biconical or broad-elliptical, constricted and septate in the middle, brown, 80–100 x 40–55 μ (108–140 x 50–65 μ, Winter). The prolongation at the ends is often obscurely 1–3-septate, but these additional septa are easily overlooked and are not always present. The sporidia are often surrounded by a broad hyaline envelope.

On old peach pits lying on the ground. Common in the peach region.

The asci are very evanescent, so that it is difficult to find an entire ascus containing mature sporidia. Those we have seen were 150–200 x 55–60 μ, but they were immature. Occasionally a peach pit is met with, on which all the perithecia produce 8-spored asci, but in this case the sporidia are smaller, 30–50 x 18–20 μ. There are generally but two of the larger sporidia in an ascus.

C. callicárpa, (Curr.)

Sphæria callicarpa, Curr. Linn. Trans. XXII, p. 321. tab. 58 fig. 62.
Caryospora callicarpa, Fckl. Symb. p. 163.

Perithecia scattered, broadly adnate at base, ¾–1 mm. diam., carbonaceous, brittle, dull black, with a prominent, obtuse, tuberculiform-conical ostiolum, which is finally deciduous, leaving the perithecium broadly perforated above. Asci broad-clavate, paraphysate, about 200 x 50 μ (210–260 x 60–70 μ, Winter). Sporidia subbiseriate, broad-fusoid, with ends mostly curved in opposite directions, subhyaline, 75–85 x 20–25 μ, 3–5-septate at first, finally broad-elliptical, with the ends narrowly pointed, brown, nearly opake, 100–112 x 40–50 μ.

On bark of decaying *Platanus*, Long Island, N. Y. (Zabriskie) and on dead birch limbs, Hull, Canada (Macoun).

The young sporidia are mostly distinctly constricted at the middle septum, but when mature, they are scarcely constricted, and the color becomes so dark that the septa can hardly be seen. The asci in the specc. examined did not seem well developed, and it is probable that the measurements given by Dr. Winter are none too large.

C. Langloísii, E. & E. Journ. Mycol. IV, p. 79.

Perithecia gregarious, nearly superficial, their bases slightly sunk in the matrix, depressed-conical, large (nearly 1 mm. across), dull black, with a distinct papilliform ostiolum. Asci broad-oblong or narrow-elliptical, subsessile, 120–140 x 40–45 μ, 8-spored, with abundant filiform paraphyses. Sporidia crowded in the asci, somewhat almond-shaped or acutely elliptical, uniseptate and slightly constricted at the septum, ends obtusely pointed, yellowish-hyaline at first, soon dark brown, 35–45 x 16–20 μ and 3-septate.

On old canes of *Arundinaria*, Louisiana (Langlois).

The two additional septa (one near each end of the sporidium) do not appear at first.

OHLÈRIA, Fckl. (Plate 23)

Symb. p. 163.

Perithecia adnate-superficial, carbonaceous, subglobose, papillate, ostiolum soon perforated. Often seated on a thin stroma or on the blackened surface of the wood. Asci elongated, stipitate, 8-spored, paraphysate. Sporidia biseriate, oblong, 3-septate, brown, constricted and separating at the middle septum. Differs from *Trematosphæria* principally in its sporidia separating in the middle.

O. rugulòsa, Fckl. l. c.

Exsicc. Ell. N. A. F. 694.

Stroma inconspicuous or none. Perithecia gregarious, adnate-superficial, conic-globose, about ½ mm. diam., carbonaceous, black, sub-velutinous below. Ostiolum conic-tuberculiform, soon perforated. Asci clavate-cylindrical, stipitate, 80–110 x 7–8 μ, paraphysate. Sporidia biseriate, fusoid-oblong, 3-septate, brown, constricted and separating in the middle, 16–20 x 4–4½ μ (14–16 x 4½–5 μ, Winter and Sylloge).

On the hard wood of a decaying oak stump, Newfield, N. J. Found also in Florida by W. W. Calkins.

O. modèsta, Fckl. l. c.

Exsicc. Fckl. F. Rh. 2173.—Ell N. A F. 694 (b)

Differs from the preceding in its distinct, pulverulent-velutinous subiculum and rather smaller sporidia (14–16 x 3½–4 μ).

On a decaying oak stump, Newfield, N. J.

WINTÈRIA, Rehm.

Ascom. Diag No 286.

Perithecia rather soft, membranaceous, not carbonaceous, greenish- or reddish-parenchymatous, generally collapsing to cup-shaped. Ostiolum perforated. Sporidia hyaline, 2- or more-septate, and sometimes subfusiform.

W. cœrùlea, E. & E. Journ. Mycol. I, p. 91.

Perithecia scattered, membranaceous, flattened, ⅓–½ mm. diam., covered by the thin epidermis, through which they are plainly visible. Ostiolum broad-papilliform, obtuse, collapsing when dry so that the perithecia appear umbilicate. Asci 75–114 x 16–17 μ, oblong-cylindrical, abruptly contracted below into a short, stout base, and surrounded by filiform paraphyses. Sporidia 8 in an ascus, broad fusiform or clavate-fusiform, narrowed below into an acute, awl-shaped base, yellowish, multi- (8–12-) septate and submuriform, 30–35 x 7–8 μ.

On bark of some living coniferous tree, Washington (leg. Suksdorf, com. C. J. Sprague).

W. rhòina, E. & E. Journ Mycol. I, p. 92. **(Plate 23)**

Exsicc. Ell. & Evrht. N. A. F. 2d Ser. 1669.

Perithecia erumpent, densely gregarious, subseriate, subglobose, black (⅓–½ mm.), membranaceous, thin and collapsing so as to become concave or patelliform. Ostiolum papilliform and mostly 4–5-stellate-cleft. Asci 45–60 x 7–8 μ, broadest in the middle. Paraphyses stout, linear, nucleolate. Sporidia biseriate, fusiform, yellowish, nucleolate, straight or slightly curved, sometimes strongly so, 20–25 x 2½–3 μ.

On weather-beaten wood of *Rhus copallina*, Newfield, N. J.

W. tuberculifera, E. & E. Proc. Acad. Nat. Sci. Phila., Pa. July, 1890, p. 240.

Perithecia gregarious, superficial, ½ mm. diam., depressed-globose, narrowed below, tubercular roughened, collapsed and cup-shaped when dry, black. Asci 35–40 x 5–6 μ. Sporidia crowded-biseriate, fusoid-oblong, hyaline, 2–4-nucleate, 6–8 x 2–2½ μ, (becoming 1–3-septate)?

On bark of wild plum (*Prunus*), London, Canada (Dearness).

Var. *cæspitosa*, on *Cerasus Virginiana*, has the perithecia collected in dense tufts 1½–2½ mm. across, surrounded by the ruptured epidermis, and the sporidia slightly curved, 6–10 x 2–2½ μ, 2-nucleate.

W. crustòsa, E. & E. Journ. Mycol. I, p. 149.

Perithecia membranaceous, ⅓–½ mm., depressed-hemispherical, tuberculose, and roughly laciniate-cleft above, seated on and partly connected by a thin, crustose, black subiculum more or less distinct. Asci clavate-cylindrical, 65–80 x 10 μ. Paraphyses soon resolved into a mass of granular matter. Sporidia biseriate, fusiform, 20–25 x 4–5 μ, yellowish or hyaline, with a faint, gelatinous envelope; endrochrome divided in the middle, exceptionally 2-times divided. The perithecia collapse when dry, so as to resemble a *Peziza* with an obtuse margin. Ostiolum not very conspicuous, papillose-conical, 4–5-stellate-cleft. Allied to *Winteria ordinata*, (Fr.), but differs in its shorter, mostly uniseptate sporidia, and depressed perithecia.

On decorticated oak, West Chester, Pa.

TEICHÓSPORA, Fckl.

Symb. p. 160, Sacc. Syll. II, p. 290.

Perithecia scattered or gregarious, superficial or with the base sunk in the matrix, spherical or ovoid, coriaceous or carbonaceous, frequently collapsing. Ostiolum papilliform or inconspicuous. Asci cylindrical or clavate-cylindrical, 8-spored, paraphysate. Sporidia ellipsoid, muriform, brown or (in the subgenus *Teichosporella*) hyaline. Lignicolous or corticolous.

Cooke and Winter make *Teichospora* a synonym of *Strickeria*, which latter name seems to have precedence, but we have retained the generic name *Teichospora*, which Saccardo has adopted in his Sylloge.

A. *Perithecia not collapsing, sporidia colored.*

Teich. obdùcens, (Fr.)

Sphæria obducens, Fr. S. M. II, p 456.
Sphæria plateata, Curr. Linn. Trans. XXII, p 318. tab. 57, fig. 35.
Sphæria miskibutris, De Not. Schema p. 47.
Teichospora obducens, Fckl. Symb p. 161.
Strickeria obducens, Winter Pilze, 3207.
Exsicc Fckl, F. Rh. 2024 —Rabh. F. Eur 638.—Rehm Asc. 42.

Perithecia gregarious or crowded, subsuperficial, ovoid-globose. unequal, rigid, rugulose, glabrous, with a papilliform ostiolum, black, about ⅓ mm. diam. Asci briefly pedicellate, clavate-cylindrical, 150–180 x 18–22 μ, 8-spored. Sporidia uniseriate or subbiseriate, sub-ellipsoid, 5–7-septate and muriform, constricted in the middle, brownish, 20–30 x 9–12 μ, with filiform paraphyses.

On Sassafras, mountains of New York, and on alder (Berk. in Grev. IV, p. 142), on *Viburnum*, London, Canada (Dearness), var. *pinea*, Sacc. Syll. I, p. 206, on wood of *Abies*, Idaho (Dr. Eckfeldt).

Teich. Eucalýpti, (Cke. & Hark.)

Sphæria Eucalypti, Cke. & Hark. Grev. XIII, p. 20.
Teichospora Eucalypti, Sacc. Syll 7105.

Perithecia scattered, subsuperficial, small, subglobose, black. Asci cylindrical, 8-spored. Sporidia uniseriate, elliptical, slightly constricted, 5–7-septate and muriform, brown, 20–22 x 8 μ.

On rotten wood of *Eucalyptus*, California.

Teich. interstitiàlis, (C. & P.)

Sphæria (*Denudatae*) *interstitialis*, C. & P. 29th Rep. N. Y. State Mus. p. 61.
Teichospora interstitialis, Sacc. Syll. 3885, Cke. Syn. 3501.

Perithecia gregarious, at first semiimmersed, always apparently so by nestling among the fibers of the wood, subglobose, pierced at the apex, black, 300–500 μ diam. Asci cylindrical. Sporidia uniseriate, polymorphous, triseptate, with occasional longitudinal septa, deeply constricted, brown, 30–35 x 12–16 μ.

On decorticated wood of cherry, Greenbush, N. Y. (Peck).

Teich. insecùra, (Ell.)

Cucurbitaria insecura, Ell. N. A. F. 882.

Perithecia scattered or subgregarious, globose-depressed, base sunk in the matrix, apex obtuse and minutely papillate, coriaceous, black, smooth, ½ mm. diam. Asci cylindric-clavate, paraphysate, 8-spored, 130–135 x 18 μ. Sporidia obliquely uniseriate, obovate-elliptical, constricted in the middle, narrower below, 24–26 x 8–10 μ, 3-septate-muriform, brown.

On partly decorticated and blackened limbs of willow, Pleasant Valley, Utah (S. J. Harkness).

Teich. solitària, (Ell.)

Cucurbitaria solitaris, Ell. in Bull. Torr. Bot. Club, VIII, p. 125.

Perithecia solitary, cylindric-ovate, rough, about ½ mm. diam. and ¾ mm. high, with a strongly papilliform ostiolum. Asci cylindrical, 125–130 x 17–18 μ. Sporidia uniseriate (mostly), oblong-elliptical, constricted in the middle, uniseptate and yellowish at first, soon becoming 3- or more-septate, dark brown and muriform, 25–33 x 12–13 μ.

On sage brush (*Artemisia*) Utah.

Teich. xeróphila, Sacc. Syll. 3907.

Teichospora aridophila, Pk. Bot. Gaz. VII, p. 57.

Perithecia minute, 250–300 μ diam., scattered, hemispherical or depressed, black, with a minute, papilliform ostiolum. Asci subcylindrical, 112–120 x 28–30 μ. Sporidia crowded or biseriate, oblong or obovate, slightly constricted in the middle, muriform, colored, 28–35 x 12–15 μ.

On bleached wood, Arizona.

Allied to *T. obducens*. (Name changed by Saccardo (Syll. II, p. 299) from *aridophila* to *xerophila*).

Teich. mammoìdes, E. & E. Proc. Acad. Nat. Sci. Phil., Pa., July, 1890, p. 242.

Perithecia erumpent-superficial, gregarious, depressed-hemispherical, brownish-black, ¾ mm. diam., with a prominent, nipple-like, black ostiolum. Asci clavate-cylindrical, subsessile, 100–110 x 12–15 μ, with abundant filiform paraphyses. Sporidia uniseriate, ovate-oblong, 5–7-septate, and muriform, scarcely constricted, yellow, becoming brown, 20–22 x 9–11 μ.

On dead stems of *Sarcobatus vermiculatus*, Montana (Anderson).

Teich. mycógena, E. & E. l. c.

Perithecia scattered, immersed, except the partially erumpent apex, which slightly raises the surface of the *Diatrype* stroma, rupturing it in a subradiate manner. Perithecia of medium size with an indistinct ostiolum. Asci subcylindrical about 100 x 12 μ, abruptly contracted below into a short stipe. Sporidia biseriate, ovate-oblong, with three distinct transverse septa, and a longitudinal septum across one or more of the cells, yellowish, becoming dark brown, distinctly constricted at the middle septum, and, when mature, 5–6-septate, 12–15 x 6–8 μ. This might be mistaken for *Lophiostoma Floridanum*, E. & E., but it has the perithecia more superficial and quite different sporidia.

Parasitic on old *Diatrype stigma*, Newfield, N. J.

Teich. umbonàta, E. & E. l. c.

Perithecia gregarious, discoid, about ½ mm. diam., seated on the surface of the inner bark exposed by the falling away of the epidermis. Ostiolum tuberculiform. Asci cylindrical, 75–80 x 7–8 μ, with paraphyses. Sporidia uniseriate, ovate, 3-septate, constricted at the middle septum, straw-yellow, 12–15 x 6–8 μ. Most of the sporidia show only the three transverse septa, but in some of them, one or both the inner cells are divided by a longitudinal septum. It is not improbable that the sporidia may finally become brown and acquire additional septa.

On dead branches of *Symphoricarpus occidentalis*, Montana (Kelsey).

Teich. megástega, E. & E. l. c. p. 243.

Perithecia gregarious, superficial, the base sunk in the wood or bark with about two-thirds of the upper part projecting, hemispheric-globose, ¾–1 mm. diam., rough, ostiolum inconspicuous, subpapilliform. Asci cylindrical, 175–200 x 15 μ; with a short, stipe-like base and abundant paraphyses. Sporidia uniseriate, about 7-septate and muriform, mostly constricted in the middle more or less distinctly, ends rounded or obtusely pointed, 25–36 x 12–15 μ. Closely allied to *T. obducens*, but perithecia less crowded, more depressed, larger and rougher, and sporidia rather larger.

On bark and wood of old weather-beaten willow and maple limbs (*Acer glabrum*), Montana (Kelsey).

Teich. Kansénsis, E. & E. l. c.

Perithecia scattered, minute, (120–175 μ), conic-hemispherical, base slightly sunk in the bark. Ostiolum papilliform. Asci oblong 75–80 x 12 μ, sometimes shorter and broader (45–50 x 15 μ). Sporidia biseriate, ovate-oblong, pale brown, 3-septate, finally 6-septate, and slightly constricted across the middle, lower end subacute, about 20 x 8–9 μ. *Teichospora pruniformis* (Nyl.) which is also found on bark of poplar and willow is much larger (½ mm. diam.)

On bark of cottonwood trees, Kansas. (Dr. Egeling).

Teich vetústa, (Ell.)

Sphæria vetusta, Ell. Bull. Torr. Bot. Club. VI, p. 135.
Strickeria vetusta, Cke. Syn. 3167.
Teichospora vetusta, Sacc. Syll. 3908.

Perithecia gregarious, superficial, of medium size, elongated-conical, thin, not polished, dull-black, with a depressed-hemispherical, black and shining ostiolum with a large opening. Asci broad cylindrical, obtuse, abruptly narrowed at base, 125x25 μ. Sporidia uniseriate, obtusely and broadly-elliptical, nearly colorless, uniseptate and more or less constricted at the septum when young, at length brown and fenestrate, 22–25 x 12–13 μ.

On a dead place in a living maple, Newfield, N. J.

Teich. papillòsa, E. & E. Proc. Acad. Nat. Sci. Phil., Pa., July, 1890, p. 242.

Perithecia gregarious, subsuperficial, depressed-globose, ⅓ mm. diam., strongly papillose-roughened, with a few short, weak, glandular hairs when young, finally collapsing above. Ostiolum papilliform, not conspicuous. Asci oblong, 75–85 x 20–24 μ, nearly sessile, paraphyses evanescent. Sporidia crowded-biseriate, 8 in an ascus, oblong or clavate-oblong, a little curved, obtuse at the ends. Mostly 5-septate, with one or two of the cells divided by a longitudinal septum, hyaline, becoming yellow-brown, 22–30 x 10–11 μ.

On weather-beaten, decorticated limbs of *Salix*, Helena, Montana (Kelsey).

Teich. Hélenæ, E. & E. l. c. **(Plate 23)**

Exsicc. Ell. & Evrht. N. A. F. 2d Ser. No. 2369.

Perithecia gregarious, semierumpent, ½–¾ mm. diam., granular-roughened, collapsing above. Ostiolum minute. Asci clavate-cylindrical, 112–120 x 10–12 μ, rather abruptly contracted below into a short, stipe-like base, and surrounded with abundant paraphyses. Sporidia uniseriate, ovate-oblong, brown, constricted in the middle, 5–7-

septate and with one or two of the intermediate cells divided by a longitudinal septum, 15–25 x 8–12 μ. Quite often asci are seen in which the sporidia are smaller, black and shriveled as if struck with blight before maturity. Closely allied to *T. patellarioides*, Sacc., but differs in its larger, globose-hemispherical perithecia without any fringe of hyphæ at the base, and in its 5–7-septate sporidia uniseriate in narrower asci.

On decorticated weather-beaten limbs of *Salix*, Helena, Montana (Kelsey), on wood of *Prunus Virginiana*, Montana (Anderson).

Teich. pygmæa, E. & E. Journ. Mycol. IV, p. 63.

Perithecia scattered, erumpent-superficial, depressed-globose, black, 200–225 μ diam., with a papilliform ostiolum, finally perforated. Asci clavate, 70–80 x 15–18 μ, subsessile, 8-spored, indistinctly paraphysate. Sporidia irregularly biseriate, ovate-elliptical, mostly 3-septate (exceptionally 5–6-septate), with one or two cells divided by a longitudinal septum, 15–20 x 6–8 μ, pale yellowish-brown. Seems nearly allied to *T. pruniformis*, Nyl., but smaller in all its parts and lacks the acute, conical ostiolum.

On bark of cottonwood trees, Kansas (com. Dr. J. W. Eckfeldt).

Teich. táphrina, (Fr.)

Sphæria taphrina, Fr. S. M. II, p 465
Teichospora taphrina, Fckl. Symb. Nachtr. I, p. 305.

Perithecia scattered, black, immersed, then emergent, subcompressed-elliptical, obtuse, with a simple, perforated ostiolum. Asci cylindrical, substipitate, 8-spored, 88 x 12 μ (p. sp.). Sporidia obliquely uniseriate, ovate or oblong ovate, sometimes curved, ends obtuse, generally irregularly 3-septate and muriform, slightly constricted at the septa, brown, 14 x 7–10 μ. Paraphyses filiform, abundant.

Found on old wood, Bethlehem, Pa. (Schw.).

Unfortunately, we can not find in the Schweinitzian herbarium any specimen of this species or of *T. seminuda*, (Pers.).

Teich. vìlis. (Fr.)

Sphæria vilis, Fr. S. M. II. p. 466.
Melanomma vile, Fckl. Symb. p. 160.
Strickeria vilis, Winter, Die Pilze, 3204, Cke. Syn. 3168.

Perithecia almost microscopical, punctiform, much scattered, smooth, glabrous, opake, emerging from the bleached fibers of the wood, and then almost superficial. Ostiolum very minute, deciduous.

Sporidia oblong, obtuse at each end, 3-septate, constricted at the septa, yellow, 12–15 x 4–5½ μ.

Not rare around Bethlehem, Pa. (sec. Schw.).

Dr. Winter states that the sporidia have one or both the middle cells divided by a longitudinal septum, and that the species can not, therefore, belong in *Melanomma*.

Teich. disseminàta. (B. & C.)

Sphæria disseminata, B. & C. Grev. IV, p. 142.
Teichospora disseminata, Sacc. Syll. 3884, Cke. Syn, 3500.

"Perithecia minute, scattered, semiimmersed, ovate when free, attenuated above. Sporidia biseriate, shortly fusiform, triseptate, constricted at the septa, sometimes divided vertically."

On bleached wood of *Liquidambar* and on oak posts, Carolina.

B. *Perithecia collapsing, sporidia colored.*

Teích. Kóchii, Korb. Parerga lich. p. 400.

Cucurbitaria Rabenhorstii, Awd. in Rab. F. E. 758.
Teichospora Rabenhorstii, Sacc. Syll. 3911.
Teichospora pezizoides, Sacc. & Speg. Mich. I, p. 350, Sacc. Syll. II, p. 300.
Exsicc. Sacc. M. Ven. 1270.—Ell. N. A. F. 112, partly, at least in some copies.

Perithecia scattered, superficial, or subgregarious, at first globose, soon collapsing to concave or cup-shaped, 250–300 μ diam., minutely rugulose, black. Ostiolum papilliform, minute. Texture of the perithecia rather soft, parenchymatous, nearly black. Asci cylindrical, attenuate-stipitate, rounded and thickened at the apex, 110–115 x 10 μ, with filiform paraphyses, 8-spored. Sporidia obliquely uniseriate, oblong-elliptical, 18–24 x 6–7 μ, rather obtuse at the ends, 3-septate and constricted at the septa, sparingly muriform, olivaceous, with the extreme cells paler.

On outer bark of living *Robinia*, New Jersey.

Teich. muricàta. E. & E. Bull. Wash. Coll. vol. I, p. 5.

Perithecia superficial, hemispherical, ⅓–½ mm. diam., scattered or subgregarious, olivaceous when fresh, dull black and collapsing above when dry, muricately roughened, and often obscurely radiate-sulcate around the prominent and very slightly compressed ostiolum. Asci 125 x 15 μ, very evanescent, with abundant, stout, granular paraphyses 2½–3 μ thick. Sporidia oblong-elliptical, 18–25 x 10–11 μ, 3-septate, and at length muriform and brown.

On bark of some tree, San Diego, Cal. (Orcutt).

Differs from *T. Kochii* in its rougher and larger perithecia less deeply collapsing above, and its rather larger asci and sporidia. When the perithecia fall away, or, on removing them with the point of a knife, a pale reddish spot is left on the bark where they stood.

C. *Perithecia not collapsing; sporidia hyaline*, (*Teichosporella*).

Teich. inflàta, (Ell.)

Sphæria inflata, Ell. Bull. Torr Bot. Club, VI, p 135.
Zignoella inflata, Sacc. Syll. 3633
Exsicc. Ell N. A. F No 98.

Perithecia scattered, carbonaceo-coriaceous, erumpent, subglobose, 200–300 μ diam., clothed, especially below, with coarse, strigose-spreading, grayish-brown hairs, finally collapsing above. Ostiolum papilliform, finally irregularly perforated. Asci at first oblong, finally obovate, 70–80 x 25–30 μ, nearly sessile, paraphysate. Sporidia 8 in an ascus, irregularly arranged, elliptical, obtuse, 3-septate with a longitudinal septum running through one or more of the cells, hyaline. 20–25 x 10 μ.

On wood of oak railroad ties, Newfield, N. J.

The sporidia at first are simply 3-septate, but when mature they are as described above. In the original description this fact and the presence of the strigose hairs was overlooked. This differs from *Rhamphoria* in the absence of any beak on the perithecia.

Teich. phellógena, (B. & C.)

Sphæria phellogena, B. & C. Grev. IV, p 144.
Teichospora phellogena, Sacc. Syll 3921, Cke Syn. 3521.

"Perithecia half immersed, subglobose, with an obtuse, papilliform ostiolum. Sporidia biseriate, shortly fusiform, fenestrate, 30–34 μ long, nearly hyaline."

On corky bark of oak, Carolina?

D. *Perithecia bristly*, (*Pleosphæria*).

Teich. modésta, (Hark.)

Pleosphæria modesta, Hark Bull Cal Acad., 1884, p. 46.

Perithecia superficial, orbicular, black, 135 μ diam., with stout, black spines 40–50 μ long. Asci 8-spored, ovate or broadly fusiform, 45–50 x 15–20 μ. Paraphyses none. Sporidia elliptical, 3–4-septate, with two longitudinal septa, pale olive-brown, 10 x 6 μ.

On decorticated rotting branches of *Eucalyptus*, California.

Teich, chlorόspora, E. & E. (in Herb.)

Perithecia superficial, scattered, submembranaceous, minute (150–200 μ diam.), clothed with spreading, black bristles 50–80 μ long,

Asci obovate, 65–75 x 15–20 μ (p. sp. 40–50 x 15–20 μ), without paraphyses. Sporidia crowded, elliptical, 16–20 x 8–10 μ, about 5-septate with a more or less perfect longitudinal septum, greenish or olivaceous brown.

On decorticated oak limbs lying on the ground, and on the bark from which the epidermis had fallen off, Newfield, N. J., May, 1890.

Teich. microlóncha, (B. & C.)

Sphæria microloncha, B. & C. Grev. IV, p. 144.
Pleosphæria microloncha, Sacc. Syll. 3926.

Perithecia scattered, superficial, ovate, with a short neck, sprinkled with short setæ. Asci clavate. Sporidia biseriate, with about four horizontal septa and a few oblique or vertical ones.

On the inside of bark of *Liriodendron*, South Carolina.

The following species, reported by Schweinitz as having been found in this country, have not been met with here since his time.

Sphæria notha, Fr. S. M. II, p. 458. (*Sphæria abnormis*, Schw. Syn. Car. 235).—Crowded, broadly effused, black. Perithecia cylindrical, very small, shining, with a large pezizoid ostiolum.

On fallen wood, Carolina.

Sphæria nigro-brunnea, Schw. Syn. N. Am. 1563.—Perithecia scattered, dark brown, on dark brown spots, ovate, rugose, attenuated into an indistinct ostiolum, some very large mixed with smaller ones, at length often collapsing. At first partly covered by the fibers of the wood, but finally bare.

On decorticated posts of *Robinia*, Bethlehem, Pa.

Sphæria Hydrangeæ, Schw. Syn. N. Am. 1562.—Perithecia scattered, seated on the epidermis, depressed-globose, black, at length collapsing, with a persistent, papilliform ostiolum. On the same limbs are found other perithecia, apparently not specifically distinct, but erumpent from the inner bark, with ostiola one-third as large as the perithecia themselves, the flattened orifice piercing the epidermis, but otherwise entirely covered.

Rare, on dead limbs of *Hydrangea*, Bethlehem, Pa. (Schw.)

Sphæria aggregata, Schw. Syn. N. Am. 1561.—Perithecia twice as large as in *Sphæria transversalis*, densely crowded, but not confluent, flattened-globose, rather smooth, brown-black, generally ash-color around the short-cylindrical or conical, deciduous ostiolum.

On very rotten wood, in extensive patches. Rare. Bethlehem, Pa.

Sphæria inconstans, Schw. Syn. N. Am. 1564.—Larger than

Sph. nigro-brunnea and *Sph. Hydrangeæ*, bursting out in crowded patches through the fibers of the inner bark, very variable in shape, at length confluent. Perithecia globose or cylindrical, sometimes ventricose above, black, rugose. Ostiola more or less prominent, cylindrical, umbilicate. When young covered by the epidermis, and then, only the ostiola are visible; at length bare and often broken away, leaving only the pezizoid base.

Under the epidermis of *Rhododendron maximum*, Bethlehem, Pa. (Schw.)

This and the preceding species were mentioned on p. 163. We give here a translation of the diagnoses published by Schweinitz, but it is impossible to tell whether they belong here or in the *Fam. Trichosphærieæ.*

FAMILY. LOPHIOSTÒMEÆ.

Stroma none. Perithecia generally more or less buried at first, sometimes with only the ostiolum projecting, finally more or less emergent or even superficial, carbonaceous or subcoriaceous. Ostiolum laterally compressed, obtuse, sometimes forming an elevated ridge across the top of the perithecium and opening by an elongated slit instead of a round pore as in the other families of the *Sphæriaceæ.*

In this family the number of septa in the sporidia, at different stages of their growth, is unusually variable; the color also runs from hyaline, through various shades of brown, so that these two characters, unaccompanied as they are, by any distinctive vegetative peculiarities, do not seem to be of generic value; we have therefore followed Winter & Cooke, in reducing *Lophiosphæra* and *Lophiotrema* to mere subgenera.

LOPHIÓSTOMA. Ces & De Not.

Schema di Class. Sferiacei, p. 45.

Perithecia and ostiola as noted in the characters of the Family. Asci paraphysate. Sporidia oblong or subfusoid, 1-multiseptate, hyaline or brown.

A. *Sporidia 5–9-septate, brown.*

L. macróstomum, (Tode).

Sphæria macrostoma, Tode. Fungi Meckl. II, p. 12.
Lophiostoma macrostomum, Ces. & De Not. Schema. p. 45.
Exsicc. Fckl. F. Rh. 923.—Rab. F. E. 2040.

Perithecia scattered, at first buried, at length more or less emergent, or even nearly superficial, sphæroid, about ¾ mm. diam., with a broad, elliptical ostiolum, which is finally cleft with an elongated opening. Asci clavate, short-stipitate, 90–110 x 15–18 μ, 8-spored, paraphysate. Sporidia subbiseriate, broad-fusoid, brown, narrowed towards each end, sometimes curved, 5–9-septate, 40–55 x 9–12 μ.

On bark of dead oak, N. J., New York (Peck), Carolina and Pennsylvania (Schw.)

L. turrìtum, C. & P. Peck's 29th Rep. p. 64.—Grev. tab. 68. fig. 2.

Perithecia subgregarious, emergent, prominent, subglobose, black, with broad, compressed, truncate necks and elongated ostiola. Asci cylindrical or clavate. Sporidia oblong-elliptical, 5-septate, brown, 20–23 μ long.

On dead branches of *Salix*, New York.

L. excipulifórme, (Fr.)

Sphæria excipuliforme, Fr. S. M. II, p. 469.
Lophiostoma excipuliforme, Ces. & De Not Schema. p. 45.
Exsicc. Rehm. Asc. 238.

Perithecia scattered or gregarious, partly sunk in the matrix or nearly superficial, spherical, black, about 1 mm. diam., with a slightly compressed, prominent ostiolum obtuse and enlarged above. Asci cylindrical, short stipitate, 8-spored, 280–300 x 20–26 μ, with filiform paraphyses. Sporidia subbiseriate or uniseriate, oblong, 6–9-septate, not constricted, brown, with the end cells subhyaline, 44–66 x 15–18 μ, (Winter). The specimens in Rehm's Asc. 238, have asci (p. sp.) about 150 x 12 μ, sporidia 30–50 x 8–10 μ.

Var. *Abietis*, E. & E., Journ. Mycol. IV, p. 64, on bark of *Abies*, Cazenovia, New York (Underwood & Cook), differs from the usual type, on bark of deciduous trees, in its larger (60–75 x 12–16 μ) sporidia and its rather narrower ostiolum.

We have seen no American specimens of the normal form, on bark of deciduous trees.

L. macrósporum, Speg. Mich. I, p. 466, F. Ital. tab. 607.

Perithecia scattered, erumpent, sphæroid, 1 mm. diam., globose, dull black. Ostiolum compressed, extending $\frac{1}{3}$–$\frac{1}{2}$ way across the vertex of the perithecium. Asci broad clavate-cylindrical, 225–300 x 40–50 μ, with a short stipe and abundant, conglutinated paraphyses. Sporidia crowded-biseriate, oblong, slightly curved, obtuse, 7–9-septate and often slightly constricted in the middle, brown, the cells filled with large nuclei, so as to appear muriform, though not really so, 80–100 x 20–25 μ.

On outer bark of white oak, Newfield, N. J.

We have referred this to *L. macrosporum*, Speg., though we have not seen any appendages on the sporidia or any hyaline envelope, but as the specimens are mature, these may have disappeared.

L. eròsum, E. & E. Journ. Mycol. II, p. 44.

Perithecia buried in the substance of the wood, globose, ½–¾ mm. diam. Ostiola erumpent, narrow, only slightly prominent. Asci clavate-cylindrical, 90–100 x 12–15 μ, with abundant filiform paraphyses. Sporidia subbiseriate above, oblong-fusoid or subnavicular, about 5-septate, hyaline at first, then yellowish and finally nearly opake, mostly 20–25 x 7–8 μ, with a shriveled appearance. The perithecia are greedily eaten out by a small beetle. Allied to *L. scelestum*, C. & E., but with smaller sporidia. Differs from *L. macrostomoides*, De Not, in its immersed and smaller perithecia, and its somewhat smaller sporidia.

On decaying wood of *Salix*, Vineland, N. J., Canada (Dearness).

L. congregàtum, Hark. Bull. Cal. Acad., Feb., 1884, p. 47.

Perithecia semiimmersed, black, broad (¾–1 mm.). Ostiolum prominent, extending ¼–½ way across the perithecium. Asci clavate. 8-spored, 70 x 12 μ. Sporidia biseriate, fusiform, widest above the middle, curved, 6–8-septate, constricted, yellow-brown, cells minutely binucleate, end cells paler, 30 x 9 μ.

On decorticated branches of *Sambucus racemosa*, Sierra Nevada Mts., Cala.

L. scelèstum, C. & E.

Lophiostoma microstomum, C. & E. Grev. IV. p. 179 (non Niessl).
Sphæria pachyascus, C. & E. Grev. IV, p. 179, tab. 68, fig 1.
Lophiostoma scelestum, C. & E. (Cke. in literis), Sacc. Syll. 5479

Exsicc. Ell. N. A. F. No 96.

Perithecia scattered, immersed, coriaceous, globose or elliptical, ½–¾ mm. diam. Ostiolum alone erumpent, small, narrow and compressed. Asci clavate-cylindrical, paraphysate, 100–120 x 15 μ. Sporidia obliquely biseriate, oblong-fusoid, 5–7-septate, brown, 25–45 x 8–10 μ (40–50 x 10 μ, Cke.)

On decorticated, weather-beaten oak, maple, *Viburnum* and pear, Newfield, N. J., on decaying wood of pear tree, Helderberg Mts., N. Y. (Peck).

The sporidia are mostly 25–35 μ long and about 8 μ wide, more or less constricted, especially at the middle septum.

L. próminens, Pk. 31st Rep. p. 50.

Perithecia very prominent, hemispherical, adnate at the base, 180–200 μ broad, smooth, black. Ostiola distinct, compressed, black and shining. Asci subclavate. Sporidia biseriate, oblong or sub-fusiform, straight or slightly curved, colored, 5-septate, 20–25 μ long.

On dead twigs of *Cephalanthus occidentalis*, New York.

L. caùlium, (Fr.)

Sphæria caulium, Fr. S. M. II, p. 509.
Lophiostoma caulium, De Not, Sfer Ital. p 68, tab. 70.
Exsicc. Rehm Ascom. 181, 484, 749.—Sydow M March. 257, 985.

Perithecia scattered, immersed, black, minute, (generally on stems from which the epidermis has fallen away), spherical or subelliptical, with a slightly prominent, bare, ellipsoid or linear ostiolum. Asci clavate-cylindrical, p. sp. 70–80 x 10–12 μ, 8-spored, with abundant paraphyses. Sporidia biseriate, fusoid-oblong, uniseptate, acute and yellowish, finally 5–7-septate and darker, with the ends more obtuse, 20–30 x 5–8 μ (mostly 20–22 x 5–6 μ), straight or slightly curved, and without appendages.

On dead herbaceous stems, Louisiana and Pennsylvania, probably common throughout.

Distinguished from *L. insidiosum* by the absence of appendages on the sporidia, but certainly very closely allied to that species.

L pileàtum, (Tode.)

Sphæria pileata, Tode Fungi Meckl. II, p. 13 fig. 78.
Lophiostoma pileatum, Sacc. Syll. 5493, Cke. Syn. 3681.

Perithecia scattered, emergent, subglobose, black, with a flat, obconical ostiolum Sporidia oblong, attenuated at each end, 7-septate, dark brown, 46–50 x 10–12 μ (Sacc. l. c.)

Found in Carolina and Pennsylvania (sec. Schw.)

B. *Sporidia* 3–5-*septate, brown.*

L. triseptàtum, Pk. 28th Rep. p. 76.

Sphæria surrecta, Cke. Grev V, p. 94.

Perithecia scattered, globose, ¾ mm. diam., about ⅔ sunk in the wood, with the upper third projecting. Ostiolum narrow, compressed. Asci linear. Sporidia uniseriate, rarely crowded, oblong-elliptical, triseptate, colored, 15–18 μ long, constricted at the septa.

On decaying wood, Buffalo, N. Y. (Clinton), Lyndonville, N. Y. (Fairman, 184), Canada (Dearness).

Differs from *L. quadrinucleatum*, Karst., in its smaller sporidia constricted at the septa.

Var. *pluriseptatum*, E. & E., on decorticated maple limbs, Lyndonville, N. Y. (Fairman, 134) has asci (p. sp.) 70–75 x 10–12 μ. Sporidia irregularly biseriate or oblique, oblong or clavate-oblong, 3-septate and constricted at the septa, obtuse, brown, 15–20 x 5–6 μ. In well developed specimens, one or both the terminal cells are again divided by a septum, making the sporidia 3–5-septate.

The spec. of *Sphæria surrecta*, Cke., on poplar board in our Herb. is certainly the same as *L. triseptatum*, Pk., sec. specc. from Peck. The ostiola are narrow and at first often not at all compressed. The sporidia are a little larger, 15–20 x 7–8 μ (25 x 8 μ, Cke.), often a little narrower below, clavate-oblong, sometimes constricted at all the septa (torulose), but more distinctly so at the middle septum. *L. triseptatum*, Pk., should probably be referred to *L. quadrinucleatum*, Karst, as a small-spored var.

L. Prùni, E. & E. Journ. Mycol. IV, p. 64. and Trans. Roch. Acad. Sci. I, p. 49. **(Plate 25)**

Perithecia gregarious, sunk in the wood, ¼–⅓ mm. diam. Ostiola erumpent through the bark and strongly compressed. Asci clavate-cylindrical, 80–91 x 8–9 μ, with abundant paraphyses. Sporidia uniseriate, mostly 4 in an ascus, oblong, rounded at the ends, brown, 3-septate and constricted at the septa, a little narrower at the lower one, 18–22 x 6–8 μ. This can hardly be *L. acervatum*, Karst, which has the perithecia erumpent in small clusters (cæspites minutos) and sporidia scarcely constricted.

On *Prunus serotina*, Lyndonville, N. Y. (Fairman).

L. rhizóphilum, (B. & C.)

Sphæria rhizophila, B. & C. Grev. IV, p. 143.
Lophiostoma rhizophilum, Sacc. Syll. 5465.

"Perithecia scattered, ovate. Ostiolum short, transverse, linear. Asci clavate. Sporidia shortly fusiform, curved, pointed, 15 μ long, triseptate, at length brown."

On exposed roots, Pennsylvania (Michener).

L. pseudomacróstomum, Sacc. Mich. I, 339.—Berlese Icon. tab. 7. fig. 1.—id. F. Moricoli, fasc. II, No. 6.

Exsicc. Ell. & Evrht. N. A. F. 1629.

Perithecia gregarious, subimmersed, globose, black, ¾ mm. diam. Ostiolum compressed, emergent, rather broad. Asci cylindric-clavate, briefly stipitate, 110–115 x 14 μ, paraphysate, 8-spored. Sporidia biseriate or obliquely uniseriate, fusoid, 28–30 x 8–10 μ, straight or slightly curved, 5-septate, 6-nucleate, deep brown, occasionally one cell divided by a longitudinal septum.

On dead branches of *Lonicera Tatarica*, Lyndonville, N. Y. (Fairman 179), and on loose hanging bark of grape vines, Newfield, N. J. (issued in N. A. F. 1695 as *L. sexnucleatum*, Cke.)

The specimens from both these localities have the perithecia smaller than in the type and correspond to *L. vagans*, H. Fabr., which can hardly be more than a var. of *L. pseudomacrostomum.*

L. Thùjæ, E. & E. (in Herb.)

Perithecia scattered, buried, except the erumpent, rounded apex, with the narrow, compressed ostiolum extending ⅓ to ½ way across. Asci clavate-oblong, 50–70 x 12–15 μ. Sporidia biseriate, oblong-fusoid, 3-septate, slightly curved, obtuse, yellowish-brown, 14–16 x 3–4 μ.

On foliage of *Thuja occidentalis* partly dead but not yet fallen, Potsdam, N. Y. and London, Canada.

L. triseptatum, Pk. seems to differ in its linear asci and obtuse sporidia.

L. magnàtum, C. & P. 29th Rep. N. Y. State Mus. p. 64.

Perithecia subgregarious, semiimmersed, globose, rather large, somewhat thin and fragile, pitchy-black. Ostiola short, compressed. Asci cylindrical or clavate. Sporidia biseriate, lanceolate, constricted in the middle, 3–5-septate, 50–60 μ long.

On decaying wood, New York (Peck). Resembles *L. macrostomum* in habit.

L. Langloìsii, E. & E. l. c.

Perithecia gregarious, subconical, ¾ mm. diam., about half buried in the bark, the projecting part dull black, roughish, with a narrow, more or less compressed, prominent ostiolum. Asci subcylindrical, 110–120 x 12–15 μ, with abundant paraphyses. Sporidia mostly biseriate, fusoid, 3-septate and slightly constricted at the septa, brown, slightly curved, each cell with a large nucleus, 34–40 x 7–8 μ.

On bark of decaying *Salix*, Louisiana (Langlois).

L. stenóstomum, E. & E. Bull. Torr. Bot. Club. X, p. 89.

Exsicc. Ell. N. A. F. 1345.

Perithecia scattered, globose, 350 μ diam., covered by the fibers of the bark which is pierced by the narrow, slightly compressed ostiolum. Asci clavate-cylindrical, 75–90 x 6–7 μ, with abundant filiform paraphyses. Sporidia biseriate, fusiform, slightly curved, yellowish, nucleate and uniseptate at first, becoming 3-septate and more or less constricted at the septa, 18–22 x 3–3½ μ. Accompanied by a *Phoma* with small subglobose sporules in perithecia scarcely different from the ascigerous perithecia, except in wanting the prominent ostiolum.

On the inner surface of loose hanging bark of grape vines, Newfield, N. J. The perithecia are more deeply buried than in *L. triseptatum*, Pk.

L. Arúndinis, (Fr.)

Sphæria Arundinis, Fr, S. M. II, p. 510.
Lophiostoma Arundinis, Ces. & De Not. Schema, p. 46.
Exsicc. Rab. Herb. Mycol. 641.—Rehm. Ascom. 692.

Perithecia scattered or gregarious, sometimes confluent, at first immersed, finally more or less erumpent, but seldom superficial, spherical, black, solid, subrugose, with a rather small, thick, obtuse or sometimes subacute ostiolum with an elongated opening. Asci clavate-cylindrical, substipitate, 100–130 x 12–15 μ. Sporidia biseriate, fusoid, mostly curved, 3–5-septate, and slightly constricted at the septa, brown, 30–37 x 6 μ.

On reeds and grasses, South Carolina (Cke. Hndbk. p. 852).

In the following species the fructification is unknown.

L. Thápsi, (Schw.)

Sphæria Thapsi, Schw. Syn. N. Am. 1607.
Lophiostoma Thapsi, Cke. Syn. 3714.

Perithecia globose-conical, not smooth, flattened at the base and innate or subimmersed, scattered on black spots, on parts of the stem from which the woolly covering has fallen off, at first only the ostiolum erumpent, finally also the perithecium itself. Ostiolum thick, conical, compressed, less than half as high as the breadth of the perithecium, at length dehiscent.

Common on decaying mullein stalks around Bethlehem, Pa.

L. variábile, (Schw.)

Sphæria variabilis, Schw. Syn. N. Am. 1608.
Lophiostoma variabile, Cke. Syn 3715.

Perithecia elliptical, flattened, rugulose, generally seated on a black spot caused by the contents of the perithecia oozing out, almost always covered by the fibers of the stem, which are raised in a pustuliform manner over it, only the compressed, very variable ostiolum, which is confluent with the perithecium, being visible. Similar to *L. Thapsi.*

On dead stalks of *Pastinaca*, &c., Bethlehem, Pa.

L. abbreviàtum, (Schw.)

Sphæria abbreviata, Schw. Syn. N. Am. 1606.
Lophiostoma abbreviatum, Cke. Syn. 3716.

Perithecia subaggregated, erumpent among the fibers of the wood by which they are covered, except the dilated-cylindrical, very black, central, subelongated ostiolum, globose-conical, somewhat shining, but not glabrous, finally becoming ash-color or brown.

On oak branches, Bethlehem, Pa.

L. subrugòsum, (Schw.)

Sphæria subrugosa, Schw. Syn. N. Am. 1609.
Lophiostoma subrugosum, Cke. Syn. 3712.

Perithecia scattered or gregarious, subimmersed, finally erumpent, spherical or elongated, crowned by a laterally elongated ostiolum reaching half way across the apex of the perithecium and opening like a *Hysterium*. Perithecia with a light colored nucleus, at length empty and black, outside minutely rugose-punctate.

Rarely met with on decaying wood of *Catalpa*, Bethlehem, Pa.

L. hysterioìdes, (Schw.)

Sphæria hysterioides, Schw. Syn. N. Am. 1610.
Lophiostoma hysterioides, Sacc. Syll. 5523.

Perithecia scattered or gregarious and aggregated, hemispherical or elliptical, or conic-globose, prominent, very black and glabrous, truncate and subcarnose above. Ostiolum forming an elongated crack across the truncate apex. The most distinct of the whole lot.

Bursting out from the bark and wood of *Kalmia*, Bethlehem, Pa. (Schw.).

C. *Sporidia hyaline, two- or more-septate* (*Lophiotrema*).

L. parasíticum, Pk. 40th Rep. p. 71.

Perithecia crowded, subsuperficial, 350–500 μ diam., clothed with minute, subcervine, pulverulent tomentum, becoming blackish-brown with age, the ostiola prominent, subterete or compressed, clothed like the perithecia. Asci subclavate, 120–150 x 15–18 μ. Sporidia crowded, at first biconical, then triseptate, constricted in the middle, hyaline, 30–40 x 7–8 μ.

On old *Hypoxylon Morsei*, New York.

L. vagabúndum, Sacc. Mich. I, p. 447.—id. II, p. 72.—F. Ital. tab. 246.

L. radicans, E. & E. Journ. Mycol. III, p. 44.
Lophiotrema vagabundum, Sacc. Syll. 5435.
Exsicc. M. March, 747

Perithecia scattered, small (160–170 μ), buried in the wood, except the narrow, compressed, erumpent ostiolum. Asci subcylindrical, 90–100 x 10–12 μ (spore-bearing part 75–80 μ long), surrounded and overtopped by abundant, thread-like, paraphyses. Sporidia biseriate, fusiform, 3-septate, hyaline, slightly curved and mostly constricted at the middle septum, 15–20 x 4–5 μ.

On decorticated, decaying stems of *Rhus radicans*, Newfield, N. J.

Var. *stenocarpum*, E. & E., on decorticated oak limb, Newfield, N. J., has sporidia about 20 x 3 μ. Var. *Asteris*, on dead stems of *Aster*, Louisiana (Langlois 1440) has sporidia 25–30 x 5–6 μ, uniseptate, becoming tardily 3-septate.

L. vestìtum, Pk. 40th Rep. p. 71.

Perithecia closely gregarious, small (350–500 μ), sunk in the wood, erumpent, conical, clothed with a slight, tawny-ferruginous, pulverulent tomentum. Ostiola naked, black, subterete or compressed. Asci clavate, 150–190 x 15–18 μ. Sporidia crowded, subfusiform, at first biconical and uniseptate, then triseptate, or quadrinucleate, constricted in the middle, hyaline, 30–40 x 7½–10 μ.

On decorticated wood of *Populus tremuloides*, New York (Peck).

L. hysterioìdes, Ell. & Langlois. Journ. Mycol. IV, p. 76 (not Schw).

Lophiotrema hysterioides, Berlese Icones Fasc. I, p 4, tab. III, fig. 4.

Perithecia gregarious, subglobose with their bases slightly sunk in the wood, mostly a little less than ⅓ mm. diam., ostiolum forming a narrow ridge entirely across the perithecia, and thus giving them the appearance of a *Hysterium*. Asci cylindrical, 60–70 x 5–6 μ, with abundant paraphyses. Sporidia biseriate, fusoid, nearly straight, 3-septate, smoky-hyaline, 14–16 x 3 μ, the next to the upper cell swollen.

On rotten oak stumps, St. Martinsville, La.

L. æquívocum, E. & E. Journ. Mycol. III, p. 118.

Perithecia gregarious, erumpent-superficial, black, nearly smooth, depressed-conical or subglobose, about ⅓ mm. diam. Ostiolum subconical, slightly compressed. Asci subcylindrical, about 80 x 5 μ, narrowed below into a short, stipitate base. Paraphyses filiform. Sporidia uniseriate, oblong-fusoid, subobtuse, yellowish-hyaline, three-septate and constricted at the middle septum, sometimes also at the other two, 12–14 x 3–3½ μ. The ostiolum varies considerably, being sometimes distinctly compressed, sometimes regularly conical, and occasionally, imperfectly radiately three-cleft.

Differs from *L. præmorsum* in its smaller perithecia and sporidia, and from *L. hysterioides*, E. & E., in its very different ostiolum.

L. angustílabrum, (B. & Br.)

Sphæria angustilabra, B. & Br. Not. of Brit. Fungi, No. 881, taf. XI, fig. 27.
Lophiostoma angustilabrum, Cke. Hndbk. p. 850.
Lophiotrema angustilabrum, Sacc. Mich. I, p. 338.
Exsicc. Vize Micr. Fungi, 280.

Perithecia semiimmersed, rugulose, sometimes elongated. Ostiolum compressed, narrow. Asci clavate. Sporidia biseriate, fusiform, curved, uniseptate and constricted, 4–6-nucleate, (becoming 3-septate)? 40–43 μ long, with a short, hyaline appendage at each end.

On a decorticated stick, New York State (Peck).

Sporidia in Vize Exsiccati, 25–30 x 5 μ, in Plowright's F. Brit. II, 49, 28–32 x 7 μ.

This is placed by Berlese as a synonym of *L. præmorsum* (Lasch), but on account of the longer, appendiculate sporidia, it seems properly distinct.

L. sexnucleàtum, Cke. Hndbk. 2543.

Lophiotrema sexnucleatum, Sacc. Syll. 5432.

Perithecia scattered, elongated, black, slightly rugose, at first immersed, then emergent. Ostiola compressed. Sporidia biseriate, fusiform, hyaline, slightly curved, 5-septate, constricted in the middle, and but little at the other septa, 35 μ long, each cell with a single nucleus.

The typical form, on dead stems of *Urtica*, does not appear to have been found in this country. Var. *Peckiana*, Sacc. Syll. II, p. 583 (*Lophiostoma sexnucleatum*, Cke, in Peck's 27th Rep. p. 110), has the perithecia subgregarious, and sporidia smaller (23–30 μ long).

On peduncles of grape vines, North Greenbush, N. Y. (Peck).

D. *Spor. uniseptate, hyaline, mostly appendiculate* (*Lophiosphæra*).

L. pulveràceum, Sacc. Mich. I, p. 336.—F. Ital. 225.—Berl. Icones. Fasc. I, tab. II, fig. 8.

L. conchoides, E. & E. (in Herb.)
Lophiosphæra pulveracea, Sacc. Syll. 5414.

Perithecia gregarious, immersed or semiimmersed in the wood, subglobose, small (200–250 μ). Ostiolum compressed, truncate, emergent. Asci clavate, short-stipitate, 70–75 x 7 μ (80 x 10–12 μ, Sacc.), obtuse and with abundant paraphyses, 8-spored. Sporidia biseriate, biconic-fusoid, 12–18 x 4–5 μ (18–20 x 4½–5 μ, Sacc.), nearly straight, uniseptate and constricted, hyaline, with a very short, hyaline appendage at each end.

On rotten wood, Louisiana (Langlois, 1770).

The appendages are very faint and easily overlooked.

L. heteróstomum, E. & E. Journ. Mycol. II, p. 99.

Lophiosphæra heterostoma, Berlese Icones Fasc. I, tab. III, fig. 2.

Perithecia erumpent, scattered or subgregarious, ⅓–¾ mm. diam., depressed-spherical, the lower half sunk in the wood, the upper half emerging. Ostiolum large, compressed, extending at first nearly or quite across the perithecium, at length deciduous, leaving the perithecium pierced above with a small, round opening. Asci clavate-cylindrical, rounded above and contracted below into a slender, stipe-like base, 70–80 x 6–7 μ, surrounded with abundant, rather stout paraphyses and containing eight fusoid, uniseptate, hyaline, slightly curved, appendiculate sporidia, 18–20 x 4½–5 μ, 3–4-nucleate at first, but the nuclei and stout, 8–10 μ long, hyaline appendages at length disappear, and the sporidia become constricted in the middle with the ends rounded (fusoid-oblong), often with a distinct yellow-brown shade. The wood just below the surface assumes a uniform purplish-red color. This might, perhaps, be considered a var. of *Lophiostoma pulveraceum*, Sacc., but differs in the stained matrix, larger perithecia and deciduous ostiola.

On an oak barrel-bottom, Louisiana, and on decorticated wood in Northern New Jersey.

L. Floridànum. E. & E. Journ. Mycol. II, p. 40.

Perithecia gregarious or scattered, hemispherical, black and rough, about ¼ mm. diam. Ostiolum linear, extending quite across the apex of the perithecium, but scarcely prominent. Asci clavate-cylindrical, 80 x 8 μ, with an elongated, slender base and surrounded with abundant, filiform paraphyses. Sporidia biseriate, fusiform, slightly curved, olivaceous, 4-nucleate, uniseptate, constricted at the septum, just above which they are slightly swollen, 18–20 x 4–4½ μ. The matrix is overrun with a hyphomycetous growth, which also embraces the base of the perithecia, but we cannot say whether this is accidental.

Parasitic on old *Diatrype stigma*, Florida (Calkins).

This seems to be a distinct species, but, unfortunately, the specimens are very meager.

L. meridionàle, E. & E. Journ. Mycol. IV, p. 76.

Perithecia scattered, minute ($\frac{1}{4}$–$\frac{1}{3}$ mm.), compressed, buried in the wood, the projecting, flattened, black ostiolum alone being visible. Asci clavate-cylindrical, 75–90 x 8–9 μ, with abundant paraphyses. Sporidia biseriate, fusoid, hyaline, slightly curved, uniseptate and slightly constricted, 30–35 x 4–5 μ, appendiculate at each end.

On rotten wood, Louisiana.

This differs from *L. angustilabrum*, B. & Br. in its smaller, buried perithecia. In the specimens of *L. angustilabrum*, issued by Vize, the perithecia are $\frac{2}{3}$–$\frac{3}{4}$ mm. diam., and nearly superficial. In the original description in Journ. Mycol., the appendages on the sporidia were overlooked. They are conical, hyaline and about 3 μ long.

E. *Sporidia appendiculate, 3–7-septate.*

L. insidiòsum, (Desm.)

Sphæria insidiosa, Desm. Ann. Sci. Nat. XV, p. 144
Lophiostoma insidiosum, Ces. & De Not. Schema, p 46
L. caulium, Fckl. Symb. p. 156.
Lophiotrema appendiculatum, Niessl. (MS)
Exsicc. Fckl. F. Rh. 927,—Kze. F. Sel. 98.—Rab. F. E. 1871.—Rehm. Ascom 88, Mycoth. March, 851, 1354.—Ell. N. A. F. 887, (in part).

Perithecia scattered or gregarious, immersed, at first with only the small, obtuse ostiolum projecting, finally suberumpent, 400–500 μ diam. Asci clavate, contracted below into a stipe-like base, 8-spored, 70–90 x 10–14 μ, with abundant filiform paraphyses. Sporidia biseriate, oblong-fusiform, slightly curved, 5–7-septate, often slightly constricted at some of the middle septa, yellow-brown, with a short, subconical (about 4 μ long), hyaline appendage at each end, 20–22 x 4–5½ μ.

On dead herbaceous stems, of *Oenothera*, *Trifolium*, &c. not uncommon.

L. Spiræ̀æ, Pk. 28th Rep. p. 76.

L. crenatum, Pers. var. *Spirææ*, Schw. Syn. N Am. 1599.
Lophiotrema Spirææ, Sacc. Syll. 5443.

Perithecia scattered, sunk to the wood, $\frac{1}{2}$–1 mm. diam., closely covered by the bark which is pierced by the compressed ostiola. Asci

100–120 x 12–15 μ (p. sp. 75–80 μ), paraphysate, 8-spored. Sporidia crowded-biseriate, elongated-fusoid, attenuate-acuminate at each end, straight or somewhat curved, yellowish-hyaline, at first 3–4-septate, finally about 7-septate, scarcely constricted in the middle, 35–55 x 6–8 μ.

On *Spiræa opulifolia*, Bethlehem, Pa. (Schweinitz), New York State (Peck & Fairman).

L. aúctum, Sacc. M. Ven. Spec. p. 110, tab. XI, figs. 5–10, F. Ital. tab. 250.

Perithecia scattered, semiimmersed, sphæroid, $\frac{1}{3}$–$\frac{1}{2}$ mm. diam., with a narrow, compressed ostiolum. Asci clavate, stipitate, 90–100 x 16 μ, 8-spored, paraphysate. Sporidia biseriate, fusiform, somewhat curved, 40 x 8 μ, 4–7-septate, deeply constricted at the septa, cells nucleate, yellowish, with a linear, obtuse, hyaline appendage at each end, especially when young.

On dead stems of wild rose, Lyndonville, N. Y. (Fairman).

The New York spece. differ from the above diagnosis (from Sacc. Syll.), in having the asci and sporidia smaller; asci 75–90 x 12–15 μ; sporidia 25–35 x 6–7 μ, 5–7-septate, only slightly constricted and obscurely appendiculate.

L. bicuspidàtum, Cke. Handbook, p. 848.

Perithecia scattered, black, immersed, elevating and pushing through the matrix with their narrow, elongated ostiola. Asci clavate. Sporidia biseriate, 5-septate, with occasional longitudinal septa, constricted, brown, each end at first furnished with a hyaline beak bent at both ends in the same direction, so as to give a curved appearance to the sporidia.

On decorticated twigs.

The above description is from Cooke's Handbook. Peck reports the species as found in New York.

L. præmórsum, (Lasch.)

Sphæria præmorsa, Lasch. in Klotzsch-Rabh, Herb. Mycol 1249.
Sphæria Jerdoni, B. & Br. Not. Brit. Fungi, No 975.
Lophiostoma præmorsum, Fckl. Symb. p. 157.
Lophiotrema præmorsum, Sacc. Mich. I, p. 338.
Lophiostoma Scrophulariæ, Pk 28th Rep p 76.
Exsicc. Fckl. F. Rh. 928.—Rabh F. E. 1239.

Perithecia scattered or gregarious, at first entirely buried, or with only the obtuse, linear ostiolum projecting, finally more or less erumpent and even superficial, black, about $\frac{3}{4}$ mm. diam. Asci elongated-clavate, contracted below into a stipe-like base, 8-spored. paraphysate, 80–105 x 10–12 μ. Sporidia biseriate, fusiform, slightly curved, hyaline, 22–30 x 4–6 μ, sometimes with a short, hyaline appendage at each end.

On dead herbaceous stems, Louisiana (Langlois, 1437).

The sporidia are for some time only 1-septate, but finally become 3–5-septate and yellowish hyaline. The figs. 7–9, tab. III, Berlese Icones, represent the sporidia very accurately.

Of the two Nos. in Sydow's Mycotheca Marchica issued as *L. præmorsum*, 25 7is a *Diplodia* and 985 a brown-spored *Lophiostoma*, (at least in our copy).

LOPHÍDIUM, Sacc.

Michelia I, p. 40.

Perithecia as in *Lophiostoma*. Sporidia ovate or oblong, brown, muriform.

L. dimínuens, (Pers.)

Sphæria diminuens, Pers. Syn. p. 57? Fr. S. M. II, 471.
Lophiostoma diminuens, De Not. Schema, p. 229.
Lophidium diminuens, Ces. & De Not. Schema, p. 220.

Perithecia scattered, prominent, rounded, subdepressed, black. Ostiolum narrow, compressed, small, sometimes subconical. Sporidia elongated-oblong, 6–7-septate, with one or more cells divided by a longitudinal septum.

On branches of *Cratægus*, *Cornus*, &c., (Europe), Carolina and Pennsylvania (Schweinitz).

There is no black spot. The perithecia on decorticated branches are bare; on limbs with the bark on, emergent, unequal, soon empty, and the ostiolum broken, truncate, sometimes conical.

L. compréssum, (Pers.)

Sphæria compressa, Pers. Syn. p. 56.
Sphæria angustata, Pers. Syn. p. 55.
Lophiostoma compressum, Ces. & De Not. Schema, p. 45.
Lophiostoma angustatum, Fckl. Symb. p. 158.
Lophidium compressum, Sacc. Mich. I, p. 340.
Exsicc. Fckl F. Rh. 924, 925.—Kze. F. Sel. 341.—Rehm Ascom. 182.—Thum. M. U. 1457, 1548.

Perithecia scattered or gregarious, mostly immersed except the long, thin, toothed or entire-margined ostiolum, sometimes finally erumpent, spherical or somewhat compressed and elliptical. Asci clavate-cylindrical, stipitate, 8-spored, 110–130 x 16–19 μ. Sporidia obliquely uniseriate or partly biseriate, oblong or clavate-oblong, straight or slightly curved, 5-septate, with one or two cells divided by a longitudinal septum, constricted at the septa, of a fine golden yellow-brown color, 18–32 x 8–10 μ. Paraphyses numerous and slender.

On wood, Carolina and Pennsylvania (Schweinitz), New York (Peek), Canada (Dearness).

Specimens on dead stems of *Spiræa* from New York State, have asci (p. sp.) 75–80 x 12 μ, with abundant paraphyses. Sporidia obliquely uniseriate, ovate-elliptical, yellow-brown, 3-septate and constricted in the middle, finally darker and mostly 5- (exceptionally 6–7-) septate and muriform 18–22 x 8–10 μ. Berlese (l. c.) notices the same variability in the number of septa. The Canada specc. on *Cratægus*, have asci (p. sp.) 110 x 18–20 μ, sporidia oblong-fusoid, 5–7-septate with one or two of the cells divided by a longitudinal septum, 25–35 x 10 μ.

L. tíngens, (Ell.) **(Plate 25)**

Lophiostoma tingens, Ell. Bull. Torr. Bot Club, VIII, p 90.
Lophidium tingens, Sacc. Syll. 5527
Exsicc. Ell. N. A. F. 693.

Perithecia buried in the wood, mostly compressed, thick and leathery, of medium size. Ostiolum barely visible on the surface of the wood, not projecting and only slightly compressed. Asci cylindrical, 80–90 x 10–11 μ, with filiform paraphyses. Sporidia uniseriate, oblong, obtuse, brownish, 3-septate, with a longitudinal septum more or less distinct, often slightly curved, variable in length, mostly about 16–18 x 7 μ.

On dry, decorticated maple limbs, Newfield, N. J.

L. obtéctum, (Pk.)

Lophiostoma obtectum, Pk. 30th Rep. p. 65.
Lophidium obtectum, Sacc. Syll. 5537.

Perithecia numerous, immersed, slightly prominent, covered by the epidermis which is pierced by the narrow, compressed ostiola. Asci cylindrical or clavate. Sporidia variable, crowded or biseriate, rarely uniseriate, at first pale, subacute and 1–3-septate, then obtuse, oblong or subfusiform, 5–6-septate, colored, 25–33 μ long, usually constricted at the septa and occasionally with longitudinal septa.

On dead branches of *Zanthoxylum Americanum*, Bethlehem, N. Y.

L. mìnus, (Ell.)

Lophiostoma minus, Ell. in Am. Nat 1883, p 316,
Lophidium minus, Sacc. Syll. 6179.
Exsicc. Ell. & Evrht. N. A. F. 2d Ser. 1690.

Perithecia erumpent, conical, ⅓ mm. diam. and in height, with a narrow and not very distinctly prominent ostiolum. Asci subcylindrical, 50–60 x 7 μ, with abundant, filiform paraphyses. Sporidia uniseriate or crowded above, oblong-elliptical, 3-septate, with occasionally a partial longitudinal septum, greenish-brown, 9–13 x 5–7 μ, sometimes with a short, obtuse, cylindrical appendage below.

On dead limbs of *Nyssa multiflora*, Newfield, N. J.

L. fenestràle, (C. & E.)

Lophiostoma fenestrale, C. & E, Grev. VI, p. 12.
Lophidium fenestrale, Sacc. Syll 5546, Berlese Icones, tab. XI. fig. 2

Immersed, scattered, subglobose. Ostiola laterally compressed, hysteriiform. Asci cylindrical? Sporidia obtuse, fusiform, constricted in the middle, fenestrate, olive-brown.

On decaying *Morus*, New Jersey.

Not in good condition. Asci dissolved, but free sporidia found which were elongated-elliptical, 50–60 x 20 μ (Cooke in Grev. l. c.) Unfortunately our specimens are lost.

Berlese (l. c.) gives the following measurements from Italian specimens. Perithecia ⅔–1 mm., scattered. Asci 210–220 x 24–27 μ. Sporidia 58–60 x 17–19 μ.

L. cùrtum, (Fr.)

Sphæria curta, Fr. S. M. II, p 470.
Lophidium curtum, Sacc. Mich. I, p. 340.

Perithecia gregarious, subprominent, rugulose, conic-globose, black. Ostiolum short, linear, compressed, extending only partly across the apex of the perithecium, sometimes conical, lips linear, closed. Asci cylindrical, with a very short stipe, 120–130 x 12 μ, paraphysate, 8-spored. Sporidia fusoid, biconical, 28–30 x 10–12 μ, 7–9-septate and muriform, constricted at the middle septum, brown (Sacc. in Syll. II, p. 714).

On very rotten limbs, Bethlehem, Pa. (Schweinitz).

This species and *L. diminuens* have not been found in this country since Schweinitz' time, and may be considered doubtful.

LOPHIONÈMA, Sacc.

Sylloge II, p. 717.

Perithecia as in *Lophiostoma*. Sporidia vermicular or filiform, septate, subhyaline.

L. vermísporum, (Ell.) (Plate 25)

Lophiostoma vermisporum, Ell. Bull Torr Bot. Club, IX, p 19
Lophionema vermisporum, Sacc. Syll. II, p. 717.
Exsicc. Ell. N. A. F. 885

Perithecia scattered or subgregarious, depressed-spherical, 150–200 μ diam., seated under the epidermis which is pierced by the stout, black, narrowly compressed ostiola. Asci clavate-cylindrical, 150–200 x 12–15 μ. Sporidia vermiform, gradually tapering towards the base, about 7-septate and slightly constricted at the septa, yellowish or nearly hyaline, each of the cells with 1–2 large nuclei, 75–88 x 3½–5 μ.

On old stems of *Oenothera biennis*, Newfield, N. J.

L. crenàtum, (Schw.)

Sphæria crenata, Pers var. *cristata*, Schw Syn. Car. 136.
Lophiostoma (*Lophionema*) *crenatum*, Cke. Syn. 3591.

Perithecia scattered or gregarious, immersed in the wood or bark, finally more or less emergent, subsphæroid, black, minute. Ostiolum compressed, very broad, subcrenate.

North Carolina (Schw.)

Cooke (l. c.) places this in his subgenus *Lophionema*. We have no knowledge of it otherwise.

Species to be rejected.

L. roseotínctum, E. & E. Journ. Mycol. I, p. 149.

This is only the diagnosis of *Leptosphæria roseotinctum*, E. & E., repeated by mistake from the preceding page.

L. mínimum, E. & E. Journ. Mycol. IV, p. 75.

The specimens are so poor and meager that we abandon this species.

L. pállidum, E. & E. Bull. Torr. Bot. Club, X, p. 52.

See *Melanomma pallidum*.

FAMILY. CUCURBITARÌEÆ.

Stroma imperfectly developed or none. Perithecia cespitose, erumpent-superficial, seated on the stroma, when that is present, sometimes superficial from the first. Substance of the perithecia hard, woody or leathery, black or dark brown. Asci elongated, paraphysate. Sporidia various, mostly brown.

CUCURBITÀRIA, Gray.

Nat. Arrangement of Brit. Plants, I, p 519

Perithecia cespitose or more rarely gregarious, generally more or less connected at the base by a thin, dematiaceous stroma, spherical, glabrous, black, coriaceous, generally rugulose, perforated above. Asci cylindrical, 8-spored. Sporidia uniseriate, oblong or elliptical, muriform, yellow, becoming brown. Paraphyses abundant.

C. elongàta, (Fr.)

Sphæria elongata, Fr. S. M. II, p. 422.
Cucurbitaria elongata, Grev. Scott. Crypt. Flora tab. 195.
Exsicc, Fckl. F. Rh. 970.—Rab. Herb. Myc. 727.—id F. Eur. 822, 1441.—Rehm Asc. 185, 336.—Thum. F. Austr. 252.—id. M. U. 565, 1362.—M. March. 146.—Plowr Sph. Brit. 54.—Ell. & Evrht. N. A. F. 2d Ser. 1955.

Stroma black, extensively effused and often surrounding the branch. Perithecia at first immersed, then erumpent, subcespitose, globose, annulate-depressed around the papilliform ostiolum. Asci cylindrical, short-stipitate, 120–150 x 12–14 μ, 8-spored, with filiform paraphyses. Sporidia uniseriate, ovate-oblong, 5–7-septate and muriform, 25–28 x 10–12 μ, yellow, then brown, constricted at the middle septum. Macropycnidial state, *Hendersonia Robiniæ*, West; pycnidia, *Diplodia Robiniæ*, stylospores ovate, brown, uniseptate, 20 x 12 μ; perithecia globose, papillate.

On *Robinia pseudacacia*, Pennsylvania (Everhart).

C. Berbéridis, (Pers).

Sphæria Berberidis, Pers. Syn. p. 52
Cucurbitaria Berberidis, Gray l. c.
Exsicc. Fckl. F. Rh. 969.—Kunze. F. Sel 103.—Rab Herb Mycol 653.—id F. E. 2230. Rehm. Asc. 281.—Thum. F. Austr 171.—id. M. U. 360.—Lin. F. Hung 281.—M. March. 161, 1716, 2943.—Plow. S. Brit. 56.—Vize. M. F. 160.—Romell, F. Scand. 76. —Ell. N. A. F 578

Perithecia seated on a thin, brownish-black, dematiaceous stroma, mostly in elongated groups of various extent, erumpent through the epidermis, crowded, subspherical, rimose-rugose, and finally subrimose-dehiscent, black, glabrous, about $\frac{3}{4}$ mm. diam. Asci cylindrical from a narrow base, 150–160 x 14–16 μ, 8-spored, with abundant paraphyses. Sporidia uniseriate, oblong-elliptical, somewhat attenuated at the ends, 8–9-septate and muriform, constricted in the middle, golden-yellow becoming brown, with the ends subhyaline, 26–30 x 12–13 μ. Spermogonium; spermatia very small, cylindrical, 4 x $\frac{1}{2}$ μ.

On dead limbs of *Berberis vulgaris*, Carolina (Schw.)

C. conglobàta, (Fr.)

Sphæria conglobata, Fr. S. M. II, p, 414
Cucurbitaria conglobata, Ces. & De Not. Schema, p. 214.

Perithecia cespitose, erumpent, subglobose, with a papilliform ostiolum. Asci cylindrical, stipitate, 8-spored, 130–155 x 14–18 μ. Sporidia uniseriate, at first 3-septate, then 5–7-septate and longitudinally divided, slightly constricted at the septa, brown, 24–30 x 11–12 μ. Stylosporous receptacles mixed with the perithecia. Stylospores spirally bent, 2–3 x $\frac{1}{2}$ μ.

On birch limbs, Itivnek, Greenland, on Corylus, Penna. (Schw.)

C. Kélseyi, E. & E. Proc. Acad. Nat. Sci. Phil., July, 1890, p. 240.

Perithecia large, ($\frac{3}{4}$–1 mm.), rough, subglobose, regularly rounded above, with a papilliform ostiolum, bursting through cracks in the bark in elongated tufts, crowded and subconfluent, connected below by a scanty grayish-black stroma. Asci cylindrical 170–190 x 15–20 μ, contracted below into a short, stipe-like base. Paraphyses filiform, abundant. Sporidia uniseriate, oblong-elliptical, straw-yellow, becoming dark brown, contracted in the middle, with three principal septa and several fainter ones (7–9 in all), muriform, 25–30 x 14–16 μ, ends at first obtusely pointed, finally rounded. The three main transverse septa are the only distinct and decided ones, the others, both transverse and longitudinal, being more or less indistinct and interrupted.

On *Philadelphus Lewisii?* Helena, Montana.

This is with difficulty distinguished from *C. Berberidis*, Gray, and might perhaps better be considered a var. of that species, from which it seems to differ in its broader sporidia. The ascigerous perithecia were accompanied by others inclining more to ovate, with a short, conic-cylindrical ostiolum, and filled with very minute (1$\frac{1}{2}$–2 x $\frac{1}{2}$ μ) sporules.

C. Fráxini. E. & E. l. c. **(Plate 26)**

Perithecia globose, rough, black (white inside), $\frac{1}{2}$ mm. diam., flattened above, with a papilliform ostiolum, seated on the surface of the inner bark in compact clusters of about 8–12, and surrounded by the ruptured epidermis. Asci cylindrical, p. sp. about 150 x 12–15 μ, with a short, stipe-like base and surrounded by numerous paraphyses. Sporidia uniseriate or subbiseriate, ovate-oblong, constricted in the middle, 5–6-septate and muriform, yellowish-brown, 25–30 x 10–14 μ.

On bark of dead *Fraxinus*, London, Canada (Dearness).

C. echinàta, E. & E.

Cucurbitaria setosa, E & E. l c, p 241.

Stroma black, carnose, 2–3 mm. across, its convex surface thickly covered with the minute, parasitic? perithecia, which are subglobose, about 165 μ diam., clothed laterally with stout, black, spreading spines 25–40 x 7 μ. Ostiolum either smooth and subpapilliform or more or less distinctly radiate-sulcate. Asci oblong clavate, 45–55 x 12 μ (p. sp.), contracted below into a narrow, stipe-like base. Paraphyses longer than the asci, evanescent. Sporidia biseriate, subrhomboidal-oblong, hyaline and multinucleate at first, becoming yellowish, 3–5-septate and submuriform, 12–15 x 4–5 μ. Accompanied by *Cornularia Persicæ*, (Schw.) The crowded perithecia bear a general resemblance to those of *Otthia morbosa*, (Schw.), from which, however, this is quite distinct.

Parasitic? on the tubercular, erumpent stroma of some *Diatrype?* on dead limbs of wild plum (*Prunus*), London, Canada (Dearness).

The stroma is carnose and black inside and out, 2–3 mm. diam., and in the bark beneath it are buried the abortive perithecia of the *Diatrype*. The specific name has been changed on account of the homonomous species *C. setosa*, Winter, which was overlooked.

C. naucòsa, (Fr.)

Sphæria naucosa, Fr in Schm & Kze. Myc Hefte, 2, p 36 and Fr. S. M II, p. 416.
Cucurbitaria naucosa, Fckl. Symb. p 173, tab I, fig. 10
Cenangium naucosum, Fr. Summa Veg. Scand. p. 364
Exsicc. Fckl F. Rh 2042.

Perithecia crowded-cespitose, globose, smooth, brownish-black, subastomous. Asci broad-cylindrical, with a narrow base, 8-spored, 110–125 x 14–16 μ. Sporidia uniseriate or partly biseriate, oblong-clavate, narrowed below, constricted in the middle, 3–6-septate, with an imperfect, longitudinal septum, honey-yellow, 19–23 x 7–9 μ.

On (elm)? limbs, Bethlehem, Pa. (Schweinitz).

The specimen in Herb. Schw. is too poor to allow one to express any opinion as to what it may be.

C. Cratægi, (Schw.)

Sphæria Cratægi, Schw. Syn. N Am. 1445

Perithecia erumpent, generally in elliptical tufts, rather large (about ½ mm. diam.), black or blackish, of irregular shape and subcircinately arranged, the lower half sunk in a furfuraceous, dark tomentum. Asci 112–120 x 15–20 μ, subsessile, paraphysate, 8-spored. Sporidia uniseriate, clavate-elliptical, hyaline and with a broad, hyaline envelope at first, becoming 6-or more-septate and yellow, strongly constricted in the middle, 35–40 x 15 μ, upper half broadest.

On various species of *Cratægus*, around Bethlehem, Pa. (Schweinitz).

The measurements of the asci and sporidia were taken from the specimen in Herb. Schw. *C. Cratægi*, Niessl. seems to be different from this.

C. Shephérdiæ, (E. & E.)

Curreya Shepherdiæ, E. & E. Proc Acad. Nat Sci Phil. July, 1890, p 249

Perithecia 4–6 together, connected at the base by a loose, brown stroma, ovate, whitish inside, covered at first, their short, obtuse ostiola soon rupturing the epidermis and revealing the clusters of brownish-black perithecia. Asci clavate-cylindrical, 100–110 x 15 μ. with obscure paraphyses. Sporidia 8 in an ascus, uniseriate, obovate, 3–5-septate and muriform, slightly constricted in the middle, yellow, 18–22 x 10–12 μ.

On dead limbs of *Shepherdia argentea*, Valley of the Teton. Northern Montana (Anderson, 539).

C. cónfluens, Plowr. Grev. V, p. 74.

Perithecia at first immersed, becoming prominent, conical, frequently confluent, covered by a very fine, black down, connected by an inconspicuous stroma. Sporidia brown, muriform, constricted in the middle, about 30 x 12 μ.

On oak bark, Sierra Nevada Mts., California.

C. Corèmæ, E. & E. Bull. Torr. Bot. Club. X, p. 89.

Exsicc. Ell. N. A. F. 1189.

Perithecia scattered, at first subcuticular, but finally throwing off the epidermis, ovate, black, rough, 260–330 μ diam., with broadly papilliform, obtusely conical ostiola. Asci cylindrical, 150 x 15 μ. Sporidia uniseriate and oblique or partly biseriate, elliptical or oblong-elliptical, yellow at first, becoming deep brown, about 7-septate, with a few longitudinal septa, mostly constricted in the middle, 20–25 x 7–9 μ.

On *Hudsonia tomentosa*, Newfield, N. J.

The host was at first taken to be *Corema Conradii*, hence the specific name.

C. umbilicàta, Ell. Bull. Torr. Bot. Club. X, p. 53.

Exsicc. Ell. & Evrht N. A F 2d Ser 1654

Perithecia scattered or subaggregated, depressed-hemispherical.

black, rough, $\frac{1}{3}$–$\frac{1}{2}$ mm. diam., collapsing above when dry. Asci cylindrical, 114 x 11 μ, 8-spored, paraphysate. Sporidia uniseriate, elliptical, constricted in the middle, 3-septate, straw color at first, becoming muriform and brown, 22–28 x 9–13 μ.

On decorticated sage brush (*Artemisia*), Utah.

C. Ribis, Niessl. Beitrag. p. 48, tab. V. fig. 34.

Exsicc Ell. & Evrht. N A F 2d Ser 2370.

Perithecia rather large, gregarious or crowded, subglobose, papillate, at length depressed, umbilicate, perforated, shining, 400 μ diam. Asci ample, oblong-clavate, short-stipitate, rounded above, paraphysate, 100–116 x 14–17 μ. Sporidia obliquely uniseriate or irregularly biseriate, obovate, constricted in the middle, 3–7-septate and muriform, dark olivaceous, 18–20 x 7–8 μ.

On decorticated stems of *Ribes*, sp. Helena, Montana (Kelsey).

C. longitudinàlis, Pk. 33rd Rep. p. 34, pl. 2, figs. 23–26.

Perithecia 500–750 μ diam., globose, arranged in short lines, erumpent through cracks in the bark, black, pierced at the apex. Asci cylindrical, 8-spored, paraphysate, 110–120 x 12 μ. Sporidia mostly uniseriate, oblong, 4–5-septate, with one or two longitudinal septa, brown, 20–32 x 10–12$\frac{1}{2}$ μ.

On dead stems of *Andromeda ligustrina*, Center, N. Y., and Newfield, N. J.

C. Comptòniæ, C. & E. Grev. VI, p. 12.

Exsicc Ell. N. A. F 95

Perithecia cespitose or scattered, often transversely erumpent through cracks in the epidermis, but also in small (1–1$\frac{1}{2}$ mm.) tufts closely surrounded by the epidermis or, when on the bare wood, scattered, ovate-globose, black, papillate, about $\frac{1}{3}$ mm. diam. Asci cylindrical, 100–112 x 12 μ, paraphysate, 8-spored. Sporidia uniseriate, oblong-elliptical, yellow-brown, 3-septate and slightly constricted in the middle, becoming 5- or more-septate and muriform, 18–23 x 10 μ.

On *Comptonia asplenifolia*, Newfield, N. J.

Asci longer and narrower and sporidia smaller than in *C congesta*.

C. congésta, C. & E. Grev. VI, p. 12.

Perithecia erumpent-superficial, crowded, ovate-globose, about $\frac{1}{3}$ mm. diam., rough, black, with a distinct papilliform ostiolum. Asci

(p. sp.) about 75 x 12 μ, paraphysate, 8-spored. Sporidia uniseriate or subbiseriate above, oblong-elliptical, yellow-brown, with three main septa, slightly constricted at the middle septum, finally multiseptate and muriform, 18–25 x 11–13 μ.

On dead limbs of *Magnolia*, Newfield, N. J.

The perithecia are mostly thickly gregarious or crowded, but not cespitose. Accompanied by perithecia with brown, *Diplodia* spores rather smaller than the ascospores.

C. Spártii, (Nees.)

Sphæria Spartii, Nees in Fr S M. II, p 424.
Cucurbitaria Spartii, Ces. & De Not Schema, p 40, Sacc. F Ital tab 532
Exsicc. Fckl. F. Rh. 974 —Rab F E 1440 —Rehm Asc 527 —Thum F Austr 253 —M. March. 34.

Perithecia erumpent, crowded and confluent or cespitose, with obtuse ostiola and sometimes with an imperfectly developed stroma, globose, with the apex finally umbilicate-depressed, blackish. Asci cylindrical, short-stipitate, 160–210 x 10–12 μ, with filiform paraphyses. Sporidia uniseriate, oblong, slightly constricted in the middle, 5–7-septate and submuriform, golden yellow-brown, 24–30 x 8–10 μ.

On dead branches of *Sarothamnus scoparius* (cult.), Newfield, N. J.

C. Labúrni, (Pers.)

Sphæria Laburni, Pers Syn. p 50.
Cucurbitaria Laburni, De Not Erb Crit. Ital No 875
Exsicc. Fckl F. Rh 965.—Kze F. Sel. 104 —Rab. Herb Myc 42.—id F. E 1014 —Rehm Asc 146.—Thum F Austr. 489 —id M. U 272

Cespitose on a subcompact stroma. Perithecia globose, rugulose, black, with a papilliform ostiolum. Asci cylindrical, 110–170 x 11–14 μ, with branching paraphyses. Sporidia elliptic-fusoid, 26–36 x 9–12 μ, subuniseriate, 5–7-septate and muriform, slightly constricted in the middle, golden-brown.

On dead branches of *Laburnum*. Common in Europe and will probably be found here.

C. tumòrum, (Schw.)

Sphæria tumorum, Schw Syn. N. Am. 1417

The specimens of this species in Herb. Schw. afford only stylospores, *Diplodia* sp.

C. radicàlis, Cke. Grev. VII, p. 51.

Exsicc. Rav. F. Am. 344

Perithecia ovate, about ⅓ mm. diam., papillate, black, erumpent-superficial, in small (1 mm.), valsoid clusters, with their bases stromatically connected. Asci clavate. Sporidia linear, slightly curved, obtuse, 8 μ long.

On bark of oak roots, South Carolina.

In the specimens in our copy of Rav. F. Am. some of the perithecia contain minute, oblong, hyaline spermatia and others, large hyaline, oblong-elliptical stylospores, 20–26 x 12–14 μ, with a thick, hyaline epispore (*Dothiorella* sp.) The perithecia are not collapsed. This must be considered a doubtful species.

FRACCHIÆA, Sacc.

Myc. Ven. Spec p. 115, tab XII, figs. 3–7.

Perithecia aggregated or subscattered, superficial, on a stromatic crust, globose or depressed, black or brown, coriaceous or coriaceo-carbonaceous. Asci clavate, polysporous. Sporidia oblong or allantoid, subhyaline.

Fr. subcongregàta, (B. & C.)

Sphæria subcongregata, B. & C. in Rav Fungi Car IV, No. 57 (1855).
Fracchiæa heterogenea, Sacc. M. Ven Spec. p 115 (1873)
Sphæria subconnata, B. & C. Grev. IV p 141 (1876).
Sphæria subconvexa, B. & Rav. (ubi)?
Gibbera moricarpa, Cke. Grev VII, p. 51 (1878).
Sphæria polycocca, B. & Rav in Rav Car IV, No. 62 (sec. Cke. in Grev. XV, p 83).
Sphæria botryosa, Tode, Scler. No. 122 (sec. specc. det by Cke. in Grev. V, p. 93).
Exsicc, Rav F. Car. IV, No 57—Speg Dec. Myc. Arg. No 41 —Rav. F. Am. 343 —Ell. N. A F 692.

Perithecia cespitose or partly scattered, seated on a thin, stromatic crust on the surface of the wood or bark, depressed-spherical, $\frac{1}{3}$ to nearly $\frac{1}{2}$ mm. diam., black and verruculose. Ostiolum minute, scarcely prominent, becoming umbilicate. Asci elongated-clavate, attenuated below, 60–70 x 12–15 μ, polysporous, (paraphysate, sec. Speg.) Sporidia allantoid, yellowish in the mass, 6–9 x $1\frac{1}{2}$ μ (12 x 2 μ, Sacc.)

On bark of *Myrica* and *Liquidambar*, Georgia (Ravenel), on *Morus*, New Jersey.

We make the sporidia in Speg. So. Am. specimens only 6–9 x $1\frac{1}{2}$ μ, mostly 6–8 x $1\frac{1}{2}$ μ, the same as in our N. Am. specc.

Fr. callista, (B. & C.)

Cucurbitaria callista, B & C. Grev IV, p 47.
Fracchiæa callista, Sacc Syll. 388.
Exsicc Rav. Car. V, No. 67.—Ell. N. A. F 1188.

Perithecia closely aggregated, seated on a patch of tobacco-brown, felt-like subiculum 2–4 mm. across, and composed of closely interwoven, pale brown, branched and sparingly septate hyphæ, globose, soon deeply collapsed, about ⅓ mm. diam., coriaceous, strigose below, glabrous above. Asci clavate, 70–75 x 10–12 μ, (paraphysate)? Sporidia densely packed, very numerous, hyaline, oblong, 6–8 x 1½–2½ μ, with a nucleus in each end.

On bark of *Cornus*, Carolina, Pennsylvania and Canada.

Fr. brevibarbàta, (B. & C.)

Cucurbitaria brevibarbata, B. & C. Grev. IV, p. 47
Fracchiæa brevibarbata, Sacc. Syll. 386.

"Cespitose, globose, not collapsing, minutely tomentose. Asci clavate, stuffed with the allantoid sporidia."

No habitat, locality or measurements are given.

NÍTSCHKIA, Otth.

In Fckl. Symb p. 165

Perithecia cespitose-erumpent, or, when on decorticated wood, superficial, spherical, collapsing to cup-shaped, bald and black; texture subcoriaceous. Asci clavate. Sporidia short-cylindrical or rod-shaped, continuous, hyaline.

N. cupulàris, (Pers.) **(Plate 26)**

Sphæria cupularis, Pers. Syn. p. 53
Sphæria cucurbitula, b. nigrescens, Tode Fungi Meckl. p 39.
Cucurbitaria cupularis, Cke. Hndbk. p. 842.
Nitschkia Fuckelii, Nits. in Fckl. Symb. p. 165.
Nitschkia cupularis, Karst. Myc. Fenn II, p. 81
Cœlosphæria Fuckelii, Sacc. M. Ven. Spec. p 115.
Cœlosphæria cupularis, Sacc. Syll. I. p. 91.
Exsicc. Fckl. F. Rh. 968.—Thum. Myc. Univ. 1947

Perithecia cespitose, erumpent in small (2 mm.), dense clusters closely surrounded by the ruptured epidermis, spherical, but collapsing to cup-shaped, 200–300 μ diam. Asci 40–60 x 7–8 μ, contracted below into a stipe-like base, and surrounded by filiform paraphyses, 8-spored. Sporidia subbiseriate, allantoid, slightly curved, hyaline, with a nucleus in each end, 9–10½ x 2–3 μ.

On dead branches of various deciduous trees, *Tilia, Acer, Prunus*, &c., on bark of *Negundo Aceroides*, Montana (Anderson, 276).

The Montana specimens are the only American specimens we have seen. They agree in all respects with the specc. in Thum. M. U. and with the description of this species in Winter's Pilze and Saccardo's Sylloge.

N. tristis, (Pers.)

Sphæria tristis, Pers Syn p 87
Nitschkia tristis, Fckl Symb. p 165.
Cœlosphæria, tristis, Sacc. Syll. 378.
Exsicc. Rab. F. E 632 —Plowr Sph. Brit 63 —Cke. F Brit 2d Ser 269

Perithecia densely gregarious, superficial, depressed-globose, soon collapsing to flat cup-shaped, coarsely wrinkled, black, without any distinct ostiolum, ½–¾ mm. diam., seated on a scanty, black, pilose subiculum. Asci clavate with a slender stipe, 8-spored, 40–45 x 8–9 μ. Sporidia lying irregularly in the asci, oblong or clavate-oblong, mostly straight, 4-nucleate, hyaline, 9–11 x 2–2½ μ.

On bark, Carolina (Schw.), also Carolina and Maine (Berkeley).

The specimen in Herb. Schw., 1413, labeled "*Sphæria cupularis*," is this species, having the perithecia ½–¾ mm. diam., and asci shorter and broader than in *S. cupularis*. Whether this is the *Sphæria tristis*, Tode, is doubtful. Persoon gives *S. tristis*, Tode, interrogatively as a synonym of his *S. tristis*. Both speak of the thin, tomentose subiculum, but Tode says the perithecia are so small as to be scarcely visible, which certainly does not apply to the species here described as *Nitschkia tristis*, (Pers.)

The specimen in Fungi Gallici, 1485, labeled *Sphæria tristis*, Tode, is (in our copy) only a patch of black tomentum without any perithecia. In the specc. issued by Cooke and Plowright the sporidia are 7–9 x 2–2½ μ, the contents often divided by a pseudoseptum across the middle.

Specc. on rotten maple from Ohio (Morgan, 976) have abundant, narrow-elliptical, hyaline sporidia? 6–8 x 3 μ, but no asci, and may be the spermogonial stage of this species.

N. euómphala, (B. & C.)

Sphæria euomphala, B. & C Grev. IV, p 141.
Sphæria craterella, B & Rav in Herb. Berk
Byssosphæria euomphala, in Cooke's Synopsis, 2603
Botryosphæria euomphala, Sacc Syll. 1784
Exsicc. Rav. Fung Car IV, 54

Perithecia densely gregarious, about ½ mm. diam., or a little less, spherical, minutely tubercular-roughened, soon collapsing. Asci clavate. Sporidia elliptical, continuous, smoky-hyaline, or pale brown, 6–8 x 3½–4½ μ.

On bark of dead trunks of *Fraxinus*, So. Carolina (Ravenel).

The perithecia are seated on a thin subiculum of creeping, septate hyphæ, with which they are also mostly fringed at base.

CŒLOSPHÆRIA, Sacc.

Mycotheca Ven. Spec p 115 (emended)

Perithecia scattered, superficial or at first covered by the epidermis, membranaceo-coriaceous, collapsing, smooth or bristly. Asci and sporidia as in *Nitschkia*.

We have modified Saccardo's generic characters so as to embrace only species with *scattered* perithecia. The evidently close relationship to *Nitschkia* forbids their removal to another family.

C. corticàta, E. & E. Proc. Acad. Nat. Sci. Phil., July, 1890, p. 222.

Perithecia scattered, globose, about ½ mm. diam., closely enveloped, except the papilliform ostiolum and the apex, by the adherent epidermis, and clothed with a thin coat of brown, branching, sparingly septate hairs about 3 μ thick. Asci (p. sp.) about 35 x 7 μ, 8-spored. Sporidia crowded-biseriate, 2-nucleate, hyaline, moderately curved, obtuse, 10–14 x 3 μ. The perithecia soon collapse down to, or a little beyond the part embraced by the epidermis, and become strongly concave.

On bark of dead *Maclura aurantiaca*, Missouri (Demetrio, 272)

C. fusariόspora, E. & E. Journ. Mycol. IV, p. 65.

Exsicc. Ell. & Evrht. N A F 2d Ser. 1957

Perithecia scattered, erumpent, about 165 μ diam., collapsing to cup-shaped, black, with a papilliform ostiolum. Asci clavate-cylindrical, 50 x 7 μ, with filiform paraphyses. Sporidia biseriate, arcuate-fusiform, continuous, 18–22 x 2½ μ, ends acute.

On bark of cottonwood, Kansas (Egeling).

C. (?) chiliopýxis, (B. & C.)

Sphæria chiliopyxis, B. & C. Grev IV, p 141.

"Extremely minute, gregarious, globose, bright. Sporidia hyaline, slightly sausage-shaped."

On rotten logs, Carolina.

C. exìlis, (A. & S.)

Sphæria exilis, A & S. Consp. p 44
Nitschkia exilis, Fckl. Symb. p. 44.
Cœlosphæria exilis, Sacc. Syll. 379

Perithecia gregarious, superficial, seated on a thin, scanty, whitish, cottony subiculum, membranaceo-coriaceous, 150–250 μ diam., globose, soon collapsing to shallow cup-shaped, sparingly clothed with black, continuous bristles, 60–80 x 6 μ. Ostiolum inconspicuous. Asci (p. sp.) 25–30 x 4 μ long. Sporidia subbiseriate, hyaline, cylindrical, curved, 3–5 x 1 μ.

On *Prunus serotina*, *Quercus Montana*, and *Cornus florida*, in

Carolina (Berk), on decaying poplar, Carolina (Fries, in S. M. II, p. 452).

The above characters are from a Finland specimen, on rotten wood, from Karsten. They certainly agree well with the description in A. & S. Conspectus. Whether this or *Melanomma exile*, (Schw.) (see p. 184) is the true *Sphæria exilis*, A. & S., we can not say, but as they are evidently two distinct things, we have given descriptions of both.

GÍBBERA, Fr.

Summ. Veg. Scand. p. 402.

Perithecia cespitose, on a superficial, thick, dematiaceous, conidia-bearing stroma, carbonaceous, fragile, bristly, obsoletely papillate. Asci cylindrical, 8-spored. Sporidia uniseriate, oblong-elliptical, uniseptate, pale yellow.

Fries takes this genus in a more comprehensive sense, including also *Gibberella* (see *Hypocreaceæ*).

G. Vaccínii, (Sow). **(Plate 26)**

Sphæria Vaccinii, Sow Eng. Fung, tab 373, fig. 1
Gibbera Vaccinii, Fr. l c.
Exsicc. Fckl. F. Rh 1757.—Rab Herb. Myc. 435.—Rehm Asc 636—Thum. F Austr 959. id. M. U. 1851.—Plowr. Sph. Brit 296.—Krieger F Sax. 168.—Roum. F. G. 565

Perithecia crowded on a black, felt-like stroma, sphæroid, slightly attenuated below, obsoletely papillate, about $\frac{1}{3}$ mm. diam., clothed with short (60–70 x 6–7 μ), rigid, black, continuous bristles. Asci cylindrical, short-stipitate, 8-spored, 90 x 8–9 μ. Sporidia obliquely uniseriate, oblong-elliptical, uniseptate, scarcely constricted at the septum, hyaline or pale olivaceous, 15–18 x 7–8 μ.

The conidial stage is *Helminthosporium Vaccinii*, Fr.

On branches of *Vaccinium Vitis Idæa*. Common in Europe but thus far not found here. This species matures only on dead branches.

ÓTTHIA, Nitschke.

Fckl. Symb Mycol p 169

Perithecia erumpent, cespitose or gregarious, subspherical, minutely papillate, black, glabrous, carbonaceo-coriaceous. Asci cylindrical, 8-spored, paraphysate. Sporidia elliptical or fusoid, brown or brownish-yellow, uniseptate. The pycnidial stage is a *Diplodia*.

A. *Sporidia yellow or brownish-yellow.*

O. hypoxyloides, (E. & E.) **(Plate 26)**

Amphisphæria Hypoxylon, E & E. Journ Mycol II, p 41.

Perithecia minute (80–90 μ), ovate-globose, white inside, rough and black outside, mostly radiate-sulcate around the ostiolum, which is not prominent and finally rather broadly perforated, densely crowded and partly sunk in a flat, blackish-brown, subcarbonaceous stroma $\frac{1}{2}$–1 cm. across or, by confluence, more, much resembling the sterile stroma of some *Hypoxylon*. Asci clavate-cylindrical, 55–60 x 10–12 μ, with abundant paraphyses. Sporidia obliquely uniseriate, ovate, uniseptate, brown, 6–9 x $3\frac{1}{2}$–$4\frac{1}{2}$ μ.

On rotten wood, Louisiana (Langlois), Delaware (Commons).

O. Symphoricárpi, (E. & E.)

Plowrightia Symphoricarpi, E. & E Proc Acad Nat Sci. Phil. July 1890, p 249. Exsicc. Ell. & Evrht N. A. F. 2d Ser 2374

Stroma convex, penetrating to the wood, but not limited by any black circumscribing line, brownish-black, whitish within, $1\frac{1}{2}$–2 mm. diam. Perithecia 10–15 in a stroma, $\frac{1}{3}$ mm. diam., the upper part mostly prominent and free. Sometimes the stroma is wanting, the perithecia being then simply cespitose or subsolitary. Ostiola obtusely conical, nearly smooth or indistinctly radiate-sulcate. Asci clavate-cylindrical, subsessile, 75–80 x 12 μ, with paraphyses. Sporidia uniseriate or subbiseriate above, ovate-elliptical, uniseptate and constricted, hyaline and granular at first, becoming yellow-brown, 15–18 x 10 μ.

On dead branches of *Symphoricarpus occidentalis*, Sand Coulee, Cascade Co., Montana (Anderson, No. 210).

O. Àceris, Winter, Hedw. 1871, p. 162.

Perithecia cespitose, depressed-globose, brownish-black, papillate, thin, $\frac{1}{2}$ mm. diam., seated on the surface of the inner bark in groups of 3–10, soon erumpent through the ruptured epidermis, and then almost superficial. Asci cylindrical, sessile, paraphysate, 180–200 x 35 μ. Sporidia overlapping-uniseriate or biseriate, oblong, uniseptate and constricted at the septum, hyaline, with a broad, hyaline envelope, becoming brown, 50–60 x 15–18 μ, each cell with a large nucleus.

On bark of dead maple limbs, Lyndonville, N. Y. (Fairman).

The sporidia exceed the measurements given by Winter, but otherwise the specimens agree with his diagnosis. The specc. in Roum. F. G. 5636, are only *Diplodia*.

O. quercícola, E. & E. (in Herb.)

Perithecia densely crowded, subglobose, 150–200 μ diam., brownish-black, granular-roughened, collapsing above, white inside, seated on a brown, felt-like subiculum of matted, septate, sparingly branched hairs, 5–6 μ diam., forming patches $\frac{1}{4}$–$\frac{1}{2}$ cm. across and generally surrounding the twig or limb, which is there swollen as in *O. morbosa*. Asci oblong, subventricose, contracted below into a short stipe, 8-spored, paraphysate, 75–80 x 18–20 μ. Sporidia crowded-biseriate, ovate-elliptical, subinequilateral, uniseptate, scarcely constricted, hyaline, becoming yellow-brown, 18–22 x 8–10 μ.

On dead twigs of white oak, Newfield, N. J.

O. fruticola, E. & E.

Parodiella fruticola, E. & E. Journ. Mycol. IV, p. 97.
Dothidea insculpta, Wallr. in Roum. F. Gall. 547, and 4955.
Exsicc. Ell. & Evrht. N. A. F. 2d Ser. 2129.

Perithecia obovate, astomous at first, about 1 mm. diam., black, flattened above and finally umbilicate-collapsed and irregularly or sublacinately ruptured above, seriately erumpent through cracks in the bark, often densely crowded, but not confluent. Asci subcylindrical, 100–150 x 15–20 μ. Paraphyses cylindrical, often branching below, faintly septate, evanescent. Sporidia uniseriate or occasionally more or less perfectly biseriate, broad fusoid-oblong, subinequilateral, uniseptate, straw-yellow, 30–35 x 8–15 μ. The perithecia are often subangular from mutual pressure and are at first filled with a whitish, grumous mass, but finally become empty.

On dead stems of *Clematis ligusticifolia*, Sand Coulee, Montana (Anderson).

The Montana specc. are certainly the same as those issued in F. Gall., above quoted, under the name of *Dothidea insculpta*, Wallr., but that species (sec. Sacc. in Syll.) has cells peripherical and minute (loculis periphericis, exiguis), which is not the case either with the Montana or F. G. specc., in which the perithecia are as described above. Specc. in our Herb. labeled *Dothidea insculpta*, Wallr., collected by Dr. Morthier in Switzerland, agree with the F. G. and Montana specc. only the sporidia are 3-septate.

B. *Sporidia subhyaline* (*Otthiella, Sacc.*)

O. álnea, (Pk.)

Cucurbitaria alnea, Pk. 28th Rep. p. 75
Otthia alnea, Sacc Syll. 2804, Cke Syn. 2490.

Perithecia cespitose, erumpent, astomous, black, white within, the tufts closely surrounded by the transversely ruptured epidermis. Sporidia uniseriate, uniseptate, subacuminate, constricted at the septum, nearly colorless, with one or two nuclei in each cell, 20–25 μ long.

On dead alder branches, Center, N. Y., in company with *Torula alnea*.

Var. *carnosa*, Cke. Grev. XV, p. 84, has the sporidia 3-septate, (*Montagnella*)? Specimens from Peck (in our Herb.) have the perithecia ovate-globose, rough, ¼–⅓ mm. diam. Ostiolum indistinct. Clusters of closely packed, subconfluent perithecia elliptical, 1–1½ x 2 mm. diam. Asci cylindrical, 75–80 x 6–7 μ. Sporidia 20–25 x 4 μ. Differs from *O. Alni*, Winter, in its smaller, hyaline sporidia.

O. staphýlina, E. & E. **(Plate 41)**

Plowrightia staphylina, E. & E. Proc Acad Nat. Sci. Phil July 1890, p. 248

Cespitose, clusters of perithecia about 1 mm. diam., mostly seriately confluent for several centimeters in length, erumpent through cracks in the bark and only slightly prominent. Perithecia black, small, 150–200 μ diam., ovate or obovate, narrowed below into a substipitate base. Ostiolum conic-papilliform, soon broadly and somewhat irregularly perforated. Asci clavate-cylindrical, subsessile, paraphysate, 60–65 x 8 μ. Sporidia mostly biseriate, oblong or clavate-oblong, hyaline, uniseptate and slightly constricted, 12–15 x 4–5 μ, ends obtuse.

On bark of *Staphylea trifolia*, London, Canada (Dearness).

O. seriàta, (Pk.)

Cucurbitaria seriata, Pk. 28th Rep p. 75.
Otthia seriata, Sacc. Syll. 2801, Cke. Syn. 2487.

Perithecia cespitose, erumpent in long, flexuous, interrupted lines, small, nearly globose, black, white within, sometimes collapsing, the stroma, if present, merely cortical and subferruginous. Asci cylindrical or subclavate. Sporidia uniseriate or rarely crowded, uniseptate, oblong-elliptical, slightly constricted at the septum, hyaline. 10–13 μ long.

On dead bark of *Euonymus*, Albany, N. Y.

O. morbósa, (Schw.)

Sphæria morbosa, Schw. Syn. Car. 134, Fr. S. M. II, p 417, Farlow in Bull. Buss. Inst 1876, p. 449, tab. IV—VI.
Plowrightia morbosa, Sacc. Syll. 5295.
Exsicc. Roum. Fungi Gall. 4450.—Ell. N. A. F. 691.

Perithecia globose, ¼ mm. diam., smooth, black, collapsing, densely crowded and forming a continuous layer on the surface of elongated swellings 4–8 cm. long and 1–2 cm. thick, formed from the scarcely altered substance of the bark. Asci clavate-cylindrical, with a short stipe, 8-spored, paraphysate, 120 x 18–20 μ. Sporidia obliquely monostichous, ovate-oblong, and rounded at the apex, subacute below, uniseptate, the lower cell narrower, hyaline, 16–20 x 8–10 μ.

Stylospores in similar perithecia, ovoid, 3-septate, yellowish, stipitate, 12 x 6 μ. Spermatia minute on slender basidia. Conidia produced on the surface of the young stroma, ovoid subolivaceous, 16 μ long, borne on filiform, simple hyphæ 40–60 x 4 μ.

Common on limbs of plum and cherry trees, to which it is very destructive.

The genus *Plowrightia*, in Sacc. Syll., is unsatisfactory, embracing species with true perithecia, such as *Sphæria morbosa*, Schw., and others with mere ascigerous cells, such as *Dothidea ribesia*, (Pers.), which is a genuine *Dothidea*. The former are here included in the subgenus *Otthiella*, as has already been done by Cooke in Grev. XV, p. 84.

MONTAGNÉLLA, Speg.

Fungi Arg. Pugill IV, p. 70.

Perithecia densely gregarious, on an effused stroma, free above, adnate below. Asci 8-spored, subcylindrical. Sporidia finally 3-septate and yellowish or brownish.

M. Heliópsidis, (Schw.) **(Plate 41)**

Dothidea Heliopsidis, Schw. Syn. Car. No 69.
Exsicc. Ell. N. A. F 682.

Perithecia depressed-globose, black (white within), connected at the base by a stromatic crust surrounding the stem on which it grows and extending longitudinally for 1 cm. or more. Asci cylindrical, short-stipitate, 90–100 x 8–10 μ. Sporidia obliquely uniseriate or biseriate, fusoid-oblong, frequently curved, 3-septate, 28–30 x 4–5 μ, yellowish-hyaline.

On dead stems of *Helianthus divaricatus?* Newfield, N. J., on *Heliopsis*, Carolina (Schweinitz).

M. platýplaca, (Berk.)

Dothidea platyplaca, Berk. & Curt in N. Pac. Expl. Exp. p 129, No 163.
Montagnella platyplaca, Sacc Syll. 5332

Stroma thin, black, penetrating the matrix, shining above, opake and fertile below, perithecia (cells)? superficial. Sporidia (sec. Cke. Grev. XIII, p. 70) 1–3-septate, becoming brownish.

On leaves, Nicaragua.

M. tumefàciens. (Ell. & Hark.)

Sphæria (Montagnella) tumefaciens, Ell. & Hark. Journ Mycol II. p 41.
Montagnella tumefaciens, Sacc. Syll. 7315
Exsicc Ell. & Evrht N. A. F. 2d Ser. 1667

Perithecia hemispherical, 200–333 μ diam., black. rough, crowded, united below in a crustose stroma, bursting out through longitudinal cracks in the bark and forming densely-compacted series, continuous or interrupted, 3–5 cm. long, on swollen portions of the limb or stem. Asci 100–120 x 10–12 μ, subcylindrical, with abundant paraphyses. Sporidia mostly biseriate, cylindric-fusiform, slightly curved, 3-septate, yellowish, 25–30 x 6–7 μ. Bears a general resemblance to *Dothidea morbosa*, Schw., but the perithecia are arranged in series and the sporidia are quite different.

On dead limbs of *Artemisia Californica*, Mt. Diabolo, Cala. (Harkness).

PARODIÉLLA, Speg.

Fungi Arg. Pugill. I, p 178.

Perithecia gregarious, superficial, astomous, black, base adnate to the leaf. Asci cylindric-clavate, 8-spored, paraphysate. Sporidia oblong or subelliptical, brown, uniseptate. Foliicolous.

P. grammódes, (Kze.) **(Plate 41)**

Sphæria grammodes, Kze. in Weigand's Exs
Dothidea grammodes, Kze in Berk Cuban Fungi, No. 868
Actidium Crotalariæ, Schw. MS.
Dothidea perisporioides, B & C. Grev. IV, p 103
Dothidea seminata, B & Rav. Grev IV, p 104.
Dothidea grammodes, Sacc. Syll. 5279.
Parodiella grammodes, (Kze.) Cke Syn, 1375.
Exsicc Ell N. A. F. 685.—Rab. F E. 3251.

Perithecia globose, astomous, black, 150–200 μ diam., base adnate, finally more or less wrinkled and umbilicate-collapsed above, often covering the whole upper surface of the leaf. Asci clavate, short-stipitate, 75 x 20 μ, paraphysate, 8-spored. Sporidia biseriate, oblong-elliptical or subbiconical, uniseptate and slightly constricted in the middle, hyaline, becoming brown, upper cell mostly a little broader, nearly straight, ends obtuse, 20–24 x 7–8 μ (28–30 x 10–11 μ, Sacc.)

On living leaves of various leguminous plants. *Psoralea*, *Rhyncosia*, *Desmodium* and *Indigofera*, Southern and Western States, common.

Dothidea seminata, B. and Rav., is given as a synonym on the authority of Cooke in Grevillea, XIII, p. 106.

P. simíllima, (B. & Rav.)

Dothidea simillima, B. & Rav. Grev. IV, p. 104.
Dothidea Desmodii, Curtis, (fide Sacc. 5096)
Parodiella simillima, Cke. Syn. 1374.

Perithecia scattered, smaller than in *P. grammodes*, granulated. Asci narrow. Sporidia hyaline, arcuate, suddenly attenuated at each end as in many *Vermiculariæ*, 15 μ long, uniseptate.

On leaves of *Desmodium*, Carolina (Ravenel).

WALLROTHIÉLLA, Sacc.

Syll. I, p. 455.

Perithecia superficial or subsuperficial, globulose, glabrous, black. Asci subcylindrical, 8-spored. Sporidia ovoid, elliptical or suboblong, continuous, hyaline.

W. Arceuthòbii, (Pk.)

Sphæria Arceuthobii, Pk. 27th Rep. p. 111.
Wallrothiella Arceuthobii, Sacc. Syll. 1756

Perithecia small, densely cespitose, oblong or cylindrical, very obtuse, shining black. Asci subclavate, fugacious. Sporida crowded, globose, colorless, about 4 μ diam.

On capsules of *Arceuthobium pusillum*, Forestburg, N. Y.

It forms little black tufts crowning the fruit at the tips of the stems and branches.

W. eunotiæ̀spora, (Cke. & Hark.)

Sphæria eunotiæspora, Cke. & Hark. Grev. XIII, p 18
Wallrothiella eunotiæspora, Sacc. Syll. 6399.

Superficial, gregarious. Perithecia obpyriform (½ mm.), black, subshining, smooth, fragile. Asci clavate. Sporidia biseriate, elliptical, inflated in the middle, continuous, hyaline, 30–35 x 12–14 μ, with granular plasma.

On decorticated Australian *Acacia*, mixed with *Diplodia*, California (Harkness).

W. melanostígma, (C. & E.)

Sphæria melanostigma, (C. & E.) Grev. VI, p. 13. tab 95, fig. 16.
Wallrothiella melanostigma, Sacc. Syll. 1760.
Exsicc. Ell. N. A. F. 782.

Gregarious, very small, black. Perithecia subglobose, 150 μ diam. papillate, emerging from the fibers of the wood, submembranaceous. Asci cylindric-clavate. Sporidia elliptical, small, hyaline, nucleate. 10 x 8 μ.

On dead places in living oak limbs, Newfield, N. J.

W. minutissima, (Crouan).

Sphæria minutissima, Crouan, Finist. p. 23
Wallrothiella minutissima, Sacc. Syll. 1761

Perithecia superficial or nearly so, depressed-globose, strongly papillate, black, scattered, 100–150 μ diam. Asci cylindrical, p. sp. 35–40 x 3½ μ. Sporidia uniseriate (end to end), oblong-elliptical, hyaline, 3½–4 x 2½ μ. Paraphyses obscure.

On dry cow dung, Newfield, N. J.

W. macilénta, (Cke.)

Cucurbitaria macilenta, Cke. Grev. VII. p 4
Wallrothiella macilenta, Sacc. Syll. 1758.

Tufts very small, erumpent. Perithecia black, papillate, subshining, collapsing when dry. Asci cylindrical. Sporidia uniseriate, elliptical, attenuated at each end, binucleate, 15 x 6 μ.

On *Abies* and *Libocedrus*, California.

W. squalídula, (C. & P.)

Sphæria squalidula, C. & P. 29th Rep. N. Y. State Mus. p. 61.
Wallrothiella squalidula, Sacc. Syll. 1759.

Perithecia gregarious, globose, semiimersed, pierced at the apex, about 300 μ diam., black. Asci cylindrical. Sporidia uniseriate, elliptical, simple, binucleate, hyaline, 12–17 μ long.

On decaying chestnut wood, New York.

W. consociàta, (Ell. & Hark.)

Sphæria consociata, Ell & Hark, Bull. Torr. Bot. Club. VIII, p. 52

Perithecia scattered, erumpent-superficial, depressed-hemispherical, about 200 μ diam., with a short, nipple-like, black ostiolum. Asci subcylindrical, mostly curved at least when young, attenuated above, about 40 x 5 μ, sessile, or nearly so. Sporidia subbiseriate, oblong-fusoid, nucleolate, slightly curved, 6–8 x 1½ μ, yellowish-hyaline.

On foliage of *Sequoia gigantea*, California.

We have drawn these characters from a reexamination of the original specc. and find the sporidia shorter than stated in the original description. In some specc. paraphyses were seen, but they were very rare and very evanescent.

FAMILY. SPHÆRELLOÌDEÆ.

Perithecia membranaceous, small, mostly perforated with a simple pore at the apex, without any distinct ostiolum, at first covered by the epidermis, finally more or less erumpent or even superficial. Asci mostly fasciculate. Paraphyses wanting or obscure.

LÆSTÁDIA, Awd.

Hedwigia 1869. p 177

Perithecia innate, membranaceous, typically pierced at the apex with a simple pore. Asci 8-spored, without paraphyses, mostly clavate. Sporidia ovoid or suboblong, continuous, hyaline.

Minute fungi growing mostly on leaves.

A. *On leaves of dicotyledonous trees and shrubs.*

L. rhytismoìdes, (Berk.)

Sphæria rhytismoides, Berk. Brit. Fungi, 178.
Sphæria Dryadis, Fckl. Symb. p. 108, tab. II, fig. 41.
Sphærella rhytismoides, De Not. Recl. Pir p. 12.
Læstadia rhytismoides, Sacc. Syll 1611, Cke. Syn. 5211.

Perithecia epiphyllous, rather large, semiimersed on large, dark colored, indeterminate spots, globose, very black, perforated, at length depressed. Asci fasciculate, stipitate, clavate, 8-spored, 50–55 x 14–16 μ. Sporidia inordinate, or distichous, oblong-ovate, subinequilateral, ends obtuse, guttulate, 14–16 x 5–6 μ, hyaline.

On leaves of *Dryas integrifolia*, Greenland.

L. carpínea, (Fr.)

Sphæria carpinea, Fr. S. M. II p. 523.
Ascospora carpinea, Fr. Summa. Veg. Sc. p. 425.
Sphærella carpinea, Awd. Myc Eur. l. c. p. 2.
Læstadia carpinea, Sacc. Syll. 1619, Cke. Syn. 5223.
Exsicc Fckl. F. Rh. 466.—Rab F. E. 365—Cke. F. Brit Ser. 1. 165.—Desm. Pl Crypt. Ed. 1. 981.—Ed. 2 285

Perithecia amphigenous, sunk in the substance of the leaf and covered by the blackened epidermis, scattered, globose, with a simple pore at the apex, black, 120–150 μ broad. Asci oblong-clavate, narrowed into a short stipe below, 8-spored, 50–60 x 8 μ. Sporidia biseriate, oblong-elliptical, inequilateral, subobtuse, one-celled, hyaline, 14–15 x 4–5 μ.

On leaves of *Carpinus*, Bethlehem, Pa. (Schweinitz).

L. albocrustàta, (Schw.)

Sphæria albocrustata, Schw Syn. N Am. 1791.
Læstadia albocrustata, Sacc. Syll. 6000, Cke. Syn. 5192.

Seated on a light cinereous, pulverulent, indeterminate, vaguely effused crust. Perithecia scattered, black, variously crowded on the crust, which in this way has a sculptured appearance, not however, confluent, at first convex, sub-rugose, at length collapsing. Asci clavate, numerous, 16 μ long. Sporidia linear, obtuse, very small, 4 x 1½ μ, hyaline.

On the lower surface of leaves of *Platanus*, Bethlehem, Pa.

L. orientàlis. E. & E. Proc. Acad. Nat. Sci. Phil. July, 1890.

Perithecia amphigenous, depressed-hemispherical. 180–200 μ diam., scattered, erumpent. Asci clavate-cylindrical, p. sp. 45–50 x 12 μ or, including the slender base, 70–75 μ long. Paraphyses none. Sporidia crowded-biseriate, inequilaterally elliptical, hyaline, granular and nucleolate, 12–14 x 5–6 μ. There is also on the same leaves a *Septoria* with gregarious, subglobose 80–100 μ perithecia, and hyaline, nucleate, subundulate, 12–25 x 1–1½ μ sporules—agreeing with the description of S. *Gilletiana* Sacc. in all but its smaller continuous sporules.

On dead leaves of "Japan Chestnut" *Castanea japonica* (cult.), LaFayette, La. (Langlois.)

L. gregària, (Cke.)

Stigmatea gregaria, Cke. Texas Fungi, No 144.

Epiphyllous. Perithecia gregarious, black, erumpent, globose, somewhat shining. Asci cylindrical. Sporidia subglobose, hyaline. 10–12 x 9 μ, continuous.

On unknown leaves, Meskat Bay, Texas (Ravenel).

L. auripúnctum, Hark. Bull. Cal. Acad. March, 1884, p. 43.

Hypophyllous. Perithecia scattered, 5–10 on yellowish, orbicular spots, 2–5 mm. diam., bordered by a dark brown line. Nucleus orange. Asci 8-spored, fusiform, long-stipitate, 54 x 10 μ. Sporidia ovate or clavate, hyaline, 10 x 6 μ.

On living leaves of *Quercus Wislizeni*, Folsom, Cala.

L. cœlàta, Harkness, l. c.

Perithecia hypophyllous, entirely concealed, scattered. Asci mucoid, clavate, long-stipitate, 45 x 15 μ. Paraphyses none. Sporidia 8, hyaline, turbinate, "shining," 12 x 6 μ, endochrome divided very near the pointed end of the sporidium.

On dead leaves of *Quercus densiflora*, California (Harkness).

L. polystígma, (E. & E.)

Sphærella (Læstadia) polystigma, E. & E Bull Torr Bot. Club, X, p 117.
Læstadia polystigma, Sacc. Syll 6371, Cke. Syn 5203

Exsicc. Ell. N. A. F 1353

Perithecia scattered thickly over the lower surface of the leaf, subhemispherical, 120–150 μ diam., covered by the blackened cuticle, finally collapsing. Ostiolum papilliform, at length perforated. Asci oblong, sessile, 35–40 x 8 μ, 8-spored. Sporidia biseriate, ovate-elliptical, continuous, subhyaline, 10–12 x 3–4 μ, in shape very much like apple seeds.

On fallen leaves of *Quercus coccinea*. Ohio and New Jersey.

L. Leucóthoes, (Cke.)

Sphærella Leucothoes, Cke. Journ Bot 1883
Læstadia Leucothoes, Sacc. Syll 6009, Cke Syn. 5247
Exsicc. Rav F Am 687.

Epiphyllous. Spots whitish, suborbicular, confluent, with a red margin. Perithecia very small, immersed, with the punctiform ostiola emergent, black. Asci clavate-cylindrical. Sporidia elliptical, continuous, hyaline, 13–15 x 4½ μ.

On leaves of *Leucothoe*, South Carolina.

L. Æsculi, Pk. 39th Rep. p. 51.

Perithecia small (150 μ), lenticular, covered by the epidermis, erumpent, opening by a minute pore, black. Asci subclavate. Sporidia crowded, subelliptical, colorless, 8½–10 x 5–6¼ μ.

On fallen petioles of horse chestnut, Albany, N. Y.

(Immature *Sphærella petiolicola*)?

L. hæmatòdes, (B. & C.)

Sphærella hæmatodes, B. & C. in Herb Berk (Cke in Journ. Bot. 1883, p 70)
Læstadia hæmatodes, Cke. Syn 5245

Epiphyllous. Spots orbicular, scattered or confluent, dark red, broadly margined. Perithecia very small, black, semiinnate, punctiform. Asci cylindric-clavate. Sporidia narrow-elliptical, continuous, hyaline, 8–10 x 2½ μ.

On *Kalmia glauca*, United States.

Cooke in Journ. Bot. 1883, p. 70, says. "Very similar externally to *Sphærella colorata*, but asci and sporidia are little more than half as long and sporidia not septate as far as seen, but immature, as they failed to leave the asci."

L. stigmatòdes, (B. & C.)

Sphærella stigmatodes, B. & C. (Cooke in Journ Bot. 1883, p. 68).
Læstadia stigmatodes, Sacc. Syll. 5999. Cke. Syn 5179

Perithecia scattered, punctiform. (resembling *S. punctiformis*).

Asci clavate-cylindrical. Sporidia narrowly elliptical, obtuse, continuous, hyaline, 8–10 x 2–2½ μ.

On leaves, Maine.

L. castanicola, E. & E. (in Herb.)

Perithecia gregarious on indefinite, pale spots; mostly hypophyllous, erumpent, depressed-hemispherical, black, pierced above, 200–250 μ diam. Asci clavate, p. sp. 60 x 10 μ, aparaphysate, subsessile. Sporidia biseriate above, uniseriate below, inequilaterally elliptical, hyaline, continuous, 10–12 x 5–6 μ.

On decaying chestnut leaves still hanging on twigs broken off last fall, Newfield, N. J., April, 1891.

Differs from *L. echinophila*, (Schw.), in its foliicolous growth and larger asci and sporidia.

L. rubéscens, E. & E. (in Herb.)

Perithecia densely gregarious on pale spots on the upper side of the leaf, minute (75–80 μ), buried in the parenchyma of the leaf, but visible through the raised epidermis, perforated above. The pale spots in some of the spece. finally assume a reddish hue. Asci cylindrical, about 40 x 6 μ, sessile. Sporidia uniseriate, closely packed, inequilaterally elliptical, hyaline, continuous, 5–6 x 3½–4 μ.

On decaying chestnut leaves with *L. castanicola*, from which it differs in its smaller, more densely gregarious, epiphyllous perithecia and smaller sporidia.

L. Magnòliæ, (Schw.)

Sphæria Magnoliæ, Schw Syn N Am 1808
Læstadia Magnoliæ, Sacc Syll 6004, Cke. Syn. 5230
Sphærella Magnoliæ, Ell Bull. Torr Bot. Club, IX p 74?
Exsicc. Ell N. A. F. 800

Perithecia aggregated in orbicular or variously shaped groups, finally confluent and occupying the whole lower surface of the leaf, punctiform, covered, flattened, astomous, rising with the epidermis, at length collapsing, black, but with the disk whitening out. Makes the upper surface of the leaf bullate. Asci sublanceolate. Sporidia biseriate, clavate-oblong, hyaline, 7 x 2½ μ.

On leaves of *Magnolia glauca*, Kaign's Point and Newfield, N. J.

There is no spec. of this species in Herb. Schw., so that we can not be sure that the specc. in N. A. F. are the *S. Magnoliæ*, Schw.

L. fraxinícola, (Curtis & Pk.)

Depazea fraxinicola, C & P. in Peck's 23d Rep p 64
Læstadia fraxinicola, Sacc. Syll 1626, Cke. Syn 5234

"Spots arid, suborbicular, sometimes with a brownish border. ¼–½ an inch diam. Perithecia black, those well developed are concave above with a slight elevation in the center. Asci subfusiform, 4–8-spored, spores densely packed in the asci, oblong, or narrow-elliptical, having a little nucleus near each end, 12–15 x 4–6 μ, about ⅓ as long as wide."

On leaves of *Fraxinus Americana*, Albany, N. Y. (Peck).

Apparently different from *Sphærella fraxinicola*, (Schw.)

L. spinícola, (E. & E.)

Sphærella spinicola, E. & E. Proc. Acad. Nat. Sci. Phil. July, 1890, p. 231.

Perithecia scattered or 3–4 together, minute 110–120 μ, collapsing, visible through the translucent epidermis as minute black specks. fringed around the base with scanty mycelium. Ostiolum papilliform. Asci oblong, sessile, 45 x 12 μ (p. sp.) Sporidia biseriate, oblong-elliptical, continuous, granular, rounded at the ends, 12–15 x 5–6 μ.

On spines of *Rosa rubiginosa*, West Chester, Pa.

B. *On leaves and stems of dicotyledonous, herbaceous plants.*

L. cineráscens, (Schw.)

Sphæria cinerascens, Schw. Syn. N. Am. 1795
Sphærella cinerascens, Cke in Journ. Bot. 1883
Læstadia cinerascens, in Cooke's Syn Grev. 18, p. 65 (5227).

Spots amphigenous, very large, irregular and indeterminate, cinereous below, black above. Perithecia innumerable, crowded on the spots, very small, black, innate, subacuminate, astomous or at length perforated, scattered or collected and arranged so as to represent various engraved or sculptured figures. Asci clavate. Sporidia narrow-elliptical, hyaline, 8 x 2¼ μ.

On leaves of *Asclepias syriaca*, Bethlehem, Pa.

L. Apócyni, E. & E. Proc. Phil. Acad. l. c.

Perithecia gregarious, depressed-spherical, 150–200 μ diam., perforated above, covered by the cuticle through which they are visible by translucence. Asci clavate-cylindrical, 40–50 x 10–12 μ. Sporidia crowded-biseriate, oblong, 2-nucleate, obtuse, mostly a little curved, 10–15 x 5–6 μ.

On dead stems of *Apocynum*, London, Canada (Dearness).

L. depréssa. (Pk.)

Sphærella depressa, Pk. 33d Rep. p. 34, (not *Sphærella depressa*, Cke. Journ Bot. 1883, *Physalospora depressa*, Sacc. Syll. 1709).

Perithecia numerous, minute, depressed, or even concave when dry, black. Asci oblong-clavate. Sporidia simple, oblong-elliptical or subfusiform, hyaline, 12–15 μ long.

On dead stems of *Mulgedium*, Center, N. Y.

The perithecia are slightly papillate and are so much depressed that they resemble a minute *Peziza*.

L. Cóptis, (Schw.).

Sphæria Coptis, Schw. Syn. N Am. 1783.
Sphærella Coptis, Farlow Cryp. Fl. of the White Mts p 247.
Exsicc. Ell. & Evrht. N. A. F. 2d Ser. 2358.

Perithecia scattered, covered by the epidermis, never denuded, convex, depressed, black, rugose, umbilicate by collapsion, often forming rather distinct whitish or yellowish spots on the leaves. Asci 38–53 x $7\frac{1}{2}$–$8\frac{1}{2}$ μ. Sporidia inordinate, hyaline or with numerous granules, navicular, sometimes curved, 15–20 x $1\frac{1}{2}$–$2\frac{1}{2}$ μ.

On leaves of *Coptis trifolia*. New Hampshire, Massachusetts and New York.

L. Epilòbii, (Wallr.)

Sphæria Epilobii, Wallr. Fl Crypt. Germ II, p. 771.
Sphærella Epilobii, Awd. Mycol Eur. Pyr. p. 14, fig. 59.
Læstadia Epilobii, Sacc. Syll 1645, Cke. Syn. 5265.

Perithecia thickly scattered, adnate with the epidermis, lens-shaped, when dry collapsing to concave, perforated above, black, 130–140 μ diam. Asci fasciculate, clavate, short-stipitate, 8-spored, 45–50 x 10 μ. Sporidia imperfectly biseriate, elliptic-oblong, attenuated toward each end but not acute, inequilateral, one-celled, hyaline, 13–17 x 3-4 μ.

On dead stems of *Epilobium angustifolium*, Greenland.

L. circumtégens, Rostr. Fungi Grönl. p. 547.

Perithecia densely gregarious, lens-shaped, perforated above. Asci cylindric-clavate, curved, 40–42 x 10 μ, 8-spored. Sporidia fusoid-oblong, guttulate, 12–14 x 3–5 μ.

On dry stems of *Draba hirta and Erigeron uniflorus*, which are thickly covered their whole length, by the perithecia, Greenland (Rostrup.)

L. árctica, Rostr. l. c.

Perithecia minute, scattered, depressed-sphaeroid, Asci obliquely oblong-ovate, 45–65 x 14–16 μ. Sporidia fusoid-oblong, hyaline, mostly 2-nucleate, 20–25 x 5–6 μ.

On leaves of *Helianthus peploides*, Godhavn, Greenland.

L. Archangélicae, Rostr. l. c.

Spots elliptical, large, 1–2 cm. diam., cinereous. Perithecia numerons, densely gregarious, depressed-sphæroid, when dry collapsing to cup-shaped, black, glabrous. Asci ovate-cylindrical, 8-spored, 32–38 x 6–8 μ. Sporidia biseriate, fusoid-elongate, 2–3-nucleate, 14–18 x 2–3 μ hyaline.

On dry stems of *Archangelica officinalis*, Greenland.

L. caryophýllea. (Cke. & Hark.)

Sphærella caryophyllea, Cke & Hark. Grev XIV, p. 9
Læstadia caryophyllea, Sacc Syll 6375, Cke Syn 5220

Perithecia minute, punctiform, densely gregarious, very black, convex, emergent. Asci obclavate, 8-spored, sessile. Sporidia obtusely lanceolate, biseriate, continuous, hyaline, 20 x 7½ μ.

On stems of *Dianthus*, California.

L. asarifòlia, (Cke.)

Sphærella asarifolia, Cke Journ Bot. 1883, p 138
Læstadia asarifolia, Sacc Syll 6012, Cke Syn 5252.

Epiphyllous. Spots orbicular and confluent, fuliginous. Perithecia small, globose, black, perforated, crowded, circinnately arranged. Asci subcylindrical, 30 μ long. Sporidia elliptical, continuous, hyaline, 7 x 2½ μ.

On leaves of *Asarum arifolium*, South Carolina.

On monocotyledonous plants, &c.

L. junipérina, (Ell.)

Sphærella juniperina, Ell. Am. Nat. 1883, p. 317
Læstadia juniperina, Sacc Syll. 6008. Cke Syn. 5254

Perithecia at first covered by the cuticle, soon erumpent, scattered or oftener subconflnent in the direction of the longer axis of the leaf, two or three together, appearing then like a minute *Hysterium*. Asci fasciculate, 35–40 x 7–8 μ. Sporidia crowded or biseriate, clavate-oblong, granular, (becoming uniseptate)? hyaline 8–10 x 2½–3½ μ. The single perithecia are conic-globose, perforated above, and about 75 μ diam.

On dead leaves of *Juniperus Virginiana*, Iowa. Found also by Karsten, in Finland.

L. Bidwéllii, (Ell.)

Sphærella Bidwellii, Ell Bull Torr Bot. Club, VII, p 90
Physalospora Bidwellii, Sacc Syll 1690.

Perithecia minute, globose, covered by the epidermis, finally suberumpent, perforated above. Asci clavate-cylindrical, obtuse, 60–70 x 10–13 μ, without paraphyses, 8-spored. Sporidia elliptical or oblong, continuous, 12–17 x 4½–5 μ, granular, hyaline.

On dried up grapes, with *Phoma uvicola*, B. & C., common.

L. gramínicola, Rostr. l. c. p. 548.

Perithecia very minute and abundant, gregarious, arranged in parallel lines. Asci fasciculate, oblong-clavate, commonly with a thick membrane, 35–45 x 12–14 μ. Sporidia fusoid-oblong, one-celled, 12–14 x 4 μ.

On culms and dry sheaths of *Colpodium latifolium* and *Agrostis rubra*, Greenland.

L. echinóphila, (Schw.)

Sphæria echinophila, Schw. Syn. N. Am. 1755
Sphæria echinophila, Ces Unio Ital Crypt, No. XXI
Sphærella echinophila, Awd. Myc. Eur Pyr p. 3 fig. 103.
Læstadia echinophila, Sacc Syll 1614
Exsicc. Ell N A F. 758.

Perithecia erumpent-superficial, equaling or exceeding the diameter of the spine on which they grow, depressed-globose, glabrous, black, papillate and finally perforated. Asci clavate-cylindrical, sessile, 8-spored, 27 x 4 μ. Sporidia biseriate, cylindric-oblong, ends rounded, scarcely crowded, continuous, 4 x 1½ μ (4 x 1 μ, Sacc.)

On spines of old chestnut burrs, Pennsylvania and New Jersey, and probably wherever the chestnut tree is found.

The species described in Sacc. Syll. I, p. 425, is evidently the same as that described by Schweinitz, to whom the species should be credited, and not to Cesati or Auerswald.

Species not well known and therefore doubtful.

L. ? brúnnea, (B. & C.)

Depazea brunnea, B. & C. Grev IV, p. 155
Læstadia brunnea, Sacc. Syll. 1636, Cke. Syn. 5256.

Spots dark brown, orbicular, about ¼ of an inch across. Asci clavate. Sporidia biseriate, narrow, fusiform, sometimes slightly curved, 4–nucleate.

On leaves of *Acer rubrum*, South Carolina.

L. glaucéscens, Cke. Grev. XVIII, p. 65.

Sphærella glaucescens, Cke. Grev. VII. p. 54.
Læstadia glaucescens, Sacc. Syll. 1637, Cke. Syn. 5257.
Exsicc. Rav. F. Am. 381.

Hypophyllous. Perithecia semiimmersed, crowded on orbicular

spots, very small. Asci clavate, 30 x 10 μ. Sporidia elliptical, 6 x 3 μ.

On leaves of *Acer rubrum*, South Carolina.

L. Cucurbitaceàrum, (Schw.)

Sphæria Cucurbitacearum, Schw. Syn. N. Am. 1699
Sphærella Cucurbitacearum, Cke l. c.
Læstadia Cucurbitacearum, Sacc. Syll 6014, Cke Syn 5269.

Perithecia emersed-innate, hemispherical, smooth, very minute. shining, membranaceous, covered by the epidermis. Asci clavate. short. Sporidia elliptical, continuous, hyaline, 7½ x 3 μ.

On gourds, Bethlehem, Pa.

The specimen in Herb. Schw. is without fruit, immature or sterile.

L. Polygonàti, (Schw.)

Sphæria Polygonati, Schw Syn. N Am 1793
Sphærella Polygonati, Cke l c.
Læstadia Polygonati, Sacc Syll 6010, Cke. Syn 5251

Perithecia scattered, innate, prominent on both sides of the leaf. hemispherical, astomous, black, nucleus white, without any fibrous subiculum.

On *Polygonatum*, Bethlehem, Pa.

SPHÆRÉLLA, Ces. & De Not.

Schema di Cassif. Sferiacei, p. 62.

Perithecia thin-membranaceous, globose-lenticular, covered by the epidermis, or suberumpent. Ostiolum impressed or short-papilliform, perforated. Asci without paraphyses, 8-spored. Sporidia elliptical or oblong, 2-celled, hyaline or subhyaline. Spermogonia and conidia of many species known. An extensive genus. Perithecia small, punctiform, generally foliicolous, rarely on twigs and fruit.

A. *Parasitic on leaves of dicotyledonous trees and shrubs.*

S. aquática. Cke. Journ. Bot. 1883, p. 106.

Exsicc. Rav. F Am 690.

Perithecia hypophyllous, globose, dark brown, densely crowded on orbicular spots, at first covered by the cuticle, then emergent. Asci clavate-cylindrical. Sporidia elongated-elliptical, uniseptate, hyaline, 20 x 4 μ.

On leaves of *Quercus aquatica*, Darien, Ga.

S. punctifórmis, (Pers.)

Sphæria punctiformis, Pers Syn. p. 90.
Sphæria sparsa, Wallr. Fl. Crypt. Germ. II, p. 772.
Sphæria insularis, Wallr. l. c. p. 814
Sphæria perexigua, Lev. Ann. Sci. Nat. Ser. III, tom. IX, p. 144.
Sphæria acerina, Fckl. and *Sphæria salicicola*, Fckl. F. Rh. 845 and 836.
Sphæria Artocreas, Rab. in Kl-Rab. Herb. Myc. Ed. I, p 360.
Sphærella punctiformis, var. *perexigua*, Rab. Herb. Myc. Ed. II, p 264.
Sphærella maculæformis, Cke. Journ. Bot. Aug 1866.
Sphærella acerina, Fckl. Symb. p. 99.
Sphærella corylaria, Fckl. l. c.
Sphærella sparsa, *paraneura*, *amphigena*, *æqualis*, Awd. Myc. Eur. V, VI, Heft. figs. 27, 39, 48.
Sphærella punctiformis, Sacc. Syll. 1819, Cke. Syn. 5307.
Exsicc. Fckl. F. Rh. 822, 836, 845, 847.—Rab. Herb. Mycol 264.—id. F. E 2441, 2945, 2548. Kze. F. Sel. 243.—Thum M. U. 264, 348.—M. March, 156.

Perithecia innate-punctiform, smooth, subshining, black, prominent, umbilicate by collapsion, perforated above, 60–120 μ diam. Asci clavate-cylindrical, subsessile, attenuated below, 28–45 x 7–9 μ. Sporidia biseriate or obliquely uniseriate, obovate-oblong, uniseptate and constricted, 6–9 x 2–3½ μ, hyaline, lower cell more acute.

On the lower surface of leaves of *Quercus*, more rarely *Castanea*. *Fagus*, *Æsculus*, *Carya*, *Cornus* and some others, common.

The synonymy is from Winter's Pilze.

S. maculifórmis, (Pers.)

Sphæria maculiformis, Pers. Syn. p. 90.
Sphæria acerina, Wallr. Fl. Crypt. II, p. 770.
Sphærella oblivia, *arcana* and *simulans*, Cke. in Journ. Bot. Aug. 1866.
Sphærella maculiformis, Awd. Myc. Eur. Pyr. V, p. 5.
Exsicc Kze F. Sel. 244 —Rab. F. E. 1219, &c., &c.—Ell. N. A. F. 1349.

Perithecia hypophyllous, thickly gregarious, forming small, black. angular spots ½–1 mm. across, and limited by the nerves of the leaf, innate-prominent, punctiform, globose, 70–80 μ broad, black and lusterless, pierced above. Asci clavate-cylindrical, sessile, 50–60 x 7–8 μ. Sporidia biseriate, obovate-oblong, uniseptate, more or less constricted at the septum, hyaline, 9–14 x 3–4 μ (or sometimes only 2 μ wide).

On leaves of *Quercus*, *Castanea*, *Fraxinus*, *Æsculus* and other deciduous trees, common.

Winter (in Die Pilze, p. 383) remarks that this species is usually gathered in the autumn, while the perithecia are still immature, so that the specimens in the various Exsiccati are, for the most part, unreliable.

S. Grossulàriæ, (Fr.)

Sphæria Grossulariæ, Fr. S. M. II, p. 521.
Sphærella Grossulariæ, Awd. Syn. Pyr. Eur. p. 11, tab. IV, fig. 44.
Exsicc. M. March, 2525.

Perithecia hypophyllous, gregarious or nearly evenly scattered over the lower surface of the leaf, globose, perforated above, 80–100 μ diam., covered by the epidermis, but distinctly prominent. Asci oblong-clavate, sessile or nearly so, 45–60 x 8–10 μ (8-spored)? Sporidia crowded, 2–3-seriate, fusoid, slightly curved, 18–22 x 2¼–3 μ, (26 x 3 μ, Winter; 26–35 x 3–4 μ, Saccardo).

On fallen leaves of *Ribes nigrum*, Ames, Iowa (Pammel).

The Iowa spece. were immature, but apparently this species. The measurements of asci and sporidia are from spec. in Sydow's M. Marchica.

S. Ravenélii, Cke. Grev. VII, p. 53.

Exsicc. Rav. F. Am. 384.

Perithecia hypophyllous, occupying the whole lower surface of the leaf, very small, scarcely visible, brown. Asci clavate. Sporidia elliptical, uniseptate, slightly thickened above, 8 x 4 μ.

On oak leaves, Aiken, South Carolina.

S. Ailánthi, Cke. Grev. VI, p. 146.

Exsicc. Rav. F. Am. 377.

Gregarious. Perithecia crowded in irregular spots, dark brown, small. Asci clavate. Sporidia elliptical, uniseptate, 6 x 3 μ.

On leaves of *Ailanthus*, Aiken, South Carolina.

S. ilicélla, Cke. Grev. VIII, p. 119.

Perithecia epiphyllous, punctiform, black, scattered on orbicular white spots. Asci elavate. Sporidia fusoid, hyaline, uniseptate, constricted, 20 x 6 μ, cells binucleate.

On leaves of *Ilex opaca*, New York State (Gerard).

S. Ílicis, Ell. Am. Nat. 1883, p. 317.

Exsicc. Ell. N. A. F. 1351.

Amphigenous. Spots round, 3–4 mm. diam., white above, with a raised, purple margin, reddish-brown below. Perithecia punctiform, black, ovate-globose, semiimmersed, often collapsing above. Asci oblong-cylindrical, 40–55 x 7½–11 μ. Sporidia biseriate, clavate-oblong, subhyaline, uniseptate, slightly constricted, 13–15 x 3 μ.

On leaves of *Ilex glabra*, Newfield, N. J.

We have not seen *S. ilicella*, Cke., but the Newfield specimens seem distinct in their smaller sporidia and amphigenous growth.

S. pardalòta, C. & E. Grev. VI, p. 16, tab, 96, fig. 23.

Exsicc. Ell. & Evrht. N. A. F. 2d Ser. 2136.

Perithecia minute, globose, black, semiimmersed, crowded in

irregular patches limited by the veinlets of the leaf. Asci cylindrical or clavate-cylindrical, 40 x 6 μ. Sporidia biseriate, narrow-elliptical, uniseptate, 8 x 3–3½ μ.

On leaves of *Myrica cerifera*, N. Jersey.

S. incanéscens, (Schw.)

Sphæria incanescens, Schw. Syn. N. Am. 1796.
Sphærella, incanescens, Cke in Journ. Bot March, 1883, p. 107.

Spots becoming hoary, broadly effused, indeterminate, appearing as if covered with frost. Perithecia punctiform, subglobose, very minute, black, often appearing as if truncate or collapsed. Asci cylindrical. Sporidia elliptical, uniseptate, hyaline, 8 x 3 μ.

On *Tilia Americana*, Pennsylvania (Schw.)

S. populifòlia, Cke. l. c. p. 197.

Exsicc Rav. F. Am. 689.

Hypophyllous. Perithecia innate-prominent, punctiform, globose, black, subconfluent-aggregated, 3–6 together, in numerous, small, black, subangular groups, or also scattered. Asci cylindrical. Sporidia sublanceolate, uniseptate, hyaline, 16–18 x 3½–4 μ.

On leaves of *Populus angulata*, South Carolina.

In our copy of Rav. F. Am., the specimen shows no fruit.

S. orbiculàris, Pk. 28th Rep. p. 81.

Perithecia minute, innate, covered by the epidermis, which is at length pierced or ruptured, occupying distinct or subconfluent, brownish spots. Asci subcylindrical. Sporidia oblong, uniseptate, colored, 10–13 μ long.

On the upper surface of fallen poplar leaves, Center and North Greenbush, N. Y. (Peck).

The spots on the leaves resemble those of *Venturia orbicularis* on oak leaves. Sometimes the epidermis peels off revealing the perithecia beneath. These are often more numerous near the margin of the spot.

S. Wistàriæ, Cke. Grev. VII, p. 54.

Punctiform, scattered. Perithecia semiimmersed, black, very small. Asci clavate. Sporidia elliptical, uniseptate, 8 x 3 μ.

On leaves of *Wistaria*, South Carolina (Ravenel).

S. fraxínea, Pk. 35th Rep. p. 145.

Perithecia numerous, minute (75 μ), black, amphigenous, generally collected in groups forming suborbicular spots. Asci oblong, often slightly narrowed above, 35–40 μ long. Sporidia crowded, oblong-ovate, uniseptate, colorless, 10–12 x 4–5 μ, divided by the septum into two very unequal parts, the smaller part ⅓–¼ the length of the larger.

On fallen leaves of *Fraxinus Americana*, Helderberg Mts., New York.

Whether this is really specifically distinct from *S. effigurata*, Schw., is doubtful, but we have no authentic spece. of this latter species. Specimens of *S. fraxinea*, Pk., in our Herb. certainly agree well with Schweinitz' description of his *S. effigurata*.

S. effiguràta, (Schw.)

Sphæria effigurata, Schw. Syn. N. Am. 1790.
Sphærella effigurata, Cke. Journ. Bot. 1883.

Spots widely effused, dark cinereous, determinate, often occupying the entire lower surface of the leaf, and appearing as if bounded by a black line, on account of the perithecia being more crowded around the margin. Perithecia very numerous, minute, crowded on the spots, subinnate, subconvex, black, much crowded, so as to form, as it were, a black, cinereous crust. Asci clavate. Sporidia elliptical, uniseptate, scarcely constricted, hyaline, 15 x 4 μ.

On leaves of *Fraxinus acuminata*, Bethlehem, Pa. Sec. Schw. not found on any other species of ash.

S. fraxinícola, (Schw.)

Sphæria fraxicola, Schw. Syn. N. Am. 1787.
Sphærella fraxinicola, Cooke Journ. Bot. 1883, p. 107.

Hypophyllous. Perithecia subinnate, black, at length rimose-dehiscent, subconnate, forming small, black, subconfluent patches. Asci clavate, short. Sporidia irregularly arranged, subelliptical, uniseptate, hyaline, lower cell narrower 7½ x 3 μ.

On ash leaves, Pennsylvania (Schweinitz), Georgia (Ravenel). Often sterile.

S. Fraxini, Niessl, has the same external appearance as this; the sporidia, however, are given as 26–28 x 4 μ, but in the specc. in Linhart's Fungi Hungarici, 162, and Rab-Winter F. Eur. 3255, they are 20–22 x 3 μ, and specc. from France (F. Fautrey) have sporidia only 15–17 x 2½–4 μ. The asci in all these are 40–60 x 7–9 μ, with the sporidia crowded-biseriate. *S. fraxinicola*, (Schw.), (sec. Cke.) has sporidia only 7½ x 3 μ, and if so, the two can hardly be the same species.

S. Asíminæ, E. & K. (in Herb.)

Perithecia on pale yellowish-white spots 1 cm. diam., scattered, convex, suberumpent, pierced above, 75 μ diam., black. Asci oblong, rather broader below, sessile, 24–27 x 12 μ. Sporidia biseriate, short-fusoid, subobtuse, slightly curved, uniseptate, hyaline, about 12 x 3½ μ, slightly constricted.

On living leaves of *Asimina triloba*, Ohio (Kellerman).

S. Opúntiæ, E. & E. Journ. Mycol. IV, p. 97.

Perithecia gregarious, 100–112 μ diam., in patches 2–10 mm. across. Ostiola erumpent, globose, imperfectly quadrisulcate-cleft. Asci oblong-cylindrical, 60 x 8–9 μ, sessile, without paraphyses. Sporidia biseriate, clavate-fusoid, uniseptate, slightly bent at the septum, nucleate, yellowish, 20–22 x 3½ μ. Remarkable for its peculiar ostiola.

On dead leaves of *Opuntia*, Louisiana (Langlois).

S. rosígena, E. & E. Journ. Mycol. III, p. 45.

Maculicolous. Spots amphigenous, reddish-brown, with a purplish border, definite, 3–4 mm. diam. Perithecia epiphyllous, thickly scattered over the spots, minute (60–75 μ), partly erumpent, subastomous, black. Asci subclavate-oblong, 25–30 x 8–10 μ. Sporidia biseriate, clavate-oblong, hyaline, uniseptate, 10–12 x 2 μ, ends subacute. Not to be confounded with *Læstadia Rosæ*, Auersw.

On living leaves of cultivated roses, Louisiana (Langlois).

S. olèina, Cke. Journ. Bot. 1883, p. 107.

Exsicc. Rav. F. Am. 754.

Epiphyllous. Spots white, suborbicular, 2–3 mm. diam., with a reddish margin, sometimes confluent. Perithecia depressed, 75–80 μ diam., perforated above, often 2–3-confluent, subcircinnate, with a slight mycelial fringe around the base. Asci clavate-cylindrical, 25–30 x 6 μ. Sporidia subbiseriate, oblong-elliptical, 12 x 4 μ. Spermogonium *Phyllosticta oleina*, Cke.

On leaves of *Olea Americana*, Aiken, South Carolina.

The measurements of perithecia and asci were taken from specc. in Rav. F. Am.

S. staphýlina, E. & E. Journ. Mycol. III, p. 128.

Maculicolous. Spots amphigenous, irregular, often narrow and elongated, mostly few on a leaf, 2–4 mm. or sometimes 1–1½ cm. diam., and occasionally occupying an entire half of the leaf, causing the affected part to dry up and fall away. Perithecia minute, visible on both sides, sublenticular and subastomous, membranaceous, black,

75–100 μ diam. Asci oblong, sessile, 40–60 x 12 μ, without paraphyses. Sporidia crowded-biseriate, oblong-elliptical, subinequilateral, uniseptate and constricted, yellowish-hyaline, 12–15 x 5 μ; accompanied by a *Macrosporium* and by smaller stylosporiferous perithecia containing elliptical, subfuscous, continuous sporules about 5 x 2½ μ.

On living leaves of *Staphylea*, Kansas.

S. exùtans, Cke. Texas Fungi, No. 141.

Spots minute, dark brown. Perithecia few, immersed, at length casting off the cuticle above them in little operculoid disks. Asci clavate-cylindrical. Sporidia elongated-elliptical, unequally uniseptate, hyaline, 12 x 4 μ.

On the upper surface of leaves of *Persea Carolinensis*, Texas.

Our specc. of this species from Ravenel are sterile or immature.

S. cerásina, Cke. Grev. VII, p. 54.

Exsicc. Ell. N. A. F. 794.

"Hypophyllous. Perithecia semiimmersed, black, crowded in suborbicular patches. Asci cylindrical. Sporidia narrow-elliptical, uniseptate, 10 x 2½ μ."

On leaves of *Prunus Lauro-cerasus*, Carolina.

The specimens sent by Ravenel and issued in N. A. F. 794, seem to be immature, and show no asci, though there are plenty of oblong, 12 x 2½ μ spores (stylospores)? The perithecia are 100 μ diam., scattered and amphigenous.

S. platanifòlia, Cke. l. c. p. 106.

Exsicc. Rav. F. Am. 756.

Hypophyllous, scattered. Perithecia small (70–80 μ), black, globose, perforated, semiimmersed. Asci oblong-cylindrical ("clavate," Cke.), sessile, 22–25 x 6 μ. Sporidia biseriate, hyaline, oblong-clavate, 6–7 x 2½ μ ("subelliptical, 8 x 4 μ," Cke.)

On leaves of *Platanus occidentalis*, Georgia.

S. Plátani, E. & M. Am. Nat. Jan. 1885, p. 77.

On round (2–4 mm.), reddish-brown spots with a narrow, dark, slightly raised border. Perithecia epiphyllous, innate-erumpent, 90–120 μ diam. Asci oblong, 8-spored, 40–60 x 12–15 μ, nearly sessile. Sporidia biseriate, subhyaline, ovate-oblong, uniseptate and constricted, nucleate, 14–16 x 4–6 μ. Quite distinct from *S. platanifolia*, Cke. There is a Phyllosticta on the same spots. Perithecia 100 μ diam. Sporules oblong-elliptical, 5–6 x 2½–3 μ.

On living leaves of *Platanus occidentalis*, Kansas (Kellerman).

S. Liriodéndri, Cke. l. c.

Epiphyllous. Spots orbicular, brown, 1 cm. diam. Perithecia subinnate, punctiform, black. Asci clavate-cylindrical. Sporidia elliptical, uniseptate, hyaline, 16 x 5 μ. Spermogonium *Phyllosticta Liriodendri*, Cke.

On leaves of *Liriodendron*, Darien, Ga. (Ravenel).

S. cornifòlia, (Schw.)

Sphæria Corni, Schw. Syn. N Am. 1792.
Sphærella cornifolia, Cke. l. c.
Exsicc. Rav. F. Am. 688.

Hypophyllous, forming large, orbicular, indeterminate spots, after the manner of *S. effigurata*. On these spots are densely crowded, compact clusters of 3–4 punctiform perithecia, not connected by any crust, but thickly scattered over the spots. The perithecia themselves are astomous, innate, with the surface punctate. Asci elavate, short. Sporidia biseriate, elliptical, hyaline, uniseptate, 5 x 2 μ.

On leaves of *Cornus florida*, Mt. Pocono, Pa. (Schweinitz).

S. lentícula, Cke. Journ. Bot. 1883, p. 107.

Exsicc. Rav. F. Am. 800

Hypophyllous. Perithecia globose, black, densely crowded in elevated lenticular pustules about 1 mm diam. Asci short-clavate, 22–25 x 8–10 μ. Sporidia irregularly biseriate 8–9 x 3½ μ, clavate-oblong ("elliptical," Cke.)

On leaves of *Cerasus Caroliniana*, South Carolina (Ravenel).

The asci we should call ovate-oblong.

S. dendroìdes, (Schw.)

Sphæria dendroides, Schw. Syn. Car. 221.
Sphærella dendroides, Cooke, l. c. p. 108.

Epiphyllous, aggregated, astomous, forming very large, cinereous, dendroid spots. Perithecia black, erumpent. Asci saccate, or broad-clavate. Sporidia lanceolate, uniseptate, hyaline, lower cell a little narrower, slightly constricted, 24–25 x 4 μ.

On leaves of *Carya*, Carolina.

S. Prini, Cke. l. c. p. 106.

Exsicc. Rav. F. Am. 753.

Epiphyllous. Scattered or occupying the whole surface of the leaf. Perithecia semiinnate, subprominent, black. Asci clavate, sessile. Sporidia narrow-elliptical, obtuse at the ends, uniseptate, hyaline, $5 \times 1\frac{3}{4}$ μ.

On leaves of *Prinos glaber*. South Carolina.

The specc. in our copy of Rav. F. Am. are sterile.

S. Gardèniæ, Cke. Journ. Bot. 1883, p. 108.

Hypophyllous. Perithecia scattered, punctiform, semiinnate, black. Asci clavate. Sporidia inordinate, elongated-elliptical, uniseptate, hyaline, $12 \times 3\frac{1}{2}$ μ. Spermogonium, *Phyllosticta Gardeniæ*. Cke.

On leaves of *Gardenia florida*, South Carolina.

S. Gordòniæ, Cke. l. c.

Hypophyllous. Perithecia scattered, covered, scarcely visible. Asci subclavate. Sporidia inordinate, elliptical, uniseptate, hyaline, 10×4 μ, scarcely constricted.

On leaves of *Gordonia lasianthus*, Darien, Ga.

Inadvertently published in Rav. F. Am. No. 799 under the name of *Sphærella Gardeniæ* on *Gardenia* instead of *Sphærella Gordoniæ*, on *Gordonia*.

S. Nigrèdo, (Schw.)

Sphæria Nigredo, Schw. Syn. N. Am. 1799.
Sphærella Nigredo, Cke. l. c. p. 109.

Hypophyllous, scarcely innate, aggregated or solitary. When aggregated, the perithecia are often covered with a pulverulent-cinereous ernst arising from the parenchyma of the leaf. Perithecia comparatively large, rugose, papillate, subperforated. Asci clavate. Sporidia sublanceolate, uniseptate, hyaline, cells subconical, $8–10 \times 3$ μ.

On leaves of *Rhus glabra*, Bethlehem, Pa. Allied to *S. maculiformis*.

S. hyperícina, Ell. Bull. Torr. Bot. Club, IX, p. 74.

Exsicc. Ell. N. A. F. 797.

Amphigenous. Perithecia minute, erumpent in little clusters of two to six. Asci oblong, $25–30 \times 4–5$ μ. Sporidia crowded, clavate-oblong, uniseptate, slightly curved, yellowish-hyaline, $10–11 \times 2\frac{1}{2}$ μ.

On fallen leaves of *Hypericum prolificum*, Newfield, N. J.

S. arbutícola, Pk. Bull. Torr. Bot. Club, X, p. 75 (July, 1883).

Sphærella Umbellulariæ, Cke. & Hark. Grev. XIII, p. 21 (1884).
Exsicc. Ell. & Evrht. N. A. F. 2d Ser. 1682.

Maculicolous. Spots suborbicular, blackish or subcinereous,

brown below. Perithecia minute, black, epiphyllous, scattered or collected in small groups, piercing the whitened epidermis. Asci oblong, somewhat narrowed above, 42–125 μ long. Sporidia crowded, narrow, hyaline, uniseptate, 12–15 x 3¾ μ.

On dead leaves of *Arbutus Menziesii* and *Umbellularia*, California.

S. cercidícola, E. & K. Bull. Torr. Bot. Club, XI, p. 123.

Perithecia scattered, erumpent, 100 μ diam., mostly on the upper side of the leaves, at length broadly perforated above. Asci oblong-cylindrical, 35 x 5 μ. Sporidia closely packed, overlapping and sub-biseriate, oblong-pyriform, uniseptate, 11–13 x 2½–3 μ, slightly curved and constricted.

On fallen leaves of *Cercis Canadensis*, Kansas.

This seems distinct from *S. Cercidis*, Pass. in its smaller sporidia.

S. nyssæcola, Cke. Hedw. 1868, p. 40, and Journ. Bot. March, 1883.

(sec. Cke.) *Asterina erysiphoides*, B. & C. in Herb. Berk, at least in part. Exsicc. Rav. F. Am. 96.

Hypophyllous. Perithecia very numerous, semiimmersed, brown, punctiform. Asci clavate, 20–25 μ long. Sporidia uniseptate, 8 x 2½ μ.

On leaves of *Nyssa multiflora*, Carolina and Florida.

S. minutíssima, Pk. 40th Rep. p. 68.

Perithecia very numerous, occupying the whole lower surface of the leaf, very minute, 50–55 μ diam., veiled by the epidermis, black. Asci oblong or slightly narrowed towards the apex, 40–50 x 8–10 μ. Sporidia crowded, oblong, straight, obscurely-septate in the middle, 15–17½ x 4 μ.

On dead leaves of *Alnus incana*, Adirondack Mts., N. Y.

The perithecia are scarcely visible to the naked eye. The affected leaves remain on the branches all winter.

S. alnícola, Pk. l. c.

Perithecia small (80–105 μ), hypophyllous, clustered or scattered, naked, black. Asci oblong or subclavate, 50–62 x 13–15 μ. Sporidia crowded, lanceolate, uniseptate, often slightly curved, 23–28 x 4½ μ.

On dead leaves of *Alnus viridis*, Mt. Marcy, N. Y.

The sporidia are narrowed towards one end and septate in the middle.

S. polifòlia, E. & E. Proc. Acad. Nat. Sci. Phil. July, 1890, p. 231.

Perithecia epiphyllous on grayish-black, indefinite spots 2 mm. or more diam., erumpent, rough, minute, broadly pierced above. Asci oblong, 35–40 x 6–8 μ, without paraphyses Sporidia biseriate, clavate-oblong, uniseptate, 10–12 x 2½–3 μ.

On living or partly dead leaves of *Andromeda polifolia*, London, Canada (J. Dearness).

S. Maclùræ, E. & E. l. c.

Spots red-brown with a definite, darker border, 3–10 mm. diam. or by confluence more, very brittle, the central part paler and soon falling out. Perithecia innate with their vertices erumpent, small (75 μ). Asci oblong-cylindrical, 50 x 8–10 μ, without paraphyses. Sporidia biseriate, oblong-pyriform, constricted, slightly curved, 12–14 x 5 μ, ends subacute.

On leaves of *Maclura aurantiaca*, Missouri (Demetrio).

S. quadrangulàta, E. & E.

S. Sapindi, E. & E. Journ. Mycol. II, p. 101.

Perithecia epiphyllous, globose, prominent, ⅓ mm. diam., scattered, on roughish, definitely margined, white spots, 3–4 mm. diam. Asci oblong-cylindrical, 60 x 12 μ. Sporidia biseriate, subelliptical (a little narrower at one end), uniseptate and slightly constricted, 16 x 4 μ. The conidial stage is a *Macrosporium* on the same spots, with long, stipitate conidia and slender, septate hyphæ.

On *Fraxinus quadrangulata*, Missouri. This was at first supposed to be on *Sapindus*.

S. salicícola, (Fr.)

Sphæria salicicola, Fr. S. M. II, p. 530.
Sphærella salicicola, Fckl. Symb. Mycol. p. 106 and Sacc. Rel. Myc. Lib. Ser. II, No. 169. tab. XX, fig. 9.

Perithecia epiphyllous, here and there in densely gregarious groups, covered by the epidermis, globose-depressed, perforated above. Asci clavate-cylindrical, very short-stipitate, 40–45 x 5–6 μ. Sporidia clavulate, uniseptate, not constricted, hyaline, 6 x 2 μ.

On leaves of *Salix herbacea*, Greenland.

S. maculàris, (Fr.)

Sphæria macularis, Fr. S. M. II, p. 502.
Pirostoma maculare, Fr. Summ. Veg. Scand. p. 395.
Sphærella macularis, Karst. Myc. Fen. II, p. 174.
Sphærella maculosa, Sacc. Syll. I, p. 487 (sec. Winter).
Exsicc. Kunze. F. Sel. 245.

Perithecia amphigenous, but mostly hypophyllous, sunk in the parenchyma of the leaf and seated on a filamentose, brown subiculum which, covered by the shining-gray epidermis, forms indefinitely limited spots, finally erumpent, gregarious, globose, black, 60–70 μ diam., with a small ostiolum. Asci sessile, elongated, inflated below, 8-spored, 50–65 x 11–13 μ. Sporidia biseriate below, uniseriate above, broadly rounded at the ends, somewhat constricted at the septum, the lower cell mostly somewhat longer and narrower than the upper one, yellowish-green, 11–15 x 3–5½ μ.

On fallen leaves of poplar, Adirondack Mts., New York.

S. pyrina, E. & E, (in Herb.)

Sphærella sentina, Fr. in Ell. N. A. F. 597.

Perithecia suberumpent, scattered or collected in irregular and indefinitely limited groups, 75–100 μ diam., pierced above, globose, black. Asci 22–30 x 5–6 μ. Sporidia biseriate, clavate-oblong, uniseptate, yellowish-hyaline, 6–9 x 2 μ, not constricted.

On fallen pear leaves in the spring of the year, Newfield, N. J.

This cannot be *S. sentina* (Fr.) which (sec. description and spece. in Thüm. M. U. 1058 and F. Sax. 372) has asci 75 x 10 μ, and sporidia 15 x 5 μ.

S. Sássafras, E. & E. Bull. Torr. Bot. Club. X, p. 98.

Exsicc. Ell. & Evrht. N. A. F. 1683.

Perithecia minute, semiimmersed, scattered over the lower surface of the leaf, or collected in groups, but not on any distinct spots. Asci subcylindrical, 35 x 5 μ. Sporidia biseriate, oblong-elliptical. uniseptate, 4 x 1½–2 μ, hyaline.

On fallen leaves of *Sassafras*, Kansas and New Jersey.

S. septorioìdes, (Desm.)

Sphæria septorioides, Desm. Ann. Sci. Nat. III. Ser. tom. VI, p. 81.
Sphærella septorioides, Niessl. in Kunze's F. Sel. 242, Rab. F. E. 1241.

Perithecia very small, innate-erumpent, globose, black, perforated above, hypophyllous, forming numerous, very small, greenish spots, which, later on, become larger, suborbicular and olive-brown. Asci cylindrical, rounded above, narrowed below, 8-spored, 30–35 x 5 μ. Sporidia subbiseriate, elliptic-oblong, rather narrower below, the upper cell broader and shorter, scarcely constricted, 8–9 x 2½ μ.

Peek cites this as found in New York. (On maple leaves)?

S. spleniàta, C. & P. Peck's 25th Rep. p. 105.

Sphærella nigrita, Cke. Grev. VII, p. 13.

Perithecia innate, densely crowded in broad (½–1 cm.), black, or bicular patches, globose, black, partly hidden by the tomentum of the leaf, subprominent. Asci linear, 40–50 x 7 μ. Sporidia oblong, hyaline, uniseptate, 12–15 x 3–4 μ.

On fallen leaves of *Quercus bicolor*, New York, on leaves of *Q. macrocarpa*, Manhattan, Ks. (Kellerman).

S. Catesbèyi, Cke. Grev. VII, p. 53.

Exsicc. Rav. F. Am. 383.

Perithecia hypophyllous, immersed, scattered, often 2–4 together, covered by the cuticle, at length rimose-erumpent. Sporidia lanceolate, uniseptate, 20–22 x 4 μ.

On leaves of *Quercus Catesbeyi*, Darien, Ga. (Ravenel).

S. molleriàna, Thüm. Contr. Mycol. Lusit. No. 537.

Perithecia amphigenous, but mostly hypophyllous, densely gregarions, punctiform, conie-subprominent, semiimmersed, shining-black, minute, on an irregular shaped spot which dries up to a dirty brown, surrounded with a narrow, darker border. Asci fasciculate, broad-clavate, narrowed at each end, 30–40 x 12–15 μ, subsessile, 8-spored. Sporidia 2–3-seriate, fusoid, acute at each end, septate in the middle, hyaline, 7–9 x 2½ μ.

On leaves of *Eucalyptus*, California (Harkness).

S. Pittóspori, Cke. Grev. VII, p. 53.

Gregarious. Perithecia black, semiimmersed, crowded in angular patches limited by the veinlets of the leaf. Asci clavate-cylindrical. Sporidia elliptical, uniseptate, rounded at the ends, slightly constricted, 10 x 4 μ.

On leaves of *Pittosporum*, Aiken, So. Carolina (Ravenel).

S. Gaulthèriae, C. & E. Grev. VII, p. 42.

Exsicc. Ell. N. A. F. 799.

Epiphyllous. Spots white, orbicular, 2–3 mm. across, with a purple margin. Perithecia punctiform, circinate, 75–85 μ diam. Asci clavate, 35–40 x 7–8 μ. Sporidia biseriate, lanceolate, uniseptate, 15–18 x 4 μ, slightly constricted.

On leaves of *Gaultheria procumbens*, Newfield, N. J.

The sporidia in our specimens are only 12–14 x 3–3½ μ.

S. Vaccínii, Cke. Journ. Bot. 1866, p. 249.

Sphærella Myrtilli, Awd. Mycol. Eur. V, Heft VI, p. 10, fig. 46.
Exsicc. Rehm Asc. 540.—Rav. F. Am. 376.—F. Sax. 373.—Fungi Gall. 3768.

Perithecia hypophyllous, thickly gregarious, forming gray, angular patches limited by the veinlets of the leaf, spherical, black, perforated above, shining, 80–100 μ diam. Asci cylindric-clavate, sessile, 40–50 x 6–7 μ. Sporidia biseriate, cylindric-oblong, obtuse at the ends, slightly curved, uniseptate, not constricted, hyaline, 12–18 x 1½–2 μ

On leaves of *Vaccinium arboreum*, Georgia, Florida, New York.

S. Bumèliæ, Cke. Grev. VII, p. 54.

Exsicc. Rav. F. Am. 375.

Epiphyllous. Perithecia black, semiimmersed, crowded in suborbicular, dark-colored patches. Asci clavate-cylindrical. Sporidia uniseptate, elliptical, 12 x 4 μ.

On leaves of *Bumelia*, Darien, Georgia (Ravenel).

S. myrtíllina, Pass. Micr. Ital. diag. No. 3.

Perithecia scattered or subgregarious, minute, pustuliform, covered by the blackened epidermis. Asci cylindric-fusoid, straight, 8-spored. Sporidia subbiseriate, fusiform, straight or curved, mostly only 2–4-nucleate, but sometimes (when mature) faintly uniseptate in the middle, hyaline, 18–20 x 3–3½ μ.

On branches of *Vaccinium uliginosum*, Kakatsiak, Greenland.

S. coloràta, Pk. 29th Rep. p. 62.

Exsicc. Ell. N. A. F. 899.

Spots small, round, 2–3 mm. diam., or by confluence more, reddish-brown and slightly raised, usually with a darker margin. Perithecia minute, black, epiphyllous. Asci cylindrical or subattenuated above, 45–50 x 7 μ. Sporidia biseriate, oblong-fusoid, uniseptate, hyaline, 14–18 x 2½ μ.

On leaves of *Kalmia angustifolia*, New York, New Jersey, and Delaware.

Depazea kalmicola, Schw. Syn. N. Am. 1812, is probably the stylosporous stage of this, and seems to include both a *Septoria* and a *Phyllosticta*. The spec. in Herb. Schw. is a *Septoria* (*S. kalmicola*, B. & C., *S. Kalmiæ*, C. & E.), while specc. of *Depazea kalmicola*, Schw., examined by Cooke (Journ. Bot. 1883) were a *Phyllosticta*.

S. Acàciæ, Cke. & Hark. Grev. IX, p. 9.

Epiphyllous. Perithecia membranaceous, covered by the epidermis, subgregarious. Asci clavate. Sporidia biseriate, hyaline, sublanceolate, constricted in the middle, uniseptate, binucleate, 25 x 7 μ.

On leaves of *Acacia*, California.

S. annulàta, Cke. Grev. VI, p. 146.

Exsicc. Rav. F. Am 378.

Scattered, immersed. Perithecia depressed, subannulate. Asci clavate. Sporidia elliptical, at length uniseptate, 8 x 3 μ.

On leaves of *Magnolia*, Aiken, South Carolina (Ravenel).

S. glaùca, Cke. Hedw. 1878, p. 39.

Exsicc. Rav. F. Am. 94.

Epiphyllous. Spots pale with a rose-colored tint, suborbicular, margin darker, 2–4 mm. diam. Perithecia semiimmersed, brown. Asci clavate. Sporidia short-lanceolate, uniseptate, hyaline, 12–14 x 4 μ.

On leaves of *Magnolia glauca*, South Carolina (Ravenel).

S. Phéllos, (Schw.)

Sphæria Phellos, Schw. Syn. N. Am 1805.
Sphærella Phellos, Cke. Journ. Bot. 1883, p. 106.

Perithecia hypophyllous, collected a few together, in small, scattered clusters, on a grayish-brown spot, penetrating to the opposite side of the leaf, globose, subprominent, minute, black, perforated. Asci clavate. Sporidia narrow-elliptical, uniseptate, hyaline, 8–10 x 2½ μ.

On leaves of *Quercus Phellos*, Carolina (Schweinitz).

S. convéxula, (Schw.)

Sphæria convexula, Schw. Syn. N. Am. 1788
Exsicc. Thum. M. U. 1149.—Rav. F. Am. 755.—Ell. & Evrht. N. A. F. 2d Ser. 1674.

Perithecia scattered over the whole lower surface of the leaf, innate, rugulose, but somewhat shining, black, subastomous, strongly convex, not collapsing.

On leaves of *Carya*, Carolina, Pennsylvania, New York and Ohio.

The specimens in all the Exsiccati above referred to, are unsatisfactory, immature or sterile. Those in N. A. F. 1674, have asci 25–30 x 6 μ, and sporidia, as near as can be made out, 6 x 2½–3 μ, but they are too immature to show an accurate outline. The perithecia in these spece. are mostly in small patches as in *S. maculiformis*, and in this respect do not agree with the description given by Schweinitz.

S. petiolícola, (Desm.)

Sphæria petiolicola, Desm. Ann. Sci Nat. III, Ser tom. XI, p 354
Sphærella petiolicola, Awd Myc Eur. V, VI, Heft, p. 8, fig. 38.
Exsicc Ell. N. A. F. 795.—Desm. Pl. Cr. Ed. 1, 2077.

Perithecia scattered or subgregarious, at first covered by the epidermis, which is either pierced or raised in a pustuliform manner. spherical, black, perforated above, about 110 μ diam. Asci cylindrical, sessile, 8-spored, 40 x 6 μ. Sporidia biseriate, oblong-clavate, rounded at the ends, two-celled, scarcely constricted at the septum, hyaline, 8 x 1½ μ.

On fallen petioles of *Robinia* and *Juglans regia*, Newfield, N. J.

S. applanàta, E. & E. Journ. Mycol. IV, p. 98.

Scattered, punctiform, flattened, covered by the thin epidermis through which the perithecia are plainly visible. Asci oblong, 40–50 x 15 μ. Sporidia biseriate, oblong-elliptical or pyriform, uniseptate, hyaline, 18–20 x 6–8 μ. On the same stems was another *Sphærella*, with sporidia continuous (*Læstadia*) and perithecia not flattened, but the material was insufficient for a satisfactory description.

On dead stems of *Clematis ligusticifolia*, Montana (Anderson).

S. alàrum, Ell. & Halst. (in Herb.)

Perithecia amphigenous, gregarious, on grayish spots bounded by the nerves of the leaf, small (60–70 μ), subglobose, perforated above. Asci fasciculate, aparaphysate, sessile, oblong, subinequilateral, about 40 x 10 μ. Sporidia biseriate, fusoid-oblong, slightly curved, subobtuse, hyaline, 12–14 x 3–3½ μ, uniseptate.

On samaræ of seedling maples, New Brunswick, N. J. (Halsted).

B. *On leaves and cones of coniferous trees.*

S. conígena, Pk. 33d Rep. p. 34.

Perithecia minute, erumpent, black. Asci broad, obovate or subclavate, somewhat pointed at the apex. Sporidia oblong or subcylindrical, when mature uniseptate, 25–40 μ long.

On old cones of *arbor-vitæ*, Helderberg Mts., New York (Peek).

S. conícola. Sacc. Add to Syll. p. 75.

Sphærella conigena, Pk. 38th Rep. p. 104.

Perithecia small, scattered or gregarious, erumpent and slightly prominent, black. Asci subcylindrical, 60–90 x 12 μ. Sporidia crowded, oblong-clavate, uniseptate, constricted at the septum, 8–10 x 2½–3 μ, (10–12½ x 4¼–5 μ, Pk.), the cells unequal, the lower one narrower than the subglobose or elliptical upper one.

On fallen cones of *Abies Canadensis*, Helderberg Mts., New York.

Differs from *S. Pinsapo* in its longer and differently shaped sporidia.

S. acícola, Cke. & Hark. Grev. XIII, p. 20.

Scattered, minute, subinnate. Perithecia globose-depressed, scattered, membranaceous, brownish-black. Asci short-clavate, 8-spored. Sporidia clavate-elliptical, uniseptate, hyaline, yellowish, one cell globose, the other narrower and subconical, 7 x 3 μ.

On pine leaves, California.

S. Canadénsis, E. & E. (in Herb.)

Perithecia scattered, minute, buried, with the apex and subpapilliform, perforated ostiolum erumpent. Asci clavate-oblong, 40–50 x 12–15 μ. Sporidia biseriate, clavate-oblong, hyaline, uniseptate, rounded at the ends, straight, slightly constricted at the septum, 20–22 x 5–7 μ.

On foliage of *Thuja occidentalis*, London, Canada (Dearness).

S. Taxòdii, Cke. Journ. Bot. March, 1883, p. 106.

Exsicc. Rav. F. Am. 686.—Ell. & Evrht. N. A. F. 2d Ser. 1676.

Perithecia amphigenous, scattered, with a mycelial fringe around the base, subprominent, black, 112–130 μ diam., perforated above. Asci cylindrical, 30–35 x 6 μ, sessile. Sporidia biseriate, clavate-oblong ("arete ellipticis," Cke.), 8 x 2½ μ.

On leaves of *Taxodium distichum*, South Carolina (Ravenel).

S. Pinsàpo, Thüm. Contr. Fungh Lit. No. 301.—Sacc. Syll. I, p. 480.

Perithecia epiphyllous, scattered or subgregarious, minute, at first covered by the epidermis, at length suberumpent, subconical, black, not on any spots. Asci sessile, narrow-cylindrical, obtusely rounded at the apex, narrowed at the base, straight or slightly curved, colorless, 38–44 x 7 μ, 8-spored. Sporidia biseriate, elliptic-cylindrical, subacute at the ends, septate in the middle, but not constricted, 2–4-nucleate, straight, colorless, 10 x 4 μ.

On fallen leaves of *arbor vitæ*, Port Henry, N. Y. (Peck).

C. *On stems and leaves of dicotyledonous, herbaceous plants.*

S. Krígiæ, E. & E. (in Herb.)

Spots reddish-brown, suborbicular, 2–4 mm. diam., definitely limited but without any distinctly raised or colored margin. Perithecia amphigenous, globose, pierced above, about 72 μ diam., erumpent. Asci oblong or clavate-oblong, 20–25 x 6–7 μ, without paraphyses. Sporidia biseriate (not well matured but apparently fusoid-oblong, 10–12 x 1½ μ).

On leaves of *Krigia amplexicaulis*, Racine, Wisconsin (Davis).

Accompanied on the same spots by a Septoria (*S. Krigiæ*, E. & E.), with bacillate sporules 20–25 x 1–1¼ μ.

S. asterinoides, E. & E. Journ. Mycol. IV, p. 98.

Perithecia scattered, lenticular, rather broadly pierced above, prominent, but covered with the cuticle, their bases sunk in the matrix, and more or less distinctly fringed with brown, branching mycelium. Asci without paraphyses, clavate-cylindrical, with a short, narrow base, 80–90 x 18–20 μ. Sporidia crowded, acutely elliptical, uniseptate and constricted, upper cell mostly broader, smoky-hyaline, 20–28 x 8–12 μ.

On dead stems of *Dipsacus*, Clyde, N. Y. (O. F. Cooke, Jr.)

S. decídua, E. & K. Bull. Torr. Bot. Club, XI, p. 122.

Perithecia visible on both sides of the leaf, of coarse, cellular structure, depressed-globose, 100 μ diam. Asci oblong, sessile, 50 x 15 μ. Sporidia crowded or biseriate, oblong or oblong-pyriform, uniseptate and constricted, mostly a little curved, 11–12 x 4–5 μ. Perithecia in small (1–1½ mm.), round, dull-white, translucent spots with a narrow, raised border; these spots are on dead, discolored parts of the leaf, which finally fall out and leave irregularly shaped holes, as if the leaf had been eaten out by insects.

On living leaves of *Vernonia*. Also on *Scrophularia nodosa*, with the spots whiter and more confluent and mostly not on dead, but on living, green parts of the leaf, Manhattan, Kansas (Kellerman).

S. Aràliæ, Cke. & Hark. Grev. IX, p. 9.

Perithecia gregarious, brown, membranaceous, covered by the epidermis, crowded in orbicular patches. Asci cylindrical. Sporidia elliptical, hyaline, uniseptate, 10 x 4 μ.

On stems of *Aralia Californica*, in California.

S. Dendromecònis, Cke. & Hark. Grev. IX, p. 9.

Exsicc. Ell. N. A. F. 798.

Perithecia scattered, covered by the epidermis, black, globose, about 100 μ diam., membranaceous. Asci oblong, sessile, mostly broader at the base, 35 x 12–14 μ. Sporidia biseriate, clavate-oblong ("elliptical," Cke.), uniseptate, hyaline, lower cell narrower, about 12 x 3 μ.

On dead stems of *Dendromecon rigidus*, California.

S. Salicórniæ, Awd. Myc. Eur. Pyr. p. 16, fig. 68.

Perithecia crowded, enveloping and blackening the stem between the upper internodes, covered by the epidermis, minute (60–80 μ), perforated above. Asci obovate or obovate-cuneate, sessile, aparaphysate, 8-spored, 25–40 x 15–20 μ. Sporidia crowded in 2–3 series, clavate-oblong, uniseptate, scarcely constricted, rounded at the ends, 12–15 x 3–3½ μ.

On *Salicornia*, Connecticut (Thaxter).

The perithecia are smaller than stated in Sacc. Syll., and the sporidia larger, agreeing better with *S Peruviana*, Speg., which, however, can hardly be more than a var. of *S. Salicorniæ*.

S. Pýrolæ. Rostr. Fungi Groenl. p. 551.

Perithecia amphigenous, subprominent, hemispherical, on large, subzonate, reddish spots, which sometimes cover the whole leaf. Asci cylindric-clavate, 50–60 x 8–10 μ. Sporidia oblong-fusoid, 15 x 4 μ.

On living leaves of *Pyrola grandiflora*, Ritenbenk, Greenland.

S. inconspícua, Schroet. Nord. Pilze, p. 12.

Perithecia sphæroid, slightly prominent, dark rufescent, 60–90 μ diam. Asci cylindric-elliptical, 40–50 x 13–16 μ. Sporidia biseriate, ovate-clavulate, 16–19 x 5–5⅓ μ, uniseptate in the middle, upper cell broader, lower attenuated downwards, hyaline.

On leaves and flower stalks of *Cassiope tetragona* and *Loiseleuria procumbens*, Greenland.

S. Astrágali, (Currey).

Sphæria Astragali, Curr. in Linn. Trans. XXII, No. 362.
Sphærella Astragali, Cke. in Journ. Bot. 1883

Perithecia scattered, membranaceous, covered, globose, dark brown, minute, here and there gregarious. Asci clavate. Sporidia subfusiform, uniseptate, hyaline, 15–18 x 4 μ.

On stems of *Astragalus*, Arctic America.

S. erióphila, Niessl. Neue Kernp. p. 86.

Perithecia scattered, erumpent, minute, subglobose, black, shining, in diffused light bright chestnut color, coriaceous, with a papilliform ostiolum. Asci obovate or ovoid-oblong, sessile, 74–83 x 30–45 μ, 8-spored. Sporidia irregularly 3-seriate or crowded, oblong-cuneate, a little attenuated below, straight, septate in the middle and slightly constricted, with a greenish tint, 20–30 x 8–9 μ.

On *Erigeron compositus*, Greenland.

S. confínis, Karst. Fungi Spetsb. et Beer. Eil. coll. p. 106. et Myc. Fenn. II, p. 179.

Perithecia amphigenous, covered by the epidermis, then protuberant, scattered or gregarious, sphæroid, perforated above, black, 100–120 μ diam. Asci short-pedicellate or subsessile, elongated or elongate-clavate, rarely somewhat inflated below, 45–52 x 10–13 μ. Sporidia 8 in an ascus, biseriate, acicular-elongated, uniseptate in the middle, not at all or only slightly constricted in the middle, hyaline or greenish-hyaline, 14–19 x 3–5 μ.

On *Antennaria alpina*, Greenland.

S. Compositàrum, Awd. Myc. Eur. Pyr. p. 15, fig. 105.

Exsicc. Rab. F. Eur. 1558.

Perithecia gregarious, innate, ovate, subacute, black, 100 μ diam. Asci gradually attenuated above from an ovate base, 8-spored, very short-stipitate, 60–70 x 20–28 μ. Sporidia 2–4-seriate-crowded, oblong. rounded at the ends, uniseptate, constricted at the septum, hyaline. 24–27 x 7 μ.

On dry stems of *Taraxacum ceratophorum*, Greenland.

S. verbascícola, (Schw.)

Sphæria verbascicola, Schw Syn N. Am. 1726.
Exsicc. Rav. F. Am. 141 (spermogonia).—Ell. N. A. F. 591

Perithecia thickly scattered, minute (150–175 μ), covered by the blackened cuticle, which is raised into distinct pustules over them. Asci oblong or clavate, or sometimes attenuated above, sessile, 8-spored, about 40 x 6 μ, without paraphyses. Sporidia biseriate, fusoid-oblong, hyaline, uniseptate, constricted at the septum, 10–12 x 5–6 μ.

On dead stems of *Verbascum Thapsus*, common.

S. Pediculàris, Karst. Fungi Spetsb. No. 53.

Perithecia scattered or gregarious, at first covered by the epidermis, then protuberant, sphæroid, subconoid-attenuated above, glabrous, black, 150 μ diam. Asci subsessile, elongated, 30–35 x 7 μ, 8-spored. Sporidia biseriate, fusoid or acicular-elongated, straight or slightly curved, guttulate or uniseptate, hyaline or yellowish-hyaline, 12–18 x 2½–8 μ.

On *Pedicularis hirsuta* and *P. lanata*, Greenland.

S. trichóphila, Karst. Myc. Fenn. II, p. 171.

Perithecia gregarious, at first covered by the epidermis, then erumpent, seated on a dark, filamentose subiculum, sphæroid, often conoid at the vertex, and armed with rather long, black, spine-like bristles, pierced above, black, scarcely papillate. Asci and sporidia as in *S. Tassiana*, viz. 40–80 x 18–26 μ, and 17–26 x 5–9 μ, respectively.

On *Pedicularis flammea, euphrasioides* and *hirsuta*, in Greenland.

S. pachyásca, Rostr. Fungi of Greenland, p, 552 ("Særtryk af Meddelelser om Grönland," III).

Perithecia scattered or gregarious, growing on either leaves or stems. Asci thick ovoid-oblong, inequilateral, 40–50 x 15–24 μ, mostly entire at the apex. Sporidia conglobate, conoid or ovoid-oblong, 16–20 x 5–6 μ.

On leaves and stems of many dicotyledonous plants, viz. *Chamænerium, Draba, Arabis, Ranunculus, Saxifraga, Pyrola, Pleurogyne, Campanula, Thymus, Stenhammaria, Plantago* and *Diapensia*, in Greenland, and on leaves of *Phlox cæspitosa*, in Montana.

This species "plays the same role among the *Dicotyledonous* plants as *S. Tassiana* among the *Monocotyledonous*."

S. Polygonòrum, (Crié).

Depazea Polygonorum, Crie sur les Dep. p. 41, tab. 8, fig 4.

Spots epiphyllous, round, surrounded by a prominent, dark red border. Perithecia epiphyllous, subprominent, black. Asci cylindrical, 8-spored. Sporidia ellipsoid, obtuse at the ends, uniseptate, 17–18 x 6 μ, pale greenish.

On dead stems and leaves of *Oxyria digyna, Polygonum viviparum* and *Koenigia Islandica*, Greenland.

S. arthopyrenioìdes, Awd. Myc. Eur. p. 15, fig. 55.

Perithecia thickly scattered, covered by the epidermis, depressed-globose, pierced above, black, 60–120 μ diam. Asci obovate, sessile, 8-spored, 38 x 18 μ. Sporidia imperfectly biseriate, ovate-oblong, rounded at the ends, uniseptate, not constricted, hyaline, 14 x 4–5 μ.

On *Papaver nudicaule*, Greenland.

S. mìnor, Karst. Myc. Fenn. II, p. 171.

Perithecia innate, covered by the epidermis, joined by innate, subradiate, creeping, brown fibers, forming cinereous, round spots, black, pierced above, 60–70 μ diam. Asci sessile, ovoid-elliptical or ovoid-sphæroid, 18–25 x 12–14 μ, 8-spored. Sporidia conglobate, ovoid-oblong, uniseptate in the middle, scarcely constricted, greenish-hyaline, 11–12 x 4 μ.

On *Linnæa borealis, Saxifraga oppositifolia*, and *S. Aizoon*, in Greenland.

S. Stellarineàrun, (Rab.)

Sphæria Stellarinearum, Rab. Herb. Mycol. 975?
Sphærella Stellarinearum, Karst. Fungi Spets. No. 48.

Perithecia somewhat scattered, amphigenous, covered, punctiform, perforated above. Asci sessile, ovoid, 60 x 20 μ, 8-spored. Sporidia collected in a mass, ovoid-oblong, straight or slightly curved, uniseptate, hyaline, not constricted, 20–24 x 6 μ.

On leaves of *Stellaria longipes*, *S. humifusa*, *Cerastium arvense*, *Alsine verna* and *Alsine biflora* and *Groenlandica*, in Greenland.

S. Cruciferàrum, (Fr.)

Sphæria Cruciferarum, Fr. S. M. II, p. 525.
Sphærella Cruciferarum, Sacc. Mich. II, p. 315.
Exsicc. Rab. F. Eur. 1853?

Perithecia scattered, innate, hemispherical, pierced at the apex, black, shining, smooth, 70–80 μ diam. Asci clavate, 40 x 12 μ, with a short, thick stipe, 8-spored. Sporidia biseriate, oblong-fusoid, rather obtuse at the ends, uniseptate and slightly constricted, curved, 15 x $3\frac{1}{2}$ μ, with 2–4 nuclei.

On *Vesicaria*, *Draba*, *Eutrema*, *Cardamine* and *Sisymbrium*, in Greenland, on pods of *Cheiranthus*, Bethlehem, Pa. (Schweinitz).

S. micróspila. (B. & Br.)

Sphæria microspila, B. & Br. Ann. Nat. Hist. No. 984, tab. 17, fig 36.

Perithecia scattered, globose, one or more innate on a small, brown spot, surrounded by a thin mycelium. Asci cylindrical. Sporidia oblong-ellipsoid, uniseptate, hyaline, 8–11 μ long.

On leaves of *Chamænerium angustifolium*, Sukkertoppen, Greenland.

S. Sibírica, Thüm. Beitr. Pilz. Sibir. No. 766.

Perithecia hypophyllous, sometimes extending through to the other side of the leaf, densely gregarious, subglobose, semiimmersed, free, shining-black. Asci broad-clavulate, narrowed at each end, with a thin membrane, 8-spored, subsessile. Sporidia long-elliptical, acutely rounded at the ends, septate in the middle, but not constricted, homogeneous, enucleate, hyaline, 20–22 x 5–6 μ.

On *Silene acaulis*, *Viscaria alpina*, *Melandrium apetalum* and *M. triflorum*, Greenland.

S. innumerélla, Karst. Mycol. Fen. II, p. 182.

Sphærella maculiformis, forma Comari, Rab. F. E 1042.
Exsicc. Rab. F. E. 1042.—Karst Fungi Fenn. 965.

Perithecia hypophyllous, gregarious, erumpent, globose, pierced above, black, 75--100 μ diam. Asci clavate-cylindrical, very short-stipitate, sometimes obliquely inflated below, 42–51 x 8–9 μ. Sporidia about 6 in an ascus, imperfectly biseriate, fusoid or rod-shaped, septate in the middle, not constricted, straight or slightly curved, hyaline, 17–24 x 3–4 μ (seldom 5 μ thick).

On leaves of *Sibbaldia procumbens*, Greenland.

S. oothèca, Sacc. Mich. II, p. 160.

Perithecia epiphyllous, subgregarious, innate-erumpent, depressed conoid-globose, black, thick-membranaceous, 100 μ diam. Asci thick-obovate. 40 x 28 μ, rounded at the apex, short-apiculate-stipitate, 8-spored, without paraphyses. Sporidia 2–3-seriate, oblong, uniseptate, very slightly constricted, 15–16 x 6–7 μ, subobtuse at the ends, hyaline.

On leaves of *Dryas integrifolia*, Disco, &c., Greenland.

S. melanoplàca, (Desm.)

Sphæria melanoplaca, Desm. XX, Not. 10.
Sphærella melanoplaca, Awd. M Eur. Pyr p. 13 *fig.* 108.
Exsicc. Desm Pl Crypt. Ed. 1. 2097

Perithecia epiphyllous, more rarely amphigenous, very minute, numerous, innate-subprominent, black, subshining, pierced above, 50–80 μ diam., with a whitish nucleus, gregarious, on large, sooty-black spots. Asci short-cylindrical, 40–42 x 8 μ, without paraphyses. Sporidia 3–4-seriate, lanceolate-fusoid, obtuse at the ends, uniseptate, hyaline, 22 x 3 μ.

On leaves of *Alchemilla alpina*, Kobbefiord, Greenland (asci 48–52 x 13–15 μ, sporidia 18–22 x 4–6 μ).

S. Impatiéntis, P. & C. 30th Rep. p. 67.

Perithecia abundant, minute, black. Asci subcylindrical. Sporidia crowded, oblong or lanceolate, uniseptate, usually quadrinucleate, 12 μ long.

On living leaves of *Impatiens fulva*, Adirondack Mts., N. Y.

S. phlògina, E. & E. Journ. Mycol. IV, p. 65.

Perithecia scattered, erumpent-superficial, conic-hemispherical, about 165 μ diam., epiphyllous. Asci ovate-oblong, 65–75 x 20–22 μ. Paraphyses none. Sporidia clavate-oblong, uniseptate and slightly constricted at the septum, rounded at the ends, yellowish-hyaline, with several small nuclei, 20–30 x 7–9 μ. Allied to *S. Stella-*

rinearum (Rabh.), but differs in its host plant, its more prominent perithecia and its straight, slightly constricted sporidia.

On dead leaves of *Phlox longifolia*, Belt Mts., Montana (Anderson).

S. Harknéssii, Sacc. Syll. I, p. 511.

Sphærella brachytheca, Cke. & Hark. Grev. IX, p. 8 (not Cke Grev. VII, p. 88).

Perithecia scattered, covered, small (60 μ), membranaceous. Asci ovate, 20 x 16 μ. Sporidia elliptical, hyaline, uniseptate, 8 x 4 μ.

On stems of *Convolvulus*, California.

S. leucophæa, E. & K. Journ. Mycol. II, p. 3.

Perithecia punctiform, minute (70–100 μ), buried in the substance of the leaf, except their slightly projecting, perforated apices. Asci oblong, sessile, or nearly so, 75–80 x 7–8 μ. Paraphyses none. Sporidia biseriate, fusiform, hyaline, uniseptate, and very slightly constricted at the septum, 18–20 x 3½ μ.

On dead leaves of *Baptisia leucophæa*, Manhattan, Kansas (Kellerman).

This is different from *Sphærella baptisiicola*, Cke., which has shorter asci, mostly not over 55 or 60 μ long, surrounded at first by paraphyses, and which has also shorter (12–15 x 3½–4 μ), yellowish sporidia.

S. baptisiícola, (Cke.)

Sphæria baptisiæcola, Cke. Grev. XI, p 110.
Exsicc. Rav. F. Am. 680.—Ell. & Evrht. N. A. F. 2d Ser 1799.

Perithecia scattered, punctiform (100 μ), covered by the epidermis, subprominent. Asci oblong, subinequilateral, mostly 35–55 x 7 μ, imperfectly paraphysate at first. Sporidia biseriate, oblong-fusoid, uniseptate, yellowish-hyaline, 15 x 3½ μ.

On dead stems of *Baptisia tinctoria*, Newfield, N. J., and on stems of *B. leucantha*, South Carolina.

S. granulàta, E. & E. Journ. Mycol. II, p. 102.

Exsicc. Ell. N. A. F. 2d Ser. 1798.

Densely gregarious, occupying a definitely limited area of the stem. Perithecia minute (165 μ), covered by the epidermis, which is raised and fissured over them, but not blackened, though the black perithecia are visible through it. Ostiolum papilliform, only slightly prominent; ascigerous nucleus, white. Asci nearly cylindrical, about 70 x 7 μ, subsessile, without paraphyses, and containing eight biseriate, fusiform, slightly curved, hyaline, granular at first, then uniseptate sporidia 20–23 x 3–3½ μ.

On dead stems of *Baptisia tictoria*, with *Sphærella baptisiicola*, Cke., Newfield, N. J.

The specimens found grew around the base of the stem, extending up for about two inches, where they were abruptly succeeded by *Sphærella baptisiicola*, Cke., in which the perithecia are more scattered and the sporidia shorter (12–16 μ), and broader (4–5 μ). The specimens of this latter species, in Rav. F. Am., are not well developed and have the sporidia narrower (3–3½ μ), and more acute, as described by Cooke.

S. depazeæfórmis, (Awd.)

Sphæria depazeæformis, Awd. in Rab. Herb. Myc 1641.
Sphærella depazeæformis, Ces. & De Not. Schema, p. 64.
Karlia Oxalidis, Rab. Herb. Myc. Ed. II, No. 567.
Sphærella Karlii, Fckl. Symb. p. 103.
Læstadia Oxalidis, Sacc. Syll. 1635.
Exsicc. Krieg. F. Sax. 332.—M. March. 1445, 2341.

Perithecia amphigenous, globose, 75–80 μ (40–46 μ, Winter & Saee.), broadly perforated at the apex, base sunk in the substance of the leaf, generally seated in the center of round whitish spots 1–3 mm. diam., with a brown border. Asci ovate-oblong, nearly sessile, 8-spored, 35–40 x 12-15 μ (34–36 x 8–10 μ, Winter). Sporidia biseriate, oblong, obtuse, septate in the middle but not constricted, 10–14 x 3–3½ μ, hyaline.

On leaves of *Oxalis corniculata*, var. *stricta*, Columbia, So. Carolina (Atkinson).

The Carolina specc. agree in all respects, size of perithecia, asci and sporidia, with the specc. in Kriegers Fungi Saxonici. The synonymy is from Winter's Pilze.

S. Thalíctri, E. & E. Journ. Mycol. I. p. 44.

Perithecia hypophyllous, 90 μ diam., of coarse cellular structure, with a rather large opening above, hemispherical (flattened when dry), scattered on small (2–3 mm.). round. white spots with a dark border. Asci sessile, about 36 x 7 μ. Sporidia crowded, ovate-oblong, granular and nucleate, (becoming uniseptate ?) 12–16 x 3–4½ μ.

On leaves of *Thalictrum dioicum*, New Jersey and Iowa.

S. Earliàna, Winter. Journ. Mycol. I, p. 101.

Perithecia amphigenous, densely crowded in small, black, angular, roundish patches about 1–2 mm. broad, globose, very small, (60–70)μ, perforated above, black. Asci fasciculate, slightly attenuated upwards from the subventricose base, very short-stipitate, 8-spored, 26–30 x 7 μ.

Sporidia inordinate, clavate, uniseptate in the middle, not constricted, hyaline, 8 x 2 μ. Paraphyses wanting.

On leaves of cultivated strawberries, Illinois. Differs from *S. Fragariæ*, Tul. especially in its small asci and sporidia.

S. xanthícola, Cke. & Hark. Grev. XIV, p. 9.

Perithecia punctiform, subprominent, black, crowded in elliptical spots, at first covered by the thin epidermis. Asci obclavate, sessile, 8-spored. Sporidia elliptical, uniseptate, scarcely constricted, rounded at the ends, hyaline, 12–15 x 5–6 μ.

On stems of *Xanthium*, California.

S. Campánulæ, E. & K. Am. Nat. 1883, p. 1166.

Exsicc. Ell. & Evrht. N. A. F. 2d Ser. 1673.

Perithecia minute (¼ mm.), scattered, covered by the epidermis which is pierced by the papilliform ostiolum. Asci subcylindrical, 35–40 x 5–7 μ. Sporidia crowded-biseriate, ovate-oblong, uniseptate and constricted, 10–13 x 3–3½ μ, ends subacute.

On dead stems of *Campanula Americana*, Ohio (Kellerman).

S. sicyícola, E. & E. Journ. Mycol. III, p. 45.

Spots amphigenous, small (1–2 mm.), dirty white, suborbicular or partly limited by the veinlets of the leaf, thin and transparent in the center. Perithecia few, 1–3, often only one in the center of a spot, epiphyllous, black and subshining, about 100 μ diam., sublenticular, with a rather broad perforation above, structure coarsely cellular. Asci cylindric-oblong, 40–50 x 6–7 μ, sessile. Sporidia biseriate, ovate-oblong, hyaline, uniseptate and somewhat constricted, 8–11 x 3–3½ μ, ends rounded or subacute.

On living leaves of *Sicyos angulata*, Missouri (Galloway).

S. solidagínea, E. & K. Journ. Mycol. III, p. 127.

Perithecia erumpent-superficial, 80–100 μ diam., subglobose, of rather coarse, cellular structure, pierced above, scattered or collected in groups. Asci 35 x 9–10 μ. Sporidia biseriate, clavate-oblong, nucleate, slightly constricted near the middle, 20 x 3½ μ, hyaline.

On dead leaves of *Solidago Canadensis*, Kansas (Kellerman).

S. Desmòdii, Winter, Journ. Mycol. I, p. 121.

Epiphyllous on large and very irregular spots, which become of a dirty gray color, with the margin more obscure but well defined, flex-nous or sinuous. Perithecia gregarious, minute, hemispherical, perforated above, black, 70–90 μ diam. Asci oblong-cylindrical, slightly attenuated above, rounded at the apex, sessile, 35–40 x 6–7 μ. Sporidia biseriate, cymbiform or fusoid, often more or less curved, uniseptate in the middle but not constricted, hyaline, 10¼–12 x 2½–3½ μ.

On living leaves of *Desmodium canescens*, Missouri.

S. Sesbániæ, E. & E. Journ. Mycol. IV, p. 98.

Densely gregarious; erumpent, perithecia 80–100 μ diam., pierced above. Asci 35–40 x 7–8 μ; paraphyses none. Sporidia biseriate, oblong-cylindrical, uniseptate, but not constricted or curved, 10–12 x 3½–4 μ, ends obtuse. Preceded by a *Macrosporium* with short, crooked, septate hyphæ, and oblong, 2–3-septate, muriform conidia, which are soon opake. In general appearance resembles *S. granulata*, E. & E., but has much smaller sporidia.

On dead stems of *Sesbania macrocarpa*, Louisiana (Langlois).

S. subcongregàta, E. & E. Journ. Mycol. II, p. 101.

Perithecia gregarious or occasionally 3–6 collected in a cluster, erumpent and superficial, ovate-globose, about ¼ mm. diam., ostiolum acute. Asci oblong, 40–45 x 12–15 μ, without paraphyses. Sporidia biseriate, oblong-cylindrical or clavate-oblong, subhyaline (yellowish), 18–23 x 3–4 μ, or, in the clavate form, 4–5 μ wide. *Pleospora permunda*, Cke., which appears to be common in the Rocky Mountain region, occurred on the same stem.

On peduncles of *Erigeron salsuginosus*, Mt. Paddo, Washington, alt. 6000 to 7000 ft.

S. Dáhliæ, C. & E. Grev. VII, p. 42.

Perithecia scattered, membranaceous, brown, covered by the epidermis, pierced above. Asci clavate. Sporidia biseriate, elliptical, uniseptate, hyaline, 12 x 4 μ.

On dead stems of *Dahlia*, Newfield, N. J.

S. Aquilègiæ, Ell. & Galw. Journ. Mycol. V, p. 66.

Perithecia scattered on the leaves and petioles, erumpent, rather acutely hemispherical, black, 100–120 μ diam., pierced above and more or less distinctly fringed at the base with brown, creeping hyphæ, texture coarsely cellular. Asci obovate-oblong, sessile, 50–60 x 22–25 μ, inequilateral, without paraphyses. Sporidia crowded-biseriate, subclavate-oblong, hyaline, straight, obtuse, slightly constricted, 20–22 x 9–11 μ, each cell 1–3-nucleate.

On *Aquilegia Jonesii*, Belt Mts., Montana (Anderson).

Differs from *S. pachyasca*, Rost., which is also found in Montana on *Phlox cæspitosa*, principally in its broader sporidia.

S. Œnotheræ, E. & E. Journ. Mycol. I, p. 151. **(Plate 27)**

Exsicc. Ell. & Evrht. N. A. F. 2d Ser. 1681.

Perithecia erumpent, hemispherical, 90–100 μ diam., broadly perforated above, densely gregarious and mostly in broad strips or series on the valves of the capsules. Asci oblong, slightly narrower above and abruptly contracted below into a short, stipe-like base. Sporidia biseriate, variable from oblong or ovate-oblong, 10–15 x 3 μ and nucleate, to oblong-fusiform, slightly curved, faintly uniseptate and 15–20 x 3–3½ μ. The smaller ones are apparently immature, being without septa.

On old capsules of *Œnothera biennis*, Newfield, N. J.

S. fuscàta, Ell. Bull. Torr. Bot. Club, VIII, p. 125.

Perithecia minute, densely crowded, forming elongated, dark colored patches 1½–2 cm. or more long, and about half as wide. Asci obovate, 40–50 x 15–18 μ. Sporidia inordinate, oblong, uniseptate, mostly narrower at one end, hyaline, 15–16 x 4–5 μ. Perithecia ovate-conical, rough, subcuticular, but finally throwing off the epidermis and blackening the matrix.

On dead herbaceous stems, Utah (S. J. Harkness).

S. Sarracèniæ, (Schw.)

Sphæria Sarraceniæ, Schw. Syn. N. Am 1759.
Sphærella Sarraceniæ, Sacc. Syll. 1941.

Perithecia scattered, depressed-hemispherical, brownish-black, covered by the blackened cuticle, pierced above, about 100 μ diam. Asci clavate, 35–40 x 6–8 μ, without paraphyses. Sporidia biseriate, fusoid-oblong, yellowish-hyaline, about 10 x 3 μ. 3–4-nucleate.

On species of *Sarracenia*, Carolina, on *Sarracenia purpurea*, Northern U. S. and Canada.

S. ciliàta, E. & E. Proc. Acad. Nat. Sci. Phil. July, 1890, p. 231.

Thickly gregarious. Perithecia subglobose, 150 μ diam., covered by the cuticle, but not sunk in the matrix, ostiolum papilliform. Asci clavate-cylindrical, 40 x 7 μ. Sporidia biseriate, clavate-oblong, hyaline, uniseptate and constricted, 10–12 x 3 μ.

On dead stems of *Steironema ciliatum*, London, Canada (Dearness).

S. Angélicæ, E. & E. l. c.

Perithecia scattered, growing under the bark and attached to it, so that when the bark is peeled off, they come with it, globose, ¼ mm. diam., collapsing below. Ostiola papilliform, barely piercing the cuticle and only slightly raising it. Asci clavate-cylindrical, with abundant paraphyses, 65–70 x 7–9 μ. Sporidia biseriate, oblong-elliptical, 2-nucleate, 12–15 x 3½ μ, becoming ovate-oblong and uniseptate.

On dead stems of *Angelica atropurpurea*, London, Canada.

Judging from the specimens in F. Eur. and F. G, this is different from *S. rubella*, Niessl.

S. Lactúcæ, E. & K. Bull. Torr. Bot. Club, XI, p. 123.

Epiphyllous on dark brown, (3–4 mm.), concentrically wrinkled spots with a distinct, raised border. Perithecia erumpent, subglobose, 120–150 μ diam., of coarse, cellular structure. Asci oblong, sessile, 40–75 x 12–14 μ. Sporidia biseriate, ovate-oblong, uniseptate and constricted, 14–15 x 5 μ, ends obtuse.

On leaves of *Lactuca Canadensis*, Kansas (Kellerman).

D. *On monocotyledonous plants.*

S. sabalígena, E. & E. Journ. Mycol. II, p. 101.

Exsicc. Ell. & Evrht. N. A. F. 2d Ser. 1800.

Perithecia gregarious, 100–125 μ diam., covered by the cinereous cuticle, which is scarcely ruptured over them. Asci subovate, 22–25 x 12–15 μ. Sporidia crowded, oblong-clavate, uniseptate, constricted at the septum, subhyaline, 10–12 x 3½–4 μ.

On dead tips of leaves of *Sabal Palmetto*, Louisiana (Langlois).

S. smilácina, E. & E. Journ. Mycol. II, p. 101.

Perithecia scattered, depressed-globose, 165 μ diam., covered by the cuticle, which is blackened directly over them and barely pierced by the minute ostiola. Asci oblong, sessile, about 35 x 7 μ. Sporidia biseriate or crowded, hyaline, fusiform-oblong or clavate-oblong, uniseptate, 8–11 x 2½ μ. The pycnidial perithecia are entirely similar to the ascigerous perithecia, only a little more prominent and mostly on bleached stems, and contain stylospores oblong or elliptical-oblong, hyaline, 2-nucleate, uniseptate and slightly constricted, 7–8 x 3–3½ μ. (*Diplodina Smilacis*, E. & E.)

On dead stems of *Smilax*. Newfield, N. J.

S. Tassiàna, De Not. Sfer. Ital. p. 87, tab. XCVIII.

Exsicc. Rehm Asc. 295?

Perithecia scattered or gregarious, either barely covered by the epidermis, or sunk in the parenchyma, more or less emergent, sphæroid, often subconical with a small opening, black, 60–150 μ diam. Asci sessile or very short-pedicellate, obliquely ovoid-oblong, inequilateral, 40–80 x 18–26 μ, 8-spored. Sporidia inordinate, crowded, ovate-oblong, uniseptate near the middle, scarcely constricted at the septum, hyaline or greenish-hyaline, 17–26 x 5–9 μ.

On leaves and culms of various species of *Gramineæ* and *Cyperaceæ; Luzula, Carex, Agropyrum, Agrostis, Aira, Trisetum, Poa, Glyceria, Calamagrostis* &c., in Greenland (Rostrup), and on leaves of *Typha latifolia*, in Delaware.

S. minimæpúncta, Cke. Journ. Bot. 1883.

Exsicc. Rav. F. Am. 681.

Scattered or aggregated. Perithecia punctiform, emergent, black, 125–140 μ diam. Asci clavate, short-stipitate. Sporidia narrow-elliptical, continuous, hyaline, 8 x 3 μ.

On stems of *Gladiolus*, South Carolina.

On the specimens in our copy of Rav. F. Am. are a *Diplodia*. perithecia 130–150 μ diam., sporules 18–20 x 8–10 μ, and a *Phoma* with smaller (110 μ) perithecia, and sporules 18–20 x 3–4 μ, but no ascigerous fungus.

S. serrulàta, E. & E. Journ. Mycol, III, p. 45.

Perithecia minute, covered by the cuticle, which is blackened over them, rather prominent, with an acute, papilliform ostiolum, mostly collected in groups of 6–12 or arranged in a seriate manner. Asci oblong-lanceolate, 35 x 7–8 μ. Sporidia biseriate, oblong-fusoid, 2-nucleate, hyaline, 6–8 x 2½ μ, ends rather obtuse. Differs from *S. sabaligena*, E. & E., in its smaller, grouped perithecia and in its smaller, obscurely septate sporidia.

On dead stems of *Sabal serrulata*, Florida (Calkins).

S. Tỳphæ, (Lasch.)

Sphæria Typhæ, Lasch. in Rab. Herb. Mycol. Ed. I, No. 660.
Sphærella Typhæ, Awd. Myc. Eur. Pyr. p. 18, fig. 110.
Exsicc. Fckl. F. Rh. 906.—Ell. & Evrht. N. A. F. 2d Ser. 1678, (forma minor).

Perithecia amphigenous, sunk in the parenchyma of the leaf, crowded and often connate in dense, sublanceolate, convex groups ¼–1 mm. long and 165 μ wide, subglobose, perforated above, 70 μ diam. Asci elongated-oblong, very short-stipitate, 8-spored, 50 x 7–8 μ. Sporidia biseriate, wedge-shaped-oblong, rounded at the ends, straight, uniseptate, not constricted at the septum, smoky-hyaline, 14 x 5 μ.

On leaves of *Typha latifolia*, West Chester, Pa.

The West Chester specc. have asci 25–38 x 10–12 μ, sporidia 11–12 x 3–3½ μ, and seem to be the *forma minor* of Fckl. Symb. p. 107, with asci 36 μ long and sporidia 10 x 4 μ. The specc. issued under this name in Mycotheca Marchica 2136, and Roum. F. Gall. 5354, afford two species of *Pleospora*, and one of *Leptosphæria*, but no *Sphærella*.

S. Pontedèriæ, Pk. 40th Rep. p. 69.

Sphærella paludosa, E. & E. in Ell. & Evrht. N. A. F. 2d Ser. 2357.

Spots rather large (1¼–2 cm.), oblong or elliptical, sometimes confluent, brown above, blackish-brown or grayish-brown below. Perithecia minute, (75 μ), hypophyllous, black. Asci oblong or subfusiform, 50–62 x 12½ μ. Sporidia crowded or biseriate, oblong-clavate, uniseptate, sometimes quadrinucleate, 15–20 x 5–6 μ.

On languishing leaves of *Pontederia cordata*, Whitehall, N. Y., and Newfield, N. J., and on leaves of *Nuphar advena*, London, Canada.

S. incìsa, E. & M. Journ. Mycol. I, p. 99.

Perithecia membranaceous, gregarious, globose or depressed-globose, ¼ mm. diam., covered by the blackened epidermis. Asci lanceolate, 100–120 x 8–10 μ, without paraphyses. Sporidia fusiform, attenuated to a bristle-like point at each end, endochrome distinctly divided in the middle, pale yellowish; length, including the bristle-pointed ends, 40–50 μ, width, 3–4 μ. The walls of the perithecia are closely adnate to the matrix, and with difficulty separable from it.

On dead petioles of *Sabal serrulata*, Florida (Martin).

S. Oróntii, E. & E. Journ. Mycol. I, p. 151.

Perithecia epiphyllous, scattered, minute (100 μ), purplish-black, membranaceous, coarsely cellular, perforated above, partly erumpent. Asci oblong, a little narrower above and abruptly contracted below into a very short, stipe-like base, 35–40 x 10–12 μ, without paraphyses. Sporidia biseriate, oblong, hyaline, nucleate and faintly uniseptate, about 14 x 4–5 μ, a little narrower at one end. *Physalospora Orontii* has larger perithecia and rather larger, regularly elliptical, continuous sporidia.

On yellowish spots on living leaves of *Orontium aquaticum*, Newfield, N. J.

S. paùlula, Cke. Grev. VI, p. 146.

Perithecia scattered, minute, globose, half free, 70 μ diam. Asci clavate, 20 μ long. Sporidia elliptical, at length uniseptate. 5 μ long.

On sheaths of *Zea Mays*, Gainesville, Florida (Ravenel). One of the smallest representatives of the genus, masked by the *Cladosporium* with which it is mixed.

We have seen no specimens, and take the diagnosis and remarks from Grevillea.

S. ignóbilis, Awd. l. c. p. 17.

Exsicc. Rab. F. Eur. 1645.

Perithecia hypophyllous, sunk in the substance of the leaf, scarcely piercing the epidermis, scattered, globose, black, pierced above, 100–125 μ diam. Asci very numerous, clavate-cylindrical, sessile, 8-spored, 50 x 6–8 μ. Sporidia overlapping-uniseriate or subbiseriate, pale, oblong-clavate, ends rounded, uniseptate, scarcely constricted, 10–12 x 4 μ.

On *Aira alpina*, Greenland.

S. lineolàta, (Rob. & Desm.)

Sphæria lineolata, Rob. & Desm. in Ann. Sci. Nat II, Ser. tom. XIX, p. 351.
Sphærella lineolata, De Not. Schema, p. 63.
Exsicc. Desm. Pl. Crypt. de France Ed. I, 1203.

Perithecia amphigenous, sunk in the somewhat blackened substance of the leaf, seriate, with the perforated apex rupturing the epidermis, more or less erumpent, globose, black, 60–70 μ diam. Asci obovate or oblong-clavate, sessile, 8-spored, 40–50 x 12–14 μ. Sporidia biseriate, oblong or obovate, ends rounded, uniseptate, not at all or only slightly constricted, hyaline, 16–18 x 3–4 μ.

On *Alopecurus alpinus* and *Poa filipes*, Greenland.

S. Lùzulæ, Cke. Grev. VI, p, 31.

Perithecia sunk in the substance of the leaf, punctiform. Asci clavate-cylindrical. Sporidia biseriate, 4-nucleate, becoming uniseptate, hyaline. No measurements given.

This species was found by Cooke on De Thümen's specc. of *Hendersonia Typhoidearum*, (Austrian Fungi, 785).

The specc. in our copy of this collection afford an abundance of cylindrical, yellowish-hyaline, 2–4-nucleate spores 10–20 x 3 μ, but no asci. Rostrup, in his Greenland Fungi, reports *S. Luzulæ*, Cke., on *Luzula confusa*, but gives no measurements.

S. Scírpi lacústris, Awd. Myc. Eur. Pyr. p. 18, fig. 73.

Perithecia scattered, sunk in the surface of the leaf, carbonaceo-coriaceous, somewhat brittle, depressed-spherical, pierced above, black, 150–180 μ diam. Asci clavate-oblong, attenuated into a short stipe below, 8-spored, 55–68 x 14 μ. Sporidia 3-seriate, elongated-fusoid, obtuse at the ends, slightly curved, uniseptate, not constricted, hyaline, 24–27 x 4–5 μ.

On *Scirpus cæspitosus*, Sukkertoppen, Greenland.

S. Wichuriàna, Schröt. Nord. Pilze, p. 12.

Perithecia gregarious, innate, very small, sphæroid, 50–65 μ diam., glabrous, texture loosely areolate, ostiolum obsolete. Asci few, aparaphysate, ovoid, 22–26 x 14–16 μ, sessile, 8-spored. Sporidia conglobate, fusoid, 11–13 x 3–3½ μ, hyaline, uniseptate, not constricted.

On leaves of various species of *Carex*, and on *Eriophorum angustifolium* and *Alopecurus alpinus*, Greenland.

S. pusílla, Awd. l. c. p. 17, fig. 115.

Exsicc. Rab. F. Eur. 2440.—Thum. M. U. 1355.

Perithecia epiphyllous, sunk in the surface of the leaf, thickly scattered, spherical, pierced above, black, 50–70 μ diam. Asci ovate-oblong, sessile, 8-spored, 40 x 12–14 μ. Sporidia 3–4-seriate, oblong-fusoid, somewhat thickened above, ends obtuse, nearly straight, constricted at the septum, with 4–5 oil globules, hyaline, 22–24 x 3–4 μ.

On various species of *Cyperaceæ* and *Gramineæ*, viz., *Kobresia caricina*, *Carex* (several species), *Festuca rubra*, *Phleum alpinum*, *Alopecurus alpinus* and *Nardus strictus*, Greenland.

S. leptopleùra, De Not. Recl. Pyr. Comm. II, p. 488.

Perithecia gregarious, often seriate between the nerves of the sheath, membranaceous, thin, black, broadly perforated above, about 200 μ diam., covered by the epidermis through which they are so plainly visible as to appear entirely bare, not strongly prominent, and more or less collapsing. Asci cylindrical, subsessile, 45–50 x 6 μ, without paraphyses. Sporidia biseriate, oblong-cylindrical, straight or slightly curved, about 15 x2½–3 μ.

On *Secale cereale* (sheaths of the leaves), Alabama (Atkinson).

There can not be much doubt that this is the species named, though we have no specimen for comparison. No septa could be seen in the sporidia but the specc. were not fully mature.

S. philochórta, Cooke, Journ. Bot. 1883, p. 137.

Epiphyllous, scattered. Perithecia minute, globose, subprominent, black, covered by the epidermis. Ostiolum papillate, perforated. Asci clavate. Sporidia narrow, cylindric-elliptical, obtuse at the ends, uniseptate, scarcely constricted, hyaline, 13–17 x 3 μ.

On leaves of some grass, Maine.

S. Muhlenbérgiæ, Ell. Am. Nat. 1883, p. 317.

Exsicc. Ell. N. A. F 1352.

Perithecia erumpent, minute, generally elongated-seriate. Asci oblong, 35 x 8–10 μ. Sporidia biseriate, elliptical, uniseptate, 11–15 x 3–3½ μ. Stylospores in larger perithecia, 13–19 x 4 μ (*Phoma*).

On leaves of *Muhlenbergia*, Newfield, N. J.

This and the preceding species, published about the same time, may be the same, but we have no specc. of Cooke's species for comparison.

S. Califórnica, Cke. & Hark. Journ. Bot. 1883, p. 136.

Perithecia small, scattered, subspherical, innate-prominent, dark brown, perforated above. Asci clavate-cylindrical. Sporidia narrow-elliptical, uniseptate, hyaline, not constricted, 8 x 2 μ.

On leaves of some native grass.

On specimens in our Herb. sent by Dr. Harkness as *Sphærella Californica*, we find only a *Septoria* with perithecia 75–110 μ diam., and sporules cylindrical, curved, hyaline, continuous, 12–14 x 1½ μ. Cooke (l. c.) is inclined to regard this as the mature state of *Læstadia perpusilla*, Sacc.

S. Spartìnæ, E. & E. Journ. Mycol. IV, p. 97.

Perithecia of coarse, cellular structure, subastomous, elliptical, 100–112 x 170–190 μ, buried in the parenchyma of the leaf, but plainly visible above or on the outside, and faintly so below, quite evenly and thickly scattered. Asci mostly immature in the specc. examined, but evidently present. Free sporidia (which we believe to be ascospores) oblong-elliptical, yellowish, constricted and faintly uniseptate in the middle, with about 4 small nuclei, 12–16 x 4½–5½ μ. This can not be the *Ascochyta Spartinæ*, Trelease, Journ. Mycol. I, p. 14, on account of the absence of any spots and the quite different sporidia. We are also confident that the Nebraska specc. are ascigerous.

On dead leaves and sheaths of *Spartina cynosuroides*, Nebraska (Webber).

S. crus-gàlli, E. & K. Journ. Mycol. III, p. 105.

Perithecia buried in the substance of the leaf and visible on both sides, but more prominent above, evenly scattered or in small groups with scattering perithecia intermediate, globose, 100–115 μ diam., with a broad, round opening above. Asci oblong, 50–55 x 10–12 μ, without paraphyses. Sporidia crowded, oblong-fusoid, subinequilateral, uniseptate and mostly constricted at the septum, hyaline, 14–16 x 4 μ. This is quite different from *Sphærella Maydis*, Pass., which has rather larger perithecia more distinctly grouped, and (sec. specc. in Rab. Fungi Europæi, No. 1851) sporidia fusoid, 16–20 μ long. It is also different from *S. Muhlenbergiæ*, Ell., which, by the way, is a good species and quite distinct from either *S. graminicola*, Fckl., which has larger asci and sporidia, or *S. pusilla*, Awd. *S. graminicola*, Fckl., (sec. specc. in Rehm's Ascom., 794, and F. Eur. 3446) has asci 75 x 12 μ, and sporidia 15–20 x 3½–4½ μ.

On withered leaves of *Panicum crus-galli*, Kansas (Swingle).

S. Zizàniæ, (Schw.)

Sphæria Zizaniæ, Schw. Syn. N. Am. 1750.
Sphærella Zizaniæ, Cke. Syn. 5673.
Exsicc. Ell. & Evrht. N. A. F 2d Ser. 1797.

Perithecia immersed and quite evenly scattered, not seriate. Asci oblong-cylindrical, turgid, 45–55 x 10–12 μ. Sporidia biseriate, clavate-oblong, uniseptate and constricted, hyaline, straight or slightly curved, 15–20 x 5–6 μ.

On languishing leaves of *Zizania aquatica*, Delaware (Commons).

The specimens of *Sphæria Zizaniæ*, Schw., in Herb. Schw., are sterile and poor, but the Delaware specimens appear to be that species.

E. *On cryptogamous plants, &c.*

S. gàllæ, E. & E. Bull. Torr. Bot. Club, XI, p. 75.

Perithecia minute, scattered or 2–3 together, rupturing and loosening the epidermis. Asci oblong-cylindrical, 40–45 x 10 μ, sessile. Sporidia crowded in 2–3 series, slender-clavate, uniseptate, 12–15 x 3 μ (at the broad end).

On galls on branches of *Vaccinium corymbosum*, and on rose bushes, Newfield, N. J.

S. indistíncta, Pk. 28th Rep. p. 81.

Perithecia minute, innate, slightly prominent, scattered or subgregarious, globose, black. Asci subcylindrical, 35–40 μ long. Sporidia crowded, elongated, hyaline, simple or obscurely uniseptate, 25–28 μ long, generally slightly curved.

On dead leaves of *Pteris aquilina*, New York State (Peck).

The perithecia are so small as to be easily overlooked. The sporidia are unlike those of *S. Pteridis*, being twice as long and not distinctly septate.

S. Lycopòdii, Pk. 39th Rep. p. 51.

Perithecia minute (100 μ), blackish. Asci oblong or subcylindrical, often slightly narrowed toward the apex, 30–40 x 10 μ. Sporidia oblong, 12–15 x 3½–4 μ.

On scales of dead spikes of *Lycopodium clavatum*, Adirondack Mts., New York.

This differs from *S. lycopodina* in its place of growth and in its smaller asci and sporidia.

S. supérflua, Fckl. in Ell. & Evrht.'s N. A. F. 2d Ser. 2134.

A reexamination of the specc. in the copies now accessible shows only a *Phoma* with oblong 5–7 x 2–2½ μ sporules. Sec. Fckl. the *Phoma* accompanying *Sphærella superflua* has sporules "cylindrical, curved, minute," indicating a different thing from this.

S. pandurata, E. & E. Bull. Torr. Bot. Club, X, p. 117, is *Didymella lophospora*, Sacc.

In the following Schweinitzian species, apparently referable to the *Sphærelloideæ*, the fructification is unknown.

Sphæria tigrinans, Schw. Syn. N. Am. 1804.—Spots round or subirregular, about 4 mm. diam., darker in the center, on account of the abundant, minute, astomous, black perithecia.

On the under side of oak leaves. Allied to *S. myriadea.*

Sphæria subbullans, Schw. Syn. N. Am. 1797.—Spots bullate-elevated, orbicular, black, covered with the numerous, innate, minute perithecia, with only their papillate ostiola visible.

On the upper side of leaves of *Pyrus Malus*, Bethlehem, Pa.

Sphæria Staphyleæ, Schw. Syn. N. Am. 1800.—Perithecia scattered or aggregated, innate, minute, irregular, flattened, black, astomous, furnished with a distinct, included sack, finally empty.

Very common on capsules of *Staphylea trifolia*, Bethlehem, Pa.

Sphæria plantaginicola, Schw. Syn. N. Am. 1806.—Perithecia minute, punctiform, very black, innate on both sides of the leaf, subconically elevated, astomous, thickly but irregularly scattered, shining, becoming empty, but hard, so as to be with difficulty cut.

Frequent on dead leaves of *Plantago lanceolata* of last year's growth, Bethlehem, Pa.

Sphæria coccineo-maculata, Schw. Syn. N. Am. 1801.—Perithecia black, seriate or inordinate, astomous, oblong, finally empty, seated on a scarlet spot slightly elevated in the center.

On leaves of grasses, Bethlehem, Pa.

Sphæria Andromedarum, Schw. Syn. N. Am. 1702.—Perithecia minute, innate, scattered or aggregated, convex-globose finally, as it were, circumscissile, the upper part deciduous.

On the lower side of leaves of *Andromeda axillaris*, Salem, N. C.

Sphæria apertiuscula, Schw. Syn. N. Am. 1785.—Perithecia scattered, brownish-black, minute, arising from the bullate parenchyma, at first innate, with a large opening above, finally empty. Resembles a minute *Peziza.*

Rarely met with, on the lower side of leaves of *Ulmus fulva*, New York State (Torrey).

Sphæria perigynicola, Schw. Syn. N. Am. 1782.—Perithecia scattered, very minute, astomous, black, erumpent, subconic-compressed, at length empty, often distantly seriate-erumpent.

Met with occasionally on the perigynia of *Carices*, especially of *C. xanthophysa*, Mt. Pocono, Pa. Allied to *S. recutita.*

Sphæria Angelicæ lucidæ, Schw. Syn. N. Am. 1781.—Perithecia scattered, innate, subprominent, dark, astomous, collapsing, occupying the lower side of languishing leaves, effused on subçrustaceous, yellow spots.

On leaves of *Angelica lucida*, Bethlehem, Pa.

Sphæria herbicola, Schw. Syn. N. Am. 1794.—Perithecia at first concealed under an effused, subhimantoid, white crust, which finally disappears (except around the ostiolum), leaving the scattered, black perithecia bare. Ostiola conical, equal in length to the diameter of the perithecia, finally deciduous; then the perithecia appear globose and perforated.

On leaves of various herbaceous plants, Bethlehem, Pa.

PHARCÍDIA, Körber.

Parerga lichenol. p. 469.

Perithecia more or less sunk in the matrix, finally erumpent, small, membranaceous, with a simple, perforated ostiolum. Paraphyses united in a shiny mass or wanting. Sporidia 1–3-septate, hyaline.

The species of this genus grow exclusively on *Lichens*.

P. boreàlis, (Sacc.)

Epicymatia borealis, Sacc. Syll. 2236.

Perithecia small, black, 200 μ diam., adnate or semiimmersed, at first subglobose, then variously wrinkled and, as it were, split open at the apex. Asci cylindric-clavate, with slender, anastomosely branched (genuine)? paraphyses. Sporidia 4–6, uniseriate, elliptical, constricted in the middle, obtuse at the ends, with two nuclei, 15–17 x 7–9 μ.

On some sterile *lichen thallus*, Greenland.

Collected during the English North Polar Exp. 1875–76.

Of the following genera included by Winter in the Family *Sphærelloideæ*, no representatives have yet been reported as found in this country, as far as we know.

TICHOTHÈCIUM, Flotow.

In Korber Krypt. Kunde (1848) p. 199.

Perithecia more or less sunk in the matrix, finally erumpent, small, of tolerably hard, horn-like, carbonaceous substance, black, with a simply perforated ostiolum. Paraphyses united in a slimy mass, obscure or wanting. Sporidia 8 or more in an ascus, 1–3-septate, brown (Winter, Die Pilze, II, p. 348.)

As in the preceding genus, all the species are lichenicolous.

MÜLLERÉLLA, Hepp.

In Muller Principes de Classif. de Lichens, XVI, Part 2d.

Perithecia more or less sunk in the matrix, globose, black, perforated above. Asci polysporous. Sporidia continuous, brown. Paraphyses obscure. Lichenicolous.

LIZÒNIA, Ces. & De Not.

Schema Sferiacei p. 41.

Perithecia subaggregated, ovoid, glabrous, coriaceo-membranaceous, erumpent-superficial, ostiolum minute. Asci clavate. Sporidia biseriate, oblong, often inequilateral, uniseptate, hyaline, becoming darker, rather large. Paraphyses none.

Foliicolous fungi, differing from *Sphærella* in their subcoriaceous perithecia crowded-erumpent, and mostly larger sporidia.

Placed by Cooke as a subgenus of *Sphærella* and by Winter as supplementary to the *Fam. Cucurbitarieæ*.

L. Sphágni, Cke. Grev. XVIII, p. 86.

"Perithecia scattered, subglobose, black, rather prominent, with a mammillate ostiolum, seated on the decayed leaves and soon becoming superficial. Asci clavate. Sporidia cylindrical, slightly curved at one or both ends, uniseptate (then probably triseptate), hyaline, 40–50 x 8 μ."

On dead *Sphagnum*, Maine.

PHYSALÓSPORA, Niessl.

Notiz. ueber neue und Krit. Pyr. p. 10.

Perithecia subglobose, covered, membranaceous or coriaceo-membranaceous, black, with the ostiolum (typically papilliform) erumpent. Asci clavate-cylindrical, paraphysate. Sporidia ovoid or oblong, continuous, hyaline or subhyaline.

Differs from *Læstadia* in the presence of paraphyses. The perithecia are also for the most part larger and of firmer texture. Included by Winter in the next Family.

P. Wrightii, (B. & C.)

Sphæria Wrightii, B. & C. Grev. IV, p. 154.
Physalospora Wrightii, Sacc. Syll. 1661.
Læstadia Wrightii, Cke. Syn. 5271.

Perithecia covered by the cuticle, collapsing and then cup-shaped. Asci lanceolate, obtuse. Sporidia biseriate, oblong, slightly curved, 25–33 μ long, hyaline, with a narrow, gelatinous border.

On leaves of *Statice Limonium*, California (Wright).

P. Arthuriàna, Sacc. Mich. II, p. 569.

Exsicc. Ell. & Evrht. N. A. F. 2d Ser. 1665.

Perithecia epiphyllous, gregarious, on yellowish spots, innate-superficial, globose-depressed, black, 125 μ diam. Ostiolum obsolete. Texture of perithecium parenchymatic, yellowish-brown. Asci cylindrical, short-stipitate, 90 x 12 μ, paraphysate, rounded at the apex, 8-spored (rarely clavate, 60 x 18 μ). Sporidia obliquely uniseriate, rarely biseriate, ellipsoid, obtusely rounded at the ends, 11–13 x 7 μ, 2–4-nucleate.

On leaves of *Iva xanthifolia*, partly dead, Iowa (Arthur).

P. Ílicis, (Sch.)

Sphæria Ilicis, Schleich. Fr. S. M. II, p. 501.
Diplodia ilicicola, Desm. Plantes Crypt. Ed. I, 988.
Læstadia Ilicis, Cke. Syn. 5298.
Physalospora Ilicis, Sacc. Syll. 6390.
Exsicc. Ell. N. A. F. 196.—Desm. Exsicc. Ed. I, 988.—Rav. Fungi Car. IV, No. 63.

Perithecia scattered, covered by the epidermis which is white in the center and finally rimosely ruptured, subprominent, epiphyllous, globose or globose-conoid, ⅓ mm. diam. Asci cylindrical, rounded above, attenuated below into a narrow stipe-like base, 80 x 12 μ, 8-spored. Sporidia biseriate, boat-shaped, rounded at the ends, 14 x 5 μ, hyaline.

On fallen leaves of *Ilex opaca*, So. Carolina and New Jersey.

P. philoprìna, (B. & C.)

Sphæria philoprina, B. & C. Grev. IV, p 154.
Læstadia philoprina, Cke. Syn. 5283.
Physalospora philoprina, Sacc. Syll. 1685.

Perithecia shining-black, white in the center from the cuticle. Asci oblong. Sporidia biseriate, hyaline, cymbæform.

On leaves of *Ilex*, North Corolina.

Doubtfully distinct from thc preceding species.

P. rhòdina, B. & C. in Curt. Cat. p. 148, Grev. XVII, p. 92.

Perithecia gregarious, covered, subglobose, very small, black, with erumpent ostiola. Asci clavate, 8-spored. Sporidia sublanceolate, continuous, hyaline, 30–35 x 10 μ.

On branches of *Rosa rubiginosa*, Carolina.

P. Gerànii, Cke. & Hark. Grev. XIV, p. 9.

Perithecia densely gregarious, convex, black, covered by the epidermis, at length erumpent. Asci clavate, ample, 8-spored. Sporidia biseriate, elliptical, continuous, granular, hyaline, 25–28 x 10 μ.

On stems of *Geranium*, California.

P. gelsemiàta, (Cke.)

Sphæria gelsemiata, Cke. Grev. VI, p. 145.
Physalospora, gelsemiata, Cke. Syn. 4092, Sacc. Syll. 1680.

Covered, scattered or subgregarious. Perithecia subglobose, subprominent. Ostiola papillate. Asci clavate. Sporidia short-lanceolate, hyaline, 30–35 x 10–12 μ.

On twigs of *Gelsemium*, Aiken, So. Carolina.

P. minutélla, (Pk.)

Sphæria minutella, Pk, 29th Rep. p. 62.
Physalospora minutella, Sacc. Syll. 1686.
Phomatospora minutella, Cke. Syn. 4347.

Perithecia minute, somewhat flattened, black, the upper part at length breaking away and leaving the base attached to the matrix.

Asci sublanceolate. Sporidia oblong, simple, colorless, 7½ μ long.

On dead stems of herbs, Greenbush, N. Y. (Peck).

P. Cupréssi, (B. & C.

Sphæria Cupressi, B. & C. Grev. IV, p.148.
Physalospora, Cupressi, Sacc. Syll. 1679, Cke. Syn. 4091.

Perithecia covered by the cuticle, but prominent, with a distinct ostiolum. Asci clavate. p. sp. 75 x 20 μ. Sporidia inordinate, inequilaterally elliptical, hyaline, 20–23 x 10 μ.

On limbs of *Cupressus thyoides*, Carolina (Ravenel).

The measurements of asci and sporidia are taken from a spec. com. by Farlow from Herb. Curtis.

P. ceanòthina, (Pk.)

Sphæria ceanothina, Pk. 29th Rep. p. 62.
Physalospora ceanothina, Sacc. Syll. 1692, Cke. Syn. 4093.

Perithecia scattered or rarely two or three crowded together, smooth, subglobose. Ostiola piercing the epidermis, somewhat rugged, often curved or deformed. Sporidia crowded or biseriate, oblong, obtuse, sometimes curved, hyaline, 12–15 μ long.

On dead stems of *Ceanothus Americanus*, New York.

P. Ludwígiæ, (Cke.)

Sphæria Ludwigiæ, Cke. Hedw. 1868, p. 39.
Physalospora Ludwigiæ, Sacc. Syll. 1720.
Phomatospora Ludwigiæ, Cke. Syn. 4357.

Gregarious. Perithecia covered by the blackened and raised epidermis. Asci clavate. Sporidia biseriate, elliptical or pyriform, hyaline, continuous, 25 x 10 μ.

On stems of *Ludwigia*, So. Carolina.

P. aurántia, E. & E. in Ell. & Evrht.'s N. A. F. 2d Ser. 2355.

(Plate 27)

Perithecia scattered, buried in the parenchyma of the leaf, suboblong, 175–200 x 130–150 μ, of a fine orange color, which is readily seen in a section of the leaf, the surface of which is uniformly blackened and slightly pustulate from the scarcely erumpent ostiola. Asci clavate-cylindrical, 100–110 x 12–15 μ, with paraphyses. Sporidia uniseriate, oblong-elliptical, granular, hyaline, continuous, 12–22 x 8–9 μ.

On dead leaves of *Astragalus pectinatus*, Sand Coulee, Montana (Anderson).

P. polàris, Rostr. Fungi Groenland. p. 548.

Perithecia scattered. Asci ovate-oblong, paraphysate, 35–40 x 6–8 μ. Sporidia elongated, guttulate, 10–12 x 2 μ.

On dried up stems of *Papaver nudicaule*, Kangerdlugsuak, Greenland.

P. Potentíllæ, Rostr. l. c.

Perithecia scattered, globose. Asci ovate-oblong, paraphysate, fasciculate, 32–34 x 8-9 μ. Sporidia elongated, guttulate, 10 x 1–2 μ.

On dry stems and petioles of *Potentilla maculata*, Upernivik, Greenland.

P. megástoma, (Pk.)

Sphærella megastoma, Pk. Bot. Gaz. IV, p. 231.
Physalospora megastoma, Sacc. Syll. 1669.
Læstadia megastoma, Cke Syn. 5273.

Perithecia minute, numerous, amphigenous, at first covered by the epidermis, then erumpent, black, with a large, circular opening at the apex. Asci subcylindrical, 100–150 x 15–18 μ, paraphysate. Sporidia uniseriate or crowded-biseriate, oblong-elliptical, granular, greenish-yellow., 10–15 x 6–7 μ.

On living or languishing leaves of *Astragalus bisulcatus*, Colorado (Jones), and on some leguminous plant (*Astragalus*), Valley City, Dakota (Seymour).

P. bìna, Hark. Bull. Cal. Acad. Sci. No. I, p. 43.

Epiphyllous, shining-black, hemispherical, half free, numerous, on irregular, whitish spots, covering a third or more of the leaf and bordered by a dark line. Asci fusoid, very delicate, 2-spored, attenuated below to a slender pedicel, paraphysate. Sporidia oblong-elliptical or slightly boat-shaped, rounded at one end and slightly pointed at the other, with several vacuoles, 15 x 6 μ.

On living leaves of *Quercus agrifolia*, California (Harkness).

P. quercifòlia, E. & E. Journ. Mycol. I, p. 92.

Perithecia ¼–⅓ mm. diam., globose with a light colored nucleus, buried in the substance of the leaf, but prominent, so as to show distinctly on both sides, covered by the epidermis which is slightly blackened and closely adherent to the perithecia. Ostiolum papilliform, barely visible through the ruptured epidermis. The perithecia finally collapse more or less distinctly. Asci oblong, 75–80 x 12 μ, with a short, abruptly contracted base. Sporidia biseriate, narrow-elliptical or broad-fusiform, granular, hyaline, 15–25 x 6–8 μ.

With *Harknessia hyalina*, E. & E., which is probably its stylosporous stage. On dry, dead oak leaves (*Q. coccinea*) still hanging on limbs cut off last season, Newfield, N. J., June, 1885.

P. phomópsis, (C. & E.)

Sphæria phomopsis, C. & E. Grev. VII, p. 41.
Physalospora phomopsis, Sacc. Syll. 1704.

Perithecia gregarious, flattened, covered by the epidermis, membranaceous, brown, 100-120 μ diam. Asci oblong or clavate-oblong, sessile, rounded at the apex, 25–30 x 8–10 μ, with rudimentary paraphyses. Sporidia ovate-oblong or subnavicular, hyaline, 10–12 x 3 μ (12–14 x 6 μ, Cke.)

On dead stems of *Desmodium*, Newfield, N. J.

The sporidia even in specc. that have lain in the herbarium for ten or twelve years, are mostly continuous, though a few were uniseptate. Some perithecia contain only stylospores (*Phoma*) 10–15 x 4–5 μ.

P. errática, (C. & E.)

Sphæria erratica, C. & E. Grev. VI, p. 95, tab. 100, fig. 35.
Physalospora, erratica Sacc. Syll. 1696, Cke. Syn. 4094.

Perithecia gregarious, globose-depressed, black, at length erumpent, often crowded. Asci clavate. Sporidia biseriate, elliptical, continuous, hyaline, granular within, 20–26 x 12–13 μ.

On branches of *Rosa*, *Lonicera* and on stems of *Solidago*, Newfield, N. J.

P. oxýstoma, Sacc. & Ell. Mich. II, p. 568.

Exsicc. Ell. N. A. F. 784.

Perithecia gregarious and often 2–4 erumpent in a cluster together, but not connate, globose-ovoid, minute, with a white nucleus. Ostiolum acutely papilliform, exserted. Asci oblong-clavate, sessile, with a thick membrane, 70–80 x 15–17 μ (90–100 x 17–18 μ, Sacc.), rounded at the apex and surrounded by jointed paraphyses. Sporidia clavate-oblong, crowded, yellowish-hyaline, continuous, 15–20 x 5 μ.

On dead culms of *Phleum pratense*, Newfield, N. J., and on dead culms of *Panicum Curtisii*, in Louisiana.

In the Louisiana specc. the perithecia are erumpent in oblong-hysteriiform tufts about 1 mm. long, 5–8 perithecia together, connected by an imperfectly developed, subpulverulent, black stroma.

P. leptospèrma, Rostr. Fungi Groenl. p. 548.

Perithecia gregarious, covered by the blackened epidermis, globose. Asci fasciculate, paraphysate, cylindrical, 8-spored. Sporidia biseriate, fusoid-elongated, guttulate, 12 x 2 μ.

On dry culms and sheaths of *Calamagrostis purpurascens*, Ikertok, Greenland.

P. Oróntii, E. & E. Journ. Mycol. I, p. 150.

Spots elongated, dead and dry, 3–4 x 1 cm. Perithecia erumpent, orbicular, flattened, 150–180 μ diam., pierced in the center with a small, round opening, texture membranaceous. Asci oblong-cylindrical, with an acute, sessile base and rather abruptly narrowed, truncate apex. Paraphyses? Sporidia biseriate and closely packed, granular, subhyaline, 14–16 x 6–7 μ. *Ramularia Orontii*, E. & M., and *Phyllosticta Orontii*, E. & M., are not improbably connected with this as conidia and spermogonia.

On dead spots in living leaves of *Orontium aquaticum*, Newfield, N. J.

P. uvæspora, (Cke.)

Sphæria uvæspora, Cke. Tex. Fungi No 144.
Physalospora uvispora, Sacc. Syll. 1689.

Gregarious, covered by the cuticle. Perithecia almost globose, rather small and not numerous, brown, pierced at the apex. Asci clavate. Sporidia shortly clavate, simple, hyaline, in form resembling grape stones, 12–15 x 5 μ.

On flower stalk of *Yucca*, Texas.

P. Pandàni, E. & E. Proc. Acad. Nat. Sci. Phil. July, 1890, p. 230.

Perithecia amphigenous, on large, dull white spots with a purplish-red border, covered by the epidermis, subglobose, 150–200 μ diam., membranaceous, of coarse, cellular structure, the apex and papilliform ostiolum erumpent. Asci cylindrical, about 200 x 10 μ, with faint, rudimentary paraphyses. Sporidia uniseriate or biseriate, hyaline, granular, oblong-elliptical, 18–20 x 7–8 μ.

On leaf of *Pandanus*, in a greenhouse, Knoxville, Tenn. (Scribner).

P. zeícola, E. & E. l. c p. 229.

Perithecia gregarious, minute ($\frac{1}{4}$–$\frac{1}{3}$ mm.), covered by the cuticle which is slightly raised and pierced by the obtusely conical, black and shining ostiola. Asci clavate-cylindrical, 75–80 x 12–15 μ, nearly sessile, with abundant paraphyses. Sporidia crowded-biseriate, elliptical or almond shaped, hyaline, granular, 18–20 x 8–10 μ.

On dead stalks of *Zea Mays* exposed to the weather through the winter, Newfield, N. J.

P. cónica, E. & E. l. c.

Perithecia gregarious, erumpent and superficial, conical, about ½ mm. broad and high. Asci oblong, about 75 x 20 μ, with abundant paraphyses. Sporidia biseriate, elliptical, hyaline, granular, often bulging on one side, 20–22 x 10–12 μ. The asci and sporidia are the same as in *Botryosphæria fuliginosa* (in N. A. F. 475–481) but its scattered, conical perithecia are quite different from any of the forms included under that name.

On old canes of *Arundinaria*, St. Martinsville, La. (Langlois).

P. entáxia, (C. & E.)

Sphæria entaxia, C. &. E. Grev. VI, p. 14, tab. 95, fig. 14.
Physalospora entaxia, Sacc. Syll. 1703, Cke. Syn. 4097.
Exsicc. Ell. N. A. F. 481.

Perithecia seriately erumpent, globose, black, papillate. Asci clavate. Sporidia narrow-elliptical, 30–32 x 12 μ. The accompanying pycnidia (*Diplodia* sp.) have stylospores elliptical, brown, uniseptate, 22 x 10 μ.

On dead branches of *Andromeda ligustrina*, Newfield, N. J.

P. citríspora, (B. & C.)

Sphæria citrispora, B. & C. Grev. IV, p. 147.
Physalospora citrispora, Sacc. Syll. 1677, Cke. Syn. 4089.

"Quite covered by the cuticle. Asci clavate. Sporidia biseriate, either lemon-shaped or narrower, 33 μ long."

On *Tilia glabra*, Virginian Mountains.

P. Eunòtiæ, (B. & C.)

Sphæria Eunotiæ, B. & C. l. c.
Phomatospora, *Eunotiæ*, Cke. Syn. 4350.
Physalospora Eunotiæ, Sacc. Syll. 1693.

"Bursting through the cuticle. Asci clavate. Sporidia oblong, hyaline, swelling in the middle on either side, like the frustules of *Eunotia*, 20–25 μ long."

On stems of Ivy, Carolina.

P. phlyctænoìdes, (B. & C.)

Sphæria phlyctænoides, B. & C. Grev. IV, p. 151.
Physalospora phlyctænoides, Sacc. Syll. 1705.
Phomatospora phlyctænoides, Cke. Syn. 4356.

"Forming little, brown, irregular specks on a white ground. Asci short, lanceolate. Sporidia cymbæform with the endochrome retracted to either end, 12–14 μ long."

On stems of *Dolichos*, Alabama (Beaumont).

P. hypericina, (B. & C.)

Sphæria hypericina, B. & C. l. c. p. 149.
Physalospora hypericina, Sacc. Syll. 1682.
Phomatospora hypericina, Cke. Syn. 4346.

"Perithecia depressed, surrounded by the cuticle. Asci clavate. Sporidia cymbæform, hyaline, 25 μ long."

On *Hypericum*, South Carolina.

P. Œnotheræ, (B. & C.)

Sphæria Œnotheræ, B. & C. l. c. p. 151.
Phomatospora Œnotheræ, Cke. Syn. 4348
Physalospora Œnotheræ, Sacc. Syll. 1687.

"Very minute black. Ostiola rather prominent. Asci clavate. Sporidia shortly fusiform, ends obtuse, hyaline.

On *Œnothera sinuata*, So. Carolina.

We have seen no specimens of this and the four preceding species. They are all placed in *Phomatospora* by Sacc. in Syll., and Cooke, in his Syn. Pyr. in Grev., places them partly in *Physalospora*, and partly in the paraphysate sections of *Phomatospora* and *Læstadia*.

Species imperfectly known.

P. subsolitària, (Schw.)

Sphæria subsolitaria, Schw. in Fr. Elench. II, p. 86.
Physalospora subsolitaria, Sacc. Syll. 1701.

Stroma consisting only of a thin, effused, black stratum under the epidermis and penetrating the inner bark. Perithecia erumpent, subsolitary, white inside. Ostiolum subprominent, white.

On bark of *Rhus*, Carolina and Pennsylvania.

Perithecia connected in pairs or standing singly, globose, black. Asci elongated, suberect.

P. subsimplex, (Schw.)

Sphæria subsimplex, Schw. Syn. N. Am. 1679.
Physalospora subsimplex, Sacc. Syll. 1718.

At first very thin, covered by the whitened epidermis, then partially denuded as the epidermis falls away. Ostiola translucent. Perithecia densely crowded, partly immersed in the bark, often confluent, oblong-hemispherical, depressed, rugose, black, papillate, the papilla deciduous and perithecium perforated. Sporidia (fide Berk.) ellipsoid, hyaline, 14–15 μ long.

On branches of *Rhus glabra*, Pennsylvania and New York.

PHOMATÓSPORA, Sacc.

F. Ven. Ser. II, p. 306.

Perithecia minute, covered or erumpent, scattered, membranaceous. Ostiolum papillate. Asci cylindric-filiform, aparaphysate. Sporidia uniseriate, 8 in an ascus, hyaline, continuous, 2-nucleate. Differs from *Physalospora* in the absence of paraphyses.

P. Datíscæ, Hark. Bull. Cal. Acad. Feb., 1884, p. 43.

Perithecia scattered, subepidermal. Ostiolum conical. Asci oblong or obclavate, 8-spored, 50–60 x 15–20 μ, paraphyses none. Sporidia oblong-oval, hyaline or minutely granular, with a small vacuole at each end, 20–30 x 8–10 μ.

On dead stems of *Datisca glomerata*, California.

P. argyrostígma, (Berk.)

Sphæria argyrostigma, Berk. Hook. Lond. Journ. Bot. VI, p. 326.
Phomatospora argyrostigma, Sacc. Syll. 1654, Cke. Syn. 4339.

Perithecia widely scattered, rather small, depressed-globose, astomous, covered by the epidermis. Spots epidermal, black, with a white center, punctiform. Sporidia boat-shaped, hyaline. *Phoma dispersa*, Berk., occurs on the opposite side of the leaf.

On decaying leaves of *Yucca filamentosa*, Ohio.

P. disrúpta, (B. & C.)

Sphæria disrupta, B. & C. Grev. IV, p. 149.
Physalospora disrupta, Sacc. Syll. 1672.
Phomatospora disrupta, Cke. Syn. 4345.

Perithecia causing little swellings in the unaltered cuticle, which are pierced in the center by the ostiola. Asci broad, obovate, with two rows of hyaline, elliptical sporidia accompanied by brown, elliptical, binucleate stylospores in different perithecia.

On *Smilax*, South Carolina.

P. Berkelèyi, Sacc. F. Ven. Ser. III, p. 306, F. Ital. tab. 605.

Sphæria phomatospora, B. & Br. Ann. Nat. Hist. No. 647, tab. 11, fig. 38.
Phomatospora Berkeleyi, Sacc. Syll. 1650, Cke. Syn. 4335.

Perithecia minute, 140–200 μ diam., sphæroid, immersed or erumpent. Ostiolum subpapillate, punctiform; nucleus with a rose-colored tint. Asci cylindric-filiform, rather long attenuate-stipitate below, 100–115 x $3\frac{1}{2}$ μ (p. sp. 50–60 x 3–$3\frac{1}{2}$ μ), 8-spored, paraphyses none. Sporidia directly or obliquely uniseriate, oblong-ellipsoid or subcylindrical, 6–8 x 2–$2\frac{1}{2}$ μ, straight, with a nucleus in each end, hyaline. Spermogonia in similar perithecia, but more superficial; spermatia cylindrical, hyaline, 8–10 x 2 μ.

APIÓSPORA, Sacc.

Consp. Gen. Pyr. p. 9, Syll. I, p. 539.

Perithecia covered, membranaceous, globulose, connected by a dark colored pseudostroma. Asci 8-spored, with paraphyses more or less distinct. Sporidia clavate-subpyriform, attenuated and often curved below, faintly uniseptate near the lower end, hyaline or yellowish.

Mostly growing on grasses.

A. Montágnei, Sacc. F. Ven. Ser. II, p. 306.

Sphæria apiospora, Dur. & Mont. Fl. Alg. I, p. 492, tab. 25, fig. 1.
Hypopteris apiospora, Berk. Dec. Fungi, 485.
Exsicc. Rab. F. E. 3157.

Seriate-erumpent black, 300 μ diam. Perithecia comparatively large, globose, uniseriate in longitudinal cracks in the epidermis, white inside, connected by a dark-colored pseudostroma. Ostiola hemispherical, umbilicate. Asci clavate or oblong, 70–90 x 15–20 μ. Sporidia biseriate, ovate-oblong, appendiculate-pyriform, 20–25 x 8–10 μ.

On culms of *Arundinaria*, in the Southern States.

Specc. from Louisiana have asci 100–110 x 25–30 μ; sporidia 35–40 x 10 μ, but specc. from Alabama have asci and sporidia agreeing with the measurements in Sacc. Syll.

A. Polýpori, E. & E. (in Herb.)

Perithecia scattered, erumpent, convex-hemispherical, about 200 μ diam., black, carbonaceo-membranaceous. Ostiolum papilliform, soon perforated. Asci oblong, obtuse, nearly sessile, 40–45 x 10–12 μ, overtopped by the abundant linear paraphyses. Sporidia crowded-biseriate, ovate-oblong, slightly curved or subinequilateral, about 12 x 4 μ, hyaline and at length with a single septum nearer the narrow end.

On old *Fomes applanatus*, Newark, Delaware (Commons).

SPHÆRÙLINA, Sacc.

Michelia I, p. 399.

Perithecia as in *Sphærella*. Sporidia hyaline, 2-or more-septate.

S. sambùcina, Pk. 38th Rep. p. 106.

Perithecia minute, numerous, closely gregarious, unequal and irregular, orbicular, oblong or even flexuous, covered by the epidermis, then erumpent, opening by a pore or a narrow chink, black. Asci clavate or subcylindrical, 75–125 x 12 μ, without paraphyses. Sporidia crowded or biseriate, oblong-clavate, 5–7-septate, constricted at the middle septum, colorless, 23–30 x $7\frac{1}{2}$–$8\frac{3}{4}$ μ; the lower half narrower than the upper.

On dead branches of *Sambucus Canadensis*, New York State (Peck).

Apparently related to *S. intermixta* (B. & Br.), but differs in its longer asci and its sporidia.

S. Péckii, (Speg.)

Sphærella Peckii, Speg. in Thum. M. U. 1356.
Metasphæria Peckii, Sacc. Syll. II, p. 172.
Sphærulina Peckii, Cke. Syn. No. 5799.

Perithecia epiphyllous, gregarious, at first innate in the epidermis, then protuberant, sphæroid-lenticular, texture subcarbonaceous, pierced above, black, 100–120 μ diam. Asci cylindrical, attenuated below into a short, thick stipe, rather acutely rounded at the apex, 8-spored, without paraphyses, 50–60 x 6–$7\frac{1}{2}$ μ. Sporidia biseriate, hyaline, fusoid or subcylindrical, obtusely acuminate at each end, mostly a little curved, protoplasm at first, 2-parted and spuriously septate, then multiseptate but never constricted, 15–18 x 3–$3\frac{1}{2}$ μ.

On dry leaves of *Amelanchier Canadensis*, Albany, N. Y.

S. dryóphila, (Cke. & Hark.)

Sphærella dryophila, Cke. & Hark. Grev. IX, p. 86.
Leptosphæria dryophila, Sacc. Syll. 3036.
Sphærulina dryophila, Cke. Syn. 5816.

Epiphyllous. Spots orbicular, reddish-brown. Perithecia brown, subimmersed. Asci clavate, sessile. Sporidia lanceolate, triseptate, pale brown, 20 x $3\frac{1}{2}$–4 μ.

On leaves of *Quercus*, California.

S. myriàdea, (DC.) **(Plate 27)**

Sphæria myriadea, DC. Fl. Fr. VI. p. 148.
Sphærella myriadea, Rab. in F. Eur. 149.
Sphærulina myriadea, Sacc. Syll. 3524.

Exsicc. Rab. F. Eur. 149.—Thum. M. U. 2157.—Cke. F. Brit. 1st Ser. 172.—id. 2d Ser. 269. Rav. F. Am. 156.—Kriegr. F. Sax. 279.—Sydow. M. March. 1931.—Rehm Asc. 947.

Perithecia densely gregarious, in tolerably large (1–10 mm.), indefinitely limited, dendroid or variously shaped groups or patches, at first buried, finally erumpent, spherical, perforated above, black, 90–100 μ diam. Asci clavate, sessile, 8-spored, 60 x 6 μ. Sporidia 2–3-seriate, elongated-fusoid, somewhat pointed at the ends, slightly curved, 3-septate, hyaline, 30–35 x 2–3 μ.

On fallen oak leaves, Carolina, Ohio, Pennsylvania, New York, New Jersey and New England.

STIGMÁTEA, Fr.

Summa Veg. Scand p. 421, (Sacc. Syll. I, p. 541).

Perithecia innate-subprominent, glabrous, thin-walled. Ostiolum minute, nucleus rather compact, white. Asci oblong, subsessile, 8-spored, paraphysate or pseudoparaphysate. Sporidia ovoid-ellipsoid, unequally uniseptate, yellowish-hyaline. Foliicolous, minute.

St. Robertiàni, Fr. Summa Veg. Sc. p. 421. (Plate 27)

Dothidea Robertiani, Fr. S. M. II, p. 564.
Cryptosphæria nitida, Grev. Fl. Ed. p. 363.
Hormotheca Geranii, Bon. Abb. p. 149.
Exsicc. Fckl. F. Rh. 419.—Rab F. E. 963, 2129.—Rehm Asc 246.—Thum. F. Austr. 151. id. M. U. 160.—Cke. F. Brit. Ser. I, 133.—Ell & Evrht. N. A. F. 2d Ser. 2362. Krieg F. Sax 230.—Erikss. F. Scand. 90.—M. March. 253.—F. Gall. 2085.

Perithecia scattered, gregarious or collected in loose groups, superficial, sessile, hemispherical, smooth and shining, about 140 μ broad and 50–60 μ high. Asci oblong, sessile, 8-spored, 33–45 x 11–14 μ, mostly broader below, with scanty paraphyses. Sporidia sub-biseriate, ovate-elliptical, unequally two-celled, upper cell broader, greenish, 12–14 x $4\frac{1}{2}$–5 μ.

On living leaves of *Geranium Robertianum*, Massachusetts (Sturgis), Canada (Dearness).

St. Arundinàriæ, Cke. Grev. VII, p. 50.

Epiphyllous, gregarious. Perithecia subglobose or depressed, black, astomous, superficial. Asci clavate. Sporidia fusoid, 4-nucleate, at length triseptate, hyaline, 26–30 x 8–10 μ.

On *Arundinaria*, Georgia (Ravenel).

St. Gerànii, Fr. Summa Veg. Scand. p. 421.

Dothidea Geranii, Fr. S. M. II, p. 558.

Epiphyllous. Perithecia gregarious on a purplish spot, ovate, minute, black, glabrous. Asci elongated, 8-spored, 40 x 8 μ. Sporidia biseriate, subclavate, didymous, hyaline, 8 x 4 μ.

On living leaves of *Geranium Carolinianum*, California (Harkness).

St. Sequoiæ, (Cke. & Hark.)

Dothidea Sequoiæ, Cke & Hark. Grev. IX, p. 87.
Stigmatea Sequoiæ, Sacc. Syll. 2110, Cke. Syn. 1384.

Scattered, convex, black, shining, minute, unicellular. Asci broad-clavate, sessile. Sporidia biseriate, lanceolate, obtuse, constricted, uniseptate, binucleate, hyaline, 23 x 7½ μ.

On leaves of *Cupressus*, California (Harkness).

From spece sent by Dr. Harkness (on *Libocedrus decurrens*), we add the following notes.—Asei 65–70 x 20 μ; sporidia narrow-ovate, unequally 2-celled, upper cell broader, 16–22 x 7–8 μ.

St. Ranúnculi, Fr. Summa Veg. Scand. p. 412.

Perithecia innate, prominent, seated on a bleached spot, crowded, more or less scattered, globose, or sometimes subconic-attenuated above, glabrous, smooth, black. Ostiola rounded, perforated. Asci oblong-ovate, inequilateral, broader below, 8-spored, 50–60 x 13–16 μ. Sporidia conglomerated, elongated-fusoid, uniseptate, straight, hyaline or yellowish, 23–30 x 4–5 μ. Paraphyses none.

On leaves and stems of *Ranunculus nivalis*, Upernivik and Disco, Greenland.

St. Juníperi, (Desm.)

Dothidea Juniperi, Desm. VIII, Not. p. 13.
Stigmatea alpina, Speg. in Thum. M. U. 1057.
Microthyrium Juniperi, Sacc. Mich. I, p. 351.
Stigmatea Juniperi, Winter Die Pilze II. p. 340.

Exsicc. Desm. Pl. Cr. de France, Ed. I, 1094.—Thum. M. U. 1057.—Ell. N. A. F. 1191. Linhart Fung. Hung 73. Sacc. M. Ven. 1269

Perithecia scattered, lenticular or subhemispherical, black, roughish, perforated above, 200–300 μ diam., superficial. Asci oblong, rounded and obtuse above, abruptly narrowed below into a short stipe, 60–70 x 20 μ. Paraphyses none. Sporidia biseriate, ovate-lanceolate, uniseptate, 16–25 x 6–8 μ, the septum nearer the lower end, and the upper cell broader yellowish-hyaline.

On living leaves of *Juniperus Virginiana*, Newfield, N. J., and on *Sequoia gigantea*, California.

A careful examination of the Newfield and California spece. shows that they are in all respects the same as the spece. in Desm. Exsicc. above referred to.

St. sclerotídea, Cke Grev. V, p. 153.

Gregarious, black. Perithecia superficial, depressed. Asci clavate. Sporidia biseriate, elliptical, hyaline, uniseptate. Externally resembling a small *Sclerotium*. Asci very soon dissolved. Sporidia 22 x 9 μ.

On leaves of *Arundinaria*, South Carolina (Ravenel).

DIDYMÉLLA, Sacc.

Mich. I, p. 377, Sacc. Syll. I. p. 545.

Perithecia covered, membranaceous, globose-depressed, minutely papillate, mostly growing on stems or branches. Asci 4–8-spored, paraphysate. Sporidia ovoid, ellipsoid or suboblong, uniseptate, hyaline. Differs from *Sphærella* in the presence of paraphyses. The perithecia also are mostly larger and firmer.

D. Canadénsis, E. & E. Proc. Acad. Nat. Sci. Phila., Pa., July, 1890, p. 232.

Perithecia irregularly but thickly scattered, buried in the bark, which is slightly raised above them and pierced by the small, black, papilliform ostiola, white inside, globose, about ⅓ mm. diam. Asci clavate-cylindrical, 75–90 x 12–15 μ, with abundant paraphyses. Sporidia crowded-biseriate, cylindrical, obtuse, hyaline, 4-nucleate, constricted in the middle, and, slightly so, near each end, 25–34 x 6–7 μ.

On dead limbs of *Salix*, London, Canada (Dearness).

D. Màli, E. & E. l. c.

Perithecia scattered, about ¼ mm. diam., buried in the substance of the bark, except the emergent rather acutely conical ostiolum. Asci clavate-cylindrical, about 70 x 7 μ, with abundant paraphyses. Sporidia biseriate, fusoid, slightly curved, about 4-nucleate, not constricted, 20–22 x 3 μ, ends acute.

On the inner surface of loose hanging bark of living apple trees, Newfield, N. J.

D. recèdens, (Cke. & Hark.)

Sphæria recedens, Cke. & Hark. Grev. IX, p. 130 (non Niessl),
Didymella recedens, Sacc. Syll. 2140,
Endophlœa recedens, Cke. Syn. 4126.
Exsicc. Thum. M. U. 1748 (in part).

Perithecia gregarious, minute, covered by the pustulately raised epidermis, papillate. Asci elongated, 100 μ long. Sporidia narrowly fusoid, uniseptate, hyaline, 18–20 x 3 μ, the two cells easily separating.

On bark of *Eucalyptus*, California.

D. sphæréllula, (Pk.)

Sphæria sphærellula, Pk. 30th Rep p. 66.
Endophlæa sphærellula, Cke. Syn 4117.
Didymella sphærellula, Sacc. Syll. 2131.

Perithecia minute, scattered or seriately placed, covered by the epidermis which is at length ruptured. Asci broad, obtuse, gradually narrowed above, suddenly contracted at the base. Sporidia crowded, fusiform, uniseptate, hyaline 12–15 μ long.

On dead, bleached twigs of *Acer Pennsylvanicum*, Catskill Mts., New York.

D. ségna, (C. & E.)

Sphæria segna, C. & E. Grev. VI, p. 95.
Didymella segna, Sacc. Syll. 2141.
Endophlæa segna, Cke. Syn. 4127.

Perithecia gregarious, covered, perforated above, raising the epidermis into pustules which are soon ruptured above. Asci oblong-clavate, 35 x 6–7 μ, paraphysate. Sporidia biseriate, fusoid-oblong, uniseptate and often slightly constricted, 12–13 x 3 μ (15 x 7 μ, Cke.)

On dead branches of *Nyssa multiflora*, Newfield, N. J.

D. Céltidis, (B. & C.)

Sphæria Celtidis, B. & C. Grev. IV, p. 146.
Didymella Celtidis, Sacc. Syll. 2144.
Endophlæa Celtidis, Cke. Syn. 4129.

"Perithecia closely packed but distinct, covered by the bark, which is slightly raised, so that the whole looks like fine shagreen. Ostiola rather prominent. Asci short and with four sporidia which are oblong, obtuse, uniseptate."

On branches of *Celtis*, South Carolina.

D. castanélla, (C. & E.)

Sphæria castanella, C. & E. Grev. VII, p. 10.
Didymella castanella, Sacc. Syll. 2142.
Endophlæa castanella, Cke. Syn. 4128.

Perithecia loosely embraced by the ruptured cuticle, here and there gregarious. Asci clavate. Sporidia biseriate, cylindrical, uniseptate, hyaline, 16 x 3 μ.

On dead twigs of *Castanea*, Newfield, N. J.

D. Raùii, (E. & E.)

Sphæria (Didymella) Rauii, E. & E. Bull. Torr. Bot. Club, X, p. 90.
Endophlæa Rauii, Cke. Syn. 4138.
Didymella Rauii, Sacc. Syll. 6476.
Exsicc. Ell. N. A. F. 1336.

Perithecia scattered or 2–3 together on the surface of the inner bark, and covered by the thin, loosened cuticle which is pierced by the papilliform ostiola. Asci 35–40 x 6–7 μ, with slender paraphyses. Sporidia biseriate, oblong-fusiform, uniseptate, constricted at the septum and slightly curved, yellowish-hyaline, with a faint, bristle-like appendage at each end, and a nucleus in each cell, 7–8 x 1½–2 μ.

On dead branches of cultivated roses, Bethlehem, Pa. (Rau).

D. lophóspora, Sacc. & Speg. Mich. I, p. 376, F. Ital. tab. 3676.

Exsicc. N. A. F. 588.

Perithecia gregarious, covered by the cuticle, subprominent and subglobose, scarcely papillate, ⅛ mm. diam., texture compact, dark. Asci thick-clavate, 65–75 x 12 μ (p. sp. 50 x 12 μ), stipitate, with abundant filiform paraphyses, 8-spored. Sporidia biseriate, cylindric-fusoid, straight or curved, 15–18 x 3–4 μ (16–17 x 5–6 μ Sacc.), uniseptate and obtuse, constricted in the middle, hyaline, with a broad, ill-defined appendage at each end. Var. *Acetosellæ*, Ell. & Sacc. has asci 80–90 x 8–10 μ, sporidia 10–12 x 4 μ.

On fallen oak leaves, Ohio. On peduncles of *Rhus copallina*, bark of grape vines, twigs of *Sassafras*, and capsules of *Œnothera*. Newfield, N. J., and on *Staphylea*, in Pennsylvania.

D. Cornùti, E. & E. l. c.

Perithecia scattered, minute, 175–200 μ diam., attached to the blackened surface of the stem just beneath the thin epidermis which is barely pierced by the prominent ostiola. Asci cylindrical, nearly sessile, 65–70 x 7 μ. Paraphyses present. Sporidia biseriate, fusiform, yellowish, very slightly curved, uniseptate and constricted, becoming 3-septate, 20–25 x 3–3½ μ, ends subobtuse.

On dead stems of *Asclepias Cornuti*, London, Canada (Dearness).

D. Andropògonis, E. & E. l. c.

Perithecia hypophyllous, subgregarious, about ½ mm. diam., buried in the substance of the leaf, with the apex and short, conic-cylindrical ostiolum projecting. Asci cylindrical, narrow, 80–90 x 5 μ, with filiform paraphyses. Sporidia overlapping-uniseriate, fusoid-oblong, 3-nucleate, becoming 1–2-septate and constricted, acute at first, but finally obtuse, hyaline or yellowish-hyaline, 12–15 x 3 μ, some of them very slightly curved. Near *D. subgemina*, B. & C.

On dead leaves of *Andropogon muricatus*, St. Martinsville, La.

D. Megarrhizæ, Cke. & Hark. Grev. XIV, p. 9.

Sphæria Megarrhizæ, Cke & Hark. Grev. XIII, p. 18?
Didymella Cookeana, Sacc. Syll. 6479.
Exsicc. Ell. & Evrht. N. A. F. 2d Ser. 1661.

Gregarious. Perithecia subconical, black, small, shining, at first covered, the acute ostiolum at length erumpent. Asci cylindric-clavate, 8-spored. Sporidia elliptical, uniseptate, constricted in the middle, straight, hyaline, 16–18 x 7 μ.

On stems of *Megarrhiza*, California.

D. Füschiæ, Cke. & Hark. l. c.

Scattered or subgregarious. Perithecia globose-depressed, black, covered by the epidermis, the conical ostiolum emergent. Asci cylindric-clavate, 8-spored. Sporidia biseriate, straight or curved, uniseptate, not constricted, binucleate, hyaline, 15 x 5 μ.

On stems of *Fuschia*, California.

D. Lupìni, (Cke. & Hark.)

Sphæria (Didymella) Lupini, Cke. & Hark. Grev. XIII, p 18.
Didymella Lupini, Sacc Syll. 6482, Cke. Syn. 4392.

Perithecia scattered, punctiform, covered, depressed-globose, slightly papillate. Asci clavate. Sporidia uniseptate, elliptical, yellowish-hyaline (immature)? granulose, 15–17 x 6–7 μ.

On stems of *Lupinus*, California.

D. nivàlis, (Fckl.)

Sphæria nivalis, Fckl. in Die zweite deutsch Nordpolarf. p. 93, tab. 1, fig. 5.
Didymella nivalis, Sacc. Syll. 6483

Perithecia lying under the dry, whitened epidermis of the leaves and stems, gregarious, minute, globose, very black. Ostiolum black, acute, subprominent. Asci oblong-ovate, attenuated towards each end, with a thick membrane, 8-spored, 94 x 20 μ. Sporidia collected in a mass or subbiseriate, oblong, unequally didymous, ends obtuse, slightly constricted at the septum, hyaline, 20 x 8 μ.

On dry stems and leaves of *Epilobium latifolium*, Franz-Joseph's-Fiord, in Greenland.

D. próminens, E. & E. Journ. Mycol. II, p. 101.

Exsicc. Ell. & Evrht. N. A. F. 2d Ser. 1796.

Perithecia erumpent-superficial, scattered, ovate-hemispherical, ¼ mm. diam., black, rough, except the prominent, tubercular-conical or short-cylindrical ostiolum. Asci clavate-cylindrical, subsessile, 40–45 x 6–7 μ, with filiform paraphyses, and eight subfusoid or subcymbiform, uniseptate, hyaline sporidia 10–12 x 3–4 μ, slightly constricted at the septum.

On dead stems of *Ambrosia trifida*, Louisiana (Langlois).

D. subexsérta, C. & E.

Sphæria subexserta, C. & E. Grev. VII, p. 42.
Didymella subexserta, Sacc. Syll. 2180, Cke. Syn. Pyr. 4408.
Exsicc. Ell. N. A. F. 189.

Perithecia scattered, depressed-globose, ⅓ mm. diam., buried in the substance of the stem, with the short-cylindrical, obtuse ostiolum slightly projecting. Asci subclavate about 60 μ long. Sporidia biseriate, narrow-elliptical, constricted and uniseptate in the middle, each cell 2-nucleate, 10–12 x 3–4 μ.

On dead stems of *Lactuca Canadensis*, Newfield, N. J., and on *Lactuca Floridana*, Louisiana (Langlois).

D. fructígena, E. & E. (in Herb.)

Perithecia scattered or gregarious or 2–3 together, but not confluent, 100–150 μ diam., subglobose, prominent, but covered, except the apex and papilliform ostiolum, by the thin epidermis; when growing on the denuded cherry stones, superficial, with the base adnate. Asci oblong-cylindrical, subsessile, 34–40 x 6–7 μ, paraphysate. Sporidia inordinate, oblong-fusoid, slightly curved, uniseptate and constricted, the upper cell abruptly swollen just above the septum, the lower cell narrower and acute, 12–15 x 3–3½ μ, hyaline.

On dried up cherries (cult.), and on the bare cherry stones lying on the ground, Newfield, N. J.

D. intercellulàris, (B. & C.)

Sphæria intercellularis, B. & C. Grev. IV, p. 153.
Didymella intercellularis, Sacc. Syll. 2183, Cke. Syn. 4412.

"Perithecia contained in the larger or dilated cells, about 50 μ diam., with a few radiating threads. Asci short, thicker at the base. Sporidia shortly cymbiform, uniseptate, 12–13 μ long."

On *Typha*, Massachusetts.

D. Nebráskæ, (B. & C.)

Sphæria Nebraskæ, B. & C. Grev. IV, p. 154.
Didymella Nebraskæ, Sacc. Syll. 2186, Cke. Syn. 4415.

"Minute, shortly hysteriiform. Asci oblong, slightly swollen. Sporidia elliptical, biseriate, uniseptate, 12–16 μ long."

On leaves of grass, Nebraska.

D. Douglásii, E. & E.

Sphærella conigena, E. & E. Proc. Acad. Nat. Sci. Phil. Pa., July, 1890, p. 230.

Perithecia gregarious on the back of the exposed tip of the scale, minute (74–110 μ), buried, except the black, smooth, conic-papilliform apex. Asci narrow clavate-cylindrical, gradually attenuated below, 75–80 x 5 μ, paraphyses filiform. Sporidia uniseriate, ovate, uniseptate and constricted at the septum, hyaline, 6–7 x 3–3½ μ.

On scales of dead cones of *Abies Douglasii*, Belt Mts., Montana (Anderson).

D. onosmòdina, (Pk. & Cl.)

Sphæria onosmodina, P. & C. 30th Rep. N. Y. State Mus. p. 67.
Didymella onosmodina, Sacc. Syll. 2165, Cke. Syn. 4388.

Perithecia numerous, minute, at first covered by the epidermis, then exposed, depressed, black. Ostiola pierced. Asci cylindrical. Sporidia crowded or biseriate, oblong-elliptical, uniseptate, hyaline, 15–18½ μ long, the cells usually unequal.

On dead stems of *Onosmodium Carolinianum*, Buffalo, N. Y.

D. Dioscorèæ, (B. & C.)

Sphæria Dioscoreæ, B. & C. Grev. IV, p 152.
Didymella Dioscoreæ, Sacc. Syll. 2190, Cke. Syn. 4419.

"Very minute. Asci short, with few sporidia, which are oblong and uniseptate, about three times longer than broad," (20 μ long).

On stems of *Dioscorea*, South Carolina.

D. lathýrina, (B. & C.)

Sphæria lathyrina, B. & C. Grev. IV, p. 155.
Didymella lathyrina, Sacc. Syll. 2167, Cke. Syn. 4391.

"Very minute. Asci oblong. Sporidia 8 in an ascus, oblong, uniseptate, constricted at the septum, biseriate, 7–8 μ long."

On stems of *Lathyrus latifolius*, Pennsylvania.

D. commanípula, (B. & Br.)

Sphæria commanipula, B. & Br. Ann. Nat. Hist. No. 645, tab. 11, fig. 31.
Didymella commanipula, Sacc. Syll. 2173, Cke. Syn. 4399.

"Scattered, at first subglobose, covered, at length denuded, collapsed. Ostiola minute. Asci cylindrical. Sporidia biseriate, short elliptic-cymbiform, uniseptate, sometimes decidedly conical, with a constriction about the center."

Specc. collected in California, on stems of spikenard, were doubtfully referred to this species in Grev. VII, p. 74.

D. eumórpha, (B. & C.)

Sphæria eumorpha, B. & C. Grev. IV, p. 145.
Sphærella eumorpha, Cke Journ. Bot. 1883.
Didymella eumorpha, Sacc. Syll. 2191, Cke. Syn. 4420.

"Linear, closely surrounded by the cuticle, opening with a narrow slit. Asci linear. Sporidia uniseriate, short-oblong or cymbiform, 12–15 x 3 μ, uniseptate."

On culms of *Arundinaria*, South Carolina.

D. hyperbòrea, (Karst.)

Sphæria hyperborea, Karst. Fungi Spetsb. No. 42.
Didymella hyperborea, Sacc. Syll. 2148.

Perithecia scattered, hypophyllous, sunk in the parenchyma of the leaf and covered by the blackened epidermis which is at length ruptured, subsphæroid, collapsing to cup-shaped, pierced with a very minute orifice, black, when moistened brownish-black, glabrous, smooth, about 250 μ diam. Asci clavate-cylindrical, subsessile, 90–100 x 11–12 μ, 8-spored. Sporidia uniseriate, ellipsoid, uniseptate, hyaline, 14–17 x 7–8 μ. Paraphyses filiform, guttulate, slender.

On *Cassiope tetragona*, Ameralik, Greenland.

D. pteridícola, (B. & C.)

Sphæria pteridicola, B. & C. Grev. IV, p. 145.
Didymella pteridicola, Sacc. Syll. 2195, Cke. Syn. 4424.

"Perithecia forming little gray, parallel lines, covered with the cuticle. Asci clavate. Sporidia oblong, slightly curved, obtuse at each end, uniseptate, 15 μ long."

On stalks of *Pteris*.

D. Catàriæ, (C. & E.)

Sphæria Catariæ, C. & E. Grev. V, p. 95.
Didymella Catariæ, Sacc. Syll. 2175, Cke. Syn. 4402.

Perithecia scattered, covered by the epidermis, slightly prominent, depressed-globose, 150–200 μ diam., with a papilliform ostiolum. Asci clavate, 55–60 x 12 μ, paraphysate. Sporidia biseriate, narrow-elliptical, slightly curved, ends obtusely pointed, hyaline, uniseptate, 15–18 x 6–7 μ (20 x 8 μ, Cke.)

On dead stems of *Nepeta Cataria*, New Jersey, Ohio and Canada. Closely allied to the next species.

Species imperfectly known.

D. incommiscíbilis, (B. & C.)

Sphæria incommiscibilis, B. & C. Grev. IV, p. 151.
Didymella incommiscibilis, Sacc. Syll. 2176, Cke. Syn. 4403.

"Minute, covered by the cuticle. Ostiola projecting, papilliform. Asci linear. Sporidia fusoid, narrow, at length uniseptate."

On stalks of some herbaceous plant, Virginian Mountains.

D. combúlliens, (B. & C.)

Sphæria combulliens, B. & C. Grev. IV, p. 147.
Didymella combulliens, Sacc. Syll. 2192, Cke. Syn. 4421.

"Scattered, covered by the cuticle, with the exception of the minute ostiolum. Sporidia uniseriate, oblong, uniseptate."

On stems of *Arundinaria*, (South Carolina)?

D. júncina, (B. & Rav.)

Sphæria juncina, B. & Rav. Grev. IV, p. 145.
Didymella juncina, Sacc. Syll. 2184, Cke. Syn. 4413.

"Forming little discolored patches which are studded with the ostiola. Sporidia oblong, uniseptate, constricted slightly at the septum."

On *Juncus*, South Carolina.

FAMILY. GNOMONÌEÆ.

Perithecia as in the *Sphærelloideæ*, but with a cylindrical or subulate ostiolum. Asci mostly without paraphyses. Sporidia oblong, fusoid or filiform, continuous or uniseptate, hyaline.

GNOMÒNIA, Ces. & De Not.

Schema Sfer. p. 57, Sacc. Syll. I, p. 561.

Perithecia covered or erumpent, submembranaceous, glabrous, generally separate. Ostiolum more or less elongated, cylindrical, central or lateral. Asci mostly aparaphysate, 4–8-spored, often perforated at the apex. Sporidia oblong, elliptical or fusoid, continuous or uniseptate, hyaline. Fungi minute, foliicolous or more rarely caulicolous.

A. *Sporidia continuous, oblong or fusoid* (*Gnomoniella*).

G. curvicólla, (Pk.)

Sphæria curvicolla, Pk. 31st Rep. p. 50.
Gnomoniella curvicolla, Sacc. Syll. 1584.
Gnomonia curvicolla, Cke. Syn. 3856.

Perithecia small, 75–100 μ diam., scattered or 2–3 confluently crowded, erumpent, at length naked, hemispherical, black. Ostiola subcylindrical, slightly curved. Asci oblong. Sporidia crowded or biseriate, colorless, 15–23 x $7\frac{1}{2}$ μ.

On decaying stems of *Polygonum articulatum*, Center, N. Y.

G. tubæfòrmis, (Tode).

Sphæria tubæformis, Tode. Fungi Meckl. II, p 51.
Gnomoniella tubiformis, Sacc. Syll. 1567.
Gnomonia tubiformis, Cke. Syn. 3839.
Ceratostoma tubæforme, Ces. & De Not. Schema p. 54.
Gnomonia tubæformis, Awd. Mycol. Eur. Pyr. p. 22, tab. 8, fig. 121.
Exsicc. Fckl. F. Rh. 866.—Kze F. Sel 249,—Rab. F. E. 54, 1454.—Rehm Asc. 96.—Desm. Pl. Crypt. 442.

Perithecia in dense groups often covering the whole lower surface of the leaf, sunk in the parenchyma, covered by the epidermis and hemispherically prominent on both sides, the cylindrical, often curved ostiolum about equal in length to the diameter of the perithecium, erumpent, dark brown, about 400 μ diam. Asci oblong, with a short stipe, 8-spored. 35–70 x 14–16 μ. Sporidia imperfectly biseriate, oblong or elliptic-oblong, often inequilateral, hyaline, 14–15 x $5\frac{1}{2}$–6 μ.

On fallen alder leaves, Carolina and Pennsylvania (Schweinitz), New Jersey and New York.

G. emarginàta, Fckl. Symb. p. 122.

Gnomoniella emarginata, Sacc Syll. 1571.
Sphæria mirabilis, Pk. 28th Rep. p. 80.
Exsicc. Fckl. F. Rh. 876.—Ell. & Evrht. N. A. F. 2d Ser. 2139.—Kze. F. Sel. 252.—(Rab. Winter F. Eur. 2756)?

Perithecia scattered, covered, tolerably large, lenticular, black, with a round, slender beak 1–$1\frac{1}{2}$ lines long arising from the concave side of the perithecium. Asci elliptical, stipitate, 8-spored. Sporidia fusoid, often curved, continuous, binucleate, hyaline.

On petioles of decaying leaves of *Acer rubrum*, Newfield, N. J.

In the Newfield spece. the asci are oblong-elliptical, 70–80 x 15–20 μ. Sporidia fasciculate, broad-fusoid, 4-nucleate, 25–30 x $4\frac{1}{2}$ μ, with a broad (6–8 x 4 μ), ovate, hyaline appendage at each end, but this is soon absorbed. See Journ. Mycol. IV, p. 81.

Whether this is really the *G. emarginata*, Fckl., may be open to some doubt. The spece. distributed in Fungi Rhenani are (sec. Winter) immature, affording neither asci nor sporidia and no measurements are given in the original diagnosis. In the Newfield specc., as well as in those in the Exsiccati quoted, the perithecia can hardly be called "emarginate," though some of them are slightly so. Peck's *Sphæria mirabilis* (on fallen birch leaves) certainly belongs here.

G. tenèlla, E. & E. Journ. Mycol. IV, p. 80. **(Plate 32)**

Exsicc. Ell. & Evrht. N. A F. 2d Ser. 2140.

Perithecia amphigenous, scattered, mostly on the lamina of the leaf and not confined to the veinlets, depressed-globose, small ($\frac{1}{4}$-$\frac{1}{3}$ mm), covered by the cuticle which is raised above it. Ostiolum black, straight, bristle-like, about 1 mm. long. Asci fusoid, 50–70 x 6–7 μ. Sporidia fasciculate, narrow cylindrical, nucleate, 16–22 x 1½–2 μ, with a long, slender-pointed, hair-like appendage at each end 15–20 μ long.

On fallen and decaying leaves of *Acer rubrum*, Newfield, N. J.

The apical appendages on the sporidia are coiled in the upper part of the ascus, and when this is ruptured, protrude like the sporidia in some species of *Ophiobolus*. The perithecia occur also on the petioles of the leaf, and are more perfectly developed there, but are readily distinguished from those of *G. emarginata* by their shorter, thinner ostiola and also by the color of the petiole itself, which is of a lighter color when occupied by the last mentioned species.

Specimens on *Rubus fruticosus* in Kunze's Fungi Sel. 113. referred to *G. setacea*, Pers., are much like this, if not the same.

G. Andropògonis, E. & E. in (Herb.)

Perithecia buried in the substance of the leaf, subglobose, about ½ mm. diam., membranaceous, black, collapsing from below. Ostiolum sublateral, erumpent, cylindrical, subobtuse, projecting about ½ mm. Asci (p. sp.) oblong-lanceolate, 40 x 10 μ, aparaphysate. Sporidia fasciculately crowded, cylindric-fusoid, hyaline, septate, each cell 1–2-nucleate, slightly curved, 20–25 x 4 μ.

On decaying basal leaves of *Andropogon*, Newfield, N. J.

G. petiolóphila, (Pk.)

Sphæria petiolophila, Pk. 35th Rep. p. 144.
Gnomonia petiolophila, Sacc. Syll, 6491, Cke. Syn. 3877.

Perithecia minute, scattered, covered by the epidermis which is pierced by the prominently papillate or short-rostrate ostiola, depressed-globose, black. Asci narrow, subcylindrical, 40–45 μ long. Sporidia biseriate, narrow-fusiform, pointed at each end, hyaline, 12–15 x 2 μ, sometimes with 3–4 nuclei.

On petioles of fallen leaves of *Acer spicatam*, Helderberg Mts., N. Y.

G. Magnòliæ, Ell. Am. Nat. 1883, p. 318. **(Plate 32)**

Exsicc. Ell. N. A. F. 1354.

Perithecia rather large, buried in the parenchyma of the leaf, the short, rufous, subulate-conical ostiola alone being visible. Asci oblong-elliptical, 40 x 7–8 μ. Sporidia fusiform, acute, pale straw-color, obscurely nucleate, 11–19 x 2 μ.

On fallen leaves of *Magnolia glauca*, Newfield, N. J.

G. amœna, (Nees.), var. petiolorum, (Schw).

Sphæria petiolorum, Schw. Syn. Car. 153.
Gnomonia amœna, Nees. var. *petiolorum*, Schw. Cke. Syn. 3840.
Exsicc. Rav. Fungi Car. I, No. 64.—Ell. & Evrht. N. A. F. 2d Ser. 2543.—Rav. F. Am. 374.

Perithecia gregarious, buried, about 400 μ diam., the slender ostiolum about 1 mm. long, rising through a white, granular, superficial tubercle. Asci clavate, p. sp. 18–22 x 6 μ or, including the slender base, 30–35 μ long. Sporidia fasciculately crowded, narrow-fusiform, nearly straight, 10–12 x 1¼–1½ μ, 3–4-nucleate, hyaline, continuous.

On fallen and decaying petioles of *Liquidambar*, Carolina, Louisiana and New Jersey, (sec. Peek also on fallen petioles of ash, New York).

The typical form is found in Europe on petioles and leaves of *Corylus*, and has, sec. Sacc., asci 36 x 8 μ and sporidia 8 x 4 μ, continuous and 2–4-nucleate; sec. Winter, asei 45–50 x 9–10 μ, sporidia fusoid, acute, septate and constricted in the middle, hyaline, 12–16 x 3 –3½ μ.

G. excéntrica, (C. & P.)

Sphæria excentrica, C. & P. 25th Rep. N. Y. State Mus. p. 105.
Gnomoniella excentrica, Sacc. Syll. 1585.
Gnomonia excentrica, Cke. Syn. 3857.

Perithecia scattered, depressed, black, at first covered by the epidermis, which is pierced by the excentric or lateral, curved, acute, rostellate ostiolum; at length superficial. Asci subclavate. Sporidia crowded or biseriate, subfusiform, 4-nucleate, hyaline, 8½–9 μ long.

On dead stems of *Polygonum*, New York State.

G. vulgàris, Ces. & De Not. Schema, p. 58.

Sphæria Gnomon, Tode Fungi Meckl. II, p. 50, fig. 125.
Ceratostoma Gnomon, Fr. S. M. II, p. 497.
Cryptosphæria Gnomon, Grev. Flora Edin. p. 360.
Gnomoniella vulgaris, Sacc. Syll. 1578.
Exsicc. Rab. F. E. 1453.—Rehm Asc. 95.—Thum. F. Austr, 163.—Thum. M. U. 562.

Perithecia scattered, hypophyllous, at first sphæroid, covered, then emergent and collapsed, beak elongated, straight, often a little thickened towards the end, as long as, or a little longer, than the diameter of the perithecium. Asci elongated-fusoid, briefly stipitate, 32–42 x 5–6 μ. Sporidia conglobate, rod-shaped, subacute, 4-nucleate, straight or slightly curved, hyaline, 12–18 x 1–2 μ.

On leaves of *Ostrya Virginica*, Troy, N. Y. (Peck).

B. *Sporidia uniseptate (Eugnomonia).*

G. setàcea, (Pers.) **(Plate 32)**

Sphæria setacea, Pers. in Usteri, Ann. d. Bot. St. 11, p. 25, taf. 2, fig. 7 a.
Gnomonia setacea, Ces. & De Not. Schema, p. 58.
Gnomonia nervisequia, Fckl. Symb. p 122. (sec. Winter).
Sphæria ischnostyla, Desm. Ann. Sci. Nat. III, Ser. tom. XI, p. 375 (sec. Awd.)
Exsicc. Fckl. F. Rh. 871, 878.—Kze. F. Sel. 113, 251.—Rehm Asc. 494, 495.—Rab. F. E. 1450? 2756.—Thum. M. U. 455, 1741.—Ell. & Evrht. N. A. F. 2d Ser. 2138.

Perithecia mostly hypophyllous, scattered, covered, globose, black, 200–300 μ diam., with a very long, slender, thread-like ostiolum generally curved and twice as long as the diameter of the perithecium. Asci oblong-clavate or fusoid, contracted below into a short stipe, 4–8-spored, 30–40 x 6–9 μ. Sporidia fasciculate-crowded, fusoid or rod-shaped, subacute at the ends and often with a bristle-like appendage, straight or slightly curved, septate in the middle, but not constricted, hyaline, 12–16 x 1½–2 μ.

On fallen leaves of *Quercus*, *Castanea* and many other trees, common.

Specc. on fallen leaves of *Carya*, found at Newfield, N. J., have the sporidia 20–25 μ long, but do not differ otherwise from the usual form.

G. Myrìcæ, C. & E. Grev. VI, p. 17.

Exsicc. Ell. & Evrht. N. A. F. 2137.

Perithecia hypophyllous, gregarious, small, only the rostrate ostiolum piercing the cuticle. Asci clavate, 25–30 x 6–7 μ, 8-spored. Sporidia mostly biseriate, narrow-elliptical, hyaline, 4-nucleate, becoming unequally uniseptate, 6–8 x 2½–3 μ (10 μ long, Cke.)

On red-brown spots on living leaves of *Myrica cerifera*, or on fallen leaves without spots, Newfield, N. J.

G. Álni, Plowr. Grev. VII, p. 74.

Exsicc. Ell. & Evrht. N. A. F. 2d Ser. 2360.

Perithecia minute, in clusters, buried in the substance of the leaf. Ostiola elongated, slightly tapering at the apex. Asci ovate-elongated, 35–40 x 10–15 μ. Sporidia 8, hyaline, curved, uniseptate, nucleate, 22–25 x 3–4 μ.

On living leaves of *Alnus*, California.

The bases of the perithecia project slightly on the opposite side of the leaf, raising the surface into minute tubercles.

G. clavulàta, Ell. Am. Nat. March. 1883, p. 318.

Exsicc. Ell. & Evrht. N. A. F. 2d Ser. 1685.

Perithecia membranaceous, globose, 100–165 μ diam., rough, bedded in the substance of the leaf, their bases projecting on the lower surface and their cylindrical, obtuse, subclavate ostiola about equal in length to the diameter of the perithecia, projecting above. Asci oblong-cylindrical, p. sp. 35–40 x 5–6 μ, aparaphysate. Sporidia biseriate, narrow-elliptical, subacute, 4-nucleate at first, becoming unequally uniseptate and obtuse, 7½–9 x 2–2½ μ, yellowish-hyaline.

On fallen leaves of *Quercus nigra* and *Carya*, Newfield, N. J.

The tips of the ostiola are generally enlarged into a knob-like swelling, and are somewhat cup-shaped with a large opening.

C. *Perithecia covered by an imperfect stroma* (*Mamiania*).

G. Córyli, (Batsch).

Sphæria Coryli, Batsch Elench. Cont. II, p. 131.
Mamiania Coryli, Ces. & De Not Schema, p. 37, and Winter Die Pilze p. 670.
Gnomonia Coryli, Awd. M. Eur. Pyr. p. 23, fig. 1123, and Cke. Syn. 3863.
Gnomoniella Coryli, Sacc. Syll. 1590.

Exsicc. Fckl. F. Rh. 881.—Rab. F. E. 333, 3260.—Thum. F. Austr. 248.—id. M. U. 1453 Cke. F. Brit. 2d Ser. 278—Eriks. F. Scand. 291.—Ell. N. A F. 598—Desm. Pl Crypt. Ed. I, 1762—id. Ed. 2d 1412.

Perithecia hypophyllous, separate, arranged in a circle, sunk in the parenchyma of the leaf, covered on both sides by the blackened and inflated epidermis, depressed-sphæroid, or even lenticular, about 300 μ diam., with a stout, cylindrical beak about equal in length to the diameter of the perithecium, rising through a minute, pallid disk. Asci clavate, with a short stipe, 36–40 x 7 μ, 8-spored. Sporidia biseriate, obovoid, simple, subhyaline, granular, 7 x 3 μ. Spermogonium *Leptothyrium Coryli*, Fckl.

On leaves of *Corylus*, from Maine to Oregon; var. *spiralis*, Pk. 34th Rep. N. Y. State Mus. p. 57, differs from the ordinary form in having the ostiola spirally coiled in about two volutions. *All* the ostiola on the *same leaf* are coiled.

G. fimbriàta, (Pers.)

Sphæria fimbriata, Pers. Syn. p. 56.
Mamiania fimbriata, Ces. & De Not. Schema, p. 37.
Gnomoniella fimbriata, Sacc. Syll. 1589.
Gnomonia fimbriata, Awd. in Rab. F. E. No. 928.

Exsicc. Fckl. F. Rh. 882.—Kriegr. F. Sax. 133.—Kze. F. Sel. 107.—Erikss. F. Scand. 197. Sydow M. March. 479.—Rehm Asc. 291.—Thum. F. Austr. 867.—Cke. F. Brit. 2d Ser. 277.—Desm. Pl. Crypt. de Fr. 1st Ser. 969.—Linn. F. Hung. 252,—Ell. & Evrht. N. A. F. 2d Ser. 2361.

Stromata irregularly scattered or sometimes subconfluent, irregularly rounded, flat-pulvinate, shining-black, ½–2 mm. broad, hypophyllous. Perithecia 2–20 in a stroma, slightly prominent, globose, 400–500 μ diam., membranaceous, with long, cylindrical, mostly curved, black ostiola, which are surrounded below with a white fringe formed from the laciniately torn epidermis. Asci oblong, with a short stipe, thickened at the apex, 8-spored, 40–50 x 8–10 μ. Sporidia biseriate, ovate-elliptical, septate near the lower end, hyaline, 9–11 x 3½–4 μ.

On leaves of *Carpinus Americana*, Rhode Island, Canada and New York.

D. *Sporidia filiform* (*Cryptoderis*).

G. melanóstyla, (DC.)

Sphæria melanostyla, DC. Flore Franc. VI, p. 129.
Gnomonia melanostyla, Awd. Leipz Tausch-Verein, 1866, p. 4.
Gnomoniella, melanostyla, Sacc. Syll. 1591.
Exsicc. Fckl. F. Rh. 869.—Kze. F. Sel. 115.—Rehm Asc. 244 —Rab. F. E. 744, 2055.—Thum. M. U. 265.—Sydow M. March. 157.—Krieg. F. Sax. 333.—Desm. Pl. Crypt. Ed. I, 1788.—id. Ed. 2d 1438.

Perithecia hypophyllous, thickly scattered, sunk in the parenchyma of the leaf, finally erumpent, with an upright, straight and very long, fusiform beak, depressed-spherical, black, about 300 μ broad and 200 μ high. Asci oblong-fusoid, stipitate-attenuated below, 8-spored, 55–60 x 4–6 μ. Sporidia filiform, lying parallel in the asci, the upper part swollen, hyaline, 36–42 x 1 μ.

On leaves of *Tilia*, New York State (fide Peek) and Bethlehem, Pa. (Schweinitz).

G. Sássafras, E. & E. Bull. Torr. Bot. Club, X, p. 98.

Exsicc. Ell. & Evrht. N. A. F. 2d Ser. 1684.

Perithecia hemispherical, erumpent, about 200 μ diam., scattered over the lower surface of the leaves and along the midrib. Ostiola filiform, 200–250 μ long, of fibrous texture, subhyaline above, a little bent. Asci lanceolate, or nearly cylindrical, abundant, mostly curved, 50 x 4 μ, aparaphysate. Sporidia filiform, with a faint yellowish tinge, indistinctly multinucleate, 35–50 x ¾ μ.

On the under side of leaves of *Sassafras*, Ohio (Kellerman), and New Jersey.

Species imperfectly known.

G. pruìna, (Schw.)

Sphæria pruina, Schw. Syn. N. Am. 1770.
Gnomonia pruina, Cke. Syn. 3909.

Scattered, very minute, at first immersed, finally emergent, very black, globose, elongated into a subsetaceous, black, deciduous ostiolum. Covered at first, as well as the leaf, with a white pruinosity.

On the under side of leaves of *Vitis æstivalis*, Bethlehem, Pa.

DITOPÉLLA, De Not. (Plate 32)

Sferiacei Ital. p. 42.

Perithecia corticolous, covered. Ostiolum suberumpent. Asci subclavate, polysporous, aparaphysate. Sporidia oblong or fusoid, continuous, subhyaline.

D. Hosáckiæ, (Cke. & Hark.)

Sphærella? Hosackiæ, Cke. & Hark. Grev. IX, p. 86.
Ditopella Hosackiæ, Sacc. Syll. 1739, Cke. Syn. 4111.

Scattered, covered, punctiform. Perithecia globose-depressed. Asci clavate, sessile, containing about 16 sporidia, which are ellipsoid, hyaline, about 6 x $2\frac{1}{2}$ μ.

On twigs of *Hosackia*, California.

The specc. sent by Harkness afford only subcuticular, flattish perithecia filled with elliptic-oblong, brown stylospores, 3-septate, with one or two cells divided by a longitudinal septum (*Dichomera Hosackiæ*, E. & E. in Herb.)

FAMILY. PLEOSPÒREÆ.

Perithecia membranaceous or coriaceous, generally with a papilliform or short-conical ostiolum, or even simply pierced above, buried in the matrix with only the ostiolum projecting, finally often erumpent and nearly free and superficial by the falling away of the epidermis. Asci clavate-cylindrical, paraphysate. Sporidia 1- or more-septate, (muriform in *Pleospora*), mostly colored, elliptical, oblong, fusoid or filiform.

DIDYMOSPHÈRIA, Fckl.

Symbolæ Mycol. p. 140.

Perithecia covered by the epidermis, submembranaceous, with a minute, papilliform ostiolum. Asci paraphysate, 4–8-spored. Sporidia ovoid-oblong, uniseptate, colored (brown).

D. palmàcea, (Cke.)

Sphæria palmacea, Cke. Grev. VII, p. 12.
Anthostomella palmacea, Sacc. Syll. 1085.
Didymosphærella palmacea, Cke. Syn. 4675.

Immersed, covered. Perithecia very small, subglobose, about 100 μ diam. Ostiolum punctiform, barely erumpent. Asci cylindrical, 75 x 10 μ, paraphysate. Sporidia uniseriate, oblique, often lying crosswise in the asci, oblong-elliptical, brown, obtuse, constricted and uniseptate in the middle, 10–12 x 4–4½ μ (13 x 6 μ, Cke.).

On palm leaves, California.

Some of the perithecia are filled with curved, hyaline, 12 x 1¼–1½ μ sporules (*Phlyctæna*).

The diagnosis is from specc. sent by Dr. Harkness.

D. Parnássiæ, (Pk.)

Sphæria Parnassiæ, Pk. 27th Rep. N. Y. State Mus. p. 111.
Didymosphæria Parnassiæ, Sacc. Syll. 2647.
Didymosphærella Parnassiæ, Cke. Syn. 4657.

Perithecia scattered, convex or subhemispherical, prominent, pierced above, black. Asci cylindrical. Sporidia long, narrow, uniseptate, generally constricted at the septum, often slightly curved, colored, 30–33 μ long, with 1–2 nuclei in each cell.

On dead stems of *Parnassia Caroliniana*, Albany, N. Y. (Peck).

D. grumàta, (Cke.)

Sphæria grumata, Cke. (in literis).
Didymosphæria grumata, Sacc. Syll. 2695, Cke. Syn. 4250.
Exsicc. Ell. N. A. F. 185.—Rehm Asc. 441.—Thum. M. U. 460.

Perithecia subgregarious, covered by the blackened and swollen epidermis, globose, about ½ mm. diam., obtusely papillate, black. Asci clavate, 100–120 x 12–15 μ, with branching paraphyses, 8-spored. Sporidia subbiseriate, clavate-oblong, uniseptate, hyaline, at length becoming pale brown and constricted, 21 x 8 μ, at first with a gelatinous envelope.

On living stems of *Andromeda ligustrina*, Newfiéld, N. J.

D. accèdens, Sacc. Proc. Rochester Acad. March, 1890, p. 48.

Perithecia gregarious, covered, ¼–⅓ mm. diam., nucleus at first white. Ostiolum papilliform, erumpent. Asci clavate-cylindrical, 120 x 10 μ, paraphysate. Sporidia uniseriate, 6–8 in an ascus, elliptical, rounded at the ends, uniseptate and slightly constricted, dark brown, 20–22 x 9–11 μ. Allied to *D. Rhamni* and *D. incerta*, but with a different ostiolum and asci.

On bark of *Fraxinus*, Lyndonville, N. Y. (Fairman).

D. Andropògonis, Ell. & Lang. Proc. Acad. Nat. Sci. Phil., July, 1890, p. 235.

Stroma consisting of the nearly unchanged substance of the culm which is there whiter than the surrounding parts, 3–4 cm. long, 1 cm.

broad, surrounded by a greenish-black line which penetrates deeply, the surface also being of a uniform slaty-black. Perithecia scattered, subglobose, ½–¾ mm. diam., entirely buried except the convex-discoid, erumpent ostiola. Asci cylindrical, about 110 x 8–10 μ, with stout but evanescent paraphyses. Sporidia uniseriate, oblong-cylindrical, rounded at the ends, slightly curved, uniseptate, hyaline at first, becoming brown, 18–22 x 4½–5½ μ.

On dead culms of *Andropogon muricatus*, St. Martinsville, La. (Langlois).

D. cùpula, (Ell.) **(Plate 28)**

Sphæria cupula, Ell Am. Nat. March, 1883, p. 317, Bull. Torr Bot. Club, X, p. 90.
Didymosphæria cupula, Sacc. Syll. 6112 and 6586, Cke. Syn. 4234.
Exsicc. Ell. N. A. F. 1338

Perithecia scattered, membranaceous, about ½ mm. diam., convex-hemispherical when fresh, collapsed when dry, buried, except the papilliform ostiolum, in the parenchyma of the leaf, and covered by the blackened, slightly raised epidermis. Asci subcylindrical, 75 x 7 μ. Sporidia uniseriate elliptical, brown, uniseptate, 9½–11½ x 4½–5½ μ. *Sphæria diplospora*, Cke., has the same fruit, but the perithecia do not collapse.

On dry oak leaves (*Q. coccinea*) still hanging on the limbs, Newfield, N. J. What seems to be the same has also been found on *Phytolacca* and *Desmodium*.

D. Ceanòthi, (Cke. & Hark.)

Sphæria Ceanothi, Cke. & Hark. Grev. XIII, p. 19.
Didymosphæria Ceanothi, Sacc. Syll. 6587, and Cke. Syn. 4235.

Perithecia scattered, covered, flattened-globose, black, with a short, papilliform ostiolum. Asci ample, clavate, 8-spored. Sporidia elliptical, uniseptate, constricted in the middle, dark brown, 35 x 15 μ.

On twigs of *Ceanothus*, California (Harkness).

D. sarménti, (Cke. & Hark.)

Sphæria sarmenti, Cke. & Hark. Grev. XIII, p. 19.
Didymosphæria sarmenti, Sacc. Syll. 6574 and Cke. Syn. 4221.

Perithecia scattered, covered, subglobose, black, smooth, at length flattened. Ostiolum short, punctiform. Asci cylindric-clavate. Sporidia elliptical, uniseptate, not constricted, brown, 12 x 5 μ.

On "Canary Vine," California (Harkness).

D. phyllógena, Winter. Journ. Mycol. I, p. 121.

Perithecia on a round or subangular, determinate, brown spot surrounded by a black line, and sometimes as much as 8 mm. diam., hypophyllous, immersed, about 100 μ diam., thin-membranaceous, the apex perforating the epidermis. Asci clavate, very short-pedicellate, 8-spored, 50–70 x 9-11 μ, with thick, cylindrical, septate paraphyses. Sporidia biseriate (rarely uniseriate), oblong, inequilateral, attenuated at each end, uniseptate, constricted at the septum, brownish, 14–16 x 5 μ.

On fallen leaves of *Liriodendron tulipifera*, Missouri (Demetrio).

D. serrulàta, E. & M. Journ. Mycol. I, p. 99.

Perithecia scattered or gregarious, covered by the cuticle, lenticular, ½–⅓ mm. diam. Asci cylindrical, 100–112 x 10–12 μ, with abundant, linear paraphyses. Sporidia uniseriate, hyaline at first and 3-nucleate, soon becoming dark brown and uniseptate, 18–20 x 5–6 μ, surrounded at first with a hyaline envelope. The sporidia are much like those of *Anthostomella leucobasis*, E. & M., only longer and uniseptate, and the perithecia are larger and more prominent.

On bleached spots on dead petioles of *Sabal serrulata*, Florida (Martin).

D. Tỳphæ, Pk. 28th Rep. N. Y. State Mus. p. 104.

Perithecia minute, punctiform, subglobose, covered by the epidermis, which is pierced by the scarcely papillate ostiola. Asci cylindrical, 55–80 x 7½–10 μ, with filiform paraphyses. Sporidia oblong or elliptical, uniseriate, uniseptate, not at all, or but slightly constricted at the septum, colored, 10–15 x 5–7 μ.

On the lower part of dead leaves of *Typha latifolia*, Albany Co., N. Y. (Peck).

D. epidérmidis, (Fr.)

Sphæria epidermidis, Fr. S. M. II, p. 499.
Sphæria atomaria, Wallr. Fl. Crypt. Germ. No. 3731.
Sphæria Araucariæ, Cke. Seem. Journ. Bot. IV, tab. 45, fig. 12.
Didymosphæria epidermidis, Fckl. Symb. p. 141.

Perithecia scattered, covered, but prominent, small, smooth, at length collapsed and flattened. Ostiolum minute, erumpent. Asci cylindrical, 75–100 x 7–8 μ, with branching paraphyses, 8-spored. Sporidia uniseriate, rarely subbiseriate, biconic-elliptical, uniseptate, constricted, colored, straight or curved, 2-nucleate, 8–11 x 6 μ. Very variable, especially in the size of the sporidia. A form occurs (on bramble stems) with the asci mostly tetrasporous.

On *Sambucus* and *Persica*, North Carolina. (Curtis).

D. yuccægena, (Cke.)

Sphæria yuccægena, Cke. Grev. VII, p 12.
Didymosphæria yuccogena, Sacc. Syll. 2673.
Didymosphærella yuccogena, Cke. Syn. 4673.

Gregarious, covered. Perithecia subprominent, black. Asci cylindrical. Sporidia uniseriate, elliptical, uniseptate, constricted, brown. 28 x 10 μ.

On *Yucca*, California (Harkness).

D. tenebròsa, (B. & Br.)

Sphæria tenebrosa, B. & Br. Ann. Nat Hist. No. 649, tab. 12, fig. 35
Didymosphæria tenebrosa, Sacc. Syll. 2685.
Didymosphærella tenebrosa, Cke. Syn. 4689.

"Perithecia scattered irregularly, covered by the cuticle and subjacent cells, which are traversed by dark, cellular mycelium here and there giving rise to short, toruloid threads. Asci large, cylindrical, obtuse, slightly attenuated below. Sporidia biseriate, composed of two apposed, irregular cones, which contain at first a single large globule, but at length have two irregular endochromes. Remarkable for its curious mycelium and large sporidia." Cke. Hndbk., 2697.

On *Vernonia Noveboracensis*, Pennsylvania (Michener).

D. círcinans, Hark. Bull. Cal. Acad. No. 1 (1884), p. 44.

Epiphyllous. Perithecia gregarious in circular spots 1–6 mm. diam., covered by the epidermis and staining the matrix around the ostiolum. Asci cylindrical, sessile, 8-spored, paraphysate, 66 x 6 μ. Sporidia oval or oblong, uniseptate, slightly constricted, olive-brown. 5–7 x 3–4 μ.

On the early form of leaves of *Eucalyptus globulus*, California (Harkness).

D. pardalìna, E. & E. Journ. Mycol. II, p. 102.

Perithecia gregarious in groups of 4–6, whose position is indicated by suborbicular or elliptical black spots 2–4 mm. in diam., or by confluence more, entirely sunk in the substance of the stem, rather large (½ mm.), with thick, coriaceous walls, and minute, punctiform ostiola, not elevating the epidermis. Asci cylindrical, 150–200 x 12–15 μ. Sporidia uniseriate, oblong-cylindrical, olive-brown, uniseptate and constricted, slightly curved, ends obtuse, 22–30 x 8–10 μ.

On dead stems of *Spartina polystachya*, Louisiana (Langlois).

D. bácchans, Pass. in Thüm. Pilze des Weinst. p. 134.

Perithecia immersed in the bark, minute, seriate, subglobose, the short, conical, black ostiola erumpent. Asci oblong-clavate, attenuated below, 4–8-spored. Sporidia obliquely uniseriate, elliptic-navicular, uniseptate, not constricted, brown. Stylospores in larger perithecia.

On dead branches of grape vines, Saugerties, N. Y. (Peck).

D. polysticta, (B. & C.)

Sphæria polysticta, B. & C. Grev. IV, p. 149.
Didymosphæria polysticta, Sacc. Syll. 2670.
Didymosphærella polysticta, Cke. Syn. 4668.

"Perithecia scarcely raising the cuticle, visible chiefly from the black, dot-like ostiola. Asci linear. Sporidia oblong, uniseptate, $7\frac{1}{2}$ x $3–3\frac{1}{2}$ μ."

On stems of *Smilax*, Alabama (Beaumont).

D. Vizeàna, (Cke.)

Sphæria Vizeana, Cke. Grev. VII, p. 12.
Didymosphæria Vizeana, Sacc. Syll. 2651.
Didymosphærella Vizeana, Cke. Syn. 4662.

Perithecia gregarious or scattered, about 200 μ diam., covered by the blackened epidermis, finally collapsing. Asci oblong, sessile, 60–75 x 18–20 μ, (paraphysate)? Sporidia biseriate, oblong-elliptical, subinequilateral, yellowish. uniseptate, 18–22 x 7–10 μ.

On dead stems of *Lathyrus venosus*, Sacramento, Cal. (Harkness).

D. adélphica, (Cke.)

Sphæria adelphica, Cke. Grev. VII, p. 42.
Didymosphæria adelphica, Sacc. Syll. 2652.
Didymosphærella adelphica, Cke. Syn. 4663.

This is an unsatisfactory and doubtful thing. The brief description in Grevillea does not enable one to recognize it. The spec. in our herb. is only a stylosporous fungus (*Diplodia*), poorly developed.

PLEÓSPORA, Rabh.

in Herb. Mycol. Ed. II, No. 547.

Stroma none. Perithecia at first covered, finally more or less perfectly erumpent, mostly membranaceous, seldom of firmer texture, black. Asci mostly oblong or clavate-cylindrical, 8-spored, paraphysate. Sporidia oblong, ovate or clavate-fusoid, with both transverse and longitudinal septa (muriform), generally colored (yellow or yellow-brown).

This interesting but difficult genus has been elaborately described and beautifully illustrated by Dr. A. N. Berlese, in his valuable monograph of *Pleospora* &c. in Nuovo Giorn. Bot. Ital. Vol. XX, Nos. 1 and 2.

Pl. herbarum, (Pers.)

Sphæria herbarum, Pers. Syn. p. 79.
Sphæria Papaveris, Schum. Enum. Fung. Fl. Saell, II, p. 155.
Pleospora herbarum, Rab. Herb. Mycol. Ed. II, 547.
Sphæria Pisi, Sow. Eng. Fungi tab. 393, fig. 8.
Sphæria Armeriæ, Corda Icones, IV, p. 41, tab. VIII, fig. 119.
Sphæria Allii, Kl. Rab. Herb. Mycol. Ed. I, No. 838.
Pleospora herbarum, var. *Allii*, id. Ed. II, No. 347.
Pleospora Asparagi, id. No. 750.
Pleospora Armeriæ, Ces. & De Not. Schema, p. 44.
Pleospora Allii, id. l. c.
Pleospora Pisi and *Pleospora Samaræ*, Fckl. Symb. p. 131.
Pleospora Meliloti, Rab. F. E. 2330
Pleospora leguminum, Wallr. Fl. Cryp. Germ. 3726.
Pleospora Cepæ, Pr. Hoyersw. No. 290.
Pleospora Grossulariæ, Fckl. Symb. p. 131.
Pleospora Dianthi, De Not. Sfer. Ital. p. 74, fig 80.
Pleospora varians, Ces. Hedw. 1882, p. 9.
Pleospora Erythrinæ, id. p. 10.
Pleospora Oxyacanthæ, Pass. et Beltr. F. Sic. Nov. No. 10.
Pleospora socia, Sacc. & Pass. Herb. Critt. Ital. 1069.
Pleospora albicans, Sacc. Syll. 3736.
Pleospora Gymnocladi, Sacc. Syll. 3783.
Sphæria pellita, Roum. F. G. 1284.
Pleospora petiolorum, Thum. M. U. 760.
Pleospora Clarkeana, E. & E. Bull. Torr. Bot. Club, XI, p. 75.*

Exsicc. Fckl. F. Rh. 811, 899.—Kz. F. sel 68.—Rab. Herb. Mycol. 347, 544, 547, 750. id. F. E. 768, 951, 1018, 1635, 1636, 1929, 2330.—Rehm Asc. 145, 341, 486, 683.—Thum. M U. 1255.—Sydow M. March. 179.—Ell. & Evrht. N. A. F. 2d Ser. 1583 and 2368.

Perithecia scattered, at first buried and covered by the epidermis. finally free, depressed-spherical, of medium size, collapsing to flat or concave, with a conic-papilliform or sometimes somewhat elongated ostiolum, glabrous or subfibrillose at base, 250–450 μ diam., coriaceous. black. Asci at first ovate, becoming oblong-clavate, stipitate, 8-spored, 90–165 (mostly 120–150) x 24–40 (mostly 27–30) μ. Sporidia biseriate, ovate-oblong, rounded at the ends, or sometimes subattenuated above, 7-septate and constricted at the septa, with 2–3 longitudinal septa, yellow or yellow-brown, becoming dark brown, but not opake, 28–33 x 14–16 μ (exceptionally reaching 40 μ) long. Paraphyses jointed, thick, simple or somewhat branched.

On dead stems of various plants, more especially of the *Papilionaceæ*. Common everywhere from Greenland to Mexico.

Pl. argyróspora, Hark. Bull. Cal. Acad. Feb. 1884, p. 45.

Perithecia minute, black, scattered, erumpent. Asci elavate, sessile, thiek-walled, 8-spored, 60–65 x 20 μ. Sporidia biseriate, hyaline, elliptic-lanceolate, 3–5-septate, with 1–2 longitudinal septa, 18 x 8 μ. Paraphyses filiform.

On dead branches of *Dendromecon rigidum*, California.

*The synonymy of this species is taken from Winter's Pilze and Berl. Monograph.

Pl. Americàna, E. & E.

Pleospora hyalospora, E. & E. Proc. Acad. Nat. Sci. Phil. July, 1890, p 238, (not Speg.)

Perithecia scattered, depressed-hemispherical, 75–90 μ, of coarse, cellular structure, at first sunk in the parenchyma of the leaf, finally more or less erumpent-superficial, with a papilliform ostiolum. Asci oblong, 75–85 x 35–40 μ, 8-spored. Sporidia oblong or slightly ovate-oblong, 3–6-septate (mostly 5-septate), with one or more longitudinal septa more or less distinct, nearly hyaline, ends obtusely pointed or rounded, slightly constricted at the septa, especially at the middle one, 25–40 (mostly 25–30) x 12–15 μ. Differs from *P. Pisi* (Sow.) in its obovate asci, nearly hyaline sporidia, and more delicate, smaller perithecia.

On leaves of *Lathyrus sativus* and *Pisum sativum*, Starkville, Miss. (Tracy), and on leaves of *Trifolium*, Canada (Dearness).

Pl. comprėssa, Hark. Bull. Cal. Acad. Sci. No. I (1884), p. 45.

Perithecia scattered, covered by the blackened cuticle, concave, surrounded by radiating, brown hyphæ. Asci clavate, short-stipitate, 8-spored, paraphysate, 80–90 x 15–18 μ. Sporidia biseriate, unequally elliptical, transversely 3-septate, brown, with a longitudinal septum across the two central cells, at first surrounded by a gelatinous coat, 20–27 x 10–19 μ.

On dead stems of *Polygonum amorphum*, Blue Canon, Cal. (Harkness). Very near Pl. permunda, Cke.

Pl. vitríspora, Cke. & Hark. Grev. IX, p. 86.

Exsicc. Ell. & Evrht. N. A. F. 2d Ser. 1582.

Perithecia scattered, buried in the wood, the surface of which is raised and blackened over them, about ½ mm. diam., globose, papillate, and finally perforated, black. Asci cylindrical, about 112 x 12 μ, with abundant paraphyses. Sporidia uniseriate, elongated-elliptical, hyaline, the contents divided in a muriform manner (merenchymatic), 22–24 x 12 μ (32 x 12 μ Cke.)

On dead branches of *Lonicera*, California.

Pl. denotàta, (C. & E.)

Sphæria denotata, C. & E. Grev. VI, p. 16, Pl. 96, fig. 20.
Pleospora denotata, Sacc. Syll. 3740 and Cke. Syn. 5027.
Exsicc Ell. N. A. F. 778.

Perithecia mostly seated on pallid spots, globose, large, prominent, at first covered by the cuticle, soon exposed. Asci clavate. Sporidia muriform, 50 x 20 μ, larger than in *Pl. herbarum.*

On dead stems of *Trifolium pratense*, Newfield, N. J.

Pl. thuriodónta, (C. & E.)

Sphæria thuriodonta, C. & E. Grev. V, p 94.
Pleospora thuriodonta, Sacc. Syll. 3768.
Delacourea thuriodonta, Cke. Syn. 4328.

Perithecia covered by the epidermis, somewhat prominent, scattered. Asci cylindrical. Sporidia uniseriate, elliptical, brown, 25 x 10 μ, 5-septate and muriform, constricted at the middle septum. Accompanied by stylospores (in other perithecia) elliptical, uniseptate, brown, 40 x 20 μ.

On dead limbs of *Nyssa multiflora*, Newfield, N. J.

Apparently rare. Has only been met with once, and then only in small quantity.

Pl. Thümeniàna, Sacc. Syll. 3787, Mich. II, p. 139.

Perithecia gregarious, semiimmersed, at first covered by the epidermis, subglobose-depressed, papillate, ¼ mm. diam., rather solid. Asci cylindrical, short-stipitate, rounded at the apex, 90–100 x 15 μ, paraphysate, 8-spored. Sporidia biseriate, oblong-biconical, 18–20 x 7–8 μ, 3-septate, constricted in the middle, with one or two longitudinal septa, cribrose-guttulate, olivaceous.

On dead leaves of *Yucca aloifolia*, Carolina (Ravenel).

Pl. láxa, Ell. & Galway Journ. Mycol. V, p. 66.

Perithecia scattered, subglobose, black, 150–170 μ diam., their bases projecting on one side of the leaf and their apices on the other. Asci few (6–8 in a perithecium), inflated-oblong, broadly rounded above, 150–200 x 35–55 μ, contracted at base into a short stipe. Paraphyses obscure. Sporidia 8 in an ascus, obovate-oblong, 6–8-septate and coarsely muriform, deeply constricted near the middle, so as to easily break in two there, straw-yellow, 35–45 x 15–20 μ (mostly 15 μ wide). This seems to differ from any of the other described species on grasses and *Carices*, in its strongly constricted sporidia. This character is very distinct through all the stages of growth. The constriction is generally at the third septum from the upper end, the part above this constriction being broader and shorter than the part below it. This comes near *Pl. Islandica*, Johans., but has the sporidia much more deeply constricted.

On dead leaves and culms of some grass, Montana (Anderson).

Pl. Harknéssi, Berl. & Vogl. Sacc. Syll. 7090.

Leptosphæria straminis, Cke. & Hark. Grev. XIV, p. 10, (not Sacc. & Speg.)

Perithecia scattered or subgregarious, on effused, blackened spots, black, erumpent, convex, perforated. Asci cylindric-clavate, 8-spored, Sporidia lanceolate, 5-septate, slightly constricted, dark, the two central cells divided by a longitudinal septum, 32–35 x 6–8 μ.

On culms of *Triticum*, California.

Pl. quadriseptàta. Cke. & Hark. Grev. XIV, p. 10.

Perithecia subgregarious, convex, black, shining, at first covered by the epidermis, small. Asci clavate, 8-spored. Sporidia biseriate, subelliptical, 4-septate, scarcely constricted, one or more of the cells divided by a longitudinal septum, brown, 20–22 x 8 μ.

On pods of "gillyflower" (*Matthiola*)? California.

Pl. sarcocýstis, (B. & C.)

Sphæria sarcocystis, B. & C. Grev. IV, p. 152.
Pleospora sarcocystis, Sacc. Syll. 3793 and Cke. Syn. 5077

"Forming little, oblong, black bodies which consist of a few closely joined perithecia with fleshy walls. Asci rather short, oblong, but tumid, containing four oblong, obtuse, sometimes clavate sporidia with three horizontal articulations, and a few vertical, 50 μ long. Perhaps more properly placed in *Dothidea.*"

On wheat, Carolina (Berk. in Grev.)

Pl. Labiatàrum, Cke. & Hark. Grev. IX, p. 8.

Perithecia scattered, black, semiimmersed, smaller than those of *Pl. herbarum.* Asci cylindrical. Sporidia uniseriate, elliptical, amber-yellow, 3-septate, 25 x 10 μ, next to the last cell divided by a longitudinal septum.

On stems of *Marrubium vulgare*, California.

Pl. Sámbuci, Plowr. Grev. VII, p. 74.

Perithecia subgregarious, at first covered by the epidermis, then exposed and superficial, globose, 300–400 μ diam., at length collapsing above. Ostiolum papilliform. Asci clavate-cylindrical, 75–100 μ long. Sporidia mostly uniseriate, ovate, 3-septate, with a longitudinal septum across one or two cells, pale yellow, 12–14 x 4–5 μ.

On *Sambucus*, California (Harkness).

The sporidia resemble those of *Pleospora Bardanæ*, Niessl.

Pl. pústulans, E. & E. Journ. Mycol. IV, p. 76.

Perithecia gregarious, membranaceous, ovate-globose, ½–¾ mm. diam., raising the bark into subconical pustules, with the papilliform ostiola erumpent. Asci clavate-cylindrical, with a short-stipitate base, about 100 x 12–15 μ., with abundant paraphyses. Sporidia uniseriate or subbiseriate, varying from ovate to oblong, and oblong-elliptical, 18–20 x 8–10 μ, at first pale brown and 3–4-septate, scarcely constricted, with the ends subacute, soon 5–7-septate and darker, with the ends obtuse, one or two of the cells divided by a longitudinal septum which finally runs through all but the terminal cells, more or less distinctly.

On the exposed inner surface of bark of *Fraxinus*, Clyde, N. Y. (O. F. Cook).

This is very distinct from *Pl. velata*, Sacc. & Roum. in F. G. 1081, in its much larger perithecia, which are not flattened, and in its mostly longer, 3–7-septate, sporidia. Dr. Berlese, in his monograph of *Pleospora*, figures *Pl. Saccardiana* (of which he gives *Pl. velata* as a syn.) with 4-septate sporidia, though he says the normal number is three, as we find them in our copy of Fungi Gallici.

Pl. vulgàris, Niessl. Notiz. p. 27.

Exsicc. Rab. F. E. 824, 1545, 3146.—Lin. Fungi Hung. 275.—M. March, 1924, 2037, 2542, 2545.—F. Gall. 5251.

Perithecia scattered, covered by the epidermis, depressed-spherical, with a flattened base, soon collapsing, small (250 μ), glabrous except the slightly fibrillose base, dark brown, thin-coriaceous, with a papilliform ostiolum. Asci clavate or clavate-cylindrical, 8-spored, stipitate, 80–140 x 10–15 μ. Sporidia uniseriate or biseriate, obtusely rounded at the ends, subinequilateral, 5-septate and constricted at the septa, more decidedly so at the middle septum, the four middle cells divided by a longitudinal septum, yellow, becoming brown, 15–21 x 8–10 μ. Paraphyses jointed, simple or sparingly branched, longer than the asci.

On dead stems of *Verbascum Thapsus*, Newfield, N. J., on stems of *Nepeta Cataria*, London, Canada, and on *Potentilla*, *Artemisia* and *Pedicularis*, Greenland. Smaller throughout than *Pl. herbarum.*

Pl. Cássiæ, E. & E. Journ. Mycol. II, p. 41.—Berl. Mon. Pleosp. p. 152.

Perithecia scattered, ¼ mm. diam., covered by the epidermis which is raised in a pustuliform manner, but not blackened over them, and barely pierced by the papilliform ostiola. Asci subcylindrical, rather abruptly contracted at the base, 75–80 x 10–12 μ, paraphysate. Sporidia biseriate, oblong-elliptical, 3-septate, yellow-brown, constricted at the middle septum, slightly curved, one or both the inner cells divided by a longitudinal septum. The perithecia become finally slightly collapsed above.

On dead stems of *Cassia*, with *Leptosphæria cassiæcola*, Houston, Texas (Ravenel).

Pl. baccàta, Ell. Bull. Torr. Bot. Club, X, p. 53.

Perithecia scattered, subcuticular, becoming bare and superficial by the falling away of the cuticle, depressed-globose, strongly papilliform, thick-membranaceous, black, 300–500 μ diam., fringed around the base with spreading mycelium. Asci 75–85 (exceptionally 114) x 20–24 μ; paraphyses abundant, conglutinate. Sporidia inordinately biseriate, oblong or ovate-oblong, at first 3-septate with a longitudinal septum across one or two cells, and constricted at the septa, when mature 5-septate, 20–30 (mostly 20–25) x 10–14 μ, with 1–3 imperfect, longitudinal septa.

On dead stems of *Cirsium*, sp., Utah.

Berlese (monograph, p. 195) makes this a synonym of *Pl. permunda*, Cke., but it is a much coarser species besides the fact (not at first noticed) that the mature sporidia become 5-septate.

Pl. aùrea, Ell. Bull. Torr. Bot. Club, X, p. 53. **(Plate 28)**

Perithecia gregarious, subcuticular, 150–250 μ diam., membranaceous and collapsing when dry, of coarse, cellular structure, surrounded at base with a fringe of brown, septate, branching, mycelial hyphæ (as in *Pl. permunda*) plainly visible through the transparent cuticle. Asci broad, obtuse, mostly inequilateral or curved, ovate-oblong, 75–114 x 25 μ, with a very short stipe. Paraphyses stout, rather longer than the asci, guttulate, about 2 μ thick. Sporidia inordinate or subbiseriate, ovate-elliptical, 22–25 x 12–14 μ, and about 10 or 11 μ thick, golden yellow, becoming dark brown.

On dead herbaceous stems, Pleasant Valley, Utah (S. J. Harkness).

Berlese (monograph, p. 195) puts this, too, as a synonym of *Pl. permunda*, Cke., the spece. sent him being that species instead of *Pl. aurea*, the two being hardly distinguishable outwardly except that *Pl. permunda* has rather larger perithecia. Under the microscope the two are readily distinguished, *Pl. aurea* having 4-septate sporidia.

The measurements of these two species in the Torr. Bull. appear to have been somehow interchanged and confused. Those here given have been made very carefully and may be relied on.

Every one who has had much to do with the microscopical examination of Ascomycetes will recognize the fact (noted by Dr. Winter in Die Pilze, p. 454) that in *Leptosphæria*, *Pleospora* and many other Ascomycetes, the length of the asci, in the same perithecium, is often very variable, depending on whether the ascus has become elongated by the absorbing of water, in the course of the microscopical examination.

Pl. planíspora, Ell. Bull. Torr. Bot. Club, X, p. 53.

Clathrospora planispora, Berl. Mon. p 200, tab. IX, fig. 5.*
Exsicc. Ell. & Evrht. N. A. F. 1584.

Perithecia gregarious, covered by the scarcely discolored epidermis, 250–330 μ diam., collapsing when dry, of coarse, cellular structure. Asci 114–120 x 25–28 μ, p. sp. 80 μ long, with abundant paraphyses. Sporidia biseriate, oblong-elliptical, 5-septate, with a longitudinal septum running through all but the terminal cells, 30–40 x 15–20 μ, and 7–11 μ thick.

On dead culms and sheaths of some grass (Elymus)? Pleasant Valley, Utah (S. J. Harkness). The perithecia are mostly on the sheaths which are clouded or mottled with a darker color in those parts occupied by the fungus. The sporidia are larger than in *Pl. aurea*, and 5-septate; the perithecia also are larger and mostly lack the fringe of mycelium around the base.

Pl. oligómera. Sacc. & Speg. Mich. I, p. 408.—F. Ital. tab. 331.

Perithecia loosely gregarious, innate-erumpent, becoming nearly superficial, depressed-globose, $\frac{1}{4}$–$\frac{1}{3}$ mm. diam., black and somewhat shining, obtusely papillate, finally umbilicate-collapsed. Asci cylindrical, rounded at the apex, very short-stipitate, 90–100 x 15–16 μ, paraphysate, 8-spored. Sporidia obliquely uniseriate, oblong-ovoid, 20–22 x 9–11 μ, 3-septate, slightly constricted at the septa, with a longitudinal septum across two or more cells, yellow, becoming brown.

On *Silene Gallica*, California (Harkness).

Pl. permúnda, Cke. Grev. V, p. 111.

Clathrospora permunda, in Berl. Mon. p. 195.
Exsicc. Ell. N. A. F. 886.

Perithecia scattered, at first covered by the epidermis, at length erumpent and collapsing above, 200–250 μ diam., fringed below with a mycelium of dark brown, branching, septate threads spreading out

*The genus *Clathrospora*, characterized by its reticulate sporidia, is with difficulty distinguished from *Pleospora*, and we have included all species referred to that genus in *Pleospora*.

beneath the epidermis. Asci oblong-cylindrical, 75 x 22 μ, subsessile, paraphysate. Sporidia biseriate, straw-yellow at first, becoming dark brown, 3-septate and constricted in the middle, ends rather acute, with a longitudinal septum extending across the two middle cells, 20–24 x 10–11 μ, subinequilateral.

On dead herbaceous stems, common in the Rocky Mountain region.

The sporidia like those of *Pl. planispora*, and *Pl. aurea* are more or less distinctly flattened.

Pl. gigáspora, Karst. Hedwigia 1884, p. 37.

Pleospora amplispora, E. & E. Bull. Wash. Coll. Lab. Nat. Hist. 1884, p. 41.

Perithecia scattered, superficial, black, subglobose or hemispherical, ⅓ mm. diam., collapsing when dry. Ostiola short, cylindrical or subconoid. Asci oblong, 250 x 75 μ (paraphysate)? Sporidia oblong, obtuse, slightly constricted in the middle, 12–16-septate, muriform, 60–75 x 20–24 μ, yellow-brown becoming dark brown.

On dead stems of *Lupinus*, Mt. Paddo, Washington (Suksdorf).

Pl. diaporthoides, E. & E. (in Herb.)

Perithecia scattered or subseriate, buried, minute (¼ mm.), with rather thick, subcoriaceous walls, often in series of 3–4 lying close together. Ostiolum exserted, short-cylindrical, rough, except the smooth, black, depressed-conical tip. Asci 75–85 x 12–15 μ, substipitate, p. sp. 60–70 μ long, with paraphyses. Sporidia uniseriate or subbiseriate above, oblong-elliptical, subinequilateral, ends obtusely pointed, 3-septate, with a more or less perfect longitudinal septum, 14–18 x 7–8 μ.

On dead stems of *Pastinaca sativa* with *Ophiobolus Bardanæ*, Fckl., Newfield, N. J., July, 1890.

Differs from *Pleospora microspora*, Niessl., in its habitat, and constantly 3-septate, rather smaller sporidia, and more prominent ostiola.

Pl. lactucícola, E. & E. Journ. Mycol. IV, p. 64.

Perithecia scattered, suberumpent, depressed-hemispherical, 175–200 μ diam. Ostiolum papilliform. Asci clavate-cylindrical, 90–100 x 10–12 μ. Sporidia obliquely uniseriate or rarely subbiseriate above, oblong, slightly constricted, 3-septate, with each of the middle cells divided by a longitudinal septum, ends subacute and suboblique, 14–16 x 6–7 μ. Allied to *Pl. Bardanæ*, Niessl., but, compared with the specimens of that species in Linhart's Fungi Hungarici, No. 168, the perithecia are smaller and the sporidia also smaller and darker colored. *Sphæria Lactucarum*, Schw., is said to grow on cinereous spots.

On decaying stems of *Lactuca Canadensis*, Newfield, N. J.

Pl. lichenàlis, (Pk.)

Sphæria lichenalis, Pk. Bot. Gaz. V, p. 36.
Pleospora lichenalis, Sacc. Syll. 3769.
Delacourea lichenalis, Cke. Syn. 4529.

Perithecia scattered, minute, 200–250 μ diam., subhemispherical, erumpent, black, with a papilliform ostiolum. Asci oblong-elliptical. Sporidia crowded, oblong, multiseptate, fenestrate, greenish-yellow, 35–40 x 10–12½ μ.

On bark of birch trees, Vermont.

The perithecia occupy a discolored spot, which gives a lichenose aspect to the fungus. Sometimes 2–3 perithecia are seriately crowded or subconfluent, so as to present a hysteriiform appearance.

Pl. Shephérdiæ, Pk. 40th Rep. p. 71.

Perithecia scattered, small, 350–450 μ diam., covered by the epidermis, erumpent, black. Asci cylindrical, 125–200 x 15 μ. Sporidia uniseriate, oblong, generally 3-septate, rarely 5-septate, with 1–2 longitudinal septa, constricted in the middle, colored, 20–25 x 7½–10 μ.

On dead branches of *Shepherdia Canadensis*, New York State.

Pl. árctica, Fckl. in Die Zweite Deutsch. Nordpolarf. II, p. 52, tab. I, fig. 2.

Perithecia scattered under the cinereous, pustulate-inflated epidermis, of medium size, globose, black, with papilliform, perforated, slightly prominent ostiola. Asci oblong, mostly curved, with a short, thick stipe, and thick membrane, 8-spored, 132 x 36 μ. Paraphyses linear, septate or multinucleate, hyaline. Sporidia imbricated-biseriate, oblong-ovate, constricted in the middle, 6–7-septate and muriform, dark olive, 28 x 14 μ.

On dry stems of *Epilobium latifolium*, Greenland.

Pl. Dràbæ, Schröt. Nord. Pilz. p. 15.

Perithecia globose-depressed, 300 μ diam., surrounded at base by creeping, brown hyphæ, otherwise glabrous, collapsing, papillate. Asci cylindric-clavate, subattenuated below, short-stipitate, 60–70 x 13–17 μ. Sporidia biseriate, ellipsoid, 16–20 x 7–9 μ, dark chestnut, 5–7-septate, with 2–3 longitudinal septa.

On *Draba hirta* and *Draba Wahlenbergii*, Upernivik, Greenland.

Pl. pentámera, Karst. Fungi Spetsb. p. 99.

Perithecia scattered, at first covered by the epidermis, then emergent, subsphæroid, very obtusely conoid or depressed above, subastomous, glabrous (rarely beset with straight, jointed, black hairs), 150–250 μ diam. Asci very short-stipitate, clavate, 105–150 x 24–30 μ, 8-spored, paraphysate. Sporidia biseriate, ovoid-elliptical, or subelongated, 5-septate (rarely 5–6-septate), mostly with one cell divided by a longitudinal septum, scarcely constricted at the septa, brownish, 20–24 x 10–15 μ.

On *Carex*, *Poa*, *Agropyrum* and *Festuca*, Greenland.

Pl. vàgans, Niessl. Notiz. p. 14.

Perithecia scattered or seriate, depressed-spherical, dark brown, sparingly fibrillose below, otherwise glabrous, with a small, conical ostiolum. Asci clavate or oblong-clavate, 8-spored, short-stipitate. Sporidia biseriate, oblong or clavate-fusoid, straight or curved, often boat-shaped, 5-septate, with an imperfect longitudinal septum, honey-yellow.

Var. *Arenaria*, Niessl. l. c., has the perithecia tolerably large (250–270 μ diam.), scarcely erumpent, with a tolerably thick, obtuse, perforated ostiolum. Asci broad, 105–120 x 21–23 μ. Sporidia clavate-fusoid, straight, the third cell broader, 27–30 x 9–10 μ.

On *Elymus arenarius*, Greenland. The part of the stem occupied is often colored brownish.

Pl. platýspora, Sacc. Mich. II, p. 251.

Perithecia gregarious, at first covered by the epidermis, then nearly free, ½ mm. diam., globose-depressed, glabrous but with creeping hyphæ around the base, with a short, conoid ostiolum. Asci cylindric-clavate, 100–120 x 18 μ, short-stipitate, paraphysate, 8-spored. Sporidia biseriate, flattened, 4-septate, not constricted, cells divided by a longitudinal septum, 22–24 x 12–14 μ, and 6–7 μ thick, straw-yellow, with a gelatinous envelope.

On *Vesicaria*, *Draba*, *Arabis*, *Armeria*, *Erigeron* and *Polygonum*, Greenland.

Pl. papaveràcea, (De Not.)

Cucurbitaria papaveracea, De Not. Sfer. Ital p. 62
Pleospora papaveracea, Sacc. Syll. 3718
Exsicc Rehm Asc. 736.

Perithecia scattered or two or three together, superficial, with a flat, sessile base, globose, somewhat collapsed at the apex, coriaceous, black, obtusely papillate. Asci cylindric-clavate, attenuated below, 8-spored, about 100 x 10 μ, paraphysate. Sporidia uniseriate or partly biseriate, elliptical or oblong-fusoid, straight, 3-septate and constricted at the septa, the second and generally the third cell divided by a longitudinal septum yellow, becoming brownish, 18–25 x 7–8 μ.

On dead stems of *Papaver nudicaulis*, Tasiusak, Greenland.

Pl. macróspora, Schröt. Nord. Pilze, p. 15.

Perithecia erumpent, seriately-gregarious, globose-depressed, 250–300 μ diam., collapsing, black, smooth and globose, with minute, papilliform ostiola. Asci cylindric-clavate, attenuated below, short-stipitate, 150–170 x 24–26 μ, 4–8-spored. Sporidia uniseriate, ellipsoid; in the 4-spored asci 35–45 x 13–17 μ, in the 8-spored, 30–33 x 9–10 μ, 3-septate, one or two of the inner cells longitudinally divided, yellow-brown.

On *Hierochloa alpina*, Christianshab, &c., Greenland.

Pl. heteróspora, De Not. Sfer. Ital. p. 76, tab. 81.

Perithecia scattered, small, punctiform, sphæroid, black, with a round opening above, projecting through the fissured epidermis, papillose under the lens, texture of dark, round, turgid cells, soft when moistened. Asci thick-walled, 8-spored, attenuated upwards, mostly curved, oblong, variable in size. Sporidia oblong or fiddle-shaped, dark brown, hardly translucent, 9–11-septate, densely tessellate-muriform, 25–30 x 10–12 μ.

On *Carex supina*, Sarkak, Greenland.

Pl. Élynæ, (Rab.)

Clathrospora Elynæ, Rab. in Hedw. I, tab. XV, fig 3.
Pleospora Elynæ, Ces. & De Not. Schema, p. 44.
Exsicc Rabh-Winter F. Eur. 2861.

Perithecia gregarious, covered by the epidermis, depressed-spherical, the distinct papilliform ostiolum piercing the epidermis, black, smooth and glabrous. about 250 μ diam. Asci oblong, short-stipitate, 8-spored, thickened at the apex and broadly rounded, 140–165 x 35–40 μ, paraphysate. Sporidia 2–3-seriate, oblong, rounded at the ends,

7-septate, with usually partial longitudinal septa, golden-brown, 55–65, x 26–30 μ.

On *Carex scirpoidea*, *C. supina*, *Luzula arctica*, *L. confusa*, and *Aira alpina*, Greenland.

Pl. Frángulæ, Fckl. Symb. p. 133.

Exsicc. Fckl. F. Rh 1767.—Thum. F Austr 482?

Perithecia gregarious on bleached spots, about as large as those of *Pl. herbarum*, covered by the epidermis, black, globose, wrinkled when old, the conical, minutely perforated ostiolum erumpent. Asci cylindrical, short-stipitate, 8-spored, 120–125 x 24–26 μ. Sporidia biseriate, oblong, narrowed at the ends, constricted in the middle, generally 7–8-septate, with an imperfect, longitudinal septum, yellow-brown, 35–39 x 10–12 μ.

On twigs of *Rhamnus Californica*, California (Harkness).

Pl. Sálsolæ, Fckl. Symb. p. 131.

Exsicc Fckl. F. Rh. 814.

Perithecia scattered, at first covered, finally free, globose, black, floccose below, bare above, crowned with a broad papilla, 300–320 μ diam. Asci oblong, sessile or short-stipitate, 8-spored, 100–115 x 26 μ. Sporidia biseriate, ovate-oblong, broadest above the middle, 7-septate, constricted at the septa, more strongly so at the middle one, with 1–2 (generally imperfect) longitudinal septa, golden yellow, 28–30 x 11–12 μ.

On *Salicornia herbacea*, California (Harkness).

PYRENÓPHORA, Fr.

Summa Veg. Scand. p. 397, Emend. Sacc. Syll. II, p. 277.

Perithecia subcuticular-erumpent, globulose, black, bristly or hairy, sclerotoid or coriaceo-membranaceous, otherwise as in *Pleospora*.

Pyr. fenestràta, Pk. Bull. Torr. Bot. Club, XI, p. 28.

Perithecia 280–370 μ diam., depressed, at first covered by the epidermis, finally erumpent and free, clothed with a few straight, divergent, black setæ. Asci fugacious. Sporidia oblong, fenestrate, deeply colored, black in the mass, 40–50 x 18–22 μ, about 7-septate, generally constricted in the middle.

On dead stems of herbs, Utah.

Pyr. depréssa, Pk. l. c.

Perithecia 250–350 μ diam., depressed or collapsed, at first covered by the epidermis, then erumpent or naked, black, sometimes surrounded at base by a few appressed, black, radiate filaments, furnished above with a few short, erect or divergent, straight, black bristles. Asci cylindrical or subclavate, 112–150 μ long. Sporidia crowded or obliquely monostichous, oblong or subovate, triseptate, with 1–2 longitudinal septa, colored, 28–35 x 15–20 μ.

On dead stems of *Arabis*, California.

Pyr. hyphàsmatis, E. & E. Journ. Mycol. IV, p. 77.

Perithecia gregarious, superficial, flask-shaped, mouse-colored, 340–370 μ diam., clothed, except the broad, truncate ostiolum, with brown, sparingly branched and sparingly septate, short, soft, somewhat crisped and matted hairs. Asci clavate-cylindrical, 65–70 x 6–7 μ including the slender base, (p. sp. 50–55 μ long), with abundant paraphyses. Sporidia crowded-biseriate, brown subnavicular with the ends at first subacute, 3-septate, with occasionally one of the cells longitudinally divided, 12–15 x 5–7 (mostly 12 x 6 μ). This has the general appearance of a *Chætomium*.

On exposed cotton cloth, Louisiana (Langlois).

Pyr. phæócomes, (Reb.)

Sphæria phæocomes, Reb. Fl. Neom. p. 338.
Ceuthospora phæocomes, Rab. Deutschl. Kr. Flora I, p 144.
Pyrenophora phæocomes, Fr. Summa Veg. Scand. p. 398.
Exsicc Rab. Herb. Mycol. 747.—id. F. E. 1931.—Fckl. F. Rh. 798. 2362.

Perithecia scattered, innate, mostly prominent on both sides of the leaf, globose-hemispherical, black, subastomous, with erect-divergent, black, bristle-like hairs at the apex. Asci not abundant, oblong-clavate, short-stipitate, 8-spored, very large, over 300 μ long and about 80 μ broad. Sporidia cylindrical, subattenuated below, broadly rounded at the ends, 6-septate and slightly constricted at the septa, with one longitudinal septum running through all the cells, the second somewhat broader, yellow, 84–90 x 21–32 μ, with a thin, gelatinous envelope.

On (*Carex*)? Carolina (Berk.), on dead grass, New York (Peck), on *Archangelica, officinalis*, Greenland, (Rostrup).

Pyr. paucítricha, (Fckl.)

Pleospora paucitricha, Fckl. in the 2d Dutch North Polar Exp. II, p 32, tab 1, fig. 3.
Pyrenophora paucitricha, Sacc. Syll 7101, Cke Syn. 5154.

Perithecia very small, scattered, subsuperficial, conical, very black, clothed at the apex with a few short, rigid, black hairs. Asci ovate-oblong, obtuse at each end, somewhat curved, thick-walled, 8-spored, 112 x 24 μ. Sporidia imbricate-subbiseriate, oblong-ovate, obtuse, straight, 5-septate and muriform, constricted in the middle, brown, 34 x 13 μ.

On leaves of *Salix Groenlandica*, Franz Joseph's Fiord, Greenland.

Pyr. relícina, Fckl. Symb. p. 215.

Pleospora polytricha, Tul. Sel. Carp. II, p. 269.

Perithecia innate-erumpent, sclerotoid, pale within, clothed with scattered, rigid, cuspidate, pluriseptate, black bristles 130–150 x 10–12 μ. Asci clavate, 150–170 x 35–40 μ, with a short, thick stipe. Sporidia biseriate, rarely uniseriate, oblong-elliptical, 3–5-septate, with a partial longitudinal septum, pale yellow, 35–46 x 20–30 μ.

On leaves of *Triticum repens*, Albany, N. Y. (Peck).

Pyr. calvéscens, (Fr.)

Sphæria calvescens, Fr. Scleromyc. Suec. Exsicc. No. 401
Sphæria eriophora, Cke. Grev. V, p. 153.
Pleospora calvescens, Tul. Sel. Carp. II, p. 266.
Pyrenophora calvescens, Sacc. Syll. 3845, Cke. Syn 5142.
Leptosphæria eutypoides, Pk. 38th Rep. p. 105.
Exsicc. Fckl. F. Rh. 2152.—Rab. F. E. 2216—Rehm Asc. 439.—Thum. M. U. 266.—Ell. N. A. F. 592.

Perithecia gregarious, on extensive blackened areas of the stem, superficial, globose or depressed, 250–300 μ diam., clothed, especially below, with spreading, black, bristle-like hairs, which finally disappear; at length collapsing to plane or concave, with a distinct, papilliform ostiolum. Asci cylindrical, short-stipitate, 8-spored, 75–110 x 10–12 μ, with abundant paraphyses. Sporidia uniseriate, ovate-oblong, rounded at both ends, often rather narrower below, more or less constricted in the middle, 3-septate, with an imperfect, longitudinal septum, 18–22 x 8–9 μ.

On dead stems of *Chenopodium*, Newfield, N. J., and New York State.

Pyr. ciliàta, (Ell.)

Pleospora ciliata, Ell. Bull. Torr. Bot. Club. VIII, p. 125.
Pyrenophora ciliata Berlese Mon. p. 237, tab. XII, fig. 6.

Perithecia at first covered by the cuticle, but at length partially erumpent, 125–175 μ diam., hemispherical. Ostiolum large, surrounded by a fringe of stout, black, reflexed hairs. Asci broad, oblong-elliptical. Sporidia muriform, yellow, slightly constricted across the middle, 33–37 x 15–18 μ, transversely 7-septate, the cells divided by 2–3 longitudinal septa.

On dead stems of *Phlox*, Utah.

Pyr. Zabriskieàna, E. & E. Proc. Acad. Nat. Sci. Phil. July, 1890. p. 238.

Perithecia loosely gregarious, erumpent-superficial, ovate-globose, 175–200 μ diam., densely clothed with straight, erect, sparingly-septate, yellowish-brown hairs 100–125 μ long and about 5 μ thick at the base, tapering gradually to the subacute tip, of membranaceous texture and dark yellowish-brown color (under the microscope). Asci clavate-oblong, 100–125 x 22–25 μ, rounded above, with a short, abrupt, stipitate base. Paraphyses slender and numerous but inconspicuous. Sporidia crowded-biseriate, ovate-elliptical, densely muriform, rounded at the ends, yellowish-brown, 22–30 x 12–14 μ, constricted across the middle. The sporidia are so closely and densely muriform as to appear granular, the granular contents being arranged in transverse lines across the sporidia so that they appear 12 or more septate. Differs from *P. polyphragmia*, Sacc. to which it comes nearest, in its smaller perithecia and shorter sporidia.

On bark of *Ulmus Americana*, New Baltimore, N. Y. (Zabriskie).

Pyr. comàta, (Awd. & Niessl).

Pleospora comata, Awd. & Niessl in Niessl Beitr. Zur Kent. der Pilze, p 30.
Pyrenophora comata, Sacc. Syll. 3864, Cke. Syn. 5169.
Exsicc. Rab F. E. 1544.

Perithecia scattered, covered by the epidermis, only the short conical ostiolum projecting and surrounded by a tuft of straight, simple, black, divergent bristles, spherical, of coriaceo-membranaceous texture, black, 180–220 μ diam. Asci broad, oblong or oblong-clavate, sessile, with a broadly-rounded apex, 8-spored, 110–130 x 40 μ, paraphysate. Sporidia biseriate, oblong-ovate, often oblique, at first with 7–9, later with 11–13 (or more) septa, with 2–3 more or less perfect longitudinal septa, dark brown, opake, 32–38 x 14–16 μ.

On *Alsine*, *Cerastium*, *Melandrium*, *Viscaria*, *Pyrola*, *Pedicularis*, *Oxyria*, *Luzula*, *Draba Saxifraga*, in Western Greenland.

Pyr. chrysóspora, (Niessl.)

Pleospora chrysospora, Niessl. Hedw. 1880, p. 173.
Pyrenophora chrysospora, Sacc. Syll. 3861, Cke. Syn. 5165.
Exsicc. Kz. F. Sel. 581.—Rab. F. E. 2860.

Perithecia scattered, at first sunk in the scarcely altered substance of the host, afterwards more or less erumpent, 260–320 μ diam., spherical or slightly depressed, tipped with a tuft of straight, rigid, black bristles and surrounded at base by brown, creeping hyphæ. Asci cylindrical, subclavate, short-stipitate, 8-spored, 100–110 x 23 μ. Sporidia biseriate, becoming uniseriate by the lengthening out of the asci, oblong, obtuse, constricted in the middle, 7-septate, with one or two longitudinal septa, 24–30 x 10½–11 μ.

On *Draba*, *Saxifraga*, and *Pedicularis*, in Greenland.

LEPTOSPHÀERIA, Ces. & De Not.

Schema, p. 60, Emend. Sacc. Syll. II, p. 13

Perithecia at first covered by the epidermis, finally more or less erumpent, subglobose, coriaceo-membranaceous, glabrous, mostly with a papilliform ostiolum. Asci subcylindrical, mostly 8-spored, paraphysate. Sporidia ovoid, oblong or subfusoid, two- or more-septate, olivaceous, yellowish, or brown.

A. *On dicotyledonous plants.*

* *On trees and shrubs; sporidia* 2–3- (*rarely* 4-) *septate.*

L. Jànus, (B. & C.)

Sphæria Janus, B & C. Grev. IV, p. 154.
Leptosphæria Janus, Sacc. Syll. 3167
Heptameria Janus, Cke. Syn. 4957.
Exsicc. Ell. N. A. F. 1299.

Perithecia collected on yellowish-brown spots on the under side of the leaf, and penetrating to the upper surface. Asci oblong, 60–75 x 9–10 μ, sessile, paraphysate. Sporidia 8 in an ascus, biseriate, short-fusoid, 4-septate, 20–33 x 4–5 μ.

On leaves of *Quercus virens*, Texas (Wright), Florida (Martin).

L. Tamáricis, (Grev.)

Sphæria Tamaricis, Grev. Scot. Crypt Fl. tab. 45, Curr. Simple Sphærias, tab. 58, fig. 81
Leptosphæria Tamaricis, Sacc. Syll. 2946, Cke. Syn. 4271.

Perithecia gregarious, scattered, subcutaneous, 150–200 μ diam., raising the epidermis into pustules which are soon ruptured in a hysteriiform manner, but the perithecia themselves are not erumpent, white inside, with an obtuse, papilliform ostiolum. Asci clavate-cylindrical, subsessile, rounded above, paraphysate, 65–70 x 15–20 μ. Sporidia crowded-biseriate, ovate-oblong, 3-septate and strongly constricted, especially at the middle septum, yellowish, 25–30 x 13–15 μ, including the broad, gelatinous envelope.

On dead branches of *Tamarix* (cult.), West Chester, Pa.

In Currey's fig. the sporidia are not constricted, and are smaller. The asci often contain but two perfect sporidia. Whether the West Chester spece. are really the *Sphæria Tamaricis*, Grev., is not altogether certain.

L. ramulícola, Pk.

Sphæria ramulicola, Pk 25th Rep p 104
Leptosphæria ramulicola, Sacc Syll. 2947, Cke. Syn. 4273

Perithecia small, scattered, seated on the inner bark, erumpent by an angular or subcircular aperture, subglobose, subfibrous, black, white within. Ostiola minute, indistinct. Asci cylindrical. Sporidia elliptical, uniseriate, biseptate, slightly constricted, colored, 20–27 x 12–13 μ.

On dead twigs of elm trees, Greenbush, N. Y.

The perithecia are abundant on all sides of the smaller branches, rendering them rough to the touch.

L. inspérsa, (Schw.)

Sphæria inspersa, Schw in Curr. Simple Sph No 334, tab 59, fig 112
Leptosphæria inspersa, Sacc. Syll 2951, Cke. Syn. 4278.

Perithecia simple, corticolous, erumpent. Asci cylindric-clavate, short-stipitate, 8-spored. Sporidia biseriate, rather broad-fusoid, 4-nucleate, (triseptate)?, constricted between the nuclei, dark brown. 25–28 μ long.

On bark North America (Schw. in Herb. Hooker).

L. Tini, E. & E. Journ. Mycol. IV, p. 64.

Maculicolous and amphigenous, on large cinereous spots with a narrow, reddish-purple border. These spots, or dead places, often occupy the margin along one side, or the apex of the leaf, 2–3 cm. in diam. Perithecia punctiform, minute, innate, the black, subacute apex alone being visible. Asci oblong-cylindrical, subsessile, with filiform paraphyses, 40–50 x 7 μ. Sporidia biseriate, fusoid, yellowish, slightly curved, faintly 3-septate, 14–16 x 3 μ.

On leaves of *Viburnum Tinus*, Lafayette, La. (Langlois).

L. Andrómedæ, (Awd.)

Sphærella Andromedæ, Awd. Syn. Pyr. p. 12, tab 7, fig 102.
Leptosphæria Andromedæ, Sacc Syll. 3031.

Perithecia epiphyllous, black, innate in the epidermis of the leaf, scattered, ovoid, 160 x 130 μ, pierced with a minute pore which is scarcely visible. Asci subclavate, sessile, 8-spored, 136 x 27 μ. Sporidia biseriate, oblong, straight. 34 x 14 μ, ends rounded, 3-septate and constricted at the septa, more deeply so at the middle septum, pale, at length brownish.

On dry leaves of *Andromeda tetragona*, in Labrador and Greenland.

L. hyperbòrea, (Fckl.)

Pleospora hyperborea, Fckl. 2d Deutsch Nordpolfahr 2, p 92 and Oud. Contr. Fl. Myc. Now. Seml. p. 153, tab. II, fig. 9.
Leptosphæria hyperborea, Sacc. Syll. 6673.

Perithecia seated on the dry upper surface of the leaves, scattered, punctiform, semiimmersed, surrounded below by the slightly raised, gray substance of the leaf, the free part obtusely-conical, pierced above and very black. Asci oblong-ovate, attenuated at each end, 8-spored, 60 x 24 μ. Sporidia crowded in the middle part of the asci, oblong, ends obtuse, straight, 3-septate, constricted at the middle septum, 24 x 8 μ, yellow-brown.

On leaves of *Andromeda tetragona*, in Greenland.

L. Ceanòthi, (Cke. & Hark.)

Sphæria Ceanothi, Cke. & Hark. Grev. XIII, p 19.
Leptosphæria Ceanothi, Sacc. Syll. 6662, Cke. Syn. 4292.

Perithecia gregarious, minute, subglobose, covered by the epidermis. Ostiolum short, punctiform. Asci short-clavate, 8-spored. Sporidia sublanceolate, triseptate, yellowish-hyaline, 16–18 x 4½ μ.

On small twigs of *Ceanothus*, California (Harkness).

L. odòra, (Cke. & Hark.)

Sphæria odora, Cke. & Hark. Grev. l. c.
Leptosphæria odora, Sacc. Syll 6666, Cke. Syn 4296

Perithecia scattered, of medium size, globose, black, sunk in the bark and covered by the epidermis. Ostiolum punctiform. Asci clavate, 8-spored. Sporidia biseriate, lanceolate, at first uniseptate, then 3-septate, yellowish-hyaline, 22–25 x 5 μ.

On branches of *Umbellularia*, California (Harkness).

L. bicuspidàta, (Cke. & Hark.)

Sphæria bicuspidata, Cke. & Hark. l. c.
Leptosphæria bicuspidata, Sacc. Syll. 6649.

Perithecia scattered, covered, innate, subprominent, globose, black. Ostiolum short. Asci clavate, 8-spored. Sporidia fusiform, 3-septate, constricted, brown, with a hyaline, cuspidate appendage at each end, 25 x 8 μ without the appendages.

On twigs of *Baccharis*, California.

L. olivæspora, (B. & C.)

Sphæria olivæspora, B. & C. Grev. IV, p. 148.
Leptosphæria olivæspora, Sacc Syll 3170.
Heptameria olivæspora, Cke. Syn. 4960.

Perithecia covered by the cuticle, but rather prominent, marked in the center with the black ostiola. Asci linear. Sporidia oblong. pointed at each end, 3-septate, about four times longer than broad.

On dead limbs of *Cornus florida*, South Carolina.

L. Fráxini, E. & E. Journ. Mycol. III, p. 44.

Spots amphigenous, rusty below, dirty white above, with a rusty brown border, orbicular, small (1 mm.) or elongated (2–3 x 1 mm.). Perithecia black, epiphyllous, few, small (75 μ), erumpent, indistinctly pierced above, and with a rudimentary mycelium around the base. Asci clavate-cylindrical, 55–70 x 8–10 μ. Sporidia biseriate, fusoid, somewhat curved, nucleate, becoming 3–5-septate (mostly 4-septate), constricted at the middle septum when mature, and slightly constricted at the others, 20–25 x 3½–4 μ.

On living leaves of *Fraxinus Americana*, Columbia, Mo. (B. T. Galloway).

L. boreàlis, E. & E. (in Herb.)

Perithecia scattered, erumpent, ⅓–½ mm. diam., the lower part sunk in the wood, the apex raising the surface into flattish pustules which are pierced in the center by the papilliform ostiolum. Asci subcylindrical, 150–170 x 8 μ. Sporidia uniseriate, oblong, 3-septate. not constricted, 18–22 x 5–7 μ, pale brown, becoming darker.

On decorticated *Salix*, Helena, Montana (Kelsey).

L. fuscélla, (B. & Br.)

Sphæria fuscella, B. & Br. Brit. Fungi, No 636, tab XI, fig. 20.
Leptosphæria fuscella, Ces & De Not Schema, p 62

Perithecia scattered, covered, subglobose, slightly depressed, raising the epidermis into little pustules. Asci cylindrical, 100 x 9–10 μ, with a short, thick stipe. Sporidia obliquely uniseriate, oblong-elliptical, rounded at the ends, sometimes slightly curved, 3-septate and somewhat constricted at the septa, olivaceous, 15–17 x 9 μ.

On dead stems of *Rubus strigosus*, Greenbush, N. Y. (fide Peck).

L. platanícola, (Howe).

Sphæria platanicola, Howe Bull. Torr. Bot. Club, V, p. 43.
Leptosphæria platanicola, Sacc. Syll 6130, Cke. Syn. 4276, 4760.

Perithecia globose-conoid, erumpent. Asci cylindrical or subclavate. Sporidia uniseriate, of a deep, dark color, 3-septate.

On branches of *Platanus*, New York.

L. taxícola, (Pk.)

Sphæria taxicola, Pk 24th Rep. p. 99.
Leptosphæria taxicola, Sacc. Syll. 3169.
Heptameria taxicola, Cke. Syn 4959.

Perithecia minute, close, black, shining, slightly prominent, at first covered by the epidermis, then erumpent. Sporidia in a single series, oblong, triseptate.

Occupying the whole upper surface of dead leaves of *Taxus Canadensis*, Sandlake, N. Y. No measurements of asci and sporidia given.

L. concéntrica, E. & E. (in Herb.)

Perithecia widely scattered, innate-erumpent, on large, coneentrically marked, dull cinereous, indefinite spots occupying a large part of the leaf. Asci (p. sp.) 50–55 x 10 μ. Paraphyses not seen. Sporidia oblong-cylindrical, 2-septate, yellow-brown, 12–14 x 4–5 μ, ends obtuse.

On apple leaves, Columbia, Mo. (H. Dorsett), and Louisiana (Langlois).

This is different from *L. Lucillæ* or from *L. Pomona*, Sacc.

L. Califórnica, (Cke. & Hark.)

Sphæria Californica, Cke. & Hark. Grev. XIII, p. 20.
Leptosphæria Californica, Sacc. Syll. 6665, Cke. Syn. 4295.

Densely gregarious, covered. Perithecia hemispheric-prominent, obtuse, black, for a long time covered by the epidermis, but finally partially denuded above. Asci clavate. Sporidia biseriate, sublanceolate, ends obtuse, 4-septate, the penultimate cell somewhat swollen, pale brown, 25–30 x 8 μ.

On *Araucaria imbricata*, *Sarothamnus*, *Rhododendron*, and *Euonymus*, California.

Differs from *Metasphæria anisometra*, Cke. & Hark., in the perithecia being densely aggregated, so as sometimes to blacken the twig for some inches, and in its brown sporidia.

** *On herbaceous stems; sporidia 2–3-septate.*

L. Tephròsiæ, (C. & E.)

Sphæria Tephrosiæ, C. & E. Grev. VII, p. 10
Leptosphæria Tephrosiæ, Sacc. Syll. 2949, Cke. Syn. 4275.
Exsicc. Ell. N. A. F. 695.

Perithecia scattered, at first covered by the epidermis, finally exposed and subsuperficial, subglobose, about ¼ mm. diam., with an acutely papilliform ostiolum. Asci cylindrical, 65–70 x 6–7 μ, paraphysate. Sporidia overlapping, uniseriate, fusoid, acute, nearly hyaline and uniseptate at first, becoming subobtuse, brown and 3-septate, 18–20 x 4–5 μ, (30 x 6 μ. Cke.).

On dead stems of *Tephrosia Virginiana*, Newfield, N. J.

L. anómala, E. & E. Journ. Mycol. III, p. 117.

Perithecia gregarious, membranaceous and of rather coarse, cellular structure, about ⅓ mm. diam., black, smooth, subspherical, at length slightly collapsing above, at first covered by the epidermis, finally erumpent. Asci oblong-cylindrical, about 100 x 20 μ, subsessile, with evanescent, filiform paraphyses. Sporidia biseriate, broad oblong-fusoid, 1–2-septate, inequilateral and slightly curved, pale straw-yellow, constricted at the main septum, 22–30 x 7–9 μ, (exceptionally 30–35 x 9–12 μ), mostly 22–25 x 7–8 μ.

On dead herbaceous stems, Scofield, Utah (S. J. Harkness.

With *Cylindrocolla diffluens*, E. & E. (J. M. l. c.).

It is possible that fresh specc. might show the sporidia to be appendiculate, in which case this would be referable to *Ceriospora*, but, as far as can be seen in the specimens examined, there are no appendages on the sporidia, though while lying in the asci their ends are acute. The sporidia, as a rule, are only 1-septate, the 2-septate sporidia being only occasionally seen.

L. Thalíctri, Winter, Hedwigia, 1872, p. 140.

Metasphæria Thalictri, Sacc. Syll. II, p. 156.

Perithecia scattered or gregarious, for a long time covered by the epidermis, strongly depressed-spherical, black, with the very small, papilliform ostiolum scarcely projecting. Asci fasciculate, oblong, mostly broader below, sessile, 8-spored. Sporidia pyriform, 2-septate, the terminal cells mostly larger than the middle one, ends rounded, slightly constricted at the septa, pale yellow, 21–23 x 10 μ. Paraphyses filiform, articulated.

On stems of *Thalictrum alpinum*, Kobbefiord and Umanak, Greenland.

The asci in the Greenland specc. were 70–75 x 20–24 μ; sporidia 25 x 11 μ.

L. Doliolum, (Pers.)

Sphæria Doliolum, Pers. Syn. p. 78.
Pleospora Doliolum, Tul. Sel. Carp. II, p 276.
Leptosphæria Doliolum, Ces. & De Not. l. c. p. 61.

Exsicc. Fckl. F. Rh. 898.—Kze. F. Sel. 76, 335, 336.—Rab. F. E. 1546 —Rehm Asc. 93, 194, 343.—Thum. F. Austr. 1148, 1255.—id. M. U. 141.—Ell. N A. F. 197.—Sacc. M Ven. 1472.—Desm. Pl. Cr. Ed. I, 714.—Myc. March. 141.—Cke. F. B. 1st Ser. 489.

Perithecia scattered or gregarious, at first buried, finally free and superficial, hemispherical, generally surrounded by several concentric furrows, about 400 μ diam. with a papilliform, black, shining, perforated ostiolum. Asci cylindrical, 8-spored, 100–120 x 5½–7 μ. Sporidia uniseriate or partly biseriate, fusiform, straight or slightly curved, subinequilateral, 3-septate and slightly constricted at the septa, yellow-brown, 20–30 x 4½ μ. Paraphyses filiform, branching.

On dead herbaceous stems, *Aster*, *solidago*, &c., &c., common.

L. subcónica, (C. & P.)

Sphæria subconica, C. & P. 29th Rep. N. Y. State Mus. p 65.
Leptosphæria subconica, Sacc Syll. 2898,
Heptameria subconica, Cke. Syn. 4699.
Exsicc Ell N. A. F. 697.

Perithecia conoid, flattened at the base, somewhat collapsed when dry, black, seated beneath the epidermis, which is at length thrown off. Asci cylindrical, sessile, paraphysate, 75–80 x 6–8 μ. Sporidia overlapping-uniseriate, broad-fusoid, subinequilateral, 3-septate and sometimes slightly constricted at the septa, 20–25 x 5–7 μ.

On dead herbaceous stems, common.

It is doubtful whether this should be separated from *L. Doliolum*.

L. viridélla, (Pk.)

Sphæria viridella, Pk. 30th Rep. p 66.
Leptosphæria viridella, Sacc. Syll. 2910
Heptameria viridella, Cke. Syn. 4714.

Perithecia small, gregarious, seated on a greenish spot, covered by the epidermis which is ruptured by the minute ostiola. Asci cylindrical. Sporidia oblong-fusiform, sometimes curved, 3-septate, greenish, 25 μ long, the third cell from the base swollen.

On dead stems of *melilot*, Bethlehem, N. Y.

The marked feature of the species is the greenish color of the spot and of the sporidia. The latter resemble those of *L. subconica*, ex-except in color.

L. conséssa, (C. & E.)

Sphæria consessa, C. & E Grev. VI, p. 95
Leptosphæria consessa, Sacc. Syll. 2906.
Heptameria consessa, Cke. Syn. 4709.

Perithecia globose, black, shining, papillate, at first covered by the epidermis, at length bare. Asci cylindrical, 75 x 7 μ, paraphysate. Sporidia biseriate, narrow-lanceolate, brown, 3-septate, 18–20 x 3–4 μ, (20–22 x 5 μ, Cke.).

On dead stems of *Helianthus tuberosus*, Newfield, N. J.

L. anthelmíntica, (Cke.)

Sphæria anthelmintica, Cke Grev. VII, p 53.
Leptosphæria anthelmintica, Sacc. Syll. 2924.
Heptameria anthelmintica, Cke Syn, 4730.
Exsicc. Rav. F Am. 372.

Perithecia scattered, covered. only slightly prominent, the puncti-form ostiola piercing the epidermis. Asci cylindrical. Sporidia fusi-form, triseptate, brownish, slightly constricted, 30 x 7 μ.

On *Chenopodium anthelminticum*, South Carolina.

L. dumetòrum, Niessl, Beitrag, p. 26, tab. III, fig. 12.

Exsicc. Rab. F. E. 2238, 2853 —Rehm Asc. 678, 688.—Thum. M. U 2248 —M March, 986, 2046, 2137, 2238.—Ell. & Evrht N. A. F. 2d Ser 2614.

Perithecia scattered, covered by the epidermis, hemispherical or subglobose, papillate membranaceo-coriaceous, black, 200 μ diam. Asci clavate-oblong, short-stipitate, rounded at the apex, 8-spored, 60–66 x 6–7 μ. Sporidia biseriate, fusoid-oblong, inequilateral, straight or slightly curved, subacute, 3-septate, the next to the upper cell swollen, pale olive-yellow, 18–20 x 3 μ. Paraphyses longer than the asci. Spermatia cylindrical, very minute, 2–4 x 1½–2 μ, in small. black, membranaceous, covered perithecia.

On dead stems of *Helianthus*, Newfield, N. J., on dead stems of *Lonicera*, New York (Fairman), and on stems of *Impatiens*, Canada (Dearness).

L. olerícola, (B. & C.)

Sphæria olericola, B & C. Grev. IV, p. 150.
Leptosphæria olericola, Sacc. Syll. 2923.
Heptameria olericola, Cke. Syn. 4729.

Perithecia gregarious, conical. Sporidia oblong, slightly curved, triseptate, 25 μ long.

On stems of *Brassica*, New England.

A curious diseased state occurs, in which the sporidia turn black and lose all trace of septa.

L. fimbriàta, E. & E. (in Herb.)

Perithecia subglobose, 200–300 μ diam., subcuticular, prominent, exposed by the peeling off of the cuticle, fringed with brown, mycelial threads around the base. Asci oblong-cylindrical, short-stipitate, par-aphysate, mostly curved, 80–90 x 15–18 μ. Sporidia crowded-biseriate, oblong-fusoid, 3-septate, scarcely constricted, brown, 20–24 x 5–6 μ, slightly curved, subobtuse.

On dead herbaceous stems, Utah.

L. Brunéllæ, E. & E. Proc. Acad. Nat. Sci. Phil. July, 1890, p. 237.

Perithecia scattered, minute (200–250 μ), covered by the epidermis, which is only slightly raised, and barely pierced by the papilliform ostiolum. Asci clavate-cylindrical, 75–80 x 10–12 μ, subsessile, with filiform paraphyses. Sporidia biseriate, fusoid, slightly curved, pale yellowish-brown, 3-septate, the next to the upper cell swollen, 22–30 x 4 μ. Differs from *L. pyrenopezizoides*, Sacc., in its perithecia not collapsing, and from *L. parietariæ*, Sacc., in its paler sporidia.

On dead stems of *Brunella vulgaris*, London, Canada (Dearness).

Accompanied by perithecia containing fasciculate, acicular stylospores (*Rhabdospora*) 40–55 x 2–2½ μ. These perithecia are white inside and rather larger. Other smaller perithecia contain spores 4 x 1½ μ (*Phoma*).

L. pyrenopezizoìdes, Sacc. & Speg. Mich. I, p. 394., F.•Ital. tab. 323

Perithecia scattered, erumpent-superficial, subglobose-depressed, 165 μ diam., papillate, finally collapsing to cup-shaped, black. Texture loosely parenchymatic, dark olive, more dense around the ostiolum within a circular space about 10 μ across. Asci cylindric-clavate, 70–75 x 8 μ, very short nodulose-stipitate, subtruncate above, with a contracted opening, 8-spored. Sporidia subbiseriate, fusoid, slightly curved, 20–25 x 4 μ, 3-septate, the second cell slightly swollen, pale yellow.

On dead herbaceous stems, Manhattan, Kansas. (See Journ. Mycol. II. p. 3).

L. vagabúnda, Sacc. F. Ven. Ser. II, p. 318.

Sphæria fuscella, Sacc. M. Ven. Spec. p. 97, tab. IX, figs. 37–46 (not B. & Br.)

Perithecia nestling in bark, covered by the epidermis, scattered or crowded globose, with a small and only slightly projecting ostiolum, black, 200–500 μ diam. Asci clavate-cylindrical, short-stipitate, 8-spored, with abundant paraphyses. Sporidia biscriate, fusoid, at first hyaline and 4-guttulate, then uniseptate and constricted and finally 3-septate and brown, constricted at all the septa, and often with an oil globule in each cell, 19–25 x 6–7 μ.

On dead stems of *Hypericum prolificum*, Newfield, N. J., and on dead limbs of *Spiræa opulifolia*, London, Canada.

Saccardo in Syll, II, p. 31 enumerates 25 different trees and shrubs on which this species is found in Italy, Germany, France and So. America. The spermogonial stage is *Coniothyrium vagabundum*, Sacc.

L. Physàlidis, E. & E. Bull. Torr. Bot. Club. XI. p. 42.

Perithecia minute, depressed-globose, of coarse, cellular structure, with a rather broad opening above. Asci 35–40 x 7–8 μ, no paraphyses seen. Sporidia biseriate, fusiform, slightly curved, yellowish, 3-septate, 15–19 x 3 μ.

On dirty-white, round spots on leaves of *Physalis pubescens*, near Lexington, Ky. (Kellerman).

L. Steironèmatis, E. & E. Proc. Acad. Nat. Sci. Phil. July, 1890, p. 237.

Exsicc Ell & Evrht N. A F. 2d Ser 2615.

Perithecia gregarious around the nodes of the stem, subepidermal, conic-hemispherical, ½ mm. diam., raising the epidermis which is pierced by the obtusely conic ostiolum. Asci clavate-cylindrical, 75–100 x 15–20 μ, with abundant filiform paraphyses. Sporidia biseriate, oblong, 3-septate, sometimes slightly constricted at the septa, brown, obtuse at the ends, mostly a little curved, 15–22 x 7–8 μ.

On dead stems of *Steironema ciliatum*, London, Canada (Dearness).

L. Silènes-acaùlis, De Not. Recl. Piren. p. 485.

L. Hausmanniana, Awd in Oesterr. Bot. Zeitschr. 1868, No. 8.
Exsicc. Rab. F. E 2765.

Perithecia epiphyllous, thickly scattered, very small, punctiform, globose, black, with a small, papilliform ostiolum which pierces the epidermis. Asci oblong-clavate, sessile, 8-spored, 60–70 x 15–17 μ. Sporidia conglomerated, fusoid, mostly slightly curved or inequilateral, 3-septate, not constricted, honey-yellow, 36–42 x 6–7 μ. Paraphyses filiform, conglutinated.

On leaves of *Silene acaulis*, Egedesminde, Greenland.

L. Stellàriæ, Rostr. Fungi Grönl. p. 557, No. 161.

Closely allied to *L. Silenes-acaulis*, differing principally in its smaller asci and sporidia, which are 45 x 10 μ, and 20 x 6 μ respectively.

On leaves of *Stellaria humifusa*, Itivnek, and on *Stellaria longipes*, Godhavn, west coast of Greenland.

L. Oxýriæ, Rostr. Fungi Grönl. p. 559, No. 168.

Perithecia gregarious, dark, depressed-spherical. Asci thick, cylindric-clavate, stipitate. 50–52 x 12–14 μ. Sporidia biseriate, fusoid-oblong, 3-septate, constricted at the septa, yellow.

On dry stems of *Oxyria digyna*, Egedesminde, Greenland. No measurements of asci or sporidia given.

L. Ranúnculi, Rostr. l. c.

Perithecia covered by the epidermis, then denuded, depressed-globose. Asci thick cylindric-clavate, 95–105 x 30–33 μ, with a rather thick membrane, 8-spored. Sporidia biseriate, oblong-elongated, 30–35 x 8–10 μ, 3-septate, slightly constricted at the septa, the second cell slightly enlarged, yellow.

On stems of *Ranunculus affinis*, Isortok, Greenland.

L. rubicúnda, Rehm, Wint. Diag., p. 10.—Sacc. F. Ital. tab. 292.

Exsicc. Rehm Asc. 92.

Perithecia gregarious on a dark red spot, innate and at first covered, but finally erumpent and partly free, brownish-black, globose, collapsing and more or less flattened or even cup-shaped, 150–200 μ diam. Asci clavate-cylindrical, 45–65 x 6–7 μ, short-stipitate, paraphysate. Sporidia subbiseriate, narrow-fusoid, 3-septate, slightly curved, the second cell from the lower end slightly swollen when mature, pale yellow, 22–25 x 2½–3 μ (45 x 2½–3 μ, Winter).

On dead stems of *Phytolacca*, Bethlehem, Pa. (Rau) and on dead stems of *Ambrosia*, Louisiana (Langlois).

L. Harknessiàna, E. & E. Journ. Mycol. I, p. 91. **(Plate 28)**

Perithecia scattered or gregarious, at first covered by the epidermis, at length bare and superficial or nearly so, hemispherical, black, smooth, ⅓–½ mm. in diameter. Ostiolum short-cylindrical, with a large circular opening. Asci cylindrical, 100–114 x 10–12 μ, 8-spored, and surrounded with filiform paraphyses. Sporidia in a single series, lying end to end, elliptical, yellow-brown, 4 septate and constricted at the septa, 18–22 x 7–9 μ, obtusely pointed above and regularly rounded below. The perithecia are much like those of *Leptosphæria subconica*, C. & P., but the ostiolum is shorter.

On dead stems of "Columbo" (*Frasera*?), Emery Co., Utah (S. J. Harkness).

L. distribùta, (C. & E.)

Sphæria distributa, C. & E. Grev VII, p. 41.
Leptosphæria distributa, Sacc. Syll 2912
Heptameria distributa, Cke Syn. 4716.

Perithecia scattered, covered by the blackened epidermis, which is raised and pierced by the acutely papilliform ostiolum, mostly oblong or elliptical, in a transverse section, longer diam. about 400 μ. Asci clavate, subsessile, paraphysate, 8-spored, 75 x 12 μ, rounded at the apex. Sporidia biseriate, broad-fusoid, slightly curved, 3-septate and constricted at the middle septum, 20–27 x 7–9 μ including the hyaline envelope.

On dead stems of *Desmodium*, Newfield, N. J.

The substance of the perithecia under the microscope is purple.

L. Utahénsis, E. & E. (in Herb.)

Perithecia scattered or subgregarious, subcuticular, slightly fringed around the base with mycelial threads, small, 150–175 μ diam., closely covered by the cuticle, collapsing to concave. Ostiolum minute, papilliform. Asci oblong-cylindrical, 55–60 x 7–8 μ, sessile, paraphysate. Sporidia crowded-biseriate, oblong-fusoid, subinequilateral, 3-septate, ends subobtuse yellowish at first, finally opake, 15 x 3 μ, remaining for some time uniseptate.

On dead stems of some (umbelliferous)? plant, Pleasant Valley, Utah (S. J. Harkness).

L. Coniothýrium, (Fckl.)

Sphæria Coniothyrium, Fckl. Symb p. 115, and Nachtr. II, p. 89.
Leptosphæria Coniothyrium, Sacc. Syll. II, p. 29.
Exsicc. Rehm Asc. 388

Perithecia gregarious, nestling under and attached to the epidermis, depressed-globose, with a papilliform, erumpent ostiolum, black, ½ mm. diam. Asci cylindrical, substipitate, 8-spored, 60–90 x 5–7 μ. Sporidia oblique or subbiseriate, oblong, subobtuse, 3-septate and slightly constricted at the septa, the second cell somewhat broader, often slightly curved, brownish, 10–15 x 2½–4 μ. Paraphyses filiform.

On dry capsules of *Salix glauca*, Kobbefiord, Greenland.

L. Kálmiæ, Pk. 39th Rep. p. 53.

Perithecia subcespitose, erumpent, 350–450 μ diam., subhemispherical, thick, black, the ostiola pertuse or slightly papillate. Asci cylindrical, 100–112 x 7–8 μ. Sporidia uniseriate, oblong or subfusoid, 3-septate, sometimes slightly constricted at the middle septum, colored, 16–20 x 6–7½ μ. Paraphyses filiform.

On dead stems of *Kalmia angustifolia*, Adirondack Mts., N. Y.

Generally there are 3–4 perithecia in a cluster, but sometimes they are single, and occasionally compressed.

*** *Mostly on herbaceous stems; sporidia 5- or more-septate.*

L. stictóstoma, (B. & C.)

Sphæria stictostoma, B. & C. Grev. IV, p. 151.
Leptosphæria stictostoma, Sacc. Syll. 3162.
Heptameria stictostoma, Cke. Syn. 4952.

Perithecia covered by the cuticle, causing little corresponding prominences pierced by the ostiola, at length free, subconical. Asci oblong. Sporidia biseriate, cymbiform, 5-septate, constricted at the septa, 20 μ long.

On some herbaceous plant, Connecticut.

L. stictoìdes. (B. & C.)

Sphæria stictoides, B. & C. Grev. IV, p. 149.
Leptosphæria stictoides, Sacc. Syll. 3171.
Heptameria stictoides, Cke. Syn. 4961.

Forming little irregular, depressed spicules surrounded by the cuticle. Sporidia nearly biconical, one division larger than the other, 5-septate, constricted at the middle septum. No measurements given.

On *Liriodendron*, North Carolina.

L. Váhlii, Rostr. Fungi Grönl. p. 557 (No. 162).

Hyphæ pale brown, with short, inflated joints. Perithecia depressed-spherical, umbilicate, papillate. Asci numerous, elongated-clavate, long-pedicellate, 90–100 x 11–13 μ, with copious, filiform paraphyses. Sporidia biseriate, cymbiform-fusoid, dark, 5-septate, 20–25 x 6–7 μ, cells 1–2-nucleate.

On dry stems of *Melandrium triflorum*, collected by Vahl, at Umanak, Greenland.

L. striàta, Winter, Hedwigia 1872, p. 140.

Perithecia membranaceous, globose, with a small ostiolum, somewhat wrinkled, brown. Asci elongated-oblong, sessile, 8-spored, 90 x 12 μ, with simple, filiform paraphyses longer than the asci. Sporidia fusoid, 5-septate, each cell with a large oil globule, greenish, the epispore marked with longitudinal stripes, 35 x 5 μ.

On stems of *Veronica alpina*, Greenland.

L. Boccòniæ, (C. & E.)

Sphæria Bocconiæ, C. & E. Grev. VII, p. 10.
Leptosphæria Bocconiæ, Sacc. Syll. 2901.
Heptameria Bocconiæ, Cke. Syn. 4703.
Exsicc. Ell. N. A. F. 586.

Perithecia scattered, buried and, when the epidermis falls away, with the rounded apex projecting, membranaceous, subglobose, brownish-black. Asci clavate, 75 x 12 μ, paraphysate. Sporidia biseriate, sublanceolate in the asci, oblong-elliptical when free, straight or slightly curved, pale yellow, 20–25 x 5–6 μ, 3-septate at first, becoming 5-septate.

On dead stems of *Bocconia* (cult.), Newfield, N. J.

L. cassiæcola, E. & E. Journ. Mycol. II, p. 41.

Perithecia gregarious, minute, 125–175 μ diam., covered by the blackened epidermis, which is raised into little pustules and at length pierced by the papilliform ostiola. Asci 70–75 x 8–10 μ. Sporidia biseriate, fusiform, slightly curved, 5-septate, yellow-brown, not constricted at the septa. Has the general aspect of a *Sphærella*.

On dead stems of *Cassia*, Houston, Texas (Ravenel).

L. subcæspitòsa, Cke. & Hark. Grev. XIV, p. 10.

Perithecia erumpent, subcespitose, black, crowded, 4–8 in a cluster or arranged in elongated strips. Ostiola cylindrical, elongated, emergent. Asci clavate, 8-spored. Sporidia lanceolate, constricted in the middle, 5-septate, brown, 35–38 x 7 μ.

On stems of *Geranium*, California (Harkness).

L. Ogilviénsis, (B. & Br.)

Sphæria Ogilviensis, B. & Br Brit. Fungi, No 642, tab XI, fig. 28.
Leptosphæria Ogilviensis, Ces. & De Not Schema, p. 61.
Exsicc. Kunze F. Sel. 332, 580 —Rab. F. E. 2115.—Rehm. Asc. 35, 534.—Thum M. U. 649. Ell. N. A. F. 585 —Lin. Fungi Hung. 163 —M. March. 746, 2537.

Perithecia mostly very numerous, thickly scattered over the stems, at first covered, or with only the papilliform ostiolum projecting, afterwards naked and free, depressed-hemispherical, with flattened, sessile base, black and shining, finally umbilicate. Asci cylindrical, sessile, 60–75 x 10–12 μ, with filiform paraphyses. Sporidia fasciculate, long-fusoid, obtuse at the ends, constricted in the middle, often with one half broader than the other, 5-septate, yellow-brown, 35–40 x 3½–4 μ.

On dead herbaceous stems, especially of the *Compositæ*, *Aster Erigeron*, *Bidens*, *Solidago*, Newfield, N. J.

L. ténera, Ell. Bull. Torr. Bot. Club, VIII, p. 124.

Perithecia gregarious, seated just beneath the cuticle which closely covers them, and is raised into little pustules and blackened, 200–300 μ diam., subglobose or a little depressed, strongly fringed around the base with brown, spreading hyphæ. Ostiolum papilliform, erumpent. Asci clavate-cylindrical, slender, 65–70 x 5–6 μ, paraphysate. Sporidia overlapping and subbiseriate above, fusoid, slightly curved,

5-septate, brown, 15–20 (mostly 15) x 2½ μ, scarcely constricted at the septa, and occasionally with one joint (near the upper end) swollen.

On dead herbaceous stems (apparently of some umbelliferous plant), Utah (S. J. Harkness).

L. planiúscula, (Riess.)

Sphæria planiuscula, Riess. Hedw. I, tab. IV, fig. 7.
Sphæria Virginica, C. & E. Grev. VIII, p. 16.
Leptosphæria planiuscula, Ces. & De Not, Schema, p 61.
Exsicc. Thum. M. U. 1847.—Myc. March 2372.—Rehm Asc. 685.—Ell. N. A. F. 579.—Rab. F. E. 946.

Perithecia scattered or gregarious, at first covered by the epidermis, then bare and free, depressed-spherical, 300–400 μ diam., black, smooth, shining, glabrous, with a small, flattened, perforated ostiolum. Asci clavate, 105–115 x 14–16 μ, stipitate, paraphysate. Sporidia biseriate, fusoid, straight, at first uniseptate and nucleate, becoming 5-septate and brown, 40–55 x 6–7 μ.

On dead stems of *Lepidium Virginicum*, Newfield, N. J., and on *L. intermedium*, Manhattan, Kansas.

The American spece. agree with the form on Cruciferous plants in Rehm's Asc. l. c. and can hardly be specifically distinct from the usual European form on *Solidago Virgaurea.*

L. comatélla, (C. & E.)

Sphæria comatella, C. & E. Grev. V, p. 52.
Leptosphæria Asparagi, Pk. 40th Rep. p. 70.
Leptosphæria comatella, Sacc. Syll. 2965.
Heptameria comatella, Cke. Syn. 4762.
Exsicc. Ell. N. A. F. 190.

Gregarious or scattered. Perithecia ovate, dark brown, covered. Ostiola erumpent, surrounded by short, erect, rigid, black hairs. Asci clavate, subsessile, paraphysate, 8-spored, 75–80 x 15 μ. Sporidia crowded-biseriate, abruptly fusiform, 5-septate, constricted at the septa, hyaline, at length brown, 20–25 x 7–8 μ (35–40 x 9 μ, Cke.).

On various dead herbaceous stems, *Asparagus*, *Verbascum*, *Gerardia*, *Desmodium*, *Baptisia*, *Phaseolus* (cult.), *Daucus*, *Tanacetum*, and *Aralia*, Newfield, N. J.

Var. *Lespedezæ*, E. & E., has the sporidia 6–7-septate, 35 x 8 μ, slightly curved, with one of the joints swollen. The bristles around the ostiolum are sometimes wanting.

L. Merténsiæ, (Ell.)

Sphæria Mertensiæ, Ell. Bull. Torr. Bot. Club, VIII, p. 90.
Leptosphæria Mertensiæ, Sacc. Syll. 2972.
Heptameria Mertensiæ, Cke. Syn. 4772.

Perithecia subcuticular, at length erumpent, about ⅓ mm. diam., black, glabrous, subspherical. Ostiolum papilliform. Asci clavate-cylindrical, 75–100 x 15–18 μ, paraphysate. Sporidia crowded, linear-lanceolate, yellowish, nucleate and uniseptate at first, finally 5- or more-septate, constricted and more or less swollen in the middle, 37–50 x 3½-5 μ.

On dead leaves of *Mertensia*, Utah (Jones). Closely allied to *L. Ogilviensis*, but sporidia rather larger, and perithecia not collapsing.

L. Lophánthi, (B. & C.)

Sphæria Lophanthi, B. & C. Grev. IV, p. 152.
Leptosphæria Lophanthi, Sacc. Syll. 3163.
Heptameria Lophanthi, Cke. Syn. 4953.

Perithecia minute, covered by the cuticle. Asci oblong. Sporidia fusiform, slightly curved, 6-septate, 50 μ long.

On *Lophanthus*, Pennsylvania (Michener).

The specc. distributed under this name in Roum. F. Gall. 3954, are some sterile, worthless thing and give no assistance.

L. mesœ̀dema, (B. & C.)

Sphæria mesœdema, B. & C. Grev. IV, p. 151.
Heptameria mesœdema, Sacc Syll. 3187, Cke. Syn. 4979.

Perithecia at length liberated, conical. Sporidia elongated, acuminate at each end, with about 8 septa, the central joint swelling, 52 μ long.

On *Eupatorium coronopifolium* and *Cirsium muticum*, South Carolina.

L. Artemísiæ, (Fckl.)

Sphæria Artemisiæ, Fckl. F. Rh. 896.
Pleospora helminthospora, Fckl. Symb. p. 138.
Leptosphæria Artemisiæ, Awd. in Herb. and in Hedw. 1878, p. 46.
Exsicc. Rab. F. E 1725.

Perithecia somewhat scattered, covered, not erumpent, obtusely globulose, and papillate. Asci elongated-clavate, 120–130 x 16–18 μ, paraphysate, 8-spored. Sporidia subbiseriate, fusoid, straight or curved, 35–38 x 6–7 μ, yellowish-olive, 5- (rarely 7-) septate, slightly constricted at the septa.

On dead stems of *Artemisia cana*, Helena, Montana (Kelsey).

The pycnidial stage, *Hendersonia Artemisiæ*, Sacc., was not observed. This has erumpent-superficial perithecia of medium size with cylindrical ostiola as long as the diameter of the perithecia, and fusiform sporules 40 x 8 μ, 5–6-septate, yellowish and slightly constricted at the septa. In the Montana specc. the sporidia are 20–26 x

8–11 μ, 3–5-septate, cylindric-oblong, agreeing with specc. from Dr. Winter.

L. ágnita, (Desm.)

Sphæria agnita, Desm. Ann. Sci. Nat. III, Ser. 16 vol. p 313.
Leptosphæria agnita, Ces. & De Not. Schema, p. 62.
Exsicc. Desm. Pl. Cryp. Ed. I, 713.—Cke. F. Br. 1st Ser. 277, 2d do. 255 —Rab. F. E. 2042. Rehm Asc. 144.—Thum. M. U. 1257.

Perithecia scattered, erumpent, becoming free, base applanate, black and subshining, about ⅓ mm. diam., seated on gray, effused spots. Ostiolum papilliform, pierced, sometimes short-cylindrical. Asci subclavate-cylindrical, narrowed slightly above and more so below, 8-spored, 120–125 x 8 μ. Sporidia inordinate, elongated-fusoid, strongly constricted in the middle, 8-celled (7-septate), the fourth cell from the upper end generally somewhat swollen, yellow-brown, 30–35 x 3½ μ. Paraphyses filiform.

On dead stems of *Impatiens*, West Chester, Pa. (Everhart), on dry stems of *Hieracium vulgatum*, at various localities in Greenland (Rostrup).

L. acùta, (M. & N.)

Sphæria acuta, M. & N. Stirp. Crypt. Vosges, No. 181.
Pleospora acuta, Fckl. Symb. p. 135.
Leptosphæria acuta, Karst. Mycol. Fenn. II, p. 98.
Sphæria coniformis, Fr. Syst. Myc. II, p. 508, sec. Karst.
Exsicc. Fckl F. Rh. 900.—Rab. Herb. Mycol 44, 729.—Thum. M. U. 1287.—M. March 45. Plowr. Sph. Brit. 89.—Cke. Fungi Brit. Ser. I, 265, Ser. II, 254.—Rehm Asc. 783. Desm. Pl. Crypt. Ed. I, 36.

Perithecia scattered, at first covered by the epidermis, afterwards bare and free, conical, running up into a thick, obtuse, perforated ostiolum, black, glabrous, shining, about 350 μ diam. Asci cylindrical, elongated, 8-spored, 130–140 x 10 μ. Sporidia biseriate, fusoid, straight or a little curved, subinequilateral, 6–10-septate, yellowish, 36–50 x 5–6 μ. Paraphyses slender, filiform.

On dead herbaceous stems, Carolina and Virginia (Berk. in Grev.). Probably not uncommon.

L. torulæspora, (Cke.)

Sphæria torulæspora, Cke. Texas Fungi, 136.
Leptosphæria torulispora, Sacc. Syll. 3000.
Heptameria torulispora, Cke. Syn. 4804.

Perithecia gregarious, erumpent-superficial, conic-hemispherical, about ½ mm. diam., glabrous, black, flattened at base, ostiolum conic-papilliform, distinct. Asci cylindrical, sessile or nearly so, 75–80 x 12 μ, with abundant paraphyses. Sporidia fasciculate, fusoid-cylindrical, brown, multiseptate, at length slightly constricted at the septa, nearly as long as the asci, and 3–3½ μ thick.

On dead herbaceous stems, Texas (Ravenel).

The foregoing diagnosis is from specc. in our Herb. from Ravenel.

L. clavígera, (C. & E.)

Sphæria clavigera, C. & E. Grev. VI, p. 16.
Leptosphæria clavigera, Sacc. Syll. 3008.
Heptameria, clavigera, Cke. Syn. 4812.

Perithecia scattered or subgregarious, covered, black, subprominent, about ⅓ mm. diam. Asci clavate, 90–100 x 12–15 μ, paraphysate and stipitate. Sporidia crowded, clavate-fusoid, slightly curved, reddish-brown, 7–9-septate, one joint (near the middle) sometimes swollen, 25–35 x 6–8 μ (40 x 8 μ, Cooke).

On dead stems of *Phytolacca*, Newfield, N. J., and on dead herbaceous stems, Utah (S. J. Harkness).

In the New Jersey specc. of this species, the sporidia remain for a long time hyaline and uniseptate, only a few being seen with 4–5 septa and pale reddish-brown, but the Utah specc. agree well with the diagnosis in Grevillea.

L. rubrotíncta, E. & E. Journ. Mycol. I, p. 150.

Perithecia scattered, membranaceous, depressed-globose, 200–250 μ diam., perforated above, covered by the slightly elevated epidermis, which is slightly ruptured, and stained pale blood-red. Asci clavate-cylindrical, sessile, 100–110 x 10–13 μ, sessile and accompanied by paraphyses. Sporidia biseriate, cylindrical, straight or slightly curved, deep straw-yellow, 8–10-septate with one joint (a little below the middle) slightly swollen, 25–40 x 4–5 μ, ends subobtuse. Distinguished by its sporidia from all other red-tinged species.

On dead twigs of *Staphylea trifolia*, West Chester, Pa. (Everhart.)

L. consímilis, E. & E. Journ. Mycol. II, p. 41.

Perithecia scattered, carbonaceo-coriaceous, ½ mm. diam., at first covered and raising the fibers of the wood or bark into little pustules. Ostiola papilliform, soon erumpent. Asci subcylindrical, 80–100 x 15–20 μ, with abundant paraphyses. Sporidia 8 in an ascus, subbiseriate, cylindric-fusiform, 3-septate and subhyaline at first, becoming yellow, and finally dark brown, and about 7-septate, more or less constricted at the septa, 28–35 x 8–10 μ. Apparently allied to *Sphæria Baggei*, Auersw., which is said to have greenish-yellow, 3–5-septate sporidia. In the Dakota specimens, the mature sporidia are quite constantly 7-

septate, exceptionally 8-septate. The perithecia occur both on decorticated limbs and on those still covered with the bark, but in the latter case, the bark is old and somewhat decayed.

On dead willow limbs, near Huron, Dakota, autumn, 1885 (Miss Nellie E. Crouch).

L. Maclùræ, E. & E. Proc. Acad. Nat. Sci. Phil. July, 1890, p. 236.

Spots as in *Sphærella Macluræ*, E. & E. (which occurs on the same leaves), suborbicular, reddish-brown, 4–10 mm. diam., with a darker margin and deciduous center. Perithecia mostly hypophyllous, innate-erumpent, small (75 μ), black. Asci oblong-cylindrical, 50–60 x 8–10 μ. Paraphyses? Sporidia biseriate, fusoid, about 6-nucleate, becoming 5-septate, slightly curved, nearly hyaline, 20–22 x 3 μ.

On leaves of *Maclura aurantiaca*, Saline Co., Mo. (Demetrio).

L. puteàna, E. & K. (in Herb.)

Perithecia gregarious, erumpent-superficial, subglobose or depressed-globose, about 200 μ diam., black, glabrous, broadly perforated above, rough. Asci subcylindrical, short-stipitate, rounded above, about 75 x 15–18 μ, paraphysate. Sporidia crowded, sometimes oblique, oblong-fusoid, 6–8-septate, sometimes a little curved, dark brown, 20–27 x 6–8 μ.

On the inside and outside of an oaken well-bucket, Manhattan, Kansas (May Varney).

B. *On monocotyledonous plants.*

* *Sporidia 3-septate.*

L. sabalígera, (B. & C.)

Sphæria sabaligera, B. & C. Grev. IV, p. 147.
Clypeosphæria sabaligera, Sacc. Syll. 3193.
Leptosphæria sabaligera, Cke. Syn. 4310.

Perithecia scattered, minute, covered by the blackened cuticle which is slightly raised. Asci (p. sp.) oblong, 40–50 x 7–9 μ. Sporidia biseriate, fusiform, slightly curved, 3-septate, straw-yellow, sometimes constricted at the septa, 15–22 x 4–5 μ (25 μ long, sec. Berk.).

On leaves of *Sabal*, Louisiana (Langlois).

L. sabalícola, (E. & M.)

Sphæria sabalicola, E. & M. Am. Nat. 1882, p. 810.
Leptosphæria sabalicola, Sacc. Syll. 6135.
Exsicc. Ell. & Evrht. N. A. F. 2d Ser. 1963.

Perithecia gregarious or scattered, with rather thick, coriaceous walls, about ⅓ mm. diam., covered by the epidermis, which is raised into little obtusely conical projections, around which the surface of the matrix is of a tawny color. Asci clavate-cylindrical, 50–60 x 7–8 μ. Sporidia biseriate, oblong, 3-septate and slightly constricted at the septa, brown, 11–15 x 3–3½ μ, ends obtuse.

On petioles of *Sabal serrulata*, Florida (Martin and Calkins).

This is closely allied to *L. sabaligera*, B. & C., but (sec. Louisiana specc. from Langlois), differs in its rather larger, less prominent perithecia and shorter, darker, constricted and obtuse sporidia.

L. gigáspora, Niessl. in Rab. F. E. 2998.

Perithecia somewhat scattered, globulose, sunk in the parenchyma of the leaf, ½ mm. diam., with a broad, flattened, scarcely erumpent ostiolum, black. Asci broad cylindric-clavate, sessile, rounded at the apex, 8-spored, 144–150 x 22–24 μ thick. Sporidia 2–3-seriate, elongated-fusoid, at first hyaline and 2-celled, finally 3-septate and somewhat constricted at the septa, honey-yellow, 50–55 x 10–12 μ.

On leaves of *Carex microglochin*, Igaliko, Greenland.

L. culmòrum, Awd. Verzeichn. des Leipz Tausch Vereins (1866), p. 4.

Leptosphæria microscopica, Karst. Fungi in Ins. Spetsb. collecti, in Oefvers, af Vetensk. Akad. Forhandl (1872), p 102.

Exsicc Rab. F E 1553, 54, 2759.—Rehm Asc 240.—Thum F. Austr. 1254.—id. M. U. 761

Perithecia scattered, buried except the very small papilliform, projecting ostiolum, finally more or less erumpent, black, glabrous, about 100 μ diam. Asci elongated-clavate, very short-stipitate, 8-spored, 60–80 x 12–16 μ, with scanty, filiform paraphyses. Sporidia 2–3-seriate, oblong-fusoid, subinequilateral or slightly curved, 3-septate, the penultimate cell often a little enlarged, honey-yellow, 17–22 (or even 27) x 6–9 μ.

On *Luzula arcuata*, *spicata*, and *multiflora*, *Poa glauca*, and *flexuosa*, and on *Alopecurus alpinus*, in western Greenland.

L. júncina, (Awd.)

Sphæria juncina, Awd. in Rab. F. E. 748.
Sphærella juncina, Awd. Myc. Eur. p. 18, tab. VI, fig 74.
Heptameria juncina, Cke. Syn. 4865.
Leptosphæria juncina, Sacc. Syll. 3094.

Perithecia very numerous, thickly scattered, covered by the epidermis, globose, pierced above, black, 50–60 μ diam. Asci elongated-ovate, very short-stipitate, 8-spored, 48 x 10–12 μ. Sporidia 3–4-seriate, fusoid, subobtuse, slightly curved, 3-septate, not constricted, brownish, 24 x 3–4 μ, often surrounded by a gelatinous envelope.

On *Juncus biglumis*, Disco, Greenland.

L. latebròsa, (Ell.)

Sphæria latebrosa, Ell. Bull. Torr. Bot. Club, VIII, p. 90.
Leptosphæria latebrosa, Sacc. Syll. 3176.

Perithecia gregarious, minute, subelliptical, 150 x 200 μ, subcuticular, but prominent and covered by the blackened cuticle. Ostiolum papilliform, erumpent. Asci clavate-cylindrical, 60–70 x 8 μ, sessile. Paraphyses filiform, interwoven, longer than the asci. Sporidia crowded, fusiform, slightly curved, 3-septate, pale brown, subconstricted at the middle septum, 20–25 x 3 μ.

On basal sheaths of old *Andropogon*, Newfield, N. J.

L. epicarécta, (Cke.)

Sphæria epicarecta, Cke. Grev. V, p. 120.
Leptosphæria epicarecta, Sacc. Syll. 3090.
Heptameria epicarecta, Cke. Syn. 4859

Perithecia scattered, covered by the cuticle, which appears darker over them. Asci clavate. Sporidia biseriate, broadly lanceolate, 3-septate, yellow, second cell rather larger, 30 x 10 μ. Perithecia when dry, scarcely visible.

On leaves of *Carex pulla*, Egedesminde, Greenland.

In the Greenland specimens the asci were 66–70 x 20–22 μ; sporidia 28–30 x 9–10 μ.

L. orthográmma, (B. & C.)

Sphæria orthogramma, B. & C. Grev. IV, p. 144.
Leptosphæria orthogramma, Sacc. Syll. 3071.
Heptameria orthogramma, Cke. Syn. 4839.
Exsicc. Ell. N. A. F. 195.

Forming parallel black lines, surrounded on each side by the cuticle, and 1 mm. or more long. Perithecia small, with only the apex erumpent. Asci clavate, about 80 x 10–12 μ, paraphysate. Sporidia biseriate, oblong-fusoid, yellow, 3–5-septate, with the middle joint swollen, 25–35 x 7 μ.

On *Zea* and *Erianthus*, Pennsylvania, New Jersey and Carolina.

L. sorgóphila, (Pk.)

Sphæria sorgophila, Pk. 31st Rep. p. 51.
Leptosphæria sorgophila, Sacc. Syll. 3072.
Heptameria sorgophila, Cke. Syn. 4840.

Perithecia very minute, immersed, erumpent through a longitudinal chink, elliptical, black. Asci elongated-clavate. Sporidia biseriate, oblong-cylindrical, triseptate, constricted at the septa, pale when young, then colored, 27–30 μ long.

On brush of an old broom-corn broom, Greenbush, N. Y.

The ostiola are so obscure that they can with difficulty be seen.

L. Michóttii, (West.)

Sphæria Michottii, West. Bull Soc. roy. bot. Belg. II. Ser. tom. VII, No. 52.
Sphærella Michottii, Awd Myc. Eur. p. 18, tab. VI, fig. 75.
Leptosphæria trimera, Sacc. F. Ven. Ser. II, p. 319.
Leptosphæria Michottii, Sacc. F. Ital. tab. 279.
Exsicc. Plowr. Sph. Brit. 267, 268.—Rav. F. Am. 750.

Perithecia scattered or gregarious, at first covered, afterwards erumpent, depressed-spherical, with a short, sharp, conical ostiolum, black, 100–120 μ diam. Asci cylindric-clavate, rounded at the apex, with a short, thick stipe, 8-spored, 50–55 x 12–15 μ, with scanty, filiform paraphyses. Sporidia biseriate, oblong-cylindrical, straight, rounded at the ends, biseptate and constricted at the septa, olive-brown, 15–18 x 3½–4½ μ, cells mostly nucleate.

On dead stems of *Juncus*, New England (Blake), on *Carex*, New Jersey and Georgia.

L. arundinàcea, (Sow.)

Sphæria arundinacea, Sow. Eng. Fung. tab. 336.
Sphæria Godini, Desm. Ann. Sci. Nat. III. Ser tom. V, p 49.
Leptosphæria Godini, Awd. Tausch-verein, 1866, p. 4.
Pleospora arundinacea, Fckl. Symb. p 137.
Melogramma arundinacea, Niessl. in Rab. F. E. 1840.
Leptosphæria arundinacea, Sacc. Syll. 3081, F. Ven. Ser. II, p. 320.
Heptameria arundinacea, Cke. Syn. 4849.
Exsicc. Rab. F. E. 1840.—Rehm Asc. 193.—Thum. M. U. 1256.—M. March. 259.

Erumpent, linear, black, with scarcely any stroma. Perithecia in one or two series, connate, minutely papillate, black inside. Asci elongated-clavate, with branching paraphyses, 8-spored, 80–88 x 10–12 μ. Sporidia subbiseriate, fusoid, 3-septate, mostly straight, 24–30 x 6 μ, at first hyaline, then honey-colored. Spermatia, in perithecia like the ascigerous perithecia, oblong, uniseptate, hyaline. Pycnidial perithecia the same, stylospores fusiform, curved, 3–5-septate, cells nucleate, yellow, 26–56 x 6 μ.

On *Arundinaria*, South Carolina.

L. hysterioìdes, E. & E. Bull. Wash. Coll. 1884, p. 4.

Perithecia globose, 165 μ diam., covered by the blackened cuticle which is pierced by the obtusely-conical ostiolum. Often several perithecia are confluent in the direction of the longer axis of the leaf so as to resemble closely some minute *Hysterium* and this resemblance is increased when, as often happens, the epidermis splits along the line of the ostiola. Asci oblong or oblong-cylindrical, 85–112 x 12-15 μ:

paraphyses matted together at their tips. Sporidia biseriate, oblong-fusiform, hyaline, 4-nucleate, becoming 3-septate and brown, 20–24 x 7–9 μ, constricted at the septa, ends obtuse.

On dead leaves of *Xerophyllum tenax*, Mt. Paddo, Washington (Suksdorf).

L. Typhàrum, (Desm.)

Sphæria scirpicola, var. *Typharum*, Desm. Cryp. de France, Ed. II, No 1428
Pleospora Typharum, Fckl. Symb p. 137.
Leptosphæria Typharum, Karst. Myc. Fenn. II, p. 100.
Exsicc. Desm. Pl. Cr. 1 c.—Fckl. F. Rh. 858—Kunze F. Sel. 256.—Rab. Herb. Mycol. 731.—id. F. E. 1040, 1448, 2552.—Rehm Asc. 142.—Thum. M. U. 352.—Myc. March. 2350.

Perithecia scattered, buried, globose or elliptical, with the broad-conoid apex projecting, black, glabrous, 160–200 μ in the longer diameter. Asci oblong, short-stipitate, 8-spored, 60–80 x 16–20 μ, paraphysate. Sporidia 2–3-seriate, oblong, obtuse, 3-septate, scarcely constricted, slightly curved, clear yellow, 21–24 x 7–9 μ.

On dead leaves and culms of *Typha latifolia*, Montana (Kelsey), Delaware (Commons).

L. filamentòsa, E. & E. Journ. Mycol. IV, p. 76.

Perithecia immersed, scattered, small (200–225 μ), depressed-globose, the upper part slightly raising and barely rupturing the cuticle. Ostiolum not prominent. Asci subcylindrical, 75–80 x 7–8 μ, with abundant paraphyses. Sporidia biseriate, oblong-cylindrical, 3-septate, yellow, constricted at the septa, not curved, 12–15 x 4–5 μ, ends obtuse. The spermogonial stage is a *Coniothyrium* (*C. concentricum*)? with small (4 μ), subglobose, brown sporules. The parts of the leaf (mostly the sides or tips) occupied by the ascigerous stage of this species are quite dead, and soon become decayed and brittle.

On dead places in living leaves of *Yucca filamentosa*, Newfield, N. J.

L. marìna, E. & E. Journ. Mycol. I, p. 43.

Perithecia irregularly scattered, subelliptical, ($\frac{1}{2}$–$\frac{3}{4}$ mm.), upper part hemispheric-conical, projecting, closely covered by the blackened epidermis, apex subtruncate with a papilliform ostiolum, which is soon deciduous, leaving a broad opening. Asci 112–150 x 25–35 μ, narrowed into a substipitate base. Sporidia 2–3-seriate, fusiform or clavate-fusiform, yellowish, 1–3-septate (mostly uniseptate), slightly constricted at the middle septum, 50–70 x 10–12 μ, ends subobtuse. In this case, the greater number of septa does not seem to indicate maturity. The perithecia are more prominent and the sporidia much longer than in *L. discors*, S. & E.

On dead culms of *Spartina* lying on the beach at Cape May, N. J. (Mrs. Caroline Treat).

L. díscors, Sacc. & Ell. Mich. II, p. 567.

Metasphæria discors, S. & E. Sacc. Syll. 3473.
Exsicc. Ell. N. A. F. 1341.

Perithecia loosely gregarious, innate, globose-papillate, black, the ostiola piercing the blackened epidermis. Asci cylindric-clavate, 120 x 20 μ, short-stipitate; paraphyses filiform, generally dichotomous. Sporidia obliquely uniseriate or biseriate, oblong-fusoid, ends rounded, 25–30 x 9–10 μ, 3-septate, constricted at the septa, 4-nucleate, the two inner cells darker, the end cells subhyaline.

On culms of *Spartina*, Cape May, N. J.

L. álgida, Rostr. Fungi Grönl. p. 560 (No. 175).

Perithecia globulose, scattered, 160 μ diam. Asci cylindric-clavate, somewhat curved, 50–54 x 12 μ, without paraphyses. Sporidia obliquely uniseriate, oblong, often somewhat curved, yellow, 3-septate. 16–20 x 6–7 μ.

On leaves of *Catabrosa algida*, Maneetsok, Greenland.

L. Xerophýlli, Ell. Am. Nat. 1883, p. 316.

Exsicc. Ell. N. A. F. 1340.

Perithecia scattered, subglobose, 150–190 μ diam., semiimmersed. Ostiolum obtuse with a rather large opening, elevating and splitting the epidermis, by which it is partly covered. Asci oblong or oblong-clavate, sessile, 55–60 x 15 μ, with obscure paraphyses. Sporidia biseriate, broad-fusoid, slightly curved, hyaline at first with 4 large nuclei, becoming 3-septate, slightly constricted at the septa, 19–25 x 5–6½ μ. Doubtfully distinct from *L. hysterioides*.

On dead leaves of *Xerophyllum asphodeloides*, Willow Grove, N. J.

Accompanied by *Hendersonia Xerophylli*, Ell. (Torr. Bull. IX, p. 74), and a *Pestalozzia*. On the same leaves are also superficial perithecia containing oblong-elliptical, brownish spores 4 x 2 μ.

L. phormícola, Cke. & Hark. Grev. XIV, p. 10.

Perithecia gregarious, punctiform, erumpent, emergent through cracks in the epidermis, black, subconvex, papillate. Asci clavate, 8-spored. Sporidia sublanceolate, rounded at the ends, biseriate, triseptate, scarcely constricted, pale brown, 22 x 6 μ.

On *Phormium*, California (Harkness).

L. folliculàta, E. & E. Proc. Phil. Acad. July, 1890, p. 237.

On pale white, elliptical spots 2–4 x 1–1½ mm. Perithecia buried in the substance of the leaf with their apices slightly prominent. few on a spot (1–6), small 60–75 μ diam. Asci clavate-cylindrical, 50 x 10–12 μ. Sporidia biseriate, oblong or clavate-oblong, 2-septate and slightly constricted at the septa, 12–15 x 3 μ, yellowish-brown, ends obtuse. Differs from the other species on *Carex* in its distinct spots and smaller sporidia.

On leaves of *Carex folliculata*, London, Canada (Dearness).

L. Corallorhìzæ, Pk. 38th Rep. p. 105.

Perithecia scattered, subcuticular, 112–125 μ diam, pierced above, raising the epidermis in a papilliform manner so that the stem is slightly roughened. Asci abundant, oblong-cylindrical, 75–80 x 15 μ, paraphysate. Sporidia biseriate, fusoid, curved, subhyaline, nucleate, then uniseptate. becoming 3-septate and constricted at the middle septum, 24–27 x 4–5 μ.

On dead stems of *Corallorhiza*, Caroga, N. Y. (Peck), London, Canada (Dearness).

The diagnosis is drawn from .the Canada specimens, which were not fully mature, but there seems no doubt that they represent the species described by Peck.

** *Sporidia 5-septate.*

L. incarceràta, (B. & C.)

Sphæria incarcerata, B. & C. Grev. IV, p. 152.
Leptosphæria incarcerata, Sacc. Syll. 3174
Heptameria incarcerata, Cke. Syn. 4968.

Perithecia under a little clouded speck of the cuticle opening by a little longitudinal fissure. Sporidia torulose, straight, 5-septate, 50 μ long.

On *Spartina*, South Carolina.

This cannot differ much from *L. albopunctata* (West).

L. Stràminis, Cke. & Hark. Grev. XIV, p. 10.

Culmicolous. Perithecia scattered or subgregarious on dark-colored, effused spots, erumpent, black, convex, perforated. Asci cylindric-clavate, 8-spored. Sporidia lanceolate, 5-septate, slightly constricted, brown, the two central cells longitudinally divided, 32–35 x 6–8 μ.

On straw, California.

L. Spartinæ, E. & E. Journ. Mycol. I, p. 43.

Perithecia gregarious (about 1 mm.), covered by the blackened cuticle which is raised into little prominences over them and pierced by the papilliform and at length broadly perforated ostiolum. Asci 115–120 x 12–15 μ, surrounded by abundant paraphyses, and containing eight two-ranked, broadly fusiform, pale yellowish, 5-septate, 35–45 x 8–10 μ sporidia.

On culms of *Spartina* decaying on the beach, Cape May and Atlantic City, N. J.

This must be very near *L. incarcerata*, B. & C., but in that species the perithecia are said to lie under "a little clouded speck in the cuticle opening by a little longitudinal fissure," while in *L. Spartinæ* the surface of the culm is uniformly blackened and raised into little pustules by the subjacent perithecia, and these pustules are not longitudinally cleft, but merely perforated by the papilliform ostiola.

L. albopunctàta, (West.)

Sphæria albopunctata, West in Kickx, Fl. Crypt. Fl I, p. 355.
Leptosphæria sticta, E. & E. Journ. Mycol. I, p. 43.
Leptosphæria albopunctata, Sacc. Syll 3116.
Heptameria albopunctata, Cke. Syn. 4892.
Exsicc. Ell. & Evrht. N. A. F. 2d Ser. 2616.

Perithecia scattered, subovate, membranaceous, buried under the blackened epidermis, which is pierced but not raised by the punctiform ostiola. Asci 100–125 x 15–20 μ, with abundant paraphyses. Sporidia 1 or 2-seriate, fusiform, pale yellow, 5-septate, very slightly constricted at the septa, ends obtuse, 30-40 x 7–9 μ. Distinguished by its punctiform ostiola, which are the only outward indication of the buried perithecia. The fruit is scarcely different from that of the preceeding species. The part of the culm occupied by the perithecia is uniformly blackened.

On decaying culms of *Spartina*, Cape May, N. J., and on culms of *Juncus maritimus*, Ocean Springs, Miss. (Tracy).

We have compared *L. sticta*, E. & E. with an authentic spec. of *L. albopunctata*, West., sent us by Prof. Oudemans, and it is undoubtedly the same.

L. culmícola, (Fr.)

Sphæria culmicola, Fr. S. M. II, p. 440.
Leptosphæria culmicola, Awd. Tausch-Verein, 1866, p. 4.
Exsicc. Rehm. Asc. 143.—Thum. M. U. 457.

Perithecia scattered or subgregarious, sunk in the substance of the leaf with the papilliform ostiolum erumpent, globose or subelliptical, black, with branched, creeping mycelial threads around the base, 200–250 μ diam. Asci clavate, substipitate, 8-spored, 70–80 x 8–9 μ, with filiform paraphyses. Sporidia fusoid, subinequilateral and often slightly curved, 5-septate, the second cell from the upper one thicker than the others, honey-yellow, 19–22 x 4–5 μ.

On culms of *Phragmites*, Iowa (Arthur), on *Andropogon muricatus*, Louisiana (Langlois), on culms of some grass, Long Island, N. Y. (Zabriskie).

L. Nárdi, (Fr.)

Sphæria Nardi, Fr. S. M. II, p. 520.
Leptosphæria Nardi, Ces. & De Not. Schema, p. 62.
Pleospora Nardi, Fckl. Symb. p. 137.
Exsicc. Rab. Herb. Myc. 640.—Fckl. F. Rh. 855.

Perithecia scattered, buried, finally erumpent, globose, black, flat and perforated above, 170–190 μ diam. Asci cylindrical, subattenuated below, sessile, 8-spored, 70–80 x 10½–12 μ, with filiform paraphyses. Sporidia biseriate, fusoid, subinequilateral or slightly curved, 5-septate, the third cell mostly a little swollen, yellow, 23–25 x 4–5 μ

On *Nardus stricta*, Nanortalik, Greenland.

L. heloniæfòlia, (C. & E.)

Sphæria heloniæfolia, C. & E. Grev. VII, p 16
Leptosphæria heloniæfolia, Sacc. Syll. 3102.

This is a stylosporous fungus (*Hendersonia*) and not ascigerous, sporules 1–2-septate, brown, 20 x 8 μ.

L. Rousseliàna, (Desm.)

Sphæria Rousseliana, Desm. XVII, Not. in Ann. Sci Nat. III. Ser tom 11, p. 355.
Leptosphæria Rousseliana, De Not. Schema, p. 236.
Exsicc. Rab. F. E. 2048.—Desm. Pl. Crypt. Ed. I, 2082.—id. Ed. 2d, 1782.

Perithecia minute, scattered or subapproximated, globose, ¼ mm. diam., immersed, covered by the blackened cuticle, gray inside. Ostiolum erumpent, punctiform. Asci elongated, subclavate, 80–100 μ long. Sporidia subfusiform, 20–25 μ long, ends obtuse, straight or curved, olivaceous, 4–5-septate.

On leaves of *Colpodium latifolium*, Greenland.

*** *Sporidia* 6–12-*septate*.

L. Sporoboli, Ell. & **Galw.** Journ. Mycol. V, p. 66.

Perithecia scattered, erumpent-superficial, subhemispherical, glabrous, black, 200–250 μ diam., with a short, thick, nipple-like ostiolum. Asci clavate-cylindrical, 75–80 x 16–18 μ, with abundant paraphyses. Sporidia crowded-biseriate, overlapping each other, oblong-fusoid, sub-obtuse, 6-septate, and not at all or finally slightly constricted at the septa, about 22 x 7 μ, straight or nearly so.

On dead culms of *Sporobolus depauperatus*, Montana (Anderson).

Differs from *L. culmifraga* in its more scattered mode of growth and quite constantly only 6-septate sporidia, and from *L. culmicola* in its superficial growth and broader sporidia without any protuberant cell.

L. scapóphila, (Pk.)

Sphæria scapophila, Pk. 30th Rep. p. 66.
Leptosphæria scapophila, Sacc Syll 3007.
Heptameria scapophila, Cke. Syn 4811.

Perithecia minute, subglobose, scattered, covered by the epidermis which is ruptured by the minute, perforated ostiola. Asci cylindrical. Sporidia crowded or biseriate, subcylindrical, yellowish, 25–30 μ long, 7-septate, one apical and three basal cells longer than the others.

On dead scapes of *Sarracenia purpurea*, Adirondack Mts., N. Y. (Peck), Minnesota (McMillan).

L. Beaumóntii, (B. & C.)

Sphæria Beaumontii, B. & C. Grev. IV, p 145
Leptosphæria Beaumontii, Sacc. Syll 3179.
Heptameria Beaumontii, Cke. Syn. 4973.

Forming little short, black lines, bursting through the cuticle. Asci elongated, clavate. Sporidia linear, sometimes oblique, with about 9 septa and a nucleus in each joint, 50 μ long.

On stalks of some grass, Alabama.

L. ceratíspora. (B. & C.)

Sphæria ceratispora, B. & C. Grev. IV, p. 150.
Leptosphæria ceratispora, Sacc. Syll. 3178.
Heptameria ceratispora, Cke Syn. 4972.

Perithecia gregarious, conical. Sporidia long, curved, acuminate at both ends, with 12 or more septa, projecting at the septa like the horns of some antelope.

On some thick, herbaceous stem, possibly *Zea*, South Carolina.

L. interspérsa, (Cke.)

Sphæria interspersa, Cke. Grev. VI, p. 146.
Leptosphæria interspersa, Sacc. Syll. 3134.

Scattered. Perithecia globose, subprominent, brown, with papilliform ostiola. Asci clavate-cylindrical. Sporidia biseriate, fusiform, 7-septate, bright brown, at first nucleate.

On husks of *Zea Mays*, Gainsville, Fla.

No measurements are given.

L. zizaniæcola, (B. & C.)

Sphæria zizaniæcola, B. & C. Grev. IV, p. 145.
Sphæria rimosa, Schw. (Herb.)
Leptosphæria zizaniæcola, Sacc. Syll. 3175
Heptameria zizaniæcola, Cke. Syn. 4969.

Forming little swollen patches, slightly discoloring the cuticle. Asci elliptical, containing four linear, straight or sigmoid, 5-6-septate sporidia 33–50 μ long.

On *Zizania*, (Carolina)?

The perithecia are entirely buried. We have seen no authentic specimens.

L. culmífraga, (Fr.)

Sphæria culmifraga, Fr. S. M. II, p. 510.
Leptosphæria culmifraga, Ces. & De Not. Schema, p. 61.
Pleospora culmifraga, Fckl. Symb. p. 137.
Exsicc. Fckl. F. Rh. 2245—Rehm Asc. 195, 700, 784, 785.—Rab. F. E. 1552.—Lin. Fungi, Hung. 74, 366.—M. March. 2354, 3034.

Perithecia scattered or seriate, at first covered, afterwards erumpent through cracks in the epidermis, subcompressed, mostly elliptical, sometimes spherical, with a short-conical ostiolum, black, glabrous or with a few brown, mycelial threads around the base, 300–400 x 200-250 μ. Asci clavate, 8-spored, 80–100 μ long, 12–14 μ broad, paraphysate. Sporidia 2–3-seriate above, fusoid-elongated, mostly curved, 7–9-septate, the third cell generally a little swollen, slightly constricted at the septa, yellow, 35–46 x 5–7 μ.

On *Panicum*, New England (Berk. in Grev.), on culms of *Phleum pratense*, New Jersey, on stems of grass, New York (Peck), and on *Panicum Curtisii* (culms), Louisiana (Langlois).

C. *On cryptogamous plants.*

L. Marcyénsis, (Pk.)

Sphæria Marcyensis, Pk 31st Rep. p. 51.
Leptosphæria Marcyensis, Sacc. Syll. 3143.
Heptameria Marcyensis, Cke. Syn. 4930.

Perithecia minute, punctiform, covered by the epidermis, which is ruptured by the distinct, slightly prominent, blunt ostiola. Asci oblong-cylindrical, sessile. Sporidia crowded, subfusiform, blunt, slightly colored, 3-septate, 25–28 x 7½ μ, the cells generally nucleate.

On leaves of *Lycopodium annotinum* and *L. Selago*, Mt. Marcy, New York.

L. lycopodiicola, (Pk.) 38th Rep. p. 105.

Leptosphæria lycopodiicola, Sacc. Syll 6690
Heptameria lycopodiicola, Cke. Syn. 4928.

Perithecia small (125–150 μ), sphæroid or elliptical, erumpent, black. Asci subcylindrical, nearly sessile, 55–75 x 7–10 μ. Sporidia oblong or subfusiform, slightly colored, 3–5-septate, 20–25 x 4½–5 μ.

On dead peduncles of *Lycopodium clavatum*, Adirondack Mts., New York.

The perithecia are associated with a minutely tufted *Cladosporium*. Some of them are laterally compressed. The sporidia are much narrower than in *L. Crepini* and *L. Marcyensis*.

L. Crépini, (West).

Sphæria Crepini, West, in Bull. de la Soc. de botan. de Belg. II, tab. VII.
Leptosphæria Crepini, De Not. Nuov. Recl. Piren. Ital. p. 10
Exsicc. Rab F. E. 444 —Thum M. U 354.

Perithecia mostly hypophyllous, gregarious, buried, covered by the epidermis, globose or subdepressed, broadly perforated above, black, 100–150 μ diam. Asci cylindrical or elongated-subclavate, sessile, 8-spored, 70–75 x 13–14 μ. Sporidia 2- or partly 3-seriate, oblong, obtuse, slightly curved, 3-septate, yellowish, 20–26 x 7–10 μ. Paraphyses filiform.

On spikes of *Lycopodium annotinum*, Mt. Marcy, N. Y. (Peck), Igaliko, &c., in Greenland (Rostrup).

L. polàris, Sacc. Syll. II, p. 83, (Sphæria No. 8, Th. Fries in Hedw. 1881, p. 59).

Perithecia small, immersed, black. Asci cylindric-clavate, with abundant branching paraphyses, 4-spored. Sporidia uniseriate, oblong, obtuse, 3-septate, somewhat constricted at the septa, especially the middle one, cells 1–2-nucleate, obscurely clouded, 28–32 x 9–10 μ.

On the lichen *Rhizocarpum geographicum*, Greenland.

METASPHÆRIA, Sacc. Syll. II, p. 156.

We have reluctantly accepted this as a genus distinct from *Leptosphæria*, from which it differs only in its permanently hyaline sporidia. Practically it makes little difference whether these hyaline-spored species are recognized under a separate generic name, or considered a section or subgenus of *Leptosphæria*. In either case they will be separated from the brown-spored species, so that it seems more convenient to give them a separate generic name. It may be noted that quite a number of the species grow on limbs, wood or bark of trees or shrubs.

A. *On dicotyledonous plants.*

* *Sporidia* 3- (*rarely* 4-) *septate.*

M. sepíncola, (Fr.)

Sphæria sepincola, Fr. S. M. II, p. 498.
Metasphæria sepincola, Sacc. Syll. 3433.
Exsicc. Cke. F. Brit. 1st Ser. 263 —Fckl. F. Rh. 2026.—Roum. F. G. 585.

Gregarious. Perithecia covered, globose, subrugose, white within, with a central nucleus, pierced with a simple ostiolum. Asci slender-clavate. Sporidia biseriate, fusiform, hyaline, 3-septate, 10 μ long, (Berk.), 20 x 8 μ (Fckl.).

On *Spiræa opulifolia*, Mountains of New York (Berk. in Grev.).

The spec. in Cke. F. Brit. is sterile, and in Roum. F. Gall. there is only some stylosporous fungus, so that we can only give the published diagnosis, which we have taken from Cooke's Handbook. The species seems not to be well understood.

M. sèmen, (C. & P.)

Sphæria (*Caulicolæ*) *semen*, C. & P. 29th Rep. N. Y. State Mus. p. 65.
Metasphæria semen, Sacc. Syll. 3460.
Psilosphæria semen, Cke. Syn. 2996.

Perithecia soon free, globose, clustered, pierced at the apex, black. Asci clavate or cylindrical. Sporidia biseriate, lanceolate, straight or curved, 3-septate, deeply constricted in the middle, hyaline, 30–35 μ long.

On fallen petioles of *Pyrus Americana*, New York State (Peck).

M. rùbida, E. & E. Proc. Acad. Nat. Sci. Phil. July, 1890, p. 237.

Perithecia gregarious, globose, minute (¼ mm.), sunk in the surface of the wood with their apices and obtusely-conic ostiola projecting. On carefully shaving off the ostiola, the upper part of the perithecium is seen to be filled with carnose, bright flesh-red material which is also often visible through the broadly perforated ostiola. The lower part of the perithecia is white inside. Asci clavate-cylindrical, 75–80 μ long (p. sp. about 40 x 12 μ). Paraphyses abundant, longer than the asci. Sporidia crowded-biseriate, oblong-fusoid, slightly curved, 3-septate, the next to the upper cell swollen, hyaline, 20–22 x 3½–4½ μ. The upper part of the perithecia seems to be covered (as in *Clypeo-*

sphæria) with a more or less distinct cap of black, carbonaceous matter which is irregularly ruptured by the emergent ostiolum.

On a decaying log of *Platanus occidentalis*, Flatbush, Long Island, N. Y. (Zabriskie).

M. Arábidis, Johans. Svamp. Island, p. 169, tab. XXIX, fig. 11.

Perithecia scattered or gregarious, depressed-spherical, black or cinereous-black, pierced with a round pore above, 180–200 μ diam. Asci cylindric-clavate, short-pedicellate, paraphysate, 8-spored, 54–60 x 8–10 μ. Sporidia biseriate, cylindric-fusoid, hyaline, 3-septate, not constricted, second cell scarcely or only slightly enlarged, straight or curved, 22–28 x 4–5 μ.

On leaves of *Arabis alpina*, Kerortusok, Greenland.

The Greenland spece. have asci 75–90 x 10 μ, sporidia 25–28 x 3–4 μ.

M. Cassiopes, Rostr. Fungi Grönl. p. 561 (No. 181).

Perithecia scattered, semiimmersed. Asci cylindric-clavate, 37–40 x 10 μ, paraphysate, 8-spored. Sporidia biseriate, fusoid-oblong, obtuse, 3-septate, hyaline, 12–15 x 5 μ.

On dry leaves of *Cassiope tetragona*, Isortok Kingua, Greenland.

M. anisómetra, (Cke.& Hark.)

Sphæria anisometra, Cke. & Hark. Grev IX, p. 86.
Metasphæria anisometra, Sacc. Syll. II. p. 163.
Endophlœa anisometra. Cke. Grev. XVII, p. 89
Exsicc Ell. N. A. F. 890.

Perithecia evenly and thickly scattered, erumpent, minute (150 μ), subglobose, black, rough, with a papilliform ostiolum soon perforated. Asci oblong-clavate, sessile, 70–75 x 12 μ, with abundant paraphyses. Sporidia lanceolate, 1–4-septate, hyaline, ends acute while lying in the asci, obtuse when free, 19–22 x 4–5 μ (26 x 8 μ, Cooke). The mature sporidia are slightly olivaceous and constricted above the middle. In the specc. distributed in N. A. F. the sporidia are mostly less than 5 μ thick.

On twigs of *Mimulus glutinosus*, *Lonicera*, *Cupressus*, *Eucalyptus*, *Rubus*, *Dracæna*, and on pods of *Robinia*, California.

M. subcutánea, (E. & E.)

Sphæria subcutanea, C. & E. Grev. VII, p 41
Metasphæria subcutanea, Sacc. Syll. II, p. 167, Cke. Syn. 4163.

Perithecia scattered, semierumpent, ovate-globose, black, thin-carbonaceous, about 200 μ diam. Ostiolum not prominent. Asci clavate-cylindrical, narrowed below, paraphysate, 70–75 x 8–10 μ. Sporidia biseriate, fusiform, 3-septate, acute, 20–22 x 2½–3 μ.

On decorticated limbs of pear or apple, Newfield, N. J.

In Grevillea (l. c.) the sporidia are said to be 5-septate, constricted in the middle, with the third joint slightly swollen, 40 x 5 μ, but our spece. are as above stated, sporidia not constricted, and none of the cells swollen.

M. boùcera, (C. & E.)

Sphæria boucera, C. & E. Grev. VIII, p 15.
Metasphæria boucera, Sacc. Syll. II, p. 161, and Cke. Syn. 4456.
Exsicc. Ell. N. A. F. 887, (in part).

Perithecia scattered, covered by the epidermis, subglobose, prominent, black. Asci cylindric-clavate. Sporidia biseriate, fusoid, triseptate, hyaline, slightly curved, with a horn-shaped appendage at each end, 30–32 x 7½ μ without the appendages (which finally disappear).

On dead herbaceous stems, Newfield, N. J.

M. plagàrum, (Cke. & Hark.)

Sphæria plagarum, Cke. & Hark. Grev. XIII, p. 19.
Endophlœa plagarum. Cke. & Hark. in Cke. Syn. 4160.
Metasphæria plagarum, Sacc. Syll. 7025.

Perithecia gregarious, covered, elevated, subglobose, black, carbonaceous, collected in patches covered by the convexly elevated cuticle. Asci clavate, sessile, 8-spored. Sporidia lanceolate, inordinate, triseptate, hyaline, 18–20 x 4 μ, ends acute.

On bark of *Eucalyptus*, California.

M. Cattànei, Sacc. Syll, 3482.

Perithecia membranaceous, scattered, black, immersed in the parenchyma of the leaf, covered by the slightly pustulate-elevated epidermis, subglobose, perforated above, 200–300 μ diam. Asci numerous, subsessile, 8-spored, 150 μ long, p. sp. 50 x 8 μ. Sporidia subbiseriate, oblong-fusoid, slightly curved, 1–3-septate, slightly constricted at the septa, 20–22 x 4–5 μ (27 x 6 μ, Sacc.), hyaline.

On withered leaves of rice, South Carolina and Louisiana.

M. Myrìcæ. Pk. 38th Rep. p. 105.

Perithecia numerous, broadly conical, 400–500 μ diam., covered by the thin, closely adhering epidermis, black, white within. Ostiola perforated. Asci clavate, obtuse, 100–125 x 16–20 μ, with abundant, conglutinated paraphyses. Sporidia crowded or biseriate, oblong or subfusoid, straight or slightly curved, at first uniseptate, quadrinucleate, strongly constricted at the septum, finally 3-septate, hyaline, 30–40 x 10–12 μ.

On dead branches of *Myrica Gale*, lying partly in water, New York State.

The epidermis is so closely adherent that the perithecia appear as if superficial or merely innate at the base. The nuclei of the sporidia are large. Sporidia with 3 septa were rare, but this may have been due to immaturity.

M. staphýlina, (Pk.)

Sphæria, staphylina, Pk. 26th Rep. p. 86.
Metasphæria staphylina, Sacc. Syll. 3447.
Endophlœa staphylina, Cke. Syn 4161,

Perithecia minute, black, covered by the epidermis which at length ruptures in a stellate manner or irregularly. Asci? Sporidia biseriate, colorless, constricted in the middle, 3–5-septate, 20–25 μ long, the two parts formed by the central septum unequal in diameter.

On dead twigs of *Staphylea trifolia*, Helderberg Mts., N. Y.

M. leióstega, (Ell.)

Sphæria leiostega, Ell. Bull. Torr, Bot. Club, VIII, p. 91.
Metasphæria leiostega, Sacc. Syll. 3432.
Exsicc. Ell N. A. F. 888

Perithecia gregarious, pustuliform, entirely covered by the epidermis which is usually not ruptured or blackened over them, subprominent, of medium size. Asci cylindrical, 80–100 x 10–12 μ, abruptly narrowed below into a short, stipe-like base, not paraphysate. Sporidia uniseriate, elliptical, nearly hyaline, 3-septate, 12–18 x 7–8 μ.

On various dead twigs, *Carya*, *Rosa*, *Vaccinium* &c., Newfield, N. J., and on *Ribes*, London, Canada (Dearness).

This is certainly very near *M. corticola*, (Fckl.), but we have no spece. of that species for comparison. It is also closely allied to *Leptosphæria fuscella*, (B. & Br.), but that has olivaceous sporidia (see p. 353).

M. helicícola, (Desm.)

Sphæria helicicola, Desm. 16. Not. 1849, p. 30.
Leptosphæria helicicola, Desm. Niessl, Beitr. p. 25, tab. III, fig. 18.
Metasphæria helicicola, (Desm.) Sacc. Syll. II, p. 169.
Exsicc. Desm. Pl. Crypt. Ed. I, 2085, Ed. II, 1785.

Amphigenous. Perithecia scattered, erumpent, at length partially free, 120–130 μ diam., globose, obtusely papillate, collapsing in the center, marginate, coriaceo-membranaceous, black. Asci large, oblong, 8-spored, 68–76 x 10–12 μ, short-stipitate, obtusely flattened at the apex, with scanty, filiform paraphyses. Sporidia biseriate, oblong or fusoid-oblong, 3-septate and constricted at the septa, 18–20 x 4 μ, ends obtuse, nucleus subolivaceous.

On dry leaves of *Hedera Helix*.

This is credited to North America by Saccardo in Syll. l. c.

M. hederæfòlia, (Cke.)

Sphæria hederæfolia, Cke. Grev. XI, p. 110.
Metasphæria hederæfolia, Sacc. Syll. 6148.
Exsicc Rav. F. Am. 683.—Ell. N. A. F. 699.

Foliicolous, gregarious. Perithecia globose, semiïmmersed, black. Asci clavate. Sporidia elliptic-lanceolate, or clavate, triseptate, hyaline, 20 x 8 μ.

On leaves of *Hedera Helix*, Aiken, South Carolina (Ravenel).

This appears to differ from the preceding species in its narrower sporidia.

M. complanàta, (Tode).

Sphæria complanata, Tode Meckl. fig. 88, and Fr. S. M. II, p. 508.
Leptosphæria complanata, De Not. Schema, p. 62.
Metasphæria complanata, Sacc Syll. II, p. 161, and Cke. Syn. 4454.

Perithecia scattered, subglobose, black, soon collapsing and flattened. Ostiolum papilliform, persistent. Asci cylindrical. Sporidia 1–2-seriate, subfusoid, curved, 4-celled, second cell subinflated, 30 x 5 μ, hyaline or greenish-hyaline.

On dead herbaceous stems, South Carolina and Virginia.

M. àcuum, (Cke. & Hark.)

Sphæria acuum, Cke. & Hark. Grev. IX, p. 86.
Metasphæria acuum, Sacc. Syll. 3459, Cke. Syn. 5786.

Perithecia black, scarcely papillate, erumpent, hemispherical, subprominent. Asci clavate. Sporidia biseriate, sublanceolate, rounded at the ends, constricted in the middle, 1–3-septate, sometimes quadrinucleate, hyaline, 23–24 x 6 μ.

On fir leaves.

Closely allied to *M. anisometra*, (C. & H.)

M. sqamàta, (C. & E.)

Sphæria squamata, C. & E. Grev. VII, p. 10.
Metasphæria squamata, Sacc. Syll. 3445.
Endophlœa squamata, Cke. Syn. 4158.

Perithecia scattered, suberumpent, depressed-hemispherical, $\frac{1}{2}$–$\frac{3}{4}$ mm. diam., membranaceous, black, ostiolum not prominent. Asci oblong or ovate-oblong, sessile, mostly subattenuated above, but obtuse, 70–76 x 15–20 μ. Paraphyses not abundant. Sporidia mostly biseriate or irregularly crowded, oblong-fusoid, 1–3-septate, hyaline, 22–25 x 6–8 μ.

On dead limbs of *Pinus rigida*, Newfield, N. J.

M. cavernòsa, (E. & E.)

Sphæria (Metasphæria) cavernosa, E. & E. Journ Mycol. I, p. 91.
Metasphæria cavernosa, Sacc. Syll 7030.

Perithecia coriaceo-carbonaceous, black, rather thin-walled, $\frac{1}{2}$–$\frac{3}{4}$ mm. diam., sometimes 2–3 united, at first covered by the fibers of the bark, the upper half at length projecting and nearly bare. Ostiolum subtuberculiform, obtuse, broad. Asci clavate-cylindrical, 80–115 x 12–15 μ, with filiform paraphyses. Sporidia uniseriate or partly biseriate above, rather acutely elliptical, endochrome, 3-times divided, hyaline, 18–22 x 7–9 μ. The upper part of the perithecium at length falls away, leaving the black, cup-shaped, hemispherical base bedded in the bark. Closely allied to *M. leiostega*, Ell., which is scarcely distinct from *M. corticola*, Fckl. It differs however in its denuded perithecia, longer and broader asci, and rather longer sporidia. The sporidia of *M. leiostega*, are mostly 14–18 x 7–8 μ, very few reaching 20 μ long.

On bark of *Taxodium distichum*, Darien, Ga. (H. W. Ravenel, 703).

** *Sporidia 5- or more-septate.*

M. brúnnea, (Cke.)

Pleospora brunnea, Cke in Rav. F. Am. 684, and Syn 5013.
Metasphæria brunnea, Sacc. Syll. 3427, Cke. Syn. 4459.

Perithecia covered by the epidermis, densely and widely gregarious, so as to blacken the stem, globose-depressed, small, 120–180 μ diam., papillulate, finally collapsing. Asci cylindric-clavate, 90–100 x 16 μ, sparingly paraphysate, very short-stipitate, 8-spored. Sporidia biseriate or obliquely uniseriate, oblong-fusoid, at first broader and surrounded with a hyaline envelope, 6-nucleate, becoming 5-septate, the third cell thicker, constricted in the middle, 28–30 x 6$\frac{1}{2}$–7$\frac{1}{2}$ μ, frequently curved.

On dead stems of *Fœniculum*, Aiken, South Carolina.

M. aùlica, (C. & E.)

Sphæria aulica, C. & E. Grev. VI, p 95.
Metasphæria aulica, (C. & E.) Sacc. Syll. II, p. 168.
Endophlœa aulica, Cke. Syn. 4164.

Perithecia somewhat scattered, covered, globose, black, subprominent, papillate, about 200 μ diam. Asci clavate, rounded above, paraphysate, about 70 x 12–14 μ. Sporidia biseriate, lanceolate, endochrome 5-parted, hyaline, narrower below, 22–25 x 4 μ (35–40 x 9 μ, Cke.), constricted at the septa.

On *Lonicera* and *Solidago*, Newfield, N. J.

M. rimulàrum, (Cke.)

Sphæria rimularum, Cke. Grev. VI, p 146.
Metasphæria rimularum, Sacc. Syll. 3502, Cke. Syn. 4500

Perithecia covered, globose, crowded in short lines, covered by the longitudinally fissured cuticle. Asci clavate. Sporidia fusoid, hyaline, 5-septate, constricted, nucleate, 40–50 x 5–6 μ.

On reeds (*Arundinaria*), Gainesville, Fla.

"The perithecia are collected in little, elongated clusters, the cuticle cracked above them in parallel lines, but the ostiola do not penetrate."

M. brachythèca, (B. & C.)

Sphæria brachytheca, B. & C. Grev. IV, p. 146.
Metasphæria brachytheca, Sacc. Syll. 3451.
Endophlœa brachytheca, Cke. Syn. 4165.

Perithecia minute, surrounded by the cuticle. Asci obovate, very short. Sporidia clavate, with about 6 septa, 25 μ long, resembling those of *Patellaria atrata.*

On *Rosa*, New England.

M. dissíliens, (C. & E.)

Sphæria dissiliens, C. & E. Grev. V, p. 51.
Metasphæria dissiliens, Sacc. Syll. II, p. 163, and Cke. Syn 4463.

Perithecia scattered, black, at length erumpent, subglobose, with punctiform ostiola. Asci clavate. Sporidia biseriate, fusiform, 8-septate, 70 x 9 μ, constricted and divided into two unequal parts, one of which is 3-septate and the other 4-septate, readily dividing at the constriction, hyaline.

On stems of *Desmodium strictum*, Newfield, N. J.

B. *On monocotyledonous plants.*

* *Sporidia 3-septate.*

M. Palmétta, (Cke.)

Sphæria Palmetta, Cke. Grev. VII, p. 53.
Metasphæria Palmetta, Sacc. Syll. 3489.
Exsicc. Rav. F. Am. 369.

Perithecia scattered, covered by the blackened epidermis, which is raised into little pustules. Ostiola erumpent. Asci cylindrical or clavate, 60–70 x 10–12 μ (30 x 7 μ Cke.). Sporidia biseriate, fusoid, hyaline, uniseptate at first, becoming 3-septate and constricted, 20–25 x 6–8 μ. Var. *foliicola*, E. & E., is on dead spots in the leaves, and has the perithecia subelongated or hysteriiform, but does not differ materially in other respects.

On dead petioles of *Sabal Palmetto*, Georgia (Ravenel).

We have retained this species on the authority of Dr. Cooke, but it is very doubtful whether it is specifically distinct from *Leptosphæria sabaligera*, B. & C. The only appreciable difference lies in the rather larger perithecia and subhyaline sporidia.

M. macrothèca, Rostrup, Fungi Grönl. p. 561 (No. 183).

Perithecia gregarious, globose-depressed. Asci very large, ovate-oblong, contracted just below the apex, stipitate, 130–135 x 30–38 μ. Sporidia irregularly 3-seriate, 8-in an ascus, hyaline, 3-septate, each cell with a cubical nucleus, 32–35 x 12–13 μ.

On dead leaves of *Carex hyperborea*, Sukkertoppen, Greenland.

M. punctulàta, E. & E. Journ. Mycol. IV, p. 76.

Perithecia scattered, immersed, the surface of the culm remaining quite even but blackened around the small, erumpent, black ostiola, or finally more or less uniformly blackened. Perithecia globose, ¼–⅓ mm. diam., with a white, rather firm nucleus. Asci clavate-cylindrical, 80–110 x 20 μ, with indistinct paraphyses. Sporidia fusoid, slightly curved, 3-septate, hyaline, 40–50 x 6–7 μ.

On dead culms of *Panicum Curtisii*, St. Martinsville, La.

M. stenothèca, (E. & E.)

Sphæria (Metasphæria) stenotheca, E. & E. Journ. Mycol. III, p. 127.
Metasphæria stenotheca, Sacc Syll. IX, p. 844.

Perithecia scattered, membranaceous, subovoid, ¼ mm. diam., buried in the matrix, except the rather prominent, depressed-conoid apex, which is covered by the blackened epidermis, with only the papilliform ostiolum erumpent. Asci linear, 70–80 x 4–5 μ, with indistinct paraphyses. Sporidia uniseriate, overlapping, oblong-fusoid, 3–4-nucleate, becoming 3-septate, subhyaline, 12–16 x 3 μ.

On sheaths of dead *Panicum Curtisii*, Louisiana (Langlois).

M. lacústris, (Fckl.)

Sphæria lacustris, Fckl. Symb. Nachtr. II, p. 22.
Metasphæria lacustris, Sacc. Syll. 3470, Cke. Syn. 4466
Exsicc. Fckl. F. Rh. 2436.

Perithecia gregarious, covered by the epidermis, finally erumpent, globose or somewhat depressed, with the apex somewhat shining and crowned with a small, perforated, papilliform ostiolum, black, 180–210 μ diam. Asci elongated-oblong, sessile, 8-spored, 70–90 x 12–14 μ, with abundant paraphyses. Sporidia biseriate, oblong, at first uniseptate, then 3-septate, constricted at the septa, rounded at the ends, hyaline, 20–22 x 5 μ.

On *Phragmites communis*, Manhattan, Kansas (Kellerman).

The sporidia are somewhat longer than in the original specc. examined by Dr. Winter.

M. infúscans, E. & E. (in Herb.)

Perithecia gregarious, on the inner surface of the sheath, depressed-globose, black, ¼–⅓ mm. diam., the subconical apex raising the epidermis into little pustules which become umbilicate at the apex by the collapsing of the papilliform ostiola. Asci oblong-cylindrical, sessile, paraphysate, 70–80 x 10–12 μ. Sporidia crowded-biseriate, clavate-oblong, 3-septate, but not constricted, yellowish-hyaline, 15–20 x 3½–4½ μ.

On the inner surface of dead, blackened, outer sheaths enclosing the spikes of *Andropogon Virginicus*, Alabama (Atkinson).

M. sabalénsis, (Cke.)

Sphæria sabalensis, Cke. Grev. VII, p 53.
Dilophia sabalensis, Sacc. Syll. 4105.
Metasphæria sabalensis, Cke. Syn. 4523.
Exsicc. Ell. & Evrht. N. A. F. 2d Ser. 1962.

Perithecia numerous, small, covered by the blackened epidermis which is raised into slight pustules. Asci clavate, 150 μ long. Sporidia biseriate, fusoid, 45–50 x 4–4½ μ, hyaline, prolonged at each end into a bristle-like appendage, uniseptate at first, then 3-septate, the extreme septa near the ends.

On dead petioles of *Sabal serrulata*, Georgia (Ravenel), Florida (Calkins).

M. ceratothèca, (Cke.)

Sphæria ceratotheca, Cke. Grev. XI, p. 109.
Metasphæria ceratotheca, Sacc. Syll. 6150, and Cke. Syn. 4487.
Exsicc. Rav. F. Am. 677

. Superficial, nestling in a black, conidiiferous subiculum. Perithecia very small, black, opake, hemispherical. Asci lanceolate, with an acute, horn-shaped apiculus above. Sporidia lanceolate, 3-septate, hyaline, 25 x 5 μ. Conidia pluriseptate, muriform, brown, 45–50 μ long.

On culms of *Zea Mays*, Aiken, South Carolina.

M. boreàlis, Rostr. l. c.

Perithecia gregarious, above the medium size, sphæroid, with a conoid papilla. Asci cylindric-clavate, 70–75 x 14–16 μ. Sporidia biseriate, cuneate-oblong, 1–3-septate, hyaline, obtuse at both ends, 22–26 x 5–6 μ.

On dry stems of *Tofielda borealis*, Umanarsuk, Greenland.

M. Panicòrum, (Cke.)

Sphærella Panicum, Cke. Grev. V, p. 153.
Metasphæria Panicorum, Sacc. Syll. 3483, and Cke. Syn. 4479.

Perithecia epiphyllous, scattered, covered, on purple spots. Asci clavate. Sporidia biseriate, fusiform, hyaline, triseptate, 25 x 5 μ.

On fading leaves of *Panicum*, South Carolina.

M. carectòrum, (B. & C.)

Sphæria carectorum, B. & C. Grev. IV, p. 153.
Metasphæria carectorum, Sacc. Syll. 3487, and Cke. Syn. 4482.

Minute, punctiform, scattered, subprominent. Asci clavate. Sporidia short-fusiform, 4-nucleate. No measurements given.

On leaves of *Carex folliculata*, United States.

M. recutìta, (Fr.)

Sphæria recutita, Fr. S. M. II, p. 524.
Sphærella recutita, Fckl. Symb. Nachtr. II, p. 21.
Metasphæria recutita, Sacc. Syll. 3484, and Cke. Syn. 4480.
Exsicc. Fckl. F. Rh. 2434.

Perithecia very small, spherical, perforated above, black, innate-erumpent, crowded in long, parallel lines, so that often the whole leaf appears gray. Asci pyriform or ovate-elliptical, sessile, 8-spored, 27–30 x 12 μ. Sporidia conglomerated, elongated-clavate, somewhat narrowed below, uniseptate, slightly constricted at the septum, hyaline, 12–14 x 3½ μ.

On dead leaves of *Carices*, Troy, N. Y. (sec. Peck).

** *Sporidia 4–8-septate.*

M. defòdiens, (Ell.)

Sphæria (*Leptosphæria*) *defodiens*, Ell. Bull. Torr. Bot. Club, VIII, p. 90.
Metasphæria defodiens, Sacc. Syll. 3505, Cke. Syn. 4503.
Exsicc. Ell. N. A F. 889.

Perithecia scattered, depressed-globose, 150–300 μ diam., buried, but raising the epidermis into strong, hemispherical protuberances, finally deciduous. Ostiolum subglobose-papillate, black, rough, erumpent. Asci clavate, attenuate-stipitate, 90–100 x 12 μ, paraphysate. Sporidia biseriate, fusoid or clavate-fusoid, 20–25 x 4–5 μ, endochrome 4–6-parted, mostly 5-parted, pale yellow.

On dead stems of *Juncus*, Iona, N. J.

The sporidia are at first surrounded with a broad, gelatinous envelope which disappears together with the bristle-like apical appendage. The measurements here given apply to the body of the sporidium and do not include the envelope. Spece. of *L. apogon*, S. & S. in Kriegers Sax. Fungi, 130, are quite different from this. *L. juncina*, Awd. (Myc. March. 2140) has much smaller and less prominent perithecia.

M. hyalóspora, (Sacc.)

Leptosphæria hyalospora, Sacc F. Ven. Ser. II, p. 323, F. Ital. tab. 273.
Metasphæria hyalospora, Sacc. Syll. 3497, Cke. Syn. 4495.
Exsicc. Ell. N. A. F. 587.

Perithecia gregarious but separate, erumpent-superficial, rather less than ½ mm. diam., globose, black. Ostiolum rather acutely conical, becoming narrowly perforated. Asci densely fasciculate, cylindric-subclavate, 90–100 x 10–12 μ, often flexuous, 8 spored, with a short, nodulose stipe, and with filiform paraphyses. Sporidia fusoid, 28–32 x 5½–6½ μ, obtusely acuminate at each end, somewhat curved, 8-guttulate, then torulose, 7-septate, hyaline.

On decaying stalks of *Zea Mays* and on culms of *Panicum crus-galli*, Newfield, N. J., and on *Sabal Palmetto*· Louisiana (Langlois).

Both the New Jersey and Louisiana specc. have the sporidia 35–45 x 7–9 μ, 9–12-septate and slightly curved.

C. *On cryptogamous plants.*

M. epipterídea, (Cke. & Hark.)

Sphæria epipteridis, Cke. & Hark. Grev. IX, p. 8.
Metasphæria epipteridea, Sacc. Syll. 3513, Cke. Syn. 4513.

Scattered, covered or erumpent and semiimmersed, black. Asci clavate, sessile. Sporidia fusoid, hyaline, 3–5-septate, 22–25 x 5 μ.

On stipes of *Pteris aquilina*, California.

M. Lycopòdii, (B. & C.)

Sphæria Lycopodii, B. & C. Grev. IV, p. 155.
Metasphæria Lycopodii, Sacc. Syll. 3511, Cke. Syn 4511.

Punctiform, quite covered by the cuticle, not the least projecting.

Asci clavate. Sporidia biseriate, shortly fusiform, hyaline, biseptate. No measurements given.

On *Lycopodium*, New Jersey.

CERIÓSPORA, Niessl. (Plate 32)

Not. Pyr. p. 9.

Perithecia scattered, sunk in the matrix, with the ostiolum erumpent, and the perithecia themselves finally suberumpent or exposed. Asci 8-spored. Sporidia fusoid, 1–3-septate, yellow or yellowish-brown, with a hyaline, mucronate appendage at each end. Paraphyses evanescent.

In the original diagnosis of the genus, only the ostiola are erumpent, but in both the species here described the perithecia themselves are finally erumpent-superficial. Dr. Winter includes this in Fam. *Gnomonieæ*, but it seems to us more closely allied to *Pleosporeæ*.

C. Montaniénsis, (E. & E.)

Lophiostoma Montaniensis, E & E. Journ. Mycol. IV, p. 64.

Perithecia scattered or oftener seriate in longitudinal cracks in the bark, erumpent-superficial, depressed-globose, $\frac{1}{2}$–$\frac{3}{4}$ mm. diam., smooth, with a small, tubercular-papilliform ostiolum pierced with a slightly elongated opening. Asci cylindrical, p. sp. 80–90 x 7–8 μ, with filiform paraphyses. Sporidia uniseriate, oblong-elliptical, 3-septate, the two middle cells brown, the terminal cells hyaline and acute, prolonged into a filiform appendage 6–8 μ long. Colored part of the sporidia 12–14 x 6–7 μ. The sporidia are exactly like those of a single-crested *Pestalozzia*, but they are produced in asci.

On dead stems of *Clematis ligusticifolia*, Montana (Anderson).

A more careful examination shows that this can not be a *Lophiostoma*. The description in Journ. Mycol. is faulty. "cm." should be "mm.," and the measurements of the sporidia are omitted.

C. Alabamiénsis, E. & E. (in Herb.)

Perithecia scattered, tubercular-hemispherical, but covered by the epidermis, obtuse and perforated above, brown, $\frac{1}{3}$–$\frac{1}{2}$ mm. diam., base broadly adnate and slightly sunk. Asci cylindrical, short-stipitate 90–100 x 6–8 μ, paraphysate. Sporidia uniseriate, ends overlapping, fusoid-oblong, yellowish-hyaline, subinequilateral, 12–15 x 4–5 μ, with a straight, hyaline bristle or awn 10–15 μ long at each end, with a large nucleus at first, becoming uniseptate.

On dead herbaceous stems, Alabama (Atkinson).

SACCARDOÉLLA, Speg.

Mich. I, p. 461.

Perithecia large, immersed, coriaceo-carbonaceous. Ostiolum papilliform. Asci cylindrical, 8-spored, paraphysate. Sporidia elongated-fusoid, setigerous-appendiculate, multiseptate, hyaline.

S. Canadénsis, E. & E. (in Herb.)

Perithecia solitary, ovate, $\frac{3}{4}$ x $\frac{1}{2}$ mm. sunk in the inner bark, of light-colored, waxy consistence inside, the apex raising the epidermis into slight pustules barely perforated by the papilliform ostiolum. Asci cylindrical, short-stipitate, 200–280 x 10 μ, with abundant filiform paraphyses. Sporidia overlapping-uniseriate, fusoid-cylindrical, about 15-septate, 40–60 x 8–9 μ, at first much smaller, uniseptate and setigerons at each end.

On bark of *Cratægus*, London, Canada (Dearness).

JULÉLLA, H. Fab.

Spher. Vaucluse, p. 113.

Perithecia simple, subglobose, typically covered, but (in the American species) suberumpent. Asci 1–2-spored. Sporidia large, clathrate-reticulate, yellowish.

J. monospérma, (Pk.)

Sphæria monosperma, Pk. 28th Rep. p. 79, pl. 2, figs. 36–39.
Julella monosperma, Sacc. Syll. 3874.

Perithecia scattered, semiimmersed, ovate-hemispherical, black, about $\frac{1}{2}$ mm. diam. Ostiolum tubercular-papilliform, flattened above and soon pierced with a large, round opening. Asei clavate-oblong, 75–150 x 30–40 μ, with abundant, filiform paraphyses. Sporidia only one in an ascus, densely clathrate-fenestrate, and nearly filling the ascus, yellowish-hyaline.

On decorticated birch wood, Forestburg, N. Y. (Peek), on decorticated bleached wood, Washington (Suksdorf).

In the Washington specc. the perithecia are rather smaller ($\frac{1}{2}$ mm. diam.), and the sporidia also are mostly not over 100 x 25 μ, but otherwise they do not seem to differ from the New York specc. sent by Peek. The Washington specc., however, are rather old, and the upper part of the perithecia is broken away so that the ostiolum can not be seen. *J. Kellermanni*, mentioned in Cooke's Synopsis Pyrenomycetum (5137 bis), is *Kellermannia yuccægena*, E. & E., which belongs in the *Sphæropsideæ* (see Journ. Mycol. I, p. 153).

OPHIÓBOLUS, Riess.

Hedw. 1853, p. 27.

Perithecia scattered, subsphæroid, submembranaceous, covered or suberumpent. Ostiolum papillate or elongated. Asci cylindrical. Sporidia filiform, guttulate or septate, hyaline or yellowish. Caulicolous or culmicolous.

O. acuminàtus, (Sow.)

Sphæria acuminata, Sow. Eng. Fungi, tab 394, fig. 3.
Sphæria Carduorum, Wallr. Fl Crypt. Germ. II. 805
Ophiobolus disseminatus, Riess, Hedw. I, p. 27
Ophiobolus acuminatus, Duby, in Rab. Herb. Mycol. Ed II. No. 57.
Exsicc. Fckl. F. Rh. 780.—Rab. F. E. 1156, 1437 —Rehm Asc. 50 —Thum. F. Austr. 476. id. M. U. 358.—Roum. F. Gall. 1849.

Perithecia scattered, at first sunk in the matrix, with only the conical or short-cylindrical ostiolum projecting, finally erumpent, and by the falling away of the epidermis, superficial, about ⅓ mm. diam., often with mycelial hyphæ around the base. Asci cylindric-clavate, 8-spored, stipitate, 120–150 x 8–10 μ, paraphysate. Sporidia filiform, multinucleate, then multiseptate (15–20-septate), usually with one joint (near the middle mostly) swollen, yellowish, 75–110 x 2½–3 μ.

On herbaceous stems: *Erigeron, Campanula, Polygonum, Cirsium, Solanum* and *Lappa*.

Specimens on *Cnicus lanceolatus*, sent from London, Canada (Dearness), have two of the joints swollen, dividing the sporidium into three subequal parts.

O. porphyrógonus, (Tode).

Sphæria porphyrogona, Tode Fungi Meckl. II, p. 12, tab. IX, fig. 72.
Sphæria rubella, Pers. Syn. p. 63.
Rhaphidophora rubella, De Not. Sfer. Ital p. 80.
Leptospora rubella, Rab. Herb. Mycol. 532.
Leptospora porphyrogona, Rab. Hedw. I. p. 116.
Rhaphidospora rubella, Fckl. Symb. p. 125.
Ophiobolus porphyrogonus, Sacc Syll. 4017.
Exsicc. Fckl. F. Rh. 787.—Kze. F. Sel. 79, 254.—Rab Herb Mycol 532.—Rehm Asc. 94. Thum. M. U. 561.—Ell. N. A. F. 191.—Sydow M. March. 1358.—Roum. F. Gall. 288.

Perithecia scattered or oftener gregarious, and mostly on purplish-red stains, at first buried, finally more or less erumpent, globose-conical, sometimes slightly depressed, with a flattened base, black, brittle, glabrous, with projecting, conical or cylindrical ostiolum, about 300 μ diam. Asci cylindrical, long and narrow, substipitate, 140–160 x 4½–6 μ, 8-spored, paraphysate. Sporidia filiform, lying parallel, about as long as the asei, multinucleate, then multiseptate, yellowish, about 1 μ thick.

On dead herbaceous stems: *Solanum, Zea, Lactuca*, &c., common.

O. consimilis, E. & E. (in Herb.)

Perithecia scattered, subcuticular, finally exposed by the falling away of the cuticle, ovate-globose, small, about 200 μ diam., with a short-cylindrical, projecting ostiolum. Asci cylindrical, stipitate, 75–100 x 8 μ, with paraphyses crisped and more or less matted together above. Sporidia filiform, multiseptate, yellow, nearly as long as the asci, without any swollen joint, about 2 μ thick.

On old tomato stems, Newfield, N. J., and on okra stems, Louisiana.

In the Louisiana specc. the perithecia are often on red stains, just as in *O. porphyrogonus*, but this is easily distinguished from that species by its smaller perithecia, shorter and broader asci and sporidia.

O. filísporus, (C. & E.)

Sphæria filispora, C. & E. Grev. VII, p. 10.
Ophiobolus filisporus, Sacc. Syll. 4074.
Raphidospora filispora, Cke. Syn. 4581.

Perithecia lenticular, ⅓ mm. diam., scattered, covered by the epidermis which is slightly raised and perforated, and stained olive-black. Asci linear, 112–150 x 5–6 μ. Sporidia filiform, nearly as long as the asci.

On dead stems of *Smilax*, Newfield, N. J. Apparently rare. Closely allied to *O. stictisporus*, (C. & E.)

O. trichísporus, E. & E. (in Herb.)

Perithecia scattered, erumpent-superficial, ovate-conical, about 400 μ high and 300 μ broad, narrowed gradually above into a stout, obtuse, cylindric-conical apex, glabrous, black, membranaceous. Asci linear, paraphysate, 180–220 x 3–3½ μ. Sporidia about as long as the asci, very slender filiform or capillary, nucleate, scarcely ½ μ thick, yellowish-hyaline.

On dead culms of some grass, London, Canada (Dearness).

The asci and sporidia scarcely differ from those of *O. stictisporus*, (C. & E.), but the erumpent-superficial perithecia easily separate this from that species.

O. impléxus, (E. & E.)

Lophiostoma (*Lophionema*) *implexum*, E. & E. Journ. Mycol. IV, p. 75.

Perithecia gregarious, brown strigose, ovate, about ⅓ mm. diam., subcuticular, the obtuse-conical, slightly compressed ostiolum and upper part of the perithecia erumpent. Asci 150–160 x 8–10 μ, clavate-cylindrical, with abundant filiform paraphyses. Sporidia filiform, closely braided or twisted together, and about as long as the asci. Well characterized by its perithecia clothed with brown strigose hairs. and its braided sporidia.

On dead adventitious roots of *Sorghum Halapense*, and on the lower part of sheathing leaves of (*Andropogon*)? Pointe a la Hache, La. (Langlois).

This evidently belongs here rather than among the *Lophiostomeæ*.

O. stictísporus, (C. & E.)

Sphæria stictispora, C. & E. Grev. VI, p. 96, tab. 100, fig. 36.
Ophiobolus stictisporus, Sacc. Syll 4067.
Raphidospora stictispora, Cke. Syn. 4574.

Perithecia scattered, immersed, and covered by the blackened cuticle which is only slightly elevated, $\frac{1}{3}$–$\frac{1}{2}$ mm. diam., with a rather large opening above. Asci linear, 150-210 x 3-4 μ, paraphysate. Sporidia capillary, nearly as long as the asci and about $\frac{1}{2}$ μ thick.

On dead culms and leaves of grasses, Newfield, N. J.

The general appearance is that of a *Stictis*.

O. oliváceus, (Ell.) (Plate 28)

Leptosphæria olivacea, Ell. Bull. Torr Bot Club, X, p. 53.
Ophiobolus olivaceus, Sacc Syll. 7127.

Perithecia submembranaceous, about 250 μ diam., buried in the substance of the stem and covered by the cuticle, which is slightly elevated, stained olive-brown and pierced by the broad, rough, obtuse ostiolum. Asci clavate, 75–85 x 15–18 μ. Sporidia fasciculate, yellow-brown, vermiform, 6–7-septate, and when mature, slightly constricted at the septa, the third joint from the tip slightly swollen, 75 x $3\frac{1}{2}$–4 μ.

On dead herbaceous stems, Utah (S. J. Harkness).

The sporidia are generally slightly bent just below the swollen joint.

O. Medúsa, E. & E. Journ. Mycol. I, p. 150.

Perithecia membranaceous, scattered, depressed-globose, $\frac{1}{3}$–$\frac{1}{2}$ mm. diam., covered by the epidermis which is not discolored or raised, but merely pierced by the black, punctiform ostiolum. Asci very long (400 μ and over by 43–15 μ broad), containing 8 filiform, curved sporidia nearly as long as the asci, and 3–$3\frac{1}{2}$ μ thick in the middle, gradually tapering to each end, yellowish or nearly hyaline, with endochrome multipartite. The perithecia lie in the furrowed cavities of the culm, attached above to the inner surface of the cuticle and covered with loose, spreading, weak, brown, septate hairs 200–300 μ long by about 3 μ thick. On culms of *Spartina*, lying partly buried in the sand on the beach at Cape May, N. J., and on culms and sheaths of *Andropogon muricatus*, Louisiana (Langlois).

O. anguíllides, (Cke.)

Sphæria anguillida, Cke. Grev. VI, p. 15.
Ophiobolus anguillides, Sacc. Syll. 4029.
Raphidospora anguillida, Cke. Syn. 4542.
Exsicc. Ell. N. A. F. 582.

Perithecia gregarious, soon exposed, ovate, black, smooth, hard, shining, papillate, ½ mm. diam. Asci cylindrical, 80–110 x 10–12 μ. Sporidia filiform, multiseptate, yellowish, 80–100 x 2½–3 μ (120 μ long Cke.).

On dead stems of *Bidens*, Newfield, N. J., and on dead stems of *Aster*, Texas (Ravenel).

The sporidia, when mature, are slightly enlarged at the upper end, which is a little curved to one side and bears a striking resemblance to the head of a serpent. The paraphyses are abundant and a little longer than the asci.

O. hamásporus, E. & E. Journ. Mycol. III, p. 117.

Perithecia scattered, globose, membranaceo-carbonaceous, ¼–⅓ mm. diam., black, buried in the substance of the leaf, except the erumpent, convex-flattened apex. Asci 70 x 8–10 μ, narrowed above, but obtuse. Paraphyses (?). Sporidia 8 in an ascus, filiform, multinucleate, yellowish-hyaline, 30–35 x 1½ μ, narrowed to a point below, and about one third of the lower part bent almost to a right angle, or even curved into a hook (*i. e.*, after the sporidia have escaped from the asci). The general aspect is that of *Didymosphæria cupula*, Ell., only the perithecia are not collapsed. The ostiolum is indistinctly papilliform.

On fallen leaves of *Quercus tinctoria* (?), Manhattan, Ks., July, 1887 (W. T. Swingle).

The leaf is sometimes blackened around the perithecia, indicating the presence of an imperfect stroma.

O? bacillàtus, Cke.

Sphæria bacillata, Cke. Hndbk. No. 2636.
Ophioceras bacillatum, Sacc. Syll. 4111.
Ceratostomella bacillata, Cke. Syn. 3786.

This (sec. spece. in our Herb. det. by Cooke) belongs to the *Discomycetes*. Asci 150 x 12 μ. Sporidia filiform, about as long as the asci and $1\frac{1}{2}$–2 μ thick, multiseptate, soon separating into joints 3–4 μ long, (*Stictis* or *Schizoxylon*).

O. staphýlinus, E. & E. l. c.

Perithecia small, covered by the fibers of the wood through which project the short, straight, roughish, black, rostellate ostiola. Asci linear, 120–150 x 4 μ, accompanied by filiform paraphyses. Sporidia 8 in an ascus, filiform, yellowish, nucleolate, and about as long as the asci.

On the same stems is a *Sphæropsis* with oblong, depressed perithecia, and sporules 18–20 x 8–9 μ; also other small perithecia partly covered by the fibers of the wood, and containing numerous elliptical, subfuscous, 3 x 2 μ sporules.

On decorticated stems of *Staphylea trifolia* West Chester, Pa.

The spece. are scanty and poor, so that we can not verify the original diagnosis, and the species must be considered doubtful. Possibly not distinct from *O. fruticum*, (Rab. & Desm.).

O. fúlgidus, (C. & P.)

Sphæria fulgida, C. & P. Peck's 29th Rep. p. 62.
Ophiobolus fulgidus, Sacc. Syll 4054.
Raphidospora fulgida, Cke. Syn. 4569.
Exsicc. Ell. N. A. F. 583.

Perithecia gregarious, sometimes disposed in lines, soon free, globose, black, smooth, shining, scarcely papillate, 250–270 μ diam., at length collapsed. Asci clavate-cylindrical, p. sp. 80–90 x 12 μ, paraphysate, stipitate, 8-spored. Sporidia cylindrical, subattenuated at each end, nearly straight, yellowish-brown, multiseptate and often constricted at the septa, 75–80 x 4–5 μ.

On dead stems of *Ambrosia trifida*, New York, New Jersey, Pennsylvania, Ohio and Kansas.

Distinguished from *O. anguillides* by its much coarser sporidia.

O. cláviger, Hark. Bull. Cal. Acad. Feb. 1884, p. 46.

Perithecia globular, with a papillate ostiolum, gregarious or scattered, erumpent, then free, $\frac{1}{3}$–1 mm. diam., at first filled with minute spermatia. Asci linear-clavate, 8-spored, 210 x 14 μ. Sporidia filiform, nucleate, pale brown, 20–25-septate, the upper third somewhat swollen and constricted, 140 x 7 μ.

On creeping stems of *Audibertia humilis*, California.

O. byssícola, Hark. l. c.

Perithecia globose, with prominent ostiola, superficial, $\frac{3}{4}$–1 mm.

diam., nestling in a dirty-brown subiculum. Asci cylindric-clavate, tapering to a slender pedicel which terminates in a bulbous base. 170 x 16 μ, 8-spored. Sporidia pale brown, tapering, obtuse at the ends, 20–30-septate, constricted, the upper cell, and 1–3 other cells, at irregular intervals, enlarged and globular, 120–140 x 4–6 μ.

On decorticated branches of *Sambucus glauca*, California.

O. collápsus, Sacc. & Ell. Mich. II, p. 374.

Exsicc. Ell. N. A F. 584.

Perithecia gregarious, covered by the epidermis, then erumpent, depressed-globose, ½ mm. diam., collapsing to cup-shaped, becoming black, not seated on any spots. Asci cylindrical, short-stipitate, 80–110 x 8–10 μ, 8-spored. Sporidia rod-shaped, slightly curved, nodulose-thickened in the middle, 70 x 2 μ, 12–15-guttulate, hyaline.

On dead stems of *Trifolium pratense*, Newfield, N. J.

O. versísporus, E. & M. Journ. Mycol. I, p. 99.

Exsicc. Ell. & Evrht. N. A. F. 2d Ser. 1961.

Perithecia scattered or gregarious, covered by the cuticle, lenticular, ¼–⅓ mm. diam., covered by the blackened epidermis which is whitened just around the short, obtuse, barely erumpent ostiolum. Asci 70–80 x 8–9 μ. Paraphyses? Sporidia filiform, curved, multinucleate at first, but at length of a uniform pale yellow color, without nuclei or septa, 60–70 x 2–2½ μ.

On dead petioles of *Sabal serrulata*, Florida (Martin).

Melanconium Sabal, Cke. is usually associated with this.

Species imperfectly known.

O? glòmus, (B. & C.)

Sphæria glomus, B. & C. Grev. IV, p. 152.
Ophiobolus? glomus, Sacc. Syll. 4055.
Raphidospora glomus, Cke. Syn. 4570.

"Perithecia convex, perforated. Sporidia linear, sigmoid, 25–50 μ long. Stylospores are produced within flat, dark specks seated on forked threads, at first joined in pairs, so as to make an obovate mass, then separating and still obovate, but narrow, 25 μ long. On *Ambrosia*, Alabama."

O. Solidáginis, (Fr.)?

Sphæria Solidaginis, Fr. Elench. II, p. 106. See also Grev. VI, p. 16.
Ophiobolus Solidaginis, Sacc. Syll. 4034, Cke. Syn. 4547.

Perithecia scattered, depressed-globose, 300 μ diam., covered by the epidermis which is raised into pustules blackened and pierced by the globose-papilliform, perforated ostiolum. Asci clavate-lanceolate, substipitate, 75–90 x 7 μ, paraphysate. Sporidia fasciculate, filiform, multinucleate, curved when free, yellowish, 50–60 x 2 μ.

On dead stems of *Solidago*, Louisiana (Langlois).

Whether this is really the *Sphæria Solidaginis*, Fr., we can not say, but it agrees fairly well with the imperfect diagnosis in Elench. l. c. There was not, however, any "white disk" observed, but this. it is said, becomes brown ("*demum fuscescens*").

Sphæria (Dothidea) Solidaginis, Schw., as shown by spece. in Herb. Schw., is a foliicolous species and quite distinct from this.

FAMILY. MASSARÌEÆ.

Stroma wanting. Perithecia mostly permanently covered by the epidermis, very seldom erumpent, typically with only the small, papilliform ostiolum piercing the epidermis; texture firm-coriaceous. Asci paraphysate. Sporidia hyaline or brown, 1- or more-septate.

MASSÀRIA, De Not.

Giorn. Bot Ital I, p. 333.

Perithecia immersed, coriaceous, sphæroid, with erumpent, papillate ostiolum. Asci ample, mostly 8-spored. Sporidia subbiseriate, oblong, 1-pluriseptate, hyaline or brown, mostly large, and surrounded by a gelatinous envelope.

Many of the species, especially those with 3-septate sporidia, are with difficulty distinguished from each other, and it is not improbable that a more careful and thorough examination will reduce some of these to mere varieties.

M. Árgus, (B. & Br.)

Sphæria Argus, B. & Br. Not. of Brit. Fungi, No. 626.
Massaria Argus, Fres. Beitr. p. 59.
Massaria Niessleana, Rehm Asc. 645.
Exsicc. Fckl. F. Rh. 802.—Rab. F. E. 259, 3057.

Perithecia gregarious, permanently covered by the scarcely raised epidermis, subdepressed-spherical, finally umbilicate, tolerably hard, black, 600–800 μ diam., with a small, conical ostiolum, piercing the epidermis in a punctiform manner. Asci cylindric-clavate, short-stipitate, 8-spored, 200–220 x 38–44 μ. Sporidia obscurely biseriate, cylindrical or oblong-clavate, ends rounded, subattenuated below, unequally divided, the upper thicker half mostly 4- (sometimes 3-)celled, the lower, smaller half 3-celled, with a gelatinous envelope which is constricted in the middle, brown, 50–74 x 14–20 μ.

On dead branches of birch trees, Portville, N. Y. (Peck).

Myxocyclus confluens, Riess is considered to be the pycnidial form of this species.

M. inquinans, (Tode).

Sphæria inquinans, Tode Fungi Meckl. II, p. 17.
Sphæria gigaspora, Desm. Pl. Crypt. Ed. I, 3065, id. Ed. II, 1765.
Massaria gigaspora, Ces. & De Not. Schema, p. 43.
Massaria Bulliardi, Tul. Sel. Carp. II, p. 236.
Massaria inquinans, Fr. Summa, p. 369.
Exsicc. Fckl. F. Rh. 803.—Rab. F. E. 1237, 1526.—Thum. M. U. 1950.—Sacc. M. Ven. 1189. Roum. F. G. 1387.—M. March. 1735.—Rehm. Asc. 989.

Perithecia thickly scattered, buried in the inner bark and penetrating to the wood, globose, 1–1½ mm. diam., raising the epidermis into distinct pustules, with the short-cylindrical ostiola erumpent. Asci oblong-clavate, short-stipitate, 180–220 x 25–32 μ, with abundant paraphyses, 8-spored. Sporidia inordinate or subbiseriate, oblong-cylindrical, hyaline and uniseptate at first, and with a broad, hyaline envelope, then brown and 3-septate, straight or only slightly curved, 70–90 x 15–20 μ.

On *Viburnum prunifolium* and *V. dentatum*, Pennsylvania (Michener), on dead maple limbs, Orono, Maine, and Lyndonville, N. Y., on dead *Cratægus*, London, Canada, and on dead *Pyrus arbutifolia*, Bethlehem, Pa. (Rau).

The perithecia often lie 2–3 close together and the bark is whitened in a narrow stratum around them. Ambiguous forms occur between this and *M. vomitoria* which may with equal propriety be referred to either species. From the absence of the circular, discoid area around the ostiola, the American specc. might be referable to *M. gigaspora*, Fckl., if that species is really distinct from *M. inquinans*, which is very doubtful. Winter gives the measurements of the sporidia as 80–103 x 21–23 μ, Sacc. 75–90 x 20–32 μ. The specc. in Rehm's Asc. have the sporidia 60–70 x 15–18 μ, and in Thüm. M. U. they are 75–80 x 18–20 μ. The specc. of *Sphæria gigaspora*, in Desm. Pl. Crypt. Ed. I, 2065, have the sporidia 75–80 x 18–20 μ.

M. vomitòria, B. & C. Grev. IV, p. 155. **(Plate 29)**

Exsicc. Ell. N. A. F. 97.—Ell. & Evrht. N. A. F. 2d Ser. 1954.

Perithecia scattered, or 2–3 lying close together, coriaceous, thick-walled, 1 mm. and over diam., buried in the bark which is scarcely raised above them, but merely pierced by the inconspicuous, scarcely projecting ostiolum. Asci broad clavate-fusoid, 150–200 x 25 μ, 8-

spored. Sporidia oblong-cylindrical, straight or slightly curved, hyaline at first, then brown and 3-septate, 55–70 x 10–15 μ, (mostly 55–65 x 12 μ).

On *Acer, Robinia, Fraxinus, Amelanchier*, and *Pyrus Malus* from New England and Canada to Carolina.

In the typical form on *Acer rubrum*, the perithecia scarcely raise the epidermis at all, but in the specc. on *Amelanchier* and *Pyrus Malus* (*M. Pyri*, Otth)? the epidermis is more or less pustuliform-elevated. All the forms here included in *M. vomitoria* certainly are very closely allied to *M. inquinans*, (Tode), and might with good reason be considered as mere varieties, or forms of that species.

M. conspurcàta, (Wallr.)

Sphæria conspurcata, Wallr. Fl. Crypt. Germ. II, p. 782.
Massaria conspurcata, Sacc. Syll. 2888, Cke. Syn 4043
Exsicc. Rehm Asc. 882 —Ell. & Evrht. N A. F. 2d Ser. 2613.

Perithecia scattered or 2–3 together, buried in the inner bark, depressed-globose, about 1 mm. diam., slightly raising and rupturing the epidermis. Asci elongated, clavate-cylindrical, short-stipitate, paraphysate, 150–200 x 20–22 μ. Sporidia subbiseriate, oblong-cylindrical, slightly curved or straight, 3-septate, scarcely constricted, hyaline, becoming brown, 40–60 x 10–12 μ.

On dead limbs of wild plum, London, Canada (Dearness).

The specc. in Rehm's Asc. have the asci a little broader and the sporidia 60–70 x 12–14 μ, but do not differ otherwise from these. *M. vomitoria*, B. & C., scarcely differs from this except in having the epidermis less distinctly pustuliform-elevated, and the ostiola smaller and less prominent.

M. distíncta, (Schw.)

Sphæria distincta, Schw. Syn. N. Am 1634.
Massaria distincta, Cke. Grev. XVII, p 92.

Scattered, covered by the thin epidermis, rather large, buried in the whitened substance of the inner bark. Perithecia black, orbicular, depressed, glabrous, persistent in the bark when the epidermis is peeled off, with a large, round opening above. Ostiola perforating the epidermis, short-cylindrical, not prominent, umbilicate.

Under the epidermis of *Sambucus pubens*, Bethlehem, Pa. (Schweinitz).

Sporidia (sec. Cke. in Grev. l. c.), biseriate, 5-septate, brown, 70–80 x 16–18 μ, constricted in the middle and surrounded at first by a hyaline envelope.

M. olivàceo-hirta, (Schw.)

Sphæria olivaceo-hirta, Schw. Syn. N. Am. 1656.
Massaria olivaceo-hirta, Cke Grev. XVII, p. 92

Perithecia scattered, rather large, with the thick, cylindrical, perforated, persistent ostiola penetrating the epidermis; when this is peeled off, the large, flattened perithecia are disclosed, clothed with an olivaceous, hairy coat, and tinging the bark in which they are buried with an olive-black color.

Under the epidermis of the larger branches of *Morus alba*, Bethlehem, Pa.

Sporidia (sec. Cke. Grev. l. c.) biseriate, lanceolate, 3–5-septate, brown, 50–60 x 12–16 μ, constricted in the middle, at first ocellate-nucleate.

M. epileùca, B. & C. Grev. IV, p. 156.

Perithecia gregarious, covered by the epidermis, globose-depressed, $\frac{3}{4}$ mm. diam., dark villose, papillate, then with a large, round opening above. Ascigerous nucleus black. Asci clavate, 130 x 30 μ, with a short, thick stipe, paraphysate, 8-spored. Sporidia biseriate, fusoid, 3–5-septate, straight or slightly curved, 65–70 x 18–20 μ, slightly constricted at the septa, surrounded by a gelatinous layer, the inner cells dark brown, and sometimes with a large nucleus, the terminal cells much smaller, paler and subapiculate.

On decaying branches of *Morus alba*, Pennsylvania and New Jersey.

M. púlchra, Hark. Bull. Cal. Acad. Feb. 1884, p. 44.

Perithecia scattered, covered, 1–1½ mm. diam., contents white. Asci broadly clavate, 126 x 36 μ, 8-spored. Sporidia fusiform-navicular, of two irregular, unequal cones united by their bases, and surrounded by a gelatinous stratum, at first uniseptate and hyaline, slowly becoming brown and unequally 3–5-septate, 58–60 x 20–22 μ.

On dead branches of *Umbellularia Californica*, California.

M. semitécta, (B. & C.)

Sphæria semitecta, B & C. Grev. IV, p. 147.
Massaria semitecta, Sacc. Syll. 2872, Cke Syn. 4021.

Perithecia half covered, subprominent, surrounded by the annular-ruptured epidermis. Sporidia clavate, triseptate, slightly constricted, 35 μ long, clothed at first with a thick, gelatinous coat.

On *Platanus*, Virginian Mountains (Berk in Grev.).

M. Ùlmi, Fckl. Symb. p. 153.

Exsicc. Fckl. F. Rh. 2008.—Ell & Evrht. N A. F. 2d Ser 2611 —Thum. M. U. 1852.

Perithecia gregarious, nestling in the inner bark, covered by the slightly blackened epidermis which is raised into slight pustules and pierced by the papilliform ostiola, about 1 mm. diam., depressed-spherical, coriaceous. Asci oblong, 8-spored, 250–300 x 30–35 μ, with abundant paraphyses. Sporidia biseriate, broad cylindric-fusoid, 3-septate, strongly constricted in the middle, brown, 50–70 x 15–20 μ, each cell with a large nucleus.

On bark of elm, London, Canada.

Differs from *M. inquinans*, in its smaller sporidia.

M. Dryàdis, Rostr. Fungi Grönl. p. 560.

Perithecia scattered, sphæroid-depressed, black. Ostiola snow-white. Asci thick-cylindrical, 90–115 x 32–38 μ, very short-stipitate, 8-spored. Sporidia biseriate, oblong, 3-septate, constricted at the septa, especially at the middle one, hyaline, surrounded by a rather broad, hyaline stratum. No measurements of sporidia given.

On the upper surface of dead leaves of *Dryas octopetala*, Western Greenland.

M. Plátani, Ces. in Rab. F. Eur. 323 (1842), and Comm. Soc. Crit. I, p. 217.

Massaria atroinquinans, B. & C. Grev IV. p 156 (1876).
Exsicc. Rav. F. Am 669 —Ell. & Evrht. N. A. F 2d. Ser. Cent. XXVIII.

Perithecia gregarious, often in subcircinate groups of 4–8, lying between the loosened laminæ of the bark, depressed-globose, ½–¾ mm. diam., finally collapsing beneath, the sporidia oozing out and staining the surface of the bark as in *Melanconium*. Asci broad clavate-cylindrical, 150–190 x 25–35 μ, subsessile, paraphysate. Sporidia irregularly biseriate, oblong-elliptical, olive-brown, 3–6- (mostly 3–5-) septate, slightly constricted at the septa, with a gelatinous envelope at first, finally opake so that the septa can with difficulty be seen, 35–55 x 14–20 μ.

In bark of *Platanus*, Carolina (Ravenel), Canada (Dearness).

The perithecia are entirely concealed, their presence being indicated only by slight, pustuliform elevations in the bark. The 3-septate sporidia are shorter and broader and scarcely constricted at the septa and are not usually mixed with the longer, narrower, mostly 5-septate sporidia in other asci in the same perithecium. The Canada specc. do not differ essentially from those in our Herb. sent from Carolina by Dr. Ravenel. We have not seen the pycnidial stage (*Hendersonia Desmazieri*).

M. atroinquinans is given as a synonym of *M. Platani*, on the authority of Berlese who has figured this species in his Icones (tab. XIV, fig. 2).

M. plumígera, E. & E. (in Herb.)

Perithecia scattered, depressed-globose, about $\frac{3}{4}$ mm. diam., slightly raising the epidermis which is pierced by the subprominent, short-conical or short-cylindrical ostiolum. Asci oblong-clavate, short-stipitate, 130–150 x 22–25 μ, 8-spored, paraphysate. Sporidia inordinate, oblong-cylindrical, hyaline, 3-septate, 55–60 x 12 μ.

On dead limbs of *Viburnum lentago*, Newfield, N. J.

The sporidia exude from the ostiola in little white, brush-like cirrhi. This is different from *M. Corni* (Fr. & Mont.), Sacc. Syll. 2859, which has brown sporidia 75–90 x 20–25 μ. It is not probable that the sporidia in *M. plumigera* ever become brown as they are perfectly hyaline when they issue from the ostiolum.

M. cleistothèca, Hark. l. c.

Perithecia minute, covered. Asci pyriform or obovate, 8-spored, thick-walled, without any stipe or point of attachment, 48 x 30 μ. Paraphyses agglutinate. Sporidia hyaline, of two opposed, rather long, equal cones, occasionally each of these divided so as to make the sporidium 3-septate, surrounded by a gelatinous stratum, 32–40 x 8–10 μ.

On dead stems of *Dendromecon rigidum*, California. Apparently an anomalous species. We have seen no specimens.

M. gigáspora, Fckl. Symb. Nachtr. II, p. 28.

Perithecia subcuticular, raising the epidermis into pustules, scattered or 2–3 together, rather large, globose, black, with a dirty-colored nucleus. Ostiolum very minute, papilliform, in a small, black disc. Asci saccate, sessile, 8-spored, 272 x 68 μ. Sporidia generally 4 in the upper part, and 4 in the lower part of the ascus, conglobate or uniseriate, very large (96 x 28 μ), oblong-ovate, obtuse at the ends, slightly curved, 3-septate, not constricted, cells uninucleate, with a narrow, hyaline margin, pale umber; paraphyses filiform, shorter than the asci.

On branches of *Viburnum Lentago*, Albany, N. Y. (Peck).

Peck gives the sporidia as 75 μ long, 4-celled, the two middle cells shorter than the terminal ones. We have not seen the specimens, and take the diagnosis above from Fckl. l. c. There may be some doubt whether the New York specc. are the genuine *M. gigas-*

pora. Winter (Die Pilze, II, p. 547) is of the opinion that *M. gigaspora*, Fckl., is only an immature state of *M. inquinans*, (Tode).

M. Gerárdi, Cke. (pro tem.) Grev, VIII, p. 118.

Sporidia very large, 90–120 x 30 μ, brown, 3–5-septate.

On bark, New York State (Gerard). Specimen too imperfect for a full description.

MASSARIÉLLA, Speg.

Fungi Arg. Pug. I, p. 2.

Perithecia and asci as in *Massaria*. Sporidia uniseptate, brown, surrounded by a hyaline stratum.

M. bufònia, (B. & Br.) **(Plate 30)**

Sphæria bufonia, B. & Br. Ann N. Hist. No. 629, tab. 10, fig. 13.
Massaria bufonia, Tul. Sel. Carp. II, p. 237.
Massariella bufonia, Speg. F. Arg. Pug. I, p. 2.
Massaria atrogrisea, C. & P. Grev. XVII, p 92.
Didymosphæria atrogrisea, C. & P. Cke. Syn. 4264.
Exsicc. Ell. & Evrht. N. A. F. 2d Ser. 3612.

Perithecia scattered or subgregarious, globose, $\frac{3}{4}$ mm. diam., coriaceous, raising the epidermis into pustules which are blackened and pierced by the papilliform ostiola. Asci cylindrical, 150 x 12–15 μ, paraphysate. Sporidia uniseriate, oblong-elliptical, uniseptate and constricted, hyaline at first with a gelatinous border, becoming dark brown, 15–20 x 8–10 μ.

On outer bark of living *Quercus alba*, New York, New Jersey, New England and Canada.

The asci and sporidia in the American spece. (*M. atrogrisea*, C. & P.) are constantly smaller than in the European specc., which have asci 150–200 x 15–20 μ, sporidia 25–30 x 12–15 μ, but there is no other difference.

M. Currèyi, Tul. Sel. Carp. II, p. 231.

Sphæria Tiliæ, Curr. Linn. Trans. XXII, tab. 59, fig. 104.
Massariella Curreyi, Sacc. Syll. 2709.

Perithecia scattered, covered, $\frac{1}{3}$–$\frac{1}{2}$ mm. diam., black, globulose, the papilliform ostiolum scarcely perforating the epidermis. Asci broad-clavate, 80–90 x 25 μ, paraphysate, 8-spored. Sporidia subbiseriate, obclavate, uniseptate-constricted, dark brown, upper cell thicker, 35 x 12–14 μ, with a gelatinous border.

What appears to be this species has been found on *Tilia*, at West Chester, Pa., but the specc. are too imperfectly developed to be decided with certainty.

M. seriàta, (Cke.)

Massaria (Massariella) seriata, Cke. Grev. XVII, p. 92.
Massariella seriata, Sacc Syll. IX, p. 739. No. 3025

Perithecia subdepressed, rather large, seriately arranged, covered by the epidermis which is finally fissured. Asci clavate. Sporidia elliptical, 60 x 18–20 μ, uniseptate, constricted in the middle, brown, cells equal, with a thick, hyaline epispore.

On branches of *Carya*, South Carolina (Ravenel).

M. bìspora, (Curtis).

Massaria (Massariella) bispora, Cke. Grev. XVII, p. 93
Massariella bispora, Sacc. Syll. IX, p. 740, No. 3027.

Perithecia corticolous, subglobose-depressed, covered, subscattered. Ostiola perforating the epidermis which is blackened by the sporidia. Asci clavate. Sporidia elliptical, uniseptate, brown, 45 x 18–20 μ, cells equal, constricted in the middle, with a hyaline envelope.

On bark of *Acer* (Dr. Curtis).

M. scoriàdea, (Fr.)

Sphæria scoriadea, Fr. El. II, p. 87
Anthostoma scoriadeum, Sacc. Syll. 1127
Massaria (Massariella) scoriadea, (Fr.) Cke. Grev. XVII, p 93.
Massariella scoriadea, Sacc. Syll. IX, p. 739.

Innate. Stroma widely effused, black, entirely hidden under the epidermis, surrounding the branches and penetrating the inner bark. Perithecia of a horn-like consistence, hemispheric-subprominent, shining, perforated, white inside, crowded but not confluent. Sporidia (sec. Cke. l. c.) elliptical, uniseptate, 70 x 23 μ, the upper cell rather larger, constricted in the middle, with a thick, hyaline epispore.

On bark of *Betula lenta*, Pennsylvania, Arctic America (Drummond).

PLEOMASSÀRIA, Speg.

Fungi Argentini, Pug. 1st.

Perithecia as in *Massaria*. Sporidia more or less distinctly muriform.

Pl. rhodóstoma, (A. & S.) **(Plate 30)**

Sphæria rhodostoma, Alb & Schw. Consp. p. 43.
Hercospora rhodostoma, Fr. Summa Veg. Scand. p. 397.
Massaria rhodostoma, Tul. Sel. Carp. II, p 238, tab. XXV, figs. 1-4.
Karstenula rhodostoma, Sacc. Syll. 3711.
Pleomassaria rhodostoma, Winter, Die Pilze, 3842

Exsicc. Fckl. F. Rh. 801.—Rab F. E. 3058.—Rehm. Asc. 236 —Thum. M. U. 862.—Krieger F. Sax. 78.

Perithecia mostly thickly scattered, gregarious, or occasionally standing singly, permanently covered by the slightly raised epidermis, depressed-globose, mostly concentrically furrowed or zoned, and umbilicate, the apex reddish, perforated and slightly erumpent. Ostiolum tolerably large, black, surrounded by a black, crustaceous mass. Asci cylindrical, attenuate-stipitate below, obtuse above, 8-spored, paraphysate, 150–170 x 10–12 μ. Sporidia uniseriate, oblong, slightly attenuated and rounded at the ends, mostly 3-septate, constricted at the septa, brown, generally with one or two of the inner cells divided by a longitudinal septum, 18–27 x 7–9 μ.

On *Rhamnus frangula*, in Sweden, Germany, England and Italy.

We find no record of this species having been met with in this country, and when the drawing on plate 30 was made, we were not aware that any species of *Pleomassaria* had been found here, but had the sporidium (from spece. in Krieger's Fungi Saxonici) figured to illustrate that genus. Since then, *Pl. siparia* and *Pl. Carpini* have been found in Iowa and New York, and the diagnosis of *Pl. rhodostoma* has been added in anticipation of that species also being yet found in America. The genus *Karstenula*, Sacc., distinguished from *Pleomassaria* only by the absence of any gelatinous envelope around the sporidia, can hardly be worthy of generic distinction.

Pl. sipària, (B. & Br.)

Sphæria siparia, B & Br. Not Brit. Fungi No. 625
Massaria siparia, Ces & De Not. Schema. p 43.
Pleomassaria siparia, Sacc. Syll. 3708.
Exsicc. Fck!. F. Rh. 2011. Rab. F. E. 260

Perithecia scattered or oftener crowded and subconnate, attached to the inner bark and covered by the pustuliform-elevated epidermis, depressed-spherical, ½–¾ mm. diam. (exceptionally 1 mm.), black, the minute ostiolum piercing the epidermis, finally umbilicate. Asci clavate, very large, attenuate-stipitate, (190–210 x 38–44 μ Winter), 8-spored, paraphyses very long and filiform. Sporidia elliptic-oblong, gradually and only slightly narrowed towards each end, obtuse, 7–8-septate and constricted at the septa, the middle cells divided by a longitudinal septum, golden-brown, 56–65 x 15–17 μ, with a gelatinous envelope.

On birch, Decorah, Iowa (Holway).

The Iowa specc. were immature, the sporidia being still hyaline, 35–85 x 5 6 μ (mostly 40–60 x 5–6 μ), vermiform-cylindrical, 5–6-nucleate and granular.

Pl. Carpini, (Fckl.)

Massaria Carpini, Fckl Symb. p. 153, tab VI, fig. 35.
Pleomassaria Carpini, Sacc. Syll. 3710.
Exsicc. M. March. 1928 —Krieg. F. Sax. 234.

Perithecia scattered, subcuticular, slightly rupturing but scarcely raising the epidermis, flattened, seated on the surface of the inner bark, broadly perforated above, brownish-black, $\frac{1}{2}$–$\frac{3}{4}$ mm. diam. Asci ventricose-clavate, 150 x 25–30 μ (p. sp. 110–120 μ), paraphysate. Sporidia biseriate, clavate-oblong, 3–5-septate and muriform, nearly hyaline and with a hyaline envelope and uniseptate at first, becoming yellowish-brown, 30–40 x 12–15 μ.

On *Carpinus Americana*, Lyndonville, N. Y. (Fairman).

At the main septum, which appears first, the sporidia are distinctly constricted. The upper and larger cell soon acquires two additional septa and the lower cell one. Most of the cells are divided by one or two longitudinal septa so that the sporidia appear to be filled with large nuclei. The asci and sporidia in the New York specc. are smaller than the measurements given by Dr. Winter (170–220 x 35–42 μ, and 45–65 x 17–21 μ). Fckl. gives 208 x 36 μ, and 48 x 16 μ for asci and sporidia respectively.

MASSARIOVÁLSA, Sacc.

Mich. II, p. 569

Perithecia buried in the surface of the inner bark, circinate. Asci and sporidia as in *Massariella*.

M. sùdans, (B. & C.) **(Plate 30)**

Massaria sudans, B. & C. Grev. IV, p 156.
Massariovalsa, sudans, Sacc. Syll. Add. vol. II, p. LV.
Massariella sudans, Sacc. Syll. I, p. 717.
Exsicc. Ell. N. A. F. 1190.

Perithecia circinate, sunk in the surface of the inner bark, ovate-globose, $\frac{1}{3}$ mm. diam., 4–8 together, their slender ostiola converging and united in a black, convex, erumpent disk. Asci clavate-cylindrical, 180–200 x 20–25 μ, stipitate, obtuse, 8-spored, paraphysate. Sporidia uniseriate, oblong-elliptical, uniseptate and constricted, olive-brown, 30–40 x 15–16 μ, with a thick, hyaline envelope.

On dead branches of *Acer*, *Carya* and *Quercus*, New Jersey and Pennsylvania.

M. caudàta, E. & E. (in Herb.)

Stroma cortical, 1–1$\frac{1}{2}$ mm. diam. Perithecia circinate, 6–10 in a pustule, about $\frac{1}{3}$ mm. diam., buried in the bark, covered by the epidermis which is raised into a flat pustule ruptured in the center by the compact cluster of black, subpapilliform ostiola. Asci varying from clavate, 75–80 x 25–35 μ, to obovate, 80–100 x 40–60 μ, narrowed below into an acute base, rounded and obtuse above, obscurely paraphysate, 4–8-spored. Sporidia inordinate, clavate-oblong or simply oblong, slightly curved and subinequilateral, 25–50 x 18–22 μ, 2–3-(mostly 2-) septate, hyaline at first, then olive-brown, ends obtuse, each with a cylindrical, hyaline, straight or curved, subpersistent appendage 12–20 x 5–6 μ. The perithecia do not penetrate so as to be visible on the inner surface of the bark.

On bark of dead *Platanus*. London, Canada (Dearness).

FAMILY. CLYPEOSPHÆRIEÆ.

Perithecia buried, without any proper stroma, but covered by a blackened, shield-like layer, which is sometimes sharply limited, and sometimes with an indefinite outline, and consists of the slightly altered and more or less blackened outer layer of the bark, leaf, or wood. This dark layer is sometimes also found under the perithecia as well as over them, or even enveloping them on all sides. Asci cylindrical or clavate-cylindrical, 8 spored, mostly paraphysate. Sporidia variable, oblong or filiform, hyaline or brown, continuous or 1–3-septate.

CLYPEOSPHÆRIA, Fckl.

Symb. p. 117.

Perithecia scattered, rarely confluent, covered by the epidermis, submembranaceous, covered above by a thin, epidermal, stromatic shield. Ostiolum erumpent, papilliform, short. Asci elongated, 8-spored. Sporidia uniseriate, oblong or oblong-cylindrical, triseptate, obtuse, often curved, brown.

Cl. sanguinea, E. & E. (in Herb.)

Perithecia gregarious on red, indefinite spots, minute (100 μ), buried in the red-stained surface of the wood, the minute, erumpent ostiolum barely visible. Asci clavate-cylindrical, 40–50 x 7–8 μ, paraphysate, 8-spored. Sporidia overlapping-uniseriate, oblong-fusoid, 3-septate and constricted at the middle septum, 12–15 x 3½–4½ μ, pale brown finally dark brown.

On exposed, weather-beaten wood of deciduous trees, Pennsylvania (Eckfeldt), Kansas (Cragin).

The stromatic shield is very obscure, so that this might perhaps go in *Leptosphæria*.

Cl. mamillàna, (Fr.)

Sphæria mamillana, Fr. S M II, p. 487.
(*Clypeosphæria mamillana*, *Lambotte*, Fl. Myc. Belg II, p 247)?
Clypeosphæria limitata, Fckl. Symb. p. 117.
Exsicc. Fckl. F. Rh. 915.

Perithecia gregarious, partly sunk in the matrix, globose, with a flat base, 400–450 μ diam., with an obtusely conical ostiolum erumpent through the black, epidermal shield, and surrounded with a whitish zone. Asci narrow-cylindrical, short-stipitate, 8-spored, 150–160 x 8–9 μ. Sporidia uniseriate, oblong-lanceolate, subattenuated and rounded at the ends, subinequilateral, 3-septate, brown 19–24 x 5–6 μ.

On branches of *Celastrus*, Bethlehem, Pa., (Schw.). Diagnosis from Winter's Pilze.

Cl. imperfécta, E. & E. (in Herb.)

Perithecia gregarious, globose, ½ mm. diam., covered by the epidermis which is raised into strong pustules blackened and pierced by the erumpent, papilliform ostiolum. Asci broad clavate-cylindrical, 80–100 x 20 μ. Sporidia biseriate, elliptical, 3-septate, brown, 20–30 x 10–12 μ, subinequilateral, slightly constricted at the septa.

On bark of living birch, Syracuse, N. Y. (Underwood).

This has the aspect of a lichen (***Pyrenula***) but seems really more closely allied to *Clypeosphæria mamillana* (Fr.).

Cl. aliquánta, (C. & E.)

Sphæria aliquanta, C. & E. Grev. V, p. 94.
Clypeosphæria ? *aliquanta*, Sacc. Syll. 3198.
Heptameria aliquanta, Cke. Syn 4984

Perithecia scattered, covered by the blackened epidermis, 200 μ diam., or less. Asci oblong-clavate, p. sp. 65–70 x 12–15 μ, paraphysate. Sporidia crowded-biseriate, oblong-fusoid, slightly curved, hyaline and uniseptate, becoming yellowish-hyaline, 3-septate and slightly constricted at the septa, 20–25 x 8–10 μ including the broad, hyaline envelope, the body of the sporidium being mostly only 18-20 x 6–7 μ.

On dead stems of *Smilax*, Newfield, N. J.

Cooke makes the sporidia 30–35 x 10 μ, but we do not find them as large as that, nor do they ever appear to become brown, only yellowish-hyaline.

Cl. Hendersònia, (Ell.)

Sphæria Hendersonia, Ell. Grev. VI, p. 14. tab. 95, fig. 8.
Leptosphæria Hendersonia, Cke. Syn. 4311.
Clypeosphæria Hendersonia, Sacc. Syll 3194.
Sphæria melantera, Pk. 29th Rep. p. 62.
Exsicc. Ell. N. A F. 581.

Perithecia gregarious, covered by the blackened cuticle which is slightly raised, but not fissured. Asci cylindrical, 75 x 5 μ. Sporidia uniseriate, oblong-elliptical, brown, 3-septate, 12–16 x 4–4½ μ (18 x 4 μ Cke.)

On dead canes of black and red raspberry, and on dead limbs of Sassafras, Newfield, N. J.

LINÓSPORA, Fckl.

Symb p. 123

Perithecia appearing late in the season, buried in the shield-shaped, black, phyllogenous pseudostroma (which is sometimes wanting), generally solitary, beak subprominent, black, more or less elongated. Asci cylindrical, 8-spored. Sporidia filiform, lying parallel in the asci, hyaline or yellowish-hyaline.

L. conflícta, (Cke.)

Sphæria conflicta, Cke. Grev. VII, p. 13.
Linospora conflicta, Sacc Syll 4095, Cke Syn. 5842.
Exsicc. Rab F. E. 3759

Spots amphigenous, pale reddish-brown, definitely limited by a narrow, darker border, suborbicular, 2–4 mm. diam., or often larger (1–2 cm.), with an irregular, subsinuous outline. Perithecia buried in the parenchyma of the leaf, globose, about 150 μ diam., with a papilliform ostiolum, the apex slightly erumpent and rupturing the epidermis in an operculoid manner. Asci oblong-cylindrical, sessile, 8-spored, aparaphysate, 75–85 x 12 μ. Sporidia subfasciculate, four above and four below, clavate-cylindrical, 40–45 x 2½–3 μ, (60 μ long, Cke.), 3–6-septate, yellowish-hyaline.

On leaves of *Quercus densiflora*, Tamalpais, California (Harkness).

The description is drawn from specc. sent by Dr. Harkness.

L. ferrugínea, E. & M. Am. Nat. Dec. 1884, p. 69.

Spots light yellowish-brown, border darker, narrow and slightly raised, 1½–2 mm. diam. Perithecia black, subglobose, 150 μ diam., immersed and covered by the blackened cuticle which is perforated by the scarcely prominent ostiolum. Asci cylindrical, 75–80 x 7 μ, sessile or nearly so, with abundant filiform paraphyses, 8-spored. Sporidia vermiform, yellowish, faintly nucleate, acute at each end, 35–45 x 1½ μ.

On leaves of *Andromeda ferruginea*, Florida (Dr. Martin).

The perithecia are solitary, one in the center of each spot, but the spots are often sterile.

L. leucóspila, (B. & C.)

Sphæria leucospila, B. & C Grev. IV, p. 153.
Linospora leucospila, Sacc. Syll. 4101, Cke. Syn 5848.

"On narrow pallid spots parallel with the nerves on the underside of the leaf. Asci linear. Sporidia filiform."

On leaves of *Platanus*, South Carolina.

We have seen no specimens.

L. Palmétto, E. & E. Journ. Mycol. III, p. 45. **(Plate 31)**

Perithecia globose, about ⅓ mm. diam., immersed, with the papillose ostiolum erumpent and included in a superficial, depressed-conic. cap-like stroma nearly as broad as the perithecia, and around which the epidermis of the leaf is blackened, as is also the parenchyma of the leaf around the perithecia. Asci lanceolate, 75–80 x 8–10 μ, with abundant paraphyses. Sporidia 8 in an ascus, linear, fusoid, yellowish, nucleate, acute, 40–50 x 2–2½ μ. The perithecia are mostly in sub-elongated spots of a paler color than the surrounding part of the leaf.

On dead places in living leaves of *Sabal Palmetto*. Point à la Hache, La., (Langlois).

ISÓTHEA, Fr.

Summa Veg. Scand. p. 421.

Perithecia covered by a phyllogenous, maculiform pseudostroma. Asci oblong. Sporidia clathrate-septate (muriform).

I. Nýssæ, B. & C. Grev. IV, p. 157.

"Shining, penetrating the leaf, seated on a little brown spot not much wider. Asci oblong. Sporidia shortly fusiform, not three times longer than broad, at length fenestrate."

On leaves of *Nyssa aquatica*. (Carolina)?

HYPÓSPILA, Fr.

Summa Veg. Scand. p. 421.

Perithecia immersed in the parenchyma of the leaf, very delicate, covered above by a black, phyllogenous, stromatic shield, beak (ostiolum) lateral, at length barely perforating the epidermis and appearing as a black speck. Asci fusoid-clavate, 8-spored, aparaphysate. Sporidia biseriate, 1- finally 3-septate, hyaline. Minute fungi covered by the epidermis which is tinged with a dark red color and swollen or inflated.

H. Groenlándica, Rostr. Fungi Grönl. p. 561.

Perithecia immersed in the parenchyma of the leaf, covered by the epidermis which is bullate-inflated on both sides, gregarious, with a lateral, black, cylindrical beak. Asci exactly fusiform, 8-spored, 95–115 x 10–12 μ. Sporidia narrow-fusiform, straight, multinucleate, 2-septate, 48–52 x 4–5 μ.

On fallen leaves of *Salix glauca*, Sukkertoppen and Sermersut, Greenland.

Sporidia about three times as long as in *H. pustula*.

H. pústula, (Pers.) **(Plate 31)**

Sphæria pustula, Pers. Syn p 91.
Phoma pustula, Fr. S M. II, p. 547.
Sphæria pleuronervia, De Not. Micr. Ital. Dec. IX, No. 9.
Isothea pustula, Berk. Outl. Brit. Fung. p 392.
Sphæria oleipara, Sollm. Hedw. V, p 65.
Gnomonia pustula, Awd Mycol. Eur. Pyr. p. 21, tab. VIII
Hypospila pustula, Karst. Mycol. Fenn. II, p. 127.
Exsicc. Fckl. F. Rh. 842.—Kze. F. Sel. 106—Rab. F. E. 1452.—Thum F Austr 472 Linht. F. Hung. 467.—Kriegr. F. Sax. 285.—Roum. F. G. 4945.—Rehm Asc. 793

Perithecia sunk in the parenchyma of the leaf, covered by the epidermis which is bullate-inflated on both sides, and is of a dark reddish tint below, scattered or gregarious, often following the course of the nerves of the leaf, 200–300 μ diam., sometimes 2–3 confluent, depressed-globose, with a short, lateral, tardily erumpent, beak-like ostiolum. Asci clavate, attenuate-stipitate, thickened above, 70 x 8–10 μ (75–100 x 7–10 μ, Sacc.). Sporidia 8 in an ascus, biseriate, oblong-fusoid, rounded at the ends, straight or subinequilateral, becoming 3-septate, but not constricted at the septa, 17–23 x 4 μ.

On fallen oak leaves. Credited to America by Saccardo, in Sylloge.

We have seen no American specimens. The figs. 12–15, on Plate 31, are from the spece. in Linhart's Fungi Hungarici. The synonymy and diagnosis are taken from Winter's Pilze. The young sporidia have (sec. Winter) a button-shaped appendage at the ends.

TRABÙTIA, Sacc. & Roum.

Revue Mycol. 1881, p. 27, tab. XIV, fig. 2.

Stroma phyllogenous, black, radiose-asteromatoid, flattened. Perithecia adnate with the stroma, separate, protuberant, with the ostiola perforated. Asci 8-spored, obsoletely paraphysate, evanescent. Sporidia ovoid-oblong, continuous, subhyaline. The genus has the habit of *Phyllachora* or *Rhytisma*.

T. quércina, (Fr. & Rud.) **(Plate 31)**

Rhytisma quercinum, Fr. & Rud. in Linn. Trans. 1830, p. 551.
Asteroma parmelioides, Desm. Pl. Crypt. Ed. I, 1737.
Rhytisma riccioides, Letellier Champ. V, tab. 629, fig. 4.
Sphaeropsis riccioides, Lev. Ann. Sci. Nat. III. Ser. tom. p 257.
Trabutia quercina, Sacc. & Roum. l. c.
Exsicc. Thum. M. U. 2271.—Ell. N. A. F. 1288

Perithecia hemispherical, subcarbonaceous, shining-black, with a minute, round, perforated ostiolum, finally collapsing and umbilicate or subplicate, but never opening as in *Rhytisma*. Asci cylindric-clavate, 100–110 x 18–21 μ, paraphysate, membrane entire at the apex, short-stipitate, soon disappearing. Sporidia biseriate, oblong-navicular, ends subobtuse, 28–30 x 8–10 μ.

On leaves of *Quercus laurifolia* and *Q. virens*, Florida and Mississippi.

T. tósta, (B. & C.)

Rhytisma tostum, B. & C Grev. IV, p. 9.
Trabutia tosta, Cke Syn. 1372

"Seated on yellow spots, thin, gyrose, only here and there producing fruit-bearing perithecia which soon shell off. Undoubtedly distinct, but the specimens are imperfect."

On leaves of *Quercus laurifolia*, Alabama.

T. erythróspora, (B. & C.)

Rhytisma erythrosporum, B. & C. Proc. Am. Acad IV, p 128 and Grev. IV, p 9.
Trabutia erythrospora, Cke. Syn. 1371.

"Minute, opening with two or three laciniae. Asci swollen. Sporidia subfusiform, salmon-colored, apiculate at each end, 33 μ long.

On leaves of *Quercus virens*, California.

The manner of dehiscence indicates *Rhytisma*.

THYRÍDIUM, Sacc.

Mich. I, p. 50.

Perithecia scattered or gregarious, immersed in the more or less altered substance of the wood or bark, and covered above by a prominent, black, stromatic shield. Ostiola papilliform. Sporidia subelliptical, brown or hyaline, muriform.

Th. lívidum, (Pers.)

Sphaeria livida, Pers. Syn. p 80.
Teichospora livida, Karst. Myc. Fenn. II, p 68.
(*Fenestella*) ? *livida*, Winter Die Pilze No 4251.
Thyridium lividum, Sacc. Syll. 3991, Cke Syn 3981.

Perithecia scattered, enclosed in a rather large, woody, elliptical, gray or blackish, nearly superficial tubercle, umbilicate-perforated above, coriaceous, thiek-walled, ovoid or subsphæroid, rather less than 1 mm. diam. Asci cylindrical, subsessile, 100–110 x 12 μ, 8-spored, paraphysate. Sporidia uniseriate, elliptical, yellowish or greenish-brown, 3–5-septate, with a more or less perfect longitudinal septum. 14–20 x 8–9 μ.

On decorticated wood of *Thuja*, Vermont (Pringle), on *Juniperus Virginianus*, Iowa (Holway), on *Rhus*, Carolina and Pennsylvania (Schweinitz), on bleached wood, Texas (Wright), on wood, Mountains of New York (sporidia 25 μ long, Berk. in Grev. IV, p. 146).

Th. antiquum, (E. & E.)

Sphæria (Thyridium) antiqua, E. & E. Bull. Torr. Bot. Club, X, p. 90
Thyridium antiquum, Sacc. Syll. 7123, Cke. Syn. 3987

Perithecia gregarious, globose, about ⅓ mm. diam., buried in the substance of the bark, which is blackened above them, and raised into little pustules. Ostiola papilliform, at length perforated. Asci cylindrical, 75–80 x 10 μ. Paraphyses filiform, abundant. Sporidia uniseriate, oblong-elliptical, at length 3-septate, and submuriform, 17–19 x 7 μ, brown.

On the inner surface of loose, hanging bark of grape vines, Newfield, N. J.

Sometimes as in *T. Garryæ*, 2–3 perithecia are covered by the same shield which may be entirely shaved away without cutting the subjacent perithecia.

Th. Gárryæ, (Cke. & Hark.)

Sphæria (Thyridium) Garryæ, Cke. & Hark. Grev. XIII, p. 20.
Thyridium Garryæ, Sacc. Syll. 7122, Cke. Syn. 3985

Gregarious or scattered. Perithecia immersed, subglobose, black, ½ mm. diam., covered above by the black, prominent, convex stromatic shield which is perforated by the tube of the papilliform ostiolum. Asci cylindrical, 4–8-spored. Sporidia elliptical, 7-septate and muriform, hyaline, becoming yellow-brown, 40–45 x 15–18 μ. Epispore thick, hyaline.

On bleached, decorticated twigs of *Garrya*, California (Harkness).

Our spece. of this species from Harkness are old and without fruit. The general appearance is the same as that of *Th. antiquum*. Occasionally two perithecia are covered by the same shield.

Th. ambleium, (C. & E.)

Sphæria amblera, C. & E. Grev. VII, p 16.
Thyridium ambleium, Sacc. Syll. 3993, Cke Syn 3983.

Perithecia scattered, black, subprominent, covered. Asci clavate-cylindrical. Sporidia broad-lanceolate or acutely elliptical, constricted in the middle, about 5-septate, yellow-brown, 15–20 x 8–10 μ (25 x 10 μ. Cke.).

On dead limbs of *Carya* and *Azalea*, Newfield, N. J.

The specc. of this species are poor and unsatisfactory.

Th. personàtum, (Cke. & Hark.)

Sphæria personata, Cke & Hark. Grev. XIII, p. 20.
Thyridium personatum, Sacc. Syll. 7124, Cke. Syn. 3986.

Lignicolous. Perithecia scattered, included in an elliptical, elevated, gray or black tubercle pierced by the short, inconspicuous ostiolum. Asci cylindrical, 8-spored. Sporidia uniseriate, elliptical, constricted in the middle, 3-septate, with 1–2 longitudinal septa, bright brown, 18–20 x 10 μ.

On decorticated *Acacia*, California (Harkness).

We have seen no specc. of *Th. personatum*, Cke. & Hark., but the description agrees in all respects with that of *Th. lividum*.

Th. Canadénse, E. & E. (in Herb.)

Perithecia scattered or gregarious, on the more or less bleached surface of the wood, minute (¼ mm. or less), covered by a thin, oblong or lanceolate, black shield 1–1½ mm. long and pierced in the center by the papilliform ostiolum. Asci cylindrical, subsessile, paraphysate, 100–120 x 12–15 μ. Sporidia uniseriate, obovate-oblong, 5–7-septate and muriform, hyaline, 20–27 x 10–13 μ.

On old (spruce)? logs, Lake Nipigon, Ontario, Canada (Macoun).

Outwardly this much resembles *Th. lividum*, but the shield is thinner. Perithecia much smaller, sporidia larger and hyaline.

Th. cingulàtum, (Mont.)

Sphæria cingulata, Mont. & Fr. in Mont. Syll. No. 833.
Thyridium cingulatum, Sacc. Syll. 3992, Cke. Syn. 3982

Covered, blackening the surface with the discharged sporidia, scattered, black. Stroma pulverulent, cortical, covered above by a conceptacle resembling a rough, hemispheric-conical pseudo-perithecium. Perithecia globose, one in each stroma, with a moderately long neck terminating in a shining, papilliform, deciduous ostiolum. Asci clavate, stipitate. Sporidia oblong, 28 x 9 μ, 5-septate and muriform, constricted in the middle, brown.

On dead branches of *Symphoricarpus racemosus*, California (Harkness).

We have not seen this, and take the diagnosis from Saccardo's Sylloge.

ANTHOSTOMÉLLA, Sacc.

Syll. I, p. 278.

Perithecia submembranaceous, globose-depressed, typically covered by the epidermis, which is generally somewhat blackened around the scarcely erumpent ostiola. Asci 8-spored (rarely 4-spored), paraphysate. Sporidia ovoid or oblong, continuous, dark, sometimes hyaline-appendiculate.

A. nigro-annulàta, (B. & C.)

Sphæria nigro-annulata, B & C. Cuban Fungi, No. 859.
Anthostomella nigro-annulata, Sacc. Syll. 1032, and Cke. Syn. 4610
Sphæria Yuccæ, Schw. Syn Car. No. 88?
Exsicc. Ell. N. A. F. 2d Ser 1672.—Rav. Car. V, 73.

Perithecia subgregarious, covered by the epidermis which is blackened (except a white spot in the center) around the slightly erumpent ostiola. Asci cylindrical, p. sp. 75–80 μ long. Sporidia obliquely uniseriate, oblong or subelliptical, brown, 12–18 (mostly 14–15) x 7½ μ.

On leaves of *Yucca*, Carolina and Florida.

A. minor, E. & M. Journ. Mycol. III, p. 43.

Exsicc. Ell. & Evrht. N. A. F. 2d Ser. 1965.

Perithecia scattered, ⅓ mm. diam., subglobose, with the upper part subconical and prominent, with a rather acute, papilliform ostiolum. Asci linear, 65–75 x 5 μ. Sporidia uniseriate, opake, 2–3-nucleate, subinequilateral, 7–8 x 2½–3 μ. The surface of the matrix, in the specimens seen, was covered with a thin, black crust, but whether this had any connection with the perithecia, we could not say.

On petioles of *Sabal serrulata*, Florida (Calkins).

A. sepelíbilis, (B. & C.)

Sphæria sepelibilis, B. & C. Grev. IV, p. 146.
Anthostomella sepelibilis, Sacc. Syll 1042, and Cke. Syn. 4623
Exsicc. Ell. N. A. F 1200

Perithecia scattered, depressed-globose, about ⅓ mm. diam., covered by the blackened cuticle which is raised into pustules and pierced by the papilliform, erumpent ostiola. Asci clavate-cylindrical, paraphysate, about 75 x 7 μ. Sporidia uniseriate or partly biseriate, oblong-elliptical, brown, 10–12 x 5–6 μ.

On dead stems of *Smilax*, Carolina and New Jersey.

A. eliminàta, (B. & C.)

Sphæria eliminata, B. & C. Grev. IV, p. 148.
Anthostomella eliminata, Sacc. Syll. 1040, and Cke. Syn. 4621.

Perithecia covered by the jet-black cuticle, which is the more conspicuous from the unoccupied parts being white, marked in the center with white above the ostiolum. Asci linear. Sporidia uniseriate, oblong, 14 x 3½–4 μ.

On stems of *Smilax*, Alabama.

We have seen no specimens. From the brief diagnosis it seems too near the preceding species.

A. leucóbasis, (E. & M.)

Sphæria (Anthostomella) leucobasis, E. & M. Am. Nat. Oct. 1882, p. 809.
Anthostomella leucobasis, Sacc. Syll. 5926, Cke. Syn. 4634.
Exsicc. Ell. N. A. F. 1199.

Perithecia globose or subelliptical, about ½ mm. diam., buried in the matrix in definite groups, above which the epidermis is generally more or less blackened, the blackened areas mostly limited by a well defined line, which does not, however, penetrate deeply. Ostiola obtuse, barely piercing the epidermis. Asci cylindrical, 75–80 x 7–8 μ. Sporidia uniseriate, elliptical, brown, 11–14 x 5½–7 μ. The substance of the matrix is partially bleached so that a horizontal section shows dull white blotches, which indicate the presence of the fungus.

On dead petioles of *Sabal serrulata*, Florida (Martin).

A. melanostícta, E. & E. Journ. Mycol. III, p. 44.

Perithecia gregarious or scattered, buried in the parenchyma of the leaf, with their black, dot-like ostiola barely projecting through the epidermis, which is not at all blackened or discolored. Asci 80–110 x 12–15 μ. Sporidia subbiseriate, elongated-elliptical and subinequilateral, brown, continuous, 18–22 x 7–9 μ, with a thin envelope at first.

On dead leaves of *Sabal Palmetto*, Louisiana.

A. Magnòliæ, E. & E. Journ. Mycol. IV, p. 122. **(Plate 31)**

Perithecia gregarious, hypophyllous, immersed, ⅓–½ mm. diam., slightly prominent and covered by the blackened cuticle, which is pierced by the papilliform ostiolum. Asci cylindrical, 75–85 x 5–6 μ, without paraphyses. Sporidia uniseriate, oblong-elliptical, pale brown.

2-3-nucleate, 7-8 x 3-4 μ, with a faint, obtuse, hyaline apiculus about 1½ μ long at the lower end, and a rather shorter one at the upper end.

On fallen leaves of *Magnolia*, Louisiana (Langlois).

A. Oreodáphnes, (Cke. & Hark.)

Sphæria (Anthostomella) Oreodaphnes, Cke. & Hark. Grev. XIII, p. 18.
Sphæria Oreodaphnes Berl & Vogl. Sacc. Syll. 6321.
Anthostoma Oreodaphnes, Cke. Syn. 4193.

Scattered, innate, covered. Perithecia globose, scarcely papillate, covered by the slightly raised, partially blackened epidermis. Asci cylindrical, 8-spored. Sporidia biseriate, elliptical, inflated in the middle, continuous, brown, 30-35 x 12-14 μ, with granular contents.

On leaves of *Umbellularia*, California (Harkness).

A. Ludoviciàna, Ell. & Langlois. Proc. Acad. Nat. Sci. Phil. July, 1890, p. 228.

Perithecia gregarious, covered by the blackened cuticle which is pierced by the papilliform, minutely perforated ostiola, 140-170 μ diam. Asci 50-55 x 3-3½ μ, cylindrical, paraphysate. Sporidia oblong-elliptical, brown, mostly 2-nucleate, 4-6 (mostly 4-5) x 2-2½ μ, uniseriate. The perithecia are often in subseriate patches, lying so near as to touch each other, but hardly confluent, and are buried in the substance of the bark, or even in the denuded wood, which is then continuously and uniformly blackened on the surface, but not within. Distinguished from other allied species by its small sporidia.

On dead stems of *Smilax*, Louisiana (Langlois).

A. smilacìnina, (Pk.)

Sphæria smilacinina, Pk. 29th Rep p. 62.
Anthostomella smilacinina, Sacc. Syll. 1043, Cke. Syn. 4624.

Perithecia abundant, slightly prominent, minute, at first covered by the thin, often blackened epidermis. Asci cylindrical or subclavate. Sporidia ovate or unequally elliptical, pale greenish-yellow, 12½-15 μ long, usually with a single large nucleus.

On dead stems of *Smilacina stellata*, New York State (Peck).

A. brachýstoma, E. & E. Bull. Wash. Coll. Vol. I (1884), p. 5.

Perithecia globose, ⅓ mm. diam., buried in the wood, their short, stout, obtuse, broadly perforated ostiola slightly projecting. Asci? Sporidia oblong-elliptical or subnavicular, brown, almost opake, 22-25 x 11-12 μ.

On rotten wood of *Tsuga Pattoniana*, Mt. Paddo, Washington. The spece. were old and the asci had mostly disappeared.

A. erúctans, E. & E. Proc. Roch. Acad. 1890, p. 50, pl. 4, figs. 7–8.

Perithecia gregarious, globose, $\frac{1}{2}$–$\frac{3}{4}$ mm. diam., with thick coriaceous walls, buried in the wood, abruptly contracted above into a short neck with an obtuse-conical, erumpent ostiolum. Asci cylindrical, 75–80 x 7–8 μ (p. sp.), with abundant paraphyses. Sporidia uniseriate, brown, continuous, rather acutely elliptical, 10–15 x 5–7 μ (mostly 12 x 5 μ). The surface of the wood is uniformly blackened. and the sporidia when mature are discharged as in *Massaria*.

On decorticated (maple?) limb, Lyndonville, N. Y. (Fairman).

A. pholidígena, (Ell.)

Sphæria (Anthostomella) pholidigena, Ell. Bull. Torr Bot. Club, X. p. 54.
Anthostomella pholidigena, Sacc. Syll. 6320. *Anthostoma*, Cke. Syn. 4181.
Exsicc Ell & Evrht. N. A F. 2d Ser 1664.

Perithecia subcuticular, erumpent, hemispherical, rough, $\frac{1}{3}$–$\frac{1}{2}$ mm. diam., with the ostiolum slightly prominent and broadly perforated. Asci linear, 114 x 7 μ, with abundant paraphyses. Sporidia uniseriate. narrow-elliptical, continuous, brown, 7–10 x 5–6 μ. Some of the perithecia contain stylospores which are much like the ascospores, only a little shorter.

On cones of red pine, Utah (S. J. Harkness). Mostly on the back of the scales, and covered by the overlapping point of the next scale below.

Differs from *A. conorum*, Fckl., in its smaller perithecia and sporidia and different stylospores.

A. ostioláta, Ell. Bull. Torr. Bot. Club XI, p. 42.

Perithecia single or 2–4 together, $\frac{1}{3}$–$\frac{1}{2}$ mm. diam., nearly buried in the unchanged inner bark, but with about one-third of their upper part projecting and closely covered by the blackened epidermis which is pierced by the short, stout ostiolum. Asci cylindrical, 80–85 x 7–8 μ. with filiform paraphyses. Sporidia obliquely uniseriate, oblong-elliptical, brown, 1-2-nucleate, 10–13 x 4–5 μ.

On dead twigs of *Laurus Benzoin*, Newfield, N. J.

The stroma is formed from the unaltered substance of the bark and not limited by any circumscribing line.

A. Baptísiæ, (Cke.)

Sphæria Baptisiæ, Cke. Grev. VI, p. 145.
Anthostomella Baptisiæ, Sacc. Syll 1061, Cke. Syn. 4603.
Exsicc. Rav F. Am. 200.

Scattered or subgregarious. Perithecia depressed, covered by the blackened epidermis. Asci obclavate. Sporidia elongated-elliptical, binucleate, brown, 14 x 4 μ.

On stems of *Baptisia perfoliata*. Aiken. South Carolina, with *Phoma Baptisiæ*, Cke.

A. nigrotécta, (B. & Rav.)

Sphæria nigrotecta, B. & Rav. Grev IV, p 155.
Anthostomella nigrotecta, Sacc Syll. 1054, Cke. Syn. 4601.

Perithecia shining black, white in the center around the ostiolum. Asci linear. Sporidia uniseriate, brown, elliptical.

On leaves of *Ilex*, Carolina.

Externally like *Physalospora philoprina*, (B. & C.) but smaller.

** *Sporidia with a hyaline appendage at one or both ends. (Entosordaria).*

A. confùsa, Sacc. Syll. 1065.

Sphæria appendiculosa, B & C Grev. IV, p 153, (not B & Br), Cke Syn. Pyr 4636.

Perithecia collected two or three together, closely surrounded at the base by the cuticle. Asci oblong. Sporidia biseriate, fusiform, quadrinucleate, 12–13 μ long, with a straight, hyaline, filiform appendage at each end.

On leaves of *Sapindus*, Texas.

A. rostríspora, (Ger.)

Sphæria rostrispora, Ger. Bull. Torr. Bot. Club, V, p. 26
Anthostomella rostrispora, Sacc. Syll. 1068, Cke. Syn. 4637.

Perithecia orbicular, densely crowded, seated on a blackish, compact, fibrous stroma. Asci cylindrical. Sporidia ovate, binucleate, brown, 15 x 5 μ, with a hyaline beak at each end.

Encircling the base of a stalk of *Inula Helenium*, New Paltz, N. Y.

A. sabalensioìdes, (E. & M.)

Sphæria sabalensioides, E. & M. Am. Nat. Oct. 1882, p. 810
Anthostomella sabalensioides, Sacc. Syll. 5932, Cke. Syn. 4635.
Exsicc. Ell. & Evrht. N. A. F. 2d Ser. 1694.

Perithecia scattered, minute, ¼ mm. diam., globose, covered, the short ostiolum barely piercing the epidermis and visible under the lens as a small black dot. Asci 75–80 x 7½–9 μ, without paraphyses? Sporidia biseriate, elliptic-fusoid, appendiculate, yellowish, surrounded by a gelatinous stratum, 13–15 x 3½–4 μ. The short, filiform appendages at each end of the sporidia are soon absorbed.

On petioles of *Sabal serrulata*, Florida.

A. clostérium, (B. & C.)

Sphæria closterium, B. & C. Grev. IV, p. 147.
Anthostoma closterium, Cke. Syn. 4198.
Anthostomella closterium, Sacc. Syll. 1067.

Minute, bursting through the cuticle which at first is closed, black and shining. Asci lanceolate. Sporidia elliptical in the center, with a long, attenuated, curved appendage at each end, 50 μ long. Sometimes the elliptical part is divided into two elliptical joints.

On *Spiræa opulifolia*, mountains of New York.

A. Cácti, (Schw.)

Sphæria Cacti, Schw. Syn. Car. 227.
Anthostomella Cacti, Sacc. Syll. IX, p. 512.

Perithecia gregarious, erumpent, subastomous, shining black, minute, mostly on round, yellowish-brown spots 3–4 mm. diam., with a definite, slightly raised border, and often confluent. Asci clavate-cylindrical, 40 x 10 μ (p. sp.), with a short, slender, pedicellate base. Paraphyses not seen. Sporidia biseriate, elliptical or obovate-elliptical, opake with a light-colored band (pseudo-septum) across the middle, the lower end subhyaline (and slightly appendiculate ?), 12–15 x 4–4½ μ.

On *Opuntia Engelmanni*, Los Angeles, Cala. (Scribner).

FAMILY. VÁLSEÆ.

Ascigerous stroma effused (diatrypoid), or subglobose, conical, or pulvinate (valsoid); often obscurely defined, or only indicated by a black, circumscribing line penetrating more or less deeply. Perithecia buried in the stroma. collected in groups or effused.

The ascigerous stromata are often preceded or accompanied by spermogonia producing sporules (mostly minute) borne on basidia which line the inner surface of the spermogonial cavity. The spermogonia accompanying the perithecia in the effused form of the ascigerous stroma, are generally simple, *i. e.* having the spermatiiferous cavity undivided, and generally pierced above with a single pore, but in the valsoid form of stroma, the cavity of the spermogonium is mostly divided into several cells or chambers stellately arranged, and inclosed by thin walls, or partitions extending in from the circumference towards the center, all these cells opening above through a single pore, or sometimes through several pores. The perithecia are either formed in the substratum beneath the spermogonia, or quite as often lie in a circle around it.

The members of this family grow in the bark or decorticated wood of dead limbs or on dead herbaceous stems.

DIAPÓRTHE, Nitschke.

Pyr. Germ. p. 240.

Stromata cortical, subvalsoid, separate (*Chorostate*), or effused, indeterminate, formed from the slightly altered substance of the bark and usually limited by a black line (*Tetrastaga*), or evenly effused, but with the surface of the matrix finally blackened, its substance otherwise unchanged, only in most cases limited by a black, circumscribing line visible on a horizontal section (*Euporthe*). Perithecia membranaceous or subcoriaceous, generally pale cinereous within, with a cylindrical or filiform exserted beak. Asci typically aparaphysate, fusoid, 8-spored. Sporidia fusoid or subelliptical, uniseptate,* generally constricted in the middle, 2–4-nucleate, hyaline, with or without appendages. The spermogonial stage is represented by species of *Phoma*.

A. Stroma valsoid; perithecia subcircinate; ostiola fasciculate. (Chorostate).

** Sporidia not appendiculate.*

D. oncóstoma, (Duby).

Sphæria oncostoma, Duby in Rab. Herb. Mycol. No. 205
Valsa personata, C. & E. Grev. VI, p 9
Diaporthe oncostoma, Fckl. Symb. p 205
Sphæria enteroleuca, Fr. Schw. Syn N. Am. 1314?
Exsicc. Fckl. F. Rh. 1730—Rab. l. c.—Ell. & Evrht. N. A. F. 2d Ser. 1952.—Thum. M. U. 1855.—Kze. F. Sel. 582.—Sydow. M. March. 551.—Roum. F. G. 2391.

Stromata scattered, globose-conical, innate-erumpent, with a black, circumscribing line which penetrates the wood. Perithecia subcircinate, globose, $\frac{1}{2}$–1 mm. diam., penetrating to the wood or partly immersed in it. Ostiola elongated, subconvergent, cylindrical or irregular, erumpent in a rather compact tuft. Asci oblong-clavate, p. sp. 45–50 x 8–9 μ. Sporidia biseriate, oblong-fusoid, 12–16 x 3–3$\frac{1}{2}$ μ, uniseptate and finally slightly constricted.

On dead branches of *Robinia pseudacacia*, Newfield, N. J.

Cke. in Grev. makes the sporidia 25–28 x 6 μ. We find them as above stated.

*Often each of the cells is divided by a faint septum but without any constriction, the division being only a separation of the cell contents and not a true septum.

D. rhòina. (C. & E.)

Diatrype rhoina, C. & E. Grev. VII, p. 8, Cke Syn. 1623.
Calospora rhoina, Sacc. Syll. 3707
Exsicc. Ell & Evrht. N. A. F. 2d Ser. 1953.

Perithecia subcircinate, 8–20 in a group, globose, $\frac{1}{3}$–$\frac{1}{2}$ mm. diam., the lower part sunk in the wood, necks convergent and erumpent in a compact fascicle of stout but short, cylindrical ostiola, which perforate the epidermis, but do not rise much above it, their tips rounded or obtusely conical, and finally with a large irregular opening. Stroma variable, orbicular, irregular in outline, often widely confluent and continuous for several inches, faintly circumscribed. Asci oblong-clavate, p. sp. 60 x 8–10 μ. Sporidia biseriate, oblong-fusoid, 2–4-nucleate, becoming uniseptate and slightly constricted, 12–15 x $2\frac{1}{2}$–3 μ.

On dead *Rhus venenata*, Newfield, N. J.

On a section exposing the perithecia they seem to lie almost evenly scattered. Sec. Cooke, the sporidia are 1–5-septate, 40 x 4 μ, but we can only find them as stated above.

D. acérina. (Pk.)

Valsa acerina, Pk. 28th Rep. p. 74.
Diaporthe acerina, Sacc. Syll 2365.
Valsa albocincta, C. & P. in Cke. Valsei of the U. S. No. 64, (fide Peck).

Perithecia sunk in the wood, $\frac{1}{2}$–$\frac{3}{4}$ mm. diam., in subcircinate groups of 6–10, circumscribed by a black line which penetrates deeply into the wood and encloses a space considerably larger than that occupied by the perithecia. Ostiola erumpent through the ruptured epidermis (which is also slightly raised), stout, short-cylindrical, obtuse, perforated, enveloped in a sulphur-yellow, grumous substance which finally disappears. The wood of the stroma within the circumscribing line is also tinged sulphur-yellow. Asci oblong-fusoid, p. sp. about 60 x 8–10 μ. Sporidia biseriate, oblong or subelliptical, subobtuse, 12–15 x 4 μ, uniseptate and constricted.

On dead *Acer spicatum*, Indian Lake, N. Y. (Peck), London, Canada (Dearness).

D. eùsticha, E. & E. (in Herb.)

Perithecia small ($\frac{1}{4}$ mm. diam.), buried in the unaltered substance of the inner bark in groups of 8–12, the ostiola converging and seriate-erumpent in a black, convex-hemispherical disc which is loosely surrounded by the ruptured epidermis. Asci clavate, 40–50 x 6–7 μ, aparaphysate, 8-spored. Sporidia biseriate, cylindric-fusoid, 4-nucleate, hyaline, straight, uniseptate, scarcely constricted, 12–15 x 3–$3\frac{1}{2}$ μ.

On dead hickory limbs, Newfield, N. J.

The ostiola do not project at all, but are slightly umbilicate-depressed.

D. Carpini, (Pers.)

Sphæria Carpini, Pers. Syn. p. 39.
Valsa Carpini, Fr. Summa Veg. Sc. p. 411.
Diaporthe Carpini, Fckl. Symb p. 205.
Exsicc. Thum. M. U. 2169.—Rehm Asc. 376.—Kunze F. Sel. 121 —Rav. F. Am. 746 —Krgr. F. Sax. 138.—Sydow M. March. 298, 1257.

Stroma cortical, pale. Perithecia circinate, numerous, crowded, closely circumscribed by a narrow line. Ostiola erumpent through the stellately torn epidermis, not confluent, at first papillate, finally umbilicate, and sometimes rostellate. Asci oblong, sessile, 8-spored, 30–60 x 7–8 μ. Sporidia subbiseriate, oblong-cylindrical, uniseptate, obtuse, hyaline, 4-nucleate, 13–18 x 3–4 μ. Spermogonia erumpent, 1–2 mm. across, with an irregular central opening; spermatia lance-fusoid, 2-nucleate, hyaline, 12–15 x $3\frac{1}{2}$–4 μ.

On *Carpinus Americana*, South Carolina (Ravenel); Canada (Dearness), New York (Peck and Fairman).

The specc. from Dr. Fairman have asci and sporidia corresponding to the smaller dimensions in the foregoing diagnosis, which is taken from Saccardo's Sylloge.

D. strumélla, (Fr.)

Sphæria strumella, Fr S. M. II, p. 365.
Diatrype strumella, Fr. Summa Veg. Scand. p. 385.
Diaporthe strumella, Fckl. Symb. p. 205
Exsicc Fckl. F. Rh. 598.—Rab. Herb. Myc. 49, 255.—id. F. E. 2431.—Rehm Asc. 429. Sydow M. March, 196.—Thum M. U. 1160, 1857.—Desm. Pl. Crypt. Ed. I, 1752, Ed. II, 1402.—Ell. & Evrht. N. A. F. 2d Ser. 2524.

Stromata thickly and evenly scattered, orbicular, depressed, formed of the unaltered substance of the inner bark, and covered by the epidermis. Perithecia 10–15 in a stroma, circinate, small, spherical, penetrating to the wood, and often adnate to it. Ostiola cylindrical, closely packed together, and erumpent in a black, elliptical, transverse disk which is finally obliterated; mostly short and obtuse, but sometimes elongated. Asci 50–60 x 7–10 μ, subsessile, clavate. Sporidia biseriate, fusoid, uniseptate and constricted at the septum, 13–22 x 3–4 μ, greenish-hyaline, slightly curved.

On dead stems of *Ribes lacustre*, Canada (Dearness), New York (Peck).

D. subcóngrua, E. & E. (in Herb.)

Perithecia ovate-globose, $\frac{1}{3}$ mm. diam., 6–12 together, enclosed in a light colored, oblong (3–4 x 2–3 mm.) stroma and penetrating to the wood, which is marked with a black, circumscribing line; necks subconvergent, their large, conic-globose ostiola connate at the base, but without any distinct disk, erumpent through cracks in the slightly raised epidermis. Asci oblong-fusoid, about 50 x 8–10 μ, subsessile, 8-spored. Sporidia biseriate, oblong-elliptical, uniseptate and slightly constricted, 12–14 x 3½–4½ μ, ends subobtuse. The ostiola resemble those of *Eutypella cerviculata*, only they are smaller and not sulcate.

On dead maple limbs, London, Canada (Dearness).

Closely allied to *D. pustulata*, (Desm.), but the specimen of that species in Desm. Pl. Crypt. has a distinct, nearly round, almost plane disk, with the ostiola only slightly prominent and narrower, and rather longer, acute spóridia.

D. cóngener, E. & E. (in Herb.)

Exsicc. Ell. & Evrht. N. A. F. 2d Ser. 2532.

Stromata evenly scattered, depressed-conical, formed of the slightly altered substance of the bark which is lighter than the adjacent parts, limited by a black, circumscribing line which penetrates the wood, apex brownish-black, erumpent through the ruptured epidermis which, however, is only slightly raised. Perithecia subcircinate, adnate to the surface of the wood but scarcely penetrating it, about 6–8 in number, ½ mm. diam., rather abruptly contracted into slender, slightly converging necks, piercing the disk with their rounded, obtuse, soon umbilicate, slightly projecting ostiola. Asci clavate-cylindrical, 70-75 x 10–12 μ. Sporidia uniseriate, direct or oblique, sometimes partly biseriate above, elliptical, hyaline, uniseptate and strongly constricted, each cell with a large nucleus, 12–14 x 6 μ.

On dead limbs of *Fraxinus*, London, Canada (Dearness).

In *D. fibrosa*, (Pers.) which is credited to this country by Schweinitz & Berkeley, the asci and sporidia are scarcely distinguishable from those of our *D. congener*, but in that species the inner bark is blackened and subcarbonized with long, coarse, light colored fibers bedded in its surface just under the epidermis, and there is, in the European specc. we have seen, no black circumscribing line penetrating the wood. The perithecia also are larger. Berkeley (Grev. IV, p. 99.) states that in specc. from the mountains of New York, the sporidia are at length "fenestrate," a character not applicable to *D. fibrosa*. *Sphæria extensa* Fr. Schw. Syn. N. Am. 1315, is apparently not distinct from *D. fibrosa*.

D. técta, (Cke.)

Valsa tecta, Cke. Grev. XI, p. 109, Syn Pyr. 2000.
Diaporthe tecta, Sacc Syll. 6091.

Exsicc Rav. F. Am 747.

Perithecia circinate, few, nestling in the unaltered substance of the bark and raising the epidermis into little pustules, small (150 μ), about four together, contracted above into slender necks terminating in the short, inconspicuous ostiola, which barely rupture the epidermis without projecting above it. Asci oblong 25–30 x 5–6 μ. Sporidia biseriate, oblong-sublanceolate, slightly curved 2–4-nucleate, uniseptate, hyaline, 6–7 x 1½–2 μ (18 x 5 μ, Cke.).

On dead limbs of *Myrica*, Darien, Georgia (Ravenel).

This discrepancy between Cooke's measurements and ours is remarkable. We have carefully examined the specc. in Rav. F. Am., and find the sporidia as stated above.

D. Woolwórthii, (Pk.)

Valsa Woolworthii, Pk. 28th Rep p 73.
Diaporthe Woolworthii, Sacc. Syll. 2383

Minute, erumpent. Perithecia 2–6 together, nestling in the inner bark. Ostiola stout, becoming umbilicate, crowded, slightly prominent, barely exserted through the ruptured epidermis. Asci p. sp. 30–35 x 7 μ. Sporidia crowded or biseriate, oblong-fusoid, uniseptate, scarcely constricted, nearly colorless, 10–12 x 2½–3½ μ.

On dead oak and hickory branches, Greenbush, N. Y. (Peck), on oak and *Tilia*, Canada (Dearness).

The clusters of perithecia are very numerous and often seriate, the epidermis being ruptured from one to another. What appears to be the same has been sent by Mr. Langlois from Louisiana, on dead limbs of white oak, with the ostiola at first erumpent through a pale disk which at length disappears.

D. farinósa, Pk. 40th Rep. p. 69.

Stroma subpulverulent or mealy, dull buff color, formed of the slightly changed inner bark, erumpent in a minute, slightly exserted disk. Perithecia irregularly circinating, generally 4–10 together, the clusters subconfluent. Ostiola black, dotting the prominent, pulverulent, buff-colored or, at length, brownish disk. Asci subcylindrical, 55–75 x 7½–10 μ. Sporidia crowded or biseriate, oblong or subfusiform, uniseptate, generally 4-nucleate, 15–20 x 3½–4 μ.

On dead branches of *Tilia Americana*, Argusville, N. Y. (Peck).

This species approaches *D. furfuracea* in its pulverulent stroma, but differs in its prominent disk, making the branches rough to the touch, and in its smaller sporidia. From *D. velata* it is easily separated by the entire absence of any black, circumscribing line or blackened surface. It evidently belongs to the subgenus *Chorostate*, but the clusters of perithecia are so numerous that they form an almost continuous stratum, which surrounds the branch and extends for a long distance under the epidermis.

We have not seen this species, and take the foregoing from the report cited.

D. Saccardiàna, Kze. Sacc. Syll. 2430.

Exsicc. Kze. F. Sel 123.

Perithecia 2–6 together, subcircinate, buried in the unaltered inner bark without any circumscribing line, about 500 μ diam., with their stout, conical or conic-cylindrical ostiola erumpent in a loose, fascicle together. Asci p. sp. oblong-fusoid, 40–45 x 8–9 μ. Sporidia biseriate, oblong-fusoid, 4-nucleate, uniseptate and very slightly constricted, 10–12 x 3–3½ μ (15 x 4½–5 μ, Sacc.).

On small, dead limbs of elm, London, Canada (Dearness).

The small groups of perithecia lie close together, almost filling the bark. There is no distinct apiculus at the ends of the sporidia, nor do we see any in Kunze's specimens in which the sporidia are a trifle larger, 10–14 x 3½-4 μ. The latter also show a distinct, circumscribing line in the older and thicker bark, but none on the smaller limbs.

D. impúlsa, (C. & P.)

Valsa impulsa, C. & P. 27th Rep. N. Y. State Mus. p. 109.
Diaporthe impulsa, Sacc. Syll. 2395.

Stroma depressed-subconical, ⅓–½ cm. diam., formed of the scarcely altered substance of the bark, with a black, limiting line penetrating the wood. Perithecia 8–12, subcircinate, ⅓–½ mm. diam., with slender necks, the obtusely-conical, finally umbilicate-collapsing ostiola erumpent through a small, grayish-black disk rather loosely surrounded by the ruptured epidermis. Asci clavate-cylindrical, p. sp. about 40 x 6 μ. Sporidia biseriate, fusoid, uniseptate, nucleate, hyaline, hardly constricted, 16–20 x 3–3½ μ.

On dead branches of *Pyrus Americana*, Adirondack Mts., N. Y. (Peck), on *Pyrus sambucifolia*, California (Harkness).

Described from spece. sent by Mr. Peck.

D. Ailánthi, Sacc. M. Ven. spec. p. 137, tab. XIII, figs. 40–43.

Stroma valsoid. Acervuli scattered, small, consisting of 5–6 perithecia lying in subcircinate groups buried in the surface of the wood and covered by the slightly raised epidermis, which is finally ruptured. Perithecia subglobose, about ⅓ mm. diam. Ostiola erumpent in a fascicle together, for some time covered by the epidermis, but finally more or less exserted. Asci fusoid, 50 x 6–7 μ. Sporidia subbiseriate, oblong-fusoid, 4-nucleate, slightly constricted in the middle, 12–15 x 3–3½ μ.

On dead limbs of *Ailanthus glandulosa*, Delaware (Commons).

The specc. sent by Mr. Commons are quite variable as to the ostiola which are sometimes ½ mm. long and then clavate. The stroma is limited by a black circumscribing line.

D. bicíncta, (C. & P.)

Valsa bicincta, C. & P. 29th Rep. N. Y. State Mus. p. 64.
Diaporthe bicincta, Sacc. Syll. 2411.

Stromata thickly scattered, depressed-hemispherical, about 3 mm. diam., seated on the inner bark without any circumscribing line, dirty white inside. Perithecia 6–20 in a stroma, of medium size, sunk in the bark, their cylindrical necks terminating in a compact fascicle of stout, black, obtusely conical ostiola piercing the epidermis but not rising much above it, and finally umbilicate-collapsed. Asci subcylindrical, 40–45 x 6 μ. Sporidia biseriate, narrow-elliptical, 4-nucleate. not constricted, 10–13 x 3 μ.

On dead branches of *Juglans cinerea*, New York (Peck).

When the epidermis is peeled off, the fascicle of ostiola is surrounded by a pale whitish zone at base, and this, again, by a dark-colored line.

D. leiphæmia, (Fr.)

Sphæria leiphæmia, Fr. S. M. II, p. 399.
Valsa leiphæmia, Fr. Summa, Veg. Sc. p. p 412
Diaporthe leiphæma, Sacc M. V. spec. p. 135, tab. XIII, figs. 26–28
Exsicc. Fckl F Rh. 611 —Kze F. Sel. 348.—Rab. Herb. Mycol 732, id. F. E 1015, 2225 Rehm Asc. 476.—Cke. F. Brit. Ser. I, 255. 2d Ser. 225 —Desm. Pl. Cr. Ed. I, 1256. Ed. II, 756.—Ell. N. A. F. 93

Stromata numerous and thickly scattered but not confluent, cortical, with a pale, erumpent, convex-hemispherical disk. Perithecia numerous, lying in the bottom of the stroma, pale, soft. Ostiola not strongly exserted. Asci fusoid, subsessile, 70–80 x 9 μ, 8-spored. Sporidia biseriate, oblong-elliptical, uniseptate, slightly constricted, 15–20 x 4 μ.

On decaying oak limbs, common. Mostly on *Quercus alba* around Newfield.

D. Raveneliana, Thüm. & Rehm in Thüm. M. U. 865, is *D. leiphœmia*, Fr. (at least in our copy).

D. crinígera, E. & E. Proc. Acad. Nat. Sci. Phil. July, 1891.

Stroma cortical. Perithecia buried in the substance of the inner bark, subcircinate, 7–20 together, ovate-globose, $\frac{1}{3}$–$\frac{1}{2}$ mm. diam., contracted above into short, slender, convergent necks with the ostiola smooth and rounded or distinctly quadrisulcate and erumpent in a small, capitate fascicle. In well developed specimens the ostiola are cylindrical, 1–2 mm. long, but quite as often they project only slightly above the epidermis. There is not a separate circumscribing line around each cluster of the perithecia, but one continuous, black layer extends along just under the surface of the inner bark over the entire space occupied by the fungus. In the early stage of growth, and where there are only a few perithecia in a cluster, the surface of the inner bark is smooth and even, but where the perithecia are more numerous and well developed, they raise the bark into little flat pustules about 2 mm. diam. Asci 45 x 7–8 μ (p. sp.), with paraphyses. Sporidia biseriate, oblong-fusoid, 4-nucleate, slightly constricted in the middle, 10–13 x 3–3$\frac{1}{2}$ μ, ends subobtuse. This was at first referred to *Diaporthe Woolworthii*, Pk., but, having compared it with a specimen of that species from Mr. Peck, we find it to differ in its larger and more numerous perithecia with long, cylindrical ostiola, and its broader sporidia; nor is there any seriate arrangement in the clusters of perithecia, or any circumscribing line. Mr. Commons sends the same from Delaware (No. 1266), differing only in the clusters of perithecia being more or less longitudinally confluent.

On dead oak limbs, London, Canada (Dearness).

D. magníspora, (E. & E.)

Valsa magnispora, E. & E. Journ. Mycol. III, p 42.
Diaporthe magnispora, Sacc Syll. IX, p. 707

Perithecia buried in the inner bark, not penetrating to the wood, or circumscribed by any black line, 6–10 in a cluster, globose-ovate, about $\frac{1}{4}$ mm. diam., contracted above into short necks which burst in a cluster through the epidermis, but project only slightly above it, their apices (ostiola) hemispherical, black, smooth and shining, with a minute, central pore, and sometimes slightly umbilicate. Asci subsessile, oblong-cylindrical, 100–120 x 18–22 μ. Sporidia biseriate, oblong-fusoid, hyaline, uniseptate, slightly curved, 25–35 x 7–11 μ. *Diaporthe Aceris*, Fckl., has asci only 60 x 8 μ and sporidia 14 x 4 μ.

On dead maple limbs, Plainfield, N. J. (G. F. Meschutt).

D. Àceris, Fckl. Symb. p. 204.

Valsa myinda, C. & E. Grev VI, p. 93.
Diaporthe myinda, Sacc. Syll. 2368.
Exsicc. Ell. N. A. F. 180.

Perithecia small, about ¼ μ diam., buried in the inner bark in subcircinate groups of 8–12, covered above by a thin, black crust lying just beneath the epidermis. Ostiola slightly convergent, rising separately through the overlying crust and raising and rupturing the epidermis, short-cylindrical, subcompressed or otherwise irregular. Asci clavate-cylindrical, subsessile, aparaphysate, 8-spored, 50–60 x 6–7 μ. Sporidia biseriate, biconical, or acutely elliptical, uniseptate and constricted, subobtuse, hyaline, 12–14 x 4 μ.

On bark of *Acer rubrum*, Newfield, N. J.

The groups of perithecia almost entirely fill the bark for a foot or more in length and surround the limb, but the bark does not seem to be in any way discolored, nor is there any circumscribing line, but the surface is roughened by the ruptured epidermis around the slightly projecting ostiola.

D. Eucalýpti, Hark. New Cal. Fungi, p. 44.

"Perithecia aggregated in valsiform groups bordered by a raised black line. Ostiola 1 mm. or more long, lax, somewhat agglutinated. Asci oblong-fusoid, 33 x 7 μ. Sporidia fusiform, slightly curved. hyaline, uniseptate, 4-nucleate, acuminate, 15 x 4 μ."

On dead leaves of *Eucalyptus globulus*, San Francisco, Cala.

D. Columbiénsis, E. & E. Proc. Acad. Nat. Sci. Phil. July, 1890. p. 233.

Perithecia in subcircinate clusters of 3–6 (occasionally only one). buried in the inner bark, their bases penetrating to the subjacent wood, large (¾–1 mm.), collapsing below, abruptly contracted above into short necks terminating in subtubercular, quadrisulcate-cleft ostiola, erumpent (but not strongly prominent) through the thin, black, superficial, convex crust that covers the stroma. The substance of the stroma consists entirely (except the black, circumscribing layer) of the bleached substance of the bark. Stroma elliptical, 2–5 mm. diam., with a distinct, black, circumscribing line, which does not penetrate deeply into the wood. Asci oblong-lanceolate, about 100 x 12 μ. Sporidia biseriate, hyaline, oblong, 20–22 x 7–8 μ, uniseptate and constricted, each cell with a large nucleus.

On dead limbs of some undetermined tree, British Columbia (Macoun).

D. oculària, (C. & E.)

Valsa ocularia, C. & E. Grev. VI, p. 11.
Diaporthe ocularia, Sacc. Syll. 2389.

Perithecia 3–12, seated on the wood, collected in a pallid stroma, and surrounded by a faint black line, so that when the bark is removed, they have an ocellate appearance, about ⅓ mm. diam., the short, obtuse, umbilicate ostiola barely piercing the bark. Asci about 40 x 6 μ (p. sp.). Sporidia subbiseriate, fusoid, 4-nucleate, becoming faintly uniseptate (faintly 5-septate, Cke.), not constricted, 15–20 x 3 μ (25 x 4 μ, Cke.).

On dead branches of *Ilex* (*glabra*)? Newfield, N. J.

This must not be confounded with *D. binoculata*, Var. *Ilicis*, which has much larger, elliptical, constricted sporidia.

D. binoculàta, (Ell.)

Valsa binoculata, Ell. Bull. Torr. Bot. Club, IX, p. 111.
Diaporthe binoculata, Sacc. Syll. 6093
Exsicc. Ell. N. A. F. 879.

Perithecia 3–6, rather large, deeply imbedded in a depressed-hemispherical stroma formed of the slightly altered substance of the bark and enclosed in a tolerably thick, hard, carbonaceous crust, which penetrates to the wood and shows on a horizontal section as a black, circumscribing line. Ostiola subglobose, with a large, irregular opening, slightly erumpent through cracks in the slightly raised epidermis. Asci clavate-cylindrical, 100–150 (mostly 100–112) x 12–20 μ, p. sp. 75–80 μ, with granular paraphyses. Sporidia uniseriate or sometimes subbiseriate, elliptical, uniseptate and constricted. with a large nucleus in each cell, 20–25 x 12–14 μ, but often smaller 15–20 x 10–12 μ, constricted at the septum.

On dead trunks of *Magnolia glauca*, Newfield, N. J., and on *Ilex verticillata*, London, Canada (Dearness).

The ostiola often fail to rise through the epidermis, so that outwardly there is no trace of the fungus. The ascigerous nucleus is soft and pale, and the perithecia themselves are almost colorless.

This differs from *D. ocularia*, C. & E., in its much longer asci and very different sporidia.

D. stictóstoma, (Ell.)

Valsa punctostoma, Ell. Am. Nat. March, 1883, p. 316.
Diaporthe stictostoma, Sacc. Syll. 6096.

Stroma cortical, formed of the unaltered substance of the inner bark. Perithecia 8–12, about ½ mm. diam., in a single layer, their short-cylindrical beaks joined in a small (¾ mm.), olivaceous, slightly elevated disk which is closely girt by the epidermis, and pierced around its circumference by the black, obtuse, slightly prominent, rather broadly perforated ostiola. Asci clavate-cylindrical, 55 x 8–9 μ.

Sporidia biseriate, oblong-elliptical, 4-nucleate, slightly constricted in the middle, hyaline, 11–13 x 4–4½ μ.

On dead limbs of *Amelanchier Canadensis*, Iowa (Holway).

D. Magnòliæ, E. & E. (in Herb.)

Stroma irregular in outline, elongated for 2–4 (or more) cm. in extent, and consisting of the unaltered substance of the bark and wood, enclosed in a thin, black layer which appears, on a horizontal section, as a narrow, black line penetrating the wood to the depth of 2–3 mm. Perithecia 4–8 together, buried in the bark and partly sunk in the subjacent wood, about ½ mm. diam., whitish or horu-color inside, with converging necks erumpent in a fascicle together, their tubercular-cylindrical ostiola, rounded and obtuse, and more or less distinctly quadrisulcate, soon perforating or rupturing the epidermis and raising it into small pustules, but hardly rising above it. Asci clavate-cylindrical, aparaphysate, subsessile, 8-spored, 40–50 x 6–7 μ. Sporidia subbiseriate, oblong-fusoid, 2–3-nucleate, then uniseptate and slightly constricted, 10–12 x 3 μ.

On dead limbs of *Magnolia glauca*, Newfield, N. J.

This was published in the Proc. Acad. Nat. Sci. Phil. July, 1890, p. 235, as *D. Americana*, Speg., but the perithecia are too large, and too distinctly clustered to allow it to be referred to that species. It belongs evidently in the subgenus *Chorostate*, where we now place it.

D. Ellísii, Rehm. in literis and in Sacc. Syll. 6554.

Exsicc. Ell. & Evrht. N. A. F. 2d Ser. 1567.

Stroma cortical without any circumscribing line. Perithecia numerous (15–20), subcircinate, about ½ mm. diam., raising the bark into small pustules, through the center of which, bursts the brown, convex disk, pierced around its margin by the black, obtuse, dot-like ostiola. Asci subclavate, 57–62 x 7–9 μ. Sporidia biseriate or overlapping, narrowly elliptical, 4-nucleate, often slightly constricted in the middle, 11–15 x 3½–5½ μ.

On dead branches of *Carpinus Americana*, West Chester, Pa.

With a *Cytispora* having oblong, hyaline, 4-nucleate sporules, 8–11 x 2–2½ μ, issuing in an orange colored mass.

D. tuberculòsa, (Ell.) **(Plate 33)**

Valsa tuberculosa, Ell. Bull. Torr. Bot. Club, VIII, p. 89.
Diaporthe tuberculosa, Sacc. Syll. 2404.
Exsicc. Ell. N. A. F. 880.

Perithecia 8–10, about 400–500 μ diam., subcircinating, and buried in a stroma formed entirely of the substance of the bark, which is not discolored, though rendered more compact and surrounded by a thin, black layer which penetrates the wood beneath. Ostiola short-cylindrical, thick, stout, obtuse, with an irregular opening. Asci broad-lanceolate; sessile, 75 x 15 μ. Sporidia biseriate, oblong-elliptical, uniseptate and slightly constricted, hyaline, 12 x 5–6 μ, with two large nuclei, at length easily separating in the midddle.

On dead limbs of *Amelanchier Canadensis*, Newfield, N. J., and on dead maple limbs, London, Canada (Dearness).

When the bark decays and falls away, the black, tuberculiform stroma remains attached to the wood and then, much resembles the stroma of some *Diatrype*; this at least is the case with the specc. on *Amelanchier*.

D. Ampelópsidis, (Ell.)

Valsa Ampelopsidis, Ell. Bull. Torr. Bot. Club, IX, p 112.
Cryptosporella Ampelopsidis, Sacc. Syll. 6025.
Valsa (Cryptosporella) Ampelopsidis, Cke. Syn. 1940.
Exsicc. Ell. N. A. F. 881.

Perithecia subcircinate, few, seated on the surface of the wood and enveloped in the bark without any distinct stroma. Ostiola cylindrical, subacute, their tips united in an elliptical, plane disk erumpent through longitudinal fissures in the bark, but at length obliterated. Asci clavate-cylindrical, about 70 x 12 μ. Sporidia biseriate, oblong-elliptical, subacute, hyaline, 1–4-nucleate, becoming uniseptate, 18–22 x 7½ μ.

On dead stems of *Ampelopsis quinquefolia*, Newfield, N. J.

D. apocrýpta, (C. & E.)

Valsa apocrypta, C. & E. Grev. VIII, p. 15.
Diaporthe apocrypta, Sacc. Syll. 2409.

Perithecia in groups of 3–8, subcircinate in the unchanged inner bark, ⅓–½ mm. diam., their short, obtuse, subfasciculate ostiola raising the epidermis into slight pustules which are finally ruptured, but the ostiola are not exserted. Asci clavate-cylindrical, 75–80 x 15–18 μ. Sporidia biseriate, oblong, uniseptate and constricted, obtuse, hyaline, 18–20 x 6 μ (25–28 x 8–9 μ, Cke.). The groups of perithecia are thickly scattered through the bark, more or less irregularly disposed, often contiguous or confluent, and scarcely visible externally.

On decaying hickory limbs, Newfield, N. J.

D. corymbòsa, (C. & E.)

Valsa corymbosa, C. & E. Grev VIII, p. 15.
Diaporthe corymbosa, Sacc. Syll. 2412.

Perithecia immersed in the wood, in an elliptical (2–3 x 1–2 μ) stroma bounded by a black, circumscribing line, 4–10 in a group, their slender, fasciculate ostiola barely erumpent or entirely concealed by the epidermis which is only very slightly or not at all elevated over them. Asci about 70 x 8–10 μ, cylindrical. Sporidia obliquely uniseriate, elliptical, uniseptate, 18–20 x 7–8 μ.

On dead branches of *Vaccinium corymbosum*, Newfield, N. J.

Much like *D. apocrypta*, but the perithecia are smaller and sunk in the wood, only their apices rising into the bark, and in that species there is no black, circumscribing line, either in the wood or bark.

D. salicélla, (Fr.)

Sphæria salicella, Fr. S M. II, p 377.
Sphæria salicina, Curr. Linn. Trans. XXII, tab. 48, fig. 149.
Halonia salicella, Fr. Summa Veg. Scand. p. 397.
Diaporthe Salicis, Nitschke, in Feckl. F. Rh. 1987.
Cryptospora salicella, Fckl. Symb 193.
Diaporthe salicella, Sacc. Mycol. Ven. spec. p. 135.
Exsicc. Fckl. F Rh. 800, 1987 —Rab. F. E. 2046.—Thum. M. U. 170.

Perithecia mostly in groups of 4–8, buried in the inner bark without any stroma, the short, conic-cylindrical ostiola breaking through the epidermis in a small fascicle projecting a little above the slightly pustuliform-elevated epidermis, and appearing like small black specks thickly scattered over the limb. Mixed with the groups are also many perithecia standing singly. Asci 65–70 x 14–16 μ, elongated-clavate. Sporidia biseriate, oblong-fusoid, inequilateral, uniseptate, not constricted, yellowish-hyaline, 15–20 x $4\frac{1}{2}$–$5\frac{1}{2}$ μ, ends subobtuse.

On dead limbs of *Salix*, New York (Fairman), Canada (Dearness), Pennsylvania (Berk. in Grev. IV, p. 147).

The perithecia are 400–500 μ diam., thin and membranaceous, collapsing from below.

D. sociàta, (C. & E.)

Valsa sociata, C. & E Grev VI, p 11,
Diaporthe sociata, Sacc Syll 2378.

Perithecia 3–5 together, buried in the unaltered substance of the bark, minute (200 μ), globose, slightly raising the epidermis, which is ruptured by the short, obtuse, rather broadly perforated ostiola. Asci oblong-clavate, about 4 x 8 μ. Sporidia oblique or biseriate, oblong-fusoid, uniseptate, hyaline, 10–12 x 3 μ (12 x 4 μ, Cke.).

A horizontal section shows a circular or elliptical stromatic area about 1 mm. diam., bounded by a faint black line which hardly penetrates to the wood.

On dead limbs of *Laurus Benzoin*, Newfield, N. J. With a tufted *Helminthosporium* rising from the pustules and bearing clavate, multiseptate conidia.

D. paùlula, (C. & E.)

Valsa paulula, C. & E. Grev. VII, p. 9.
Diaporthe paulula, Sacc. Syll. 2391.

Perithecia few, 2–4 together, subglobose, about 200 μ diam., buried in the unaltered substance of the bark, which is raised into small pustules ruptured by the short, obtuse, subumbilicate ostiola, which scarcely project. Asci clavate, p. sp. about 40 x 6 μ. Sporidia biseriate, oblong-fusoid, uniseptate and constricted, hyaline, 16–18 x 3–4 μ. Spermogonial stroma radiate-multilocular, about 1 mm. diam., rupturing the epidermis like the ascigerous stroma, from which it can scarcely be distinguished outwardly. Spermatia allantoid, hyaline, 4 x 1 μ.

On dead twigs of *Nyssa*, Newfield, N. J., with *Cornularia hispidula*, (Ell.).

Differs from *D. sociata*, C. & E., in the absence of any circumscribing line.

D. nivòsa, Ell. & Holw. Proc. Acad. Nat. Sci. Phil. Pa. July, 1890.

Perithecia mostly 8–12, about ¾ mm. diam., subcircinate, buried in the unaltered substance of the bark which is raised in a pustulate manner over them, contracted above into short necks with black, subhemispherical, papilliform ostiola erumpent around the margin of a snow-white disk rather less than 1 mm. diam., having the same general appearance as *V. nivea*, Fr. Asci (p. sp.) about 60 x 12 μ. Sporidia subbiseriate, oblong, 4-nucleate, uniseptate, constricted, 12–16 x 3–4 μ, straight or very slightly curved. There is no black, circumscribing line around the stroma.

On dead alders, Isle Royale, Lake Superior (Holway).

D. conjùncta, (Nees.)

Sphæria conjuncta, Nees. Syst. p. 305, fig. 337
Diaporthe conjuncta, Fckl. Symb. p. 206.

Perithecia circinate, globose, crowded, with short necks and with elongated-cylindrical, smooth ostiola thickened at the tips and joined in a round or transversely elongated, slightly projecting disk. Asci clavate, sessile, 8-spored, 63–78 x 10 12 μ. Sporidia biseriate, fusoid, slightly attenuated at the ends, inequilateral, septate in the middle, not constricted, hyaline, 17–20 x 3½–4 μ.

D. Comptòniæ, (Schw.)

Sphæria Comptoniæ, Schw. Syn. N. Am. 1353.
Diaporthe Comptoniæ, E. &. E. Proc. Acad. Nat. Sci. Phil. July, 1890, p 234.
Exsicc. Ell & Evrht. N. A. F 2d Ser. 2364

Perithecia immersed in pustules formed of the scarcely altered substance of the inner bark, 4–12 together, globose, black, 200 μ or over diam., contracted above into slender necks with the ostiola barely piercing the pustulate-raised epidermis. The pustules are often thickly scattered and subconfluent. Asci about 50 x 7 μ, clavate, aparaphysate. Sporidia biseriate, fusoid, 4-nucleate, then uniseptate, scarcely constricted, straight or slightly curved, 16–20 x $2\frac{1}{2}$–3 μ, attenuate-acuminate at each end.

On *Comptonia asplenifolia*, Pennsylvania and New Jersey.

On old dead stems of *Comptonia*, the cracks in the outer bark often extend entirely around the stem, and in these cracks the necks of the perithecia, which were at first enclosed in the bark, are exposed, projecting in fascicles, or often in a continuous series half way round the stem, but their tips scarcely rise above the surface of the bark. The specc. of *S. Comptoniæ*, Schw., in Herb. Schw., do not afford any fruit, so that we cannot certainly say that what we have here described as *D. Comptoniæ*, (Schw.), is really what he designated under that name. The outward appearance, however, of the specc. in the Schweinitzian collection is the same as that of the specc. we have referred to that species.

D. phomáspora, (C. & E.)

Valsa phomaspora, C. & E. Grev. VI, p. 10.
Valsa (*Cryptosporella*) *phomaspora*, Cke. Syn. 1938.
Exsicc. Ell. N. A. F. 179.

Pustules covered by the slightly raised epidermis, scattered, small, about 1 mm. diam. Perithecia 4–8 in a stroma, minute (150 μ), subcircinate in a cortical stroma which is a little paler than the surrounding portions of the bark, ovate-conical, necks short, converging, with the fascicle of minute ostiola barely visible through short, transverse cracks in the thick epidermis. Asci oblong-clavate, rounded above at first, then fusoid, 40 x 6–7 μ, sessile, 8-spored, aparaphysate. Sporidia biseriate, fusoid-elliptical, 4-nucleate, becoming uniseptate, hyaline, 8–10 x $2\frac{1}{2}$–$3\frac{1}{2}$ μ.

On dead stems of *Myrica cerifera*, Newfield, N. J.

D. Ontariénsis, E. & E. (in Herb.)

Perithecia 2–6 together, about $\frac{1}{2}$ mm. diam., in a depressed-hemispherical, cortical stroma with a black outer stratum penetrating the wood. Ostiola at first conical, then umbilicate, thick and stout, erumpent in a black, carbonaceous disk which raises the epidermis into distinct pustules. Asci lanceolate, p. sp. about 65 x 12 μ. Sporidia biseriate, elliptical, uniseptate and constricted, obtuse, hyaline, 12–15 x –$6\frac{1}{2}$ μ.

On maple limbs, London, Canada (Dearness).

This is closely allied to *D. pustulata*, Desm., but the specimens of that species in Desm. Pl. Crypt. 1755, have the sporidia fusiform, 4-nucleate, acute, and only 3–3½ μ thick, and the black stratum enclosing the stroma is thicker and harder, both perithecia and stroma coming off with the bark, which is not the case with the Canada specimens.

D. Robergeàna, (Desm.)

Sphæria Robergeana, Desm. Not. 19, 1851, p. 11.
Diaporthe Robergeana, Niessl, Rab. F. E. 2222.
Exsicc. Rab. F. E. l. c.—Desm. Pl. Crypt. Ed. I, 2053, Ed. II, 1755.

Corticolous, immersed, minute. Perithecia 6–15 together, covered by the adnate epidermis, irregularly circinate, black, collapsing to concave. Ostiola cylindrical, converging, erumpent, prominent. Asci subfusoid, 75 μ long. Sporidia biseriate, subellipsoid, uniseptate, or 2-nucleate, greenish-hyaline, 15 x 5 μ.

On branches of *Staphylea trifolia*, Albany, N. Y. (Peck).

D. Cratægi, (Curr.)

Valsa Cratægi, Curr. Linn. Trans. XXII, p. 278, tab. 48, fig. 135*a*.
Diaporthe (*Chorostate*) *Cratægi*, Fckl. Symb. p. 204.

Perithecia subvalsoid, tolerably compact. Ostiola slightly emergent, rather obtuse. Asci fusoid, 90–100 x 9–12 μ, aparaphysate, 8-spored. Sporidia biseriate, cylindrical, subobtuse, 17–18 x 5–6 μ, uniseptate, slightly constricted, 4-nucleate, hyaline.

On dead ash branches, Catskill Mts. N. Y. (sec. Peck 31st Rep. p. 50).

** *Sporidia appendiculate.*

D. galericulàta, (Tul.)

Valsa galericulata, Tul. Sel. Carp. II, p. 263.
Diaporthe galericulata, Sacc. Syll. 2435, Cke. Syn. 2028.

Stroma cortical, disk obtuse, whitish, erumpent, dotted with the 8–10, round, black ostiola. Asci ovate-oblong, 8-spored. Sporidia biseriate, lanceolate, slightly curved, uniseptate, with a short, bristle-like appendage at each end; hyaline, 19–23 x 6½ μ. Spermogonium (*Fusicoccum*)—spermatia fusoid, 6–10 x 3½–4 μ.

On *Fagus sylvatica*, New York State (fide Peck).

D. decèdens, (Fr.)

Sphæria decedens, Fr. in Kze. & Schm. Mycol. Hefte II, p. 49.
(*Diaporthe decedens*, Fckl. Symb. Nachtr. I, p. 318)?

Perithecia in subcircinate groups of 8–12, semiimmersed in the wood and covered by the bark, which is only slightly raised, about ⅓ mm. diam. Ostiola stout, short, obtuse, papilliform at the apex and soon perforated, not joined in a disk but erumpent singly, and only slightly convergent. Asci (sec. Fckl.) 70–80 x 9 μ. Sporidia lanceolate-oblong, septate in the middle, hyaline, acutely appendiculate at each end, 12 x 4 μ.

On dead limbs of *Corylus*, Iowa (Holway).

We have no authentic spec. of *D. decedens*, but the Iowa specc. agree so well with the description of that species that we have referred them to it. The sporidia were 12–15 x 4–4½ μ, uniseptate and distinctly constricted in the middle, ends obtuse. There were no appendages visible, but as the specc. were rather old, these may have disappeared; nor is it unusual in this genus for sporidia at first acute to become obtuse at maturity.

D. marginàlis, Pk. 39th Rep. p. 52.

Pustules numerous, covered by the epidermis which is somewhat elevated. Perithecia valsoid, 8–15 in a pustule, nestling in the inner bark, with no circumscribing line, the ostiola slightly emergent, black, surrounding the margin of the whitish pulverulent, erumpent disk. Asci subcylindrical, 65–75 x 10–12 μ. Sporidia crowded or biseriate, uniseptate, obscurely apiculate at each end, 20–23 x 5–7 μ.

On dead branches of *Alnus viridis*, New York.

Externally like *Valsa ambiens*. In the larger pustules the ostiola form a marginal circle around the disk, but in the smaller ones they sometimes emerge centrally and obliterate the disk.

D. epimícta, E. & E. (in Herb.)

Exsicc. Ell. N. A. F. 495.—Rab. F. E. 3154.

Stromata scattered, cortical, orbicular, about 3 mm. diam., convex, brownish-black outside, subcinereous within, with a faint, circumscribing line, which does not penetrate the wood. Perithecia 6–10 in a stroma, subglobose, about ½ mm. diam., subcircinate, necks short, ending in the black, roughish, hemispherical ostiola erumpent in a black disk surrounded by the ruptured epidermis. Asci (p. sp.) about 50 x 8 μ, clavate-cylindrical. Sporidia biseriate, fusoid, hyaline, 3–4-nucleate, becoming uniseptate, 12 x 8 μ, with a bristle-like appendage at each end about half as long as the sporidium.

On dead stems of *Ilex verticillata*, Newfield, N. J., also sent from Plainfield, N. J., by Mr. G. F. Meschutt.

Issued in N. A. F. as *Diatrype Badhami*, Curr., from which (sec. authentic specc.) it is quite distinct.

D. cercóphora, (Ell.)

Valsa cercophora, Ell. Bull. Torr. Bot. Club, IX, p. 99.
Diaporthe cercophora, Sacc. Syll. 6097.
Exsicc. Ell. N. A. F. 1187.

Perithecia few, 4–6, rather large, imbedded in a subcarbonaceous stroma, which is limited by a black line penetrating the wood. Ostiola stout, obtuse, with a large opening, united in a subconical disk, which pierces the epidermis and rises slightly above it. Asci clavate-cylindrical, 75–85 x 10–12 μ. Sporidia biseriate, oblong-elliptical, constricted in the middle and uniseptate, appendiculate at each end, 10–15 x 2¾–4 μ, hyaline.

On dead limbs of *Ilex opaca*, Newfield, N. J.

The appendages, which are finally absorbed, are as long as or longer than the sporidium itself, and the upper one generally recurved at the end.

D. talèola, (Fr.)

Sphæria taleola, Fr. Syst. Mycol. II, p. 391.
Valsa taleola, Fr. Summa Veg. Sc. p. 411.
Aglaospora taleola, Tul. Sel. Carp II, p. 168.
Diaporthe taleola, Sacc. Syll. 2426
Exsicc. Desm. Pl. Crypt. Ed. I, 2054, Ed. II, 1754.—Rab. F. E. 821.—Cke. F. Brit. Ser. I, 252, Ser. II, 231.

Stroma cortical, with a black, circumscribing line which does not penetrate the wood, only slightly raising the bark, depressed-pulvinate, 2–4 mm. across, closely covered by the epidermis. Perithecia not numerous (4–10), buried in the inner bark, with their ostiola converging and commonly erumpent in a small, light-colored disk, but not projecting, and often scarcely visible. Asci cylindrical, 120–140 x 10–12 μ, (paraphysate)? Sporidia uniseriate, elliptical, uniseptate, constricted, with a setaceous appendage about as long as the sporidium at each end, and often with 2–3 similar ones at the septum, 15–22 x 8–9 μ (20–25 x 7–10 μ, Winter).

On dead oak limbs, Pennsylvania (Schw.), Iowa (Holway). Allied to *Melanconis.*

D. oxýspora, (Pk.)

Valsa oxyspora, Pk. 28th Rep. p. 73, pl. 2, figs. 26–29.
Diaporthe oxyspora, Sacc. Syll. 2427.

Stromata scattered, subconical, cortical, enclosed in a thin, black sheet which penetrates to the wood, surrounded above by the triangularly or stellately ruptured epidermis. Perithecia sunk to the wood. Ostiola few and short, moderately exserted. Asci clavate, about 40 x 8 μ (p. sp.). Sporidia biseriate, oblong-fusoid, uniseptate and constricted, 4-nucleate, 12–15 x 3 μ, with a bristle-like appendage at each end.

On dead oak limbs, New York. Sent also from Louisiana by Mr. Langlois.

Cooke, in Valsei U. S. p. 121, makes this a Syn. of *V. taleola*, Fr., but we find it very different.

D. tessélla, (Pers.)

Sphæria tessella, Pers. Syn. p. 48.
Valsa tessella, Fr. Summa Veg. Sc. p. 411.
Cryptospora tessella, Karst. Myc. Fenn. II, p. 78.
Valsa mucronata, Pk. 28th Rep. p. 74.
Diaporthe tessella, Rehm Asc. 176.

Stroma depressed-hemispherical, formed of the scarcely altered substance of the bark, but surrounded by a black, enveloping crust visible through the epidermis as a small, grayish-black circle 2–3 mm. diam. Perithecia few (4–6), deeply bedded in the stroma, sunk to the subjacent wood which is slightly blackened on the surface, but not penetrated by any circumscribing line. Ostiola separately erumpent, but not projecting, perforated and often umbilicate. Asci oblong-fusoid, attenuated at each end, stipitate, 100–130 x 18–22 μ. Sporidia mostly biseriate, oblong-cylindrical, slightly curved, with a short, slender appendage at each end, 48–52 x 7–8 μ.

On dead willow limbs, Carolina and Pennsylvania (Schw.), Iowa (Holway). Described from the Iowa specc.

D. tèssera, (Fr.)

Sphæria tessera, Fr. S. M. II, p. 405.
Diaporthe tessera, Fckl. Symb. Nachtr. I, p. 318
Exsicc. Fckl. F. Rh. 592.—Kze F. Sel. 119.—Thum. M. U. 1261.

Perithecia 3–10 together, subcircinate, buried in the unchanged substance of the inner bark which is raised into indistinct pustules a little paler in the center and barely pierced by the minute, black ostiola. The groups of perithecia are very numerous and lie close together almost entirely filling the bark, but there is no discoloration. Asci oblong-fusoid, 8-spored, 60–70 x 10–12 μ. Sporidia biseriate, oblong-fusoid, uniseptate and constricted (finally 3-septate), 15–22 x 5 –7 μ.

On dead limbs of *Corylus*, New York (Peck), Iowa (Holway), on *Corylus rostrata*, California (Harkness).

D. obscùra, (Pk.)

Valsa obscura, Pk. 28th Rep. p. 73
Diaporthe obscura, Sacc. Syll 2429
Exsicc. Ell. N. A. F. 877.

Stromata cortical, not discolored, minute, about 1 mm. diam., sunk to the wood. Perithecia 3–8. Ostiola short-cylindrical, obtuse, perforated, erumpent in a short, compact fascicle rising from the center of the pulvinulate stroma, and rupturing the closely enveloping epidermis in a stellate manner. Asci subcylindrical. Sporidia crowded or biseriate, oblong, a little narrower at one end, obscurely uniseptate, hyaline, with a minute bristle at each end, 7–8 x 2 μ, 2- or more-nucleate.

On dead stems of *Rubus strigosus*, New York and New Jersey.

D. syngenésia, (Fr.)

Sphæria syngenesia, Fr. S. M. II, p. 382.
Valsa syngenesia, Fr. Summa, p. 411.
Diaporthe syngenesia, Fckl. Symb. p. 204.
Diatrype Frangulæ, (Pers.), Cke. Hndbk. II, p. 816
Exsicc. Fckl. F. Rh 601.—Kze. F. Sel. 120, 349 —Rab. F. E 1249, 2525 —Sydow, M. March. 264.—Thum. F. Austr. 697.—id. M. U 2171.—Cke. F. Brit. Ser.1, 238, Ser. 2, 222.

Stroma mostly conical, adnate to the wood, concentrically substriate, black, dusky inside, about 2 mm. diam. Perithecia 5–9 in a stroma, sphæroid, black. Ostiola bound together in a compact disk which raises the epidermis in a pustuliform manner and ruptures it, but scarcely rises above it. Asci sessile, cylindric-clavate, 54–58 x 7–9 μ, 8-spored. Sporidia biseriate, clavate-oblong, 4-nucleate, greenish-hyaline, 13–15 x 3½ μ, with a short, bristle-like appendage at each end.

On dead limbs, Carolina and Pennsylvania (Schw.). (Syn. N. Am. 1321.)

D. sulphúrea, Fckl. Symb. p. 205.

Melanconis ? umbonata, Sacc Mycol. Ven. Spec. p. 126, tab. XII, figs. 39–41 (fide Winter).
Exsicc. Fckl. F Rh. 2539.—Kunze F. Sel. 350.

Stromata scattered, slightly convex, orbicular or elliptical, seated on the inner bark and covered by the epidermis, which is raised into slight pustules, but not discolored, 2–3 mm. diam., sulphur-yellow. Perithecia 4–10 in a stroma, subcircinate, bedded in the bark, globose, at length flattened, tolerably large, their ostiola united in a small, round, or subelliptical, erumpent disk which rises but slightly above the epidermis. Asci subclavate or fusoid, sessile, 80–90 x 12–14 μ. Sporidia biseriate, oblong, narrowed and rounded at the ends, septate in the middle, 20–24 x 7 μ, with a small, globose appendage at each end.

On dead branches of *Corylus rostrata*, Ganesvort, N. Y. (Peek).

D. decipiens, Sacc. F. Ven. Ser. IV, p. 6.

Exsicc. Kze. F. Sel. 350.—Thum. M. U. 469.—M. March, 984.—Rab. F. E. 2421.

Perithecia 5–15 in a pustule, circinating in a sulphur-yellow stroma formed of the altered substance of the bark, depressed-globose, ½ mm. or a little more in diam., contracted into slender, convergent necks, with their papilliform, minute, black ostiola erumpent mostly around the margin of a small, flat, circular, yellowish disk which pierces the epidermis, but scarcely rises above it. Asci (p. sp.) 70–75 x 12–15 μ, 8-spored. Sporidia biseriate, fusoid-oblong, uniseptate, each cell nucleate, hyaline, 15–20 x 5–7 μ, with an obscure apiculus at each end.

On bark of dead *Carpinus*, London, Canada (Dearness).

The yellow color of the stroma is sometimes very distinct, and again scarcely perceptible. It is doubtful whether this is more than a form of *D. sulphurea*, Fckl.

D. Hýstrix, (Tode).

Sphæria Hystrix, Tode Fungi Meckl. II, p. 53, tab. XVI, fig. 127.
Diaporthe Hystrix, Sacc. F. Ven. IV, p. 6.
Exsicc. Roum. F. Gall. 76.—Ell. N. A. F. 89?

Perithecia collected in valsoid groups lightly covered by the pustulate, superficial layer of the inner bark, ovate-globose, 12–20 in a stroma. Ostiola erumpent together, but not confluent, about ½ mm. long, obtusely pointed and mostly smooth at the apex, black and brittle. Asci subfusoid, 8-spored, p. sp. 40–45 x 8–9 μ. Sporidia inordinate or subbiseriate, cylindric-fusoid, very slightly curved, uniseptate and slightly constricted in the middle, with a short appendage at each end, hyaline, 10 x 3 μ.

On bark of *Acer rubrum*, Newfield, N. J.

The specc. in N. A. F. are in poor condition and uncertain, but probably this species.

D. glýptica, (Berk. & Currey.)

Valsa glyptica, Berk. & Currey, Grev. IV, p 100.
Diaporthe glyptica, B. & Curr. Sacc. Syll. 2433. Cke. Syn. 2026.

"Quite covered by the bark, which is merely pierced by the ostiola, surrounded more or less evidently by a black line. Sporidia fusiform, sometimes sigmoid, uniseptate, 45–50 μ long."

On willow, South Carolina (Berk.)

Cooke, in the Valsei of the U. S., doubts whether this is distinct from *D. tessera*, (Fr.), as the slight appendages may have been overlooked.

**** Species imperfectly known.*

D. carpinígera, (B. & C.)

Diatrype carpinigera, B. & C. Grev. IV, p. 96, and Cke. Syn. 1617.
Diaporthe carpinigera, Sacc. Syll. 6092.

"Pustules small, black. Perithecia hidden. Sporidia oblong, uniseptate, 14–15 μ long."

On hornbeam, Pennsylvania (Berk. in Grev. l. c.).

D. innàta, (B. & C.)

Valsa innata, B. & C. Grev. IV, p. 102.
Diaporthe innata, Sacc. Syll. 2440, Cke. Syn. 2032.

"Perithecia forming a flat, annular ring around the raised disk within which the ostiola are concealed. Asci filiform. Sporidia in a single row, oblong, slightly narrowed each way from the center, 7–8 μ long."

On *Castanea vesca*, mountains of New York.

D. ciliàta, (Pers.)

Sphæria ciliata, Pers. Syn. p. 35.
Diaporthe ciliata, Sacc. Syll. 2442.

Perithecia (sec. Winter) 10 or less together, circinating in the inner bark, and covered by the pustuliform-raised epidermis, ovate, converging. Ostiola long, slender, flaccid, diverging, smooth, 1–2 lines long. Sporidia (sec. Stevenson Add. to Cooke's Valsei of the U. S.), naviculoid, biseptate, 12½ x 5 μ.

On bark of dead limbs, Carolina and Pennsylvania (Schw.).

D. sphendámnina, (B. & C.)

Diatrype sphendamnina, B. & C. Grev. IV, p. 96, and Cke. Syn. 1608.
Diaporthe sphendamnina, Sacc. Syll. 2367.

"Pustules splitting the bark longitudinally, and overtopped by its fragments. Ostiola short, cylindrical. Asci clavate. Sporidia uniseriate, narrow, hyaline."

On *Acer rubrum*, Pennsylvania (Michener).

D. lixívia, (Fr.)

(*Sphæria lixivia*, Fr. S. M. II, p. 385)?
Valsa lixivia, Cke. Vals. U. S. No. 124, Sacc. Syll. 2410.

Stromata small, yellowish. Perithecia small, subglobose, black, loose, ("laxis"). Ostiola fasciculate, granulose, becoming relaxed ("relaxatis"). Sporidia lanceolate, uniseptate and constricted, 4-nucleate, 12 μ long.

On bark of *Juglans cinerea*, Bethlehem, Pa. (Schw.)

D. tortuòsa, (Fr.)

Sphæria tortuosa, Fr. S. M II, p. 395.
Valsa tortuosa, Cke. Valsei of the U. S. 125, Cke Syn. 2033.
Diaporthe tortuosa, Sacc Syll. 2441.

Perithecia globose, glabrous, smooth, circinate-crowded in the substance of the inner bark, and covered by a subprominent, cortical pustule. Ostiola converging within the pustule, then fasciculate-erumpent, short, subdivergent, nearly smooth or elongated, deflexed, subnodose, with the habit of *Calosphæria pulchella*.

In Schw. Syn. N. Am. No. 1450, this species is said to have been found in New Jersey, on a pine limb infested with *Peridermium Pini;* very rare.

The spec. in Herb. Schw. labeled *Sphæria tortuosa*, Fr., is the *Caliciopsis pinea*, Pk., which is certainly not the *Sphæria tortuosa*, Fr. Whether Schweinitz was mistaken in his determination, or the error has resulted from some confusion of labels, we can not say, but incline to the former alternative, as the name (in Schweinitz' handwriting apparently) is written on a piece of paper to which the specimen is glued.

B. *Perithecia gregarious, more or less sunk in the wood, which is often blackened on the surface and circumscribed by a black line within* (*Euporthe*).

* *On dicotyledonous plants.*

(*a*) *Arboricolæ.*

D. concréscens, (Schw.)

Sphæria concrescens, Schw. Syn. N. Am. 1301.
Diaporthe concrescens, Cke. Grev. XIII, p 38.

Covered by a kind of black crust formed from the bark, transversely erumpent, orbicular or elliptical, with the disk concave, surrounded by the substellate-ruptured epidermis and an elevated margin. 1–1½ cm. long. In this crust the perithecia are partly sunk, globose-depressed, minute, attenuated above into a cylindrical ostiolum rather thick and somewhat exserted. Perithecia with a light-colored nucleus, at length changing their form and becoming conic-cylindrical. Sporidia (sec. Cke. l. c.) fusiform, 4-nucleate, then uniseptate, 12 μ long.

On dead stems of *Ribes aureum* (cult.), Bethlehem, Pa. (Schw.).

Allied to *D. recondita*, Schw., but differs in its longer sporidia, as well as in some other characters. Considered by Schweinitz as a good species and quite distinct.

D. Péckii, Sacc. Syll. IX, p. 713.

Diaporthe sparsa, Pk. 39th Rep. p 52.

Perithecia few, minute, scattered, immersed in the wood, which is blackened on the surface. Asci clavate or subcylindrical, 75–87 x 7½–10 μ. Sporidia crowded, oblong or subfusoid, hyaline, uniseptate and constricted at the septum, 4-nucleate, 20–27 x 5–8 μ.

On dead branches of *Rhus Toxicodendron*, New York.

D. Megalóspora, E. & E. Proc. Acad. Nat. Sci. Phil. July, 1890, p. 235.

Perithecia globose, ½–¾ mm. diam., scattered, buried in the wood, which is blackened on the surface but remains white within, abruptly contracted above, and prolonged into a long (2–3 mm.), rough, subflexuous ostiolum. Asci (p. sp.) 70–90 x 10–12 μ. Sporidia biseriate, oblong-fusoid, slightly curved, uniseptate and constricted at the septum, each cell with 1 or 2 large nuclei, acute at the ends, 25–35 x 4½–5½ μ. Narrower than in *D. leucosarca.*

On dead wood of *Sambucus Canadensis*, Manchester, Mass. (W. C. Sturgis).

D. gorgonoìdea, Cke. & Hark. Grev. XIII, p. 18.

Stroma effused, covered by the easily separable bark, blackening the surface of the wood, or forming an interrupted crust, limited within by a black line. Perithecia globose, immersed in the wood, mostly much crowded. Ostiola cylindrical, slender, flexuous, much elongated. Asci cylindric-clavate, 8-spored. Sporidia biseriate, fusiform, straight, 4-nucleate, at length 2-celled, hyaline, 15–17 x 3 μ.

On Australian *Acacia*, California.

Closely allied to *D. medusæa* and *D. fasciculata.*

D. crýptica, Nitschke, Pyr. Germ. p. 265.

Exsicc. Sydow, M. March. 2239.

Stroma variously effused, covered by the unaltered bark, and when this is removed, the surface of the wood is seen to be marked and spotted with variously shaped areas bounded by a black line which penetrates the wood. Perithecia scattered, or here and there in small groups or clusters of 2–4 together, sunk more or less deeply into the wood, globose or angular from mutual pressure, ⅓–½ mm. diam. Ostiola thick, nodulose or ventricose, more or less elongated, and mostly curved or decumbent. Asci clavate, p. sp. 45–50 x

6–7 μ, oblong-clavate. Sporidia biseriate, fusiform, 3–4-nucleate, hyaline, obtuse, slightly curved, 12–15 x 3½–4 μ.

On dead stems of *Lonicera* (cult.), Newfield, N. J.

D. biglobòsa, (C. & E.)

Sphæria biglobosa, C. & E. Grev. VII, p. 9.
Diaporthe biglobosa, Sacc. Syll. 2478, Cke. Syn. 2250.

Covered by the blackened epidermis. Perithecia depressed. Asci cylindrical. Sporidia uniseptate, strongly constricted, hyaline. 14 x 7 μ, cells globose.

On *Sassafras*, Newfield, N. J.

Our specc. are lost or mislaid.

D. spiculòsa, (A. & S.)

Sphæria spiculosa, A. & S. Consp p. 16.
Valsa circumscripta, Mont. Syll. p. 220.
Valsa tortuosa, Fckl Enum. Fungi Nas. p. 55.
Diaporthe spiculosa, Nitsch. Pyr. Germ. p. 256.
Exsicc. Rab. F. E 2045.—Rehm Asc. 430.—Thum M. U. 868.

Stroma widely effused, surrounding the branch or stem, or occupying numerous small spots, or areas, mostly covered by the bark, but sometimes exposed, surface of the wood blackened with a circumscribing line penetrating deeply. Perithecia loosely scattered or here and there crowded in subvalsiform groups, small, globose, partly or wholly sunk in the wood. Ostiola cylindrical, erumpent, spine-like or, by the pressure of the enclosing bark, deformed and bent. Asci cylindrical or subclavate, sessile, 8-spored, 50–60 x 7–9 μ. Sporidia biseriate or oblique, narrow-fusoid, 12–15 x 3 μ, scarcely constricted.

On dead stems of *Sambucus*, New York to Louisiana.

D. tumulàta, (C. & E.)

Sphæria tumulata, C. & E. Grev. V, p. 49.
Diaporthe tumulata, Sacc. Syll. 2452, Cke. Syn. 2223.

Perithecia scattered irregularly or subcircinate, 2–8 together buried in the surface of the wood, about ½ mm. diam., globose, abruptly contracted above, their slender necks converging and emerging from the bark in a little fascicle of short (½–1 mm.), black ostiola, with abruptly swollen tips, the bark being raised into slight pustules around them. Stroma variable in extent, 1–3 or more centimeters long and ½–1 cm. wide, consisting of the scarcely altered substance of the wood, and surrounded by a black line penetrating deeply. Asci p. sp. 45–55 x 7 μ, fusoid-oblong. Sporidia biseriate or obliquely uniseriate, fusoid-oblong, 4-nucleate, hyaline, subacute, slightly constricted in the middle, straight, 11–14 x 3–4 μ.

On dead stems of *Corylus Americana*, Newfield, N. J.

D. recóndita, (Schw.)

Sphæria recondita, Schw. Syn. N. Am. 1300.

Perithecia sunk in the wood, depressed-globose, with very long, rough ostiola protruding here and there through the epidermis which is finally thrown off, revealing the black, indefinitely effused, crustaceous stroma.

On dead stems of *Ribes floridum*, Bethlehem, Pa. (Schw.).

The specimen in Herb. Schw. has fusoid-oblong, hyaline sporidia 5–7 x $1\frac{1}{2}$ μ.

D. subpyramidàta, (B. & C.)

Eutypa subpyramidata, B. & C. Grev. IV, p. 97.
Diaporthe subpyramidata, Sacc. Syll. 2456, Cke. Syn. 2227.

"Effused; perithecia of a somewhat pyramidal shape, forming a rasp-like stratum. Asci lanceolate, much attenuated below. Sporidia biconical, each division much attenuated."

On oak, North Carolina.

No measurements of asci and sporidia given.

D. grìseo-tíngens, (B. & C.)

Sphæria griseo-tingens, B. & C. Grev. IV, p. 148.
Diaporthe griseo-tingens, Sacc. Syll. 2490, Cke. Syn. 2262.

"Gregarious, forming little detatched or continuous, short, linear prominences, marked with the ostiola. Asci clavate. Sporidia obliquely fusiform, 20 μ long."

On *Juniperus Virginiana*, Pennsylvania.

** *On dicotyledonous, herbaceous plants.*

D. apiculòsa, Ell. Bull. Torr. Bot. Club. IX, p. 19.

Exsicc. Ell. N. A. F. 787.

Perithecia scattered, small, buried in the substance of the stem, which is uniformly blackened on the surface. Ostiola rather stout, slightly projecting. Asci oblong-clavate, 40–50 x 7 μ, 8-spored. Sporidia biseriate, elliptical, 2–3-nucleate, with a faint apiculus at each end, $8\frac{1}{2}$–10 x $3\frac{1}{2}$ μ.

On the lower part of the stems and the thick part of the roots of old *Erigeron Canadensis*, decaying on the ground, Newfield, N. J.

D. díscrepans, Sacc. Mich. II, p. 568.

Perithecia gregarious, immersed in the more or less blackened stem, globose. $\frac{1}{3}$ mm. diam. Ostiola erumpent, short-rostellate. Asci fusoid-cylindrical, subobtuse at each end, 65 x 8–9 μ, 8-spored, aparaphysate. Sporidia biseriate or oblique, cylindrical or subfusoid, obtuse at each end, straight or curved, very slightly constricted in the middle. 10–12 x $3\frac{1}{2}$–4 μ, 4-nucleate, hyaline.

On stems of *Rumex acetosella*, New Jersey.

Var. *Polygoni*, E. & E., on dead stems of *Polygonum acre*, Louisiana (Langlois 1351), differs only in the stems not being blackened.

D. Eburénsis, Sacc. Mich. II, p. 60.

Stroma broadly effused, blackening the surface of the stem, but without any black line within. Perithecia densely gregarious, not deeply buried, $\frac{1}{3}$ mm. or a little over in diam., subglobose. Ostiola subconical, shortly ($\frac{1}{4}$–$\frac{1}{3}$ mm.) exserted and roughening the stem. Asci fusoid, 70 x 10–11 μ, p. sp. 40 x 10 μ), 8-spored. Sporidia biseriate, fusoid, subobtuse, slightly curved, 15–18 x 3–$3\frac{1}{2}$ μ (15–16 x 4 μ, Sacc.), constricted, uniseptate, 4-nucleate, hyaline.

On some large herbaceous stem, Iowa (Holway).

Differs from *D. Arctii* in its larger, constricted sporidia and shorter, obtuse ostiola; from *D. immersa* and *D. orthoceras*, in its curved, constricted and larger sporidia.

D. Phacèliæ, Cke. & Hark. Grev. IX, p. 86.

Subeffused. Stroma blackening the inner surface of the bark. Perithecia subglobose, immersed. Ostiola cylindrical, elongated, flexuous. Asci clavate, sessile. Sporidia straight, sublanceolate, 4-nucleate, then uniseptate, 18 x $3\frac{1}{2}$ μ.

On branches of *Phacelia*, California (Harkness).

D. euspina, (C. & E.)

Sphæria euspina, C. & E. Grev. V, p. 93.
Diaporthe euspina, Sacc. Syll. 2535, Cke. Syn. 2311.

Thickly scattered, small, $\frac{1}{4}$ mm. diam., buried in the surface of the woody stem (just under the bark, which is blackened on the inner surface, as is the surface of the stem when the bark falls away, but there is no penetrating line or discoloration within. Ostiola erumpent, short-cylindrical, obtuse. Asci clavate, p. sp. about 40 x 6 μ. Sporidia biseriate, fusoid, 4-nucleate, uniseptate, scarcely constricted, hyaline, 10–13 x $2\frac{1}{2}$–3 μ (18 x 3 μ, Cke.). Resembling *D. spiculosa*.

On dead stems of *Chenopodium album*; mostly near the base, Newfield, N. J.

D. salviicola, (C. & E.)

Sphæria salviæcola, C. & E. Grev. V, p. 93.
Diaporthe salviicola, Sacc. Syll. 2525, Cke. Syn. 2301.

Perithecia scattered or here and there gathered in groups of 3–5 or more, but not crowded, small, sunk more or less in the woody stem, which is sometimes blackened on the surface, but is not penetrated by any black line or otherwise discolored within. Ostiola erumpent, conical or short-cylindrical, rather stout. Asci clavate-oblong, about 60 x 6 μ. Sporidia biseriate, fusoid, 4-nucleate, curved, hyaline, 15–20 x 2½–3 μ.

On dead stems of *Salvia officinalis*, Newfield, N. J.

D. Desmazièrii, Niessl, Beitrag, p. 53.

Sphæria inquilina, Desm. Pl. Crypt. Ed. II, No. 1766.
Exsicc. Ell. N. A. F. 1197.—Sacc. M. Ven. 1262.—Sydow, M. March. 2056.

Stroma effused, at length blackening the surface of the stem, which remains unchanged within. Perithecia scattered, small, pale at first, buried more or less deeply. Ostiola acute and, when well developed, ½–¾ mm. long, cylindrical, slender. Asci oblong, sessile, 40–55 x 6–7 μ. Sporidia mostly biseriate, navicular-fusoid, mostly acute, both ends usually slightly curved in opposite directions, yellowish-hyaline, 12–16 x 3–3½ μ, uniseptate, scarcely or only slightly constricted.

On dead stems of *Brunella vulgaris*, Newfield, N. J.

Found mostly near the base of the stem. Higher up are perithecia filled with stylospores scarcely distinguishable from the ascospores.

D. Árctii, (Lasch.)

Sphæria Arctii, Lasch. in Rab. Herb. Mycol. 1046.
Sphæria orthoceras, id. 1435.
Diaporthe Arctii, Nits. Pyr. Germ. p. 268.
Exsicc. Fckl. F. Rh. 2337.—Kze. F. Sel. 133.—Rab. F. E. 2116, 2869.—Rehm Asc. 332, 668. Thum. M. U. 173.—Ell. & Evrht N. A. F. 2d Ser. 1793.—Krieger F. Sax. 236, 237. Lin F. Hung. 462 —Plowr. Sph. Brit. 234.—(Ell. N. A. F. 1194) ?

Stroma more or less effused, often covering the stem for a considerable extent, surface mostly blackened, sometimes not discolored or only gray, internally composed of the unaltered substance of the stem surrounded by a thin, black layer. Perithecia scattered, or here and there approximated, entirely buried, mostly a little depressed, about ⅓ mm. diam., contracted into a slender neck, erumpent in a cylindrical or conic-cylindrical ostiolum, about as in *D. orthoceras*. Asci oblong-clavate, sessile, 8-spored, 40–50 μ long, 6–8 μ wide. Sporidia biseriate, fusoid, ends subacute, uniseptate, mostly slightly curved, scarcely constricted, 10–14 x 3–3½ μ (mostly about 12 μ long).

On dead stems of *Lappa major*, New Jersey, and on *Ambrosia trifida*, Louisiana, on wild parsnip, California (Harkness). Found in Europe also on *Cirsium*, *Tanacetum*, and other species of *Compositæ*.

This is with difficulty distinguished from *D. orthoceras*, but differs from that species in its rather acute, subinequilateral or slightly curved sporidia. Winter and Saccardo give the sporidia as 10–14 μ long. We find the American specc. of *D. orthoceras* to have the sporidia mostly shorter than in *D. Arctii*, about 8–10 μ.

D. adúnca, (Desm.)

Sphæria adunca, Rob. & Desm. XIX. Not. p. 14.
Diaporthe adunca, Sacc. Syll. 2514.
Exsicc. Desm. Pl. Crypt. Ed. I, 2071.—Kunze F. Sel. 134.

Perithecia buried, with the apex slightly pustuliform-prominent, small (150–200 μ), often slightly elongated, scattered or sometimes two lying close together. Ostiola short-cylindrical, exserted, 150–200 x 75–80 μ. Asci oblong-clavate, sessile, 35–40 x 6–7 μ, aparaphysate. Sporidia biseriate, oblong-fusoid, mostly straight, uniseptate, but not constricted, 12–15 x 3–4 μ, ends subacute, hyaline.

On dead scapes of *Plantago lanceolata*, Newfield, N. J.

The part of the scape occupied by the fungus, soon becomes uniformly blackened on the surface, but is not discolored within. The foregoing diagnosis is drawn from the specc. in Desm. Exsicc., the Newfield specc. not being well matured.

D. placoides, E. & E. (in Herb.)

Stromata elliptical, 1–3 x 1 mm. diam., or about that, black on the surface, white within, but limited by a distinct black line, gregarious, at first covered by the epidermis and not prominent, but when the epidermis falls away, standing out like little black, elliptical shields or disks. Perithecia often only one or two in a stroma, very small, 100–150 μ diam., buried in the surface of the stroma, with the papilliform or conic-papilliform ostiola projecting. Asci oblong-fusoid, 35–40 x 6 μ, 8-spored, aparaphysate. Sporidia biseriate, narrow-elliptical, 3–4-nucleate, hyaline, not constricted, 7–8 x 3½ μ.

On decaying stems of *Lactuca Canadensis*, Newfield, N. J.

The stromata, especially near the base of the stem, are sometimes confluent. Outwardly there is a strong resemblance between this and *D. Gladioli*.

D. orthóceras, (Fr.)

Sphæria orthoceras, Fr. Elench. II, p. 97.
Diaporthe orthoceras, Nits. Pyr. Germ. p. 270.
Exsicc. Fckl. F. Rh. 897.—Kze. F. Sel. 130, 131, 132, 353.—Rehm Asc. 331, 523.—Rab. F. E. 534.—Thum. M. U. 974.

Stroma effused, short and narrow or continuous and surrounding the entire stem, which is finally blackened on the surface and penetrated by the black, stromatic lines. Perithecia globose or subdepressed, scattered irregularly or oftener in closely packed groups, buried in the matrix, attenuated above into short necks. Ostiola more or less elongated, with a conical base, more or less nodulose, generally straight, moderately exserted. Asci oblong-clavate, 8-spored, 40–50 x 6–8 μ. Sporidia biseriate, fusoid, mostly straight, 2-celled, not constricted, subobtuse, 3–4 nucleate, hyaline, 10–14 x 3–4 μ.

On dead stems of *Solidago*, *Aster*, and *Achillea*, New Jersey.

Var. *Lactucæ*, E. & E., has the surface of the stem uniformly blackened, and the sporidia only 8–10 μ long. On dead stems of *Lactuca Canadensis*, New Jersey and Louisiana.

D. aculeàta, (Schw.)

Sphæria aculeata, Schw. Syn. N. Am. 1287.
Diaporthe aculeata, Sacc. Syll. 2534, Cke. Syn. 2310.
Exsicc. Ell. N. A. F. 589 (as *D. spiculosa*).

Extensively and indefinitely effused. Perithecia deeply buried in the substance of the stem, depressed-globose, black, scattered, their long, cylindrical ostiola erumpent through a thin, black crust, lying just beneath the thin cuticle, and at length exposed, or where more scattered, rising through a small, elongated, black, superficial tubercle. Asci p. sp. about 85 x 7 μ. Sporidia biseriate, fusiform, with 4 large nuclei which, by crowding against each other, causethe sporidium to appear 3-septate, not constricted, ends subacute, 11–13 x 2½–3 μ.

On dead stems of *Phytolacca decandra*. Common wherever this plant is found.

D. semiinscúlpta, Sacc. Syll. 2528.

Sphæria semiimmersa, B. & C. Grev. IV, p. 146. (non Nits.)

"Perithecia immersed below, above subcylindrical or subconical. Sporidia linear, oblong, curved, 15 μ long."

On dead herbaceous stems, Connecticut (Wright).

*** *On monocotyledonous plants.*

D. Màydis, (Berk.)

Sphæria Maydis, Berk. Hook. London Journ. Bot. Vol. 6, p. 326.

"Spots minute, elevated, often purple-brown, punctiform or subelliptical, rarely linear, containing very few perithecia with a single, broad-conical ostiolum. Sporidia oblong, slightly curved, uniseptate.

Habit that of *Leptosphæria arundinacea.* Very different from *Sphæria* (*Diplodia*) *Zeæ*, Schw."

On dead culms of *Zea Mays*, Cincinnati, Ohio (Lea).

On account of the raised spots, and broad-conical ostiola, this seems distinct from *D. Kellermanniana*, Winter, and from *D. incongrua*, E. & E.

D. incóngrua, E. & E. (in Herb.)

Diaporthe Kellermanniana, Winter, in Journ. Mycol. II, p 100 (not Bull. Torr. Bot. Club, X, p. 49).

Stroma broadly effused, 5 or more centimeters long and nearly surrounding the culm, which it penetrates and blackens on the inner surface; the outside is also finely mottled with narrow-elliptical, dark-colored spots about 1 mm. long, lighter in the center, and so numerous and closely confluent as to cause the surface of the culm to appear, at first sight, as if uniformly blackened. The whole area is limited by a distinct black line, visible on the surface of the culm, especially at the ends, where the stroma is often prolonged in narrow strips. Perithecia scattered or subcespitose, sunk in the substance of the culm. $\frac{1}{4}$–$\frac{1}{3}$ mm. diam., their long (1 mm.) rather crooked, black ostiola projecting, either singly or in little tufts of two or three together. Asci lanceolate, about 40 x 7–8 μ. Sporidia biseriate, oblong-fusoid, 4-nucleate and yellowish, becoming constricted and uniseptate, ends rather obtusely pointed, 7–10 x 3 μ.

On decaying culms of *Zea Mays*, Kentucky and Louisiana.

The distinct stroma will distinguish this from *D. Kellermanniana*, Winter, and the cylindrical ostiola, from *D. Maydis*, Berk.

D. Kellermanniàna, Winter, Bull. Torr. Bot. Club, X, p. 49.

Stroma none. Perithecia deeply immersed, depressed-globose, membranaceous, black, 210–260 μ diam., erumpent. Asci oblong-fusoid, 8-spored, 35–40 x 5–7 μ. Sporidia oblong, often inequilateral, rounded or subacute at the ends, uniseptate in the middle, not constricted, 4-guttulate, 9–11 x 3–3$\frac{1}{2}$ μ.

On decaying culms of *Zea Mays*, Lexington, Ky. (Kellerman).

This seems quite distinct from either of the two preceding species.

D. Gladìoli, E. & E. Journ. Mycol. II, p. 101.

Exsicc. Ell. & Evrht. N. A. F. 2d Ser. 1794.

Perithecia sunk in the substance of the stem just below the epidermis, which is blackened above them, forming elliptical, definitely limited spots 2–3 mm. long or, by confluence, 1 cm. or more. Perithecia about ¼ mm. diam., few, often only one or two in a spot, sometimes 6–8. Asci (p. sp.) about 40 x 6–7 μ, with a substipitate base. Sporidia biseriate, subfusoid, 7–10 x 2½ μ, 2-nucleate, becoming uniseptate, hyaline. The ostiola project like slender, black bristles about 1 mm. long, but are easily broken off. This is very different from *Sphærella minimæpuncta*, Cke., also on *Gladiolus*.

On dead stems of *Gladiolus*, Louisiana.

C. *Perithecia gregarious, buried in the bark, which is mostly blackened on the surface, circumscribed* (*Tetrastaga*).

* *On dicotyledonous plants.*

(*a*) *Arboricolæ.*

D. Neilliæ, Pk. 39th Rep. p. 52.

Perithecia numerous, 250–270 μ diam., loosely and irregularly aggregated in extensive patches, immersed in the inner bark, with their bases often slightly sunk in the wood, covered by the epidermis, which is pierced by the black, conical or rostellate ostiola. Asci subcylindrical, p. sp. 55–75 x 7½–10 μ. Sporidia crowded or biseriate, oblong or subfusoid, uniseptate and slightly constricted, 2–4-nucleate, 12–17 x 5–7 μ.

On dead branches of *Neillia opulifolia*, Albany, N. Y.

The surface of the branch is rough to the touch from the projecting ostiola. The perithecia are sometimes clustered as in *Valsa*, and often collapse below.

D. Conrádii, Ell. Am. Nat. March, 1883, p. 816.

Exsicc. Ell. N. A. F. 1193.

Perithecia scattered, minute, depressed-spherical, barely covered by the epidermis, not penetrating the wood or limited by any black line. Ostiolum cylindrical, straight, rough, black, abruptly pointed above. Asci subcylindrical, 35–40 x 6–7 μ. Sporidia biseriate, ovate-elliptical, uniseptate, hyaline, scarcely constricted, 6–8 x 2½–3 μ.

On dead stems and branches of *Hudsonia tomentosa*, New Jersey.

When the species was published in the Am. Nat., the host was supposed to be *Corema Conradii*, hence the specific name.

D. densíssima, Ell. Am. Nat. March, 1883, p. 316. **(Plate 33)**

Exsicc. Ell. N. A. F. 1192.

Perithecia minute (¼ mm.), black, buried in the unchanged substance of the inner bark mostly in groups of 15–30, their short, subulate ostiola slightly converging, but not united, and barely penetrating the pallid, loosened epidermis, which soon disappears around them, so that the bark of the affected shoots appears thickly dotted with little circular openings. Asci clavate-cylindrical, 40–45 x 5–6 μ. Sporidia biseriate, fusiform, at first 4-nucleate, becoming 1–3-septate, 11–15 x 1½–2 μ. There is a faint, bristle-like appendage at each end of the young sporidium.

On dead shoots of *Quercus coccinea*, Newfield, N. J.

The upper part of the dead shoots for a foot or more is entirely occupied by the fungus, which is definitely limited, but not surrounded by any black line.

D. Æsculi, Cke. & Hark. Grev. IX, p. 86.

Cortical, collected in elongated groups. Perithecia globose-depressed. Asci lanceolate, sessile. Sporidia sublanceolate, straight, 4-nucleate, 18 x 3½ μ.

On *Æsculus Californica*, in California (Harkness).

D. spina, Fckl. Symb. p. 210 (not Schw.)

Exsicc. Fckl. F. Rh. 2257.—Krieger F. Sax. 139.—Kze. F. Sel. 136, 357.—Rab. F. E. 1715 Rehm Asc. 330.—Thum. M. U. 67.

Perithecia scattered or gregarious, covered, immersed in the unaltered substance of the bark, of medium size (⅓ mm.), globose, black, with a short, conical ostiolum about equal in length to the diameter of the perithecium piercing the epidermis and slightly prominent. Asci oblong-elliptical, 8-spored, sessile, p. sp. 40–50 x 5–10 μ, 35–40 x 15–16 μ (Sacc.), 38–48 x 5–7 μ (Winter). Sporidia overlapping, 2–3-seriate, fusoid, curved, 4-nucleate, uniseptate, hyaline, about 20 x 2½ μ.

On dead limbs of *Salix*, London, Canada (Dearness).

Sphæria spina, Schw. is (sec. specc. in Herb. Schw.), *Sphærographium Fraxini*, (Pk.), and not an ascigerous fungus. There is (sec. Winter) a small appendage on each end of the sporidia.

D. velàta, (Pers.)

Sphæria velata, Pers. Syn. p. 32.
Diaporthe velata, Nitschke Pyr. Germ. p. 287.

Stroma widely effused, enveloping the limbs and twigs, the limiting lines penetrating the wood, enclosing areas of various size and shape, and blackening the inner surface of the bark. Perithecia evenly scattered, buried in the surface of the inner bark, here and there 2–4 together, small, subglobose, soon depressed and even lenticular, abruptly contracted into a short neck, with thick, short, cylindrical or subconical ostiola scarcely or only slightly exserted, or under the loosened epidermis decumbent and longer. Asci narrow-fusoid, subsessile. 56–64 x 8–10 μ. Sporidia biseriate or oblique, narrow-fusoid, 4-nucleate, subobtuse, straight or a little curved, uniseptate, rarely a little constricted, hyaline, 10–15 x 3 μ.

On dead limbs of *Tilia*, Carolina and Pennsylvania (Schw.)

D. leucosárca, E. & E. Proc. Acad. Nat. Sci. Phil. July, 1891, p. 234.

Perithecia gregarious, thickly scattered and enveloping the limb for several inches, depressed-spherical, about ½ mm. diam., white inside, closely covered by the epidermis which is raised into small, lead-colored pustules pierced by the minute, punctiform ostiola. Asci elongated-clavate, with abundant paraphyses, 8-spored, 80–114 x 20 μ. Sporidia oblique and overlapping or biseriate, broad fusiform, slightly curved, uniseptate and constricted, with 1–2 large nuclei in each cell, 22–30 x 8–10 μ.

On dead limbs of *Carpinus Americana*, London, Canada (Dearness).

The bark is unaltered and there is no limiting line.

D. rostellàta, (Fr.)

Sphæria rostellata, Fr. S. M. II, p. 476.
Diaporthe rostellata, Nitschke, Pyr. Germ. p. 298.
Exsicc. Fckl. F. Rh. 930.

Perithecia thickly and evenly scattered over the stems, without any distinct stroma, small (150–200 μ diam.), seated on and slightly sunk into the surface of the inner bark, abruptly contracted into a nearly cylindrical beak ⅓–½ mm. long, erumpent through the epidermis and making the dead canes rough and prickly to the touch. Asci clavate, subsessile, 8-spored, 40–45 x 6–7 μ. Sporidia biseriate, fusoid or oblong-fusoid, 2–4-nucleate, often slightly constricted in the middle, mostly a little curved, 12–15 x 3–3½ μ, ends at first acute and faintly appendiculate, finally obtuse and without any appendages.

On dead canes of *Rubus villosus*, common around Newfield, N. J., but often sterile. On stems of rose and *Rubus*, Bethlehem, Pa. (Schw), on *Rubus Nutkanus*, California (Harkness).

The appendages on the sporidia are faint and easily overlooked, but in the young sporidia, at least, they are certainly present. *D. vepris*, (de Laer.) differs only in its smaller perithecia and sporidia, the latter only 8 x 12 μ long.

D. ophìtes, Sacc. Syll. I, p. 679.

Exsicc. Ell. & Evrht. N. A. F. 2d Ser. 1657, Sacc. M. V. No. 214.

Stroma broadly effused, mottling the surface of the bark and the wood with oblong or variously shaped spots from 1 mm. to 1 cm. or more long, with the black, limiting line penetrating the wood. Perithecia gregarious, subglobose, of medium size, covered by the bark, and more or less sunk in the wood. Ostiola erumpent, straight, slender, spine-like, often 1 mm. long, but quite as often barely perforating the bark. Asci fusoid, 8-spored, 50–60 x 9–10 μ. Sporidia biseriate, short-fusoid, uniseptate and constricted, 12–13 x $4\frac{1}{2}$–5 μ, 4-nucleate, subobtuse, hyaline.

On dead trunks and limbs of *Hibiscus*, New Jersey and Louisiana.

D. Wíbbei, Nits. Pyr. Germ. p. 305.

Stroma mostly broadly effused, covering the entire stem and branches, closely covered by the unaltered epidermis, through which is visible the black, circumscribing line of the stroma. Perithecia tolerably large, irregularly scattered or gregarious, or even collected 2–4 together in valsoid groups, at first globose, but soon depressed, nestling in the bark, their apices more or less prominent, and raising the bark into little pustules, their very short ostiola erumpent through cracks in the epidermis, but scarcely projecting. Asci narrow-clavate, sessile, 8-spored, 52–60 x 8 μ. Sporidia biseriate or oblique, narrow-fusoid, subobtuse at the ends, cylindrical, straight, hyaline, 1–3-septate, 4-nucleate, not constricted, 16–18 μ long, 3 μ thick.

On dead branches of *Myrica Gale*, Adirondack Mts., N. Y.(Peck).

D. gallóphila, Ell. Bull. Torr. Bot. Club, VIII, p. 90.

Densely gregarious, perithecia subcuticular, depressed-hemispherical, 200–250 μ diam., rugose. Ostiola cylindrical, obtuse, minutely roughened, 150–200 μ long. Sporidia biseriate, oblong-fusoid, 2–4-nucleate, and mostly constricted, hyaline, slightly curved, when young faintly appendiculate at each end, variable in length, 12–18 μ long.

On galls on dead canes of *Rubus villosus*, and on the canes themselves, Newfield, N. J.

The parts occupied by the fungus appear to the naked eye as if covered with a black pubescence, so thickly are they covered with the hair-like ostiola.

D. Lupìni, Hark. Bull. Cal. Acad. Feb. 1884, p. 441.

Exsicc. Ell. & Evrht. N. A. F. 2d Ser. 1655.

Perithecia gregarious, about $\frac{1}{3}$ mm. diam., buried in the unaltered bark and sunk to the wood, depressed-globose. Ostiola erumpent singly, short-conic-cylindrical, black, rendering the branches rough to the touch. Asci (p. sp.) oblong-fusoid, 50–60 x 9 μ. Sporidia biseriate, oblong-fusoid, 4-nucleate, uniseptate and constricted and easily separating at the septum, obtuse, hyaline, 12–16 x 4–4$\frac{1}{2}$ μ.

On branches of *Lupinus arboreus*, California.

Diagnosis drawn from specc. sent by Dr. Harkness.

D. Baccháridis, (Cke.)

Sphæria (Diaporthe) Baccharidis, Cke. Grev. VII, p. 53.
Diaporthe Baccharidis, Sacc. Syll. 2636, Cke. Syn. 2424.
Exsicc. Rav. F. Am. 370.

Perithecia scattered, punctiform, covered by the epidermis, which is blackened, slightly raised, and pierced by the short ostiola. Asci clavate-oblong, sessile, 35 x 12 μ. Sporidia biseriate, oblong-fusoid, 15 x 3–3$\frac{1}{2}$ μ, 3–4-nucleate (18–20 x 3 μ Cke.)

On *Baccharis*, Darien, Ga. (Ravenel).

D. Murràyi, (B. & C.)

Sphæria Murrayi, B. & C. Grev. IV, p. 147.
Diaporthe Murrayi, Sacc. Syll. 2564.

"Perithecia covered by the cuticle, rather prominent. Asci lanceolate. Sporidia oblong, constricted in the middle, with 4 nuclei, probably septate when older. Each perithecium is surrounded externally with short, white hairs, but it is uncertain whether they belong to the plant."

On apple, New England.

** *On dicotyledonous, herbaceous plants.*

D. mucronulàta, Sacc. Mich. II, p. 568.

Exsicc. Ell. N. A. F. 1196.

Perithecia scattered or 2–3 lying close together in a line, covered by the bark which is slightly raised and blackened, about $\frac{1}{4}$ mm. diam., not circumscribed. Ostiola short-rostellate, emergent. Asci fusoid, subobtuse at each end, 60 x 10 μ, aparaphysate, 8-spored. Sporidia biseriate or obliquely uniseriate, broad-fusoid, straight or curved, ends acute and mucronulate, 14–15 x 6 μ, faintly septate and constricted in the middle, 4-nucleate, hyaline.

On dead stems of *Solidago*, not much decayed, Newfield, N. J.

D. Gerànii, Cke. & Hark. Grev. XIV, p. 8.

Perithecia gregarious, immersed in the bark, often surrounded by a black line, subglobose, black. at first covered, matrix finally rimose-fissured, and the short ostiola emergent. Asci clavate. Sporidia lanceolate, 4-nucleate, hyaline, 15–16 x 4 μ.

On stems of *Geranium*, California (Harkness).

D. elephantìna, Cke. & Hark. l. c.

Pustules valsiform, rarely scattered. Perithecia buried in the bark or attached to the wood, globose, black, with slender, cylindrical, elongated, flexuous, emergent ostiola. Asci clavate, 8-spored. Sporidia lanceolate, 4-nucleate, soon uniseptate, hyaline, 12 x 3½ μ.

On stems of *Geranium*, California (Harkness).

D. Asclepiàdis, E. & E. Bull. Torr. Bot. Club, X, p. 98.

Stroma forming black patches on the surface of the stem ½ cm. long or, by confluence, much longer, limited by a deeply penetrating, black, circumscribing line. Perithecia scattered, globose, ¼ mm. diam., buried in the substance of the stem, but not deeply. Ostiola rather stout, cylindrical, subobtuse, ¼–⅓ mm. long. Asci (p. sp.) 35–40 x 9 μ. Sporidia biseriate, elliptical, constricted in the middle, 3–4-nucleate, subobtuse, 10–12 x 3–4 μ.

On dead stems of *Asclepias tuberosa*, Newfield, N. J., and on *Asclepias Cornuti*, Iowa (Holway).

D. exercitàlis, (Pk.)

Sphæria exercitalis, Pk. 30th Rep. p. 66.
Diaporthe exercitalis, Sacc. Syll. 2641, Cke. Syn. 2429.

Perithecia minute, crowded, arranged in long strips, at first covered by the epidermis, which is at length ruptured in long cracks, through which rise the short, spine-like, subacute ostiola; or oftener the stem is more or less blackened and raised into oblong tubercles, or elongated ridges, pierced, but not ruptured by the ostiola. Asci subcylindrical, 42–50 x 6–7 μ, subsessile. Sporidia subbiseriate, oblong-fusoid, 4-nucleate, colorless, 12 x 2½–3 μ.

On dead herbaceous stems, not much decayed, Catskill Mts., N. Y. (Peck).

D. immutàbilis, Cke. & Hark. Grev. XIV, p. 9.

Stroma variously effused and interrupted, scarcely discoloring the surface of the stem, but limited by a black line within. Perithecia scattered, globose, with short, punctiform ostiola. Asci clavate, 8-spored. Sporidia biseriate, sublanceolate, straight or slightly curved, 2–4-nucleate, uniseptate, hyaline, 12–14 x 4 μ.

On dead stems of *Scrophularia*, California (Harkness).

D. Phaseolòrum, (C. & E.)

Sphæria Phaseolorum, C. & E. Grev. VI, p. 93.
Diaporthe Phaseolorum, Sacc. Syll. 2635, Cke. Syn. 2423.
Exsicc. Ell. N. A. F. 188.

Perithecia gregarious, buried, very small. Ostiola spine-like, slender, projecting for ¼–½ mm. Asci clavate, 30–35 x 6–7 μ. Sporidia biseriate, oblong-lanceolate, 4-nucleate, scarcely or only slightly constricted, 10–12 x 3 μ (16 μ long, Cke.).

On decaying bean vines left exposed through the winter, Newfield, N. J. Mostly around the nodes of the stem, the surface mostly blackened and the stroma limited within by a black line.

D. Desmòdii, (Pk.)

Sphæria Desmodii, Pk. 26th Rep. p. 87.
Sphæria desmodiana, C. & E. Grev. VI, p. 93.
Diaporthe Desmodii, Sacc. Syll. 2633, Cke. Syn. 2420.

Perithecia scattered or seriately placed, minute, black, covered by the epidermis which is pierced by the acute or narrowly conical ostiola. Asci clavate, 35 x 5 μ. Sporidia biseriate, fusiform, colorless, 2–4-nucleate, 8–10 x 2½ μ.

On dead stems of *Desmodium*, New York and New Jersey.

In Grev. l. c. the sporidia are given as 18 μ long, but we can make them only 10 μ.

D. racèmula, (C. & P.)

Sphæria racemula, C. & P. 29th Rep. N. Y. State Mus. p. 65.
Diaporthe racemula, Sacc. Syll. 2629, Cke. Syn. 2417.

Perithecia cespitose, rugose, small, flattened, black, at length collapsing, separating with the epidermis which is pierced by the elongated ostiola. Asci clavate, sessile. Sporidia narrowly lanceolate, colorless, 4-nucleate, 15 μ long.

On dead stems of *Epilobium angustifolium*, New York.

VÁLSA, Fr.

Summa Veg. Scand. p. 140, emend. Sacc. Syll. I, p. 103.

Perithecia immersed in the bark, more or less distinctly circinate and lying in a single layer. Ostiola entire, converging to the center and mostly united in a small, erumpent disk.

Stroma cortical, without any circumscribing line. Disk mostly black. (Euvalsa Nits. Pyr. Germ. p. 126).

* *Microsporæ. Sporidia not exceeding 8 μ long.*

V. ceratóphora, Tul. Sel. Fung. Carp. II, p. 191.

Sphæria ceratosperma, Fr. S. M. II, p. 364.
Valsa Rosarum, De Not. Sfer. Ital. tab. 42, (sec. Winter)
Valsa Rubi, Fckl. Symb. p. 200, *V. Rubi*, Pk. 28th Rep. p. 72.
Valsa excorians, C. & E. Grev. VIII, p. 14.
Exsicc. Fckl. F. Rh. 606, 1567, 2260.—Kze. F. Sel. 346.—Rab. F. E. 2867.—Rehm Asc. 326, 525 —Thum. M. U. 870.—Ell. N. A. F. 496, 864.—Cke. F. Brit. Ser. I, 251.

Stromata scattered, often standing close together, but not confluent, orbicular or elliptical, varying from depressed-hemispherical to conical, covered by the epidermis or erumpent, brown outside and staining the adjacent bark the same color. Perithecia 5–20 in a stroma, crowded, globose, small. Ostiola more or less elongated, slender, cylindrical, smooth, mostly united at base with their tips more or less spreading. Asci narrow-clavate, sessile, 32–40 x 4–5 μ. Sporidia biseriate, allantoid, straight or slightly curved, hyaline, 6–8 x $1\frac{1}{2}$–2 μ.

On *Betula*, *Quercus*, *Carya*, *Amelanchier*, *Viburnum*, *Cratægus*, *Acer*, *Cornus*, and *Chionanthus*, common.

The ostiola in all the American specc. we have seen, remain short, projecting but little above the epidermis. They are often clavate-thickened above. From a careful examination of the specc. of *V. excorians*, in our Herb., we cannot separate this species from *V. ceratophora*, Tul. The sporidia (6–8 x $1\frac{1}{2}$ μ), and all the other characters of *V. excorians*, agree so well with the specc. of *V. ceratophora* in the various Exsiccati, that it cannot safely be separated as specifically distinct.

V. Americàna, B. & C. Grev. IV, p. 102.

Valsa Alni, Pk. 25th Rep. p. 103.
Exsicc. Rav. F. Am. 191.

Pustules small ($1\frac{1}{2}$ mm.), flat, covered by the epidermis which is pierced by the small fascicle of short-cylindrical, black ostiola, which are rounded or subconical at the apex. Perithecia 3–8 in a stroma, small ($\frac{1}{4}$ mm.), buried in the thin, inner bark, ovate-globose or subangular from mutual pressure, necks short. Asci (p. sp.) oblong-fusoid, 20–25 x 5–6 μ, 8-spored. Sporidia biseriate, allantoid, only very slightly curved, hyaline, 7–8 x $1\frac{1}{2}$–2 μ.

On various shrubs and trees, *Maclura aurantiaca*, *Alnus*, *Clethra*, *Vitis*, &c., common.

V. coronàta, (Hoff.)

Sphæria coronata, Hoff. Veg. Crypt. I, p. 26.
Valsa coronata, Fr. Summ. Veg. Scand. p. 412.

Stromata scattered or approximated and confluent at base, orbicular, about 1 mm. broad, convex, scarcely or only slightly prominent, crowned with a very small disk, which penetrates the epidermis, but hardly rises above it. Perithecia 4–12 in a stroma, subcircinate, buried in the unaltered substance of the bark, crowded, small, globose, with very small, short, black and shining, crowded ostiola piercing the small, flat disk. Asci narrow-clavate, sessile, 24–28 x 4 μ. Sporidia biseriate, allantoid, hyaline, 6–7 x 1½ μ.

On dead twigs of *Castanea* and *Bignonia*, Carolina (Curtis), on dead limbs of *Cornus*, Canada (Dearness).

This appears to differ from *V. ceratophora*, Tul., in its smaller disk, and perithecia only 4–12 in a stroma.

V. lutéscens, Ell. Bull. Torr. Bot. Club. IX, p. 111.

Exsicc. Ell. N. A. F. 876.

Stromata cortical, dark brown inside, conical, about ½ mm. broad at base, not circumscribed, thickly scattered over the matrix. Perithecia 10–20, subcircinate in the bottom of the stroma, small (⅓ mm. or less), necks converging, slender, with their short-cylindrical, stout, obtuse, smooth or sometimes substellate-furrowed, soon broadly and irregularly perforated ostiola erumpent around the conic-convex, subcoriaceous tobacco-brown disk. Asci lanceolate-clavate, 35 x 5–6 μ. Sporidia subbiseriate, allantoid, hyaline, 5–7 x 1½ μ. Spermogonium occupying the center of the stroma, pierced above, lined inside with innumerable, slender, simple or branched basidia, bearing minute (4–5 x 1 μ), hyaline spermatia.

On dead limbs of *Quercus coccinea*, Newfield, N. J.

The wood beneath the stroma is generally tinged with yellow. This differs from *V. clausa*, C. & E. in its smaller sporidia and smaller perithecia.

V. Pini, (A. & S.)

Sphæria Pini, A. & S. Consp. p. 20.
Valsa Pini, Fr. Summa Veg. Sc. p. 412.
Exsicc. Fckl. F. Rh. 608.—Rab. F. E. 147, 634, 1013.—Rehm Asc. 432.—Sydow Myc. March. 477, 1950, 1829.—Thum. M. U. 2256 —Roum F. G. 5028.

Stromata evenly scattered or subgregarious, hemispherical, 1½–2½ mm. across, covered by the pustuliform-elevated epidermis, which is pierced or ruptured by the closely packed, short ostiola with rounded or obtuse tips, forming a black, flat or concave disk. Sometimes the marginal ones and sometimes all are elongated (½–1 mm.), but usually they remain short. Perithecia very small but numerous, 20–30 in a stroma, closely packed, globose, with short necks. Asci narrow-clavate, sessile, 25–30 x 5–6 μ. Sporidia irregularly crowded, cylindrical, curved, hyaline, 6–9 x 1½ μ.

On dead limbs of *Larix Europœa*, Newfield, N. J., and on *Pinus*, New York and New England (Cke. Valsei of the U. S. No. 25). The spermogonia are like the ascigerous stroma, only instead of the black disk formed by the tips of the crowded ostiola, they are capped by a tubercular-papilla which is at length pierced with a small round opening, whitish-gray at first, finally darker. Spermatia 4 x 1 μ, issuing in yellow cirrhi.

The Newfield specc. agree perfectly with the European, only the spermatia are about twice as long. Dr. Winter who examined the specc. was of the opinion that the spermogonia did not belong to the *Valsa*, but they are certainly very intimately associated. This species is distinguished from *V. Abietis* and *V. cenisia* by its larger hemispherical stromata and more numerous perithecia, but the three are closely allied.

V. Abiètis, (Fr.)

Sphæria Abietis, Fr. in Kze. & Schm. Mycol. Hefte. II, p. 47.
Sphæria pinastri, Grev. Scot Crypt. Fl. tab. 50.
Valsa Abietis, Fr. Summa, p. 412.
Exsicc. Fckl. F. Rh. 606 —Rab. F. E. 2324, 3554 —Ell. N. A. F. 174 —Cke F Brit 2d Ser. 484.—Rehm Asc. 776.—Krieger F. Sax. 378.

Stromata irregularly scattered, depressed-conoid, small, covered by the subpustulate-elevated epidermis, which is pierced by the dense fascicle of short, or sometimes elongated-cylindrical ostiola, which are globose, smooth, black and densely crowded. Perithecia 5–15 in a stroma, small, globose or angular from pressure, with very short necks, buried in the unaltered substance of the bark. Asci narrow-clavate, sessile, 25–32 x 5–6 μ. Sporidia irregularly crowded, allantoid, slightly curved, hyaline, 6–9 x $1\frac{1}{4}$–$1\frac{1}{2}$ μ. Spermogonia in stromata similar to the ascigerous ones, multilocular, narrowed above and erumpent, with a light-colored disk pierced by a single pore (rarely by several). Spermatia slender, 3 x 1 μ.

On dead limbs of *Pinus rigida* and *Cupressus thyoides*, New Jersey, on *Abies* and *Thuja*, in Canada and probably in other parts of the country.

V. cenísia. De Not. Sferiac. Ital. p. 38, tab. 44.

Exsicc. Fckl. F. Rh. 2139 —Thum. M. U. 571.—Ell. N. A. F. 177.

Stroma conic-truncate or depressed-hemispherical, covered by the adherent epidermis, of a dark brown color, 1–2 mm. broad. Perithecia 2–20 in a stroma, subspherical or angular from mutual pressure, densely crowded, with short necks and ostiola erumpent in a flat disk or separate, short or subelongated, subcylindrical, rather stout, rounded at the apex and at length perforated. Asci cylindric-clavate, 30–40 x 5–6 μ (p. sp. 25–30 x 5–6 μ). Sporidia biseriate, allantoid, 6–12 x 1½–2 μ. Spermogonia in similar stromata with a white disk and large central pore, surrounded by smaller ones. Spermatia 5–7 x 1–1½ μ, erumpent in whitish cirrhi (when fresh), yellowish (when dry).

On dead limbs of *Juniperus Virginiana*, New Jersey and Delaware.

Distinguished from *V. Pini* by its larger spermatia.

V. Frièsii, (Duby.)

Sphæria Friesii, Duby Botan. Gall. II, p. 610.
Valsa Friesii, Fckl. Symb. p. 198.
Exsicc. Fckl. F. Rh. 610.—Sydow M. March. 148.—Rab. F. E. 2537.

Stromata numerous, quite evenly scattered, depressed-conical, sunk in the substance of the bark and raising the epidermis more or less distinctly into pustules. Perithecia 6–10 in a stroma, subcircinate, small, globose, buried in the unaltered bark, with very short necks. Ostiola small, thickened, globose or ovate, acute or obtuse, black, with a small, round pore; mostly crowded around the margin of a grayish-brown disk, but sometimes occupying the entire disk. Asci narrow-clavate, sessile, 40 x 6 μ. Sporidia crowded, cylindrical, hyaline, slightly curved, 10–12 x 1½–2 μ.

On dead limbs of *Abies balsamea*, Adirondack Mts, N. Y. (Peck), Canada (Dearness).

This species is peculiar on account of the spermogonia being produced on the leaves. They are very small, obtuse-conical, with their apices projecting and crowned with a grayish-black disk bearing a small, perforated papilla. The interior divisions are radiately arranged. Spermatia cylindrical, hyaline, curved, 4–5 x 1½ μ. Distinguished from the two preceding species by its longer sporidia.

V. junipérina, Cke. Grev. VI, p. 144.

Exsicc. Rav. F. Am. 195.—Rab. F. E. 2950.—Krieger F. Sax. 80.

Pustules covered by the slightly raised epidermis, which is pierced by the pale, farinaceous disk. Perithecia few (3–6), buried in the unaltered substance of the bark, ⅓mm. diam. or over, contracted abruptly into slender, converging necks, with the conic-hemispherical, slightly radiate-sulcate ostiola barely erumpent through the pale disk. Asci clavate, 30–35 x 5–6 μ. Sporidia biseriate, allantoid, slightly curved, hyaline, 6–8 x 1½ μ.

On dead limbs and trunks of *Juniperus Virginiana*, Carolina and New Jersey.

The spermogonial stromata are outwardly like the ascigerous stromata, but radiate-cellular within. Spermatia minute. The sporidia are smaller than in *V. cenisia.*

The specimen in Rav. F. Am. is entirely sterile, even the spermogonia being without fruit, but the specc. in the other Exsiccati quoted, agree with this in outward appearance, and the sporidia are as stated in Cooke's diagnosis of *V. juniperina*, only there are no "elongated cylindrical ostiola," unless the necks of the perithecia inclosed in the stroma are reckoned as ostiola.

V. gossýpina, Cke. Valsei of the U. S., No. 115.

Perithecia quarternate-circinate, covered, disk erumpent. Ostiola not sulcate. Asci 8-spored. Sporidia allantoid, minute, hyaline, 5–6 μ long.

On branches of *Gossypium*, So. Carolina.

Has the habit of *V. quaternata.*

V. florifórmis, E. & E. Proc. Acad. Nat. Sci. Phil. July, 1890, p. 222. **(Plate 33)**

Stroma conic-hemispherical, about 2 mm. broad and 1½ mm. high, seated on the inner bark, and covered by the epidermis, which is either pierced or sublaciniately ruptured by the thick fascicle of cylindrical (1 mm. or more long), somewhat spreading, rather obtusely pointed ostiola swollen just below the tip, and erumpent through a yellowish disk which is soon obliterated. Perithecia numerous (25–50), packed in a single layer in the lower part of the stroma, 1 x ¾ mm. diam., ovate or irregular from compression, contracted above into slender necks which rise through the cinereous contents of the stroma, and terminate above in the cylindrical ostiola. Asci (p. sp.) 35–40 x 5 μ. Sporidia biseriate, cylindrical, hyaline, slightly curved, 6–7 x 1½ μ, 8 in an ascus.

On dead limbs of *Populus monilifera*, Missouri (Demetrio).

Differs from *V. verrucula*, Nits., in its long ostiola and smaller asci and sporidia. Has much the same general appearance as *V. scoparia*, Schw., but ostiola not sulcate and asci and sporidia larger.

V. delicátula, C. & E. Grev. VI, p. 10.

Valsa decidua, C. & E Grev. VI, p. 11, Pl. 95, fig. 11.
Exsicc. Ell N. A. F. 865

Pustules small (1 mm.) prominent, orbicular or elongated, covered by the bark, except the minute ostiola, which are clustered in a brown disk. Perithecia shaped like a Florence flask, membranaceous, the neck equal in length to the perithecium or a little more or less, slender, swollen at the apex. Sporidia sausage-shaped, 6–8 x 1½ μ, moderately curved. Asci clavate.

On dead stems of *Andromeda racemosa*, *Vaccinium corymbosum*, *V. Pennsylvanicum*, *Azalea*, *Gaylussacia resinosa*, Newfield, N. J.

We find no definite characters to separate *V. decidua* from *V. delicatula*. In both, the tips of the ostiola are more or less swollen, but usually not as abruptly as represented in the figure in Grevillea. The ostiola are united in a pale brown disk, which in very thrifty specimens becomes obliterated. The deciduous habit is common to both, and the sporidia, as we see them, do not differ appreciably in size or shape. The perithecia (10–20 in a stroma) are of the same size and shape in both.

V. Lavateræ, Cke. & Hark. Grev. XIV, p. 8.

Stromata gregarious, immersed in the blackened wood, covered by the epidermis. Perithecia 3–6 in a stroma, subglobose, black. Ostiola elongated, cylindrical, slender, emergent. Asci clavate, 8-spored. Sporidia allantoid, hyaline, 8 x 2 μ.

On stems of *Lavatera assurgentifolia*, California (Harkness).

V. ribèsia, Karst, Myc. Fenn. II, p. 138.

Valsa agnostica, Cke. & Hark. Grev. XIII, p. 17.

Erumpent, subrotund, nestling in the bark, at first covered, then, with the suborbicular disk, bare. Perithecia 8–12, subglobose, black, collected in a pallid stroma. Ostiola short, straight, obtusely rounded, shining. Asci subclavate, p. sp. 22–25 x 5–6 μ, 8-spored. Sporidia allantoid, straight or somewhat curved, biseriate, hyaline, 6–7 x $1\frac{1}{2}$ μ.

On branches of *Ribes*, California and Canada.

We have seen no authentic specimen of *V. ribesia*, Karst., but the specc. of *V. agnostica*, from Harkness, are the same as collected in Canada by Dearness, and both agree so well with the diagnosis of *V. ribesia*, Karst, that we have no hesitation in referring them to that species.

V. Lupìni, Cke. & Hark. Grev. XIV, p. 8.

Exsicc. Ell. & Evrht. N. A. F. 2d Ser. 1570.

Pustules small (1 mm.), numerous, prominent, black. Perithecia 4–8 in a stroma, collected in a subconical group. Ostiola minute, converging, globose-conical, only slightly projecting, erumpent in a black disk. Asci clavate, long-attenuated below, 75–80 x 6 μ, 8-spored. Sporidia allantoid, hyaline, nearly straight, 7 x 2 μ.

On stems of *Lupinus*, California (Harkness).

V. Eucalýpti, Cke. & Hark. Grev. IX, p. 85.

Exsicc Ell. N. A. F. 871.

"Erumpent, subrotund, convex, black. Perithecia oblong. Ostiola elongated, cylindrical, smooth, straight. Asci clavate, sessile. Sporidia straight or slightly curved, ends obtuse, hyaline, 8–9 x $1\frac{1}{2}$ μ. Pustules small, containing only 5–6 perithecia."

On twigs of *Eucalyptus*, California.

The specc. in N. A. F. show mostly spermogonia, convex, black, $1\frac{1}{2}$–2 mm. diam., with a central, perforated papilla. Cells numerous, radiate. Spermatia cylindrical, hyaline, curved, 5 x 1 μ.

V. conspurcàta, (Schw.)

Sphæria conspurcata, Schw. Syn. N. Am. 1336.
Valsa conspurcata, Stevenson Additions to Cooke's Valsei, No. 110.

Conceptacle rather large, black, but covered, as well as the effused black crust around the base, with a dirty brownish, pulverulent mass. Several conceptacles are often confluent beneath the epidermis, which is stellately ruptured with revolute laciniæ. Perithecia rather large, subcircinate, compressed in the conceptacle, surrounded by a cinereous-brown stroma, with a horn-colored nucleus, necks elongated-connate. Ostiola 6–12, short-cylindrical, rising together through the whitish or cinereous disk, rounded and nearly smooth at the tips. Sporidia 5–6 x $1\frac{1}{2}$–2 μ.

On oak firewood, Bethlehem, Pa. (Schw.)

The description of the ostiola and sporidia was made from specimen in Herb. Schw.

V. modésta, (Schw.)

Sphæria modesta, Schw. Syn. N. Am. 1337
Valsa modesta, Stevenson l. c.

Conceptacle nearly free in the inner bark, making a round, subelevated tubercle, at length erumpent through the revolute-fissured epidermis. Disk dark brown. Ostiola black, subprominent, rather long, densely crowded, cylindric-conical, at length deciduous. Stroma in which the perithecia lie, cinereous. Diameter of the conceptacle 2–3 lines.

On young branches of elm, Bethlehem, Pa. (Schw.)

Sporidia (sec. Stevenson l. c.) allantoid, hyaline, $7\frac{3}{4}$ x $3\frac{1}{4}$ μ.

V. Tóxici, (Schw.)

Sphæria Toxici, Schw. Syn. N. Am. 1330.

Conceptacles rather large, suborbicular, closed below, black, rugose, at first covered, finally exposed, often longitudinally confluent. Ostiola collected in the center, round, obtuse, sometimes apparently stellate. Perithecia oblong, crowded, narrowed into a neck, brown, not black, included in a scanty, whitish (becoming brownish) stroma. Sporidia allantoid, yellowish, 6–7 x 1½ μ.

On branches of *Rhus radicans*. Frequent around Bethlehem, Pa. (Schw.)

The *Valsa Toxici* in Cooke's Valsei of the U. S., with brown, uniseptate sporidia is something else. The specc. in Herb. Schw. have sporidia as stated above.

V. micróspora, Cke. & Plowr. Grev. VII, p. 82, (1879).

Valsa minutella, Pk. Bull. Torr. Bot. Club, XI, p. 27, (1884).

Pustules minute. Perithecia 6–20 in a pustule, nestling in the bark, crowded, black. Ostiola black, erumpent in a minute ferruginous disk, which is closely surrounded by the ruptured epidermis. Asci short, clavate or fusiform, scarcely pedicellate, 22–30 x 5–6 μ. Sporidia allantoid, crowded, 5–6 μ long.

On bark of beech (*Fagus ferruginea*), Canada (Macoun).

The Canada specimens agree so well with the description of *Valsa microspora*, in Grevillea, and with specimens from Plowright, that there can hardly be any doubt that *V. minutella*, Pk., and *V. microspora*, Cke. & Plowr., are the same.

V. defórmis, (Fr.)

Sphæria deformis, Fr. S. M. II, 398
Valsa deformis, (Fr.) Cke. Syn. & Stevenson in Add. to Cke. Valsei of the U S

Pustulate, irregular. Stroma pulverulent, ferruginous. Ostiola solitary or aggregated, globose, at length rostellate.

On oak limbs, Bethlehem, Pa. (Schw.) Sporidia (sec. Stevenson, l. c.) 6¼–8⅓ x 1½ μ.

The fungus described by Fries grew on the inner surface of bark of fir trees lying on the ground. Perithecia minute, ovate, covered by a cortical pustule, and lying in a ferruginous stroma. Ostiola erumpent in the center (of the pustule), without any distinct disk, at first papilliform, smooth, finally exserted, tentaculate. The Schweinitzian species being on oak is doubtfully synonymous with this.

V. variolària, (Schw.)

Sphæria variolaria, Schw. Syn N Am 1371
Valsa variolaria, Cke. Syn. 1785.

Subpustulate, subconfluent far and wide under the smooth, closely fitting epidermis, which is stellately ruptured by the prominent ostiola erumpent at first in a brown disk, which finally disappears. Perithecia suberect, circinate, surrounded by a scanty, light-colored stroma, but without any distinct conceptacle. On a horizontal section, a dark circumscribing line is seen in the inner stratum of the bark, surrounding many individuals or whole groups. Sporidia (sec. Stevenson) $7\frac{3}{4} \times 3\frac{1}{4}$ μ.

On branches of *Tilia*. Frequent around Bethlehem, Pa. (Schw.).

V. æquilineàris, (Schw.)

Sphæria æquilinearis, Schw. Syn. N Am. 1293.

Perithecia covered, immersed, small, ovate, subcircinate, in small, subseriate clusters in an elongated, linear, dark-colored stroma, the cylindrical ostiola bursting out in a seriate manner through cracks in the bark, but not projecting much above it. Sporidia 5 x 1 μ.

On dead stems of *Berberis Canadensis*, Carolina (Schw.).

We have taken the measurements of the sporidia and supplemented the original diagnosis from an examination of the spec. in Herb. Schw.

Species imperfectly known.

V. subscrípta, (Fr.)

Sphæria tessella, Fr *B. subscripta*, Fr. S. M. II, p. 393.
Valsa subscripta, Sacc. Syll 504.

Perithecia circinate, the black, circumscribing line subobliterated. Ostiola singly erumpent in a small disk. Asci minute. Sporidia allantoid, minute.

Mentioned in Cke. Valsei of the U. S. No. 33, and in Curtis' Cat. p. 142.

V. erinàcea, (B. & Rav.)

Hypoxylon erinaceum, B. & Rav. Grev. IV, p. 94.
Valsa erinacea, Cke. Grev. XI, p. 128.

"Perithecia ovate, pulverulent, with an elongated, often curved neck. Sporidia clavate, much attenuated below."

On *Liquidambar*, South Carolina.

V. múnda, B. & C. Grev. IV, p. 100.

"Pustules completely covered by the bark which is blackened over them, or appears black by transparence, the disk alone, which is bordered with white, being free. Asci lanceolate. Sporidia sausage-shaped.

On smooth branches of *Cornus*, Alabama.

V. fulvélla, B. & Rav. Grev. IV, p. 101.

Pustules closely covered by the bark which is raised up. Disk pale tawny, dotted with the black ostiola. Sporidia allantoid.

Of the following species apparently referable to *Valsa:*

Sphæria pugillus, Schw. Syn. N. Am. 1322 (sec. Stevenson) is a *Sphæronema.*

Sphæria frustum-coni, Schw. Syn. N. Am. 1329. The spec. of this in Herb. Schw. is some spermogonial form.

Sphæria radicum, Schw. Syn. N. Am. 1335. Specimen abortive or sterile.

Sphæria oligostoma, Schw. Syn. N. Am. 1333. The specimens in Herb. Schw. are barren.

** *Mesosporæ. Sporidia* 8–12 μ *long.*

V. rufèscens, (Schw.)

Sphæria rufescens, Schw. in Herb Schw.—(and Schw. Syn. N. Am. 1395) ?
Valsa rhuiphila, C. & E. Grev. VII, p 9.
Exsicc. Ell. N. A. F. 872.

Perithecia lying in subcircinate clusters of 15–30 in the inner bark, which is but slightly altered, only assuming a reddish-brown color which is of a more dirty shade directly over the pustules; these are orbicular, and about 2 mm. diam., convex, with the compact tuft of ostiola rising like a disk or crown through the apex, and erumpent through the epidermis, but scarcely exserted. Perithecia crowded, about $\frac{1}{3}$ mm. diam., or less. Ostiola short-cylindrical, obtuse, becoming subumbilicate. Asci 25–35 x 6–7 μ, 8-spored. Sporidia biseriate, allantoid, hyaline, slightly curved, 6–7 x $1\frac{1}{4}$–$1\frac{1}{2}$ μ (10 μ long, Cke.).

On dead branches of *Rhus glabra*, Carolina, Pennsylvania (Schw.), and on *R. copallina* and *R. venenata*, Newfield, N. J.

This is probably the *Sphæria rufescens*, Schw. Syn. N. Am. 1395, but we have never seen the ostiola as there stated "very long, decumbent, flexuous and diverging." This character applies sometimes to *Sphæria aculeans*, Schw. No. 1399. Apparently the descriptions of these two species have been in some way confused. The species here described is certainly the *Sphæria rufescens*, Schw. in Herb. Schw.

V. Línderæ, Pk. 29th. Rep. p. 59.

Pustules small, rather prominent, crowded or scattered, closely surrounded by the ruptured epidermis, circumscribed by a black line. Perithecia usually 4–6, nestling in the inner bark. Ostiola crowded, short, dull black, obliterating the blackish disk. Asci slender-clavate. Sporidia 8 in an ascus, yellowish in the mass, cylindrical, curved, obtuse, 8–12 μ long.

On dead branches of *Lindera Benzoin*, Albany, N. Y.

V. vitígera, Cke. Grev. V, p. 125, XIV, p. 46.

Valsa Vitis, Fckl. Symb p 199, Sacc Syll. 449, Sacc Myc. Ven. Sp 133, tab XIII, figs. 19–21.
Exsicc. Fckl. F. Rh. 607.—Sacc. M. Ven. 186.

Pustules subcuticular, scattered, small, slightly prominent. Perithecia 3–6, about ⅓ mm. diam., with short necks terminating in the short-cylindrical ostiola, conical at the apex, smooth or slightly sulcate, and erumpent in a compact tuft in the center of the pustule, but not much exserted. Asci clavate, p. sp. about 40 x 6–7 μ. Sporidia biseriate, allantoid, only slightly curved, hyaline, 8–10 x 1½–2 μ.

On small, dead shoots of grape vines, Newfield, N. J. Sent also from Louisiana (Langlois, 1767).

The sporidia in the Louisiana specc. average larger (8–12 μ long), but there is no other difference. The perithecia lie in the unchanged substance of the bark without any circumscribing line, and in well developed specc. each separate perithecium raises the bark above it slightly, as in *V. ambiens.*

On the Newfield specc. are also spermogonia, multilocular, gray inside, opening by a common central pore, and a little more prominent than the ascigerous pustules. Spermatia allantoid, hyaline, 4 x 1 μ.

This is evidently the *V. vitigera*, Cke., and is quite distinct from *Eutypella Vitis*, (Schw.), which has stouter, sulcate ostiola and smaller, yellowish sporidia, and is in fact only a form of *Eu. stellulata.*

V. ligústrina, Cke. Grev. VIII, p. 14.

Stromata cortical, brown within, but not circumscribed, small, elliptical, about 1½ x 1 mm., raising the surface of the inner bark into slight pustules which are not noticed till the outer layer is stripped off. Perithecia 3–8 in a pustule, ovate, small. Ostiola erumpent through longitudinal cracks in the outer bark, but not exserted, stout, cylindrical, with the apex at first conical, then broadly and irregularly perforated. Asci clavate-fusoid, about 35 x 6 μ, 8-spored. Sporidia biseriate, allantoid, hyaline, moderately curved, 7–9 x 1½–2 μ.

On bark of dead *Andromeda ligustrina*, Newfield, N. J.

The ostiola are smaller and the sporidia rather larger than in *V. delicatula*, and the fascicles of ostiola are laterally compressed.

V. Liquidámbaris, (Schw.)

Sphæria Liquidambaris, Schw. Syn. N. Am. 1352.
Valsa Liquidambaris, Cke. Valsei U. S. No. 29, Sacc. Syll. 477.

Pustulate, small. Perithecia decumbent, 10–20 in a pustule. Ostiola minute, crowded, conical-globose, rounded at the apex and smooth, or sometimes faintly sulcate-striate, erumpent in a close fascicle, mixed with yellowish-farinaceous matter. Pustules convex, $1\frac{1}{2}$–2 mm. across, formed of the scarcely changed substance of the inner bark, and not circumscribed. Asci p. sp. about 30 x 6 μ, 8-spored. Sporidia allantoid, subbiseriate, hyaline, 9–12 x $1\frac{1}{2}$–2 μ.

On dead branches of *Liquidambar styraciflua*, Carolina and New Jersey.

V. múltiplex, C. & E. Grev. VIII, p. 14.

Exsicc. Ell. N. A. F. 874.

Stromata convex-hemispherical. about 2 mm. diam., often 2–3 confluent, covered by the loosened epidermis. Perithecia numerous, 30 or more closely packed in a single layer in the bottom of the stroma, small, necks slender, with their small, black, shining ostiola erumpent in a brown, abruptly elevated disk which is soon obliterated. Asci clavate, 26–40 x 5–6 μ, 8-spored. Sporidia biseriate, cylindrical, hyaline, slightly curved, with a nucleus in each end, 6–8 x $1\frac{1}{4}$–$1\frac{1}{2}$ μ (10 μ long, Cke.).

On oak wood cut and piled for market, Newfield, N. J.

The stromata are numerous, and the tufts of erumpent, but scarcely exserted ostiola loosen the epidermis. The surface of the inner bark is of a uniform, dull black color, but there is no circumscribing line.

V. pùsio, B. & C. Grev. XIV, p. 46.

Pustules small, erumpent. Perithecia black, subglobose, nestling in the bark, surrounded by the fissured epidermis. Ostiola subrugose, truncate, emerging in an orbicular disk. Asci clavate, subsessile, short, 8-spored. Sporidia allantoid, smoky-hyaline, 8 x 2 μ.

On bark of *Morus multicaulis*, North Carolina.

V. púllula, B. & C. Grev. XIV, p. 47.

Pustules small, at first covered. Asci clavate, 8-spored. Sporidia cylindrical, slightly curved, rounded at the ends, hyaline, 8 μ long.

On twigs of *Castanea*, Pennsylvania.

V. truncàta, C. & P. 25th Rep. N. Y. State Mus. p. 103.

Erumpent, prominent, truncate. Perithecia 6–8, nestling in the inner bark, globose, black, the necks united in an orbicular or elliptical disk which is pierced by the ostiola, and generally pulverulent on the margin. Asci small, lanceolate. Sporidia minute, sausage-shaped, hyaline, 8–10 μ long. Spermogonia cytisporoid, disk erumpent, truncate, pulverulent in the center, sometimes having a bilabiate appearance. Spermatia amber in the mass, minute, linear.

On dead branches of alders, New York and Massachusetts.

V. decòrticans, (Fr.) Summa Veg. Scand. p. 412.

Exsicc. Ell. N. A. F. 496?

Stroma orbicular or oval, depressed-conical, rather large, brown or finally black, abruptly attenuated above into a disk erumpent through the epidermis, which is usually thrown off. Perithecia 6–20 in a stroma, not deeply buried, monostichous, globose or angular from pressure, attenuated into a moderately long neck. Ostiola more or less exserted, slender, cylindrical, black, shining, forming an orbicular or elliptical, erumpent disk, mostly densely crowded, with the apex rounded or truncate, pierced with a distinct pore, either attenuated-conical, or very rarely elongated-cylindrical, and considerably exserted and flexuous. Asci narrow-oblong or clavate, sessile, 45–50 x 6 μ. Sporidia conglobate, cylindrical, curved or straight, hyaline, 10–12 x $2\frac{1}{2}$ μ. Spermogonia *Cytispora* sp. Spermatia allantoid, 4–5 x 1 μ.

On bark of *Kerria Japonica* and *Syringa*, Carolina (Curtis), sec. Cooke Valsei of the U. S., No. 35.

In Europe the species is found on *Fagus* and *Carpinus*. The N. A. F. specc. are on maple, and do not agree well with the above diagnosis taken from Sacc. Sylloge. They are probably referable to *V. ceratophora*, having sporidia about 7 μ long.

V. rubíncola, (Schw.)

Sphæria rubincola, Schw. Syn. N. Am. 1331.
Valsa rubincola, Schw. in Stevenson's Add. to Cke Valsei of the U S No 105.

Growing in the inner bark, at length deciduous, leaving little cavities or pits. (Stroma) orbicular, depressed in the center, where rise the rough ostiola, gregariously confluent. At first covered by the epidermis, and buried so deeply that only the ostiola project, but finally emergent and free. Perithecia rather large, few, depressed-globose, surrounded by a white-pulverulent stroma. The conceptacle is sometimes crowned with a single ostiolum, cylindrical, with a broad opening—sometimes with several divergent ostiola.

Common on stems of *Rubus* in gardens, Bethlehem, Pa. (Schw). Sec. Stevenson, l. c., sporidia allantoid, hyaline, 9 x 2 μ.

V. rhízina, (Schw.)

Sphæria rhizina, Schw Syn. N. Am. 1398.
Valsa rhizina, Stevenson Add to Cooke's Valsei of the U S No 122

Perithecia circinate, closely covered by the inner bark. Ostiola terete, subdivergent, with a large orifice, at first sometimes papillate. Perithecia densely circinate, much depressed, numerous, olive-black, white inside, minute, nestling in pits in the bark without any stroma.

Sporidia allantoid, hyaline, $9\frac{1}{3} \times 3\frac{1}{4}$ μ. Spermatia minute, 3–4 x 1 μ.

On a root of *Rhododendron*, Bethlehem, Pa. (Schw.).

V. anómia, (Schw.)

Sphæria anomia, Schw. Syn. N. Am. 1316.
Valsa anomia, Schw. Cke. Valsei of the U. S. No. 34.

Convex, irregular, free, rugose, stroma formed of the wood, cinereous-black. Ostiola exserted, distant, black, smooth. Sporidia (sec. Berk.) 10 μ long, with uniseptate stylospores in other perithecia. Sporidia (sec. Cke.) 7–8 μ long.

On *Robinia*, North Carolina (Schw.).

Apparently different from the European species of the same name which appertains to *Pseudovalsa profusa*.

V. micróstoma, (Pers.)

Sphæria microstoma, Pers. Syn. p. 40.
Valsa microstoma, Fr. Summa, p. 411.

Stroma roundish-oval at base, convex or more rarely conical, attenuated abruptly into an oval, elliptical or orbicular disk, immersed in the outer layer of the inner bark, of which the surface assumes a chestnut color. Disk erumpent through transverse cracks in the epidermis Perithecia 6–20 in a stroma, in a single layer, minute, crowded. Ostiola somewhat thickened, rounded or almost globose, abbreviated, black and shining, crowded in a plane disk. Asci oblong or clavate, sessile, 8-spored, 40–48 x 6–8 μ. Sporidia crowded or biseriate, allantoid, hyaline, 8–10 x $2–2\frac{1}{2}$ μ. Spermogonia cytisporoid. many-celled. Spermatia allantoid, 5–6 x $1\frac{1}{2}$ μ, on long basidia.

On bark, Pennsylvania and Carolina (Schw.).

V. allóstoma, (Schw.)

Sphæria allostoma, Schw. Syn. N. Am. 1332.
Valsa allostoma, Stev. l. c., No. 106.

At first immersed in the wood, with only the globose-stellate ostiola prominent, finally emergent and almost free. Conceptacles large, and often extensively confluent. Ostiola at length becoming terete, generally dilated at the apex, rigid. Perithecia rather large, crowded, almost entirely without any stroma. The groups of confluent conceptacles finally form, as it were, a continuous crust. Sporidia (sec. Stevenson l. c.) 9 x 3 μ, allantoid, hyaline.

On a log of *Robinia*, near Lancaster, Pa. (Schw.).

V. monadélpha, Fr. Summa Veg. Sc. p. 411.

Sphæria monadelpha, Fr. S. M. II, p. 382.
Valsa monadelpha, Fr. Summa Veg. Sc p. 411.
Exsicc. Desm. Pl. Cr. Ed. I, 961, Ed. 2nd 261.

Perithecia circinating around a central one, included in a cinereous-black stroma, necks short, ostiola, densely crowded and erumpent in a convex disk. On bare wood it varies, becoming effused, with the ostiola distinct. Asci 8-spored. Sporidia allantoid, 10 μ long.

On *Prunus*, New England and Pennsylvania (Schw.).

Species imperfectly known.

V. leiphæmioides, B. & C. Grev. IV, p. 101.

Exsicc. Rav. F. Am. 192.

"Pustules, when the cuticle is stripped off, covered with the brown bark, the ostiola only exposed and mixed with white matter. Sporidia sausage-shaped, 10 μ long."

On oak, New England and Carolina. Closely resembling *Diaporthe leiphæmia*, (Fr.), but with different fruit.

The specimens in our copies of Rav. F. Am. are entirely without fruit. The specc. issued in Ell. & Evrht. N. A. F. 2d Ser. 2344, are a *Calosphæria*, apparently a compact form of *C. lasiostoma*, E. & E., which is probably the same as *C. scabriseta* Schw. This same *Calosphæria* is found on the spec. in Rav. F. Am. 192.

V. Nyssæ, Cke. Grev. VI, p. 145.

Exsicc. Rav. F. Am. 194

As nothing satisfactory can be made out of the spec. in Rav. F. Am., we can only quote the brief description in Grevillea.

"Pustules covered by the bark. Perithecia ovate, attenuated into a long neck, black. Asci clavate. Sporidia linear, curved, hyaline, 8 μ long. Often the center of the pustule is occupied by a large spermogonium around which the perithecia are clustered."

On *Nyssa*, Aiken, So. Carolina.

The specimens of *Sphæria radicum*, Schw., and *Sphæria oligostoma*, Schw., in Herb. Schw., are without fruit, and are apparently barren, *Valsas*.

V. centripeta, (Fr.)

Sphæria centripeta, Fr. S. M. II, p. 402.
Valsa centripeta, Fr. in Peck's 26th Rep. p. 86.

Pustulate, orbicular or elliptical, disk transversely erumpent, but variable. Perithecia rather large, globose-depressed, covered, the very small, crowded, exserted, semiglobose ostiola mostly arranged in two parallel lines, forming an elongated, black spot in the center of the erumpent, flat disk;

This species (sec. Peck) was found on dead alders, near Buffalo, N. Y.

We have seen no specc., and translate the diagnosis from Fries' Systema.

*** *Macrosporæ. Sporidia more than 12 μ long.*

V. ámbiens, (Pers.)

Sphæria ambiens, Pers. Syn. p. 44.
Sphæria deplanata, Nees, in Fr. Summa Veg. Sc. p. 394.
Sphæria capsularis, Pers. Syn. p. 42.
Sphæria sphinctrina, Fr. (pr. p.) S. M. II, p. 400.
Valsa corticis, Tul. in Ann. Sci. Nat. Ser. IV, tom. V, p. 117.
Sphæria tetraspora, Curr. Linn. XXII, p. 279, fig. 148.
Valsa ambiens, Fr. Summa, p. 412.
Valsa conscripta, C. & E. Grev. VII, p. 80.
(*Valsa cooperta*, Cke. Valsei of the U. S. p. 118)?

Exsicc. Fckl. F. Rh. 616, 2141.—Kze. F. Sel. 149.—Rab. F. E. 1131, 1534.—Rehm Asc 48, 171, 223.—Sydow. M. March. 33, 465, 663.—Krgr. F. Sax. 172, 431.—Roum. F. G. 1178, 4447.—Cke. F. Brit. 2d Ser. 232.—Plowr. Sph. Brit. 46.—Sacc. M. Ven. 1493.—Ell. N. A. F 868.

Stromata numerous and generally thickly scattered, surrounding and extending for some distance along the limbs, orbicular or elliptical at base, and 1½–3 mm. broad, obtusely conical above and, for the most part, distinctly pustulate-prominent, covered by the closely adherent epidermis, which is pierced by the erumpent disk or stellately cleft with the laciniae adhering. Perithecia 4–20 (mostly 6–10), subcircinate in the unaltered substance of the inner bark, about ½ mm. diam., and mostly not crowded, necks slender, decumbent, with their large, depressed-spherical, black ostiola erumpent around the margin of the small, whitish disk or irregularly arranged, crowded and obliterating the disk. Sometimes the epidermis is slightly depressed just around the disk, with a circle of slight protuberances indicating the position of the subjacent perithecia. Asci (p. sp.) 40–55 x 12–15 μ, 4–8-spored, with a slender, stipe-like base 30–35 μ long. Sporidia conglomerate or subbiseriate, allantoid, hyaline, slightly curved, obtuse, 14–24 x 3–5 μ in the 8-spored asci, in the 4-spored asci, 24–36 x 5–8 μ.

On dead limbs of various deciduous trees, common, at least throughout the northern U. S. and Canada.

We have specc. from Canada on ash, hawthorn, elm, oak, basswood and hickory; from Iowa, on *Pyrus coronaria;* from Montana, on *Shepherdia Canadensis;* from New York, on *Acer rubrum;* from Nebraska, on *Sambucus;* from New Jersey, on *Nyssa*, *Liquidambar*, and *Pyrus*, and the tetrasporous form on *Morus*, from Illinois.

V. salícina, (Pers.)

Sphæria salicina, Pers. Obs. I, p. 64
Valsa salicina, Fr. Summa Veg. Scand. p 412
Exsicc Fckl. F. Rh. 614, 615.—Kz. F. Sel. 147, 345.—Rehm Asc. 82.—Thum. M U. 468, 869.—Cke. F. Brit. Ser. I, 377.—F. Sax. 259, 432.—Roum. F G. 558.—M. March. 145, 461, 1724.—Lin. F. Hung. 173.—Ell. & Evrht. N. A. F. 2d Ser. 1951.

Stromata thickly scattered, depressed-conical, truncate, slightly prominent, pustuliform, closely covered by the epidermis, except the small, whitish disk. Perithecia 6–12 in a stroma, circinating in the unchanged substance of the inner bark, with very short and slender necks and minute, black, globose, perforated ostiola erumpent around the margin of the disk or irregularly scattered over it, scarcely exserted. Asci narrow-oblong or clavate, 4–8-spored, 40–65 x 7–8 μ, subsessile. Sporidia allantoid, hyaline, slightly curved, 12–18 x $2\frac{1}{2}$–4 μ (in the 8-spored asci), 20–30 x 5–7 μ (in the 4-spored asci).

On dead willow limbs, Carolina and Pennsylvania (Schw.), also New York, New England and Canada.

V. ambiens also occurs on *Salix*, but is larger throughout from stroma to sporidia.

V. sórdida, Nits. Pyr. Germ. p. 203.

Valsa deplanata, Fckl. Enum Fung Nassov p 55.
Exsicc Sydow, M. March. 464 —Rehm Asc. 729.—Krgr. F. Sax. 173.

Pustules numerous, thickly and evenly scattered, only slightly prominent, 2–3 mm. diam. Perithecia 6–12 together, subcircinate in the unaltered substance of the inner bark, globose or subcompressed, medium size ($\frac{1}{3}$–$\frac{1}{2}$ mm.). Ostiola short, thick, rounded at the apex and finally with a rather narrow pore, erumpent mostly around the margin of a grayish, finally brownish, disk. Asci clavate, 40–45 x 8 μ (p. sp.), or including the stipe-like base 50–60 μ long. Sporidia subbiseriate, cylindrical, hyaline, only slightly curved, 8–10 x $1\frac{1}{2}$ μ in the Iówa specc., 9–11 x $1\frac{1}{2}$ μ in Krieger's specc., and 10–12 x 2 μ mostly in Rehm Asc. Spermatia 4–1 μ, borne on long, slender basidia and oozing out in yellow cirrhi.

On dead *Populus*, Iowa (Holway), California (Harkness).

Differs from *V. ambiens* and *V. salicina* in its slenderer sporidia and spermatia.

V. boreélla, Karst. Mycol. Fenn. II, p. 141.

Exsicc. Sydow, M. Marchica, 1826.—Ell. & Evrht. N. A. F. 2d Ser. 2520

Perithecia 8–20 together, circinating in the unchanged substance of the inner bark, subglobose, about $\frac{1}{3}$ mm. diam., subdecumbent, necks slender, converging. Ostiola obtuse, rounded, finally sublaciniately dehiscent, erumpent around the margin of a grayish disk which pierces the epidermis and is rather loosely embraced by it. Asci clavate, p. sp. about 35 x 6 μ. Sporidia conglobate, allantoid, hyaline, slightly curved, 8–11 x 1½–2 μ. Spermatia (fide Karst) elongated, 4–6 x 1 μ, somewhat curved.

On bark of willow, Canada (Dearness), Montana (Kelsey).

The specc. in *M. Marchica* have the sporidia a little larger (10–12 x 2 μ mostly), but do not differ otherwise. The species scarcely differs from *V. sordida*.

V. claùsa, C. & E. Grev. VIII, p. 13.

Exsicc. Ell. N. A. F. 870.

Stroma cortical, brown inside, 1–1½ mm. diam., depressed-conical, subcoriaceous above, not circumscribed. Perithecia 4–8 (sometimes as many as 12 or 15), buried in the lower part of the stroma, ⅓–½ mm. diam., subcircinate around a central spermogonial cell, pierced above with an irregular opening. Necks converging with the obtuse, roughish, black ostiola erumpent around the opening in the central cell. Asci clavate, p. sp. 40–45 x 6–7 μ. Sporidia subbiseriate, allantoid, hyaline, 14–18 x 3–4 μ, slightly curved. Spermatia allantoid, minute, 3–4 x 1 μ.

On dead limbs of *Quercus coccinea*, Newfield, N. J.

In the smaller pustules the bark is only slightly raised and scarcely ruptured, but in the larger ones the apex of the stroma is erumpent and loosely embraced by the ruptured epidermis. Accompanying the ascigerous stromata are others containing a single irregular-shaped cell with a thick, compact, light-colored wall, and lined on the inside with basidia bearing fusoid, continuous or uniseptate, hyaline sporules, 8–12 x 2½–3 μ. The outside appearance of this species is almost the same as in *V. lutescens*, but the sporidia and perithecia are larger.

V. córnina, Pk. 38th Rep. p. 102.

Pustules small, scattered, at first covered by the epidermis, which is at length longitudinally ruptured. Perithecia 2–5 in a pustule, nestling in the inner bark, black, the ostiola scarcely exserted. Asci clavate, blunt, 50 x 10 μ. Sporidia collected in the upper part of the asci, allantoid, 15–18 x 4 μ.

On dead branches of *Cornus paniculata*, Albany, N. Y.

Differs from the other species on *Cornus* in its different habit and larger sporidia.

V. Menispérmi, Ell. & Holway, Journ. Mycol. I, p. 4.

Perithecia circinating in a cortical stroma without any circumscribing line, 6–10 in number, and about ¼ mm. diam., with membranaceous, coarsely cellular walls. Ostiola very short, united and concealed in a circular disk, ⅓–½ mm. diam., and entirely covered by the epidermis through which its outline is seen as a small, black circle, with a black dot in the center. Asci clavate, 70 x 12 μ. Sporidia crowded, cylindrical, yellowish, curved, 15–20 x 4–6 μ.

On dead stems of *Menispermum Canadense*, Iowa (Holway).

On stripping off the epidermis, the perithecia sometimes adhere to it, and sometimes remain buried in the surface of the inner bark. Closely allied to *Valsa ambiens*, Fr., but differs in the nature of its permanently covered disk.

V. opulifòlia, Pk. 38th Rep. p. 103.

Pustules subconical, or subhemispherical, erumpent. Perithecia 5–20 in a pustule, nestling in the inner bark, crowded, often angular from mutual pressure. Ostiola crowded, black, obliterating the grayish disk. Asci subclavate, p. sp. 30–45 x 6–7 μ. Sporidia allantoid, crowded above, uniseriate below, 10–12 x 2–2½ μ.

On dead branches of *Spiræa opulifolia*, West Albany, N. Y. (Peck).

Apparently related to *V. pustulata*, Awd., but the crowded ostiola are central in the disk. When the epidermis is torn away, the pustules appear much like those of *V. colliculus*, Wormsk. We have seen no specc., and take the above diagnosis and remarks from the Report cited.

V. leucostomoìdes, Pk. 38th Rep. p. 193.

Pustules numerous, minute, covered by the epidermis, which is pierced by the orbicular, white, or grayish disk. Perithecia 2–6 or more in a pustule, the ostiola punctiform, black, dotting the disk. Asci clavate or subfusoid, 49–50 x 8½–10 μ. Sporidia crowded, allantoid, colorless, 12–18 x 4–5 μ.

On branches of sugar maple, Helderberg Mts., N. Y. (Peck).

The very small size of the pustules, and the minute, white, pulverulent disk, make this species resemble *V. leucostoma*, but there is no black, circumscribing line.

V. coryneoìdes, B. & C. Grev. XIV, p. 47.

Pustules very small, at first covered, then erumpent. Perithecia few, subglobose. Ostiola crowded in a black disk. Asci clavate. Sporidia allantoid, hyaline, 10–12 μ long.

On bark of *Juniperus Virginiana*, New Jersey, (*Valsa thelebola*, Schw. in Herb. Berk)?

We have not seen either of these species.

V. quaternàta, (Pers.)

Sphæria quaternata, Pers. Syn. p. 45.
Quaternaria Persoonii, Tul. Sel. Carp. II, p. 105, tab. XII, figs. 16-25.
Valsa quaternata, Fr. Summa Veg. Scand. p. 412.
Exsicc. Fckl. F. Rh. 621.—Rab. F. E. 255, 815, 1247.—Thum. F. Austr. 182.—id. M. U. 465. M. March. 260.—Cke. F. B. 2d Ser 221, 224.—Ell. N. A. F. 175.—Desm. Pl. Cr Ed. 1st 562, Ed. 2d 1752.

Perithecia 3-6, circinating in the inner bark, decumbent, large, ½ mm. diam., ovate. Ostiola short, large, obtuse, erumpent in a small, black tubercle, but not connate. The pustules are thickly scattered, but the bark is only slightly raised above them. Asci oblong-clavate, stipitate, p. sp. 50–75 x 8–10 μ. Sporidia biseriate, allantoid, only slightly curved, smoky-hyaline, 12–18 x 3–4 μ.

On poplar and *Alnus serrulata*, Pennsylvania and Vermont (Berk. in Grev.), on *Acer rubrum*, Carolina and New Jersey, on beech, New York (Peck).

The spermogonial stage is *Libertella faginea*, Desm.

V. pauperàta, C. & E. Grev. VI, p. 93.

Exsicc. Ell. & Evrht. N. A. F. 2d Ser. 1571.

Perithecia 2–4 together (mostly 3), buried in the unaltered substance of the bark, subglobose, black, ⅓–½ mm. diam., necks decumbent. Ostiola large, obtuse, at length perforated, short, erumpent together in the form of a minute, black, flattish tubercle. Pustules numerous, subseriate, only slightly raising the bark. Asci clavate, p. sp. 60–70 x 8–10 μ. Sporidia crowded-biseriate, allantoid, not much curved, hyaline, 12–20 x 3–4½ μ, mostly 12–15 x 4 μ. Spermogonia (*Cytispora* sp.), minute, 1 mm. diam., conical, multilocular, grayish-black inside, opening by a single pore at the apex. Spermatia allantoid, hyaline, 5–6 x 1 μ, on lanceolate basidia 12–15 μ long, thickened below.

On dead limbs of *Acer rubrum*, New Jersey and New York.

This is scarcely more than a depauperate form of *V. quaternata*, unless the spermogonia warrant its separation as a species.

V. grìsea, Pk. Bull. Torr. Bot. Club, XI, p. 28.

Pustules small. Perithecia 4–15 in a pustule, nestling in the inner bark, their necks converging and united in a small (less than 1 mm.), orbicular, grayish or brownish disk. Ostiola punctiform, black.

Asci clavate, 50–55 x 7–8 μ. Sporidia subbiseriate, allantoid, colorless, 10–13 x 2–2½ μ.

On dead branches of *Fraxinus Americana*, Canada (Macoun).

The pustules are often arranged in rows and the disk becomes darker with age.

We have specimens from Dr. Macoun labeled *V. fraxinina*, Pk., but, not being certain whether they are genuine, we have not ventured to give any supplementary notes.

Valsa fraxinina, Pk. l. c., differs only in its larger sporidia (12–17 x 4–5 μ) and the absence of any gray pulverulent disk. The two are doubtfully distinct.

V. dissépta, Fr. Summa Veg. Sc. p. 411.

Sphæria dissepta, Fr. S. M. II, p. 392.
Sphæria stipata, Curr. in Linn. Trans. XXII, p. 274, fig. 197
Diatrype stipata, Berk. & Br. Ann. & Mag Nat. Hist. Ser III, vol VII, p 452.
Valsa hypodermia, B. & Br. l c., III, p. 368.
Quaternaria dissepta, Tul. Sel. Carp. II, p. 107.

Exsicc Rehm Asc. 49.—Cke. F. Brit 2d. Ser. 230 —Plowr. Sph. Brit. 44.—Roum F. G. 1476 —Sydow, M. March. 1718.—Rab. F. E. 320.

Stromata mostly crowded in patches of irregular outline, 1–3 cm. across and circumscribed by an irregular, undulate, black line. Perithecia sunk in the unaltered substance of the bark, covered by the slightly raised epidermis, 2–6 in a stroma, subcircinate or irregularly scattered or even solitary, deeply buried and separated by the parenchyma of the stroma. Necks short. Ostiola thick, convergent and piercing the epidermis together in a compact fascicle, but only slightly exserted, finally pierced with a wide, funnel-shaped opening. Asci slender-clavate or cylindrical, long-stipitate, very delicate, 8-spored, 120–140 (p. sp.) x 16 μ. Sporidia subbiseriate, allantoid, smoky-hyaline, 24–32 x 6–8 μ.

On birch bark, Carolina (Schw.), on *Ilex opaca* and *Amorpha fruticosa*, South Carolina (Berk. in Grev.).

The conidial stromata are depressed-conical, furrowed and sulcate above. Conidia sessile, filiform-cylindrical, curved, yellow, 30–40 μ long. Sec. Cke. (Valsei U. S. No. 52), it is doubtful whether the Carolina specc. belong to this species.

V. dolòsa, (Fr.)

Sphæria dolosa, Fr. S. M. II, p. 405.
Valsa dolosa, Nitschke Pyr Germ. p 200.

Stromata gregarious, often crowded and subconfluent at base, tolerably large, (1¼–2 mm. broad), strongly convex, closely covered by the epidermis which is slightly fissured or subentire. Perithecia 6–12 (rarely only 3–4) in a stroma, rather large, subcircinate or irregularly crowded, globose, buried in the unaltered bark. Ostiola globose or ovoid, or almost conical, large, black, fasciculate or irregularly erumpent, or sometimes in a circle around the margin of a whitish disk. Asci narrow-clavate, oblong, sessile, 64 x 10 μ. Sporidia biseriate, cylindrical, curved, hyaline, 14–18 x 3–5 μ.

On branches of *Celastrus*, Carolina (Schw.).

This rests only on the authority of Schweinitz, not having been found here since his time. The sporidia (sec. De Notaris) are uniseptate, and if so, this would come in *Diaporthe* (*Chorostate*.)

V. opérta, (Fr.)

Sphæria operta, Fr. S. M. II, p 407, Schw. Syn. N. Am. 1381.

Circinate, perithecia small, buried in the inner bark, irregularly aggregated, attached to a crust which covers them above. Ostiola twice as long as the perithecia, erumpent, at first convergent, then erect, perforated at the apex. Asci clavate-cylindrical. Sporidia allantoid, hyaline, 18 x 6 μ.

On branches of *Populus Italica*, Bethlehem, Pa. (Schw.).

V. Thùjæ, Pk. 40th Rep. p. 67.

Exsicc. Ell. & Evrht. N. A. F. 2d Ser. 2518, 2519.

Pustules scattered, slightly prominent, closely covered by the epidermis. Perithecia nestling in the inner bark, subcircinate, 5–10 in a pustule, about ½ mm. diam. Ostiola erumpent in a small, round, whitish disk, obtuse, black, conic-papilliform. Asci oblong-clavate, p. sp. about 40 x 6 μ, subsessile. Sporidia allantoid, 10–14 x 2–2½ μ, hyaline, not strongly curved.

On dead branches and foliage of *Thuja occidentalis*, New York State (Peck), Canada (Dearness).

Cýpri, Tul. Sel. Carp. II, p. 194, tab. XXV, figs. 10–20.

Sphæria Ligustri, Schw. Syn. N. Am. 1684.
Valsa obtecta, C. & E. Grev. VII, p. 9, (not Schw. Syn. N. Am. 1639).
Exsicc. Fckl F Rh. 1969.—Rehm Asc. 226.—Sydow, M. March. 911.

Stromata scattered, small, depressed-conical, only slightly prominent, covered by the epidermis and, when this is peeled off, coming off with it. Perithecia 3–8, medium size (⅓–½ mm.), ovate, subdecumbent, sunk in the inner bark and leaving their impress on the surface of the wood or on the surface of the inner stratum of the bark, scarcely crowded, collapsing, subcircinate, necks very short, with their large, obtuse, rounded-conical ostiola united and forming a small, black disk barely erumpent through a short fissure in the epidermis. Asci p. sp.

40–50 x 10–12 μ, clavate, substipitate. Sporidia subbiseriate, allantoid, slightly curved, yellowish-hyaline, 12–15 x 3½–4 μ.

On branches of *Ligustrum*, Bethlehem, Pa., and on dead limbs of *Clethra alnifolia*, Newfield, N. J.

We have seen no specc. of *Sphæria ligustrina*, Schw., but from the diagnosis given by Schweinitz there can not be much doubt that it belongs here. The specc. of *Valsa obtecta*, C. & E., have been carefully compared with *V. Cypri*, Tul., as represented in Rehm's Asc. 226, and seem to us to be the same. We find the ostiola united in a disk, but there is on the same specimen a *Valsa* similar in outward appearance, with the ostiola separate. *Sphæria pruinosa*, Fr. S. M. II, p. 486, is reckoned as the spermogonial stage.

V. Ceanòthi, (Schw.)

Sphæria Ceanothi, Schw Syn. N. Am. 1376.

Subpustulate, closely covered by the epidermis, only the prominent, black, round, shining, perforated ostiola visible. Perithecia suberect, circinate in a cortical stroma, pyriform or irregular, white inside. Pustules minute, 1½ lines diam. Sporidia (sec. Stevenson l. c.) allantoid, nucleated, 21½ x 6¼ μ.

Frequent on dead stems of *Ceanothus*, Bethlehem, Pa. (Schw.).

V. mesoleùca, B. & C. Grev. IV, p. 103.

"Disk white, surrounded by the black ostiola, or sometimes dotted. Sporidia hyaline, sausage-shaped, 20 μ long."

On *Viburnum dentatum*, Pennsylvania (Michener).

V. inclìnis, Sacc. Syll. I, p. 137.

Valsa acclinis, Fr. in Cooke's Valsei of the U. S. p. 116.

Perithecia circinate, about 5 together, globose, erect. Ostiola parallel, crowded, thickened, prominent. Sporidia collected in the upper part of the asci, allantoid, 20–25 μ long.

On branches of *Sassafras*, Bethlehem, Pa. (Schw.).

The *Valsa acclinis*, Fr., of the European mycologists has sporidia only 7–9 x 2 μ (Sacc.), or 12–14 μ long (De Not.), and must be different from the *V. acclinis*, of Cooke's Valsei of the U. S.

Species imperfectly known.

V. laùrina, C. & E. Grev. VII, p. 9.

Covered by the blackened epidermis. Pustules convex. Ostiola convergent. Asci clavate. Sporidia cylindrical, slightly curved, obtuse, hyaline, 20 x 4 μ.

On twigs of *Sassafras*, New Jersey.

We have never been able to find anything corresponding to the above description. Our specc. labeled *V. laurina*, C. & E., are *V. ceratophora*, Tul., with sporidia 8–10 x 1½ μ. *V. subclypeata*, C. & P., occurs on sassafras, but is still different.

V. Mahàleb, C. & E. Grev. VI, p. 11.

This is also unknown to us. Our specc. under this name are *Botryosphæria fuliginosa*.

V. quérna, Curr. Linn. Trans. XXII, p. 279, tab. 48, fig. 141.

Exsicc. Rav. F. Am. 745?

"Sporidia slightly curved, simple, linear, colorless, crowded in the apex of the ascus."

The spec. in Rav. F. Am. 745, on *Myrica*, is without perithecia.

**** *Stroma cortical, circumscribed, disk mostly white or cinereous. (Leucostoma, Nitschke).*

V. nìvea, (Hoff.)

Sphæria nivea, Hoff. Veg. Cr. I, p. 26.
Sphæria nivea, Pers. Syn. p. 38.
Valsa nivea, Fr. Summa, p. 411.

Exsicc. Fckl. F. Rh. 602.—Rab. F. E. 635, 2768.—Rehm Asc. 328.—Thum. F. Austr 255. Kriegr. F. Sax. 434.—Sydow, M. March. 283, 2959.—Ell. N. A. F. 869.—id. 2d Ser 2120 —Desm. Pl. Cr. Ed. 1, 281.

Stroma scutellate, dimidiate, 1–2 mm. across, adnate to the inner surface of the epidermis, white inside, disk erumpent, white, truncate. Perithecia 4–10 in a stroma, circinate around a central spermogonium or inordinate, globose, small, necks slender. Ostiola small, black, subglobose, erumpent in a circle around the margin of the white disk, or oftener scattered irregularly. Asci clavate, 35–45 x 5–6 μ, subsessile, 8- or 4- spored. Sporidia allantoid, hyaline, not strongly curved, 7–9 x 1¼–1½ μ. Spermogonia mostly with a single central pore, radiate-celled. Spermatia cylindrical, curved, 6 x 1 μ, oozing out in reddish cirrhi.

On *Populus grandidentata*, Ohio and Nebraska, and on *P. tremuloides*, Montana and Colorado, on poplar, Pennsylvania and Carolina. Schweinitz states that it is common on apple trees at Bethlehem, Pa.

European specimens have as stated in Syll. &c, larger sporidia 12–14 x 3 μ. The specc. issued in N. A. F. 869 are mostly only spermogonia or sterile. The only perfect ones we have seen are from Colorado and Montana. These agree with the European specc. in all but their smaller sporidia and rather smaller asci. This species seems to attain perfect development only in high latitudes or in mountainous regions.

V. leucóstoma, (Pers.)

Sphæria leucostoma, Pers Syn. p. 39.
Valsa Persoonii, Nits. Pyr. Germ. p 222.
Valsa leucostoma, Fr. Summ. Veg. Sc. p. 411
Exsicc. Ell. N. A. F. 173.—Rav. F. Am. 364. Thum F. Austr. 256

Stromata strongly convex, 2–3 mm. diam., scattered irregularly, sometimes confluent, finely granular and whitish inside, the outer layer coriaceous and darker, and closely adnate to the inner surface of the epidermis. Disk erumpent through short, transverse cracks in the epidermis, white, dotted with the black, rounded, slightly projecting ostiola. Asci fusoid-clavate, 35–45 x 7–8 μ, subsessile. Sporidia biseriate, allantoid, hyaline, not strongly curved, 9–12 x 2–2½ μ.

On peach, plum and almond trees, Carolina, Pennsylvania, New Jersey, and probably throughout the country where these trees are found.

This species closely resembles *V. nivea*, but in our American specc. the asci and sporidia are larger. The stromata also are larger. In both, the stroma with its enclosing shell (conceptacle) remains attached to the epidermis when it is peeled off, appearing like little blisters on its inner surface.

V. pállida, E. & E. (in Herb.)

Stromata numerous, crowded, but mostly not confluent, occupying definitely limited areas, small, conical, base elliptical, 1–1½ mm. diam., lighter colored inside, enclosed in a thin, black layer, penetrating to the liber, but not to the wood. Perithecia few, 4–8, lying near the bottom of the stroma, small and rather pale, ¼ mm. diam., contracted into slender necks rising together through the center of the stroma, their apices enlarged into obtuse, papilliform, finally substellately dehiscent, black ostiola erumpent through a small, orbicular, subferruginous, loosely granular or farinaceous disk, which pierces the epidermis but does not rise much above it. Asci clavate, 8-spored, 30–35 x 5 μ. Sporidia subbiseriate, allantoid, slightly curved, hyaline, 6–8 x 1¼–1½ μ.

On bark of willow, Clyde, N. Y. (O. F. Cook, No. 360).

V. subclypeàta, C. & P. 27th Rep. N. Y. State Mus. p. 109.

Exsicc. Ell. N. A. F. 91.

Stromata gregarious, minute, (1 mm.), covered by the blackened epidermis, convex-hemispherical, lighter inside, raising the epidermis into little pustules. Perithecia 3–5 in a stroma, subglobose, minute (150 μ). Ostiola erumpent (not exserted), in a small, black disk, surrounded by the whitened, torn margin of the thin epidermis, rather broadly perforated. Asci clavate-cylindrical, 35–40 x 5–6 μ, subsessile. Sporidia subbiseriate, allantoid, hyaline, slightly curved, 6–8 x 1–1½ μ.

On dead limbs of *Quercus* and *Rhododendron*, Forestburg, N. Y (Peck), and on dead limbs of *Laurus Sassafras*, and the prostrate stems of *Rubus Canadensis*, Newfield, N. J.

V. cincta, Fr. Summa Veg. Sc. p. 411.

Exsicc. Fckl. F. Rh. 2140, 2348.—Rehm Asc. 224.—M. March. 353?

Stromata rather widely scattered, strongly convex, or obtusely conical, 2–3 mm. diam., formed of the slightly altered (or a little paler) substance of the inner bark, enclosed in a very thin, black layer which shows as a faint, black circumscribing line on a cross section. Perithecia in a single layer, or generally arranged around a central spermogonium, 10–15, globose, or subangular, rather over ⅓ mm. diam., attenuated into very slender, converging necks, with the rather large, black, perforated ostiola ranged around the central pore of the spermogonium, in the convex, whitish disk, which soon becomes brown. The stroma forms a distinctly prominent pustule, covered by the longitudinally or substellately-cleft epidermis. Asci 50–60 x 9–12 μ, clavate, subsessile. Sporidia biseriate, allantoid, hyaline, curved, 12–20 (mostly 12–16) x 3½–4 μ.

On dead twigs of *Amelanchier alnifolia*, Sand Coulee, Montana (Anderson).

The perithecia are rather smaller than in Dr. Rehm's specimens, but the habit is the same. We have not found any spermatia.

V. prǽstans, B. & C. in Curtis' Cat. p. 143, and Cke. Valsei of the U. S. No. 42.

Stroma orbicular at base, 1–2 mm. diam., convex-conical, filled with whitish grumous matter, enclosed in an olivaceous-brown external layer of similar nature, but more compact; on the small limbs and twigs raising the epidermis into slight pustules pierced at the apex by the small, round, dirty-white disk. On the larger limbs with thicker bark, the pustules are less conspicuous. Perithecia 10–20 or more lying in a single layer in the bottom of the stroma, small, ¼–⅓ mm. diam., globose, oblong or otherwise irregular in shape from mutual pressure, rather abruptly contracted into long, very slender necks with their short, rounded and papillate, then perforated and umbilicate ostiola erumpent through the whitish disk (which is soon obliterated), but scarcely exserted. Asci (p. sp.) oblong-fusoid, 30–35 x 5–6 μ.

Sporidia allantoid, hyaline, not strongly curved, 7–9 x 1½ μ (10–12 x 5 μ, Cke.).

On dead limbs of *Nyssa*, Carolina (Curtis), New Jersey (Ellis).

On the larger limbs the pustules are larger, with the dense tufts of short, black ostiola, subseriately erumpent in longitudinal cracks in the bark.

The above diagnosis is from the specc. referred to by Cooke in Grev. V, p. 92, and VI, p. 10.

V. translùcens, (De Not.)

Sphæria translucens, De Not. Micr. Ital. V, No. 2.
Valsa translucens, De Not. Schema Sfer p. 34.
Exsicc. Rab. F. E. 747.—Rehm Asc. 225.

Stromata numerous, gregarious or scattered, often covering the whole limb on which they grow, conical or hemispherical from a cir. cular base, mostly small, scarcely ½ mm. diam., but sometimes larger (1 mm.), visible through the thin, closely adherent epidermis, which is raised into pustules flattened above and pierced by the disk, which is surrounded by a slight depression. Perithecia 2–8 in a stroma, irregularly monostichous, sphæroid, minute, with slender necks and punctiform ostiola, erumpent in the center of the small, whitish disk. Asci clavate or oblong, sessile, 8-spored, 40–44 x 8 μ. Sporidia conglobate, cylindrical, curved, hyaline, 9–14 x 2 μ. Spermogonia few-celled, sometimes one-celled, with a single central pore, or, more seldom, with 2–3 pores penetrating the whitish, brown-margined disk. Spermatia cylindrical, curved, 4–5 x 1 μ.

On dead willow branches, West Albany, N. Y. (Peck).

V. morígena, B. & C. Grev. XIV, p. 46.

Perithecia globose, immersed in a pale cortical stroma, which is covered at length with a black crust, and circumscribed by a black line. Ostiola subconfluent, and slightly prominent in a minute, black disk. Asci cylindric-clavate. Sporidia allantoid, hyaline, 10 x 2 μ.

On bark of *Morus multicaulis*, South Carolina.

V. caryígena, B. & C. Grev. IV, p. 102, and Cke. Valsei of the U. S. No. 32.

Exsicc. Ell. N. A. F. 867.

Stroma as in *V. præstans*, but mostly smaller, convex-conical, 1–1½ mm. diam., whitish inside, raising the bark into distinct pustules, mostly seriately arranged, and pierced at the apex by the small, dirty-white, round disk. Perithecia in a single crowded layer in the bottom of the stroma, small, mostly not over ¼ mm. diam., necks converging, their short, black, rounded, then umbilicate and open ostiola soon obliterating the disk, except a faint white ring around the margin, but not usually much exserted. Asci clavate, subsessile, 25–30 x 5 μ. Sporidia biseriate, allantoid, not strongly curved, hyaline, 6–8 x 1¼–1½ μ, (10 μ, Berk; 10–12 μ, Cke.).

On hickory saplings killed by fire, Newfield, N. J., on branches of *Carya*, Pennsylvania (Michener), New York (Cook).

Var. *chlorodisca* (*Valsa chlorodisca*, C. & E., Grev. VIII, p. 13) differs only in the yellowish disk. On the same limbs are pustules of the normal sort. *V. caryigena* and *V. præstans* are closely allied.

V. orbícula, B. & C. Grev. IV, p. 100, Cke. Valsei U. S. No. 38.

"Minute, orbicular, showing the subjacent perithecia by transparence, but not blackened, surrounded by a black line. Asci lanceolate. Sporidia sausage-shaped, 10 μ long.

On willow, South Carolina, No. 3404."

V. collículus, (Wormsk.)

Sphæria colliculus, Wormsk. Fr. S. M. II, p. 389.
Valsa colliculus, Wormsk. Cke. Valsei U. S No 23.
Exsicc. Ell. & Evrht N. A. F. 2d Ser. 1577.

Pustules prominent, convex-hemispherical, 2–3 mm. diam., closely covered by the cinereous epidermis, and uneven from the projecting perithecia which (10–20 in number) are closely packed within and covered by a scanty stroma formed from the scarcely altered substance of the bark, about ⅓ mm. diam., with short, convergent necks, and the minute, papilliform, black ostiola crowded and erumpent, but scarcely exserted. Asci slender-clavate, 35–40 x 3 μ. Sporidia subbiseriate, allantoid, hyaline, 5–6 x 1¼ μ.

On dead limbs of *Pinus strobus*, New York, New England and Pennsylvania.

EUTYPÉLLA. Nits.

Pyr. Germ. p. 163, Sacc. Syll. I, p. 145.

Stroma valsoid, immersed in the wood or bark, and surrounded by a black, circumscribing line. Perithecia in one or several layers, never simply circinate, with rather thick, subcoriaceous walls, black and shining inside when mature. Ostiola stellate-cleft. Asci long-stipitate, 8-spored. Sporidia allantoid, mostly yellowish-hyaline.

The most obvious character separating this from *Valsa* is the stellate-cleft ostiola.

Eu. capillàta, E. & E. Journ. Mycol. IV, p. 74.

Stromata pustuliform, 2–3 mm. diam., numerous and closely contiguous for 5–20 cm., blackening and carbonizing the bark, and bounded by a black line which penetrates the wood, but not deeply. Perithecia membranaceous, thick-walled, black and shining within (when dry), 6–12 in a group, not distinctly circinating, $\frac{1}{3}$–$\frac{1}{2}$ mm. diam., their bases slightly sunk in the wood. Ostiola capillary, very long ($\frac{1}{2}$–1 cm.), crooked, rough, brittle (readily breaking square off when dry), apices rounded and 4–5-sulcate, altogether resembling a mass of black, strigose, coarse hair covering the matrix with a nearly continuous coat. Asci clavate, truncate above, 15 x 3$\frac{1}{2}$ μ (p. sp.), with a slender base about 15 μ long, without paraphyses. Sporidia 8, crowded, yellowish in the mass, strongly curved, with a nucleus in each end, about 3$\frac{1}{2}$ x $\frac{1}{2}$ μ. In its smaller sporidia and very long ostiola, this appears distinct from *Eutypella Bonariensis*, Speg., and from *Valsa scoparia*, Schw. Var. *subsimplex* has the perithecia larger ($\frac{1}{2}$–$\frac{3}{4}$ mm.), more deeply buried in the wood, and only 1–2 in a stroma, which is rounded and protuberant like the perithecia of some large, suberumpent, simple *Sphæria*.

On decaying limbs lying on the ground, St. Martinsville, La. (Langlois).

Eu. deústa, (E. & E.)

Valsa deusta, E & E Journ Mycol. IV, p. 64.

Perithecia 4–6 together, sunk in the surface of the wood, with thick, membranaceous walls, shining-black inside (when dry), raising the bark into distinct pustules. Ostiola erumpent in a compact fascicle, short-cylindrical ($\frac{1}{2}$ mm.), obtuse, quadrisulcate. Asci (p. sp.) clavate, about 15 x 4 μ. Sporidia crowded, 8 in an ascus, allantoid, strongly curved, minute (3$\frac{1}{2}$–4 x $\frac{1}{2}$–$\frac{3}{4}$ μ). The cuticle is soon thrown off, leaving the exposed surface of the inner bark uniformly blackened.

On decaying limbs of *Carya*, Louisiana (Langlois).

Eu. stellulàta, (Fr.)

Sphæria stellulata, Fr S. M. II, p. 380.
Eutypella stellulata, Sacc. Syll. 571
Valsa ventriosa, C. & E. Grev VI, p. 93.
Valsa innumerabilis, Pk. 30th Rep N. Y State Mus p 65
Valsa tetraploa, B & C. Grev V, p. 55

Exsicc. Fckl F Rh 597 —Kze F. Sel 146 —Rab. F E 1535 —Ell N A F 178, 688, 689 Rav. F. Am. 190, 361, 362, 663.—Plowr F. Brit 42 —Lin F Hung. 175 —Kriegr. F Sax 430 —Sydow, M. March 760 —Rehm Asc. 730 —Vize Micr Fungi, 164. Roum. F Gall. 5348

Stromata obtusely conical or subspherical, rarely subeffused, 1–2 mm. across. Perithecia in a single stroma, few or numerous, generally crowded, rarely subconcentrically arranged, subsphæroid or angular from mutual pressure, small, necks converging. Ostiola short and small or oftener more or less elongated, rough, sulcate-cleft, 3–6-sided, connate, at least below. Asci cylindric-clavate, 40–50 x 5–6 μ (p. sp). Sporidia crowded or subbiseriate, allantoid, yellowish, 7–11 (mostly 7–9) x 1½–2 μ, (8–12 x 1½–2 μ, Sacc.). Spermogonia many-celled, larger than the ascigerous stromata. Spermatia filiform, curved, 20–25 μ long, issuing in yellow cirrhi.

Common on limbs of various deciduous trees throughout the United States.

Often when the pustules are crowded, a single circumscribing line surrounds an entire group, and the surface of the inner bark is uniformly blackened. As far as the specc. above quoted are concerned, we can find no distinctive characters to separate *Valsa ventriosa* and *V. tetraploa* from this species, and have therefore included them in *Eu. stellulata*, (see Cke. Valsei of the U. S. p. 112).

Eu. Vitis, (Schw.)

Sphæria Vitis, Schw. Syn. Car No 117, Syn N. Am. 1362.
Valsa Vitis, Cke Valsei of the U S. p. 113
Eutypa viticola, Sacc. Syll. 669.
Sphæria propagata, Plowr Grev VII, p 73
Valsa (Eutypella) Vitis, Cke. Grev XIV, p 45.
Exsicc. Rav. F. Am 363, 664.

Perithecia 3–4, rather large, nestling in the inner bark and rendering the surface bullate. Ostiola irregular, subpulverulent, black, and (sec. Cke. Grev. XIV, p. 45), distinctly sulcate, and sporidia (see Valsei U. S. p. 113) 12–14 μ long.

Rather rare on young shoots of grape vines, Carolina (Schw.), Pennsylvania (Michener), New York (Peck).

It is doubtful whether this is more than a var. of *Eutypella stellulata*, (Fr.).

Eu. fraxinicola, (C. & P.)

Valsa fraxinicola, C & P 29th Rep N. Y State Mus p. 59.

Perithecia 6–12, subcircinately arranged and buried in the unaltered inner bark, about ⅓ mm. diam., black, thick-walled, with converging necks and stout, quadrisulcate, black ostiola erumpent in a short, compact fascicle in the center of the pustule, closely embraced by the perforated, slightly pustulate epidermis, above which they project but slightly. Asci clavate, p. sp. about 35 x 5 μ. Sporidia biseriate, allantoid, yellowish, 6–8 x 2 μ, moderately curved.

On bark of dead ash, New York State (Peck). Delaware (Commons).

In specc. from Peck the pustules are subseriately arranged and here and there subconfluent, but there is no circumscribing line.

Eu. Berchèmiæ, (Cke.)

Valsa Berchemiæ, Cke. Valsei of the U S p 112.
Valsa syngenesia, var. *Berchemiæ*, Curtis l c.
Eutypella Berchemiæ, Sacc. Syll. 586.
Exsicc Rav F. Am. 659 —Ell. & Evrht, N. A. F. 2d Ser 1576

Pustules scattered or seriate, small, rather prominent. Perithecia 3–8 in a pustule, ovate-globose, mostly less than ⅓ mm. diam., buried in the surface of the inner bark, without any distinct circumscribing line, necks short, with the obtuse, quadrisulcate, short ostiola erumpent through a convex, dark brown disk, but only slightly exserted. Asci clavate, 8-spored, p. sp. 20–25 x 5 μ, with a slender stipe of about the same length. Sporidia biseriate, allantoid, moderately curved, yellowish, 7–8 x 2 μ.

On bark of dead *Berchemia volubilis*. Carolina and Louisiana.

Eu. cerviculàta, (Fr.)

Sphæria cerviculata, Fr in Kze & Schm Mycol. Hefte, II, p 43.
Valsa cerviculata, Fr Summa, p. 411.
Eutypella cerviculata, Sacc Syll. 564
Diatrype megastoma, E. & E. Journ Mycol. I, p. 141
Exsicc. Fckl. F. Rh. 2454 —Rab. F E 1612.—Ell & Evrht N A. F 2d Ser 1791, 2346 —Rav Fungi Car. III, 53.—Lin Fungi Hung 174 —Rav F. Am 662. (in our copy).

Stroma obtusely-conical or subcylindrical, ¼–½ cm. diam., sunk in the substance of the bark which is unaltered or of a paler color, soon surrounded by a strong circumscribing line, which penetrates deeply into the wood. Perithecia numerous, 15–25 or more in a stroma, crowded in a compact layer, ⅓–½ mm. diam., globose or angular from lateral pressure, with a tolerably long neck. Ostiola short, strongly thickened, subglobose, 4–6-radiate-sulcate, black, erumpent and closely packed, forming small (1–2 mm.), black, shield-like patches on the surface of the bark. Asci narrow-clavate, long-stipitate, about 30 x 4–5 μ (p. sp.). Sporidia allantoid, yellowish, 5–7 x 1½ μ.

On dead limbs of *Carpinus*, *Corylus*, *Alnus*, and *Betula;* common, Canada, New England. New York, Iowa, New Jersey and Carolina.

This was issued in Rav. Car. l. c. and in Ell. & Evrht. N. A. F 2346, as *Diatrype haustellata* and *Valsa haustellata*, Fr.

Eu. glandulòsa, (Cke.)

Valsa glandulosa, Cke Grev VII, p 52
Valsa clavulata, Cke Grev XVIII, p 86
Exsicc Rav. F Am. 661

Pustules covered by the epidermis. Perithecia 4–20, globose, with thick walls, $\frac{1}{3}$–$\frac{1}{2}$ mm. diam., buried in the inner bark with their bases often slightly sunk in the wood beneath, necks slender, converging. Ostiola erumpent in a compact fascicle, in all well-developed specc. distinctly quadrisulcate-cleft, generally not rising much above the surface of the bark, but sometimes elongated-cylindrical, even 1 mm. or more long, and slightly swollen at the apex. Asci clavate, minute, 20–25 x 5 μ (p. sp.), (30 x 10 μ, Cke.). Sporidia irregularly crowded, minute, 3–4 x 1 μ, allantoid, strongly curved, hyaline, (5 μ long, Cke.).

On dead limbs of *Ailanthus glandulosa*, Carolina (Ravenel), Staten Island, N. Y. (Mrs. Britton), Long Island, N. Y. (Underwood).

The surface of the inner bark in all the specc. is uniformly blackened. Sec. Cke. Grev. VII, p. 52, the ostiola are not sulcate, but in our copy of Rav. F. Am. and in that at the Phila. Acad. they are unmistakably so. The yellowish sporidia and black, thick-walled perithecia point directly to *Eutypella*, and we have no hesitation in placing this species there. The length of the ostiola in this and other species is often variable. On a hickory limb lying with one end on moist ground and the other end high and dry, we have seen *Valsa caryigena*, B. & C., occupying the entire length of the limb, with the ostiola at the upper, dry end of the limb, scarcely projecting above the bark, while at the lower, damp end, they were exserted, 1–2 mm. long.

Eu. prunástri, (Pers.)

Sphæria prunastri, Pers. Syn. p. 37
Valsa prunastri, Fr Summa, p 411.
Eutypella prunastri, Sacc. Syll. 566
Exsicc Fckl. F Rh 596 —Rehm Asc. 477.—Sydow, M. March. 976 —Desm. Pl. Crypt. Ed I, 478

Stroma valsiform, suborbicular or elliptical in outline, pulvinate, convex or subconical, black, adnate to the wood, at first covered by the bark, finally erumpent through transverse cracks in the epidermis. Perithecia numerous, irregularly crowded, sometimes in more than one layer, the central ones erect, the marginal ones ascending, subglobose or angular, attenuated into a neck of variable length. Ostiola thickened, 3–5- (generally 4-) sulcate, short and erect, or longer and divergent or flexuous. Asci narrow-clavate, long-stipitate, 20–30 x 3–4 μ (p. sp.). Sporidia in the upper part of the asci, subbiseriate, allantoid, curved, subhyaline, 6–8 x 1$\frac{1}{2}$ μ. Spermogonia, *Cytispora rubescens*, Fr.

On *Prunus serotina*, Carolina (Schw.).

Eu. angulòsa, Nitsch. Pyr. Germ. p. 173.

Sphæria prunastri, B Betulæ, Somm Flor Lappon p 208
Sphæria Halseyana, Schw. Syn N. Am. 1319
Eutypella angulosa, Sacc. Syll 572
Exsicc. Ell. & Evrht N. A. F 2d Ser. 1574

Stroma conical, whitish inside, disk ellipsoid or triangular, $\frac{1}{2}$–1 cm. long, 2–3 mm. wide. Perithecia numerous, small, ovoid, lying in the bottom of the stroma, walls thick, black, coriaceous, contracted into a long neck above, terminating in the short-cylindrical, deeply 4–5-cleft ostiola. Asci (p. sp.) 30–40 x 4–5 μ, with a long, filiform base. Sporidia 8 in an ascus, allantoid, brownish or yellowish, rather thicker in the middle, 6–7 x $1\frac{1}{2}$–2 μ, ends subobtuse.

On dead birch limbs, New York State.

This certainly is very closely allied to *Eu. prunastri.*

Eu. rugiélla, (C. & E.)

Valsa rugiella, C. & E. Grev. V, p 92.
Eutypella rugiella, Sacc. Syll 596
Exsicc Ell. N A. F. 176

Perithecia minute (200 μ diam.), 10–20 crowded in orbicular clusters just beneath the epidermis, which is slightly elevated. Pustules very numerous and sometimes subseriately arranged, not circumscribed, necks very short, and the short-cylindrical, faintly sulcate, finally umbilicate ostiola erumpent in a compact fascicle, but not much exserted. Asci (p. sp.) about 25 x 5 μ. Sporidia subbiseriate, cylindrical, nearly straight and nearly hyaline, 4–5 x 1 μ.

On bark of dead maple, Newfield, N. J.

Eu. venústa, (Ell.)

Valsa venusta, Ell. Bull Torr. Bot Club, IX, p. 112.
Eutypella venusta, Sacc. Syll. 591
Exsicc. Ell. N. A. F. 875

Perithecia 15–20, 250 μ diam., lying in a loose, cortical stroma, 3–4 mm. diam., and circumscribed by a black line penetrating deeply into the wood. Ostiola cylindrical, roughish, slender, slightly swollen above, convergent and erumpent through cracks in the bark, but scarcely exserted, 2–3 times as long as the diameter of the perithecium, more or less distinctly 4-sulcate at the apex, but often rounded and smooth; united at first in a black, uneven, elongated disk which is finally obliterated. Asci clavate, p. sp. 30 x 35 μ. Sporidia subbiseriate, allantoid, curved, subhyaline, 7–9 x $1\frac{1}{4}$–$1\frac{1}{2}$ μ.

On dead branches of *Robinia pseudacacia*, Newfield, N. J.

Spece. of *Valsa conseptata*, Schw. in Herb. Schw., are different from this. The diagnosis here given is from a reexamination of the original specc., and differs in some respects from that in Torr. Bull.

Eu. Plátani, (Schw.)

Sphæria Platani, Schw Syn N Am 1372.
Eutypella Platani, Sacc Syll 592

Pustules gregarious, small, numerous, lying under the epidermis which is slightly raised, and radiately fissured around the erumpent fascicle of short-cylindrical or conic-cylindrical, strongly quadrisulcate, slightly exserted ostiola. Perithecia few in a pustule (4–6 mostly), not deeply buried, small, necks short. Sporidia allantoid, nearly hyaline, 10–12 x 2 μ. The asci in the specc. examined had disappeared. The ostiola are at first joined in a brown disk which is soon obliterated. Stroma of a lighter color than the surrounding bark, and without any circumscribing line.

On loose bark of *Platanus*, New York, Pennsylvania and Louisiana.

Eu. Leaiàna, (Berk.)

Sphæria Leaiana, Berk. Hook Lond Journ. Bot IV, p 311.
Eutypella Leaiana, Sacc. Syll 585

Innate. Stroma pallid, about ½ a line diam., of a rather loose texture, and circumscribed by a black line penetrating the wood. Perithecia not numerous, circinate, elliptical. Ostiola forming a little tuft, rather elongated, umbilicate, finely grooved, granulated. Asci lanceolate. Sporidia minute (7–8 μ long), curved like those of *Diatrypella verrucæformis*. Distinguished from *Diaporthe Carpini*, by its pretty, granulated ostiola and its minute, curved, not lanceolate sporidia.

On bark of dead hornbeam (*Carpinus*), Ohio and Carolina.

Eu. gonióstoma, (Schw.)

Sphæria goniostoma, Schw Syn N. Am 1373.
Valsa goniostoma, Cke Valsei of U S No. 9
Eutypella goniostoma, Sacc Syll. 587.

Pustules closely covered by the epidermis, flat and thin, often confluent 1½–2½ mm. diam. Perithecia 8–15 in a pustule, ⅓–½ mm. diam., crowded, subcircinate, buried in the bark, without any stroma. Ostiola stout, crowded, erect, deeply quadrisulcate, short conic-cylindrical, erumpent in an orbicular or elliptical tuft closely embraced by the perforated epidermis. Asci about 25–30 x 5 μ (p. sp.). Sporidia subbiseriate, allantoid, yellowish, not strongly curved, 7–9 x 2 μ.

On various branches, Carolina (Curtis), on the younger branches of *Sassafras*, Carolina (Schw.), on branches of *Rhus venenata*, New Jersey (Ellis).

Eu. juglandícola, (Schw.)

Sphæria juglandicola, Schw Syn N Am 1328
Valsa juglandicola, Stevenson, l c.

"Innate, prominent, pustules in confluent parallel series, stroma cortical, circumscribed; easily recognized from its manner of growth, longitudinally erumpent and confluent; even the perithecia in the pustules seriately arranged. Conceptacle obliterated.

On branches of *Juglans alba*, Carolina," (Schw.).

Specimens sent under this name by Peck from Troy, N. Y., on *Juglans*, agree with the diagnosis of Schweinitz quoted above, only the perithecia are not seriately arranged in the pustules, but irregularly circinate, and there is no discoloration of the bark or any circumscribing line. The perithecia in the Troy specimens are numerous, 10–16 in a pustule, small ($\frac{1}{4}$ mm.). Ostiola short, obtuse, hemispherical, sulcate, erumpent in small, seriate fascicles rising but little above the epidermis, which is raised into little seriate pustules so close together as to form almost continuous, close-lying, longitudinal ridges. Asci clavate, about 40 x 6 μ. Sporidia crowded-biseriate, about 7 x $1\frac{1}{2}$–2 μ.

Notwithstanding the slight discrepancies, this seems to be the species described by Schweinitz.

Var. *juglandina* (*Valsa juglandina*, C. & E. Grev. V, p. 92), differs only in the pustules being scattered or only subseriate, and perithecia mostly only 3–8 in a pustule. We make the sporidia 6–8 x $1\frac{1}{2}$–2 μ (10–15 x 3 μ, Cke.).

Eu. scopària, (Schw.)

Sphæria scoparia, Schw. Syn. Car. 101, id. Syn N Am 1318.
Valsa scoparia, Stevenson, l c.
Eutypella longirostris, Pk
Exsicc. Ell. & Evrht. N. A. F. 2d Ser. 2517

Stromata cortical, orbicular, 2–4 mm. diam., often confluent, surrounded by a thin, black stratum, showing a distinct, black, circumscribing line on a horizontal section, raising the bark but slightly. Perithecia 5–12 in a stroma, buried in the scarcely altered substance of the bark, globose, about $\frac{1}{2}$ mm. diam., black and polished inside, with coriaceous walls, contracted above into slender necks terminating in a close fascicle of more or less elongated, cylindrical, deeply 4–5-cleft ostiola. Asci clavate, p. sp. 18–22 x 4 μ. Sporidia crowded, allantoid, minute, strongly curved, yellowish-hyaline, 4 x 1 μ, with a nucleus in each end.

On dead elm branches, Canada to Missouri.

Agrees with specimens in Herb. Schw.

Eu. microcárpa, E. & E. Journ. Mycol. IV, p. 122.

Perithecia in clusters of 4–12, buried in the inner bark, which is uniformly stained, of a pale slate color, their bases scarcely penetrating the wood, globose, about ¾ mm. diam., with thick, coriaceous walls, black and shining within. The surface of the bark is raised into distinct pustules over the perithecia, and is more or less cracked and pierced by the cylindrical, rough, black, 1–2 mm. long ostiola, which are distinctly quadrisulcate-cleft at their tips, and issue in a little fasciele, with their bases more or less connate, but diverging above. Asci minute, 12–14 x 4–5 μ (p. sp.), with a slender base. Sporidia crowded in the asci, yellowish in the mass, allantoid, strongly curved, with a nucleus in each end, 3–4 x 1 μ (mostly not over 3½ μ long).

On decaying limbs of (peach)? St. Martinsville, La. (Langlois 1481).

Differs from *Eutypa heteracantha*, Sacc., in the absence of bristles, the sulcate ostiola, and smaller sporidia. More closely allied to *Eu. scoparia*, from which it differs in its more scattered growth, mostly fewer (4–8) perithecia in a stroma, and the absence of any circumscribing line. In *Eu. scoparia* also the stromata are often seriately-connate.

Eu. Maclùræ, (C. & E.)

Valsa Macluræ, C. & E. Grev. VIII, p 14.
Exsicc. Ell. N. A. F 873

Pustules small (1–1½ mm.), slightly elevating the bark, numerous. Perithecia 2–8 in a pustule, globose, black, ⅓ mm. diam., necks short, erumpent in a pale, furfuraceous disk which is soon obliterated. Ostiola globose-conical, finally more or less distinctly radiate-cleft. Asci (p. sp.) about 30 x 5 μ, 8-spored. Sporidia partly biseriate, oblong-fusoid, slightly curved or subinequilateral, yellowish, 5–7 x 2–2½ μ, (5 μ long, Cke.).

On dead limbs of *Maclura aurantiaca*, Newfield, N. J.

The fasciculate ostiola rise abruptly together, but they are not elongated much above the disk. The sporidia are mostly a little thicker in the middle.

Eu. canodísca, Ell. & Holway, Proc. Acad. Nat. Sci. Phil. July, 1890, p. 223.

Stroma depressed-hemispherical, 1½–2 mm. diam., flattened above and covered by a circular blackish-gray, definitely limited disk 1–1½ mm. diam., and pierced in the center by the fascicle of deeply 4-sulcate ostiola. Perithecia 6–15 in a stroma, seated on the surface of the subjacent wood, ovoid or subangular from mutual pressure, ¼–⅓ mm.

diam., with thick, black walls, and contracted above into short, converging necks with quadrisulcate ostiola collected in a slightly erumpent fascicle in the center of the disk. The upper part of the stroma around and between the necks of the perithecia, is filled with whitish. grumous matter. Asci about 100 μ long, including the slender base, (p. sp. 50 x 10–12 μ). Paraphyses filiform, abundant. Sporidia 8 in an ascus, allantoid, yellowish, moderately curved, 12–18 x 3½–4 μ. The stromata are often confluent. The wood beneath is marked by a distinct, black, circumscribing line. The circular, flat, grayish-black disk is a distinguishing character.

On dead branches of *Salix*, Decorah, Iowa (Holway).

Eu. sabálina, Cke. Grev. VII, p. 52.

Pustules small, oblong, 1–2 mm. long, prominent. Perithecia 2–5 in a stroma composed of the unaltered or more or less blackened substance of the stem, globose, about ⅓ mm. diam., black and shining inside, walls thick. Ostiola subconvergent, erumpent but scarcely prominent, obtusely conical, distinctly quadrisulcate, stout. Asci 35–40 x 5–6 μ, lanceolate-clavate, 8-spored. Sporidia subbiseriate, allantoid, nearly hyaline, obtuse, 7–8 x 1½–2 μ.

On *Sabal*, Georgia and Florida.

The surface of the stem is generally blackened continuously where occupied by the fungus, but there is no distinct circumscribing line around each separate pustule.

Eu. tumídula, (C. & P.)

Valsa tumidula, C. & P. 29th Rep. N. Y State Mus. p 58.
Eutypella tumidula, Sacc. Syll. 593

Erumpent, piercing the elevated, discolored cuticle, ultimately exposing the blackened disk. Perithecia 4–6, semiimmersed in the wood, circumscribed by a black line. Ostiola obtuse, quadrisulcate. Asci clavate. Sporidia linear, straight or curved, obtuse, hyaline, 10–13 μ long.

On dead branches of *Cratægus*, New York State (Peck).

Eu. aleùrina, (B. & C.)

Valsa (*Entypella*) *aleurina*, B. & C. Grev. XIV, p. 46.

Pustules convex, orbicular, densely gregarious, erumpent, the epidermis stellately cleft around the ostiola. Perithecia black, nestling in the bark, their sulcate necks converging. Asci clavate, 8-spored. Sporidia allantoid, hyaline, 8 μ long.

On bark of *Platanus*, North Carolina.

Seems to differ from *Eu. Platani*, Schw., in its shorter sporidia and convex pustules.

Eu. constellàta, (B. & C.)

Valsa (Eutypella) constellata, B. & C. Grev. XIV, p. 46.

"Pustules orbicular, seriate or irregularly scattered, erumpent, densely gregarious. Perithecia black, globose, few (4–6), crowded, necks abbreviated, ostiola sulcate. Asci clavate, 8-spored. Sporidia allantoid, hyaline.

On branches of *Carya*, &c., North Carolina."

Eu. rádula, (Pers.)

Sphæria radula, Pers. Syn. p. 37, sec. Schw. Syn. N. Am. 1369.
Valsa radula, Pers. in Cke. Valsei U. S. No. 15.
Eutypella grandis, (Nits)? Sacc. Syll. 578.

Commonly in extensive tracts confluent far and wide under the closely enveloping epidermis which is roughened by the projecting tufts of ostiola more or less prominent and mostly 5-sided. Perithecia rather large, lying in a white stroma formed from the substance of the bark, without any conceptacle or circumscribing line.

On various branches, especially of poplar, Carolina (Schw.).

Sec. Cke. (l. c.), the sporidia are 10–12 μ long. This measurement, however, was from European specc.

Eu. niphoclìna, (Cke.)

Valsa niphoclina, Cke. Grev. XI, p. 109.
Exsicc. Rav. F. Am. 748.

Lineate-erumpent. Perithecia ovate, lying in a white stroma. Ostiola short, converging, sulcate, erumpent in transverse lines. Asci cylindric-clavate. Sporidia allantoid, hyaline, 8 x 2 μ.

On bark of *Betula nigra*, Florence, So. Ca. (Ravenel).

The spec. in our copy of Rav. F. Am. is worthless—not even affording perithecia.

Eu. conseptàta, (Schw.)

Sphæria conseptata, Schw Syn. N. Am. 1374.
Valsa conseptata, Stev. Add. to Cooke's Valsei, No. 114.

Not pustulate, gregariously erumpent, effused under the epidermis, disk at first brownish-black, convex, at length obliterated by the slightly exserted, stellate-angular ostiola. Perithecia nestling in the bark, variously subcircinate, without any distinct conceptacle, but several clusters surrounded by one common circumscribing line penetrating the bark and wood. Sec. Stevenson, l. c. sporidia allantoid, hyaline, $8\frac{1}{3} \times 3\frac{1}{4}$ μ.

Found under the bark of *Gleditschia*, Bethlehem, Pa. (Schw.).

Species imperfectly known.

Eu. monticulòsa, (B. & C.)

Valsa monticulosa, B. & C. Cke. Valsei U. S. No. 6.
Eutypella monticulosa, Sacc. Syll. 591.

Merely the name is given and measurement of sporidia, 8–10 μ long.

On *Magnolia glauca*, Carolina (Curtis).

Eu. corynóstoma, (B. & Rav.)

Valsa corynostoma, B. & Rav. Grev. IV, p. 102.
Eutypella corynostoma, Sacc. Syll. 597.

"Pustules small, scarcely raising the bark. Ostiola fasciculate, club shaped. Sporidia minute, sausage-shaped."

On *Acer rubrum*, Carolina (Ravenel).

Eu. indistíncta, (Schw.)

Sphæria indistincta, Schw. Syn. N. Am. 1377.

Scattered, covered closely by the slightly elevated, scarcely ruptured, not revolute epidermis. Ostiola slightly prominent, angular, black, stout, exuding a dark-colored juice. Perithecia 3–4, large, black, bedded in a cortical stroma scarcely discolored.

On young branches of *Sassafras*, Pennsylvania (Schw.).

Sporidia (sec. Stevenson l. c.), $6\frac{1}{4}$ x $4\frac{1}{2}$ μ.

Eu. quadrífida, (Schw.)

Sphæria quadrifida, Schw. Syn. N. Am. 1378.
Valsa quadrifida, Schw. Stevenson l. c.

At first covered and gregarious in elongated patches forming a tubercle in the bark two lines across and covered by the adherent epidermis moderately elevated. Ostiola thick, short, black, generally 4-cleft, obtuse. Perithecia rather large, 3–4 together, shining black, ovate, suberect, lying in a cortical stroma. Sporidia (sec. Stevenson) $9\frac{1}{3}$ x $3\frac{1}{4}$ μ.

On branches and trunks of *Vaccinium corymbosum*, Bethlehem, Pa. (Schw.).

Eu. pentágona, (Pers.)

Sphæria pentagona, Pers. Syn. p. 42, Fr. S. M. II, p. 407.

Perithecia circinate, large, covered by the bark. Ostiola shining-black, 4–5-angled.

Found (sec. Schw.) in North Carolina, on branches of *Sassafras*, but the species is a very doubtful one.

EÙTYPA, Tul.

Sel. Carp. II, p. 52.

Stroma effused, either sunk in the matrix which is blackened on the surface, but unchanged within, except in being limited by a black, circumscribing line, or raised partly above the surface of the matrix, the raised part differing in substance and color from the rest. Perithecia scattered, lying in one or more layers, mostly with short, exserted ostiola. Asci long-stipitate, 8-spored. Sporidia allantoid, hyaline or yellowish-hyaline.

From the material at our command, we are not able to add many original observations on the species of this genus, and the descriptions given are mostly taken from Winter's Pilze and Saccardo's Sylloge.

* *Ostiola 4- or more-sulcate.*

Eu. spinòsa, (Pers.)

Sphæria spinosa, Pers Syn. p. 34, tab. II, figs. 9–12.
Sphæria limæformis, Schw. Syn. Car. 47.
Diatrype Berengeriana, De Not. Sfer. Ital. p. 27, tab. 26.
Valsa spinosa, Nits. Pyr. Germ. p. 127.
Eutypa spinosa, Tul. l. c.
Exsicc. Fckl. F. Rh. 1050.—Rab. F. E. 936—Ell. N. A. F. 1183—Lin. Fungi Hung. 273 Rav. Fung. Car. III, 58.

Stroma widely effused, often for a foot or more in extent, seated in the wood or the thicker and harder parts of the bark, the outer layer of which is soon thrown off, exposing the blackened matrix which it penetrates, forming a black crust 2–4 mm. thick, clothed at first with black conidia-bearing hairs simple or branched. Conidia (*Trichosporium Berengerianum*, Sacc.) obovate, truncate at base, dark brown, 7 x 5 μ, solitary or verticillate, terminal. Perithecia buried in the surface of the wood or bark, globose or ovate, 1–1½ mm. high, thickly crowded and thus becoming angular or flattened, with very large, thick, wrinkled, pyramidal, deeply 4-cleft. exserted ostiola. Asci narrow-clavate, long-stipitate, 8-spored, p. sp. 30–40 x 5–6 μ. Sporidia subbiseriate, allantoid, slightly curved, pale brownish or yellowish, 8–10 x 2 μ.

On old logs and limbs of various deciduous trees, maple, oak, &c., common.

Eu. ludibúnda, Sacc. Michelia, I, p. 15.

Valsa referciens, Sacc. Myc. Ven. Spec. p. 128.
Exsicc. Rab. F. Eur. 2323.—Ell. & Evrht. N. A. F. 2d Ser. 2118, a. & b.—Sacc. M. Ven. 200, 1464.

Stroma broadly effused, formed of the unchanged or oftener blackened substance of the bark or wood, or also pulvinate, tuberculiform or valsoid, sometimes hardly perceptible. Perithecia subglobose, monostichous, scattered irregularly as in *Cryptosphæria* or collected in valsoid groups as in *Eutypella*, white-furfuraceous outside at first. Ostiola mostly short, but also subelongated, curved or obtuse, more or less deeply 4–5-sulcate, never smooth as in *Eu. lata*. Asci clavate, long-stipitate, p. sp. 35–55 x 6–8 μ. Sporidia biseriate, allantoid, rounded at the ends, slightly curved, yellowish, 9–15 x 2½–3 μ.

On dead wood and bark of *Ulmus*, Canada (Dearness), on *Robinia pseudacacia* and (*Genista tinctoria*)? Newfield, N. J.

The spece. in N. A. F. (a) and the Canada spece. on *Ulmus*, have the stroma pulvinate as in Diatrype, ¼–½ cm. across, and more or less confluent, white inside, circumscribed with a black line; N. A. F. 2118 (b) has the stroma much smaller and valsiform or subeffused. Saccardo enumerates 38 different hosts, and in Syll. I, p. 168 refers to this as probable synonyms, *Valsa leprosa*, (Pers.), *V. confluens* Nits., and perhaps *V. referciens*, Nits. The species is a very variable one and widely diffused.

Eu. milliària, (Fr.)

Sphæria milliaria Fr in Kze. & Schmidt Mycol. Hefte, II, p 36.
Diatrype milliaria, Fr. Summa Veg. Sc. p 385.
Valsa milliaria, Nits. Pyr. Germ. p. 149.
Eutypa milliaria, Sacc. Syll 649.

Stroma broadly effused or maculiform, roundish or elongated to as much as 3 cm., or irregularly confluent, forming narrow, parallel stripes lying close together, sunk in the wood which is pustuliform-elevated and more or less blackened, at first only on the surface, but finally also within. Perithecia entirely buried in the wood, monostichous, crowded, globose, with short necks and globose, entire or slightly sulcate ostiola, which render the smooth (at first uncolored) surface of the wood black-punctate. Asci clavate-cylindrical, long-stipitate, 8-spored, about 25 x 5 μ (p. sp.). Sporidia biseriate or uniseriate below, allantoid, slightly curved, hyaline, 7–9 x 1½–2 μ.

On decorticated wood, New Jersey (Schw.), Pennsylvania (Michener sec. Berk in Grev. We have seen no spece. Found in Europe on hard, dry wood of oak, beech, *Staphylea*, &c.

This and *Eu. leioplaca* are always on bare wood, which is raised into pustuliform swellings with the surface at first uncolored, but finally blackened through and through. The perithecia being entirely sunk in the stroma, leave the surface smooth and even, but punctate-roughened by the slightly projecting ostiola.

Eu. Achàrii, Tul. Sel. Carp. II, p. 53.

Lichen Eutypus, Achar. Lich. Prodr. p. 14.
Sphæria decomponens, Sow. Eng. Fungi, II, tab. 217.
Sphæria operculata, Pers. Syn. p. 80, (pr. p.)
Sphæria astroidea and *S. Eutypa*, Fr. S. M. II, p. 478.
Valsa Eutypa, Nits. Pyr. Germ. p. 131.
Exsicc. Rab. F. E. 1925 —Rehm Asc. 169.—Plowr. Sph. Brit. 22.

Stroma broadly effused, sunk in the wood which is blackened both on the surface and within, at first clothed with short, dark brown, thickly tufted, conidial hairs, finally bare, lusterless, black, roughened by the numerous, slightly exserted, conical or obtuse, 2–5-cleft ostiola. Perithecia monostichous, thickly and mostly quite evenly scattered, globose, small, deeply buried, with more or less elongated, slender necks. Asci narrow-clavate, long-stipitate, 8-spored, p. sp. 20–30 x 4–5 μ. Sporidia biseriate, allantoid, curved, brownish, 5–7 x 1–1¼ μ.

On the bare wood and on the bark of various deciduous trees. On poplar, New York State (Peck), on limbs of *Pyrus* and *Carya*, New Jersey. Not as common as the preceding species, or at least, not as abundant.

Eu. élevans, (Schw.)

Sphæria elevans, Schw. Syn. N. Am. 1303.

Covered by the loosened fibers of the wood, which are raised so as to make the surface uneven over a considerable extent. Perithecia covered by the black-pulverulent stroma, rather large (½ mm.), depressed-globose, subradiate-circinate, the stout, prominent ostiola bursting out in small fascicles or singly, deeply quadrisulcate, and rendering the wood rough to the touch.

On denuded wood of *Rhus glabra*, Bethlehem, Pa. (Schw.).

Sporidia allantoid, yellowish, moderately curved, 5–6 x 1½ μ. Measurements of perithecia and sporidia from specc. in Herb. Schw.

** *Ostiola not distinctly sulcate.*

Eu. heteracántha, Sacc. Mich. I, p. 504. **(Plate 33)**

Valsa heteracantha, Sacc. Mycol. Ven. Spec. p. 129, tab. XIV, figs. 35–42.
Valsa hylodes, E. & E. Journ. Mycol. II, p. 40.
Valsa atomæspora, Cke. Grev. XI, p. 109.
Eutypa echinata, E. & E. Journ. Mycol. III, p. 43.
Exsicc. Thum. M. U. 1859.—Sacc. M. V. 1460, 1461, 1462.—Roum. F. G. 1175, 3942.—Rab. F. E. 2770.—Rav. F. Car. IV, No. 43 —Rav. F. Am. 660.—Ell. N. A. F. 690

Perithecia membranaceous, globose, ½–⅔ mm. diam., in clusters of 4–6, buried in the scarcely altered, fibrous substance of the inner bark, their stout, cylindrical, roughish, black ostiola ⅓ mm. long, pierced with a small aperture at the smooth, rounded apex, bursting through the epidermis (which is not split or torn) in little fascicles, with numerous light brown, coarse, tow-like hairs as long as or a little longer than the ostiola, and causing the surface of the bark to appear as if covered with clumps of miniature, brush-like hairs. The ostiola are of a carbonaceous character, as they readily break square off, so as to appear truncate. The clusters of perithecia lie in parallel series, or lines, extending for three or more centimeters, and the epidermis, which remains closely attached, is scarcely elevated by the subjacent perithecia. Asci clavate, with a slender, thread-like base, spore-bearing part 18–22 x 5 μ, with the upper part broader and obtuse. Sporidia cylindrical, hyaline, or with a faint yellow tint, rather strongly curved, about 5 x 1 μ, with a faint nucleus near each end.

On bark of a decaying log of *Carya olivæformis*, Louisiana (Langlois), on bark of *Fraxinus*, South Carolina (Ravenel).

Saccardo finds conidia on the hyphomycetous growth around the ostiola, obovoid, 6 x 5-6 μ, dark brown, 1-nucleate. *Harpographium fasciculatum*, Sacc. (N. A. F. 2000) he considers the macroconidial stage. The species is widely diffused in Europe and probably here.

Eu. velùtina, (Wallr.)

Sphæria velutina, Wallr Fl. Cr. No. 4066
Sphæria mela, Schw. Syn. N. Am. 1308, sec. spec. in Herb. Schw.
Eutypa velutina, Sacc. F. Ven. Ser. IV, p. 16, F. Ital. tab. 472.
Exsicc. Ell. N A. F. 680.—Rehm Asc. 976.—Sydow, M. March. 2062.

Stroma broadly effused, originating under the bark, which is soon thrown off. Perithecia globose, black, thickly scattered, sunk in the wood, the surface of which is not swollen or pustulate but even, attenuated above into short necks with conical, black, shining ostiola rounded and smooth at the apex, barely erumpent but not exserted. Asci fusoid, long-stipitate, p. sp. 22–26 x 3½–4 μ. Sporidia allantoid, pale olivaceous, slightly curved, 5–6 x 1 μ.

On dead trunks and limbs of *Quercus obtusiloba*, Newfield, N. J., on oak wood, Bethlehem, Pa. (Schw.).

Spece. on *Acer campestre*, found by Saccardo in Italy, have the asci and sporidia larger, 35 x 4½ μ and 7–9 x 2 μ.

Eu. rivulòsa, (Schw.)

Sphæria rivulosa, Schw. Syn. N. Am. 1304.

Stroma elongated, reaching one foot long and one inch wide, at first subimmersed and covered by the cinerascent fibers of the wood, finally entirely emergent, with an irregular outline, surface rivulose and undulate, and when old narrowly sulcate longitudinally, and colliculose, and then quite black. Ostiola prominent on the ridges, black, subcylindrical, subelongated, apices subtruncate, rugose and perforated. Perithecia rather large, subglobose, crowded, monostichous, covered by the black stroma which is whitish above.

On wood of *Laurus*, Bethlehem, Pa. (Schw.).

The spec. of this species in Herb. Schw. agrees fairly well with the above diagnosis except the ostiola which are tuberculiform-hemispherical, distinctly papilliform at first, finally perforated or irregularly sublaciniate-dehiscent, and the perithecia small, ovate-globose, less than $\frac{1}{3}$ mm. diam. Asci clavate-fusoid, 30–35 x 5 μ. Sporidia subbiseriate, allantoid, yellowish, 5–6 x $1\frac{1}{2}$ μ. The stroma becomes at length quite superficial, forming a black crust less than 1 mm. thick, and ridged like the furrows of a plowed field. Allied to *E. maura*, sec. Schw.

Eu. làta, (Pers.)

Sphæria lata, Pers. Obs. Mycol. I, p. 66.
Diatrype lata, Fr Summa, p. 385.
Valsa lata, Nits. Pyr. Germ. p. 141
Eutypa lata, Tul. Sel. Carp. II, p. 56.
Exsicc. Fckl. F. Rh. 1046.—Rab. F. E. 935.—Cke F Brit. Ser. I, 375 —id. Ser. II, 470, 471.

Stroma widely effused, continuous, innate in the wood or bark, uneven from the irregularly emergent, scattered perithecia, brown or cinereous, becoming black. Perithecia monostichous, immersed in the wood or bark which is not discolored within, more or less protuberant, globose, about $\frac{1}{2}$ mm. diam., necks very short or almost none. Ostiola obtusely conical or subhemispherical, entire or rarely obscurely sulcate. Asci cylindric-clavate, long-pedicellate, p. sp. 40–45 x 4–5 μ. Sporidia subbiseriate, elongated, curved or nearly straight, yellowish, 8–12 x $1\frac{1}{2}$–2 μ. Spermogonia hemispherical or subconical, simple, immersed in the young stroma. Spermatia slender-cylindrical, variously curved, subsessile, hyaline, at length expelled in rose-colored cirrhi, 18–22 x 1 μ.

On dead limbs of various deciduous trees, common.

Eu. flavovirèscens, (Hoff.)

Sphæria flavovirescens, Hoff. Veg. Crypt I, p. 10.
Sphæria multiceps, Sow. Eng. Fungi. tab. 394. fig. 8.
Diatrype flavovirens, Fr Summa Veg. Sc. p. 385.
Valsa flavovirens. Nits. Pyr. Germ. p. 139.
Eutypa flavovirens, Tul. Sel. Carp. II, p. 57, tab. VII, figs. 1-7.
Eutypa flavovirescens, Sacc. Syll, 643
Exsicc. Fckl. F. Rh 1049, 1825.—Rab. Herb. Mycol. 48.—Rehm Asc. 219.—Thum. M. U. 1364 —Roum. F. G. 171.—Plowr. Sph. Brit 23 —Cke. F. Brit. Ser. I, 368. Ser. II, 469.—Desm. Pl. Cr. Ed. I, 477.

Stroma of very different form and extent, tuberculiform, flattened-pulvinate, elongated and interruptedly-confluent or continuous, extending often for several inches, with the surface colliculose and uneven, nearly superficial or sometimes sunk in the wood or bark and only slightly prominent, surface black, yellow or greenish-yellow within. Perithecia irregularly monostichous, globose, small, with very short necks, and small, obtuse, wrinkled, conical, not sulcate ostiola. Asci cylindric-clavate, long-stipitate, 8-spored, p. sp. 30–50 x 2–3 μ.

On decorticated limbs, Lyndonville, N. Y. (Fairman), Carolina (Ravenel).

Probably not uncommon, but the specc. quoted are the only ones we have seen.

Eu. maùra, (Fr.)

Sphæria maura, Fr. in Kze. and Schm. Mycol. Hefte, II, p. 50.
Eutypa maura, Sacc. Syll. 627.

Stroma widely effused, deeply immersed in the wood which is more or less elevated and blackened within and without, the surface smooth and at length paler, densely black-punctate from the numerous ostiola. Perithecia deeply buried, monostichous, globose, rather large, generally crowded, with rather long, thick necks. Ostiola not thickened, hemispherical or subglobose, rarely much exserted, but then subconical, very black, smooth, shining, at length slightly quadrisulcate. Asci narrow-clavate, very long-stipitate, 8-spored, 30 x 5–6 μ (p. sp.) Sporidia subbiseriate, allantoid, somewhat curved, yellowish, 8–10 x 2 μ.

On decorticated limbs, Pennsylvania, frequent (sec. Schw.).

Eu. leióplaca, (Fr.)

Sphæria leioplaca, Fr. S. M. II, p. 370.
Diatrype leioplaca, Fr. Summa, Veg. Sc. p. 385.
Valsa leioplaca, Nits. Pyr. Germ. p. 151.
Eutypa leioplaca, Cke. Hndbk. II, p. 800.
Exsicc. Cke. Fungi, Brit. Ser. I, 366.—Sydow, M. March. 1727, 1827.

Stroma widely effused or interrupted and maculiform, very thin, immersed in the wood and not at all or only slightly raised above it, smooth and pale brown at first, becoming at length dirty black. Perithecia monostichous, globose, very small, densely crowded, sunk deeply in the wood which becomes blackened, suddenly contracted into a tolerably long, slender neck, with the very small, punctiform ostiolum scarcely exserted. Asci clavate, long-stipitate, p. sp. 36–40 x 5–6 μ, 8-spored. Sporidia biseriate, allantoid, slightly curved, yellowish, 6–16 x 1½–3 μ (mostly 8–12 x 2 μ).

On decorticated wood, Carolina and New Jersey (sec. Berk. & Schw.).

We have seen no American specimens.

*** *Species imperfectly known.*

Eu. Mòri-rùbræ, (Schw.)

Sphæria Mori-rubræ, Schw. Syn. N. Am. 1302.

Bare, rather thick, scarcely concrescent, semiimmersed, sometimes confluent in patches ¼ of an inch across, and sometimes subsimple (subsolitary). Surface very uneven and rough. Ostiola subprominent, perforated. Perithecia rather large, monostichous, immersed in a sooty-black stroma, entirely black outside.

On decaying wood of *Morus rubra*, Bethlehem, Pa. (Schw.).

Eu. confùsa, (Schw.)

Sphæria confusa, Schw. l c. 1306.

Broadly effused, not deeply immersed, concrescent, black, irregular in outline. Perithecia polystichous, in a black, pulveraceous stroma, obovate, contracted above into a rough, subpyramidal, angular, finally perforated ostiolum.

On bark and wood, Bethlehem, Pa. (Schw.).

The spec. in Herb. Schw. looks like some crowded *Melanomma*, but old and broken down, spores all gone.

Eu. elongàto-compréssa, (Schw.)

Sphæria elongato-compressa, Schw. l. c. 1305.

Patches abbreviated, much elevated, at first subimmersed, compressed, seriately confluent longitudinally, rough, irregularly elliptical, subconically compressed and elevated in the center. Ostiola irregular, slightly prominent, subglabrous. Perithecia globose-depressed, rather large, few, stroma very scanty and black. The wood is colliculose-roughened, but does not turn black between the groups of perithecia.

On decorticated wood, Bethlehem, Pa. (Schw.).

The spec. in Herb. Schw. looks like some old *Diatrype*, but is entirely without fruit.

Eu. denigràta, (Schw.)

Sphæria denigrata, Schw. l. c. 1307.

Broadly effused, crust thin and black, blackening the wood. On this crust are seen wart-like processes longitudinally elongated, lying parallel and subconfluent, formed by a congeries of perithecia immersed in them with their minute ostiola seriately erumpent and but slightly prominent. The perithecia are white inside, depressed-globose, scattered in the stromatic crust.

On wood of *Rhododendron maximum*, Bethlehem, Pa. (Schw.).

The spec. in Herb. Schw. is a mere sterile crust.

Eu. sepùlta, (B. & C.)

Sphæria sepulta, B. & C. Grev. IV, p 151.
Eutypa sepulta, Sacc. Syll. 659.

"Perithecia buried as in *Sphœria Berkeleyi* (*Diaporthe*). Ostiola emergent. Sporidia allantoid, minute.

On *Smilax*, Car. Inf. No. 1882."

Sphœria (*Eutypa*)? *oppansa*, Fr. S. M. II, p. 374.—Covered. effused. Perithecia scattered, subcircinate, delicate, black, joined in a pale membranaceous crust, ostiola erumpent. The crust seems to be formed from a gelatinous substance. Perithecia solitary or irregularly circinate, readily collapsing. Differs from *Eu. lata* in the nature of the perithecia and color of the crust.

Found (sec. Schw.) on willow bark, Bethlehem, Pa.

Sphœria (*Eutypa*) *subcutanea*, Wahl. Fr. S. M. II, p. 371.—This is quoted by Schw. as found under the epidermis of young branches of *Ribes* at Bethlehem, Pa., but the spec. in Herb. Schw. seems to be the same as *Diatrype Dearnessii*, E. & E., in N. A. F. 2526, sporidia 5 x 1 μ. The *Sphœria subcutanea*, Wahl., is under the epidermis of willow and has (sec. Sacc. in Syll.) sporidia 12–14 x 3–3½ μ.

CALOSPHÆ̀RIA, Tul.

Sel. Carp. II, p. 108.

Stroma none. Perithecia free or seated on the inner bark, scattered or oftener collected in more or less distinctly circinate groups. Ostiola more or less elongated. Asci clavate, generally racemose-fasciculate, sessile or stipitate, mostly 8-spored. Paraphyses mostly much longer than the asci, stout, lanceolate, evanescent. Sporidia mostly small, cylindrical, curved, hyaline, continuous.

Cal. prínceps, Tul. Sel. Carp. II, p. 109, tab. XIII, figs. 17–22.

Sphæria pulchella, Pers. Disp. p. 3.
Valsa pulchella, Fr. Summa, p. 412.
Exsicc. Fckl. F. Rh. 618.—Kunze F. Sel. 272.—Rab. F. E. 525.—Thum. M. U. 1854 —Plowr Sph. Brit. 481.—Ell. N. A. F. 497.

Perithecia lying on the surface of the inner bark in orbicular or elliptical groups, generally densely crowded, globose, smooth and shining. Necks very long, decumbent, flexuous, cylindrical, with their ostiola directed towards transverse cracks in the epidermis, erumpent and more or less exserted and obtuse. Asci clavate, with long and slender pedicels, p. sp. 18–26 x 4 μ, 8-spored, overtopped by the long paraphyses. Sporidia loosely conglomerated, cylindrical, curved, hyaline, 5–6 x 1½ μ.

On dead peach and plum trees, common. Also at Newfield, N. J. on *Pyrus arbutifolia.*

Cal. pulchelloidea, (C. & E.)

Valsa pulchelloidea, C. & E. Grev. VI, p. 92.
Exsicc. Ell N. A. F. 498.

Perithecia about 300 μ diam., subglobose, brownish-black, seated on the surface of the inner bark, in loose groups of 30–50 or more, with their long, slender ostiola converging and erumpent through large, subcircular openings in the epidermis, or irregularly scattered under it, often in broad, continuous, or interrupted strips. The ostiola are very brittle, variable in length, mostly not decumbent, and their tips are not united in a disk. Asci abundant, 25–30 μ long (including the slender base), and about 5 μ thick, clavate and rounded above. Paraphyses lanceolate, stout, much longer than the asci. Sporidia allantoid, nearly straight, hyaline, 5–6 x 1 μ.

On bark of oak logs and limbs, Newfield, N. J.

Cal. ássecla, (Schw.)

Sphæria assecla, Schw. Syn. N. Am. 1622.

Very minute, brown. Perithecia globose, collected in extensive groups sometimes for half an inch around the pustules of other *Sphærias*, but seldom circinate. Ostiola three times as long as the diameter of the perithecia, inclined, terete, comparatively stout.

Under the epidermis of *Castanea* always associated with or surrounding the pustules of other *Sphærias*, Pennsylvania (Schw.).

Cooke in his Synopsis 1901, places this in his subgenus *Calosphæria.* The species is not represented in Herb. Schw. The description applies very well to *C. pulchelloidea*, C. & E., which may be the same thing.

Cal. microthèca, (C. & E.)

Sphæria microtheca, C. & E. Grev. V, p. 51, and VI, p. 14.
Calosphæria microtheca, Sacc. Syll. 398.
Exsicc Ell. N. A. F. 580.

Perithecia subgregarious or scattered, small (¼ mm. or less), globose, black, submembranaceous, covered by the epidermis at first, but when this falls away they become superficial. In some cases they appear to have been superficial from the first, as on wood of bleached limbs; in this case the base of the perithecia is slightly sunk in the wood. Ostiola variable, often short, reduced to a mere subulate point, erect, or elongated (1 mm.) and decumbent, directed towards some opening in the epidermis. Asci clavate, stipitate, 22–25 (p. sp. 15–18) x 6 μ (25 x 10 μ, Cke.), racemose-fasciculate, soon truncate above.

Paraphyses not observed. Sporidia biseriate, allantoid, hyaline, 4 x 1¼ μ.

On various dry, dead limbs, Newfield, N. J.

Cal. rimícola, (Schw.)

Sphæria rimicola, Schw. Syn. N. Am. 1397.
Valsa (microsporæ) rimicola, Cke. Syn. 1789.

Perithecia in compact circinate groups of 3–12, suberect, ovate-globose, about ⅓ mm. diam., often collapsing, seated on the surface of the inner bark and slightly sunk in it, closely covered by the epidermis, which is raised into slight pustules and ruptured in narrow, transverse cracks, through which issue the slender, cylindrical, smooth, black ostiola in a loose fascicle, not united in any disk and hardly rising above the surface of the bark. Asci (p. sp.) 22–25 x 5 μ, with stout, lanceolate paraphyses much longer than the asci. Sporidia crowded-biseriate, allantoid, 4–5 x 1 μ.

On dead limbs of *Comptonia asplenifolia*, Pennsylvania (Schw.), New Jersey (Ellis).

When the epidermis is peeled off, the perithecia sometimes adhere to it, and sometimes not. The specimens of this species in Herb. Schw. are very poor, but are apparently what we have here described. Schweinitz says the pustules are immersed in the inner bark and closely covered by it, but in our specc. the perithecia are seated on the inner bark and not covered by it.

Cal. Coókei, E. & E. (in Herb.)

Valsa parasitica, C. & E. Grev. VII, p. 9.

Perithecia circinate, decumbent, soft and pale, finally collapsing, 150–200 μ diam., lying on the surface of the inner bark, above the pustules of some *Valsa*. Ostiola slender, pale, soft (carnose), yellowish horn-color when dry, cylindrical, decumbent, converging. Asci clavate, 25–30 x 5 μ. There is no note of the paraphyses in the fresh state, and they are not visible in the dry specimen. Sporidia subbiseriate, allantoid, moderately curved, 5 x 1 μ.

On an oak log (*Quercus coccinea*), Newfield, N. J.

Readily distinguished by its soft, carnose texture, and its habitat.

Cal. subcuticulàris, (C. & E.)

Valsa subcuticularis, C. & E. Grev. VIII, p. 14.
Valsa didymospora, Ell. Bull. Torr. Bot. Club, IX, p. 98.

Perithecia small (200 μ), globose or ovate, 6–15 together in compact clusters, seated on the inner bark, with ostiola erect-converging or lying in a circle, with slender, decumbent ostiola converging to the center, and there united in an erumpent disk. Asci racemose-fasciculate, with a short but slender base, p. sp. about 18–20 x 4 μ, 8-spored. Paraphyses very long, lanceolate, evanescent. Sporidia allantoid, not strongly curved, 4–5 x $\frac{3}{4}$ μ.

On dead limbs of *Ilex opaca*, Newfield, N. J.

The epidermis is ruptured and pustuliform-elevated, and when this is stripped off, the clusters of perithecia are exposed, seated on the surface of the inner bark. The surface of the wood beneath the clusters is marked with a slight circumscribing line.

Cal. alnícola, E. & E. Proc. Acad. Nat. Sci. Phil. July, 1890, p. 221.

Subcuticular. Perithecia scattered, subglobose, $\frac{1}{2}$ mm. diam., roughish, seated on the surface of the inner bark, at length slightly collapsed above. Ostiola short-cylindrical, slightly raising and barely perforating the epidermis. Asci racemose-fasciculate, clavate-oblong, 20–22 x $3\frac{1}{2}$–4 μ, 8-spored, the upper end of the spore-mass truncate, and surmounted by the empty, transparent, dome-shaped apex of the asci. Sporidia crowded-biseriate, allantoid, curved, 5–6 x 1 μ. When the epidermis is peeled off, the perithecia either adhere to it or remain attached to the surface of the inner bark, in which respect this differs from *Cryptosphæria secreta*, C. & E., in which the perithecia always adhere to the epidermis. This latter species also has longer, distinctly clavate asci, with a long, slender base, and longer sporidia, and is, we believe, specifically distinct from the species on alder, though much resembling it.

On dead alder, Newfield, N. J.

Cal. fagícola, (E. & E.)

Valsa fagicola, E. & E. Bull. Torr. Bot. Club, X, p. 118.

Perithecia 10–15, globose, membranaceous, black and shining inside, circinating in the surface of the inner bark, and covered with a dirty yellow powder, collapsing more or less when dry, abruptly contracted into slender necks $\frac{1}{4}$–$\frac{1}{3}$ mm. long, decumbent and converging to the center, with their ostiola erumpent in a small, black, slightly elevated disk, obtusely conical, at length with a rather broad, irregular opening. Asci (p. sp.) 18–20 x 3 μ, racemose-fasciculate, 8-spored, at first with a rounded, hyaline apex which soon disappears, leaving them truncate above. Sporidia biseriate, allantoid, hyaline, curved, $2\frac{1}{2}$–$3\frac{1}{2}$ x $\frac{1}{2}$–$\frac{3}{4}$ μ.

On dead limbs of *Fagus ferruginea*, West Chester, Pa. (Everhart & Haines).

The measurement of the perithecia in Torr. Bull. is too large. They are about $\frac{1}{4}$ mm. diam. This differs from *Eutypella microspora*, Cke., with specc. of which it has been carefully compared, in its smooth ostiola and smaller perithecia, asci, and sporidia. *Cal. tumidula*, Sacc., also has the asci 40 μ, and the sporidia 7–8 μ long. The epidermis is only slightly raised, and when stripped off, the perithecia mostly come with it.

Cal. herbícola, E. & E. (in Herb.)

Perithecia scattered or 2–3 together, minute, about 100 μ diam., covered at first by the cuticle, but when this disappears, superficial, globose, black, with a short, spine-like, erect ostiolum. Asci wedge-shaped, 18–20 x 5 μ, truncate above. Paraphyses much longer than the asci, stout and tapering, but evanescent. Sporidia biseriate, oblong-elliptical, hyaline, continuous, 3–4 x $1\frac{1}{2}$–2 μ.

On decaying stems of *Lactuca Canadensis*, Newfield, N. J.

Cal. microspérma, E. & E. Proc. Acad. Nat. Sci. Phil. July, 1890, p. 221. **(Plate 33)**

Perithecia subcuticular, circinate, 6–18 together, about $\frac{1}{2}$ mm. diam., their cylindrical necks converging and erumpent in a small, compact fascicle of short ostiola projecting but slightly, and mostly 4-sulcate. Asci clavate, 22–25 x 5 μ, gradually attenuated to a slender base; paraphyses much longer than the asci. Sporidia minute, $3\frac{1}{2}$ x $\frac{3}{4}$ μ, curved into a semicircle.

On *Carpinus Americana*, London, Canada (Dearness).

Cal. Myrìcæ, (C. & E.)

Valsa Myricæ, C. & E. Grev. VII, p. 8, and Cke. Syn. 1850.
Eutypella Myricæ, Sacc. Syll. 590.

Perithecia globose, often collapsing to concave, about $\frac{1}{2}$ mm. diam., lying on the surface of the inner bark in groups of 10–20, either densely circinate or occasionally loosely aggregated, abruptly contracted into decumbent, cylindrical necks converging to a central point, with their obtusely rounded, smooth or obscurely radiate-sulcate ostiola erumpent in a loose fascicle perforating the thick epidermis, but scarcely or only slightly rising above it. Asci cylindric-clavate. Sporidia allantoid, 10–12 x $1\frac{1}{2}$–3 μ, hyaline.

On dead stems of *Myrica cerifera* (or possibly of *Ilex glabra*), Newfield, N. J.

The specc. of this species in our Herbarium are unfortunately in poor condition, sterile or old, asci entirely gone and only a few free sporidia. The habit is almost the same as that of *Cal. princeps*. The decumbent ostiola are often 1 mm. or more long, and perforate the epidermis with a circular opening.

Cal. barbiróstris, (Dufour).

Sphæria barbirostris, Dufour Fr. S. M. II. p. 473
Sphæria scabriseta, Schw. Syn. N. Am, 1394
Valsa lasiostoma, E. & E. Bull. Torr. Bot. Club, X, p. 89.
Exsicc. Ell. N. A. F. 186.—id. 2d Ser. 2122.

Perithecia circinating or scattered, about $\frac{1}{3}$ mm. diam., globose and tuberculose-roughened, lying on the surface of the inner bark or slightly bedded in it, their long (sometimes as much as 1 mm), slender, cylindrical, decumbent necks converging and piercing the epidermis in a small, compact fascicle, but not united in a disk and only slightly projecting. In the scattered forms often densely gregarious, with the ostiola erect and separate, and the perithecia more or less sunk in the bark. Asci p. sp. 32–40 x 4 μ, clavate or clavate-cylindrical. Sporidia biseriate, cylindrical, nearly straight, 5–6 x $1\frac{1}{2}$–2 μ.

On decaying white oak limbs lying on the ground, and on cast off maple bark, Newfield, N. J.

The ostiola are slightly swollen above and, except the bare, black tips, covered with an olive-brown pubescence.

Cal. ciliátula, (Fr.)

Sphæria ciliatula, Fr. S. M. II, p. 406.
Calosphæria ciliatula, Karst. Myc. Fenn. II, p. 156.

Perithecia circinate in orbicular or oblong groups, globose, glabrous, black, nearly $\frac{1}{2}$ mm. diam., 12–20 together, necks decumbent, of variable length, converging to the center with their obtuse, swollen tips barely erumpent through short, transverse cracks in the thick epidermis. Asci 25–30 x 4 μ (p. sp.), with long, stout, lanceolate paraphyses. Sporidia subbiseriate, oblong-fusoid, scarcely curved, hyaline, 4–5 x 1 μ.

On dead birch limbs, Bethlehem, Pa. (Schw.), Maine (Harvey), Iowa (Holway).

When the epidermis is peeled off, the perithecia come with it. The ostiola are said to be sometimes 1 mm. long. In the specc. we have seen they are short.

Cal. éxpers, (Schw.)

Sphæria expers, Schw. Syn. N. Am. 1396.
Valsa (Macrosporæ) expers, Cke. Syn. 1888.

Immersed, covered by the epidermis. Perithecia scarcely penetrating the substance of the bark, elegantly circinate, suberect joined into a short neck, without any disk. Ostiola very short, rather thick. black, umbilicate, looking out from cracks in the epidermis.

On tender shoots of *Rosa corymbosa*, Bethlehem, Pa. (Schw.).

Sporidia (sec. Stevenson l. c.) 19 x 6 μ.

CORONÓPHORA, Fckl.

Symb. p. 229.

Differs from *Calosphæria* in its many-spored asci, which, in some of the species, are flattened, and, as it were, coronate at the apex. The perithecia lie on the surface of the inner bark, covered only by the epidermis.

Cor. oothèca, (B. & C.)

Sphæria ootheca, B. & C. Grev. IV, p. 108.
Coronophora ootheca, Sacc. Syll. 423, Cke. Syn. 1679.

Perithecia crowded into little groups, globose, seated on a dark spot. Asci obovate, stuffed with numerous, sausage-shaped, minute sporidia.

On oak (Berk. l. c.).

This (sec. Berk.) is the *Sphæria mucida*, Fr., var. *rostellata*, of Schw. Syn. N. Am. 1515, in which the asci are 25 μ long. In the specc. on oak, they are rather larger.

CRYPTOSPHÆRIA, Nitschke.

Pyr. Germ. p. 159 (as a subgenus).

Stroma effused or wanting, not limited by any black, circumscribing line. Perithecia buried in the unaltered substance of the bark, irregularly scattered, not penetrating to the wood. Ostiola erumpent, but not exserted. Asci 8-spored, aparaphysate. Sporidia allantoid, hyaline.

Nitschke included also species with polysporous asci (*Cryptovalsa*).

Cr. popùlina, (Pers.)

Sphæria populina, Pers. Icones p. 52, tab. 21, fig. 5.
Valsa millepunctata, Nitschke. Pyr. Germ. p. 161 (not Grev.).
Cryptosphæria populina, Sacc. Syll. I, p. 183.
Exsicc. Fckl. F. Rh. 908 —Rab. F. E. 1269.—Rehm Asc. 433, 434.—Ell. N. A. F. 577.

Stroma cortical, effused in patches of greater or less extent (½–2 inches), sunk in the bark, which is blackened down to the wood, and mostly swollen and raised on the surface in the form of a broad, flat blister, the margin gradually slanting off, or abrupt, sometimes rising as much as 1 mm. above the surrounding bark, and mostly somewhat blackened on the surface. Perithecia evenly scattered, tolerably large, monostichous, buried. Ostiola erumpent, subhemispherical, rough and wrinkled or subconical at the apex, finally irregularly or sublaciniately dehiscent, often obscurely quadrisulcate. Asci narrow-clavate, 30–40 x 5–7 μ (p. sp.), narrowed above. Sporidia biseriate, allantoid, yellowish, moderately curved, 8–10 x 2 μ.

On bark of poplar logs and limbs, Newfield, N. J., Illinois (Calkins), Colorado (Cockerell), Dakota (Williams).

Eutypa subtecta (Fr.), found in Europe on *Acer campestre* and *A. pseudoplatanus* is reported by Schweinitz as found at Bethlehem, Pa. ("passim sub epiderm."), but he gives no definite host and there is no spec. in Herb. Schw. *Cryptosphæria millepunctata* Grev. is found on *Fraxinus*, and has larger sporidia (14–18 x 3 μ).

Cr. fissícola, (C. & E.)

Sphæria fissicola, C. & E. Grev. VI, p. 94.
Cryptosphæria fissicola, Sacc. Syll. 684, Cke. Syn. 4076.

Perithecia scattered or subseriate, small, about ¼ mm. diam., sunk in the surface of the inner bark, with their obtusely conical, perforated ostiola barely erumpent through short, longitudinal cracks in the epidermis; sometimes the ostiola of 3 or 4 perithecia lying near each other on the same line, split the epidermis continuously for a centimeter or more in length. Asci (p. sp.) oblong-fusoid, 20–22 x 4–5 μ, short-stipitate, aparaphysate. Sporidia crowded-biseriate, allantoid, slightly curved, hyaline, 6–8 x 1½ μ.

On dead stems of rose bushes, Newfield, N. J.

Cr. secrèta, (C. & E.)

Sphæria secreta, C. & E. Grev. V, p. 94.
Cryptosphæria secreta, Sacc. Syll. 688, Cke. Syn. 4079.

Perithecia scattered, globose, ¼–⅓ mm. diam., covered by the epidermis which is not at all pustulate, but merely pierced by the short-cylindrical, obtusely-conical, perforated ostiola which do not rise above the surface. Asci clavate, 40–45 x 7 μ, narrowed into a short, stipe-like base and obtusely rounded above. Sporidia crowded, inordinate, allantoid, only very slightly curved, hyaline, 8–10 x 1½–2 μ, (10–12 μ, Cke.).

Under the epidermis of dead *Viburnum lentago*, Newfield, N. J.

When the epidermis is peeled off, the perithecia come with it. This is a very distinct and well marked species.

Cr. vexàta, (C. & E.)

Sphæria vexata, C. & E. Grev. V, p. 94.
Cryptosphæria vexata, Sacc. Syll. 685, Cke. Syn 4077.

Gregarious, semiimmersed. Perithecia ovate, black, small, less than ¼ mm. diam., raising the epidermis into minute, thickly scattered pustules ruptured above by the slightly erumpent, conical ostiola. Asci clavate. Sporidia allantoid, hyaline, 10 μ long. Stylospores in distinct perithecia with longer necks, minute, straight, linear, 5 μ long.

On dead branches of *Vaccinium corymbosum*, Newfield, N. J.

This is a very unsatisfactory thing, for though not uncommon, we have never found it mature.

Cr. inordinàta, (B. & C.)

Sphæria inordinata, B. & C. Grev. IV, p. 146.
Cryptosphæria inordinata, Sacc. Syll. 686, Cke. Syn. 4078.

"Covered by the cuticle which is raised by the subjacent perithecia into little prominences which make the whole surface like a rasp. Asci clavate. Sporidia sausage-shaped."

On *Rosa lævigata*, Carolina.

CRYPTOVÁLSA, Ces. & De Not.

Schema. Sfer. p. 29.

Stroma effused or subvalsoid, cortical, but sometimes lignicolous, more or less blackened within. Perithecia immersed, irregularly scattered or subvalsiform-aggregated, ostiola scarcely exserted. Asci polysporous, aparaphysate. Sporidia allantoid, yellowish-hyaline. Spermogonia, when present, cytisporoid.

Cr. spársa, E. & E. Proc. Acad. Nat. Sci. Phil. July, 1890, p. 224.

Perithecia with thick, coriaceous walls, black and shining inside, ⅓–½ mm. diam., buried in the inner bark, either scattered singly or oftener in valsoid groups of 3–4, faintly circumscribed, and the bark around and under them more or less blackened, each cluster or single perithecium raising the bark into a little pustule closely embraced by the sublaciniately ruptured epidermis, attenuated above into short necks terminating in more or less distinctly quadrisulcate ostiola erumpent in a small, pustuliform disk, but scarcely projecting. Asci polysporous, p. sp. 40–50 x 8–10 μ or, including the slender base, 70–75 μ long. Sporidia allantoid, yellowish, moderately curved, with a nucleus in each end, 6–7 x 1½ μ.

On dead oak limbs, Louisiana (Langlois).

In the original diagnosis "p. sp." should have been omitted, and, on a reexamination, we find no definite paraphyses.

Cr. Nítschkei, Fckl. Symb. p. 212.

Valsa Mori, Nits. Pyr. Germ. p. 157.
Exsicc. Sacc. M. V. 1187.—Fckl. F. Rh. 955.

Stroma effused, innate in the substance of the bark, covered by the epidermis which is not discolored, but merely perforated by the ostiola, forming an irregular, often interrupted and valsiform crust more or less elevated and surrounded by an irregular, black line, the enclosed areas being a little paler. Perithecia sunk to the wood, of medium size, crowded and often angular from mutual pressure, more rarely loosely scattered, suddenly narrowed into a short, thick neck, with much thickened, rather large, subglobose, faintly quadrisulcate ostiola slightly prominent. Asci narrow-clavate, long-pedicellate, polysporous, 80–86 x 12–15 μ (60–66 x 9 μ, Sacc.). Sporidia conglobate, allantoid, 12–15 x 2–2½ μ, yellowish, (8–10 x 2½ μ, Sacc.).

On bark of *Ulmus*, London, Canada (Dearness).

This seems to agree fairly well with *C. Rabenhorstii*, Nits., but as we have no authentic specc. of that species, we have adopted the determination of Saccardo, who considers it a large-spored form of *C. Nitschkei*

Cr. pustulàta, (E. & E.)

Diatrypella pustulata, E. & E. Journ. Mycol. III, p. 116.

Perithecia gregarious, either standing singly and tuberculiform or conic-hemispherically prominent, (½–¾ mm. diam.), or 2–4 confluent in a thin, tuberculiform stroma, 1–1½ mm. diam., brownish-black outside, whitish within, closely covered by the blackened epidermis which is pierced by the short-cylindrical, stout, obtuse, mostly quadrisulcate ostiola sometimes as much as 200–300 μ long. The bases of the perithecia are slightly sunk in the wood, but when the bark becomes loosened they remain attached to it and fall away with it, leaving the wood pitted with shallow cavities. Asci clavate-fusoid, 100–110 x 10 μ, including the stipitate base, paraphysate, polysporous. Sporidia irregularly crowded, allantoid, slightly curved, yellowish-hyaline, 5–8 (mostly 5–6) x 1¼–1½ μ.

On dead stems of *Lonicera* (cult.), Newfield, N. J., and on *Symphoricarpus vulgaris*, Manhattan, Kansas (Kell. & Swingle, 1926).

With the specc. on *Lonicera* was a *Libertella* with curved spores 35–40 x 1–1½ μ, (*L. Loniceræ*, Cke. & Hark)? The part of the branch occupied by the fungus is deeply penetrated by a black, circumscribing line marking the limits of the stroma.

Cr. eutypæfórmis, Sacc. Syll. 5907, Mich. II, p. 569.

Diatrype quercina, (Pers.) var. lignicola, C. & E. Grev. V, p. 54.

Perithecia gregarious, immersed in the blackened surface of the wood, subglobose, $\frac{1}{2}$–$\frac{3}{4}$ mm. diam., black. Ostiola scarcely emergent, depressed-conical, smooth, finally perforated. Asci subfusoid, gradually narrowed below into a stipe about 50 μ long, subtruncate above, p. sp. 70–80 x 9–10 μ, polysporous. Sporidia irregularly crowded, allantoid, moderately curved, 5–7 x $1\frac{1}{2}$ μ, olivaceous.

On weather-beaten wood of maple, Newfield, N. J.

Distinguished from *Cr. Citri* and *Cr. elevata*, by its much smaller sporidia. The perithecia are mostly in dense clusters, and raise the blackened surface of the wood into little pustules more or less confluent, and varying from subhemispherical or elliptic-elongated, wart-like protuberances, with the sides abrupt, to mere convex swellings.

VALSÉLLA, Fckl.

Symb. p. 203

Stroma valsoid, cortical, generally limited by a black, circumscribing line. Ostiola entire (not sulcate). Disk generally pale. Asci sessile, polysporous. Sporidia allantoid, hyaline. Spermogonia, when present, cytisporoid.

V. melástoma, (Fr.)

Sphæria melastoma, Fr. S. M. II, p. 388.
Valsella melastoma, Sacc. Syll. 617.

Stroma conical, attenuated, more or less protuberant, about 1 mm. diam., attached to the epidermis, disk orbicular or subelliptical, minute. at first whitish-cinereous, becoming brown. Perithecia 2–6, subcircinate, subsphæroid, rather large, necks short, slender. Ostiola minute, rounded, black, crowded, rarely scattered. Asci clavate-cylindrical, 30–50 x 6 μ. Sporidia conglobate, allantoid, slightly curved or nearly straight, 4–7 x 1 μ.

On limbs of apple trees, Pennsylvania (Schw.), Syn. N. Am. 1341.

V. Láschii, (Nits.)

Valsa Laschii, Nits. Pyr. Germ. p. 235.

Stromata, minute orbicular at base, conical or depressed-conical, closely covered, except the small, punctiform, orbicular, dirty-white disk, by the thin, translucent epidermis. Perithecia 2–4 in a stroma, subcircinate, very small, globose or depressed. Ostiola only visible with a lens, collected in the center of the minute disk, scarcely prominent, rounded and shining. Asci cylindric-clavate, sessile, polysporous,

42 x 7–8 μ. Sporidia conglobate, yellowish, allantoid, slightly curved or nearly straight, 11–13 x 2½ μ.

Var. *acerina*, Pk. on branches of *Acer spicatum*, Port Henry, N. Y., has the asci broader (12–15 μ).

V. adhærens, Fckl. Symb. Nachtr. II, p. 36.

Exsicc. Fckl. F. Rh. 2538.

Conceptacles immersed, minute, 1–1½ mm. broad, depressed, irregularly orbicular, covered by the adherent epidermis. Perithecia 3–5, packed in the brown stroma, minute, black. Ostiola in a small, transversely-erumpent, elliptical disk, ovate, perforated, black. Asci oblong, sessile, polysporous, 54 x 6–7 μ. Sporidia allantoid, slightly curved, hyaline, 6 x 1 μ.

Var. *Americana*, Pk., on bark of *Betula populifolia*, Sandlake, N. Y., has 5–12 perithecia in a pustule, and the sporidia are colored in the mass.

V. clópima, (Fr.)

Valsa clopima, Fr. S. M. II, p. 401.
Valsella clopima, Sacc. Syll. 611.

Stroma hemispherical, pustuliform, depressed, almost superficial, closely covered by the thin, blackened, closely adherent epidermis which is barely pierced by the suborbicular disk. Perithecia 4–16 in a stroma, monostichous or subcircinate, subspherical, brown. Ostiola scattered, rounded, or suboblong, black, shining or opake, with a broad opening. Asci clavate-cylindrical, sessile, 40–46 x 6 μ. Sporidia cylindrical, curved or straight, hyaline, 5–8 x 1–1½ μ.

On shrubs, Bethlehem, Pa. (Schw.) Syn. N. Am. 1360.

V. nigro-annulàta, Fckl. Symb. Nachtr. I, p. 29.

Exsicc. Fckl. F. Rh. 2458.

Stromata gregarious and more or less confluent, orbicular, ½ mm. diam., white inside, depressed, dull gray, covered by the epidermis, through which the outline of the stroma appears like a dark, slightly raised ring. Perithecia about 4 in a stroma, light-colored, small (less than ¼ mm.), globose, with very short necks, and minute, punctiform, black ostiola erumpent in the center of the small, grayish disk, which is just visible through a circular opening in the epidermis. Asci clavate-oblong, sessile, 35–45 x 7–8 μ (28–35 x 7–8 μ, Winter), 16–25-spored. Sporidia conglobate, allantoid, subhyaline, slightly curved, 7–9 x 1½–2 μ (9–13 x 2 μ, Winter), (6–7 x 1 μ, Rehm).

On dead limbs of (*Salix*)? Wilmington, Del. (Commons).

The measurements of Winter and Rehm, both from Fückel's specc., show quite a variation. Sec. Dr. Rehm the Delaware specc. differ from his specimen from Fückel, in their smaller and less crowded perithecia.

V. papyríferæ, (Schw.)

Sphæria papyriferæ, Schw. Syn. N. Am. 1375.
Valsa papyriferæ, Cke. Syn. 1668.

Gregariously surrounding the branches, and strongly raising the epidermis, which is stellately ruptured, revealing the pustules with their black, rough, truncate disks protruding; the short, irregular-shaped, umbilicate ostiola at length emerging. Perithecia few, rather large, globose, enclosed in a milk-white stroma, with a black exterior. The small, prominent pustules roughen the branch.

On branches of *Morus papyrifera*, Bethlehem, Pa. (Schw.).

Placed by Cooke in his subgenus *Valsella*.

ENDÓXYLA, Fckl.

Symbolæ, Nachtr. I, p. 33.

Stroma none. Perithecia buried in the wood, scattered or crowded, with the large, sphæroid or pezizoid ostiola erumpent. Asci narrow clavate-cylindrical, 8-spored, with abundant, filiform paraphyses. Sporidia uniseriate, or biseriate above, cylindrical or 1- or more-septate. Xylogenous or corticolous.

Kalmusia is included as a subgenus.

* *Sporidia continuous.*

E. macróstoma, Fckl. Symbolæ, Nachtr. I, p. 322.

Perithecia scattered or crowded in irregular groups, entirely buried in the wood, without any visible stroma, about $\frac{1}{3}$ mm. diam., abruptly narrowed above into a stout, straight neck, about as long as the diameter of the perithecium, the swollen apex subglobose-enlarged to nearly or quite the diameter of the perithecium itself, and semi-erumpent; papilliform at first, then perforated with a round opening. Asci clavate-cylindrical, 80–110 x 6–7 μ (p. sp. about 60 μ long), surrounded with abundant, filiform paraphyses. Sporidia 8 in an ascus, overlapping-uniseriate or subbiseriate above, oblong-cylindrical, obscurely 2-nucleate, olive-brown, straight or very slightly curved, 8–11 x 3–3$\frac{1}{2}$ μ.

On decorticated, half-rotten limbs of oak and laurel (*Kalmia*), Newfield, N. J.

The specimens agree so well with the description of *E. macrostoma*, Fckl., that there can be hardly any doubt that they are referable to that species. The habitat and continuous sporidia separate this from *E. parallela*, (Fr.), and the sphæroid, not pezizoid ostiola would seem to separate it from *E. operculata*, (A. & S.), if, in fact, *E. macrostoma* is really distinct from *E. operculata*.

** *Sporidia one- or more-septate* (*Kalmusia*).

E. parállela, (Fr.)

Sphæria parallela, Fr. S. M. II, p. 373.
Sphæria uda, Schum. Enum. plant. Sæll. II, p. 161.
Eutypa parallela, Karst. Mycol. Fenn. II, p. 130.
Valsa parallela, Nits. Pyr. Germ. p. 154.
Endoxyla parallela, Fckl. Symb. Nachtr. I, p. 322.
Exsicc. Rab. F. E. 1244.—Ell. N. A. F. 99 (specc. on pine), id. 194.

Perithecia tolerably large, sunk in the unaltered substance of the wood, mostly crowded in small, subconfluent groups forming parallel series or lines, abruptly contracted into short, slender necks terminating in small, slightly thickened, globose, entire, smooth, black ostiola crowded in small, superficial groups, finally perforated. Asci narrow-clavate, long-stipitate, 8-spored, p. sp. 50–60 x 6–7 μ. Sporidia subbiseriate, allantoid, moderately curved, brownish or olivaceous, 2–3-nucleate, becoming uniseptate, 10–14 x 3 μ.

On decaying pine wood, Newfield, N. J.

In the New Jersey specc., as well as those in our Herb. from Karsten, and specc. from Fries (com. by Cooke), and the spec. in Herb. Acad. Nat. Sci. at Philada. (from Fries), the sporidia become uniseptate.

E. eutypoìdes, (E. & E.)

Thyridaria eutypoides, E. & E. Journ. Mycol. IV, p. 78.

Perithecia minute (110–120 μ diameter), immersed, scattered quite uniformly through the blackened and subcarbonized substance of the bark, but lying mostly near the surface, and here and there collected in valsiform groups. Ostiola short-cylindrical, with a round opening at the subtruncate and slightly swollen apex, and so numerous as to appear under the lens like a fine black pubescence. Asci (p. sp.) about 35 x 7 μ, or with the short, stipe-like base 40–45 μ long, surrounded with abundant, filiform paraphyses. Sporidia biseriate, oblong or clavate oblong, 3-septate and slightly constricted at the septa, olive-brown, slightly curved, ends subobtuse, 10–12 x 2½–3 μ.

On bark of decaying *Melia*, Louisiana (Langlois).

E. inústa, (Cke.)

Sphæria inusta, Cke. Grev. VII, p. 53.
Kalmusia inusta, Sacc. Syll, 3378.
Xylosphæria inusta, Cke. Syn. 3973.

Scattered, immersed, blackening the wood. Ostiola whitish. Asci cylindrical. Sporidia uniseriate, elliptical, triseptate, brown, 15–18 x 8 μ.

On *Juniperus*, Darien, Ga. (Ravenel).

E. Fráxini, (E. & E.)

Thyridaria Fraxini, E. & E. Proc. Acad. Nat. Sci. Phil. July, 1890, p. 223.

Perithecia thickly scattered, buried in and almost filling the bark which is uniformly blackened on the inner surface but otherwise unchanged; globose, $\frac{1}{2}$–$\frac{3}{4}$ mm. diam., coriaceous with thick walls, black and shining inside, contracted above into a short neck terminated by an erumpent, subhemispheric-tuberculiform, black ostiolum more or less distinctly radiate-sulcate. Asci clavate, with a slender base, 90–100 x 15–20 μ, 8-spored.* Paraphyses obscure. Sporidia conglomerate, vermiform-cylindrical, brown, 3–6-septate, moderately curved, 20–26 x 4 μ. The central septum is distinct, the others fainter. Spermogonia (*Cytisporina Fraxini*) central. Sporules filiform, curved, 40 μ long; near *Thyridaria incrustans*, Sacc.

On dead limbs of *Fraxinus* (and on maple)? London, Canada (Dearness).

FAMILY. MELANCONÍDEÆ.

Stroma valsoid, pulvinate, conical or hemispherical, often inconspicuous or wanting. Perithecia buried more or less deeply in the stroma. Conidial stroma free or covered by the epidermis.

The *Melanconideæ* are separated as a family from the *Valseæ*, only by the accompanying conidia which are produced in a stroma not unlike the ascigerous stroma, generally covered at first by the epidermis, and thickly studded on its upper surface with closely packed basidia which, by constriction of the terminal cell, produce and throw off an abundance of conidia, which soon rupture the overlying epidermis and escape. The perithecia are developed after the conidia, and are either produced in the same stroma or in a separate stroma.

MELÁNCONIS, Tul.

Sel. Carp. II, p. 115.

Stroma subconical, pulvinate or hemispherical, small, buried, with only the apex erumpent, often imperfectly developed. Perithecia buried, mostly globose, with long, cylindrical necks. Asci 8-spored, typically paraphysate. Sporidia uniseptate, hyaline or colored.

*In the original diagnosis the asci were inadvertently said to be polysporous They are only 8-spored.

Conidia (*Melanconium*, sp.) one- or more-celled, variously shaped, not superficial, for the most part dark-colored. (Winter in Die Pilze).

* *Sporidia hyaline.*

M. stilbóstoma, (Fr.)

Sphæria stilbostoma, Fr. *a*, S. M. II, p. 403
Valsa stilbostoma, Fr. Summa, p. 412.
Melanconis stilbostoma, Tul. Sel. Carp. II, p. 119.
Sphæria pulchella, Currey, Linn. Trans. XXII, p. 280.
Exsicc. Fckl. F. Rh. 590 —Rab. F. E. 933.—Rehm Asc. 675

Stromata scattered, oblong or subelliptical at base, depressed-globose, or hemispherical, 2–3 mm. diam., permanently covered by the pustuliform-elevated epidermis, which is barely pierced by the erumpent, whitish disk surrounded by a slight depression. Perithecia 3–12 in a stroma, surrounded and partly covered by a granular-pulverulent, yellowish material; sunk in the surface of the inner bark, but so slightly that, when the epidermis is peeled off, they often adhere to it; black, small, depressed and collapsing when dry, with converging, cylindrical necks bent upwards at their extremities and piercing the disk with their globose or short-cylindrical, black, perforated, slightly projecting ostiola. Asci cylindrical, substipitate, 8-spored, 90–110 x 12–16 μ. Sporidia subbiseriate, elliptical or oblong, straight, obtuse, hyaline, uniseptate and constricted, 18–25 x 7–9 μ. Conidial stroma, *Melanconium bicolor*, Nees., subcuticular, conical, producing numerous brown, continuous, ovate conidia, 13–16 x 7–10 μ.

On dead limbs of birch, *Acer* and *Melia*, Carolina and Pennsylvania (Schw. & Ravenel), on birch, New York and Iowa (Peck & Holway).

M. bitorulòsa, (B. & Br.)

Valsa bitorulosa. B. & Br. Not. Brit. Fungi, No. 861 in Ann. & Mag. 1859, III, tab. X, fig. 16.
Cryptospora bitorulosa, Niessl, in Rab. F. Eur. No. 2243.
Diaporthe bitorulosa, Sacc. Syll. 2355.
Exsicc. Rab. F. E. 2243, 2421.—Cke. F. Brit. Ser. II, 249.—Sydow, M. March. 263.

Perithecia circinate, 8 or more together covered by the slightly raised epidermis, globose, more or less collapsing, ½ mm. diam., subpulverulent, with straight, horizontal necks converging to the center, where their extremities are confluent and erumpent in a small, black disk. Asci clavate, p. sp. 65–75 x 12–15 μ. Sporidia biseriate, uniseptate, oblong-fusoid, 15–19 x 4–5 μ, hyaline.

On dead limbs of *Carpinus Americana* and *Ostrya Virginica*. Iowa (Holway).

The measurements of asci and sporidia are from spece. in the Exsiccati cited. The original diagnosis makes each of the two cells of the sporidia constricted in the middle and 2-nucleate, but in only a few of the sporidia was there any constriction except the central one, or any nuclei visible; in all of them, however, there was a distinct hyaline envelope and generally a small, subglobose apiculus at each end. The Iowa specimens are in no way distinguishable from the European except in being accompanied by a *Melanconium* (*M. zonatum*, E. & E.).

The characters agree so well with *Melanconis* that we have not hesitated to refer the specc. to that genus.

M. modònia, Tul. Sel. Carp. II, p. 141, tab. XV, figs. 1–6.

Exsicc. Fckl. F. Rh. 2006.—Rehm Asc. 379.—Thum. M. U. 1062.—Ell. & Evrht. N. A. F. 2d Ser. 1564.

Perithecia 3–12 (or more), subcircinate in the slightly blackened inner bark, globose, about ½ mm. diam., their long, converging, cylindrical necks piercing the roundish, dirty gray disk, with their short, obtuse ostiola only slightly prominent. Asci cylindrical, obtuse above, attenuated below, 8-spored, 150–200 x 12–15 μ, with abundant paraphyses. Sporidia irregularly biseriate, elliptical or elliptic-oblong, obtuse, uniseptate and constricted in the middle, hyaline (slightly brownish when mature), 27–35 x 11–13 μ.

On dead limbs of *Castanea*, Newfield, N. J.

The young stromata produce an abundance of conidia, ovate, obovate, ovate-lanceolate, or pyriform, straight or slightly curved, dark brown, 2–8-septate, 20–60 x 10–13 μ.

M. thelèbola, (Fr.)

Sphæria thelebola, Fr. S M. II, p. 408.
Aglaospora thelebola, Tul. Sel. Carp. II, p. 161, tab. XXI, figs. 1–18.
Melanconis, thelebola, Sacc. Syll. 2350, Cke. Syn. 2059.
Exsicc. Thum. M. U. 362.—Vize. Micr. Fungi, 168 —Sydow, M. March. 1722.

Perithecia circinate, ⅓–½ mm. diam., 4–5 in a stroma formed of the scarcely altered substance of the bark, and surrounded by a faint circumscribing line. Ostiola decumbent, convergent, their globose, at first papilliform, finally umbilicate ostiola erumpent in a subelliptical disk, which raises the epidermis into pustules soon ruptured at the apex. Asci clavate-cylindrical, obtuse, 110–130 x 15–18 μ, 8-spored, paraphysate. Sporidia biseriate, cylindrical, slightly curved, septate in the middle, but only slightly or not at all constricted, obtuse, hyaline, with a setaceous, hyaline, deciduous appendage at each end, 30–45 x 8–10 μ (40–60 x 10–12 μ, Winter).

On dead alders, New Jersey, New York (Peek), Canada (Macoun), on *Juniperus Virginiana* Pennsylvania (Schw.).

M. Alni, Tul. differs in its stroma light-colored or yellowish within, and its smaller sporidia. Tulasne describes spermogonia and pycnidia associated with *M. Alni*, but in the spece. we have examined, we have not yet seen them.

M. Everhártii, Ell. Bull. Torr. Bot. Club X, p. 117. **(Plate 35)**

Melanconis dasycarpa, E. & K. Journ Mycol. II, p 3
Exsicc. Ell. & Evrht. N. A. F. 2d Ser. 1561, 1565.

Stroma slightly sunk in the surface of the inner bark, scarcely exceeding 1 mm. diam., without any black circumscribing line, hemispherical or convex. Perithecia 3–18 in a stroma, small, mostly about ¼ mm. diam., their necks rising together and piercing the mostly stellate-cleft epidermis in a rather loose fascicle, without any distinct disk. Ostiola cylindrical, more or less exserted, their tips pierced with a rather large, round opening. Asci oblong- or clavate-cylindrical, 100–120 x 18–20 μ, with paraphyses more or less distinct. Sporidia biseriate, oblong-elliptical, slightly curved, uniseptate, nearly hyaline, slightly or not at all constricted at the septum, 25–38 x 8–11 μ, with a slightly oblique, stout, hyaline, horn-shaped appendage, 16–20 x 3 μ at each end.

On dead maple limbs, West Chester, Pa., on dead limbs of *Acer dasycarpum*, Kansas, (Kellerman).

The ostiola soon throw off the epidermis, and the perithecia themselves soon fall out, leaving light-colored, circular spots marking the place of their attachment. Distinguished from *M. Meschuttii*, and *M. Alni*, by its elongated ostiola and larger sporidia.

M. tiliàcea, (Ell.)

Diatrype tiliacea, Ell. in Am. Nat. Feb., 1883, p. 195
Exsicc. Ell. & Evrht. N. A. F. 2d Ser. 2521.

Perithecia subcircinate, 4–12 together in the substance of the inner bark, which is a little paler, and circumscribed by a more or less distinct black line, globose ½ mm., or ovate ½ x ⅓ mm. diam., necks rather thick, cemented together above into a slate-colored, firm, waxy mass, through which protrude the conic-cylindrical ostiola, sometimes ½–1 mm. long, at length irregularly dehiscent. Asci oblong or clavate-oblong, p. sp. mostly 75–80 x 13 μ, but sometimes 80–90 x 15–22 μ, with obscure, evanescent paraphyses. Sporidia biseriate or sometimes 3-seriate, oblong-cylindrical, straight or slightly curved, uniseptate and slightly constricted at the septum, hyaline, 22–30 x 7–8 μ.

On dead *Tilia Americana* Ames, Iowa (Arthur), Bellville, Canada (Macoun), and London, Canada (Dearness).

There may be some doubt whether this is sufficiently distinct from *Hercospora Tiliæ*, Tul., but on account of the longer, narrower sporidia, the elongated ostiola and absence of the green disk, it may claim specific rank. The pycnidial stage also (N. A. F. 2522), though not well developed, does not agree well with *Rabenhorstia Tiliæ*. The stroma is smaller and the stylospores more elongated.

M. salícina, E. & E. Proc. Acad. Nat. Sci. Phil. July, 1890, p. 236.

Exsicc. Ell. & Evrht. N. A. F. 2d Ser. 2523.

Stroma flat, thin, orbicular, about 2 mm. diam., composed of the slightly altered substance of the bark, which is not perceptibly elevated above it, surrounded by a black circumscribing line, which does not penetrate below the surface of the wood. Perithecia 3–6 (exceptionally only one) in a stroma, large ($\frac{3}{4}$ mm.), globose, membranaceous, with a light-colored nucleus, contracted above into short necks, which terminate in rather broad, round, concave ostiola piercing the epidermis, but scarcely rising above it. Asci broad-lanceolate, 90–110 x 12–16 μ (p. sp.), with abundant paraphyses. Sporidia crowded-biseriate, oblong-fusoid, uniseptate and slightly constricted, a little bent or curved, 40–60 x 8–10 μ, yellowish-hyaline, with a short, obtuse, apiculus at each end. Spermogonia in a central perithecium in the middle of the stroma. The bases of the perithecia are sunk in the surface of the subjacent wood.

On dead limbs of *Salix*, London, Canada (Dearness).

M. dolòsa, (Fr.)

Sphæria dolosa, Fr. S. M. II, p. 405.
Valsaria dolosa, De Not. Sfer. Ital. p. 57, tab. 54 A.
Melanconis dolosa, Sacc. Syll. 2348, Cke. Syn. 2055.

Perithecia 2–5 together, circinate, depressed-globose, contracted into a short, thick, often excentric neck. Ostiola converging, obtuse and perforated at the tips, only slightly projecting above the pale, roundish, erumpent disk. Asci elongated-clavate, narrowed below, 8-spored, without any distinct paraphyses. Sporidia oblong or oblong-subclavate, rounded at the ends, septate in the middle, hyaline, 25 μ long.

This diagnosis is from Winter's Pilze, translated by him from Notaris, who is supposed to have examined an original specimen from Fries. Found (sec. Schw.) on branches of *Celastrus*, in North Carolina. Apparently both this and *Valsa dolosa* (see p. 481) were included in the *Sphæria dolosa*, Fr.

*** Sporidia becoming brown (Melanconiella).*

M. acrocýstis, (Pk.)

Valsa acrocystis, Pk. 33d Rep. p. 34, pl. 2, figs. 19–22.
Melanconis biansata, E. & E. Bull Torr. Bot. Club, X, p. 118
Melanconiella acrocystis, Sacc. Syll. 6628.

Perithecia circinate, 6–12 together on the surface of the inner bark, subglobose, ½–¾ mm. diam., enveloped in and covered above by a yellowish-gray, interwoven, felt-like layer; often collapsing, their short, round or subquadrate, obtuse, black ostiola erumpent in a brownish disk bursting through the transversely fissured or subradiately-cleft epidermis, which is slightly raised. Asci 190–200 x 25–30 μ, stipitate, 8-spored. Sporidia oblong-elliptical, uniseptate, soon becoming brown and with a short, hyaline, broad, subtruncate appendage at each end, 35–50 x 15–19 μ (without the appendages).

On dead limbs of birch, Plainfield, N. J. (Meschutt), on *Betula lenta*, Greenbush, N. Y. (Peck).

The felt-like covering of the perithecia was at first overlooked, but in both the New Jersey and New York spece. this forms an important character. *Valsa acrocystis*, Pk., and *Melanconis biansata*, E. & E., were published at about the same time.

M. Meschúttii, E. & E. Bull. Torr. Bot. Club, X, p. 117.

Diatrype nigrospora, Pk. 33d Rep. p. 35.
Melanconiella Meschuttii, Sacc. Syll. 6629.

Perithecia 10–20, subglobose, ¼–⅓ mm. diam. (mostly less than ⅓), seated on the surface of the inner bark, in a thin, dark-colored, orbicular stroma 1½–2 mm. diam. Ostiola rising together in a laterally compressed tuft, united in a dirty-brown or grayish disk erumpent through short, transverse cracks in the epidermis, their tips subconical and, in well developed specc., distinctly quadrisulcate. Asci p. sp. about 75 x 12–15 μ, subsessile, obscurely paraphysate. Sporidia biseriate, oblong-elliptical, very slightly curved, uniseptate and constricted, subhyaline at first with a faint, horn-shaped, hyaline appendage at each end, but these are soon absorbed and the sporidia become brown, 14–16 x 6–8 μ.

On dead limbs of birch, Plainfield, N. J. (Meschutt).

Gelatinosporium betulinum, Pk., occurs on the same branches. Where the epidermis is thinner, the appearance of the stroma is different, it being more prominent with the epidermis sublaciniately cleft around the erumpent disk. The perithecia sometimes collapse so that on stripping off the epidermis, their position is indicated by little circular concavities around the margin of the stroma. Differs from

M. Decoraensis in its smaller perithecia and appendiculate sporidia. Spece. of *Diatrype nigrospora*. Pk. (from Peck), are certainly the same as this. Faint traces of the appendages were still visible on the sporidia quite as distinct as they now are in our spece. of *M. Meschuttii.* In the description of *D. nigrospora* the perithecia are said to be "sunk to the wood," but in the spece. from Mr. Peck we find the perithecia enclosed in a small, lenticular stroma seated on the surface of the inner bark. The sporidia are surrounded with a thin, hyaline envelope, which indicates a *Melanconis* rather than a *Valsaria*.

M. chrysostròma, (Fr.)

Valsa chrysostroma, Fr. Summa Veg. Sc. p. 412.
Sphæria xanthostroma, Mont. Ann Sci. Nat. Ser. II, tom. I, p. 301.
Valsa xanthostroma, Tul. l c. Ser. IV, tom. V, p. 117.
Melanconis chrysostroma, Tul. Sel. Carp. II, p. 125, tab. XXIV, figs. 14-20
Melanconiella chrysostroma, Sacc. Syll 2806.
Melanconis chrysostroma, Cke Syn. 2062.
Exsicc Fckl F. Rh. 1732.—Ell. & Evrht. N. A. F. 2d Ser. 1563.

Perithecia circinate, 6–15 together, decumbent, thin-walled, collapsing, slightly sunk in the surface of the inner bark, about ½ mm. diam., covered with greenish-yellow, pulverulent matter, with short, convergent, sublateral necks and subhemispherical, black, prominent ostiola erumpent in a small, tuberculiform, yellowish (becoming brown) disk which pierces the epidermis and rises above it. The clusters of perithecia are seriately placed and raise the epidermis into slight, flattish pustules. The epidermis is not ruptured, but simply pierced, and when peeled off, the perithecia mostly adhere to it. Asci clavate-cylindrical, 70–75 x 12 μ. Sporidia biseriate, narrow-elliptical, uniseptate and constricted, straight or nearly so, yellowish-hyaline, 15–20 x 5–6 μ, with a hyaline envelope and a short, hyaline appendage at each end.

On *Carpinus Americana*, West Chester, Pa, (Haines & Everhart).

The above diagnosis is from the spece. in N. A. F. l. c., or rather from spece. from the same locality since found in better condition. Whether this is really the *M. chrysostroma*, Tul., is not entirely certain. Saccardo and Fückel describe the sporidia as *brown*, but the spece. distributed in M. March. 1656 (on *Fagus sylvatica*), 1723 on (*Carpinus Betula*) and in Linhart's Fungi 266, on the last-named host, have the sporidia yellowish-hyaline. No. 350, Kze. F. Sel. labeled *Diaporthe sulphurea*, Fckl., is in no way distinguishable from the above mentioned spece., and we believe they are all the same species which, on account of the hyaline envelope of the sporidia, may well be considered as a *Melanconis*. Dr. Winter who examined the N. A. F. spece. said: "This ought properly to be *Melanconis chrysostroma*, though I find only hyaline sporidia. It is also very similar to *Dia-*

porthe sulphurea, Fckl., which differs only in its somewhat larger sporidia" As regards the color of the sporidia of *M. chrysostroma*, Tulasne calls them yellow or yellowish-green ("flavis aut luteo-virentibus").

M. apocrýpta, Ell. Am. Nat. Feb., 1883, p. 194. **(Plate 35)**

Perithecia subcircinate, ⅓ mm. diam., membranaceous, 8–12 buried in the inner bark, without any distinct stroma, entirely concealed by the epidermis, which, without being ruptured, is raised into slight, whitish pustules by the pressure of the short, fasciculate ostiola. Sporidia elliptical, 25–30 x 11–13 μ, at first surrounded with a hyaline, gelatinous envelope, and more or less perfectly biseriate in asci 114 x 22 μ, but at length becoming brown, uniseptate and uniseriate, in elongated asci 120–150 x 12 μ.

On dead poplar branches, Decorah, Iowa (Holway).

The conidial stage is probably *Melanconium populinum*, Pk. This is closely allied to *M. occulta*, (Fckl.), but differs in its smaller, brown sporidia.

M. Decoraénsis, Ell. in Am. Nat. Feb. 1883, p. 195.

Melanconiella Decoraensis, Sacc. Syll. 6123.
Exsicc. Ell. & Evrht. N. A. F. 2d Ser. 1562.

Perithecia 8–20 in a stroma, angular from pressure, coriaceous, black, circinate, ½ mm. diam., subdecumbent, with stout, converging necks, and small, black, obtuse ostiola erumpent in a light-colored, elliptical disk bursting through transverse cracks in the epidermis but scarcely rising above it. Asci cylindrical, briefly stipitate, obscurely paraphysate, p. sp. 95–115 x 10–12 μ. Sporidia uniseriate, elliptical, obtuse, uniseptate and constricted, becoming brown, 12–20 (mostly 14–16) x 8–10 μ. The accompanying *Melanconium* has spores (conidia) rather shorter and broader than the sporidia.

On dead limbs of birch, Decorah, Iowa (Holway).

Var. *major*, E. & E. Journ. Mycol. III, p. 42, on dead birch limbs, Plainfield, N. J. (Meschutt), has the sporidia larger (18–26 x 8–9 μ), but does not differ otherwise from the original specimens from Iowa. When well matured, the ostiola in both are distinctly quadrisulcate. Var. *subviridis*, Pk. 40th Rep. p. 70, on dead bark of *Betula populifolia*, Ganesvoort, N. Y., has both the disk and the stroma yellowish-green and pulverulent.

We have not been able to detect any appendages on the sporidia at any stage of growth. Otherwise this could hardly be separated from *M. spodiæa*, Tul. The perithecia are so slightly sunk in the unaltered substance of the inner bark as to be partially visible through the thin layer that covers them.

CRYPTÓSPORA, Tul.

Sel. Carp. II, p. 144.

Stroma valsoid, pustuliform, covered by the epidermis which is pierced by the ostiola, sometimes much reduced or wanting. Perithecia mostly bedded in the unaltered substance of the inner bark, subcircinate, with converging necks united in an erumpent disk. Asci 4–8-spored, mostly without paraphyses. Sporidia hyaline, ovate or elliptical, continuous (*Cryptosporella*), oblong, fusoid, or cylindrical, continuous (*Eucryptospora*), septate or pseudoseptate (*Calospora*).

* *Sporidia ovate, elliptical or oblong* (*Cryptosporella*).

Cr. paucíspora, (Pk.)

Valsa paucispora, Pk. 33d Rep. p. 33.
Cryptosporella paucispora, Sacc. Syll 6411.

Pustules covered by the slightly elevated epidermis which is at length ruptured. Perithecia 2–5, seated on the inner bark. Ostiola short, black, piercing the minute, pallid disk, smooth or rarely slightly radiate-sulcate. Asci short, 55–60 μ long, subcylindrical, tetrasporous. Sporidia continuous, uniseriate, nearly colorless, ovate-elliptical, 15–20 x 10–12 μ.

On dead alder twigs, North Greenbush, N. Y.

Allied to *Cr. aurea*, (Fckl.), from which it differs in its paler disk, shorter, four-spored asci, and uniseriate sporidia.

Cr. divérgens, (Schw.)

Sphæria divergens, Schw. Syn. Car. 123, id. Syn N. Am. 1393.
Valsa (*Cryptosporella*) *divergens*, Cke. Syn. 1941.

The following diagnosis is from Grev. XIII, p. 40:

Receptacle somewhat swollen. Groups of perithecia 2–4 lines across, surrounded by the ruptured epidermis. Perithecia black or cinereous. Ostiola very long, round, divergent, three times as long as the diameter of the perithecia. Asci 8-spored. Sporidia subelliptical, hyaline, 7 x 2 μ (7 x 1¼ μ, Stevenson).

On fallen branches of *Liquidambar*, Carolina (Schw.).

Cr. leucòpis, (Fr.)

Sphæria leucopis, Fr. Kze. & Schm. Myc. Hefte, II, p 48.
Diaporthe leucopis, Sacc. Syll. 6094.
Valsa leucopis, Stevenson, Add. to Cooke's Valsei of the U. S. No. 129.

Perithecia crowded, subcircinate, few, rather large, included in a hollow tubercle, white inside and formed from the inner bark. Necks very short, their apices united in a minute, unequal, black disk exuding a minute central globule.

On branches of *Syringa*, Bethlehem, Pa. (Schw.).

Sporidia (sec. Stevenson) subelliptical, hyaline, 22⅓ x 6¼ μ.

Saccardo (l. c.), on the authority of Quelet, gives the sporidia as lanceolate, 4-guttulate, 20 μ long, with a bristle-like appendage at each end, and places the species in *Diaporthe* (*Chorostate*).

Cr. vasculòsa, (Fr.)

Sphæria vasculosa, Fr. S. M. II, p. 408.
Valsa vasculosa, Stev. l. c.

Circinate. Perithecia subovate. Ostiola long, thickened above and joined in an opake, transversely erumpent disk. An enigmatical species, on account of the great number of species on birch, yet certainly distinct from *Sphæria* (*Diatrypella*) *melasperma*, &c. From *Calosphæria princeps*, Tul., to which it is allied, it differs in the position of the perithecia and the ostiola dilated at the apex and forming a disk.

On *Prunus*, Bethlehem, Pa. (Schw.).

Sporidia (sec. Stev.), cuneate, nucleated, 19 x 12½ μ.

Cr. lentáginis, (Rehm).

Cryptosporella lentaginis, Rehm in litteris and in Sacc. Syll. 6026.

Pustules small (1 mm. diam.). Perithecia mostly but one in a pustule, large (½ mm. diam.), soft, whitish inside, obtuse-conical from a globose base, contracted above into the stout, obtusely conical ostiolum which barely pierces the slightly elevated epidermis and is soon broadly perforated. Asci clavate, stipitate, 40–50 x 8–9 μ, aparaphysate, 8-spored. Sporidia biseriate, cylindrical, curved, or nearly straight, continuous, 2–3-nucleate, hyaline, 12 x 2 μ.

On dead *Viburnum lentago*, Iowa (Holway).

The inner bark is uniformly blackened, but the numerous small stromata are light-colored within and appear as light-colored spots in a longitudinal section through the bark. The species is anomalous in having for the most part only a single perithecium in a stroma.

Cr. aùrea, Fckl. Symb. p. 193.

Valsa aurea, Fckl Enum. Fung. Nassov. p. 53, fig. 20.
Valsa amygdalina, Cke. Seem. Journ. (1866).
Valsa rutila, Tul Sel. Carp. II, p 197.
Cryptosporella aurea, Sacc. Syll. 1803.
Exsicc. Rab. F. E. 1940

Stroma conical or convex from an orbicular base, generally more or less protuberant, pustuliform closely covered by the adherent epidermis, disk reddish or brick-colored, dotted with the black ostiola. Perithecia 4–10, circinate in the unchanged inner bark, minute, attenuated into a slender neck. Ostiola minute, cylindrical, not thickened, obtuse, scarcely exserted around the margin of the scanty disk. Asci narrow-oblong or subcylindrical, sessile, 8-spored, 108–160 x 16–18 μ. Sporidia obliquely uniseriate or subbiseriate, elliptic-ovate, straight or rarely subinequilateral, continuous, subhyaline, 24–30 x 8–12 μ (20 x 8 μ, Peck).

On dead branches of *Carpinus*, New York State (Peck), and Canada (Dearness).

Cr. umbilicàta, (Pers.)

Sphæria umbilicata, Pers. Syn. p. 45.
Valsa umbilicata, Stevenson, l. c.

"Circinate, small. Ostiola crowded, somewhat cup-shaped, rough. Among the smallest of the genus. Ostiola prominent, rough to the touch, excavated at the apex or deeply umbilicate."

On *Lonicera sempervirens*, Carolina (Schw.).

Sporidia (sec. Stevenson) oval, 6¼ x 5 μ.

Cr. anómala, (Pk.)

Diatrype anomala, Pk. 28th Rep. p. 72.
Cryptosporella anomala, Sacc. Syll. 1813.
Exsicc. Ell. N. A. F. 1185.

Pustules prominent, subrotund or elliptical, 2–5 mm. diam., erumpent, penetrating the wood, generally with a thin, black crust beneath and around them, the disk convex or slightly depressed, rough, cinereous-brown or black, sometimes whitish-pulverulent. Perithecia crowded, deeply imbedded in the stroma, often elongated. Ostiola scattered or crowded, convex, often radiate-sulcate, black. Asci short, broad, fugacious. Sporidia crowded, elliptical, simple, often with a nucleus at each end, hyaline, 7–8 μ long.

On living stems of *Corylus Americana*, Albany, N. Y. (Peck), Iowa (Holway), on *Corylus Avellana*, Newfield, N. J.

The pustules appear first on the smaller branches, and are seriately arranged along one side of the branch; afterwards they appear also on the larger branches and on the trunk itself, and in the course of two or three years, the part of the tree above ground is entirely killed. The roots, however, still retain their vitality and continue to send up each year a luxuriant growth of new shoots destined to be destroyed the succeeding year by the inexorable post. The imported trees seem to be more injuriously affected than the native species.

Cr. leptásca, (C. & P.)

Valsa leptasca, C. & P. 29th Rep. N. Y. State Mus. p. 59.
Cryptosporella leptasca, Sacc. Syll. 1811.
Valsa (*Cryptosporella*) *leptasca*, Cke. Syn. 1931.

Subpustulate, blackish, erumpent. Perithecia small, numerous, tapering above into the papillate or subconical ostiola. Asci elongated, cylindrical, slender. Sporidia uniseriate, continuous, oblong or elliptic-oblong, usually binucleate, colorless, 7½ μ long.

On dead branches of *Rhus typhina*, Buffalo, N. Y. (Peck).

Sometimes the pustules are confluent or effused.

Cr. farinòsa, (Ell.)

Valsa farinosa, Ell. Bull. Torr. Bot. Club, IX, p. 99.
Cryptosporella farinosa, Sacc. Syll. 6024.
Exsicc. Ell. & Evrht. N. A. F. 2d Ser. 1572.

Stroma cortical. Perithecia few (2–4), pale, 250–300 μ diam., raising the bark into little protuberances which indicate their position. Disk tuberculiform, yellowish-white, of a loose granular or mealy consistence. Ostiola large, pale horn-color, ovate or conical, at length disappearing and leaving a large opening. Asci clavate-cylindrical, obtuse, p. sp. 100–110 x 15–20 μ, with filiform paraphyses, narrowed below into a stipe-like base. Sporidia 1–2-seriate, narrow-elliptical and subacute, subhyaline, with a large, central nucleus, 18–22 x 10 μ. Pycnidia (*Harknessia caudata*, E. & E. Journ. Mycol. I, p. 92).

On dead oak shoots, Newfield, N. J., and on dead limbs of *Liriodendron*, Plainfield, N. J. (Meschutt).

Cr. cómpta, (Tul.)

Valsa compta, Tul. Sel. Carp. II, p. 196
Cryptospora compta, Winter Die Pilze II, p. 771.

Stromata mostly numerous, irregularly scattered, pustuliform, 2–3 mm. broad, the small, round disk piercing the epidermis, which is flattened or depressed around it. Perithecia 2–6 in a stroma, of medium size, globose, circinating in the unaltered, inner bark, abruptly contracted into slender necks. Ostiola small, scarcely or only a little thickened, hardly projecting above the very small, white-pulverulent, then ashy-gray disk. Asci 8-spored, narrow-clavate, oblong or cylindrical, sessile, 100 x 16 μ. Sporidia broad-ovate, or sometimes cylindrical, straight or slightly curved, hyaline, 16–28 x 8–16 μ.

On dead limbs of *Fagus ferruginea*, New York State (fide Peck, who finds the sporidia ovate or oblong-elliptical, hyaline, 18–24 μ long).

** *Sporidia cylindrical, continuous or nucleate.* (*Eucryptospora*).

Cr. suffūsa, (Fr.)

Sphæria suffusa, Fr. S. M. II, p. 399.
Cryptospora suffusa, Fr. Summa Veg. Sc. p. 412.
Exsicc. Fckl. F. Rh. 620, 1997.—Kze. F. Sel. 142.—Rab. F. E. 730, 1130, 2022.—Rehm Asc. 46.—Thum. M.U. 171.—Sydow, M. March. 168.—Ell. & Evrht. N. A. F. 2d Ser. 1578.

Stromata scattered, raising the epidermis into flattish pustules, but not discoloring it, 1–2½ mm. broad. Perithecia 4–12, subcircinate, depressed-globose or angular from crowding, collapsing when dry, lying in the unaltered inner bark, with long, cylindrical, converging, decumbent necks, united at their ends in a small, black, erumpent disk, or sometimes a part or all of them remaining isolated. Asci oblong, sessile, 8-spored, 70–100 x 22–30 μ. Sporidia fasciculate or interwoven, cylindrical, obtuse, hyaline, 50–65 x 3⅓–4 μ.

On dead alders, New York (Peck), New Jersey (Berk. in Grev. IV, p. 101).

Cr. Bétulæ, Tul. Sel. Carp. II, p. 149, tab. XVII, figs. 13–27.

Exsicc. Ell. & Evrht. N. A. F. 2d Ser. 1792.—Roum. F. Gall. 3944.

Perithecia 6–10, globose, small, ¼ mm. or a little over, circinate, immersed in the surface of the unaltered inner bark, which is raised into a slight pustule over them, necks convergent, decumbent, erumpent in rather prominent, punctiform, black ostiola, in the scarcely emergent, small, black disk just visible through short, transverse cracks in the epidermis. Asci cylindric-clavate, attenuated below, p. sp. 55–60 x 15–18 μ, aparaphysate. Sporidia fasciculate, cylindrical, subarcuate, obtuse, nucleate, hyaline, 30–40 x 3½–4 μ. Conidial stage *Cryptosporium Neesii*, Cda. and *C. betulinum* Sacc. Conidia 50 x 4½–5 μ, shaped like the ascospores.

On dead birch limbs, Plainfield, N. J. (Meschutt).

The N. A. F. specc. do not show the conidia.

Cr. Tíliæ, Tul. Sel. Carp. II, p. 151.

Pustules irregularly scattered, orbicular, small, slightly prominent, cortical, the erumpent disk about 1 mm. broad. Perithecia 4–6, circinate, subglobose, necks short, decumbent, with subhemispherical ostiola. Asci obovate, 65–80 x 13–16 μ, obtuse above, subacute below, 8-spored. Sporidia fasciculate, straight, cylindrical, or sometimes subclavate, 30–35 x 5–6½ μ.

On branches of *Tilia Americana*, Helderberg Mts., N. Y. (Peck).

The conidial stage (*Cryptosporium Tiliæ*) has conidia fusoid, acute, straight or curved, hyaline, 40–50 x 6½–10 μ.

Cr. femoràlis, (Pk.)

Valsa femoralis, Pk. 28th Rep. p. 74.
Cryptospora femoralis, Sacc. Syll. 4117.
Valsa (Cryptospora) femoralis, Cke. Syn. 1944.

Pustules small. Perithecia few, nestling in the inner bark. Ostiola few, short, black, erumpent through small and mostly transverse chinks in the bark, crowded or scattered. Asci lanceolate. Sporidia crowded, elongated, sublinear, straight or slightly flexuous, obtuse, slightly thickened at the ends, 35–75 μ long.

On dead alder limbs, Albany, N. Y. (Peck), and California (Harkness).

Closely related to *Cr. suffusa*, but the sporidia are thickened at the ends. The perithecia adhere to the epidermis and are torn away with it.

Cr. cínctula, (C. & P.)

Valsa cinctula, C. & P. 29th Rep. N. Y. State Mus. p. 59, pl 2, figs. 21–24.
Cryptospora cinctula, Sacc. Syll. 4121.
Valsa (Cryptospora) cinctula, Cke. Syn. 1948.
Exsicc. Ell. N. A. F. 94.

Perithecia 3–8, subcircinate, buried in the unaltered inner bark, of medium size, pale, necks convergent and erumpent through the slightly raised epidermis in stout, black, rough ostiola covered at first with white or light-colored grumous matter appearing like an efflorescence of the bark, and sometimes remaining as an obscure white ring, around the ostiola. Asci oblong-clavate, p. sp. 75–80 x 8–10 μ, stipitate and imperfectly paraphysate. Sporidia subfasciculate, 8 in an ascus, cylindrical, multinucleate, becoming 3–7-septate, 56 x 5 μ (sec. Pk.).

On dead limbs of *Castanea*, New York, New Jersey and Pennsylvania.

The perithecia are of about the same color as in *Diaporthe leiphæmia*. We do not find the sporidia as large as stated by Peck: only about 35–45 x 3–3½ μ.

Cr. tomentélla, (Pk.)

Valsa tomentella, Pk. 35th Rep. p 144.
Cryptospora tomentella, Sacc. Syll. Add. I, p. 192.

Perithecia 4–8, subcircinate, nestling in the inner bark, black, clothed below with a whitish tomentum, disk lanceolate, whitish or brownish, erumpent through a narrow, transverse chink which is acute at each end, and pierced by the smooth, black ostiola. Asci oblong, broad, subcylindrical to fusoid, sessile, 50–75 μ long. Sporidia cylindrical, crowded, more or less curved, colorless, obtuse at the ends, usually multinucleate, 50–67 x 4–5 μ.

On bark of *Betula populifolia*, New Albany, N. Y. (Peck).

Allied to *Cr. cinctula*, but differs in the character of the disk and the tomentose perithecia.

Cr. albofúsca, (C. & E.)

Valsa albofusca, C. & E. Grev. V, p. 31.
Cryptosporella albofusca, Sacc. Syll. 1815.
Valsa (Cryptosporella) albofusca, Cke. Syn. 1934.
Exsicc Ell. N. A. F. 92.

Pustules small, covered by the slightly raised epidermis, scattered. Perithecia few, 4–8 in a stroma formed of the scarcely altered substance of the inner bark, pale, necks short, the small, black, erumpent ostiola at first covered by a mealy, white pseudo-disk which soon disappears. Asci oblong-clavate, sessile, aparaphysate, 65–80 x 8–10 μ. Sporidia lying parallel in the asci, cylindrical or clavate-cylindrical, 30–40 x 3½–4 μ.

On dead limbs of *Quercus obtusiloba*, Newfield, N. J.

When the mealy white layer has disappeared, the ostiola are seen to be united in a brown disk.

*** *Sporidia 2- or more-septate, or pseudoseptate (Calospora).*

Cr. trichíspora, (C. & P.)

Valsa trichispora, C. & P. Cke. Valsei U. S. p. 119, Pk. 29th Rep. p. 58.
Cryptospora trichispora, Sacc. Syll. 4122.
Valsa (Cryptospora) trichispora, Cke. Syn. 1949.

Pustules small. Stroma cortical, pale ochraceous as well as the erumpent disk. Perithecia few, dark brown when mature. Ostiola exserted, quadrisulcate. Asci clavate. Sporidia filiform, hyaline, 5–7-septate, 55 x 2½ μ.

On dead twigs of oak, Greenbush, N. Y. (Peck).

It looks like a miniature *Diaporthe leiphæmia.*

Cr. acùleans, (Schw.)

Sphæria aculeans, Schw. Syn. N. Am. 1395.
Calospora aculeans, Sacc. Syll. 3703.
Valsa (Calospora) aculeans, Cke. Syn. 2045.
Valsa Rhois, Cke. F. Brit. Ser. I, No. 245, id. Ser. II, No. 258.
Valsa albovelata, B. & C. Grev. IV, p. 102.
Valsa stilbostoma, Fr. in Rav. Car. III, 68.
Exsicc. Cke. l. c.—Rav. F. Car. III, 68.— Ell. & Evrht. N. A. F. 2d Ser. 1569.

Stromata scattered, orbicular, convex, 2–3 mm. diam., formed of the unaltered substance of the bark, which is scarcely or only slightly discolored and not circumscribed by any black line. Perithecia generally 5–12, subcircinate, globose, necks converging, with their short-cylindrical, subconical, white-pulverulent ostiola erumpent through a whitish disk which is soon obliterated. Asci (p. sp.) 56–60 x 6–7 μ, aparaphysate, 8-spored. Sporidia biseriate, fusoid, 4-nucleate, becoming uniseptate or indistinctly 3-septate, 12–15 x 3 μ, with a slight hair-like appendage at each end, readily seen in the young sporidium, but finally absorbed.

On dead *Rhus copallina* and *R. typhina*, common.

We have seen no authentic specc. of *V. albovelata*, B. & C., but the diagnosis of that species does not enable one to separate it from the *Sphæria aculeans*, Schw., and (sec. Farlow) the specc. of *V. albovelata*, B. & C. in Herb. Curtis, are the same as *Sphæria aculeans*, Schw. *Valsa Rhois*, Cke., is certainly the same as shown by the specc. in Cooke's exsiccati cited. The two additional septa sometimes seen, are only pseudosepta, being only the dividing line between two adjacent nuclei.

Cr. Pennsylvánica, (B. & C.)

Valsa Pennsylvanica, B. & C. Grev. IV, p. 100.
Diaporthe cylindrospora, Pk. 38th Rep. p. 104.
Valsa prunicola, Pk. 33rd Rep. p. 34.
Calospora Pennsylvanica, Sacc. Syll. 3700.
Valsa (*Calospora*) *Pennsylvanica*, Cke. Syn. 2043.

Pustules valsoid, somewhat prominent, erumpent, scattered. Perithecia numerous, 15–30 or more, crowded, covered by the thin, blackened surface of the inner bark, the ostiola rather long, stout, cylindrical, rounded at the tips, crowded, exserted. Asci narrow, subfusiform, 35–50 x 5–7 μ. Sporidia subcylindrical, crowded or biseriate, 4-nucleate, hyaline, 13–19 x 3–4 μ (25 μ long, Berk.).

On dead branches of *Prunus Pennsylvanica*, Adirondack Mts., N. Y., and on *P. Americana*, Iowa.

The original description of this species in Grevillea is as follows: "Bursting transversely. Perithecia in the center of a facette. Sporidia narrow, oblong, sometimes wide at one end, sometimes slightly curved, triseptate. 25 μ long. On *Cerasus Pennsylvanica*, mountains of New York (Berk. l. c.)."

Cr. Cáryæ, Pk. 38th Rep. p. 106, tab. 2, figs. 28–31.

Pustules scattered, covered by the epidermis, then erumpent, circumscribed by a black line or at length covered by a black crust beneath the epidermis. Perithecia 4–12 in a pustule, globose or angular from mutual pressure, rather less than $\frac{1}{3}$ mm. diam., buried in the light-colored stroma in a single layer with rather short but slender necks and subglobose, black ostiola erumpent in a dark brown, convex disk loosely surrounded by the ruptured epidermis. Asci clavate, 100–125 x 12–15 μ, aparaphysate, subsessile. Sporidia biseriate above, fusoid-oblong, hyaline, subobtuse, 3–5-pseudoseptate, 40–55 x 6–8 μ.

On dead branches of *Carya*, Knowersville, N. Y. (Peck), Iowa (Holway).

We have supplemented the original description by an examination of specimens sent us by Mr. Peck.

Species imperfectly known.

Cr. gemmàta, (B. & C.)

Valsa gemmata, B. & C. Grev. IV, p. 102.
Valsa (Calospora) gemmata, Cke. Syn. 2046.
Calospora gemmata, Sacc. Syll. 3704.

"Perithecia few, circinating, necks united. Ostiola stellate. Sporidia shortly fusiform, triseptate." On *Rhus radicans*, South Carolina.

Cr. ciliàta, (Pers.)

Sphæria ciliata, Pers. Syn. p. 35.
Valsa ciliata, Stevenson's Add. to Cke. Valsei, No. 100.
Diaporthe (Chorostate)? ciliata, Sacc. Syll. 2442.

Perithecia about 10 together, erect-convergent, circinating in the inner bark, which is raised into a small, round pustule above them. Ostiola very slender, 1–2 lines long, subsetaceous, divaricate, subflaccid. Sporidia (sec. Stevenson l. c.), navicular, biseptate, 12½ x 5 μ.

On bark of elm, Carolina and Pennsylvania (Schw.).

Cr. inconspícua, (C. & E.)

Valsa inconspicua, C. & E. Grev. VI, p. 11.
Valsa (Calospora) inconspicua, Cke. Syn. 2047.
Calospora inconspicua Sacc. Syll. 3706.

Perithecia valsoid-aggregated. Sporidia narrow-fusoid, 3-septate, 4-nucleate, 15 x 4 μ.

On dead alder, Newfield, N. J.

Our specc. of this species are entirely without fruit.

PSEUDOVÁLSA, Ces. & De Not.

Schema Sferiac. p. 32.

Stroma and perithecia as in *Melanconis*. Asci 4–6 or 8-spored, typically paraphysate. Sporidia 2- or more-septate (not muriform), mostly brown.

Ps. profusa, (Fr.)

Sphæria profusa, Fr. S. M. II, p. 392.
Sphæria irregularis, DC. Fl. Fr. VI, p. 116.
Sphæria anomia, Fr. (not Schw.) l. c. p. 381.
Valsa profusa, Fr. Summa, p. 411.
Massaria setridia, B. & C. Grev. IV, p. 155.
Aglaospora profusa, De Not. Micr Ital. Dec. V, p. 5, No. 3.
Valsa Sartwellii, B. & C. (in Herb.).
Sphæria ocellata, Schw. Syn. N. Am. 1638.
Sphæria amorphostoma, Schw. Syn. N. Am 1334 (sec. specc. in Herb. Schw.).
Pseudovalsa profusa, Winter, Die Pilze, II, p. 785.
Exsicc. Fckl. F. Rh. 583.—Rab. F. E. 733, 1137, 1441, 2514.—Rehm Asc. 45.—Sydow, M. March. 176, 941.—Thum. M. U. 969.—Ell. N. A. F. 172.

Stromata scattered, numerous, subseriate and subconfluent, variable in size and shape, pyramidal, pulvinate or hemispherical, round, elliptical or elongated (on a transverse section), surface uneven or undulate, black, paler inside, of tough, horn-like consistence, sunk in the substance of the bark, enclosed in a rather thick, black layer, covered by the epidermis which is only slightly perforated, finally exposed by the falling away of the bark. Perithecia 2–4 in a stroma, deeply buried, globose or angular, contracted into a thick, irregular-shaped neck, with short, black, obtuse ostiola erumpent in a compact fascicle but scarcely exserted. Asci cylindrical, short-stipitate, 4-spored, 180–210 x 21–24 μ. Sporidia cylindrical, rounded at the ends, spuriously 4-celled, with a subtriangular nucleus in each cell, hyaline at first, then clear brown. Paraphyses filiform, abundant.

On dead limbs of *Robinia pseudacacia*, common.

Ps. Titan, (B. & Rav.)

Diatrype Titan, B. & Rav. Grev. IV, p. 97.
Pseudovalsa Titan, Cke. Syn. 2093.
Titania Berkeleyi, Berlese Icones. tab. 37, fig. 1.

Pustules irregular, black, rough with the protruding ostiola, minutely granulated. Asci (sec. Cke. l. c.) 2-spored. Sporidia fusiform, obtuse at each end, dark, with about 6 septa, 100 μ long.

On bark of hornbeam, South Carolina (Ravenel).

Ps. haplocýstis, (Berk. & Br.)

Sphæria haplocystis, B. & Br. Ann. & Mag. Nat. Hist. Ser. II, p. 317.
Haplocystis Berkeleyi, Awd. in Fckl. F. Rh. 585.
Calospora haplocystis, Fckl. Symb. p. 191.
Pseudovalsa haplocystis, Sacc. Mich. I, p. 44.
Exsicc. Rab. F. Eur. 1146.—Cke. F. Brit. Ser. I, 253, Ser. II, 229.—Roum. F. G. 4448, 4449.

Perithecia covered, subglobose, thin-walled, minutely tomentose, permanently covered by the blackened epidermis which is raised into slight pustules, depressed-globose, 4–10 together in subcircinate groups without any distinct stroma, with converging, short-cylindrical necks bent upwards at their extremities and just visible through short, narrow cracks in the epidermis, but not exserted. Asci elliptical or oblong-elliptical, with a short stipe, 8-spored, 90–105 x 35–40 μ. Sporidia conglomerate, oblong or oblong-cylindrical, ends broadly rounded, subinequilateral, often a little curved, 2- (sometimes 3-) septate, not constricted, brown, 25–35 x 12–14 μ, with a short, thick, hyaline appendage at each end.

On dead limbs of *Platanus occidentalis*, New York (Peck).

Ps. bicórnis, (Cke.)

Melanconis bicornis, Cke. in 28th Rep. N. Y. State Mus. p. 72.
Pseudovalsa bicornis, Sacc. Syl. 3364, Cke. Syn. 2110.

Perithecia circinating, 5–7, seated beneath the epidermis which is but slightly raised. Ostiola short, convergent, just piercing the epidermis, with a regular orifice. Sporidia expelled when mature, blackening the matrix around the ostiola, fasciculate, obtusely fusiform, triseptate, straight or curved, brown, 63–78 μ long, scarcely constricted, end cells smallest, each extremity tapering into a hyaline, at first straight, then curved or flexuous, horn-shaped appendage one-half to one-third the length of the sporidium.

On bark of *Platanus occidentalis*, New York State (Peck).

Allied to *Ps. haplocystis*, (B. & Br.), but distinct. When the epidermis is torn away, the perithecia come off with it. They are slightly whitish-floccose or tomentose above.

Ps. sambùcina, (Pk.)

Valsa sambucina, Pk. 28th Rep. p 75.
Pseudovalsa sambucina, Sacc. Syll. 3356, Cke. Syn. 2103.

Pustules erumpent, sometimes seriately placed. Ostiola slightly prominent, even or radiately sulcate, scattered or crowded, Asci linear. Sporidia 8, uniseriate, oblong, colored, triseptate, 12–15 μ long.

On dead stems or branches of *Sambucus*, Catskill Mts., N. Y. (Peck).

When young, the sporidia are paler. The pustules vary much in size; those on the branches being larger and more scattered than those on the main stems and trunks.

Ps. stylóspora, E. & E. Proc. Acad. Nat. Sci. Phil. Pa., July, 1890, p. 223. **(Plate 35)**

Stroma cortical, convex, 2–3 mm. diam., covered by the epidermis. Perithecia circinate, 4–8 in a stroma, globose, ¾ mm. diam., collapsing when dry, contracted above into short necks, terminated by small, globose ostiola, subseriately arranged and erumpent through a small crack in the bark. Asci (p. sp.) 80–85 x 15 μ. Sporidia biseriate, oblong-elliptical, hyaline, faintly subglobose-appendiculate and granular at first, becoming brown and 2-septate, and slightly constricted at the septa, 25–30 x 10–14 μ. Pycnidia central, bearing cylindrical 3-septate, hyaline stylospores, 40–55 x 10–12 μ, on short basidia.

On bark of dead *Acer spicatum*, London, Canada (Dearness).

The drawing was made from a young specimen and does not show the two septa in the sporidia. The appendages also were accidentally omitted. They are very obscure and easily overlooked.

Ps. lancifórmis, (Fr.)

Sphæria lanciformis, Fr. S. M. II, p. 362.
Sphæria quercina, *b*. A. & S. Consp p. 11 (sec. Fr.).
Sphæria Betulæ, Schum. Enum. Plant. Sæll. II, p. 171.
Sphæria cincta, DC. Fl. Fr. VI, p. 119.
Sphæria melasperma, Fr. S. M. II, p. 389 (sec. Tul.).
Diatrype lanciformis, Fr. Summa, p. 385.
Sphæria favacea, Tul. in Compts. Rend. tom. XXXII, p. 472 and 474.
Melanconis lanciformis, Tul. Sel. Carp. II, p. 135, tab. XIV.
Pseudovalsa lanciformis, Ces. & De Not. Schema Sfer. p. 32.
Melanconis elliptica, Pk. 25th Rep. p. 102, 28th Rep. p. 87.
Exsicc. Fckl. F. Rh. 1996—Rab. F. E. 248, 1258, 1438.—Rehm Asc. 584.—Thum. M. U. 1551.—Sydow, M. March. 752.

Stromata scattered, buried, erumpent through transverse cracks in the bark, dark brown or black above, brownish inside, laterally compressed so as to become lanceolate, not rising much above the bark, of carnose-suberose consistence, 1–3 mm. long. Perithecia 4–10 in a stroma, irregularly crowded, globose, black, rather large ½ (mm.), necks cylindrical, with their ostiola only slightly projecting through the lance-shaped disk. Asci oblong-cylindrical, narrowed below, 8-spored, 150–160 x 26–28 μ. Sporidia biseriate, oblong, ends broadly rounded, 3–6-septate, with a large nucleus in each cell, not constricted at the septa, brown, straight, 35–48 x 12½–15 μ. Paraphyses very long, filiform.

On dead limbs of birch, Bethlehem, Pa. (Schw.), New York (Peck).

Ps. Fairmani, E. & E. Proc. Roch. Acad. 1890, p. 51, Plate 3, figs. 1, 2, 3, 10, 11.

Stromata convex-pulvinate, 1–1½ mm. diam., formed of the slightly altered substance of the inner bark, the surface only subcarbonized and blackened, not surrounded by any distinct circumscribing line, covered by the epidermis, which is pierced by the stout, short-cylindrical or conical ostiola with smooth or quadrisulcate tips. Perithecia

4–8 in a stroma, closely packed, ovate or subangular from compression, about ⅓ mm. diam., with whitish, waxy contents. Asci (p. sp.) 75–85 x 20 μ, mostly only 6-spored. Sporidia oblong-cylindrical, yellowish, 3- (exceptionally 5-) septate, 30–40 x 4–7 μ, slightly constricted at the septa. The young stromata contain an abundance of *pycnidial* spores (about the size and shape of the *ascospores*), borne on stout or branching sporophores about as long as the spores themselves.

On dead hickory limbs (*Carya*), Lyndonville, N. Y. (Fairman).

Ps. sigmoidea, (C. & E.)

Melanconis sigmoidea, C. & E. Grev. VI, p. 92, pl. 100, fig. 26.
Pseudovalsa sigmoidea, Sacc. Syll. 3352, Cke. Syn. 2099.

Stromata conical or conic-hemispherical, 1–2 mm. diam., orbicular or elliptical at base, outer layer thick, black, seated in the inner bark and raising the epidermis into pustules. Perithecia rather deeply buried in the stroma, small, ¼–⅓ mm. diam., contracted above into moderately stout necks with their obtuse, subhemispherical ostiola subcircinate-erumpent in a convex, brown disk (apex of the stroma) which pierces the epidermis, but does not rise much above it. Asci broad-clavate, 120–150 x 19–22 μ (p. sp. 100–110 μ long), 8-spored, stipitate, obscurely paraphysate. Sporidia biseriate, oblong-fusoid, slightly curved, 3–7- (mostly 5-) septate, hyaline at first, finally brown, 45–75 x 10–12 μ, subacute at first, finally subobtuse.

On dead branches of *Quercus tinctoria*, *Q. ilicifolia* and *Q. alba*, Newfield, N. J., on oak limbs, Canada (Dearness).

Perithecia on the first two mentioned hosts, 3–5 in a stroma; on *Q. alba*, 3–12. The conidial stage is a *Coryneum*, forming black, erumpent tubercles about ¾–1 mm. diam. Conidia clavate-fusoid, 3–7-septate, 50–80 x 10–13 μ, on densely fasciculate, stout, brownish, septulate basidia 100–120 x 4–5 μ.

Ps. Texénsis, (E. & E.)

Diatrype Texensis, E. & E. Journ. Mycol. II, p. 40.
Thyridaria Texensis, Sacc. Syll. 7058.

Stroma pulvinate, subcarbonaceous, black, suborbicular, 2–4 mm. across, at length plane or subconcave above, seated on the surface of the inner bark and bursting through the epidermis which closely surrounds it. Perithecia coriaceous, with thick walls, globose or subangular by pressure, 6–20 in each stroma, ⅓–½ mm. diam. Ostiola subtuberculiform or hemispherical, with a rather large, though indistinct opening. Asci clavate-cylindrical, about 75 x 12 μ, with abundant paraphyses. Sporidia partly biseriate, yellowish-brown and uniseptate at first, becoming dark brown and 3-septate, ovate or oblong-elliptical, 15–20 x 6–7 μ, scarcely constricted at the septa, the terminal cells subhyaline. Has much the same general appearance as *Valsaria cincta*, (Curr.). The bark and the surface of the wood beneath the stroma are more or less blackened.

On bark of fallen limbs of *Tilia*(?), Houston, Texas (Ravenel).

Ps. Comptòniæ, (E. & E.)

Diatrype Comptoniæ, E. & E. Journ. Mycol. II, p. 89.
Thyridaria Comptoniæ, Sacc. Syll. 7491.

Stroma erumpent, subtuberculiform, small (1–3 mm.), subhemispheric or elongated, dull black outside, whitish within and consisting of the scarcely altered substance of the wood. Perithecia often single in the smaller stromata, or in the larger and more elongated ones 2–12, with thick walls, ovate or subangular from mutual pressure, $\frac{1}{4}$–$\frac{1}{3}$ mm. diam., contracted above into a short neck, with a short, cylindrical or subconical, slightly projecting, smooth ostiolum. Asci clavate, 75–85 μ long, including the slender, stipitate base, surrounded with abundant paraphyses and containing eight subfusoid, yellowish-brown, 3-septate, slightly curved, 12–15 x 4–5 μ sporidia, which are crowded into the upper half. The general appearance is much like that of *Cryptovalsa eutypæformis*, Sacc. The ostiola are not sulcate and have a smooth, round opening. The stromata arise either directly from the wood or are seated on the lower stratum of the bark denuded by the flaking off of the superficial layer.

On dead, partially decorticated stems of *Comptonia asplenifolia*, Newfield, N. J.

Species imperfectly known.

Ps. tubulòsa, (B. & C.)

Valsa tribulosa,* B. & C. Grev. IV, p. 102.
Pseudovalsa tubulosa, Sacc. Syll. 3358, Cke. Syn. 2104.
Calospora tribulosa, Sacc. Syll. 3699 (Berlese Icones, pl. XXXIV).

"Perithecia few, necks subcylindrical, projecting. Sporidia fusiform, triseptate, obtuse at the ends, brown, 30 x 10 μ, (30–32 x 10–12 μ, Berlese).

On alder, South Carolina."

Ps. convérgens, (Tode).

Sphæria convergens, Tode Meckl. II, 39, fig. 111.
Valsa convergens, Fr. (in Cke. Valsei U. S. p. 123).
Pseudovalsa convergens, Sacc. Syll. 3354, Cke. Syn. 2101.

*"Tribulosa" is a misprint for "tubulosa."

Circinate, minute. Perithecia about 6 together, entirely covered, ovate, the cylindrical, subattenuated, convergent ostiola erumpent. Asci ——? Sporidia subcymbiform, 3-septate, sometimes slightly constricted, dark brown, 50–52 x 12–14 μ, straight or curved.

On branches of *Rosa corymbosa*, Carolina and Pennsylvania (Schw.).

FENESTÉLLA, Tul.

Sel. Carp. II, p. 208, emend. Sacc. Mich. I, p. 50.

Perithecia and stroma as in Valsa. Asci cylindrical, 4–8-spored. Sporidia uniseriate, oblong or elliptical, muriform, yellow-brown.

F. prínceps, Tul. Sel. Carp. II, p. 207.

Valsa fenestrata, B. & Br. Ann. & Mag. Nat. Hist. Ser. III, tom. III, p. 366.
Thyridium Faberi, Kunze, F. Sel. 263.
Exsicc. Fckl. F. Rh 1999.—M. March. 261, 1571, 1885.—Roum. F. G. 5253.

Stroma orbicular or elliptical at base, lenticular, sunk in the parenchyma of the bark, with the epidermis slightly pustulate-elevated over it, consisting of brown, floccose matter. Perithecia 3–14 in a stroma, crowded and mostly angular from compression, collapsing when dry, black, about ½ mm. diam., with short-cylindrical, perforated ostiola, united in a roundish or lanceolate, erumpent disk, scarcely rising above the epidermis. Asci cylindrical, short-stipitate, 150–200 x 18–20 μ (210–260 x 22–27, Sacc. in Syll.), with numerous paraphyses. Sporidia uniseriate, oblong-elliptical, constricted in the middle, closely multiseptate and muriform, golden-yellow, becoming brown, with a hyaline, papillose projection at each end, 25–40 x 14–20 μ.

On dead limbs of *Cratægus*, London, Canada (Dearness), on dead limbs of *Salix*, Lyndonville, N. Y. (Fairman).

F. amórpha, E. & E. Journ. Mycol. IV, p. 58.—Proc. Acad. Nat. Sci. Phil. July, 1890, p. 239.—Proc. Roch. Acad. Aug. 1890, p. 48. **(Plate 35)**

Stroma tuberculiform, seated on the wood, variable in size from 1 mm., inclosing a single perithecium to 3 or 4 mm. with 4–6 perithecia, or sometimes confluent in a seriate manner, for 1 cm. or more. Perithecia flask-shaped, about 1 mm. high and ¾ mm. broad, black outside but the internal texture white, connate and subseriate, the short (½ mm.), cylindrical ostiola converging, but not united in a disk, bursting out through cracks in the bark, but scarcely projecting. Asci cylindrical, with a narrow base, 150–175 x 12–15 μ, with abundant filiform paraphyses. Sporidia uniseriate, oblong-elliptical, about 6-septate, with a single, longitudinal septum, dark brown, 20–22 x 12 μ, not constricted at the septa, becoming almost opake, so that the septa are hardly visible. When the bark falls away, the stroma becomes superficial.

On dead hickory limbs, Lyndonville, N. Y. (Fairman).

F. vestita, (Fr.)

Sphæria vestita, Fr. S. M. II, p. 410.
Valsa vestita, Fr. Summa Veg. Scand. p. 412.
Thyridium vestitum, Fckl. Symb. p. 195.
Cucurbitaria vagans, Sacc. M. Ven. Spec. p. 122, tab. XII, figs. 34-38.
Fenestella vestita, Sacc. Syll. 4004.
Exsicc. F. Rh. 954.—Rehm Asc. 179, 684.—Sacc. M. Ven. 1272.—Sydow, M. March. 1249, 1570, 1918, 1919.—Roum. F. Gall. 4775.

Stroma suborbicular, 2–3 mm. across, depressed, sides abrupt, penetrating the bark, and when this falls away, remaining attached to the wood, crowned above with the large, black, tuberculiform disk erumpent through the adherent epidermis, and rising but little above it. Perithecia 8–15 in a stroma, irregularly crowded, ½ mm. or more in diameter, globose or angular, with thick walls, necks cylindrical, convergent with their thick, black, perforated ostiola united in the erumpent disk, but not exserted. The stroma is covered with a granulose-floccose, pale yellow substance that is very noticeable. Asci cylindrical, 120–150 x 12–15 μ (p. sp. 80–100 μ long), with filiform paraphyses. Sporidia uniseriate, elliptical, olive-brown, 3–5-septate, and muriform, scarcely constricted at the septa, 12–20 x 10–12 μ (19–25 x 10–12 μ, Winter, 20–26 x 10–14 μ, Sacc.).

On dead limbs of *Quercus*, *Ulmus*, and *Sambucus*, Canada (Dearness).

F. superficiàlis, (P. & C.)

Melogramma superficialis, P. & C. in 29th Rep. N. Y. State Mus. p. 57.
Fenestella superficialis, Sacc. Syll. 4012.

Stroma superficial, depressed, 1–2 lines across, pale or yellowish within. Perithecia unequal, more or less irregular, crowded, depressed, blackish-brown or black. Asci very broad, varying from subglobose to oblong-clavate, fugacious. Sporidia oblong, obtuse, fenestrate, slightly constricted at the center and appearing uniseptate, 25–33 μ long.

On bark of living *Pyrus Americana*, Buffalo, N. Y. (Clinton).

F. phæóspora, Sacc. F. Ven. Nov. Ser. IV, p. 13, F. Ital. tab. 140.
Acervuli valsoid, small (2 mm.), orbicular, flattened, seated on the

slightly discolored surface of the inner bark and covered by the epidermis. Perithecia 4–6, closely packed and mostly surrounded by a thin, dark-colored stroma, rather large, nearly ½ mm. diam., attenuated above into short, cylindrical necks with ostiola convergent and perforated, piercing the epidermis, but not rising above it. Asci cylindrical, 180–250 x 18–20 μ (180 x 18 μ, Sacc.), short-stipitate, with filiform paraphyses; 6–8-spored. Sporidia uniseriate, oblong-elliptical, 35–45 x 16–18 μ, hyaline at first, becoming opake and almost black, 9–11-septate, with a few of the cells at first divided by longitudinal septa, finally densely muriform.

On decaying branches of *Quercus alba*, Newfield, N. J.

When the epidermis is peeled off, the clusters of perithecia remain attached to the surface of the inner bark. The sporidia are scarcely constricted at the septa.

F. Canadénsis, E. & E. (in Herb.)

Stroma cortical, covered by the epidermis, orbicular, 1–3 mm. diam., sunk in the surface of the inner bark which is only slightly raised above it. Perithecia 3–10 in a stroma, of medium size, circinate, buried in the stroma, their short-cylindrical, rounded and obtuse, perforated ostiola slightly raising and piercing the epidermis in a compact fascicle, but hardly rising above it. Asci cylindrical, p. sp. 100–115 x 15 μ, briefly stipitate, 8-spored. Paraphyses filiform, subgelatinous, guttulate, indistinct. Sporidia uniseriate in some asci, biseriate in others, acutely-elliptical, 3-septate, the two inner cells mostly divided by a longitudinal septum, hyaline at first, becoming olivaceous-brown, except the subapiculate end cells which remain hyaline, 22–30 x 10–12 μ.

On bark of dead *Carpinus*, London, Canada (Dearness).

The conidial stage is a *Coryneum* (issued in N. A. F. 2578 as *C. umbonatum*, Nees.) with subfusoid, 6–9-septate conidia curved at the ends, 40–60 x 8–10 μ, borne on simple, stout basidia 10–15 x 4–5 μ. The *Coryneum* grows under the epidermis which is finally perforated and slightly raised, and afterwards the circle of perithecia send up their ostiola through the same opening.

F. hormóspora. Cke. Grev. XIV, p. 48.

Valsa conjuncta, Schw. (not Fries) Syn. N. Am 1383.
Valsa hormospora, B. & C. in Herb.

Perithecia few, crowded in irregular circles, disk erumpent, whitish, pulverulent. Ostiola short, subconnate. Asci ample, clavate. Sporidia subelliptical, not appendiculate, constricted in the middle, 5-septate, at length merenchymatic, brown. Disk orbicular.

On branches, Bethlehem, Pa. No measurements given.

F. condensàta, (B. & C.)

Valsa condensata, B. & C. Grev. IV, p. 103.
Fenestella condensata, Sacc. Syll. 4007, Cke. Syn. 2123.

"Pustules small. Ostiola short. Sporidia obovate, divided horizontally and vertically, 20 μ long, 5-septate. Stroma brown.

On *Quercus montana*, Virginian mountains" (Berk. l. c.).

F. castanícola, (B. & C.)

Valsa castanicola, B. & C. Grev. IV, p. 101.
Fenestella castanicola, Sacc. Syll. 4014, Cke. Syn. 2130.

"Pustules small, rather flat. Ostiola not prominent. Asci slightly clavate. Sporidia biseriate, cymbæform, pointed, triseptate, at length vertically divided, 10 μ long."

On small twigs of *Castanea*, Virginian mountains (Berk. l. c.).

BOTRYOSPHÆRIA, Ces. & De Not.

Schema Sferiac. p. 37.

Stroma pulvinate or depressed-conical, erumpent, becoming more or less superficial, black, of pseudoparenchymatic structure, often reduced to a thin crust, connecting the bases of the perithecia, and inconspicuous or nearly wanting. Perithecia at first sunk in the stroma, often remaining so, but also finally more or less prominent, so as to present a botryoidal aspect, mostly small and globose, with an inconspicuous, papilliform ostiolum. Asci 8-spored, paraphysate. Sporidia oblong, elliptical or ovate, hyaline, continuous.

B. fuliginòsa, (M. & N.) **(Plate 36)**

Sphæria Quercuum, Schw. Syn. Car. 125.
Melogramma fuliginosum, (M. & N.) Ell. in Proc. Acad. Nat. Sci. Phil. 1879, p. 66.
Sphæria ambigua, Schw. Syn. N. Am. 1492.
Sphæria Calycanthi, Schw. l. c. 1434.
Sphæria Persimmons, Schw. l. c. 1444.
Sphæria thyoidea, C. & E. Grev. VI, p. 14.
Sphæria Syringæ, Schw. Syn. N. Am. 1667.
Sphæria viscosa, C. & E. Grev. V, p. 34.
Sphæria pyriospora, Ell. Bull. Torr. Bot. Club, V, p. 46.
Dothidea moricola, C. & E. and *Dothidea venenata*, C. & E. Grev. V, p. 95.
Dothidea Tamaricis, Cke. Grev. XI, p. 108.
Sphæria Uvæ-sarmenti, Cke. Grev. XI. p. 109.
Botryosphæria pustulata, Sacc. (in Ell. N. A. F. 480).
Dothidea Cerasi, C. & E. Grev. V, p. 34
Thuemenia Wistariæ, Rehm in Thum. M. U. 971.
Sphæria mutila, Fr. in Rav. Car. III, 62.
Melogramma Callicarpæ, Cke. in Rav. F. Am. 767.
Botryosphæria Berengeriana, De Not. Sfer. Ital. fig. 90.
Sphæria fuliginosa, M & N. Exsicc. (sec. Tul.)
Melogramma Aceris, C. & E Grev. VII, p. 8, *Sphæria eriostega*, C. & E. Grev. V, p. 34, and *Valsa Mahaleb*, C. & E. Grev. VI, p. 11, also appear to belong here, but our specc. of these are sterile or immature.

Exsicc. Ell. N. A. F 475-481.—Rav. F. Am. 666, 667, 668, 678, 740, 767.

Stroma variable, erumpent, often rudimentary, orbicular, about 1 mm. diam., or by confluence larger and variable in shape. Perithecia sunk in the stroma, finally emergent and often quite superficial, stromatically connected below, botryoidally aggregated or subcespitose, white inside. Asci inflated-clavate, paraphysate, 8-spored, 75–100 x 18–25 μ. Sporidia imperfectly biseriate, inequilaterally elliptical, hyaline, continuous, 18–38 x 8–15 μ.

On dead limbs of various deciduous trees and shrubs, throughout the U. S. and Canada. Common and variable. See paper in Proc. Acad. Nat. Sci. Phil., referred to above. The forms with scattered perithecia (var. *simplex*) have been removed to *Physalospora*.

Repeated observation convinces us that it would be simply absurd to separate the various forms above quoted as distinct species. Without knowing the host, it would be impossible to separate them. *Sphæria Hibisci*, Schw., and *S. Cratægi*, Schw., which in the paper above referred to were interrogatively included as synonyms, prove on examination to be different. Figs 1 and 2, Pl. 36, represent the two extreme forms.

B. minor, E. & E. Journ. Mycol. IV, p. 77.

Perithecia minute (150–170 μ), white inside, mostly in small, erumpent groups of 2–6 joined in an imperfectly developed stroma. Asci 75 x 20 μ. Sporidia crowded-biseriate, subelliptical (narrower at the ends), 14–16 x 6–7 μ, yellowish-hyaline. Differs from the larger forms included under *Botryosphæria*.

On *Sesbania*, Louisiana (Langlois).

B. abrùpta, B. & C. Grev. XIII, p. 101.

Stromata erumpent, closely surrounded by the fissured epidermis, subdiscoid. Perithecia globose, semiimmersed, black, at length confluent, truncate-obtuse, abrupt, white within. Asci clavate, 8-spored. Sporidia elliptic-fusoid, subattenuated at each end, obtuse, continuous, hyaline, becoming yellow, granular within, 22–24 x 10 μ.

On *Cyrilla*, Carolina.

B. meláthroa, B. & C. Grev. l. c.

Stroma longitudinally effused, innate, erumpent. Perithecia ovate, crowded, subdistinct, black, rounded at the apex, small. Asci clavate-cylindrical. Sporidia elliptical, continuous, 20 x 8 μ, hyaline.

On *Cratægus cordata*, Pennsylvania.

B. Aràliæ, Curtis, Grev. XIII, p. 101.

Stromata innate-erumpent, corticolous, subpulvinate, black, generally subseriate, (1 mm. diam.), white inside, surrounded by the epidermis. Perithecia scarcely distinct (subconfiuent), rather obtuse, opake. Asci clavate, 8-spored. Sporidia elliptical, continuous, hyaline, 20 x 8 μ.

On bark of *Aralia spinosa*, Carolina (Curtis).

B. Vibùrni, Cke. Grev. l. c.

Stromata erumpent, gregarious, very small ($\frac{1}{2}$–$\frac{3}{4}$ mm.), corticolous, depressed-pulvinate, dark brown, surface tuberculose, white inside. Perithecia 6–10, subimmersed, convex above, perforated. Asci clavate. Sporidia narrow-elliptical, continuous, turning yellow, granular, 18–21 x 8 μ.

On branches of *Viburnum opulus*, United States.

Closely allied to *B. Araliæ*, of which it may be a variety.

B. Hypericòrum, Cke. Grev. XIII, p. 102.

Subelliptical, flattish, black. Perithecia small, connate, white inside, closely embraced by the ruptured epidermis. Asci clavate, 8-spored. Sporidia narrow-elliptical, continuous, hyaline, about 20 x 5 μ (immature).

On stems of *Hypericum proliferum*, United States.

B. stomática, Schw. Grev. l. c.

Widely effused, emergent, staining the wood black. Perithecia semiimmersed, rather large. Ostiola large and prominent, cylindrical, obtuse. Asci clavate, 8-spored. Sporidia elliptical, continuous, brown, 12–13 x 5 μ.

On rotten wood, Indiana.

Differs from the others in its brown sporidia.

We have seen no specc. of the last six species, and take the descriptions from Grevillea.

B. Arctostáphyli, (Plowr.)

Sphæria Arctostaphyli, Plowr. Grev. VII, p. 73.
Botryosphæria Arctostaphyli, Sacc. Syll. 1764, Cke. Syn. 1405.

Perithecia superficial, depressed-globose, black, roughish, gregarious, $\frac{1}{2}$–1 mm. Ostiola obscurely papilliform. Asci clavate-pyriform, 70 x 15–20 μ. Sporidia continuous, ovoid, sometimes a little curved, 15–18 x 4–5 μ, hyaline. Spermogonial perithecia about the same as the ascigerous ones. Spermatia oblong, hyaline, 5–7 x 2 μ.

On bleached wood of *Arctostaphylus*, California (Harkness).

Our specc. from Dr. Harkness show only spermogonia. The stroma seems to be entirely wanting.

B. Van Vléckii, (Schw.)

Sphæria Van Vleckii, Schw. Syn. N. Am. 1427.
Botryosphæria Van Vleckii, Sacc. Syll. 1800, Cke. Syn. 1449.

Irregularly cespitose on a black, subpulverulent stroma, rising from the inner bark, or from the wood itself, irregular in shape, and easily deciduous, at first covered by the epidermis, at length erumpent, causing the trunk to appear as if pitted with small pox, and throwing off the bark. Perithecia cylindrical-globose, black (white within), rugose, scarcely confluent. Ostiola long, rostellate, mixed with shorter ones, divergent, and finally deciduous. Sporidia narrow-fusoid, hyaline, 15 μ long.

On *Bignonia*, Carolina (Schw.), Pennsylvania (Michener).

B. graphídea, (B. & Rav.)

Melogramma graphideum, B. & Rav. Grev. IV, p. 98.
Botryosphæria graphidea, Sacc. Syll. 1791, Cke. Syn. 1442.

Looks at first sight like a Lichen, as the strata are of various shapes, triangular, sinuated, &c., and surrounded by a rigid border of the bark. Sporidia short-fusiform, hyaline, 20 μ long.

On *Myrica cerifera*, (Carolina)?

B. Fìcus, Cke. Grev. XI, p. 108.

Exsicc. Rav. F. Am. 797.

Perithecia very small, obtuse, black, erumpent in tufts or lines. Asci clavate. Sporidia elliptic-lanceolate, continuous, hyaline, 25 x 10 μ.

On bark of *Ficus carica*, South Carolina (Ravenel).

Species imperfectly known.

B. Liriodéndri, (Cke.)

Melogramma Liriodendri, Cke. in Rav. F. Am. 353.
Botryosphæria Liriodendri, Sacc. Syll. 1798, Cke. Syn. 1430.

The specc. in Rav. F. Am. show only stylospores, oblong-fusoid, hyaline, about 30 x 5 μ, and are entirely without asci.

B. euómphala, (B. & C.)

Sphæria euomphala, B. & C. Grev. IV, p. 141.
Botryosphæria euomphala, Sacc. Syll. 1784.
Byssosphæria euomphala, Cke. Syn. 2603.

"Perithecia cup-shaped, rugose, crowded. Asci clavate. Sporidia biseriate, hyaline, subelliptical."

On fallen branches in damp places, South Carolina (Ravenel).

B. horizontàlis, (B. & C.)

Melogramma horizontale, B. & C. Grev. IV, p. 98.
Botryosphæria horizontalis, Sacc. Syll. 1789, Cke. Syn. 1429,

"Bursting the bark transversely. Perithecia minutely pulverulent. Asci large, clavate. Sporidia hyaline, fusiform."

On *Gossypium*, South Carolina (Ravenel).

B. rhizógena, (Berk.)

Sphæria rhizogena, Berk Hook. Lond. Journ. Bot. Vol. IV, p. 312.
Botryosphæria rhizogena, Sacc. Syll. 1786, Cke. Syn. 1440.

"Patches nearly orbicular, 4 mm. or more broad, surface rather irregular, here and there depressed. Stroma pale yellowish-brown. Perithecia minute, dull, not shining, partially immersed, pale brown when shaded from the light, nearly black above, at first pruinose, globose, with a minute and sometimes depressed papilla, filled with pale brownish jelly. Asci linear. Sporidia elliptical. Exactly the habit of *Cucurbitaria Laburni*, but differs in structure."

On roots of *Gleditschia triacanthus*, Ohio (Lea).

B. subconnàta, (Schw.)

Sphæria subconnata, Schw. Syn. N. Am. 1443.
Thuemenia valsarioides, Rehm in Thum. M. U 2166. (sec. Cke. Grev. XIII, p. 101).

"Cespitose, confluent, elliptical, erumpent through the thin epidermis, scarcely a line long, black. Perithecia few, comparatively large, flattened at the apex, or irregularly subcollapsing, papillate-ostiolate, white inside, at length deciduous, leaving white, cup-shaped cavities."

On *Gossypium*, Carolina (Schw.), Georgia (Ravenel).

The spec. in our copy of Myc. Univ. is immature. The perithecia are 80–110 μ diam., and rather numerous, instead of "comparatively large and few." The spec. 1443 in Herb. Schw. is only a *Phoma* with sporules, 5–7 x $2\frac{1}{2}$ μ, hyaline.

B. Castàneæ, (Schw.)

Sphæria Castaneæ, Schw. Syn. Car. No. 124.
Botryosphæria Castaneæ, Sacc. Syll. 1797, Cke. Syn. 1447.

Stroma ovate, tuberculose, about 6 mm. long, dark brown, subelevated. Perithecia globose-depressed, few or numerous (30–100), distinctly papillate, the upper part of the perithecia deciduous, leaving the basal cavities resembling a honeycomb; when young white inside.

On chestnut limbs, Carolina (Schw.).

B. propúllans, (Schw).

Sphæria propullans, Schw. Syn. N. Am. 1447.
Botryosphæria propullans, Cke. Syn. 1434.

Black, seated in the inner bark, and erumpent through the epidermis, by which it is closely surrounded, subrotund, base contracted, subturbinate, seated on a scanty stroma. Perithecia rather large, prominent, subdepressed, crowned with a papilliform, central ostiolum. Tufts of perithecia numerous, appearing punctate-rugose above.

On much decayed stems of *Celastrus*, Bethlehem, Pa. (Schw.).

The diagnosis (taken from Schw. Syn.) is obscure, and the species doubtful. There is no spec. in Herb. Schw.

MYRMÈCIUM, Sacc.

Syll. I, p 600.

Perithecia botryose-aggregated, erumpent-superficial, glabrous, white inside, seated on a depressed-pulvinate, rather soft stroma also whitish within. Asci cylindrical, paraphysate, 8-spored. Sporidia ovoid-oblong, uniseptate, hyaline.

M. endoleùcum, Sacc. Mich. II, p. 138.

Acervuli gregarious, subsuperficial, depressed-pulvinate, $\frac{1}{4}$–$\frac{1}{3}$ mm. diam., very black outside, but the nuclei of the perithecia and the inside of the stroma white. Perithecia slightly sunk in the stroma, subglobose, obtuse, tuberculose-prominent, texture parenchymatic, dark brown. Asci cylindrical, very short-stipitate, rounded at the apex, 90 x 5–6 μ, paraphyses grumose-deliquescent. Sporidia uniseriate, ovoid-oblong, 10–13 x 5–6 μ, 2-nucleate, uniseptate, not constricted, hyaline.

On bark of limbs, Corpus Christi, Texas (Ravenel).

M. dichænoìdes, (B. & C.)

Sphæria dichænoides, B. & C. Grev. IV, p. 98.
Myrmæcium dichænoides, Sacc. Syll. 2387, Cke. Syn. 1455 (subgenus).

"Spots bursting out transversely, but generally orbicular, very rough with the conical, rugose, pulverulent ostiola. Asci clavate. Sporidia hyaline, in one or sometimes two rows, oblong, sometimes narrower below, obtuse, 25 μ long. Looks at first sight like a *Dichæna*."

On oak, Alabama (Beaumont).

M. subáquilum, (B. & C.)

Melogramma subaquilum, B. & C. Grev. IV, p. 98.
Myrmæcium subaquilum, (subgenus) Cke. Syn. 1456.

"Perithecia few, bursting through the bark longitudinally. Asci clavate. Sporidia narrow, uniseptate, constricted more or less at the septum, sometimes almost biconical, 20–22 μ long."

On *Acer striatum*, Massachusetts.

ENDÓTHIA, Fr.

Summa Veg. Scand. p. 385.

Stroma at first covered, then erumpent, tubercular, bright-colored inside and out, subcoriaceous. Perithecia buried in the stroma, mostly in a single layer, with long necks. Asci oblong-fusoid or clavate. 8-spored, without paraphyses. Sporidia broad-fusoid or elliptical, 2-celled, hyaline (Winter in Die Pilze).

E. gyròsa, (Schw.) **(Plate 36)**

Sphæria radicalis, Schw. Syn. N. Am. 1269, Fr. Elench. II, p. 73.
Sphæria gyrosa, Schw. Syn. Car. No. 24.
Valsa radicalis, Ces. & De Not. Schema, p. 33.
Melogramma gyrosum, Tul. Sel. Carp. II, p. 87.
Endothia gyrosa, Fckl. Symb. p. 226.
Endothia radicalis, Fr Summa, p. 385.
Exsicc. Rab. Herb. Mycol. 254.—Thum. M. U. 769.—Ell. & Evrht. N. A. F. 2d Ser. 1956.

Stromata gregarious or subconfluent, tubercular-pulvinate, round, elliptical or subelongated, finally flattened above, erumpent, deep orange-yellow outside, paler yellow inside, subpulverulent, 1–2 mm. diam. Perithecia buried in the stroma, submonostichous, ⅓ mm. or less in diam., contracted into slender necks rising through the substance of the stroma, with their obtuse, black, perforated ostiola slightly erumpent. Asci attenuated-substipitate, 8-spored, 25–28 x 8–9 μ, subclavate, aparaphysate. Sporidia biseriate, oblong-fusoid, or narrow-elliptical, septate in the middle, but not constricted, hyaline, 7–9 x 2½–3 μ.

On bark of decaying oak, Florida (Calkins), on *Liquidambar*, Newfield, N. J.

Perithecia 10–30 in a stroma. Herb. Schw. 1269, is the spermogonial stage, spermatia 2 x ½ μ, hyaline.

E. Párryi, (Farlow).

Dothidea Parryi, Farlow MSS.
Endothia Parryi, Cke. Grev. XIII, p. 102.

Stromata pulvinate-erumpent. Perithecia numerous, concentrically aggregated, brown, surrounded by the orange-red epidermis. Asci linear-clavate, 76–110 x 16–20 μ, aparaphysate, 8-spored. Sporidia irregularly biseriate, hyaline, 20–26 x 5–7 μ, uniseptate, elliptical, subacute, slightly constricted.

On *Agave Shawii*, S. W. United States.

MELOGRÁMMA, Fr.

Summa Veg. Sc. p. 386

Stroma suborbicular, depressed-hemispherical, pulvinate or conical, erumpent and superficial, mostly containing numerous perithecia. Asci subcylindrical, 8-spored, paraphysate. Sporidia fusoid or cylindrical, with several septa, hyaline or colored.

M. Bulliárdi, Tul. Sel. Carp. II, p. 81. **(Plate 36, M. vagans.)**

Variolaria Melogramma, Bull. Champ. tom. I, p. 182, tab. 492, fig. 1.
Sphæria ocellata, Pers. Disp. p. 2.
Sphæria Melogramma, Pers. Syn. p. 13.
Melogramma fusisporum, and *campylosporum*, Fr. Summa, p. 386.
Melogramma vagans, De Not. Micr. Dec. IX, No. 2.
Hypocrea Ravenelii, Berk. in Rav. F. Car. V, 51.
Diatrype lateritia, Ell. Bull. Torr. Bot. Club, IX, p. 19.
Hypoxylon myriangioides, B. & C. in Ell. N. A. F. 474.

Exsicc. Fckl. F. Rh. 1033.—Rab. F. E. 937, 1923.—Thum. M. U. 366.—Roum. F. G. 173, 1174.—Ell. N. A. F. 474.—Rav. Fungi Car. V, No. 51.

Stromata gregarious, scattered or subseriate and sometimes confluent, erumpent-superficial, depressed-pulvinate, subhemispherical or obtusely conical, 1–3 mm. diam., dull brick-red at first, at length clay color or nearly black outside, brick-red or nearly blood-red inside, of subcarnose texture, more or less roughened and pitted from the collapsing of the upper part of the perithecia which are arranged in a single peripherie layer, 10–40 in a stroma, ovate, small (less than ¼ mm. diam.), at length more or less emergent and collapsing above. Asci clavate, sessile, paraphysate, 8-spored. Sporidia crowded-biseriate, fusoid, 3-septate, slightly curved, light brown, end cells paler, 38–50 x 3½–5 μ.

On dead limbs of *Carpinus Americana*, Carolina, Pennsylvania, New York (Brown), and Canada (Dearness).

In some of the Canada spece. the stroma is reduced to a mere crust on which stand the cup-shaped perithecia, exactly as in *Fracchiæa callista.* The red color of the stroma both inside and out finally disappears.

M. Méliæ, Curtis, in Curtis Cat. p. 143 and Grev. XIII, p. 103.

(Not *Sphæria Meliæ*, Schw., which is *Botryosphæria*).

Acervuli erumpent, black, subdiscoid or oblong. Perithecia prominent, at length partially free, rugose, flattened above. Asci clavate, large, 8-spored. Sporidia sublanceolate, 3-septate, not constricted, rarely 5-septate, 38–42 x 10 μ, brown.

On *Melia*, Lower and Middle Carolina (Curtis).

M. Hibísci, (Schw.)

Sphæria Hibisci, Schw. Syn. N. Am. 1444.
Botryosphæria, Hibisci, Cke. Syn. 1422.

Stromata scattered, small (1–2 mm.), suborbicular or by confluence elongated and subflexuous, surrounded by the substellate-cleft epidermis. Perithecia 3–6 in a stroma, with the apex prominent, white inside at first. Asci cylindrical, 75–80 x 8–9 μ, paraphysate, 8-spored. Sporidia uniseriate, elliptical, or ovate-elliptical, yellowish-hyaline, 3-septate and slightly constricted at the middle septum, 18–20 x 6–7 μ.

On *Hibiscus Syriacus* (cult.), Newfield, N. J.

The spece. in Herb. Schw. show only the stylosporous stage (*Diplodia* (*Dothiorella*) *hibiscina*, C. & E. Grev. VII, p. 5).

M. ferrugíneum, (Pers.)

Sphæria ferruginea, Pers. Obs. I, p. 66.
Diatrype ferruginea, Fr. Summa Veg. Sc. p. 385.
Melogramma ferrugineum, Ces. & De Not. Schema Sferíac. p. 30.
Silia ferruginea, Karst. Mycol. Fenn. II, p. 159.
Exsicc. Fckl. F. Rh. 1038.—Sydow, M. March. 1883.—Rehm Asc. 978.

Stromata mostly numerous, irregularly scattered, often confluent, pulvinate, round, elongated or sometimes 4-angled, 1–3 mm. broad, surrounded by the erect or recurved margin of the ruptured epidermis, black outside, rusty-yellow-pulverulent within. Perithecia generally numerous, crowded in a single layer, or occasionally one lying above another. globose, ovate, or irregular in shape from mutual pressure, with long, cylindrical necks terminating in the thickly crowded, more or less elongated, sometimes curved, rough, wrinkled, shining-black, erumpent ostiola. Asci narrow clavate-cylindrical, sessile, 8-spored, 80–110 x 15–17 μ. Sporidia filiform, slightly curved, with several oil globules, faintly 6–7-septate, hyaline, 60–85 x 3–4 μ.

On chestnut and oak wood, Carolina (Schw. & Ravenel), on *Populus tremuloides*, Pennsylvania (Michener).

We have no American specimens of this, and take the diagnosis from Winter's Pilze.

M. platýroum, B. & C. Grev. XIII, p. 103.

Perithecia botryoid-aggregated, erumpent, subeffused, rarely free, sprinkled below with a grayish powder. Asci clavate, 8-spored. Sporidia fusoid, triseptate, hyaline, 20–22 x 4 μ.

On bare wood, Pennsylvania.

M. Spragüei, B. & C. Grev. IV, p. 99, and XIII, p. 103.

Thyridium Spraguei, Sacc Syll 3994.

"Undulate, pulverulent, looking like the work of some burrowing larva. Perithecia entirely concealed. Asci linear. Sporidia short-clavate, with three horizontal and a few vertical septa."

Sec. Cke. (Grev. l. c.) this has no affinity with *Thyridium*. It occurs on small branches (of *Pinus strobus*), Massachusetts.

VALSÀRIA, Ces. & De Not.

Schema Sferiac. p. 31.

Stroma variable, valsiform or effused, usually covered by the epidermis. Asci mostly 8- (sometimes 4-) spored, paraphysate. Sporidia oblong or elliptical, uniseptate, brown.

V. insitìva, Ces. & De Not.

Diatrype cincta, (Curr.) B. & Br. in Ell. N. A. F. 170.
Valsa clethræcola, C. & E. Grev. V, p. 92.
Diatrype Æthiops, C. & E. Grev. VI, p. 10.
Exsicc. Fckl. F. Rh. 1821, 1822.—Rab. F. E. 1128, 2111, 2112.—Rehm. Asc. 170.
Ell. N. A. F. 170.—Sacc. M. Ven. 1185.

Stroma variable, 1 mm.–1⅓ cm. diam., valsiform, diatrypoid, pulvinate-tubercular, &c., dull black outside and subcrustaceous, light tobacco-brown, and softer (subcarnose) within, blackening the wood beneath, and the bark around it. Perithecia deeply sunk in the stroma, scarcely over ⅓ mm. diam., ovate or flask-shaped, black, coriaceous, numerous, contracted into rather long, slender necks, which rise through the stroma, and barely pierce its surface in subcircinate groups, but are not exserted, papilliform or short-conical, and then substellate-cleft, finally umbilicate. Asci cylindrical, short-stipitate, paraphysate. Sporidia uniseriate, mostly lying end to end, oblong-elliptical, uniseptate and constricted at the septum, dark brown, rounded at the ends, 12–20 x 7–10 μ.

On dead limbs of *Morus*, *Clethra*, *Quercus*, *Carya*, *Pyrus*, *Melia*, *Berberis*, and probably other deciduous trees, from Canada to Louisiana.

Whether this is the *Sphæria insitiva*, Tode, does not seem to be definitely known. The *Diatrype cincta*, in Ell. N. A. F. 170, does not differ appreciably from Rehm Asc. 170, which Dr. Winter quotes as *V. insitiva*. *V. clethræcola*, C. & E., varies with the stroma valsoid and comparatively thin and small, or subtubercular and much larger. The sporidia in all the specc. we have examined are mostly about 15 x 8 μ. South American specc. from Spegazini, on a woody species of *Solanum*, have the sporidia mostly less than 15 μ long, while most of the U. S. specc. have sporidia 15–18, and occasionally even 30 μ long.

Sydow, *M. Marchica*, 2145, is *Anthostoma*; sporidia continuous, brown. Roum. F. Gall. 5230, is only the spermogonial stage of some *Valsa*, and Thüm. M. U. 466, is a *Fenestella*; sporidia 3–4-septate and submuriform.

V. Farlowiàna, Sacc. Mich. II, p. 137. **(Plate 36)**

Acervuli (stromata) depressed-pulvinate, gregarious, 1–1½ mm. diam., black, at first subcutaneous, but when the bark falls away, denuded and seated on the blackened wood. Perithecia numerous, globulose. Ostiola erumpent in a circular, umbilicate disk. Asci cylindrical, evanescent, 8-spored. Sporidia obliquely uniseriate, oblong-ellipsoid, rounded at the ends, slightly constricted, uniseptate, dark brown, 15–18 x 7–8 μ.

On branches of *Berberis vulgaris*, Cambridge, Mass. (Farlow).

This does not seem to us specifically distinct from *V. insitiva*, but is separated from that species by Saccardo, on account of the carbonaceous stroma and umbilicate disk. Specc. from Farlow do not agree with the description published, having the stroma much larger (3 mm.–1½ cm. diam.), and where the bark has fallen away, with the margin abrupt. The outer layer in these denuded stromata is subcarbonaceous, which results apparently from exposure, the inside being of the usual color and consistence. The same subcarbonized surface has been noted in spece. found at Newfield, N. J., on decorticated limbs of apple trees.

V. exásperans, (Gerard).

Diatrype exasperans, Ger. in Bull. Torr. Bot. Club, V, p. 40, (Sept. 1874).
Diatrype obesa, B. & C. in Herb.
Diatrype quadrata, Schw. (fide Berk. in Grev. IV, p. 94.
Sphæria aspera, Fr. in Herb. Schw. 1440.
Exsicc Thum. M. U. 467, 2261.—Rehm Asc. 325.—Rav. F. Am. 665.—Rav. Car. IV, No. 47.—Ell. N. A. F. 493.

Stromata scattered or seriately confluent, orbicular, oblong or variously shaped, flattened below and enclosed in a distinct conceptacle, open above, sides rising abruptly 1½–2 mm. high, subcarnose and dirty blackish- or yellowish-brown inside, the tawny-yellow apex erumpent and closely embraced by the epidermis, seated on the surface of the inner bark. Perithecia polystichous, subglobose, about ⅓ mm. diam. or a little over, coriaceous, black, necks cylindrical, rising through the substance of the stroma, with their coarse, black, rough, perforated, conic-hemispherical ostiola (finally subcollapsing) tardily erumpent, but not exserted. Asci cylindrical, short-stipitate, paraphysate, p. sp. 80–100 x 10 μ. Sporidia uniseriate, oblong-elliptical, uniseptate and constricted, obtusely rounded at the ends, deep reddish-brown, 14–16 x 9–10 μ.

On bark of dead *Acer*, *Fagus*, *Carpinus*, *Prunus*, *Liquidambar*, *Nyssa*, *Ulmus*, &c., common, Canada to Florida.

On bark with a thick epidermis, like cherry and scarlet oak, the conceptacles adhere to the epidermis and come off with it. This does not agree with the description of *Sphæria quadrata*, Schw. or with the specc. of that species in Herb. Schw. The *Sphæria quadrata* in Herb. Schw. is a *Hypoxylon*, as indicated in the diagnosis in Schw. Syn. N. Am. 1223. This species may be recognized by the tawny-yellow color of the erumpent apex of the stroma.

V. Akèbiæ, E. & E. (in Herb.)

Perithecia globose, small, about 200 μ diam., 3–6 together, subcircinate in the scarcely altered substance of the inner bark, contracted above into very short necks, which raise the bark into little excrescence-like pustules through the pulverulent surface of which the minute, black ostiola are barely visible. Asci cylindrical, subsessile, paraphysate, 8-spored, 100–110 x 8–10 μ. Sporidia uniseriate, oblong-elliptical, straight or slightly curved, brown, uniseptate, obtuse, scarcely constricted, 12–18 x 5–6 μ.

On dead vines of *Akebia quinata* (cult.), Newfield, N. J.

V. pústulans, (E. & E.)

Diatrype pustulans, E. & E. Journ. Mycol IV, p. 80.
Valsaria pustulans, Sacc. Syll. Add. II, p. 758.

Stroma flattened, formed of the scarcely altered substance of the matrix, covered by the cuticle which is blackened and raised in a pustuliform manner, and finally pierced by the slightly projecting, papilliform ostiola. The separate stromata are ¼–½ cm. across, but are more or less confluent with each other for 2–4 cm. or more in extent; the surface of the culm being continuously blackened, and the entire area bounded by a black circumscribing line. Perithecia membranaceo-coriaceous, subglobose, or a little flattened, of medium size, 8–12 in a stroma. Asci slender, 75–85 x 6–7 μ, subsessile, with distinct, filiform paraphyses. Sporidia uniseriate, oblong, uniseptate and constricted, slightly narrowed at the ends, straight, brown, 10–12 x 3 μ. This is preceded or accompanied by a *Coniothyrium* with numerous small, immersed perithecia, and small (2 μ), brown sporules, which ooze out and stain the surface of the matrix with an olivaceous, pulverulent coat.

On dead stems of *Arundinaria*, St. Martinsville, La. (Langlois).

V. apatòsa, (C. & E.)

Valsa apatosa, C. & E. Grev. VI, p. 12, Pl. 95, fig. 1
Valsaria apatosa, Sacc. Syll. 2822.

Stroma valsoid, slightly pustuliform, 1–2 mm. diam., covered by the epidermis. Perithecia 4–6 in a stroma, circinate, small, mostly less than ⅓ mm. diam. Ostiola confluent, obtuse, perforated, piercing the epidermis in the form of a small, black disk, but scarcely rising above it. Asci oblong-cylindrical, 100–115 x 20–25 μ, paraphysate, subsessile. Sporidia obliquely uniseriate or biseriate, oblong-elliptical, uniseptate, scarcely constricted, 22–35 x 10–12 μ (35 x 5 μ Cke.).

On dead limbs of *Nyssa*, Newfield, N. J.

V. angulàris, (Pk.)

Diatrype angularis, Pk. Bot. Gaz. V, p. 36.
Valsaria angularis, Sacc. Syll. 2820.

Stromata scattered, ovate-globose, 1–2 mm. diam., the base sunk in the bark, but not penetrating to the wood, the upper part erumpent and closely embraced by the ruptured epidermis, black outside, cinereous with a reddish tint within. Perithecia 1–7 in a stroma, ovate, about ½ mm. high and ⅓ mm. broad, subcircinate near the bottom of the stroma, brownish, attenuated into rather long, stout necks (which show as light-colored points on a cross section), terminating in stout, angular, black, subpyramidal ostiola mostly converging so that the erumpent part of the stroma appears deeply sulcate. Asci cylindrical, short-stipitate, 200–230 x 18–20 μ, with stout (3–4 μ thick), jointed paraphyses. Sporidia 4–8 in an ascus, elliptical or oblong-elliptical, brown, uniseptate, scarcely constricted, 30–52 x 12–15 μ.

On bark of dead basswood (*Tilia Americana*), Vermont (Pringle).

A very distinct and curious species. Described from specc. sent by Mr. Peck.

V. moroìdes, (C. & P.)

Diatrype moroides, C. & P. 29th Rep. N. Y. State Mus. p. 63.
Valsaria moroides, Sacc. Syll. 2839.
Exsicc. Ell. N. A. F. 90.

Stroma pulvinate, buried in the scarcely altered substance of the bark, with the subtubercular apex erumpent and closely embraced by the epidermis, suborbicular or elliptical, 2–4 mm. long, a little paler than the surrounding portion of the bark, but the outer layer slightly blackened so as to show on a cross section near the base a faint circumscribing line, and when the bark is peeled off, the inner surface shows faint, whitish spots indicating the position of the stromata buried in the bark. Perithecia 6–15 crowded in the bottom of the stroma, small ($\frac{1}{3}$ mm. or less), black, coriaceous, their rather short necks terminating in hemispherical, black ostiola scattered and erumpent through the strongly convex disk. Asci cylindrical, short-stipitate, 75–80 x 4 μ, paraphysate. Sporidia uniseriate, oblong, uniseptate, brown, not constricted, straight or slightly curved, 12–15 x 3–4 μ.

On bark of dead alders, New York and New Jersey.

V. purpùrea, Pk. Bull. Torr. Bot. Club, XI, p. 28.

Pustules prominent, erumpent, covered with a purplish tomentum. Perithecia 6–20 in a pustule, crowded, subglobose, black. Ostiola piercing the tomentum, rostrate, cylindrical or elongated-conical, rugged, sometimes curved or flexuous, black. Asci cylindrical, paraphysate, 75–105 x 10 μ. Sporidia uniseriate, colored, oblong-elliptical. uniseptate, 15–18 x 7½–8 μ.

On bark of dead *Fraxinus*, Canada (Macoun).

V. Péckii, (Howe).

Valsa Peckii, Howe, 27th Rep. N. Y. State Mus. p. 109.
Valsaria Peckii, Sacc. Syll. 2825, Cke. Syn. 2076.
Exsicc. Ell. N. A. F. 171.

Stroma oblong or elliptical, 1–2 mm. long, covered by the bark and partly sunk in the wood, gray inside. Perithecia 5–15 crowded in the bottom of the stroma, rather less than $\frac{1}{3}$ mm. diam., ovate-globose, coriaceous, thick-walled. Ostiola erumpent, connate, stout-cylindrical, rough, mostly sparingly clothed with a thin, light brown, farinose tomentum, conic-hemispherical or obtuse at the apex, perforated, rising through the slightly elevated bark in an oblong, dense fascicle. Asci narrow-cylindrical, short-stipitate, 100–110 x 5–6 μ, paraphysate. Sporidia uniseriate, oblong-cylindrical, straight or slightly curved, uniseptate, brown, 12–15 x 5 μ.

On dead stems of *Kalmia latifolia* and *Vaccinium corymbosum*, New York and New Jersey.

V. Nièsslii, (Winter).

Phæosperma Niesslii, Winter, in Hedw. 1874, p. 131.
Valsaria Niesslii, Winter, Sacc. Syll. 2836, Cke. Syn. 2086.

Stroma immersed in the inner bark, thick, hemispherical from an orbicular or oval base, brown inside, with an elliptical or suborbicular disk, at first brown, at length black (from the discharged sporidia), erumpent, rugulose. Perithecia 10–40 in a stroma, sunk to the bottom of the stroma, inordinate, densely crowded, subglobose or angular from compression, black, 600–700 μ diam. Ostiola very long, thick, rugose-tuberculose, thickened at the apex, very black, piercing the disk and exserted 1 mm. above it. Asci cylindrical, very long-stipitate, 8-spored, p. sp. 70–90 x 6–7 μ. Paraphyses filiform, very slender, guttulate. Sporidia uniseriate, oblong, rounded at the ends, straight or curved, septate in the middle, not constricted at the septum, pale brownish-black, 8–15 x 4–5 μ.

On dead white birch, New York State (Peck).

We have seen no specimens and take the above diagnosis from Sacc. Syll.

V. Robíniæ, (Schw.)

Sphæria Robiniæ, Schw. Syn. Car. 61, Fr. S. M. II, p. 352.

Forming tubercles an inch long and ¼ of an inch across, sunk to the wood, the upper part narrowed, erumpent, flattish, opake. Perithecia sunk to the bottom of the stroma, crowded, ovate-oblong, rather small, shining, the long necks rising to the surface and terminating in smooth, umbonate, distant ostiola either short or 1–2 lines long. Sporidia (sec. Cke. Grev. XIII, p. 37) elliptical, uniseptate, brown, 20 μ long.

On dead limbs of *Robinia pseudacacia*, Carolina (Schw.).

V. melastròma, (Fr.)

Sphæria melastroma, Fr. S. M. II, p 399, Schw. Syn. N. Am. 1358.
Exsicc. Fr. Scl. Suec. No. 223.

Has the habit of *Diatrypella verrucæformis*, but lesss prominent, and conceptacle entirely cortical. Perithecia about 7, suberect, crowded, deeply immersed, vertically collapsing. Disk erumpent through the angularly cleft epidermis, formed from the inner bark, but the surface smooth, opake, blackened. The ostiola pierce the disk but are not exserted.

On bark of *Ulmus*, Carolina (Schw.). Sporidia (sec. Cke.) elliptical, brown, uniseptate, 22 x 14 μ.

V. salicina, E. & E. Proc. Acad. Nat. Sci. Phil. Pa. July, 1890. p. 236.

Stroma subovate, 2–2½ mm. diam., buried in the bark, the upper part light-colored within, and projecting so as to form a brownish-black, subhemispherical tubercle, 1–2 mm. across, and less than 1 mm. high, minutely papillose above from the slightly projecting ostiola. The upper, projecting part of the stroma, is of a light horn-color inside. Perithecia 10–20 irregularly crowded in the bottom of the stroma, ovate-globose, with thick, coriaceous walls, contracted above into slender necks 1 mm. or more long, terminating above in the papilliform ostiola. Asci slender, 75–80 x 5–6 μ (p. sp.). Paraphyses abundant, longer than the asci. Sporidia obliquely uniseriate, oblong, crowded, cylindrical, 2-nucleate, brown, uniseptate, 10–12 x 3½ μ. Allied to *V. anthostomoides*, Sacc.

On dead limbs of *Salix*, London, Canada (Dearness).

V. majúscula, Cke. & Hark. Grev. XIII, p. 17.

Covered. Pustules scattered, scarcely prominent, composed of 6–10 perithecia, at length perforating the cuticle. Ostiola short and obtuse. Asci ample. Sporidia elliptical, constricted in the middle, ends rounded, uniseptate, brown, 50 x 25 μ.

On branches of *Salix*, California (Harkness).

Sometimes the large sporidia are extruded and form blackened spots around the ostiola as in *Massaria*, but this is not a constant feature. The absence of any hyaline investment of the sporidia also confirms this as a *Valsaria* rather than a *Massaria*.

V. Céltidis, (Cke.)

Valsa Celtidis, Cke. Grev. V, p 55. tab 81, fig. 3.
Valsaria Celtidis, Sacc. Syll. 2808, Cke. Syn. 2070.

Perithecia irregularly circinate, subimmersed. Ostiola black, emergent in a brown, pruinose disk. Asci cylindrical, attenuated below, 8-spored. Sporidia uniseriate, elliptical, uniseptate, brown, constricted, 20 x 10 μ.

V. fulvopruinàta, (Berk.)

Sphæria fulvopruinata, Berk. Hook. Lond. Journ. Bot. IV, p. 312.
Valsaria fulvopruinata, Sacc. Syll. 2831, Cke. Syn. 2081.

Forming subangular pustules about 2 mm. broad, rather effused at base, as seen through the thin cuticle. Disk angular, tawny, pulverulent, pierced by the black, angular, punctiform ostiola. Stroma tawny like the disk. Perithecia globose. Asci linear. Sporidia elliptical, uniseptate, with a single globose nucleus in each cell, ends subacute, becoming dark brown, 15–18 μ long.

On branches of *Platanus*, Ohio (Lea), Carolina (Curtis), Connecticut (Wright).

V. Notarísii, (Mont.)

Valsa Notarisii, Mont. Syll. No. 754.
Valsaria Notarisii, Sacc Syll 2810, Cke. Syn. 2071.
Sphæria Periplocæ, De Not. Micr. VIII.

Pustulate, perforating the epidermis, at length bare. Perithecia ovoid, circinate, black, buried in a cinereous-black, cortical stroma, the ostiola connate and erumpent in a flat, black, rugulose disk. Asci cylindrical, 100 x 10 μ. Sporidia uniseriate, uniseptate, oblong, brown, constricted at the septum, 20 x 10 μ.

On *Robinia* and *Gleditschia*, Carolina (Curtis).

V. vitícola, (Schw.)

Sphæria viticola, Schw. Syn. Car. No. 64.
Diatrype viticola, Berk. & Curt. Grev. IV, p. 96.
Diatrype (*Valsaria*) *viticola*, Cke. Grev. XIV, p. 44.
Valsaria viticola, Sacc. Syll. 2712.

Pustules small, scattered or in parallel series, grayish-black, rugose. Perithecia few, 2–3, black, in a shining-brown stroma, or scattered singly in the thick, blackened bark. Sporidia oblong, uniseptate, constricted slightly at the septum, 15 μ long, brown.

On dead branches of *Vitis rotundifolia*, Carolina (Schw.).

We have seen no specimens. Perhaps a depauperate form of *V. insitiva*.

V. gregàlis, (Schw.)

Sphæria gregalis, Schw in Fr. Elench. II, p. 68.
Valsaria? gregalis, Sacc. Syll. 2847.
Melogramma (*Valsaria*) *gregale*, Cke. Syn. 1488.

Irregularly effused. Perithecia very delicate, globose, subacute, pulverulent, nearly superficial and connate, papillate, brown. Stroma very thin. Sporidia oblong, uniseptate, 12–13 μ long.

On rotten wood, Carolina (Schw.).

The stroma is very thin and indeterminate. Perithecia minute, exactly globose, sometimes connate and sometimes only crowded, very fragile, brownish-black. Ostiola minute, papilliform.

V. actídia, Berk. & Rav. in Herb. Berk. Grev. XIV, p. 14.

Stromata innate-emergent. Perithecia large, few (4–6). Ostiola cuneiform, stellately arranged. Asci ample, (4-spored)? Sporidia elliptical, uniseptate, constricted, brown, 40–42 x 16–18 μ.

On branches of *Ostrya* and *Carpinus*, South Carolina (Ravenel).

V. gemmàta, (B. & Rav.)

Hypoxylon gemmatum, B & Rav Grev IV, p. 50.
Hypoxylon Walterianum, Rav. Fung. Car. IV, No. 35.
Melogramma (*Valsaria*) *gemmatum*, Cke. Grev. XIII, p. 103.

Small, scarcely 2 mm. broad, pulvinate but flat, rust-colored, sometimes transverse, studded with the black, prominent ostiola, which are perforated in the center. Sporidia uniseptate, brown, 10–12 x 6 μ.

On branches of *Liquidambar*, South Carolina (Ravenel).

V. Bignòniæ, (Schw.)

Sphæria Bignoniæ, Schw. Syn. N. Am. 1310
Valsaria Bignoniæ, Cke. Grev. XIII, p. 39, Sacc. Syli. 4194.

Very variable. When young, pustuliform, black, small, forming, like some of the other species, a conceptacle subimmersed in the fibrous bark. Ostiola umbilicate, prominent, cylindrical. Finally the stromata become confluent for one or two inches in length, bursting out through longitudinal cracks in the bark, the apex of the conceptacles truncate above, and on this truncate surface are seen the short, thick, irregular ostiola. Perithecia of medium size, immersed in a cinereous flesh-colored stroma.

On dead limbs of *Bignonia*, Carolina (Schw.).

Sporidia (fide Cke. Grev. l. c.) olive-brown, uniseptate, 15 x 8 μ.

V. Phoradéndri, (B. & C.)

Melogramma Phoradendri, B. & C. in Curt. Cat. p. 143 and Grev. XIII, p. 103.
Valsaria Phoradendri, Sacc. Syll. 6633.

Perithecia gregarious, covered, at length erumpent through the fissured bark, black, obtuse. Asci subcylindrical. Sporidia subelliptical, uniseptate, slightly constricted, brown, 28–30 x 10–12 μ.

On bark of *Phoradendron flavescens*, seaboard of Carolina (Curtis).

V. collematoìdes, (B. & Rav.)

Botryosphæria collematoides, B. & Rav. Grev. XIII, p. 102.
Valsaria collematoides, Berl. & Vogl. in Sacc. Syll. Add I, p. 131.

Stroma effused, thin, black. Perithecia small, oval, crowded, at length flattened at the apex, opake, at length botryoid-aggregated, subconfluent. Asci cylindrical, 8-spored. Sporidia elliptical, uniseptate, scarcely constricted, brown, 15 x 7 μ.

On bark, United States (Cooke, l. c.).

Has the habit of *Botryosphæria*, with the fruit of *Valsaria*. In Grev. XIII, p. 108, this is placed in a section having hyaline sporidia. Not having seen a spec., we cannot say which is right.

V. grandínea, Berk. Grev. XIII, p. 103.

Melogramma insidens, Berk. in Grev. IV, p. 99, (not Schw., and not *Diatrype grandinea*, B. & Rav. Grev. IV, p. 95).
Valsaria grandinea, Sacc. Syll. 6634.

Stroma effused, subrotund, brownish-black. Perithecia emergent, crowded, ovate, depressed or subumbilicate at the apex. Asci clavate. Sporidia elliptical, brown, at length uniseptate, not constricted, 8 x 4 μ.

On bark of *Fraxinus*, United States.

V. Beaumóntii, (B. & C.)

Hypoxylon Beaumontii, B. & C. Grev. IV, p. 93.
Valsaria ? Beaumontii, Sacc. Syll. 2848.

Perithecia globose, connate, rather small, at first slightly brown, then black, smooth, with a distinct, papilliform ostiolum. Asci linear. Sporidia uniseriate, oblong-elliptical, uniseptate, 10 μ long.

Found in Alabama by Beaumont.

Species imperfectly known.

V. nigrofácta, C. & E. (sub *Valsa*) Grev. VI, p. 12, tab. 95, fig. 4.

Perithecia few, large, covered by the blackened, shining epidermis. Ostiola convergent. Asci cylindrical. Sporidia uniseriate, narrow-elliptical, uniseptate, strongly constricted, brown, 25 x 9 μ, cells almost globose.

On dead limbs of *Sassafras*, Newfield, N. J.

We have never seen specc. of this in fruit.

V. Diospýri, (Schw.)

Sphæria Diospyri, Schw. Syn. Car. No. 57.
Valsaria. Diospyri, Sacc. Syll. 2813.

Pustules prominent, variable, 2 lines broad, covered by the epidermis, which is finally blackened by the extruded sporidia, at length erumpent and bare. Perithecia minute, globose, monostichous on the surface of an innate tubercle, and closely covered by the epidermis. Necks short. Ostiola minute, obsoletely-prominent, at length perforated. Stroma brown.

On dead limbs of *Diospyros*, Carolina (Schw.).

Sporidia (fide De Not.) ellipsoid, two-celled, brown, slightly constricted. No measurements of sporidia given.

V. Gledítschiæ, (Schw.)

Sphæria Gleditschiæ, Schw. Syn. N. Am. 1435, id. Syn. Car. 31.
Melogramma Gleditschiæ, Berk. in Grev. IV, p. 98, Cke. Syn. 1184.
Botryosphæria Gleditschiæ, Sacc. Syll. 1792.

Variable, black, with a brown stroma. Perithecia conical, connate, stuffed, astomous.

On bark of *Gleditschia*, Carolina (Schw.), Pennsylvania (Michener).

This is placed, by Cooke in his Synopsis, in the *Valsaria* Section of *Melogramma* (sporidia uniseptate, brown). The same thing apparently is referred to by Berk. (Grev. IV, p. 47), as *Cucurbitaria Gleditschiæ*, Schw., with the note—"sporidia ovate, uniseptate."

V. nudicóllis, (B. & C.)

Hypoxylon nudicolle, B. & C. Grev. IV, p. 93.
Valsaria ? nudicollis, Sacc. Syll. 2849.

Perithecia connate, forming a continuous or slightly interrupted stratum. Ostiola papilliform, black. Perithecia crowded, covered with brownish matter. Asci linear. Sporidia uniseriate, oblong, narrow, uniseptate.

On pine wood, Carolina.

DIATRỲPE, Fr.

Summa Veg. Sc. p. 384, emend. Nitschke Pyr. Germ. p. 64.

Stroma erumpent-superficial, effused or discoid. Perithecia immersed in the stroma, necks scarcely converging. Asci 8-spored. Sporidia allantoid, small, mostly yellowish.

D. stígma, (Hoff.)

Sphæria stigma, Hoff. Veg. Crypt. I, p. 7, tab. II, fig. 2.
Sphæria decorticans, Sow. Eng. Fungi, II, tab. 137; III, tab. 371, fig. 3.
Sphæria undulata, Fr. S. M. II, p. 350.
Sphæria decorticata, DC. Fl. Fr. II, p. 289.
Diatrype Dearnessii, E. & E., Ell. & Evrht. N. A. F. 2d Ser. 2526.
Diatrype tennissima, Cke. in Rav. F. Am.
Eutypa micropuncta, Cke. Grev. VI, p. 144.
Sphæria concolor, Schw. and *Sphæria subcutanea*, Wahl. in Herb. Schw.
Diatrype undulata and *stigma*, Fr. Summa Veg. Sc. p. 385.
Stictosphæria Hoffmanni, Tul. Sel. Carp. II, p. 49, tab. VI, figs. 1-11.
Diatrype stigma, De Not. Sfer. p. 26, tab. 25.
Exsicc. Fckl. F. Rh. 1043, 1044, 1047, 2263.—Rab. F. E. 820, 2953.—Rehm Asc. 220.—Cke. F. Brit. Ser. II, 417.—Sydow, M. March. 191, 448. &c., &c.—Ell. N. A. F. 491. Rav. F. Am. 357 and 359.

Stroma widely effused, continuous or interrupted, often surrounding the limb, sometimes extending for several inches in length, at first covered by the epidermis, finally exposed, brownish or dusty-white, becoming darker or nearly black, whitish inside, definitely limited but irregular in outline, ½–1 mm. thick, sometimes undulate and of unequal thickness, the thinner parts then often sterile. Perithecia monostichous, evenly distributed, ovate, small, with short necks and punctiform, discoid, depressed-hemispherical or conical, entire or 4-cleft ostiola. Asci clavate-oblong, 8-spored, p. sp. 30–50 x 4–8 μ. Sporidia subbiseriate, allantoid, brownish, 6–9 x 1–1½ μ; 6–12 x 1½–2 μ, Sacc.; 6–12 (mostly 8) x 1—3 μ, Winter.

On dead limbs of various deciduous trees, common.

We do not find the sporidia in any American specc. over 9 μ long, mostly not over 7 μ. Specc. in Cooke's Fungi Brit. have sporidia 8–11 x $1\frac{1}{2}$–2 μ; in Rehm Asc., 6–8 x $1\frac{1}{2}$–2 μ.

Diatrype Dearnessii, E. & E., is almost exactly the form represented in Rehm Asc. 220 μ, with conical, 4-sulcate ostiola; *D. tenuissima*, Cke., is a thin form, but cannot be specifically distinct. The ostiola vary from flat-discoid, orbicular, as in Rehm Asc. 220 *b*, to conical and sulcate, R. Asc. 220 *a*; between these two extremes are various gradations. The specc. in Ell. N. A. F. 491, which are the same as in Rav. F. Am. 359, were issued by Dr. Winter (F. E. 2954) as *D. platystoma*, (Schw.), but that is a coarser species with larger perithecia and ostiola, and a thicker (1 mm.) stroma. In all the American specc. of *D. stigma*, the stroma is mostly about ½ mm. thick.

D. bullàta, (Hoff.)

Sphæria bullata, Hoff. Veg. Crypt. I, p. 5, tab. II, fig. 3.
Sphæria depressa, Bolt. Fungi Halif. III, tab. 122, fig. 1.
Sphæria placenta, Tode Fungi Meckl. II, p. 26.
Diatrype bullata, Fr. Summa, p. 385.
Exsicc. Fckl. F. Rh. 1042.—Rab. F. E. 536.—Rehm Asc. 631.—Linhart's F. Hung. 360. Cke. F. Brit. Ser. I, 485.—Desm. Pl. Crypt. Ed. I, p. 334, Ed. II, p. 961.

Stromata gregarious, often 2–3 confluent, flattish-pulvinate or shield-shaped, 2–5 mm. broad, orbicular or often with a sinuous, undulate outline, at first covered by the epidermis, then erumpent, with a smooth, dark brown surface, whitish within, surrounded at base by a black line. Perithecia numerous, monostichous, sunk in the stroma, ovate or angular from mutual pressure, rather small, with short necks and punctiform, entire, perforated ostiola only slightly prominent. Asci narrow-clavate, long-stipitate, p. sp. 40–55 x 4–6 μ. Sporidia subbiseriate, allantoid, yellowish-hyaline, 6–10 (mostly 5–8) x $1\frac{1}{2}$–2 μ.

On dead saplings, New York (Peck).

D. platýstoma, (Schw.) **(Plate 34)**

Sphæria platystoma, Schw. Syn. Car. 43
Diatrype discostoma, Cke. Grev. VI, p. 144.
Exsicc. Rav. Car. V, No. 55.—Rav. F. Am. 358.—Ell. N. A. F. 169.

Stroma effused or oftener in suborbicular patches 1–2 cm. across or elongated 2–4 x ½–1 cm., soon erumpent, about 1 mm. thick, margin abrupt, slaty-black or black. Perithecia monostichous, ovate-oblong, ½–¾ mm. high. closely packed. Ostiola prominent, hemispherical, entire or quadrisulcate. Asci (p. sp.) 25–30 x 4 μ, stipitate. Sporidia subbiseriate, allantoid, hyaline, 6–8 x $1\frac{1}{2}$ μ.

On dead trunks of *Ostrya Virginica*, Carolina (Ravenel), Ohio (Morgan), on dead *Hamamelis*, New York (Peck), on dead limbs of

Acer rubrum, Newfield, N. J. *D. discostoma*, in Rav. F. Am., is labeled "on *Carpinus*," but it is apparently on *Ostrya*.

Specc. of *Sphæria subaffixa*, Schw., in Herb. Schw., cannot be distinguished from this.

D. consobrìna, Mont. Syll. Crypt. No. 747.

Exsicc. Ell. & Evrht. N. A. F. 2d Ser. 2125.

Perithecia small, crowded in a narrow, linear, brown stroma, often continuous for 2 or more centimeters, and raising the epidermis into parallel ridges split along the top, and revealing the brown stroma roughened by the numerous, papilliform ostiola, which are finally perforated and subumbilicate. Asci clavate, stipitate, p. sp. 20–25 x 5 μ. Sporidia subbiseriate, allantoid, yellowish-hyaline, slightly curved, 5–6 x $1\frac{1}{4}$–$1\frac{1}{2}$ μ.

On dead culms of *Arundinaria*, Louisiana (Langlois).

D. tùmida, E. & E. (in Herb.)

Diatrype Daldiniana, De Not. in Ell. & Evrht. N. A. F. 2d Ser. 2525.

Stromata pulvinate, orbicular or oblong, 3–4 mm. diam., subseriate and subconfluent for 1–2 cm., broad convex-tuberculiform, nearly black outside, whitish within, buried below in the bark and circumscribed with a distinct black line. Asci sublanceolate, p. sp. 40–55 x 6–7 μ, with a long, slender stipe. Sporidia subbiseriate, allantoid, yellowish, slightly curved, obtuse, 10–12 x $2\frac{1}{2}$–3 μ. Ostiola prominent but not elongated, deeply quadrisulcate-cleft. Perithecia with thick, black, coriaceous walls.

On dead bark of elm, London, Canada (Dearness).

This seems to differ from *D. Daldiniana* in its sulcate ostiola and somewhat smaller sporidia. The stroma also is larger and more prominent than in the specimen of that species in Roumeguere's F. Gallici, No. 1078.

D. Hullénsis, E. & E. (in Herb.)

Stromata subsuperficial, oblong or elliptical, 4–5 mm. in the longer diameter, pulvinate, often confluent for 2–3 cm. or more, carbonaceous and black outside, softer and greenish-yellow within. Perithecia buried in the stroma, ovate globose, $\frac{1}{3}$ mm. diam., thin-walled, contracted above into slender necks, with their large, hemispherical, 4–5-sulcate-cleft ostiola roughening the surface of the stroma. Asci (p. sp.) clavate, 35–40 x 5–6 μ, with a slender stipe, 8-spored, aparaphysate, Sporidia biseriate above, allantoid, yellowish-hyaline, slightly curved, obtuse, 8–11 x 2–$2\frac{1}{2}$ μ.

On rotten wood, near Hull, Canada (Macoun).

D. cornùta, E. & E. (in Herb.)

Stromata scattered, erumpent, black, pulvinate-tuberculiform, 2–3 mm. diam., roughened by the short-cylindrical (1 mm. long), obtuse, scattered ostiola, not penetrating to the wood, but blackening the inner surface of the bark, and surrounded by a narrow circumscribing line. Asci (p. sp.) clavate, 25–35 x 8–10 μ, 8-spored, with a slender stipe and abundant paraphyses. Sporidia conglobate, allantoid, obtuse, yellow, about 10 x 3 μ.

On bark of dead *Ailanthus* limbs, Lyndonville, N. Y. (Fairman).

D. fibritécta, C. & E. Grev. V, p. 31.

Stromata small (1½ mm.), pustuliform, dark brown inside, seriate-confluent, forming parallel lines or strips continuous for several centimeters. Perithecia crowded, small, brown. Ostiola erumpent through the fibers of the bark which covers the pustules, short-cylindrical, black, obtusely pointed, finally irregularly perforated. Asci clavate, rounded at the apex, p. sp. 20–25 x 3–3½ μ. Sporidia biseriate, allantoid, hyaline, 5–6 x 1–1½ μ.

On bark of dead *Juniperus Virginiana*, Newfield, N. J.

D. verrucoìdes, Pk. in Sacc. Syll. IX, p. 473.

Pustules small, limited by a black, circumscribing line, at first covered by the epidermis which is longitudinally or stellately ruptured. Stromata black outside, whitish within, frequently covered above by a cinereous tomentum. Perithecia 3–8 in a pustule. Ostiola black, depressed, stellate-sulcate. Asci clavate. Sporidia continuous, cylindrical, straight or slightly curved, 20 x 4 μ.

On dead branches of birch, New York State (Peck).

This is said to have been published in 32d Rep. N. Y. State Mus.—a publication we have never seen.

D. asteróstoma, B. & C. Grev. IV, p. 96.

Exsicc. Rav. F. Am. 91, 655 and 656.—Ell. N. A. F. 88.

Stromata wart-like, erumpent, scattered or subconfluent, 1½–2½ μ diam., embraced at base by the ruptured epidermis, black (white inside). Perithecia 10–20 in a stroma, globose, black, abruptly contracted into slender necks rising through the white substance of the stroma with prominent, stellate-cleft, erumpent ostiola. Asci clavate, long-stipitate, p. sp. 25 x 4–5 μ, 8-spored. Sporidia allantoid, nearly hyaline, moderately curved, 6–10 x 1½–2 μ.

On dead limbs of *Magnolia* and of (*Nyssa*)? South Carolina (Ravenel).

The diagnosis here given is drawn from the specc. in Rav. F. Am. The species seems to differ from *D. disciformis* principally in its smaller stroma and more prominent ostiola, but is hardly more than a variety of that species.

D. discifórmis, (Hoff.)

Sphæria disciformis, Hoff. Veg Crypt. I, p. 15, tab. IV, fig. 1.
Sphæria depressa, Sow. Eng. Fung. II, tab. 216.
Sphæria grisea, DC. Flore Fr. II, p. 122.
Diatrype disciformis, Fr. Summa, 385.
Exsicc. Fckl. F. Rh. 1041, 2545, 2546.—Rab. F. E. 137.—Rehm Asc. 521.—Lin. Fung. Hung. 179.—Cke. F. Brit. 2d Ser. 218 —Krieger F. Sax. 86.

Stromata scattered or gregarious, flattened-pulvinate, round, 2–3 mm. diam., discoid, erumpent and loosely embraced by the ruptured epidermis, dark brown, whitish inside, base sunk to the wood and circumscribed by a black line. Perithecia 20–30 (or more) in a stroma, ovate, about $\frac{3}{4}$ x $\frac{1}{2}$ mm., closely packed and subangular, with short necks and small, obtusely conical, smooth or 3–5-cleft, slightly projecting ostiola. Asci clavate, long-stipitate, p. sp. 22–30 x 4–5 μ. Sporidia subbiseriate, allantoid, yellowish, slightly curved, 6–8 x 1$\frac{1}{2}$–2 μ.

On dead limbs of *Fagus*, also on other deciduous trees.

Schweinitz, in his Syn. Car., speaks of this as very common ("vulgatissima"). Berkeley also, in Grev. IV, p. 95, quotes it from North Carolina, Alabama and Pennsylvania under eight different numbers, but we have never seen any American specimens agreeing well with the European, of which the description is given above.

D. viréscens, (Schw.) **(Plate 34)**

Sphæria virescens, Schw. Syn. N. Am. 1239.
Diatrype disciformis, var. *virescens*, Berk. Grev. IV, p. 95.
Exsicc. Ell N. A. F. 776.—Rav. F. Car. IV, 48.

Stromata scattered, erumpent-superficial, orbicular, about 2 mm. diam., depressed-pulvinate, surrounded at base by the ruptured epidermis, disk greenish-yellow pulverulent, becoming darker with age. Perithecia monostichous, covered by the waxy, whitish substance of the stroma, about $\frac{1}{3}$ mm. diam., 10–15 in a stroma, necks terminating in obtusely conical, 4-cleft, black ostiola barely erumpent through the greenish disk. Asci p. sp. about 35 x 5 μ, stipitate, 8-spored. Sporidia subbiseriate, allantoid, yellowish, moderately curved, 7–9 x 1$\frac{1}{2}$–2 μ.

On dead limbs of *Fagus*, common where that tree is found.

Considered by Berkeley in Grev. IV, p. 95, as a var. of *D. disciformis*, which differs in its flatter, broader, black-brown stroma and smaller asci and sporidia.

D. albopruinòsa, (Schw.)

Sphæria albopruinosa, Schw. Syn. N. Am. 1238.
Diatrype roseola, Winter Journ. Mycol. I, p. 121.
Diatrype albopruinosa, Cke. Grev. XIII, p. 37.
Diatrype Durieui, Mont. in Herb. Curtis.
Diatrype Webberi, E. & E. in Herb.
Exsicc. Ell. & Evrht. N. A. F. 2d Ser. 2527

Stromata scattered or subgregarious, sometimes confluent, suborbicular, 1½–2½ (exceptionally 3–4 mm.) broad, slightly convex, surrounded by the ruptured epidermis, whitish when young and fresh, becoming finally dark brown, light-colored inside, surface becoming more or less rimose. Perithecia 10–30 in a stroma, crowded, ovate or elliptic-oblong. Ostiola more or less prominent, 3–5-sulcate-cleft, black. Asci oblong-clavate, long-stipitate, p. sp. 55–65 x 7 μ. Sporidia conglobate, allantoid, obtuse, slightly curved, 12–16 x 2½–4 μ, yellow-brown.

On dead limbs of *Fagus*, Pennsylvania (Schw.), on dead oak limbs, Missouri and New Jersey, on *Cratægus*, and *Salix*, London, Canada (Dearness).

This differs from *D. disciformis*, in its much larger asci and sporidia.

D. Eucalýpti, Cke. & Hark. Grev. IX, p. 85.

Suborbicular, convex, black. Ostiola conical, sulcate. Asci clavate, long-stipitate. Sporidia slightly curved, ends obtuse, hyaline, 10 x 1½ μ.

On branches of *Eucalyptus globulus*, California (Harkness).

We have not seen this, and take the foregoing brief notes from Grevillea.

D. Maclùræ, E. & E. (in Herb.)

Stroma pulvinate, oblong or linear, 2–10 mm. long by 1–2 mm. wide, almost superficial and faintly circumscribed, brownish-black outside, white within. Perithecia crowded in the stroma, ovate, about ½ mm. diam., with thick, black, coriaceous walls, necks short, with their thick, flattened, pustuliform, irregularly dehiscent ostiola slightly prominent. Asci clavate, paraphysate, 8-spored, p. sp. about 40 x 6 μ. Sporidia biseriate above, allantoid, yellowish, moderately curved, obtuse, 10–14 x 3 μ.

On decorticated dead limbs of *Maclura aurantiaca*, London, Canada (Dearness).

On account of its elongated stroma and large sporidia, this seems quite distinct.

D. próminens, Cke. & Hark. Grev. IX, p. 85.

Stromata erumpent, pulvinate, hemispherical 2–3 mm. diam. or oblong 2–3 x 1 mm., closely girt by the epidermis which also adheres in fragments to the surface, whitish or cinereous inside. Perithecia 8–15 lying in the bottom of the stroma, about ⅓ mm. diam. Ostiola rather large, stellate-cleft. Asci subsessile, p. sp. 35–40 x 6 μ. Sporidia subbiseriate, allantoid, slightly curved, yellowish, 6–8 x 2 μ.

On *Arbutus Menziesii*, and *Mimulus glutinosus*, California (Harkness).

The specimen on *Mimulus glutinosus*, sent by Harkness, has the stromata smaller (1–2 mm.), less distinctly erumpent and dark inside, and the sporidia larger (8–10 x 2 μ), and hardly seems to be the same as that on *Arbutus*. Sporidia (sec. Cke.) 12–13 x 2 μ.

D. infúscans, E. & E. (in Herb.)

Stromata gregarious, small (1–1½ mm.), conic-hemispherical or tuberculiform, closely covered (except the apex) with the adherent, blackened epidermis. Perithecia 3–6 in a stroma, globose, thin-walled, ¼–⅓ mm. diam., contracted above into short necks, with the large, tuberculiform ostiola crowded and erumpent together. Asci (p. sp.) 25 x 4 μ, with a slender stipe of about the same length, and filiform paraphyses, 8-spored. Sporidia biseriate, allantoid, slightly curved, yellowish-hyaline, 5–6 x 1¼–1½ μ.

On dead stems of *Smilax*, Houston, Texas (Ravenel 242).

The epidermis is blackened around the pustules, and where they stand near each other, it is also blackened continuously between them. This is distinct from *D. smilacicola* (Schw.), on account of its very different stroma and smaller sporidia, and from *Diatrypella prominens*, Howe, which outwardly it much resembles, by its 8-spored asci and prominent ostiola.

D. smilacícola, (Schw.)

Sphæria smilacicola, Schw. Syn. N. Am. 1251.

Subpulvinate-effused, seated on the epidermis, surrounded by a kind of sterile, sublobate margin which adheres to the wood and leaves a black line. Perithecia few, subprominent, black on the surface and black or dark brown inside. Stroma scanty, pulverulent. It occurs often 2–6 mm. long and so elevated in the center as to become conical.

Rather rare on dead stems of *Smilax rotundifolia*, Salem, N. C.

Sporidia (sec. Cke. Grev. XIII, p. 37) allantoid, 12–14 μ long.

D. subferruginea, B. & Rav. in Rav. Fungi Car. IV, No. 44.

Exsicc. Ell. & Evrht. N. A. F. 2d Ser. 1558.

Stromata scattered or subseriately confluent, conical from a suborbicular base, 1–3 mm. diam., black on the outside, composed inside of the scarcely altered substance of the bark, the apex splitting the epidermis in a sublaciniate manner, but scarcely raising it into pustules. Perithecia 8–20 in a stroma, small (about 200 μ diam.), brown, buried in the bottom of the stroma. Ostiola short-cylindrical or elongated, (50–250 μ long), smooth at the apex. Asci not well made out. Sporidia (free spores) allantoid, nearly hyaline, 5–6 x 1–1¼ μ, only slightly curved.

On bark of dead oak, South Carolina (Ravenel).

The stroma is entirely wanting below.

D. Ceanòthi, Cke. & Hark. Grev. XIII, p. 17.

Stromata wart-like, innate-erumpent, black, round, closely surrounded by the ruptured epidermis, black within. Perithecia compressed and of various shapes. Ostiola short, stellate-sulcate. Asci clavate, 8-spored. Sporidia cylindrical, curved, rounded at the ends, yellowish-hyaline, 12–14 x 2 μ.

On branches of *Ceanothus*, California (Harkness).

D. capnóstoma, B. & Rav. in Rav. Fungi Car. IV, No. 42.

Stromata scattered, small (1 mm.), erumpent, not circumscribed. Perithecia few (3–6), globose, black, not deeply buried, rather less than ⅓ mm. diam., necks short, with prominent, short-cylindrical, perforated ostiola, piercing the slightly pustulate epidermis together. Asci stipitate, p. sp. 25–30 x 4 μ. Sporidia subbiseriate, 8 in an ascus, allantoid, slightly curved, yellowish-hyaline, 5–7 x 1½ μ.

On dead limbs of *Morus*, South Carolina (Ravenel).

The stromatic material in which the perithecia lie is very scanty, and of a dark color. Differs from *D. infuscans*, E. & E., in its cylindrical ostiola.

D. Callicárpæ, B. & Rav. in Rav. Fungi Car. IV, No. 41.

Stromata thickly scattered, small, (1–1½ mm.), conic-hemispherical, erumpent, white inside, not circumscribed. Perithecia 3–8, about ⅓ mm. diam., deeply sunk in the stroma. Ostiola prominent, obtusely conical, sometimes convergent so as to cause the erumpent part of the stroma to appear quadrisulcate. Asci stipitate, p. sp. 35 x 5–6 μ. Sporidia subbiseriate, allantoid, yellow, slightly curved, with a nucleus in each end, 6–8 x 1½–2 μ.

On dead limbs of *Callicarpa*, South Carolina (Ravenel).

Closely allied to the preceding species.

D. azedaráchtæ, Cke. Grev. XI, p. 108.

Exsicc. Rav. F. Am. 744.

"Erumpent. Stroma black, convex, suborbicular, same color within. Perithecia compressed. Asci clavate. Sporidia allantoid, pale brown, 12 x 3 μ."

On branches of *Melia*, South Carolina.

The specc. in Rav. F. Am. have the stromata thickly scattered, convex-tuberculiform, 1½–2 mm. diam., roughened above by the prominent, obtusely conical ostiola. Perithecia 4–8 in a stroma, about ⅓ mm. diam., not compressed. Asci stipitate, p. sp. about 35 x 6 μ, 8-spored. Sporidia subbiseriate, allantoid, yellowish, 6–8 x 1½–2 μ. The stroma seems to have been light-colored inside and out when fresh. These notes do not agree well with the published characters quoted above. The specc. do not differ essentially (only in the rather larger stromata) from *D. Callicarpæ*, B. & Rav.

D. radiàta, Ell. Am. Nat. Feb. 1883, p. 196.

Stromata 2–3 mm. diam., orbicular, elliptical or sinuous, subhemispherically rounded above and closely embraced by the laciniately cleft epidermis, circumscribed by a black line which does not penetrate to the wood. Perithecia membranaceous, ⅓–½ mm. diam., ovate-globose, deeply sunk in a light-colored, tuberculiform stroma. Ostiola obtuse, scarcely prominent. Asci clavate, with a slender stipe, p. sp. about 35 x 6 μ. Sporidia conglomerated, yellowish, allantoid, slightly curved, 6–12 x 2–2½ μ, with a nucleus in each end.

On dead limbs of *Ulmus*, Decorah, Iowa (Holway).

Outwardly resembles *Diatrypella verrucæformis*.

D. sambucívora, (Schw.)

Sphæria sambucivora, Schw. Syn. N. Am. 1275.
Diatrype sambucivora, Cke. Grev. XIII, p. 38.

Emergent-superficial, becoming black, longitudinally aggregated-confluent, becoming superficial as the bark falls away, surface uneven, rough, but not crustose. The separate tubercles are ovate or of irregular shape. Perithecia rather large, numerous, polystichous, black, crowded in the tubercles, surrounded, when fresh, with a white stroma. Ostiola polygonal, prominent, at length irregular, deeply umbilicate. Very variable in shape as it is found on the wood or on the bark. Tubercles 4–6 mm. high, in elongated series of several inches in extent. Sporidia (Cke. l. c.) 8 μ long, allantoid.

On dead stems of *Sambucus Canadensis*, Bethlehem, Pa. (Schw.). Allied to *Sphæria scabrosa*.

D. micróstega, E. & E. (in Herb.)

Stromata gregarious, small, $\frac{3}{4}$–1 mm. diam., erumpent, superficial, black, rough, subglobose. Perithecia 1–5 in a stroma, small, black. Ostiola prominent, subpyramidal, and more or less stellate-cleft. Asci (p. sp.) 35–40 x 6–7 μ. Sporidia subbiseriate, allantoid, yellowish, slightly curved, 10–12 x 1$\frac{1}{2}$–2 μ.

On bark, California (Harkness). Sent as *Sphæria submoriformis*, Plowr.

D. Macoùnii, E. & E. Proc. Acad. Nat. Sci. Phil. July, 1890, p. 224.

Stroma discoid, gray, 3–4 mm. across, and about 1 mm. thick, suborbicular or subelliptical, seated on the surface of the inner bark, and loosely embraced by the upturned, ruptured epidermis, circumscribed by a distinct, black line, which penetrates the bark and stains the surface of the subjacent wood, but does not penetrate it. Perithecia numerous, 30–50, in a single layer, ovate-globose, $\frac{1}{4}$–$\frac{1}{3}$ mm. diam., contracted above into short necks terminating in a small, indistinctly radiate-cleft, black ostiolum, which is in a slight depression of the stroma. Asci (p. sp.) 20–30 x 3 μ or, including the thread-like base, 50–60 μ long. Sporidia biseriate above, allantoid, slightly curved, 4–6 x $\frac{3}{4}$–1 μ. Substance of the stroma dirty white inside.

On maple bark (*Acer rubrum*)?, Agassiz, British Columbia (Macoun).

D. Hochelàgæ, E. & E. Proc. Acad. Nat. Sci. Phil. July, 1890, p. 224. **(Plate 34)**

Exsicc. Ell. & Evrht. N. A. F. 2d Ser. 2528.

Stroma orbicular or elongated, 2–3 mm. long and 1–2 mm. wide, often more or less confluent, pulvinate-verrucose, with the margin abrupt or slanting off at the ends, with a faint circumscribing black line, which does not penetrate deeply into the wood, dull black outside, dirty white within. Perithecia crowded in the stroma, subglobose, about $\frac{1}{2}$ mm. diam., with thick, black, coriaceous walls. Ostiola conic-hemispherical, deeply 4–5-sulcate-cleft. Asci (p. sp.) 40–45 x 7–8 μ, with stout paraphyses and allantoid, yellowish, moderately curved, 8–12 x 2½ μ sporidia. Specific name from Hochelaga, an Indian name for the St. Lawrence River.

On decorticated elm wood, London, Canada (Dearness).

D. tremellóphora, Ell. in Am. Nat. March, 1882, p. 239.

Diatrype disciformis, B. & C.
Diatrype disciformis, var. *Magnoliæ*, Thum. in Bull. Torr. Bot. Club, VI, p. 95.
Exsicc. Rav. F. Am. 360.—Thum. M. U. 359.—Ell. N. A. F. 775.

Stroma as in *D. disciformis*, only mostly smaller and closely embraced by the laciniæ of the ruptured epidermis, at first concave and covered by a thin, circular, tremelloid, reddish orange-colored membrane which soon turns black and falls off, revealing the disk of the stroma beneath it minutely white-punctate from the incipient ostiola. The stroma finally becomes more erumpent, flattish-convex, brown, and subrimose, and the punctiform ostiola darker but not prominent. Asci and sporidia as in *D. disciformis*.

On dead trunks of *Magnolia glauca*, New Jersey and Carolina.

D. mínima, E. & E. Journ. Mycol. I, p. 91.

Stroma cortical, formed of the scarcely altered substance of the bark, elliptical, 1–2 mm. diam., limited by a black, circumscribing line which penetrates the wood beneath. Perithecia 8–12 in a stroma, lying in a single layer, globose, 150–200 μ diam., membranaceous, with black, rather thick walls and short, obtuse ostiola, their apices papilliform, black and shining at first, then distinctly perforated with a rather broad opening. Asci cylindrical, 70–80 x 2½–3 μ. Paraphyses obscure (or none?). Sporidia uniseriate, lying end to end, oblong-elliptical, 2-nucleate, yellowish, nearly hyaline, 5–7 x 2 μ. The black, scarcely projecting ostiola which dot the small, tuberculiform stroma, are visible through short, longitudinal cracks or chinks in the slightly elevated epidermis.

On dead shoots and limbs of *Magnolia glauca*, Newfield, N. J. Probably not uncommon, but easily overlooked.

D. sphæróspora, E. & E. Journ. Mycol. III, p. 42.

Stroma formed of the scarcely altered substance of the bark, erumpent, but not very prominent, surrounded by the ruptured epidermis, small (½–1 mm.). Perithecia in a single layer, 3–12, black, membranaceous, minute (166–200 μ), their smooth, black, obtusely conic ostiola dotting the surface of the stroma. Asci cylindrical, spore-bearing part 30–35 x 3 μ, with a slender, thread-like base about 20 μ long. Paraphyses not observed. Sporidia uniseriate, yellowish-hyaline, 8 in an ascus, globose, 3 μ diam. Outwardly, this is scarcely distinguishable from *D. minima*, but the marked difference in the sporidia seems to entitle it to specific rank. The stroma in some of the specimens is limited by a black line, as in *D. minima*, but in others not.

On dead shoots of *Magnolia glauca*, Newfield, N. J.

This and the preceding species vary from the usual type of *Diatrype*, in the shape of their sporidia.

Species imperfectly known or to be rejected.

D. manipulàris, B. & C. in Herb. Curtis.

This is mentioned in Grev. XIV, p. 16. No description given—as far as we know.

D. plàgia, B. & C. Grev. IV, p. 96.

"Bursting through the bark transversely. Ostiola substellate. Stroma brown, scanty. Asci clavate. Sporidia 8 in each ascus, sausage-shaped."

On *Liriodendron*, South Carolina.

D. pilulífera, (Fr.)

Sphæria pilulifera, in Herb. Schw., Syn. N. Am 1234.

Sec. Grev. XIII, p. 37, this can hardly be the *S. pilulifera*, Fr.

D. corniculàta, (Ehr.)

This is quoted by Schw. & Berk. as found in this country, but the species is not well known. The specc. in Rav. Car. IV, No. 43, are *Eutypella heteracantha*, Sacc.

D. collariàta, C. & E. Grev. IV, p. 102.

Our. spece. of this are only *Valsa caryigena*, B. & C., with the ostiola abnormally elongated.

D. bìspora, B. & C. Rav. Car. IV, 45.

This is not an ascigerous fungus, but a species of *Didymosporium.*

ANTHÓSTOMA, Nitschke.

Pyr. Germ. p. 110

Stroma more or less effused (diatrypoid), or pulvinate, conical or hemispherical (valsoid), often only partially developed. Perithecia sunk in the stroma or in the matrix, mostly with elongated necks which, in the pulvinate stroma, generally converge and are erumpent together as in *Valsa*. Asci cylindrical or elavate, with paraphyses more or less distinct. Sporidia oblong or elliptical, continuous, brown.

* *Stroma effused Euanthostoma*).

A. melanòtes, (B. & Br.)

Sphæria melanotes, B. & Br. Brit. Fungi, No. 634, tab. 9, fig 6.
Anthostoma melanotes, Sacc. Mich. I, p. 326

Stroma effused, entirely immersed in the wood, which is blackened on the surface forming elongated spots often confluent; finally blackening the wood more deeply, and with a black, circumscribing line. Perithecia small, depressed-globose, entirely buried in the wood, monostichous, loosely scattered. Ostiola exserted, minute, entire, conical or hemispherical, shining, at length perforated. Asci cylindrical, short-stipitate, 8-spored, paraphysate, 70 x 7 μ. Sporidia obliquely uniseriate, obtusely-fusiform, straight or rarely subinequilateral, becoming dark brown, 12–14 x 5–6 μ.

On decorticated oak limbs, Newfield, N J.

A. grandínea, (B. & Rav.)

Diatrype grandinea, B. & Rav Grev IV, p. 95.
Anthostoma grandinea, Sacc. Syll. 1114.
Camarops grandinea, Cke. Syn. 1469.
Exsicc. Ell. N. A. F. 494.—Rav. Car. IV, 90.

Stroma effused, thin, crustaceous, overspreading the surface of the inner bark, and throwing off the epidermis for 6 inches or more in extent, brownish at first with a rufous tint, finally darker, less than ½ mm. thick and easily overlooked, being of about the same color as the bark, surface minutely papillose-roughened by the slightly prominent, sub-hemispherical or obtusely-conical ostiola. Perithecia covered above by the stroma, small, less than ¼ mm. diam., depressed-globose, thin-walled, white inside, narrowed directly into the ostiola, which are sometimes faintly radiate-sulcate. Asci clavate, 50–60 x 10 μ (p. sp. 40 x 10 μ), short-stipitate, with imperfectly developed paraphyses. Sporidia subbiseriate above, short-elliptical or subglobose, 7–10 x 5–8 μ, becoming nearly black, 1–2-nucleate, but not septate.

On bark of dead *Quercus coccinea*, Carolina and New Jersey.

A. picàceum, (C. & E.)

Sphæria picacea, C. & E. Grev. VII, p. 9.
Anthostoma picacea, C. & E. Cke. Syn. 4207
Anthostomella? picacea, Sacc. Syll. 1093
Exsicc. Ell. N. A. F. 183.

Perithecia subgregarious, globose, deeply buried in the wood, which is blackened in patches on the surface and pierced by the minute, erumpent ostiola. Asci cylindrical, p. sp. 50–55 x 5–6 μ, paraphysate. Sporidia uniseriate or partly biseriate, oblong-cylindrical, brown, with a single nucleus, obtuse, 8–10 x 3–3½ μ (10 x 4 μ, Cke.).

On decorticated limbs of *Vaccinium* and *Acer*, Newfield, N. J.

A. gigásporum, Cke. & Hark. Grev. XIII, p. 18.

Scattered, subimmersed, covered by a raised, globose tubercle formed from the substance of the matrix. Perithecia globose, large (1½–2 mm.), black-punctate, with an obtuse ostiolum. Asci ample, saccate, 8-spored. Sporidia elongated-elliptical, slightly narrowed at the ends, continuous, dark brown 65–80 x 25–30 μ.

On decorticated twigs, California (Harkness).

In habit resembling *Sphæria cubicularis*, Fr., but with much larger sporidia, and, as in that species, the perithecia fall out, leaving cavities not unlike a large *Stictis*.

A. pulvíniceps, Pk. (in literis).

Perithecia 8–12 in a pustule, sunk to the wood, covered by the bark. Ostiola erumpent, crowded, prominent, black, forming a cushion-shaped mass. Asci clavate. Sporidia crowded, subelliptical or broadly fusiform, multinucleate, slightly colored, 10–15 μ long.

On dead stems of *Sambucus Canadensis*, New York (Peck).

A. flavovíride, Ell. & Holw. Geol. Survey of Minn. Bull. No. 3, 1886, p. 32.

Stroma effused, thin, 1–3 x ½–1 inches, more or less, covered at first with a thin coat of greenish-yellow, short, matted hyphæ, bearing small (1–1½ μ), subglobose, subhyaline conidia but finally bare and black. Perithecia membranaceous, black, globose, (⅓ mm.), sunk in the scarcely altered substance of the wood, contracted above into short, narrow necks with their black, papilliform ostiola erumpent. Asci (p. sp.) 75 x 6–7 μ, or with the slender base 120 μ long. Sporidia uniseriate, narrow-elliptical, continuous, nearly hyaline at first, becoming dark, 10–12 x 4–4½ μ. The stroma is limited by a dark, circumscribing line penetrating the wood.

On decaying poplar wood, Vermilion Lake, Minn. (Holway).

A. susténtum, (Plowr.)

Sphæria sustentum, Plowr. Grev. VII, p. 73.
Anthostoma sustentum, Sacc. Syll. 1120.

Perithecia immersed, with their ostiola erumpent between the bleached fibers of the wood. Asci cylindrical, 150 x 10–15 μ. Sporidia ovoid, dark brown, becoming black, 20–25 x 10–11 μ.

On bleached limbs of *Arctostaphylus*, California.

A. polynèsia, (B. & C.)

Sphæria polynesia, B. & C. Grev. IV, p. 146.
Anthostoma polynesia, Sacc. Syll. 1110.

"Forming little oblong, black spots which are studded with the ostiola. Asci very slender. Sporidia oblong, brown, 7 μ long."

On wood, Mountains of Virginia.

** *Stroma valsoid or tuberculiform*.

A. gástrinum, (Fr.)

Sphæria gastrina, Fr. S. M. II, p. 379.
Sphæria irregularis, Sow. Eng. Fungi, tab. 374, fig. 9.
Hypoxylon gastrinum, Fr. Summa, 383.
Melogramma gastrinum, Tul. Sel. Carp. II, p. 89.
Quaternaria Nitschkei, Fckl. Symb. p. 230.
Anthostoma gastrinum, Sacc. Syll. 1129.
Exsicc. Fckl. F. Rh. 2005.—Rab. F. E. 627.—Sacc. M. Ven. 1441.—Desm. Pl. Cr. Ed. I, 1254, Ed. II, 754.

Stroma variable; when seated on the bare wood, superficial or very nearly so, depressed-globose or pulvinate, black, smooth and shining outside, 2–4 mm. broad; when immersed in the bark, and covered by the raised and blackened epidermis, hemispherical or convex, disk erumpent; in either case whitish inside, but enclosed in a distinct, black stratum, which shows as a black, circumscribing line in a horizontal section. Perithecia 8–40 crowded in the bottom of the stroma, ½–¾ mm. diam., ovate, with thick, black, coriaceous walls, and stout, cylindrical, elongated necks, which, in the superficial stromata, on wood, rise straight through the stroma, with their obscure, papilliform ostiola erumpent over the whole of the upper surface, but when the stroma is buried in the bark, converging and erumpent in the exposed disk. Asci cylindrical, short-stipitate, 90–120 x 5 μ (p. sp. about 80 μ), paraphysate, 8-spored. Sporidia uniseriate, oblong or oblong-elliptical, ends rounded and obtuse, straight, dark brown, becoming opake or nearly black, 10–14 x 4–4½ μ (10–14 x 5–6 μ, Winter).

On bark of dead *Ulmus*, Canada (Dearness), on oak, Carolina and Georgia (Ravenel).

The Canada specc. agree perfectly with the foregoing description (which is mostly from Winter's Pilze) and with the spece. in Sacc. M. Veneta and Desm. Pl. Cr. 1254, as well as with German specc. in our Herb., only they have the perithecia rather smaller (not over ½ mm.). This is the size given them in the Syll., and in Winter's Pilze, but the specc. in the Exsiccati quoted have the perithecia ½–1 mm. diam.

A. túrgidum, (Pers.)

Sphæria turgida, Pers. Obs. Mycol. I, p. 17.
Sphæria faginea b. turgida, Pers. Syn. p. 44.
Valsa turgida, Fr. Summa, p. 412.
Anthostoma turgidum, Nits. Pyr. Germ. p. 121.
Exsicc. Fckl. F. Rh. 591.—Rab. F. E. 735, 1144.—Krieger, F. Sax. 494.—Sacc. M. V, 1442. Roum. F. G. 1947.

Stromata mostly numerous and standing close together, covered by the epidermis and forming hemispherical pustules with their bases sunk in the scarcely altered substance of the inner bark, crowned with a small, erumpent disk, surrounded by the ruptured epidermis. Perithecia 6–8 in a stroma, tolerably large, globose, crowded in a circular group, nearly erect, black, the obtuse, conical ostiola united in a small, concave, brownish-black disk but slightly prominent. Asci cylindrical, sessile, 8-spored, 100–120 x 7–8 μ, paraphysate. Sporidia uniseriate, broad elliptical, rounded at the ends, straight, black, 8–12 x 7–8 μ.

On dead beech limbs, New York State (Peck).

We have seen no American spece. and take the foregoing description from Winter's Pilze.

A. adústum, (C. & P.)

Diatrype adusta, C. & P. 29th Rep. N. Y. State Mus. p. 58.
Anthostoma adustum, Sacc. Syll. 1142.

"Pustules small, slightly elevated, subconical, blackish, covered by the epidermis which is pierced by the very small disk. Stroma white. Ostiola few, small, black. Asci cylindrical. Sporidia uniseriate, simple, elliptical, colored, 18–22 μ long."

On dead branches, New Baltimore, N. Y. (Howe).

A. amygdálinum, (Cke.)

Melanconis amygdalina, Cke. Grev. IV, p. 55, tab. 81, fig. 4.
Anthostoma amygdalinum, Sacc. Syll. 1147.

Loosely circumscribed. Stroma pale. Perithecia circinate, globose. Ostiola convergent, slightly prominent, united in a brown disk. Asci cylindrical. Sporidia almond-shaped, brown, with a large, globose nucleus, 23–25 x 12 μ, with a slender, hyaline appendage at each end 20–30 μ long.

On *Liquidambar*, South Carolina (Ravenel).

A. dryóphilum, (Curr.)

Diatrype dryophila, Curr. Linn. Trans. XXII, p. 279, fig. 75.
Anthostoma dryophilum, Sacc. Syll. 1149.
Exsicc. Ell. N. A. F. 87.

Stromata scattered, sunk to the wood, circumscribed, black inside and out, orbicular or elliptical at the base, subconical, 3–4 mm. diam., or by confluence more, apex erumpent and at first tuberculiform, finally truncate and concave, areolate from the large, obtusely flattened ostiola. Asci (p. sp.) cylindrical, about 60 x 4 μ, paraphysate, 8-spored. Sporidia uniseriate, oblong, becoming opake, continuous, 8–12 x 3–$3\frac{1}{2}$ μ, (10–15 x 3–4 μ, Curr.).

On dead white oak limbs, Newfield, N. J., Texas (Ravenel), and Iowa (Holway).

Var. minor, Cke. (Grev. V, p. 32), on black and scarlet oak, Newfield, N. J., has the stroma more effused and less prominent, covered, except the small, tuberculiform disk, by the blackened epidermis.

A. phæospérmum, (Ell.)

Diatrype phæosperma, Ell. in Am. Nat. Feb. 1883, p. 195.
Anthostoma phæospermum, Sacc. Syll. 5936.

Stroma small (1 mm.), tuberculiform, closely embraced by the sub-laciniate-cleft epidermis. Perithecia 6–8 in a stroma, $\frac{1}{3}$ mm. diam., with thick, coriaceous walls, lying in a single layer in the bottom of the white substance of the stroma, which is circumscribed by a black line which scarcely penetrates to the wood. Asci clavate-cylindrical, p. sp. about 55 x 7 μ. Sporidia subbiseriate, cylindrical, curved, continuous, brown, 10–12 x 3–$3\frac{1}{2}$ μ, ends obtuse.

On dead limbs, Decorah, Iowa (Holway).

A. Ontariénse, E. & E. Proc. Acad. Nat. Sci. Phil. July, 1890, p. 228. **(Plate 34)**

Stroma convex, $\frac{1}{4}$–$\frac{1}{2}$ mm. diam., more or less subseriately confluent often for several cm., formed of the unaltered substance of the bark and surrounded by a black, circumscribing line which penetrates the wood. Perithecia crowded in the stroma, subglobose, $\frac{1}{2}$–$\frac{3}{4}$ mm. diam., with thick, coriaceous walls, contracted above into a narrow neck terminated by the subglobose, deeply quadrisulcate, erumpent ostiolum .Asci slender, 8-spored, 90–110 μ long (p. sp. 75–80 x 8–10 μ), with abundant paraphyses. Sporidia subbiseriate, cylindrical, moderately curved, brown, 20–26 x 4–$4\frac{1}{2}$ μ. Has much the same general appearance as some compact forms of *Valsa stellulata*, Fr.

On dead limbs of *Salix*, London, Canada (Dearness).

A. cercidícolum, B. & C. 25th Rep. N. Y. State Mus. p. 101.

Diatrype cercidicola, B. & C. in Herb. Curtis.
Anthostoma cercidicolum, Sacc. Syll. 1136.

Stroma black, plane, suborbicular, 6–8 mm. diam., thin, seated on the inner bark, surrounded by the ruptured epidermis, dotted by the minute, depressed or umbilicate, at length perforated ostiola. Perithecia crowded, elliptical or ovate. Sporidia unequally ovate, colored, 10 μ long.

On bark of unknown wood, Buffalo, N. Y. (Clinton).

The inner surface of the bark is stained black.

A. micrósporum, Karst. Fungi Fenn. 860.

Phæosperma helvetica, Fckl. Symb. p. 224, tab. VI, fig. 40 (fide Karst.).
Diatrype microspora, Ell. Bull. Torr. Bot. Club, VIII, p. 74.
Anthostoma, Ellisii, Sacc. Syll. I, p. 308.

Stroma ventricose-hemispherical, swollen, black inside and out, the upper part erumpent, rounded, subviscose when fresh, appearing like a mass of black, exuding gum, 2–6 mm. broad, 2–3 mm. thick, often confluent. Perithecia deeply immersed in the stroma, oblong, leathery, closely packed. Ostiola papillate, only slightly prominent, broadly and irregularly perforated. Asci clavate-cylindrical, stipitate, p. sp. 25–30 x 5 μ, 8-spored, paraphysate. Sporidia obliquely uniseriate, oblong, obtuse, 1–2-nucleate, olive-brown, 4–5 x $1\frac{3}{4}$–2 μ. Var. *exudans*, Pk. 40th Rep. p. 67, has the perithecia collected in a thin, crowded, angular, cortical stroma, closely covered by the pustulate-elevated, irregularly ruptured epidermis. Ostiola obscure or concealed beneath the defiled epidermis. Asci very slender, cylindrical, 40 x 4 μ. Sporidia minute, oblong, straight, colored, 5 μ long, oozing out and staining the surface of the matrix.

On dead alders, Maine (Blake), New York (Peck).

Sec. Karsten ostiola exserted, 1 mm. long, but spece. sent by him have short, obtuse ostiola just like the American specc.

A. tuberculósum, (Schw.)

Sphæria tuberculosa, Schw. Syn. Car. No. 164.
Xylosphæria (*Anthostoma*) *tuberculosa*, Cke. Syn. 3936.

Perithecia ovate, immersed, glabrous. Ostiola superficial, very large, oblong-ovate, tuberculose-roughened, scattered or aggregated, large, black but without any black spot, only the sphæriiform ostiola half as large as the perithecia projecting.

On decaying birch wood, Carolina, and on *Robinia viscosa*, in Pennsylvania (Schw.).

A. sapróphilum, E. & E. Journ. Mycol. III, p. 43.

Stroma effused, blackening the surface of the wood but not discoloring it inside, circumscribed, forming black, subelongated, subconfluent, indefinite spots $\frac{1}{2}$–1 cm. or more in extent. Perithecia membranaceous, globose ($\frac{1}{3}$–$\frac{1}{2}$ mm.), buried in the wood and irregularly arranged in groups of 6–10 or more, with their hemispheric-conical ostiola distinctly prominent and finally perforated. Sporidia elliptical, pale brown, 1–2-nucleate, uniseriate, 5–6 x $2\frac{1}{2}$–3 μ. Much resembles *A. melanotes*, B. & Br., but readily distinguished by its much smaller sporidia.

On rotten maple wood, Newfield, N. J.

A. amplísporum, (Cke.)

Fuckelia amplispora, Cke. Grev. XII, p. 51.

Stroma pustulate, at first covered by the epidermis, which is finally stellate-fissured, subprominent, 1–2 mm. diam. Perithecia few, monostichous, 4–6 in a stroma, rather large, globose, thin-walled. Asci cylindrical, short-stipitate. Sporidia uniseriate, elliptical, obtuse, brown, 16 x 10 μ.

On bark, probably of *Quercus*, United States.

DIATRYPÉLLA, De Not.

Schema Sferiac. Ital. p. 28

Stroma and perithecia as in Diatrype. Asci polysporous.

D. quércina, (Pers.)

Sphæria quercina, Pers. Syn. p. 24.
Diatrype quercina, Fr. Summa Veg. Scand. p. 385.
Diatrypella Rousselii, De Not. Sfer. Ital. taf. 32.
Diatrypella quercina, Nits. Pyr. Germ. p. 71.
Exsicc. Cke. F. Brit. Ser. I, 242; II, 219.—Desm. Plantes Crypt. Ed. I, 2052.—Sacc. M. Ven. 1188.—Lin. Fung. Hung. 178.—Sydow, M. March. 544.

Stromata erumpent, surrounded by the substellate-cleft, adherent epidermis, pulvinate, orbicular or angular, rugulose, thick, disk plano-convex, becoming black, mostly solitary, but sometimes 2–3 confluent, 2–4 mm. diam. Perithecia 2–15 in a stroma, in a single or double layer, ovoid, or subangular from compression, $\frac{1}{2}$–$\frac{3}{4}$ mm. diam., attenuated into rather long necks rising through the grayish-white stroma with their conic-hemispherical, quadrisulcate ostiola distinctly erumpent. Asci clavate-fusoid, long-stipitate, p. sp. 80–100 x 9–10 μ. Sporidia numerous, allantoid, yellowish, strongly curved, 8–12 x 2–3 μ.

On oak limbs, Carolina and Pennsylvania (Schw.), Alabama (Berk.), on *Cratægus*, New York (Peck).

Specc. on dead oak limbs from Canada (Dearness, 1665), and on oak limbs from Louisiana (Langlois, 1699), have the stroma smaller (1½–2 mm.) and, except the small, black disk, closely invested by the adherent epidermis; the perithecia also are smaller (not over ½ mm.), and scarcely compressed. The asci and sporidia are almost the same as in the European specc.

D. verrucæfórmis, (Ehr.)

Sphæria verrucæformis, Ehr. in Plant. Crypt. Exsicc. 280
Sphæria avellana, Pers. Dispos Meth. p. 2.
Diatrype verrucæformis, Fr. Summa, p. 385.
Diatrypella verrucæformis, Nitsch Pyr. Germ. p 76.
Diatrypella informis, E. & E. in Ell. & Evrht. N. A. F. 2d Ser. 2530.
Diatrypella Tocciæana, var. *subeffusa*, E. & E. Journ. Mycol. IV, p. 62.
Diatrypella affinis, Cke. and *D. subglobata*, Cke. & Gerard, Grev. XIV, p. 14.
Exsicc Thum. M. U. 65.—M. March. 169, 466, 755, 1720, 2150 —Krieger, F. Sax 178, 179. Rab. F. E 135, 1022.—Ell. & Evrht. N. A. F. 2d Ser 2530.—Cke. F. Brit. Ser. I, 483. Ser. II, 220.—Roum. F. Gall. 1473

Stromata erumpent through the variously fissured and closely adherent epidermis, subrotund or irregular in shape, wart-like or flattened-pulvinate, 1–6 mm. diam., dirty drab or dark rust-color at first, finally black, white inside, thick, sometimes confluent 2 or 3 together, or forming an uneven crust of considerable extent. Perithecia 25–50 in a stroma, crowded, ovate or globose, ½–¾ mm. diam. or ¾–1 mm. high by ½–¾ mm. broad, with more or less elongated necks. Ostiola obtusely conical, entire or substellate-cleft, slightly projecting. Asci narrow-clavate, long-stipitate, paraphysate, 120–200 x 8–12 μ. Sporidia irregularly crowded, allantoid, slightly curved, yellowish, 5–8 x 1½–2 μ.

On dead limbs of *Quercus*, *Carpinus*, and various deciduous trees, common.

The stroma is broader and mostly flatter and whiter inside than in *D. quercina*, and the sporidia are smaller.

Var. *Spegazziniana*, Sacc. (*D. tuberculata*, Ell. & Calkins in Herb.), on *Sabal serrulata*, Florida, has the small (1 mm.), tuberculiform stromata subseriate. Asci (p. sp.) 75–80 x 10–12 μ. Sporidia 5–7 x 1¼ μ.

The species (sec. Dr. Winter) is specially characterized by the shape of the asci, which are broadest just below the apex.

E. Tocciæàna, De Not. Sfer. Ital. p. 30, tab. 31.

Microstoma verrucæforme, Awd. in Rab. F. E. 253.
Exsicc. Fckl. F. Rh. 2059.—Rab. F. E. 253, 2059 —Rehm Asc. 275.—Thum M. U 65

Stromata scattered or gregarious, small, pustuliform or hemispherical, often angular, with a broad base (1½–2 mm.), with a black, orbicular or elliptical disk, which pierces the epidermis, through which the black circumscribing line is sometimes visible. Perithecia 3–8 in a stroma, tolerably large, globose, with short, thick necks, submonostichous, lying close together. Ostiola either slightly prominent, globose-conical, faintly sulcate or depressed, funnel-shaped and perforated. Asci narrow-oblong-clavate, obtuse at the apex, long-stipitate, polysporous, p. sp. 100–120 x 10–12 μ, with long, filiform paraphyses. Sporidia conglomerated, cylindrical, straight or slightly curved, brownish, 5–7 x 1 μ.

On dead hazel and alder branches, Sandlake, N. Y. (fide Peck).

Characterized by the stroma abruptly enlarged at base.

D. nigro-annulàta, (Grev.)

Sphæria nigro-annulata, Grev. Flor. Ed. p. 385.
Sphæria angulata, Fr. S. M. II, p 390.
Valsa angulata, Fr. Summa, p. 411.
Diatrype angulata, Ces & De Not. Schema, p. 28.
Diatrypella angulata, Nits. Pyr. Germ. p. 81.
Exsicc. Rab. F. E. 1022.—Krieg. F. Sax. 178.—Sydow, M. March. 215b.

Stroma $1\frac{1}{2}$–3 mm. broad, suborbicular at base, narrowed above to conical or hemispherical, angular, erumpent, clothed, except the dark gray or nearly black apex, by the laciniæ of the ruptured epidermis. Perithecia 3–8 in a stroma, submonostichous, ovate or globose, or angular from mutual pressure, with more or less elongated, slender necks and small, scarcely prominent, obtuse, entire or faintly 4-sulcate ostiola, which finally become umbilicate and perforated. Asci narrow-clavate, long-stipitate, paraphysate, 100–180 x 10–12 μ. Sporidia crowded, allantoid, yellowish, slightly curved, 5–8 x $1\frac{1}{2}$ μ.

On dead limbs of *Ilex*? Mississippi (Earle), Texas (Ravenel).

D. Tocciæana and *D. nigro-annulata*, (Grev.), scarcely differ from *D. verrucæformis* except in having fewer perithecia in a stroma; but as in this genus, this character is a variable one, not much importance can be attached to it. On the larger branches and trunks, the stromata are larger, and the number of perithecia greater than in stromata on the smaller branches and twigs, so that the number of perithecia in a stroma can hardly be relied on to separate species otherwise closely allied. (Karst. Mycol. Fenn. II, p. 155).

D. favàcea, (Fr.)

Sphæria favacea, Fr. S. M. II, p. 354.
Sphæria quercina, var *betulina*, A. & S. Consp. p. 11.
Diatrype favacea, Fr Summa, p. 385.
Diatrype verrucæformis, Tul. Sel. Carp. II, p. 100.
Diatrypella favacea, Nitsch. Pyr. Germ. p. 77.
Exsicc. Fckl. F. Rh. 1040.—Thum. F. Austr. 502.—Desm. Pl. Crypt. Ed. I, 2051, Ed. II, 1751.—Ell. N. A. F. 686.

Stromata irregularly scattered, often 2 or more standing near each other, and confluent below, conical from an elliptical (seldom orbicular) base, seated on the inner bark, 3–6 mm. long, 1–2 mm. high, the elliptical disk erumpent, enclosed on each side by the transversely-ruptured epidermis, whitish inside. Perithecia 6–30 in a stroma, in one or two layers, ovate or angular from crowding, with necks more or less elongated, and rather large, rounded, faintly 4–6-sulcate, black, slightly prominent ostiola. Asci oblong-clavate, long-stipitate, p. sp. 70–100 x 9–12 μ, paraphysate, polysporous. Sporidia irregularly crowded, allantoid, yellowish, 6–8 x $1\frac{1}{2}$ μ.

On dead birch limbs, Bethlehem, Pa. (Ell. & Hark.).

D. betùlina, Pk. 25th Rep. p. 101, Pl. I, figs. 27–31.

Exsicc. Ell. & Evrht. N. A. F. 2d Ser. 2347.

Stroma transversely erumpent, prominent, elliptical, black outside, green within, $1\frac{1}{2}$–$2\frac{1}{2}$ mm. diam., loosely surrounded by the ruptured epidermis, penetrating to the wood, on which it forms a white spot surrounded by a black line, nearly plane above and dotted by the numerous, slightly prominent, stellate ostiola. Perithecia crowded in a single layer, elliptical, black. Asci polysporous, stipitate, clavate-fusoid, 100–110 (p. sp. 50–60) x 6–7 μ, paraphysate. Sporidia crowded, allantoid, yellowish, slightly curved, $4\frac{1}{2}$–$5\frac{1}{2}$ x 1 μ.

On bark of dead birches, northern U. S. and Canada.

D. decoràta, Nitschke, Pyr. Germ. p. 79.

Microstoma vulgare, Awd. in Fckl. F. Rh. 1035.

Stromata small, tuberculiform-hemispherical, 2–$2\frac{1}{2}$ mm. broad at base, embraced laterally by the lobes of the ruptured epidermis, the brownish-black apex slightly arched. Perithecia 6–12 in a stroma, monostichous, crowded, globose or ovate, abruptly contracted above into slender, short necks with their stellate-cleft, black ostiola barely erumpent. Asci (p. sp.) 40–48 x 5 μ, with a long, slender stipe and slender, filiform paraphyses. Sporidia crowded, cylindrical, slightly curved or straight, brownish, 5–6 x 1 μ.

On dead limbs of birch, London, Canada (Dearness).

D. Missouriénsis, E. & E. (in Herb.)

Stromata scattered, wart-like or tubercular-hemispherical, $1\frac{1}{2}$ mm. diam., and about 1 mm. high, seated on the inner bark, with a dark circumscribing line penetrating to the wood, erumpent and closely surrounded, except the drab-gray disk (which becomes black), by the lobes of the substellate-ruptured epidermis. Perithecia 8–12 in a stroma, closely packed, monostichous, ovate or angular, about ⅓ mm. diam., with short necks terminating in the depressed-hemispherical, sometimes obscurely radiate cleft, and finally broadlý and irregularly perforated and collapsing ostiola. Asci about 80 x 8 μ, narrow-clavate, polysporous, stipitate. Sporidia 40–50 in an ascus, yellow in the mass, nearly hyaline when seen singly, not strongly curved, 5–6 x 1 μ.

On dead stems of *Corylus*, Missouri (Demetrio).

This has the general appearance of *Diatrype albopruinosa.*

D. citrícola, E. & E. (in Herb.)

Stromata scattered, pustuliform, small (1–2 mm.), apex erumpent, black, lower part buried in the bark, faintly circumscribed, white inside. Perithecia monostichous, 5–15 in a stroma, ⅓ mm. diam., with black, coriaceous walls, sunk to the bottom of the stroma, necks short. Ostiola only slightly prominent, convex, broadly and irregularly dehiscent. Asci clavate, 110–120 x 10–12 μ. Sporidia numerous, inordinate, yellowish in the mass, nearly hyaline when free, slightly curved, obtuse, 6–7 x 1½–2 μ.

On dead twigs of orange trees, Florida (Underwood).

D. discoìdea, Cke. & Pk. 28th Rep. p. 71.

Exsicc. Thum. M. U. 864.—Ell. N. A. F. 492.—Rav. F. Am. 188.

Stroma orbicular or elliptical, transversely erumpent, surrounded by the epidermis, disk naked, plane, grayish-black. Ostiola small, scarcely exserted, nearly smooth or 4–6-sulcate. Perithecia monostichous, ovate, 6–12, or in the larger stromata 20–40, small. Asci polysporous, paraphysate. Sporidia allantoid, nearly straight, 5–6 x ¾–1 μ. When the outer bark is torn off the stroma and perithecia come with it. There are two forms, one with the stroma narrow, and transversely erumpent, the other with the stroma orbicular. Var. *Alni*, Cke., has the stroma orbicular and superficial or nearly so.

On bark of dead birch trees, Northern U. S. and Canada; the var. on alder, in South Carolina and California.

D. opàca, Cke. Texas Fungi, No. 113.

Stromata scattered, tuberculiform, about 1½–2 mm. diam., erumpent-superficial, rufous-brown, becoming black, very white inside, sometimes subconfluent. Perithecia 3–8 in a stroma, about ½ mm. diam. Asci clavate, p. sp. 75–90 x 12–15 μ, polysporous, paraphysate. Sporidia allantoid, crowded, moderately curved, yellowish-hyaline, 6–8 x 1½ μ.

On dead limbs of *Ilex opaca*, Texas (Ravenel), Florida (Calkins).

D. Sássafras, E. & E. (in Herb.)

Stromata cortical, gregarious and subconfluent, small (1–2 mm.), pustuliform, covered by the blackened epidermis, surrounded by a faint circumscribing line which does not penetrate the wood. Perithecia 6–15 in a stroma, ovate-globose, $\frac{1}{3}$–$\frac{1}{2}$ mm. diam., contents soft and gelatinous when young, crowded, with very little stromatic material between and over them, necks short and thick. Ostiola coarse, depressed-hemispherical, black, obscurely stellate-cleft, and irregularly dehiscent. Asci clavate-oblong, stipitate, 100–110 (p. sp. 75–80) x 10 μ; paraphyses? Sporidia numerous, allantoid, yellow in the mass, not strongly curved, 6–7 x 1$\frac{1}{2}$–2 μ.

On dead limbs of *Sassafras*, Newfield, N. J.

The outside appearance is almost exactly that of *Valsaria nigrofacta*, C. & E.

D. herbàcea, E. & E. Journ. Mycol. III, p. 142.

Stroma tuberculiform, 1–2 mm. diam., white inside, tinged with yellow above, but black externally. Perithecia ovate-globose, about $\frac{1}{3}$ mm. diam., rather abruptly contracted above into a short, narrow neck, expanded at the surface of the stroma into a broad, obtuse, quadrisulcate ostiolum. Asci, including the slender base, 100–120 x 10–12 μ; paraphyses soon disappearing. Sporidia crowded in the upper half of the asci, numerous, pale yellowish, cylindrical, curved, 7–8 x 1–1$\frac{1}{2}$ μ. On the same stems was a *Calosphæria* with scattered or subseriate, beaked perithecia having fasciculate asci about 20 x 3$\frac{1}{2}$–4 μ, truncate above, and sporidia 3$\frac{1}{2}$–4$\frac{1}{2}$ x $\frac{3}{4}$ μ.

On dead herbaceous stems (*Ambrosia trifida?*), Louisiana (Langlois).

D. ramulàris, E. & E. Journ. Mycol. l. c.

Stroma tuberculiform, 1–2 mm. diam., bursting out through longitudinal cracks in the bark, penetrating to the wood, which is marked with a black, circumscribing line, subtruncate above, dirty-white within. Perithecia 4–12 in each stroma, globose, with short necks, walls thick and coriaceous. Ostiola only slightly prominent, flat, 4–5-stellate-cleft, finally broadly perforated. Asci broad, clavate, 90–110 x 12–15 μ. Sporidia many, allantoid, yellowish, moderately curved, 6–10 x 1$\frac{1}{2}$ μ.

On dead branches of *Lonicera Japonica*, Pointe á la Hache, La. (Langlois).

D. Pópuli, Ell. & Hol. Journ. Mycol. I, p. 4.

Perithecia cartilaginous, ovate or subangular from mutual compression ($\frac{1}{2}$–$\frac{3}{4}$ mm.), closely packed in an orbicular, lens-shaped stroma 2–3 mm. diam., dark brown outside, dirty-white within, and seated on the surface of the inner bark. Ostiola short, stout, tips obscurely 4-cleft and united in an obscure black disk or piercing the epidermis separately. Asci long-clavate, 100–115 x 15 μ. Sporidia cylindrical, yellowish, curved, 10–12 x 1$\frac{1}{2}$–2 μ. The surface of the wood is blackened except directly under the stroma, where it retains its light color.

On dead limbs of *Populus*, Iowa (Holway).

D. decípiens, E. & E. Journ. Mycol. IV, p. 80.

Stroma erumpent, black (lighter colored at first), orbicular or oblong, 2–6 mm. across, pulvinate, convex or, in the larger specimens, almost plane, whitish inside, with a black, circumscribing line around the base. Ostiola slightly prominent, quadrisulcate, situated in slight depressions. Perithecia monostichous, oblong-ovate, about $\frac{3}{4}$ mm. long, contracted abruptly into a short neck above. Asci (p. sp.) 50–70 x 6–7 μ, polysporous. Sporidia yellowish, allantoid, moderately curved. 3$\frac{1}{2}$–4$\frac{1}{2}$ (or exceptionally 5 μ long) and less than 1 μ thick. This can not be distinguished by its external characters from *Diatrype bullata*, (Hoff.), but internally it is very different.

On bark of *Umbellularia Californica*, Oregon (Carpenter)

D. hysterioìdes, E. & E. Journ. Mycol. II, p. 99.

Exsicc. Ell. & Evrht, N. A F. 2d Ser. 1790.

Stroma erumpent, tuberculiform, prismatic, often deeply quadrisulcate and subcornute, $\frac{1}{2}$–1 mm. diam., yellow inside (about the same shade of yellow as in *Hypoxylon sassafras*, Schw.), often elongated, hysteriiform (1$\frac{1}{2}$–2 mm. long), with a longitudinal furrow above like a *Hysterium*. Perithecia 2–6 in a stroma, $\frac{1}{3}$–$\frac{1}{2}$ mm. diam., with thick, black, coriaceous walls, narrowed above into a short neck, the apex of which is visible on the surface of the stroma as a small, papilliform or sometimes conical ostiolum. Asci clavate-cylindrical, with a slender base, 100–115 x 10–12 μ (p. sp. 75–80 μ), filled with a multitude of allantoid, yellowish, 2-nucleate, 6–7 x 1$\frac{1}{2}$ μ sporidia. The species is well characterized by its peculiar stroma. The color of the young stroma is much deeper, orange-red, but the color finally disappears.

On a decorticated poplar limb, in a willow jungle, Louisiana (Langlois).

D. Vìtis, E. & E. Proc. Acad. Nat. Sci. Phil. July, 1890, p. 225.

Stroma tuberculiform-hemispherical, about 1 mm. diam., erumpent-superficial, black inside. Perithecia 1–4 in a stroma, globose, ½ mm. diam., black and shining within, contracted above into a short neck. Ostiola scarcely prominent, quadrisulcate. Asci polysporous, 75–80 x 10–12 μ, clavate-cylindrical, rather abruptly contracted below into a stipitate base, and surrounded by obscure paraphyses. Sporidia allantoid, yellowish-hyaline, 6–7 x 1½ μ, not strongly curved. The surface of the wood beneath the stroma, as well as the inner surface of the bark, is marked with a black circumscribing line.

On dead vines of *Vitis bipinnata*, Bayou Chene, La. (Langlois).

D. Demetriònis, E. & E. l. c.

Stroma pulvinate, depressed-hemispherical, orbicular, slate-black, 1½–2 mm. diam., penetrating to the wood, which is marked with a black circumscribing line, closely embraced by the superficial layer of the bark, which forms a narrow, adnate margin; inner substance whitish. Ostiola only slightly prominent, distinctly but not deeply radiate-cleft. Perithecia of medium size, globose or angular from mutual pressure. Asci slender-clavate, 75–80 μ long (p. sp. 35 x 6 μ). Sporidia crowded, pale yellowish in the mass, nearly hyaline when separated, minute, allantoid, slightly curved, about 4–5 x 1–1½ μ This comes near *D. exigua*, Winter, which is also on willow limbs, but that species is said to have the stroma very small ("minutissimis") and the sporidia 8 x 1½ μ. In the Colorado specimens we found no sporidia over 5 μ long, and mostly only about 4 μ. The general appearance is almost exactly that of *Diatrype disciformis.*

On dead limbs of *Salix chlorophylla*, Wet Mountain Valley, Colorado (Demetrio).

D. áspera, (Fr.)

Sphæria aspera, Fr. S. M. II, p. 354.
Diatrype aspera, Fr. Summa, p. 385.
Diatrypella aspera, Nits. Pyr. Germ. p. 74.

Stromata scattered or subconfluent, rising from a suborbicular base, cylindrical or prismatical, strongly erumpent, sides abrupt and closely surrounded by the adherent lobes of the ruptured epidermis, whitish inside, apex bare, black, plane or convex, and roughened by the thick, rough, conical or cylindrical ostiola. Perithecia 4–12 in a stroma, monostichous, often circinating around a central one, rather large, subglobose, or subcompressed, suddenly attenuated into a very short neck. Asci subcylindrical, with a long stipe and filiform paraphyses, polysporous, p. sp. 112–120 x 10–12 μ. Sporidia conglobate, allantoid, yellowish, 6–7 x 1½ μ.

On *Cornus*, New York State (Peck).

D. Rhòis, (Schw.)

Sphæria Rhois, Schw. Syn. Car. No. 62.

Stroma wart-like, obovate, rugose, black (cinereous within), 1½–3 mm. diam., subconfluent or densely gregarious, more or less covered by the fibers of the wood. Perithecia few in a stroma. Ostiola conical, subprominent. Sporidia numerous, hyaline, slightly curved, 5 x 1 μ.

On dead stems of *Rhus radicans*, Carolina and Pennsylvania (Schw.), and on *Rhus diversiloba*, California (Harkness).

In the specc. in Herb. Schw. no asci could be found.

D. ribèsia, (Schw.)

Sphæria ribesia, Schw. Syn. N. Am. 1257.

Effused, consisting of many small, aggregated, confluent pulvinuli: very rough above from the abundant, prominent ostiola, which are thick, short, irregular, with an elongated, hysteriiform opening. Perithecia rather large, mostly obovate, in a scanty, whitish stroma. Attached to the wood, and erumpent through the epidermis in strips reaching as much as 24 mm. long, by 6 mm. wide.

On dead stems of *Ribes rubrum*, Bethlehem, Pa. (Schw.).

Specc in Herb. Schw. have asci (p. sp.) 40–50 x 10 μ. Sporidia allantoid, hyaline, slender, 5–7 x 1–1½ μ. This is different from *Valsa ribesia* Karst (see p. 466).

D. Cephalánthi, (Schw.)

Sphæria Cephalanthi, Schw. Syn. Car. No. 59.
Diatrypella Cephalanthi, Sacc. Syll. 769.
Exsicc. Ell. N. A. F. 777.

Stromata scattered, 2–3 mm. diam., scarcely penetrating to the wood, at first small and subtubercular-erumpent, black, apex generally traversed by a single, longitudinal, deep furrow, giving the appearance of a *Hysterium*, but this furrow is soon obliterated as the apex of the stroma protrudes itself still further, assuming an orbicular or elliptical form, 1–3 mm. diam. and roughened by the protruding, black, rough, subconical ostiola which are often 3–4-stellate-cleft. Perithecia 6–20 in a stroma, closely packed, ovate or subangular, about ½ mm. diam., lying in and covered by the scanty, whitish substance of the stroma;

walls thick, black, coriaceous, necks short. Asci fusoid, 110–115 (p. sp. 75–80) x 8–10 μ, polysporous, stipitate, paraphysate. Sporidia allantoid, yellowish-hyaline, not strongly curved, 6–9 x 1½ μ.

On dead stems of *Cephalanthus occidentalis*; common.

D. spíssa, (Schw.)

Sphæria spissa, Schw. Syn. N. Am. 1253.

Crowded in longitudinal cracks in the wood, forming narrow strips often half an inch long, acuminate at each end. and becoming black, considerably elevated in the middle, surface very rough from the prominent perithecia. Stroma scarcely any. Perithecia (½–¾ mm. diam.), immersed in the wood itself. Ostiola prominent, subconical. Asci (p. sp.) 45–55 x 10–12 μ. Sporidia hyaline, slightly curved, 5 x 1½ μ.

On dry oak wood, Bethlehem, Pa. (Schw.). (Measurements from spec. in Herb. Schw.).

D. variolósa, (Schw.)

Sphæria variolosa, Schw. Syn. N. Am. 1272.

Stromata pulvinate, erumpent, about 2 mm. diam., flattish, subconfluent, closely embraced by and partly covered with fragments of the ruptured epidermis, black, rugose. Ostiola numerous, short, quadrisulcate. Perithecia obovate or pyriform, rather large, polystichous, immersed in the white substance of the stroma. Asci clavate-cylindrical, 75–80 x 12–15 μ, with abundant paraphyses. Sporidia yellowish-hyaline, allantoid only slightly curved, with a nucleus in each end, 5–6 x 1¼ μ.

On young branches of *Populus Italica*, Bethlehem, Pa.

D. próminens, (Howe).

Diatrype prominens, Howe, Bull. Torr. Bot. Club, V, p. 42.

Stroma wart-like, orbicular or angular, 1–1½ mm. diam., erumpent and closely girt by the ruptured epidermis, subprominent, black outside, whitish within. Perithecia 4–8 in a stroma, subglobose, about ⅓ mm. diam. necks short, with their ostiola obtusely conical, slightly prominent and finally broadly and irregularly perforated. Asci clavate-stipitate, 75–80 (p. sp. 55–60) x 10–12 μ, paraphysate, polysporous. Sporidia allantoid, yellowish-hyaline, moderately curved, 6–8 x 1–1½ μ.

On bark of dead *Platanus*, New York and California.

The diagnosis is from New York specc. from Peck. *Diatrype prominens*, Cke. & Hark. is distinct from this, having 8-spored asci.

D. deústa, Ell. & Martin. Am. Nat. Oct. 1882, p. 809.

Exsicc. Ell. N. A. F. 1184.

Pustules oblong or elliptical, 1–3 mm. diam., covered by the blackened and raised epidermis, scattered irregularly or often longitudinally-subconfluent, forming blackened strips several centimeters long and 1–2 mm. wide, uneven on the surface, and with the appearance of elongated swellings or flat ridges roughened by the seriately-erumpent, obtuse, imperfectly quadrisulcate, black ostiola. Perithecia in clusters of 6–15, buried in the somewhat blackened substance of the matrix but without any distinct, circumscribing line, flask-shaped, brown, about $\frac{1}{3}$ mm. diam. or rather less, necks short. Asci clavate, 75–85 (p. sp. 60–65) x $5\frac{1}{2}$–7 μ, polysporous, stipe comparatively short. Sporidia allantoid, yellowish, moderately curved, $5\frac{1}{2}$–7 x $1\frac{1}{4}$ μ.

On petioles of *Sabal serrulata*, Florida (Martin).

D. Liriodéndri, (Schw.)

Sphæria Liriodendri, Schw. Syn. Car. No. 60.

Erumpent, prominent, subconfluent, 4–6 mm. diam., seated on the bark (which is blackened to the wood) and at first surrounded by it: surface pulverulent, roughened by the indistinct ostiola, which finally become more prominent, at length black inside. Perithecia 3–5, subglobose, with short necks. Sporidia allantoid, hyaline, curved, 6 x $1\frac{1}{2}$ μ.

On dead limbs of *Liriodendron*, Salem, N. C. (Schw.).

The spec. in Herb. Schw. has the stroma smaller than stated above, only $1\frac{1}{2}$–2 mm.

D. Fróstii, Pk. Bot. Gaz. III, p. 35.

Exsicc. Ell. & Evrht. N A. F. 2d Ser. 2529.

Stromata wart-like, 1–2 mm. diam., seated on the inner bark which is blackened and marked around each stroma with a circumscribing line penetrating to the wood, erumpent, surrounded and partly covered, except the apex, by the ruptured epidermis, black outside, greenish within, forming a white spot on the wood beneath. Perithecia ovate-globose, 10–15 in a stroma, $\frac{1}{3}$–$\frac{1}{2}$ mm. diam. Asci polysporous, cylindric-clavate, stipitate, paraphysate, p. sp. 75–80 x 12–15 μ. Sporidia yellowish, allantoid, moderately curved, 7–8 x 2 μ, with a nucleus in each end.

On dead maple limbs, Vermont (Frost), Connecticut (Thaxter), California (Harkness), Iowa (Holway).

The ostiola vary from obtuse-conical to subelongated, conic-cylindrical, entire, often obscure.

D. irreguláris, C. & E. Grev. VI, p. 92.

Stromata wart-like, 2–3 mm. diam., closely covered by the epidermis, which is pierced by the apex of the stroma, often confluent, forming a flat blister-like swelling 1 cm. broad, and entirely covered by the epidermis, except where it is pierced by the small, black apices of the separate stromata, cinereous inside, not penetrating to the wood, and not circumscribed; other stromata on a decorticated part of the same limb, are seated on the wood; these are oblong and subseriate-confluent. Perithecia 8–12 in a single stroma, ovate, black, about ⅓ mm. diam., necks short, with their obtusely conical, sometimes quadrisulcate, and finally irregularly dehiscent ostiola slightly prominent. Asci oblong-clavate, stipitate, 100–115 x 10–12 μ (p. sp. 40–50 μ long), polysporous.

On dead limbs of *Pyrus communis*, Newfield, N. J.

Our specimen is not well matured, and the sporidia refuse to separate and leave the asci. Sec. Cooke they are 8 μ long. This seems more closely allied to *D. verrucæformis* than to *D. quercina*, and we are inclined to regard it as only a var. of that species.

D. subfúlva, (B. & C.)

Diatrype subfulva, B. & C. Grev. IV, p. 97.
Diatrype olivacea, C. & E. Grev. VI, p. 9.
Diatrypella subfulva, Sacc. Syll. 764.

"Erumpent, convex, at first olivaceous, then black. Perithecia large, few. Ostiola obtuse. Asci clavate, subsessile. Sporidia numerous, yellow, linear, obtuse, slightly curved, 10–12 μ long."

On dead limbs of *Nyssa*, Carolina and New Jersey.

Our spece. of *D. olivacea*, C. & E., are without fruit. The species is now considered by Cooke as synonymous with *D. subfulva*, B. & C.

D. ánnulans, (Schw.)

Sphæria annulans, Schw. Syn. N. Am. 1270.
Diatrype annulans, Cke. Grev. XIII, p. 38.

Stromata conic-tuberculiform, truncate above, 1–1½ mm. diam., seriate-subconfluent in rings surrounding the limb or extending longitudinally, white inside, rarely solitary. Ostiola prominent on the truncate apex of the stroma, quadrisulcate, stout and short. Perithecia (4–6 in a stroma), contracted into short, black necks above. Asci (p. sp.) 45–50 x 15–20 μ, with a slender base. Sporidia yellow in the mass, nearly hyaline when single, allantoid, 6–7 x 1¼–1½ μ.

On decorticated branches of various kinds of soft wood, Carolina and Pennsylvania (Schw.).

Measurements from spec. in Herb. Schw.

D. enteroleùca, (Fr.)

Sphæria enteroleuca, Fr. S. M. II, p. 381, Schw. Syn. N. Am. 1314.
Valsa enteroleuca, Cke. Syn. 1971.

"Conceptacle crustaceous, rough, nearly free, about as large as a pea. Perithecia minute, subglobose, numerous, irregularly scattered in the white substance of the stroma. Ostiola rounded, often short-cylindrical, sometimes elongated and tentacular."

On decorticated branches, Bethlehem, Pa. (Schw.).

Spece. in Herb. Schw. have the stroma pustuliform-tubercular, erumpent, 1–1½ mm. diam., closely embraced at first by the fibers of the wood, then bare, black (white inside). Perithecia crowded, small, about ⅓ mm. diam. Ostiola only slightly prominent, rather flattened, sulcate. Asci 50 x 12–15 μ (p. sp.). Sporidia allantoid, 5-6 x 1 μ.

D. prorúmpens, (Wallr.)

Sphæria prorumpens, Wallr. MS. Fr. S. M. II, p. 357, Schw. Syn. N. Am. 1247

Perithecia globose, irregularly aggregated, nestling in the bark, with short necks, covered above with a thin, black stroma, erumpent in elongated, irregular strips or series. The circumscribing line at the base surrounds not a single disk with its subjacent perithecia, but whole groups of perithecia.

On branches of *Robinia*, Bethlehem, Pa. (Schw.).

Asci and sporidia about as in *D. Rhois*, which this much resembles.

D. friábilis, (Fr.)

Sphæria friabilis, Fr. S. M. II, 361. Herb. Schw. and Schw. Syn. N. Am 1260.
Diatrype friabilis, Cke. Grev XIII, p. 38.

Stroma very prominent, orbicular or elliptical, 2–4 mm. broad, convex or depressed, subplicate, pale white within, indistinctly circumscribed around the base. Ostiola hidden. Asci (p. sp.) 45–50 x 10–12 μ. Sporidia allantoid, not much curved, 5–7 x 1½ μ.

On *Ilex prinoides* and *Viburnum prunifolium*, Carolina and Pennsylvania (Schw.).

The measurements of asci and sporidia are from spec. in Herb. Schw.

D. melaspérma, (Fr.)

Sphæria melasperma, Fr. S. M. II, p. 389.
Diatrypella melasperma, Sacc Syll. 767

In the young state the pustules are innate in the bark, and circumscribed by a dark line, stroma pale yellowish and without perithecia. Disk erumpent, linear, black. When mature, pustules elliptical, 4 mm. long, depressed, adnate to the inner bark. Conceptacle nearly free, thin, fragile, opake, black, attached to the epidermis so that when this is torn away, the scutelliform base remains. Stroma pulveraceous, olivaceous, becoming dark brown. Perithecia larger than in the allied species, few, aggregated, erect-convergent, ovate, fragile. Disk erumpent, elliptical, minute, black. Ostiola obsoletely prominent, scattered. Asci clavate, stipitate, polysporous. Sporidia numerous, crowded, allantoid, yellowish, 5–7½ μ long.

On bark of dead birch trees, Pennsylvania (Schw.).

SUBORDER. DOTHIDEÀCEÆ.

Compound. Stroma pulvinate, effused or linear, coriaceous or subcarbonaceous, black. Perithecia reduced to mere cells, (generally peripherical) in the substance of the stroma, and not separable from it, papillate or simply perforated. Asci 4–8-spored. Sporidia hyaline, yellowish or brown.

PHYLLÁCHORA, Nitschke.

in Fckl. Symb. p. 216.

Stroma variable in shape, elliptical, oblong or lanceolate, covered by the epidermis, black, more or less roughened by the apices of the small, ascigerous cells, which are buried in the stroma, and often at first contain spermatia. Sporidia ovate, elliptical or oblong, continuous and mostly hyaline. Asci subcylindrical.

* *On dicotyledonous plants.*

Ph. Dalibárdæ, Pk. 27th Rep. p. 109, Pl. I, figs 7–9.

Stromata scattered, innate and subprominent on both sides of the leaf, which is stained immediately around them reddish-brown, small, less than ½ mm. diam., subtuberculose-roughened, black. Ascigerous cells, 6–12 in a stroma, subglobose, 75–80 μ diam. Asci oblong-fusoid, 50–60 x 10 μ. Sporidia crowded-biseriate, oblong-fusoid, hyaline, continuous, 12–16 x 3–3½ μ.

On living leaves of *Dalibarda repens*, Forestburg, N. Y.

Described from a specimen sent by Mr. Peck.

Ph. Wittróckii, (Erikss.)

Dothidea Wittrockii, Erikss. Fungi Scand. No. 40.
Phyllachora Wittrockii, Sacc. Syll. 5130.

Stromata effused, surrounding 2 or 3 of the swollen, upper internodes of the stem, 5–25 mm. long, very fragile, black, nearly smooth. Ascigerous cells numerous, buried in the stroma, white. Asci cylindrical, subsessile, 70–80 x 6–10 μ, 8-spored. Sporidia uniseriate, ellipsoid or ovoid, continuous, hyaline, 12–20 x 4–7 μ. Paraphyses filiform.

On stems of *Linnœa borealis*, White Mountains, N. H. (Farlow).

Mr. Kelsey finds, on the same host, in Montana, the minute, scattered stromata of some Dothideaceous fungus on the upper side of the leaves but entirely sterile.

Ph. coccòdes, (Lev.)

Sphæria coccodes, Lev. Ann. Sci, Nat. III, (1845) p. 50.
Dothidea Cubensis, Berk. (fide Cooke).
Physalospora coccodes, Sacc. Syll. 1717.
Dothidea coccodes, Cke. Syn. 1240.
Exsicc. Ell. & Evrht. N. A. F. 2d Ser. 1950.

Stromata scattered, amphigenous, flattened-pulvinate, ½–1½ mm. diam., black, roughened by the obtuse, prominent ostiola so as to appear under the lens minutely tuberculate. Ascigerous cells subovate, immersed, white inside, about 200 μ diam., not crowded. Asci clavate, stipitate, 85–90 x 15 μ, paraphysate, 8-spored. Sporidia overlapping-uniseriate, elliptical (sometimes subinequilateral,) hyaline at first, becoming yellowish, 12–16 x 6–8 μ, continuous.

On leaves of *Olea Americana*, Florida (Calkins).

This is given as *D. coccodes* (Lev.) on the authority of Dr. Cooke. It does not agree well with Saccardo's description of *Physalospora coccodes*, though the same reference to Sci. Nat. is given by Berk. in his Cuban Fungi under *Dothidea coccodes*, Lev., as that given by Sacc. under *Physalospora coccodes*. At any rate, the spece. issued in N. A. F. can not be *Physalospora*, having the characteristic cells of the *Dothideaceæ* immersed in the stroma.

Ph. Trifòlii, (Pers.)

Sphæria Trifolii, Pers. Syn. p. 30.
Dothidea Trifolii, Fr. Summa, Veg. Scand. p. 387.
Phyllachora Trifolii, Fckl. Symb. p. 218.
Exsicc. Fckl. F. Rh. 1022 —Rab. Herb Mycol. 657.—id. F. Eur. 1331.—Kze. F. Sel. 270 Thum. M. U. 2269.—Sydow, M. March. 697.—Linhart F. Hung. 375.

Stromata hypophyllous, gregarious, collected in small, mostly elongated groups extending along the nerves of the leaf, black, subglobose or tuberculiform, prominent, often confluent. Sporidia (sec. Cke. in Grev. XIII, p. 63) elliptical, hyaline, continuous, 20 x 10 μ.

On leaves of *Trifolium repens*, Ohio (Morgan), on *Trifolium*, New York (Peck).

The spece. we have seen are immature or sterile.

Ph. Ambròsiæ, (B. & C.)

Dothidea Ambrosiæ, B. & C. Grev. IV, p. 105.
Phyllachora Ambrosiæ, Sacc. Syll. 5127.

Stroma convex, shining. Asci linear, shorter than the slender paraphyses. Sporidia uniseriate, elliptical, hyaline.

On leaves of *Ambrosia elatior*, Carolina, and on *A. artemisiæfolia*, Alabama (Beaumont).

D. Ambrosiæ, Schw. Syn. N. Am. 1935 (sec. Cke. Grev. XIII, p. 43), is without fruit and not a *Phyllachora*.

Ph. Lespedèzæ, (Schw.)

Sphæria Lespedezæ, Schw. Syn. N. Am. 1488.
Phyllachora Lespedezæ, Cke. Grev. XIII, p. 63.
Exsicc. Ell. N. A. F. 487.

Stromata gregarious or subconfluent, conic-convex, black and shining, about 1 mm. diam., epiphyllous, seated on indefinite, yellowish spots. Ascigerous cells few (1–4), small, pale. Asci? Sporidia (sec. Cke. in Grev. XIII, p. 63) elliptical, continuous, hyaline, 20 x 10 μ.

On leaves of various species of *Lespedeza*. Not uncommon, but mostly sterile.

Ph. Beaumóntii, (B. & C.) Grev. XIII, p. 63.

Epiphyllous. Stroma hemispheric-convex, opake, black, contracted at base. Asci clavate. Sporidia inordinate, elliptical, continuous, hyaline, 8–10 x 4 μ.

On leaves of *Cerasus*, Alabama.

Ph. pìcea, (B. & C.)

Dothidea picea, B. & C. Grev. IV, p. 105.
Phyllachora picea, Sacc. Syll. 5129, Cke. Syn. 1073.

Forming little orbicular, papillose patches. Asci short, clavate. Sporidia hyaline, narrow, cymbæform.

On twigs of *Vitis æstivalis*.

Ph. ténuis, (B. & C.)

Dothidea tenuis, B. & C. N. Pac. Expl. Exp. No. 162.
Phyllachora tenuis, Sacc. Syll. 5114, Cke. Syn. 1077.

Pitch-black. Stroma subpenetrating, thin, minutely granulose from the slightly prominent cells. Sporidia hyaline, oblong-clavate.

On leaves of *Bauhinia*, Nicaragua (Wright).

Ph. inclùsa, (B. & C.)

Dothidea inclusa, B. & C. N. Pac. Expl. Exp. No. 164.
Phyllachora inclusa, Sacc. Syll. 5116, Cke. Syn. 1067.

Stromata scattered, innate, amphigenous, orbicular, about ½ mm. diam., black, surface convex and slightly papillose, slightly prominent. Asci stipitate, paraphysate, clavate, p. sp. 75–100 x 18–20 μ, 8-spored. Sporidia obliquely uniseriate or biseriate, elliptical, acutely so at first, but finally more obtuse, greenish-granular, 15–22 x 9–10 μ.

On leaves of *Jacquinia*, Nicaragua (Wright).

A few sporidia were seen with the endrochrome divided in the middle. Described from specc. collected by Wright in Nicaragua.

** *On monocotyledonous plants.*

Ph. Gráminis, (Pers.) **(Plate 40)**

Sphæria Graminis, Pers. Syn. p. 30.
Dothidea Graminis, Fr. Summa, p. 387.
Phyllachora Graminis, Fckl. Symb. p. 216.
Hypopteris Luzulæ, Rab in Bad. Krypt. 335.
Phyllachora Bromi, Fckl. Symb. p. 216.

Exsicc Fckl. F Rh. 1018, 2264 —Kze. F. Sel. 269, 361.—Rab. F. E. 533 —Rehm Asc. 647. Thum. F. Austr. 176.—Sydow, M. March. 84, 86, 254.—Desm. Pl. Cr. Ed. I, 968.—id. Ed. II, 268.—Cke. F. Brit. Ser. I, 678.—Ell. N. A. F. 484.

Stromata scattered or confluent, penetrating the leaf and more or less prominent on both sides, covered by the adnate, black and shining epidermis, of various forms, mostly oblong or lanceolate, black, uneven, rugulose, about 1 mm. long. Ascigerous cells subseriate. Ostiola obscure. Asci short-stipitate, cylindrical, 75–80 x 7–8 μ, 8-spored, paraphysate. Sporidia obliquely uniseriate, ovate, simple, generally with one large nucleus, hyaline, 6–12 x 4–5 μ. Spermatia (in the young cells) subfalcate, 16 x 1½–2 μ, pluriguttulate.

On living leaves of various grasses; common.

Ph. Càricis, (Fr.)

Dothidea Caricis, Fr. Summa Veg. Sc. p. 397.
Phyllachora Caricis, Sacc. Syll. 5242.

Covered, uneven, confluent, black, tuberculose from the astomous, prominent perithecia. Allied to *P. Graminis*, but immersed in the unchanged parenchyma of the leaf, the prominent perithecia (cells) distinct, but covered by the blackened epidermis.

On leaves of *Carex Pennsylvanica*, Albany, N. Y. (Peck).

Cooke in Grev. XIII, p. 63, states that the asci are clavate and the sporidia elliptical, continuous and hyaline.

Ph. Cynodóntis, Niessl, Not. Pyr. p. 74.

Exsicc. Rehm Asc. 377.—Rab. F. E. 2241.—Thum. M. U. 1067.

Stromata scattered, suborbicular or subelongated, small, mostly not over ½ mm., not strongly prominent, black, ascigerous cells small.

Asci clavate-subfusoid, paraphysate, p. sp. 45–50 x 12–15 μ. Sporidia crowded, irregularly biseriate or sometimes obliquely uniseriate, acutely elliptical, yellowish-hyaline, continuous, 12–15 x 5–6 μ.

On leaves of some grass, Alabama (Atkinson).

Differs from *Ph. gramius* in its smaller stromata and crowded sporidia. The Alabama specc. agree with those in the Exsiccati quoted.

Ph. melanóplaca, (Desm.)

Dothidea melanoplaca, Desm. Not. 17, p. 33.
Phyllachora melanoplaca, Sacc. Syll. 5151.

Amphigenous; spots large, indeterminate, opake. Ascigerous cells numerous, small, connate, whitish within, becoming black. Asci ample, 50 μ long. Sporidia ovoid, hyaline, subolivaceous.

On languishing or dead leaves of *Veratrum viride*, Catskill and Adirondack Mts., N. Y. (Peek).

Ph. Dasylírii, (Pk.)

Dothidea Dasylirii, Pk. Bot. Gaz. VII, p. 57.
Phyllachora Dasylirii, Sacc. Syll. 5149, Cke. Syn. 1123.

Stromata amphigenous, small, narrow-elliptical, oblong or linear, black, for a long time covered by the epidermis, which is at length split longitudinally. Ascigerous cells few, white inside. Asci oblong or subcylindrical. Sporidia crowded or biseriate, oblong or clavate-oblong, colorless, 27–40 x 16–18 μ.

On leaves of some species of *Dasylirion* (D. Wheeleri)?, Arizona (Pringle).

Ph. Júnci, (Fr.)

Sphæria Junci, Fr. S. M. II, p. 428.
Dothidea Junci, Fr. Summa, p. 387.
Phyllachora Junci, Fckl. Symb. p. 216.
Exsicc Fckl. F. Rh. 1026.—Rab. F. E. 755.—Desm. Pl. Cr. Ed. I, 720.—Cke. F. Brit. 2d Ser. 243.

Stromata gregarious, often confluent, sunk in the substance of the culm, more or less elongated, thin, brown or nearly black, covered by the pustulate-raised epidermis, which finally becomes brown and splits with a narrow lanceolate opening. Ascigerous cells sunk in the stroma, finally emergent, forming globose or subcompressed tubercles, subseriate, small. Asci short-stipitate, cylindrical, 8-spored, 60–85 x 7–8 μ. Sporidia obliquely uniseriate or partly biseriate, oblong, binucleate (becoming uniseptate)? yellowish-hyaline, 9–10 x 3–3½ μ.

On various species of *Juncus*; common, but often sterile or poorly developed.

*** *On Acotyledonous plants.*

Ph. Ptéridis, (Reb.)

Sphæria Pteridis, Reb. Neom. p. 314, tab. I, fig. 3.
Dothidea Pteridis, Fr. S. M. II, p. 555.
Phyllachora Pteridis, Fckl. Symb. p. 218.
Rhopographus Pteridis, Winter Die Pilze, II, p. 915.
Exsicc. Thum. M. U. 175, 2176.—Thum. F. Austr. 1053.—Rehm Asc. 522. 581.—Romell F. Scand. 82.—Erikss. F. Scand. 39.

Stromata hypophyllous, elongated, oblong, following the nerves, cinereous-black outside, deeper black within, 2–3 mm. long, ½–¾ mm. wide, minutely punctate from the ostiola. Asci numerous, cylindrical, 8-spored, 64 x 14 μ. Sporidia biseriate, elliptical, hyaline, simple, 2-nucleate, 8–10 x 5–6 μ. Paraphyses none. On fronds of *Pteris aquilina*, common. New England to Carolina, and west to Montana.

None of the spece. in the Exsiccati quoted nor, in fact, any we have seen, are ascigerous, but those in Thüm. M. U. (collected in New Jersey) have clavate-oblong stylospores 12–20 x 2½–3 μ. The description of the asci and sporidia is taken from Sacc. Syll.

Ph. leptostromoìdea, Cke. Grev. XIII, p. 64.

Stroma irregular, minute, flattened, black, sometimes confluent. Ostiola scarcely visible. Asci clavate, 8-spored. Sporidia elliptical, continuous, hyaline, 8 x 3 μ."

On fronds of *Pteris*, South Carolina (Ravenel). Resembles a *Leptostroma.*

Ph. flabélla, (Schw.)

Sphæria flabella, Schw. Syn. N. Am. 1489.
Phyllachora flabella, Sacc. Syll. 5154, Cke. Syn. 1126.
Exsicc. Thum. M. U. 176.—Rehm Asc. 473.—Rav. F. Am. 99.—Ell. N. A. F. 486.

Stromata as in *Ph. Pteridis*, only epiphyllous, covered by the epidermis and often interrupted and tubercular. Ascigerous cells rather large, white inside. Asci cylindrical, sessile, 60–90 x 8–10 μ, paraphysate, 8-spored. Sporidia mostly overlapping-uniseriate, acutely elliptical, or ovate-elliptical, subinequilateral, yellowish-hyaline, simple, 10–12 x 5–6 μ.

On dead fronds of *Pteris aquilina*, New England, New Jersey and New York; probably also in other localities where its host is found, but, like many other species of this genus, it is often found sterile. It can, however, be easily recognized by its peculiar habit.

In the following species the fructification is unknown.

Ph. pérmeans, (B. & C.)

Dothidea permeans, B. & C. N. Pac. Expl. Exp. No. 161.

Spots irregular, substellate-orbicular, black, permeating the entire matrix, elevated-subrugose. Ostiola indistinct.

On leaves, Nicaragua.

Ph. oxálina, E. & E. (in Herb.)

Stromata amphigenous, minute ($\frac{1}{4}$–$\frac{1}{3}$ mm.), on subindefinite, pale spots, black. Stylospores fusoid-oblong, hyaline, 7–8 x 2–2½ μ, binucleate, becoming uniseptate.

On leaves of *Oxalis corniculata*, Delaware (Commons).

Ph. látitans, (Fr.)

Dothidea latitans, Fr. S. M. II, p. 522.
Phyllachora latitans, Sacc. Syll. 5164.

Stromata innate in the parenchyma of the leaf, and not readily seen until the portion of the leaf covering them is cut away. Cells numerous, subrotund, crowded, white.

On leaves of *Vaccinium Vitis-idæa*, Nugarsunguak, Greenland.

Resembles an old *Phacidium*.

Ph. Pánici, Schw. Syn. N. Am. 1925 (sub *Dothidea*).

Spots gray, thin, linear, longitudinally and laterally confluent between the nerves of the leaf, without any distinct fibrils. Cellules minute, subseriately arranged on the opposite side of the leaf. Spots numerous but minute.

On leaves of different species of *Panicum*, Bethlehem, Pa. (Schw.); also common around Newfield, N. J.

Ph? Polemònii, Hark. Bull. 7, of the Cal. Acad. p. 446.

Amphigenous, spots black, roundish, 2–6 mm. broad, papillate and shining.

On living leaves of *Polemonium humile*, Sierra Nevada Mts., California; alt. 7000 ft.

Very showy, but, as yet, only found sterile.

Ph. Eupatòrii, (B. & C.)

Dothidea Eupatorii, B. & C. Grev. IV, p. 105.
Phyllachora Eupatorii, Sacc. Syll. 5198.

Forming black, thin, irregularly erumpent, granulated patches an inch or more long. Asci short, obovate.

On stems of *Eupatorium coronopifolium*, North Carolina.

Ph. élegans, Schw. Syn. N. Am. 1914 (sub *Dothidea*).

Spots half an inch across, olive-green, at length black, often confluent, composed of branching fibrils, radiate-divergent and anastomosing; shape of the spots always elliptic-acuminate. Cells minute, abundant, shining-black, often scattered outside the limits of the spots.

On *Phytolacca* stems, Carolina and Pennsylvania (Schw.).

Ph. Phlògis, Schw. Syn. N. Am. 1931 (sub *Dothidea*).

Spots dark brown, subindeterminate, narrow, sublinear, not covered, composed of an aggregation of minute cellules.

On stems of *Phlox undulata* (cult.), Bethlehem, Pa. (Schw.).

Ph. Cheonpòdii, Schw. Syn. N. Am. 1919 (sub *Dothidea*).

Spots of medium size, formed of radiating fibrils, dark brown, tolerably thick, scarcely branching, elevated in the center, loaded with perithecia or cells, rather large, covered, globose-depressed and ostiolate. Spots often confluent.

On large stems of *Chenopodium*, Bethlehem, Pa. (Schw.).

Ph. ramòsa, Schw. Syn. N. Am. 1912 (sub *Dothidea*).

Spots oblong, black, with branching fibrils radiating only longitudinally, partially covered at first. Cellules seriate, crowded, at length ruptured. Spots not exceeding 4–6 mm. long.

On herbaceous stems, especially of *Chenopodium*, Bethlehem, Pa. (Schw.).

Ph. effùsa, Schw. Syn. N. Am. 1893 (sub *Dothidea*).

Broadly effused on determinate black spots consisting of innumerable minute perithecia, elevated, black, crowded, innate and confluent with the epidermis, but more scattered around the margin and seated on a fibrillose subiculum.

On stems of *Helianthus annuus*, Bethlehem, Pa. (Schw.).

This and the four preceding species seem to be referable to *Asteroma*.

The following numbers placed under *Dothidea* by Schweinitz, in his Synopsis of North American Fungi, are (sec. spece. in Herb. Schw.) not ascigerous.

1874. *Dothidea radicalis*, Schw., on roots of trees dug up and exposed, is a *Dothiorella*; sporules hyaline, subglobose, 15–20 μ in the longer diameter. This is quite different from *Sphæria radicalis*, Schw. Syn. N. Am. 1269.

1875. *Dothidea capreolata*, Schw., on decaying branches of *Bignonia capreolata*, is a *Haplosporella*; sporules 10–15 x 4–5 μ.

1876. *Dothidea Robiniæ*, Schw., on young branches of *Robinia*. This is a *Haplosporella*; sporules oblong, brown, 18–20 x 10–12 μ.

1878. *Dothidea Viburni dentati*, Schw., on branches of *Viburnum dentatum*, is also a *Haplosporella*; sporules oblong, brown, 12–15 x 4–5 μ.

1881. *Dothidea orbiculata*, Schw., on leaves of *Diospyros*, and

1883. *Dothidea Rhois*, Schw., on leaves of *Rhus glabra* and *R. copallina*, are both entirely sterile.

1886. *Dothidea Missouriensis*, Schw., on pecan nuts in the market, is only a thin, black, sterile crust.

1888. *Dothidea gramma*, Schw., epiphyllous, on living leaves of *Stylosanthes;* minute, crowded perithecia forming a narrow, black strip parallel with the margin of the leaf, on each side. The perithecia are filled with minute, hyaline sporules.

1889. *Dothidea Brachystemonis*, Schw., on leaves of *Pycnanthemum* (*Brachystemon*), looks like a *Phyllachora*, but is sterile.

1890. *Dothidea exasperans*, Schw., mere sterile spots. Cooke (Grev. XIII, p. 42) finds *Phyllosticta* sporules 6 x 1½ μ. *Phyllosticta exasperans*, (Schw.).

1891. *Dothidea stipata*, specimen sterile.

1892. *Dothidea culmicola*, Schw., on culms of *Andropogon;* minute, sterile perithecia so thickly crowded as to blacken the culm in irregular patches.

1894. *Dothidea Glycineos*, Schw., on living leaves of *Amphicarpæa* (*Glycine*) *monoica*.

1895. *Dothidea frigoris*, Schw., on dead twigs of *Viburnum*. Specimens of these two numbers are entirely sterile.

1896. *Dothidea fructigena*, Schw., on rotten apples; only a black, sterile crust.

1897. *Dothidea chalybea*, Schw., on dry, denuded wood of *Kalmia*, and

1898. *Dothidea denigrans*, Schw. These two are a mere sterile crust, in the latter with sterile perithecia.

1899. *Dothidea Lauri Bourboniæ*, Schw., on leaves of *Laurus Bourbonia* (Georgia), only small, black, sterile spots.

1901. *Dothidea penicillata*, Schw., on scapes of *Allium*, &c., has the aspect of a *Vermicularia*, only there are no bristles, nor are there any spores.

1909. *Dothidea pomigena*, Schw., on mature apples, apparently the fructigenous form of *Fusicladium dendriticum* (Wallr.).

1910. *Dothidea Gentianæ*, Schw., on fading leaves of *Gentiana ochroleuca;* sterile perithecia on black spots. Cooke (Grev. XIII. p. 42) finds sporules of a *Phoma* 8 x 1 μ.

From 1911–1855, inclusive, all the specimens in Herb. Schw. (except 1041, which is *Actinonema Rosæa*, Lib.) are sterile, indeterminable things, entirely without fruit of any kind, or at most only stylosporous, so that it would hardly be worth the while to go over them separately. Many of them are mere discolorations of the leaf or stem, or a mere sterile crust, with or without rudimentary perithecia. Cooke (see Grev. XIII, pp. 41–44) has examined specimens of these numbers in Herb. Berk. and finds them, as we do, very unsatisfactory, mostly without fruit.

DOTHIDÉLLA, Speg.

Fungi Argentini, Pugill. I, Appendix.

Stroma effused or discoid, thin, black; cells immersed. Asci 8-spored. Sporidia ovoid or oblong, uniseptate, hyaline. This differs from *Phyllachora* in its uniseptate sporidia.

D. betùlina, (Fr.)

Sphæria xylomoides, Fr. in Vetensk. Akad. Handl. 1816, p. 143.
Sphæria atronitens, Holl. in Fr. Scler. Suec. No. 144.
Phyllachora betulina, Fckl. Symb. p. 217.
Dothidea betulina, Fr. S. M. II, p. 554.
Dothidella betulina, Sacc. Syll. 5256.
Exsicc. Fckl. F. Rh. 1015.—Rab. F. E. 1638, 2671.—Thum. F. Austr. 970.

Stromata numerous, gregarious or thickly scattered, often covering a large part of the leaf, and more or less confluent, small, mostly less than ½ mm. broad, irregular in shape, orbicular or angular, slightly arched, roughened by the prominent perithecia, shining, black inside and out. Ascigerous cells (perithecia) thickly crowded, globose, narrowed above into the punctiform, subprominent, finally umbilicate ostiola. Asci oblong, often broader below, sessile, 38–44 x 12–12½ μ. Sporidia biseriate, elliptical, ends rounded, the septum above the middle, not constricted, greenish-hyaline, 12–14 x 5 μ.

On living leaves of *Betula nana*, *B. intermedia*, and *B. glandulosa*, Amitsuarsuk and Holstensborg, Greenland. Sent also from Maine, on leaves of *Betula populifolia*, by Rev. Jos. Blake, and reported from New York State by Mr. Peek.

D. Álni, Pk. 40th Rep. N. Y. State Mus. p. 71.

Stroma orbicular (4–12 mm.), thin, convex and black above, concave, brown or grayish-black and papillosely rugulose below. Asci cylindrical. Sporidia ovate-elliptical, obscurely uniseptate near one end, hyaline, 15–20 x 7½–8¾ μ.

On dead leaves of *Alnus viridis*, Mt. Marcy, N. Y. (Peck).

D. Kálmiæ, (Pk.)

Dothidea Kalmiæ, Pk. 25th Rep. p. 102.
Dothidella Kalmiæ, Sacc. Syll. 5264.

Thin, effused, investing the branches, black, shining, brownish within. Asci linear. Sporidia uniseptate, constricted, subhyaline, 10–12 x 5–6 μ, the eells generally nucleate and unequal.

On branches of *Kalmia angustifolia*, Sandlake, N. Y. (Peek).

The fungus forms a black crust which entirely surrounds the smaller branches and which, in fertile specimens, is dotted with the black ostiola. Within, it resembles half-charred wood. It kills the branches. In the young stage, before the development of the asci, simple, oblong, spore-like bodies 20 μ long are produced.

D. Osmúndæ, (P. & C.)

Dothidea Osmundæ, P. & C. 30th Rep. p. 64.
Dothidella Osmundæ, Sacc. Syll. 5268.

Minute, linear, erumpent through a narrow chink, scarcely emergent, black, nuclei whitish. Asci subcylindrical. Sporidia oblong, narrow, uniseptate, slightly constricted, colorless, 14–17 μ long, one cell usually a little swollen at the septum.

On dead stems of *Osmunda*, New York and New Jersey.

The N. J. specc. have the sporidia only about 12 μ long.

D. scùtula, (B. & C.)

Dothidella scutula, B & C. Grev. IV, p. 105.
Dothidella scutula, Sacc. Syll. 5270.
Exsicc. Rav. F. Am. 385.—Rab. F. E. 3559.—Ell. N. A. F. 684.

Stromata scattered or gregarious, epiphyllous, orbicular, flattened, 1–1½ mm. diam., superficial and easily deciduous, black. Cells numerous, globose, 75 μ diam. Asci oblong, 40–50 x 4–6 μ.

On living leaves of *Magnolia* and *Laurus Caroliniensis*, Southern States; common.

D. Vaccínii, Rostr. F. Groenl. p. 566.

Sphæria conferta, Fr. S. M. II, p. 435?

Stromata hypophyllous, irregularly angular, scattered, tuberculose from the projecting perithecia, black, about 1 mm. diam. Asci elongated-clavate, stipitate, 80–130 x 6–8 μ, 8-spored. Sporidia uniseriate, or biseriate above, oblong-ellipsoid, uniseptate, yellowish-hyaline, with a short appendage at each end, 16–20 x 5–6 μ.

On living leaves of *Vaccinium ulignosum*, Itivnek, Sukkertoppen, &c., Greenland.

D. Aristidæ, (Schw.)

Sphæria Aristidæ, Schw. Syn. N. Am. 1471.
Phyllachora Aristidæ, Sacc. Syll. 5230.

Exsicc. Ell. N. A. F. 86.

Stroma effused, thin, black, interruptedly continuous over both sides of the leaves and enveloping the culms for several cm. in extent, causing them to appear charred. Ascigerous cavities small, mostly less than 75 μ diam., subglobose and seriate between the parallel nerves of the leaves. Asci subglobose, 15–25 μ diam., sessile, 8-spored. Sporidia inordinate, subelliptical, or ovate-oblong, obtuse, hyaline, uniseptate, 8–12 x 5–6 μ.

On dead culms and leaves of *Aristida dichotoma*, North Carolina (Schw.), New Jersey (Ellis), on *Distichlis maritima*, Oregon, and on *Aristida purpurascens*, Mississippi (Tracy).

The Oregon specc. are well matured and show the asci and sporidia distinctly. Newfield specc. also show immature asci and in parts of the same stroma, abundant oblong-clavate, hyaline stylospores 12–15 x 3 μ, on short basidia. Schweinitz says perithecia comparatively large, but we find them as noted both in specc. in Herb. Schw. and from the other localities.

D. Úlmi, (Duv.)

Sphæria Ulmi, Duval in Hoppe. Bot. Taschenb. 1809, p. 105.
Sphæria ulmaria, Sow. Eng. Fungi, tab. 374, fig. 3.
Xyloma ulmeum, Mart. Fl. Erlang, p. 309.
Sphæria xylomoides, DC. Fl. Fr. II, p. 288.
Dothidea Ulmi, Fr. S. M. II, p. 555.
Phyllachora Ulmi, Fckl. Symb, p. 218.
Dothidella Ulmi, Winter, Pilze, II, p. 904.

Exsicc. Fckl. F. Rh. 1013.—Rab. Herb. Myc. 658.—Thum. F. Austr. 499.—id. M. U. 2064. Sydow, M. March. 256.—Linb. F. Hung. 374 —Erikss. F. Scand. 292.—Briosi and Cavara F. Parass. 73.—Desm. Pl. Cr. Ed. I, 284.—Cke. F. Brit. Ser. I, 184.

Stromata epiphyllous, scattered or approximate, or even confluent, numerous, often thickly scattered over the entire upper surface of the leaf, flattish-pulvinate, or depressed-hemispherical, suborbicular in outline, black, but covered by the thin, gray epidermis, verrucose from the projecting apices of the cells, about 1 mm. broad. Asci cylindrical, subsessile, 60–70 x 8–9 μ. Sporidia obliquely uniseriate, obovate-elliptical, greenish-hyaline, continuous at first, but when mature, with a single transverse septum near the base, 10–11 x 5 μ.

On fallen leaves of *Ulmus*.

Sent to Schweinitz by Dr. Torrey, from New York State. We do not find any other reference to this species as being found in this country, and we have seen no American specc.

D. úlmea, (Schw.)

Sphæria ulmea, Schw. Syn. Car. No. 288, Fr. S. M. II, p. 436.
Exsicc. Ell. N. A. F. 1347.—Rav. F. Am. 752.—Rav. Car. II, 67.

Perithecia buried in the parenchyma of the leaf either singly or 2–8 together, stromatically connected, their bases projecting and forming little tuberculiform heaps on the upper surface of the leaf, and their papilliform ostiola erumpent through the ruptured epidermis on the lower side. Asci oblong-cylindrical, 55–60 x 8 μ, subsessile, 8-spored. Sporidia subbiseriate, hyaline, ovate, 7–9 x 3–3½ μ, finally with a septum near the lower end.

On leaves of *Ulmus*, common.

Found mature only after the leaves have lain on the ground through the winter. This must not be confounded with *D. Ulmi*, (Duv.), which, though having sporidia almost the same as this (only a little larger), differs essentially in its other characters. This (*D. ulmea*) is anomalous on account of the ascigerous cells assuming the character of perithecia as in *Homostegia Kelseyi*.

D. sphæroìdea, (Cke.)

Dothidea Sphæroidea, Cke. Grev. VII, p. 50.
Exsicc. Rav. F. Am. 387.

Stromata scattered, solitary, superficial, 350–400 μ diam., hemispherical, rough, black. Ascigerous cells immersed in the stroma, not crowded, about 30 in number, subglobose, 100–150 μ diam. Ostiola papillose-prominent, roughening the surface of the stroma. Asci obovate, 72–80 x 30–40 μ, sessile. Paraphyses branched, thickened at their tips. Sporidia crowded, obovate, unequally uniseptate, 24–28 x 20 μ diam., upper cell of the sporidia globose, 20 μ diam., lower cell subglobose, 10–12 μ diam., hyaline, becoming yellow-brown.

On living foliage of *Juniperus Virginiana*, South Carolina (Ravenel).

A very distinct and well marked species. The general appearance is that of *Stigmatea Juniperi*, (Desm.), but the microscopical characters are very different, as above noted. The last three species differ in their thicker stroma, which approaches *Dothidea.*

D. bullulàta, (Berk.)

Dothidea bullulata, Berk. Nares Arct. Expd. II, 322.

Disks small, bullate, marked by the punctiform ostiola, arising from a filamentose base. Sporidia uniseriate, uniseptate, slightly attenuated at each end, 15 x 6–8 μ.

On leaves, Disco Island.

D. Artemisiæ, (Schw.)

Sphæria Artemisiæ, Schw. Syn. N. Am. 1227.

Small (4 mm.), ovate, subangular, subcupulate, margined by the lobes of the stellately-ruptured epidermis, erumpent, subconfluent, very black, with the disk rugose. Ostiola umbilicate, not prominent, numerous. Perithecia globose, minute, hollow, buried in the stroma which is seated on the inner bark, nearly free, black and rugose on both sides, not really confluent, but crowded. Sporidia (sec. Berk. Grev. IV, p. 104) oblong, uniseptate, constricted at the septum, 25 μ long.

On branches of *Artemisia*, Bethlehem, Pa. (Schw.).

DOTHÍDEA, Fr.

Obs. II, p. 347

Stroma erumpent, mostly pulvinate or tuberculiform, slightly arched, flat or slightly concave above, black, wrinkled or punctate from the ostiola. Perithecia (cells) sunk in the stroma. Sporidia subelliptical or elongated, 2-celled, hyaline or colored.

* *Sporidia continuous, hyaline* (*Bagnisiella*), *or colored* (*Auerswaldia*).

D. Ílicis, Cke. Texas Fungi, No. 143.

"Gregarious, erumpent. Pustules elliptical, black, the cells being enclosed in the stroma. Asci clavate. Sporidia elliptical, sometimes slightly attenuated towards each end, simple, hyaline, 30 x 10 μ."

On bark of *Ilex opaca*, Texas (Ravenel).

D. episphæria, Pk. 30th Rep. p. 64.

Stromata small, slightly prominent, scattered or subconfluent, often irregular, carbonaceous, black. Nuclei numerous, white within. Sporidia crowded or biseriate, lanceolate or subfusiform, colorless, 15–18 μ long.

On old *Diatrype stigma*, New York State.

D. Cércidis, Cke. Grev. XIII, p. 66.

Erumpent, orbicular, hemispherical, at length depressed in the center, black, cells few. Asci clavate. Sporidia biseriate, elliptic-lanceolate, continuous, hyaline, 30 x 10 μ.

On branches of *Cercis Canadensis*, Carolina.

Dothidea moricola, C. & E. Grev. V, p. 95, and *D. Tamaricis*, Cke. Grev. XI, p. 108, are only forms of *Botryosphæria fuliginosa*, (M. & N.). In the spec. of *D. Tamaricis*, in Rav. F. Am. 668, the sporidia are mostly 15–20 x 6–9 μ, but in the same perithecium are asci with sporidia 30 x 15 μ.

D. Prínglei, Pk. Bot. Gaz. VII, p. 57.

Auerswaldia Pringlei, Sacc. Syll. 5249.

Stromata amphigenous, orbicular or oblong, 2 mm. to 1 cm. diam. or by confluence two or more cm., prominent on both sides, flat-discoid, black and solid, penetrating the matrix deeply, covered by the thin epidermis which finally becomes loosened and is then white. Ascigerous, cells few (2–6) in a stroma, ovate, deeply seated, about ¼ mm. diam., narrowed above into the tuberculiform-erumpent, perforated ostiola, which are finally deciduous, leaving a round hole penetrating the stroma as if eaten out by some insect. Asci cylindrical, subsessile, paraphysate, 8-spored, 180–210 x 15–18 μ. Sporidia uniseriate, oblong-elliptical, rounded at the ends, hyaline, becoming yellow-brown, 25–36 x 12–14 μ, continuous. The surface of the stroma, except where roughened by the prominent, tuberculiform ostiola, is smooth.

On leaves of *Yucca macrocarpa*, Arizona (Pringle), San Diego, Cal. (Palmer).

** *Stroma pulvinate or tuberculiform; ramulicolous* (*Eudothidea*).

D. Sámbuci, (Pers.)

Sphæria Sambuci, Pers. Syn. p 14.
Sphæria natans, Tode F. Meckl. II, p. 27.
Dothidea Frangulæ, Fckl. Symb. p. 222.
Dothidea Sambuci, Fr. S. M. II, p 551.
Exsicc. Fckl. F. Rh. 1007, 1816.—Kze. F. Sel. 158.—Rab. F. E. 2952.—Thum. F. Austr. 764.—Roum. F. G. 1471.—Sydow, M. March. 861.

Stromata gregarious, mostly numerous, often several standing near each other, but not often confluent, orbicular or elongated, erumpent and strongly prominent, flat-pulvinate, black, at first smooth, finally shrunken and wrinkled, ½–1 mm. diam. Cells very numerous, crowded and small. Asci oblong-cylindrical, attenuated below, 8-spored, 78–96 x 14–16 μ. Sporidia obliquely uniseriate or partly biseriate, oblong-subclavate, uniseptate, the lower cell smaller, olive-brown, 19–21 x 8–9 μ.

On dead limbs of *Morus Multicaulis*, Carolina, on dead stems of *Sambucus Canadensis*, New York State (Peck).

D. Montaniénsis, E. & E. **(Plate 40)**

Dothidea Bigeloviæ, E. & E. Proc. Acad. Nat. Sci. Phil. July, 1890, p. 248.

Stromata depressed-tubercular, suborbicular, about 1 mm. diam., smooth and black, gregarious, partly sunk in the bark of the dead stems. Ascigerous cavities peripheric, minute, numerous. Asci oblong, 40–50 x 12 μ, soon disappearing. Sporidia subbiseriate, uniseptate, constricted at the septum, 15–20 x 7–10 μ.

On dead stems of *Bigelovia?* Helena, Montana (Kelsey, No. 141).

D. ribèsia, (Pers.)

Sphæria ribesia, Pers. in Usteri N. Ann. d. Bot. V, p. 24.
Dothidea ribesia, Fr. S. M. II, p. 550.
Plowrightii ribesia, Sacc. Syll. II, p. 635.
Exsicc. Fckl. F. Rh. 1005.—Kze. F. Sel. 157.—Rab. Herb. Mycol. 655.—Desm. Pl. Cr. Ed. I, 283, Ed. II, 835. Rehm Asc. 517.—Kriegr. F. Sax. 583.—Roum. F. Gall. 293.—Linht. F. Hung. 66.—Sydow. M. March. 1300, 2158.—Ell. N. A. F 483

Stromata scattered or gregarious, sometimes crowded in small dense clusters or groups, mostly transversely erumpent and surrounded by the ruptured epidermis, round or elliptical, or more or less irregular, mostly flat or depressed above, often grooved, black, lusterless, 1–3 mm. broad. Perithecia (cells) scarcely crowded, subglobose, perforated above, finally slightly prominent so that the surface of the stroma appears uneven. Asci oblong-cylindrical, stipitate, 8-spored, 75–90 x 10 μ. Sporidia oblique or biseriate, oblong-fusoid, divided unequally by the single septum, slightly constricted, hyaline, 15–22 x 5–6 μ.

On dead stems of various species of *Ribes*, common.

D. Mezèrei, Fr. S. M. II, p. 551.

Plowrightia Mezerei, Sacc. Syll. 5287.
Exsicc. Fckl. F. Rh 1818 —Rab. F. E. 1845 —Rehm. Asc. 371.

Stromata gregarious, often very numerous and thickly scattered.

erumpent and distinctly prominent, round or irregularly tuberculiform, mostly flattish-pulvinate, with an uneven, tubercular-rugose surface which is punctate from the slightly projecting ostiola, black, 1–2 mm. diam., sometimes confluent. Ascigerous cells numerous, subpyriform, 100–120 μ broad. Asci cylindric-clavate, attenuate-stipitate, 8-spored, 70–80 x 10–11 μ. Sporidia biseriate, narrow-clavate, attenuated and rounded at the ends, uniseptate and slightly constricted at the septum, hyaline, 17–20 x 4½ μ.

On a dead trunk of *Daphnis*, Bethlehem, Pa. (Schw.).

D. concaviúscula, E. & E. (in Herb.)

Stromata gregarious, irregular in shape, mostly oblong or elliptical, 1–2 x ¾–1 mm., often subseriately confluent, erumpent, but hardly rising above the bark, mostly concave above, black. Ascigerous cavities about 100 μ diam., hardly crowded. Ostiola not conspicuous, leaving the surface of the stroma nearly smooth. Asci oblong-clavate, contracted at the base into a very short stipe, paraphysate, 8-spored, 55–65 x 12–15 μ. Sporidia biseriate, oblong-fusoid, uniseptate, yellowish-hyaline, about 15 x 4 μ.

On dead branches of *Magnolia glauca*, Newfield, N. J.

D. Calystégiæ, Cke. & Hark. Grev. XIV, p. 8.

Stromata subgregarious, erumpent, elliptical, black (1–2 x ½ mm.), surrounded by the ruptured epidermis. Ascigerous cells immersed. Asci clavate, 4–8-spored. Sporidia inordinate, elongate-elliptical, uniseptate, not constricted, hyaline, 22 x 7 μ. Conidia oval, continuous, 5 x 3 μ, pale.

On stems of *Calystegia sepium*, California (Harkness).

D. Baccháridis, Cke. Grev. XI, p. 108.

Exsicc. Rav. F. Am. 738.—Ell. & Evrht. N. A. F. 2d Ser. 1559.

"Pustules erumpent, black, depressed, ovate or elliptical, cells immersed in the stroma. Asci clavate. Sporidia subelliptical, uniseptate, brown, lower cell much smaller, 20 x 12 μ."

On branches of *Baccharis*, South Carolina (Ravenel), California (Harkness).

All the specc. examined were old and without asci, but there were (sporidia)? as above described, differing from those of *D. collecta*, (Schw.), in the broader upper cell. The ascigerous cavities are numerous, subglobose, 100–110 μ diam.

D. tuberculifórmis, Ell. Bull. Torr. Bot. Club, VIII, p. 124.

Stroma discoid, contracted below, orbicular or subelongated, $\frac{3}{4}$–$1\frac{1}{4}$ mm. diam., sometimes subconfluent, black, easily deciduous. Ascigerous cells peripherical, 80–110 μ diam., crowded. Ostiola obscure or minutely punctiform, leaving the surface of the stroma nearly smooth. Asci clavate-oblong, abruptly stipitate, 55–65 x 13–15 μ, 8-spored. Sporidia biseriate, ovate-oblong, hyaline and subacute at first, finally yellow-brown and subobtuse, uniseptate near the middle, lower cell a little narrower, scarcely constricted, 13–15 x 6–7 μ, mostly a little curved.

On dead limbs of some deciduous tree or shrub, Utah (S. J. Harkness).

D. collécta, (Schw.)

Sphæria collecta, Schw. Syn. N. Am. 1271.
Dothidea tetraspora, B. & Br. Brit. Fungi, No. 899, tab. XI, fig. 39.
Dothidea crystallophora, B. & C. in Herb. Curtis.
Dothidea Loniceræ, Cke. Grev. XIII, p 66?
Exsicc. Ell. N. A. F. 168.

Stromata thickly scattered, erumpent, pulvinate, 1–3 mm. diam., orbicular or elongated, brownish-black. Ascigerous cells globose. 100 μ diam. peripherical. Ostiola punctiform, barely visible under a lens. Asci oblong, sessile, 50 x 10–12 μ. Sporidia biseriate, clavate-oblong, uniseptate, 18–24 x 6–9 μ, yellow (hyaline at first), four in an ascus.

On dead limbs of *Maclura*, *Lonicera*, *Genista*, *Andromeda*, *Viburnum*, *Salvia*, Newfield, N. J., and on *Iva frutescens* and *Maclura*, New York; on *Ulmus*, Delaware.

We have seen no spece. of *D. Loniceræ*, Cke., which differs only in its 8-spored asci, and which, if we are not mistaken, Dr. Cooke himself now considers a mere variety of this. The above diagnosis is from spece. in Herb. Schw. at the Philadelphia Acad. which are the same as *Dothidea tetraspora* as issued in N. A. F. 168.

D. Colùteæ, B. & C. Grev. IV, p. 104.

"Pulvinate, smooth, containing several cells. Asci clavate. Sporidia oblong, uniseptate, constricted at the septum, 12–20 μ long."

On twigs of *Colutea*, Pennsylvania (Michener).

Spece. in our Herb. from Farlow have the outward appearance of *D. tetraspora*, but are without fruit.

D. Berbéridis, (Wahl.)

Sphæria Berberidis, Wahl. Flor. Suec. Ed. II, p. 1060.
Dothidea Berberidis, De Not. Micr. Ital. I, p. 66.
Exsicc. Fckl. F. Rh. 1817.—Kze. Fungi Sel. 159.

Stromata solitary and scattered or 2–3 subseriately approximate, erumpent and bordered by the ruptured epidermis, orbicular-pulvinate or more or less elongated, reaching as much as 5 mm. long, rather flattened above and often depressed in the center, black-punctate from the projecting ostiola. Perithecia numerous, subglobose, 240–260 μ diam., narrowed above into the ostiola. Asci cylindrical, stipitate, gradually narrowed above, 8-spored, 50–60 x 8–11 μ. Sporidia biseriate, narrow-clavate, subinequilateral, uniseptate below the middle, hyaline, 17–20 x 4–4½ μ.

On dead stems of *Berberis*, New England.

D. Muhlenbérgiæ, Ell. Bull. Torr. Bot. Club, VIII, p. 74.

Exsicc. Ell. N. A. F. 482.

Stromata scattered or gregarious and subconfluent, subelongated-tuberculiform, or hysteriiform, minute (¼–½ mm.), not penetrating deeply, covered by the blackened epidermis which opens with a narrow, longitudinal cleft or is pierced by the short, acute ostiola. Ascigerous cells few, 150–200 μ diam., often but one in the smaller stromata, 2–4 in the larger, white inside. Asci clavate, rounded above, paraphysate, 8-spored, 50–75 x 15–20 μ, narrowed below into a stipe-like base. Sporidia biseriate or inordinate, oblong-fusoid, hyaline and granular at first, becoming yellowish and uniseptate, slightly constricted at the septum, narrowed and subacute at the ends, slightly curved, 15–20 x 4–6 μ.

On dead culms of *Muhlenbergia*, Newfield, N. J.

Prof. Niessl in Hedwigia, May, 1883, refers this to *Leptosphæria* and finds the sporidia 20–25 x 6–8 μ, 4–5-celled and hyaline, but after having carefully reexamined abundant and good material, we find the sporidia as here noted, and though the scanty stroma often gives the appearance of perithecia, a section shows the genuine ascigerous cells characteristic of the *Dothideaceæ*.

D. corýlina, Ck. & Hark. Grev. IX, p. 87.

"Erumpent, orbicular, depressed, black, same color inside. Asci ample, clavate. Sporidia obtuse, lanceolate, constricted in the middle, 1–3-septate, brown, 50 x 15 μ."

On twigs of *Corylus rostrata*, California (Harkness).

Species imperfectly known and doubtful.

D. Línderæ, Ger. Bull. Torr. Bot. Club, V, p. 40.

Exsicc. Ell. & Evrht. N. A. F. 2d Ser. 2126.

"Stroma black, erumpent, roundish or oblong, often confluent,

surface convex, papillate. Sporidia oblong, brown, uniseptate, 15 x 5 μ."
On dead branches of *Lindera Benzoin*, New York and Pennsylvania.

The spece. in N. A. F., like those from Gerard, are sterile. The general aspect is that of *D. collecta*, which perhaps this is.

D. smilacícola, Cke. & Ger. Grev. VII, p. 14.

"Scattered, black, erumpent, convex. Asci cylindrical. Sporidia biglobose, strongly constricted, brown, 18–20 x 10 μ."
On *Smilax* twigs, New York (Gerard).

D. Epilòbii, Fr. S. M. II, p. 559.

Spots amplexicaul, smooth, uniformly pitch-black, roughened by the subprominent cells.
On dead stems of *Epilobium angustifolium*, Adirondack Mts., N. Y. (Peck).

D. abnórmis, Howe, Bull. Torr. Bot. Club, V, p. 42, from the description given, should be *Hypoxylon*, perhaps a form of *H. smilacicolum*, Howe.

HOMOSTÉGIA, Fckl.

Symb. p. 223.

Stroma somewhat flattened, or hemispherical, carbonaceous black, (gray inside). Ascigerous cells immersed, pale. Ostiola minute, papillate. Asci oblong-clavate, 8-spored. Sporidia ovate-elliptical, 3-septate, brown.

H. Piggòtii, (B. & Br.)

Dothidea Piggotii, B. & Br. Not Brit. Fungi, No. 660.
Sphæria homostegia, Nyl. in Flora, 1857, p. 688.
Homostegia adusta, Fckl. Symb. p. 224.
Homostegia Piggotii, Karst. Myc. Fenn. II, p. 221.
Exsicc. Fckl. F. Rh. 953.—Ell. N. A. F. 485.

Stroma adnate to the lichen thallus, convex-pulvinate, 1–3 mm. diam., dull black, wrinkled and roughened by the projecting ostiola. Ascigerous cells ovate-globose, white, 120–150 μ diam. (260 μ, Winter), deeply sunk in the stroma. Asci oblong, short-stipitate, paraphysate, p. sp. 60–70 x 15 μ (60–70 x 21–24 μ Winter). Sporidia imperfectly biseriate, clavate-oblong, subinequilateral, slightly curved, 3-septate, brown (almost opake), 18–22 x 7–8 μ, slightly constricted at the septa.

On thallus of *Parmelia saxatilis*, Rhode Island (Farlow), Pennsylvania (Eckfeldt).

The measurements are from Farlow's specimens and are smaller (for cells and asci) than those of Dr. Winter.

H. Magnòliæ, (Cke.)

Dothidea Magnoliæ, Cke. Grev. VII, p. 50.
Homostegia Magnoliæ, Sacc. Syll. 5342.
Exsicc Rav. F. Am. 386.

Epiphyllous, scattered, black, elliptical or angular, flattened-convex, rugose. Asci clavate-cylindrical. Sporidia cylindrical, obtuse, slightly curved, triseptate, hyaline, 20 x 5 μ.

On leaves of *Magnolia*, Darien, Ga. (Ravenel).

We have not been able to find any fruit in the spec. in Ravenel's F. Am..and can only copy the description from Grevillea.

H. rugodísca, (Cke. & Hark.)

Dothidea rugodisca, Cke. & Hark. Grev. IX, p. 87.
Homostegia rugodisca, Sacc. Syll. 5343.

Hypophyllous, spots irregular, brown. Cells angular, flattened, rugose, crowded. Asci subclavate. Sporidia rounded above, attenuated below, triseptate, hyaline, 16–17 x 4 μ.

On leaves of *Arbutus Menziesii*, California (Harkness).

H. Kelsèyi, E. & E. Proc. Acad. Nat. Sci. Phil. June, 1890, p. 248. **(Plate 40)**

Perithecia gregarious or cespitose, or united 3–6 together in an imperfect stroma, often transversely seriate through cracks in the bark, ovate, ½–¾ mm. diam., with a conical or cylindric-conical, stout ostiolum which is sometimes imperfectly radiate-sulcate. Asci cylindrical, 150–190 x 8–9 μ, with paraphyses. Sporidia uniseriate, hyaline, oblong-cylindrical, 3-septate, straight, obtuse, 15–20 x 7–9 μ.

On dead stems of *Ribes rotundifolia?* Helena, Montana.

This species, on account of the scanty stroma and often nearly free perithecia, seems allied to the *Cucurbitarieæ* or *Melogrammeæ*, but in some cases the perithecia are entirely buried in the stroma, as in genuine *Dothideaceæ*.

H. gangræna, (Fr.)

Sphæria gangræna, Fr. in Duby Bot. Gall. II, p. 695.
Phyllachora gangræna, Fckl. Symb. p. 217.
Sphærella gangræna, Karst. Myc. Fenn. II, p. 185.
Homostegia gangræna, Winter, Die Pilze, II, p. 917.
Exsicc. Fckl F. Rh. 2261.—Rav. Car. II, No. 59.—Desm. Pl. Cr. Ed. I, 1267.

Stromata gregarious or scattered, irregularly elongated, 3–5 mm.

long, often confluent, only slightly prominent, black and covered by the blackened epidermis, appearing as black, slightly swollen blotches on the leaves and culms. Ascigerous cells mostly subseriate, sunk in the stroma, globose, about 100–150 μ diam. Ostiola papilliform, finally umbilicate. Asci oblong-clavate, subsessile, 8-spored, 35–40 x 10–11 μ. Sporidia biseriate, oblong-fusoid, hyaline, 1–2-septate, not constricted, 12–15 x 4–4½ μ.

On living culms and leaves of *Isolepis capillaris*, South Carolina (Ravenel).

MAZZÁNTIA, Mont.

Syll. p. 245.

Stroma erumpent, oblong or suborbicular, clypeiform, convex, tolerably thin, black, covered at first, then erumpent, white inside, texture almost like that of a *sclerotium*. Ascigerous cells sunk in the stroma, mostly few. Sporidia oblong or elliptical, continuous, hyaline.

M. Gálii, (Fr.)

Sphæria Galii, Fr. Elench. II, p. 105.
Sphæria Aparines, Cast. Cat. de Pl. de Mars. p. 171.
Mazzantia Galii, Mont. Syll. p. 246.
Exsicc. Fckl. F. Rh. 795.—Rab. F. E. 537.—Thum. M. U. 71, 1956.

Stromata scattered, covered at first, then erumpent and very prominent, elliptical or oblong, strongly arched, slightly wrinkled, black, ½–1 mm. long. Ascigerous cells 1–5, narrowed above into the slightly projecting ostiolum, depressed-spherical or lenticular, membranaceous, pale. Asci cylindrical, short-stipitate, rounded above, 8-spored, 50–52 x 5–9 μ. Sporidia biseriate, oblong-elliptical, continuous, with a nucleus in each end, 8–10 x 4–4½ μ.

On dead stems of *Galium Aparine*, California (Harkness).

M. sèpium, Sacc. & Penz. Mich. II, p. 609.

Stromata linear-oblong, subcutaneo-erumpent, shining-black, 1½–2 mm. long by scarcely ½ mm. wide, rather flat, obscurely punctulate from the ostiola, ascigerous cells few, whitish. Asci cylindric-clavate, 54–58 x 7–8 μ, without paraphyses, 8-spored, short-stipitate. Sporidia obliquely uniseriate, cylindric-fusoid, straight or curved, 2- (rarely 4-) nucleate, 10–12 x 2½–3½ μ, hyaline.

On dead stems of *Calystegia sepium*, North Greenbush, N. Y. (Peck). Sporidia a little larger than in the typical form, and trinucleate.

CURRÈYA, Sacc.

Syll. II, p. 651.

Stroma hemispherical or discoid, black. Asci elongated, 8-spored. Sporidia ovate, elliptical or oblong, pluriseptate, muriform, yellow or brown.

C. Harknéssii, E. & E. (in Herb.)

Stromata gregarious, small (less than 1 mm.), depressed-hemispherical or subdiscoid, black, at first covered, finally exposed, but scarcely rising above the bark, deciduous. Ascigerous cells few, small. Ostiola only slightly prominent. Asci oblong-cylindrical, sessile, paraphysate, 8-spored. Sporidia biseriate, oblong-ovate, 3-septate, with 1–2 partial longitudinal septa, yellow-brown, 12–15 x 5–6 μ.

On dead stems of *Convolvulus Californicus* Mt. Diablo, Cala.

Sent by Dr. Harkness as *Dothidea Calystegiæ*, Cke. & Hark., from which it is quite distinct.

C. excavàta, (C. & E.)

Dothidea excavata, C. & E. Grev. V, p. 34, tab. 75, fig. 4.
Curreya excavata, Sacc. Syll, 5348.

Gregarious, discoid, irregular, depressed, at length concave, black. Asci cylindrical. Sporidia uniseriate, elliptical, multi-septate, muriform, brown, 18–20 x 10 μ.

On dead limbs of *Magnolia glauca*, Newfield, N. J.

We know nothing of this species. Our specc. (2361) are *Dothidea concaviuscula*, E. & E., which agrees with this only in external appearance. The specc. in Roum. F. Gall. 1835, under the name of *Dothidea excavata*, C. & E., are only the spermogonial stage of some species unknown—apparently not *Dothidea*.

RHOPÓGRAPHUS, Nitschke.

in Fckl. Symb. p. 219.

Stromata subinnate, linear, irregular, variously confluent, erumpent through cracks in the epidermis, black. Ascigerous cells immersed, seriate, comparatively large. Asci oblong-ovoid, 8-spored. Sporidia suboblong, 3–5-septate, yellowish.

Rh. filícinus, (Fr.) **(Plate 40)**

Hysterium aquilinum, Schum. Enum. Plant. Sæll. III, p. 152.
Sphæria filicina, Fr. S. M. II, p. 427.
Dothidea filicina, Fr. Summa, p. 386.
Rhopographus filicinus, Nits. in Fckl. Symb. p. 219, tab. VI, fig. 31.
Rhopographus Pteridis, Winter, in Kze. F. Sel. 583.

Exsicc. Kze. l. c.—Rab. F. E. 2672.—Rehm Asc. 581.—Sydow, M. March. 163.—Thum. M. U. 2176.—Kze. F. Sel. 583.

Stromata gregarious, subinnate, parallel, irregular, variously confluent, black, shining, at first smooth, then punctate-roughened by the subprominent ostiola. Ascigerous cells seriate, connate, depressed-globose or flat-hemispherical, 180–210 μ broad and about 80 μ high. Asci oblong, short-stipitate, 8-spored, 70–75 x 15–17 μ. Sporidia biseriate, oblong-fusoid, attenuated and rounded at the ends, and with a short, subglobose, hyaline appendage, 3- (rarely 5-) septate, 28–30 x 7 μ.

On dead stems of *Pteris aquilina*, Center, N. Y. (Peck).

The N. Y. specc. have (sec. Peck) sporidia 25–27½ μ long.

The stromata are covered by the blackened epidermis which is raised into elongated, variously confluent, black blotches. We have seen no American specimens.

Rh. clavísporus, (C. & P.) **(Plate 28, figs. 6–7)**

Dothidea clavispora, C. & P. 29th Rep, N. Y. State Mus. p. 63.
Hysterium clavisporum, C. & P. 28th Rep. p. 69.
Leptosphæria clavicarpa, E. & E. Journ. Mycol. I, p. 43.

Stromata innate-erumpent, narrow, oblong or elliptical, containing 2–10 perithecia or often linear-elongated, ¼–1 cm. long, partially erumpent through cracks in the epidermis which is raised into parallel ridges. Perithecia seriate, 150–200 μ diam. Ostiola obtuse, inconspicuous. Asci broad, oblong-cylindrical, 75–100 x 22–25 μ, obtuse, with a short, narrow base. Paraphyses filiform. Sporidia crowded in several series, clavate-oblong, slightly curved, obtuse and rounded above, narrowed rather abruptly below, 7–9-septate, and when mature, deep yellow-brown, 25–40 x 8–10 μ, only slightly constricted at the septa. Spermatia in similar perithecia, oblong-cylindrical, slightly curved, hyaline, 3–5 x 1¼ μ. The general appearance is that of *Sphæria orthogramma*, B. & C.

On dead culms of *Phragmitis communis*, New York (Peck), Iowa (Arthur).

The ascigerous cells assume the character of true perithecia, at least in the mature specc., and are connate in a single series. The substance of the scanty stroma is brown inside.

OPHIÓDOTHIS, Sacc.

Syll. II, p. 652.

Stroma contracted or broadly effused, somewhat flattened, granulose, becoming black. Asci elongated, 8-spored. Sporidia filiform, continuous, hyaline, sometimes multinucleate.

O. tárda, Hark. Bull. Cal. Acad. Feb. 1884, p. 46.

"Hypophyllous, in irregular, angular spots 3–4 mm. broad, often confluent, papillate, shining black. Asci 8-spored, cylindrical, abruptly contracted, curved and bulbous at the base, 57 x 9 μ. Sporidia linear, pluriguttulate, 42 x 2 μ."

On leaves of *Rhus diversiloba*, California (Harkness).

O. Haydeni, (B. & C.)

Dothidea Haydeni, B. & C. Grev. IV, p. 104.
Ophiodothis Haydeni, Sacc. Syll. 5351.

"Forming elongated, irregular, papillose patches. Sporidia linear, attenuated at each end."

On stems of *Aster* and *Erigeron*, Nebraska.

O. álnea, (Fr.)

Dothidea alnea, Fr. S. M. II, p. 564.

Amphigenous, scattered, subrotund, black, shining, collapsing to rugose-plicate.

On *Alnus serrulata*, Pennsylvania (Michener). Sporidia (sec. Berk. Grev. IV, p. 105) linear, 10 μ long.

MYRIÁNGIUM, Mont. & Berk.

Mont. Sylloge, p. 380.

Stromata numerous, small (1 mm.), crowded, rounded or angular from mutual pressure, convex above, black, with or without an orbicular, thalloid subiculum, multilocular, each cell with a single aparaphysate ascus. Sporidia muriform, hyaline.

M. Durièui, Mont. & Berk. in Mont. Syll. p. 380.

Myriangium Curtisii, Mont. & Berk. in Mont. Syll. l. c.
Pyrenotheca Yunnanensis, Pat. Bull. Soc. Bot. France, 1886, p. 155.
Phymatosphæria Yunnanensis, Sacc. Syll. VIII, p. 847.
(*Phymatosphæria Brasiliensis*, Speg. Fungi Puigg. p. 174)?
(*Phymatosphæria Abyssinica*, Pass. Fungi Abiss. in Nuovo Giorn. Bot. Ital. VII, p. 188)?
Cenangium asterinosporum, E. & E. in Ell. N. A. F. 1279.
Exsicc. Rav. F. Am. 332.—Ell. N. A. F. l. c.

Stromata in densely crowded patches or clusters ½–1 cm. across, the single stromata ½–1 mm. diam., black, plane or convex above, suborbicular or angular from crowding. Ascigerous cells minute, scattered irregularly but abundantly through the substance of the stroma, obovate or subglobose, each containing a single ascus. Asci globose, obovate or pyriform, 50–80 μ in the longer diameter and about 40 μ in the shorter, 8-spored, without paraphyses. Sporidia ovate-oblong, hyaline, 5–7-septate, with one or more partial longitudinal septa. rounded and obtuse at the ends, 20–30 x 12–15 μ, mostly a little constricted at the middle septum and slightly curved.

On trunks and branches of living trees, Carolina (Curtis & Ravenel), Alabama (Beaumont), Massachusetts (Sprague), Pennsylvania (Michener), Michigan (Hicks), New Jersey (Ellis); on limbs of living orange trees, Florida (Hopkins).

The measurements of asci and sporidia are from the Florida specimens; those from more northern localities have the sporidia mostly smaller. The Florida specc. (*M. Yunnanensis*) also differ from those found in the northern States, in the absence of any free-margined, thalloid, effigurate subiculum. The genus *Myriangium* (from which *Phymatosphæria* can not well be separated) has been classed among the *Lichens*, but the absence of spermogones and gonimous cells, seems to exclude it from that order. Its true place in the mycological system is also doubtful, but its affinity seems to us with the *Dothideaceæ*. *Eurytheca*, De Seynes, is closely allied. We have seen no specc. of *Phymatosphæria Abyssinica* and *P. Brasiliensis*. which are given above as probable synonyms.

FAMILY. XYLARIÈÆ.

Stromata very variable in form and size, mostly free, but often more or less sunk in the matrix; either upright and often branched or horizontal, effused, crustaceous, pulvinate, globose or hemispherical, mostly black or becoming black, of woody, carbonaceous or suberose-carnose consistence. Perithecia peripherical, concentrically arranged. Asci cylindrical, 8-spored. Sporidia continuous, brown or black, often inequilateral; young stroma clothed with a conidial layer.

NUMMULÀRIA, Tul.

Sel. Carp. II, p. 42.

Stroma orbicular, cup-shaped or discoid, becoming black, marginate, the margin more or less distinctly sterile. Perithecia monostichous, peripherical, immersed. Asci cylindrical, 8-spored. Sporidia uniseriate, subelliptical, continuous, dark. The genus is too closely allied to *Hypoxylon*, especially the discoid forms.

A. *Stroma cup-shaped or concave.*

N. discrèta, (Schw.) **(Plate 39)**

Sphæria discincola, Schw. Syn. Car. No. 63
Sphæria discreta, Schw. Syn. N. Am. 1249.
Sphæria excavata, Schw. l. c. 1250, (sec. spec. in Herb. Schw.).
Nummularia discreta, Tul. Sel. Carp. II, p. 45.
Exsicc. Rav. Fungi Car. II, No. 57.—Ell. N. A. F. 489.

Stroma erumpent, orbicular, 2–4 mm. diam., cup-shaped, with a thick, raised margin, dirty cinereous, then black, the concave surface at first white-punctate from the minute, punctiform ostiola which, in the mature state, are scarcely visible. The wood beneath the stroma is marked with a black, circumscribing line. Perithecia monostichous, ovate-cylindrical, nearly 1 mm. long, rather abruptly contracted above into a short neck, their rounded bases penetrating to the bottom of the stroma. Asci cylindrical, 110–120 x 10–12 μ, with long, filiform paraphyses. Sporidia subglobose, nearly hyaline at first, finally opake, 10–12 μ diam.

On dead branches and trunks of *Pyrus Malus* and *Amelanchier Canadensis*, Newfield, N. J.; on the first-named host, New England (Farlow), New York (Peck); on *Gleditschia triacanthos*, Ohio (Morgan), found also (sec. Sacc. in Syll.) on *Sorbus*, *Ulmus*, *Cercis*, and *Magnolia*.

Sec. Cooke Grev. XII, p. 6, the specimen of *Sphæria discincola*, Schw., in the Kew Herbarium, figured by Currey in Linn. Trans. 1858, Pl. 47, fig. 105, does not differ from *S. discreta*, Schw.

N. repánda, (Fr.)

Sphæria repanda, Fr. S. M. II, p. 346, Obs. Mycol. I, p. 168,
Hypoxylon repandum, Fr. Summa Veg. Sc. p. 383.
Nummularia peziz oides, E. & E. Bull. Torr. Bot. Club, XI, p. 74.
Nummularia repanda, Nitsch. Pyr. Germ. p. 57.
Exsicc. Fckl. F. Rh 2178.—Thum. M. U. 1460.

Stroma erumpent-superficial, orbicular, or subelliptical, ½–1 cm. diam., concave, and often with a thin, erect, rather broad margin, rufo-cinereous at first, finally black; disk mammillose from the projecting ostiola. Perithecia monostichous, immersed, ovate-oblong, ½–¾ μ long, crowded, often subangular from mutual pressure. Asci cylindrical, subsessile, 8-spored, 110–120 x 8 μ, with long, filiform paraphyses. Sporidia obliquely uniseriate, narrow ovate, obtuse, subinequilateral, dark brown, 11–14 x 4–5 μ (15–16 x 6–7 μ, Sacc. in Syll.). Distinguished from *N. discreta* by its mammillose disk and differently shaped sporidia.

On bark, Ottawa, Canada (Macoun), on bark and wood, Topeka, Kans. (Cragin), and on bark of *Ulmus Americana*, Missouri (Demetrio). Found in Europe on branches and trunks of *Sorbus aucuparia*.

N. subconcàva, (Schw.)

Sphæria subconcava, Schw. Syn. N. Am. 1251.

Gregarious and often confluent, erumpent, $\frac{1}{4}$–$\frac{3}{4}$ cm. across, surrounded by the ruptured epidermis, and consisting of a black, crustaceous shell, inclosing the few, rather large, globose-depressed perithecia, connected by a very scanty stroma. Disk subconcave, subrugose and black. Ostiola globose-papillate. elevated, few, black, sometimes confluent, connected by a very short neck with the perithecia, which have the ascigerous nucleus white. Sporidia (sec. Stevenson) oblong, light brown, 15–19 x 5$\frac{3}{4}$–7$\frac{1}{2}$ μ, some of them slightly constricted in the middle, but not septate.

On branches of *Viburnum dentatum*, Bethlehem, Pa. (Schweinitz).

N. succenturiàta, (Tode).

Sphæria succenturiata, Tode Fungi, Meckl. II, p. 37.
Hypoxylon succenturiatum, Fr. Summa Veg. Scand. p. 383.
Nummularia succenturiata, Nits. Pyr. Germ. p. 58.

Stroma at first erumpent through the closely adherent epidermis; finally, after the bark has fallen away, superficial, seated on the wood and surrounded by a black, circumscribing line, 2 cm. or more long, either elongated, elliptical or orbicular at base, pulvinate, thick and flattened above, or subconstricted at the base, with the orbicular disk slightly concave, dark brown and rugose-corrugated outside, dark gray, becoming black within. Perithecia irregularly distichous or sometimes monostichous, ovate or oblong, often compressed, tolerably large, with very thick, firm walls, suberect, entirely sunk in the stroma, with necks more or less elongated according to their position, and small, not projecting, perforated ostiola. Asci cylindrical, sessile, 8-spored, 120 x 8–9 μ, with filiform paraphyses. Sporidia obliquely uniseriate, fusoid, or ovoid, obtuse at the ends, straight or subinequilateral, brown, 16–18 x 5–7 μ.

On branches, Bethlehem, Pa. (Schw.); rather rare.

The stroma varies in size, 6–9 mm. long by 4–6 mm. broad, or sometimes not more than 2 mm. broad, 2$\frac{1}{2}$–3 mm. thick, pulvinate and flat or concave above. The above diagnosis is from Winter's Pilze.

B. *Stroma convex.*

N. Bulliárdi, Tul. Sel. Carp. II, p. 43. tab. V, figs. 11–19.

Hypoxylon nummularium, Bull. Champ. tab. 468, fig. 4.
Sphæria nummularia, DC. Flore Fr. II, p. 290.
Sphæria anthracina, Schm. & Kze. Mycol. Hefte, I, p. 55.
Exsicc. Ell N. A. F. 85.—Rab. F. E. 2956.—Rehm Asc. 977.

Stroma at first covered by the epidermis, soon emergent, almost superficial and free, convex, orbicular or oval, rarely of irregular shape, sometimes broadly effused, black inside and outside, punctulate from the slightly prominent ostiola, clothed at first with the rufo-ferruginéous conidial layer. Perithecia rather large, ovate, black, loosely included in closely packed cells in the stroma, something as in *Daldinia concentrica.* Asci cylindrical, briefly pedicellate, 100–115 x 10 μ, with very long and stout paraphyses. Sporidia uniseriate, elliptical, hyaline at first, soon opake, 12–15 x 7–9 μ.

Common on dead trunks and limbs of various deciduous trees around Newfield, N. J. Mostly confined to dead oak.

The hymenium in this species, as in *Hypoxylon Petersii*, B. & C., is at first covered by a carnose-coriaceous membrane, which soon disappears, except around the margin.

N. glycyrrhìza, (B. & C.)

Hypoxylon glycyrrhiza, B. & C. Exot. Fungi, Schw. p. 285.
Nummularia glycyrrhiza, Sacc. Syll. 1541.—Cke. Syn. 822.

Suborbicular, thin (about 1 mm.), 3–5 cm. diam., convex, marked in the center by the papilliform ostiola which are depressed in the center. Perithecia oblong, ¾ mm. in height, crowded. Asci (p. sp.) 40 x 5 μ.

On bark, Ohio (Morgan, No. 284).

Differs from *N. Bulliardi*, which it resembles, in its closely packed perithecia and umbilicate ostiola, as well as in its smaller sporidia. The Ohio specimen agrees with one in Herb. Schw.

N. obulària, (Fr.)

Hypoxylon obularium, Fr. Nova Symb. p. 130.
Nummularia obularia, Sacc. Syll. 1540.

Immersed, erumpent, at length broadly effused, determinate, not polished, stroma black, perithecia immersed, oblong. Ostiola hemispheric-subprominent, umbilicate.

On dead trunks, Costa Rica (Oersted).

Closely allied to *N. Bulliardi*, Tul., but differs in having its stroma connate with the matrix and inseparable from it, at first subrotund, then concrescent in a continuous crust, generally elongated, and ostiola depressed. The specimens examined by Fries were old, and no trace of asci or sporidia remained. As far as one can judge from the diagnosis, this and the preceding species can hardly be distinct.

N. micróplaca, (B. & C.)

Diatrype microplaca, B. & C. Journ. Linn. Soc. X, p. 586.
Anthostoma microplacum, Sacc. Syll. 1112.
Nummularia microplaca, Cke. Syn. 837.

Exsicc. Rav. Fungi Car. IV, 39.—Rav. F. Am. 355.—Ell. & Evrht. N. A. F. 2d Ser. 1555.

Stroma much the same as in *N. hypophlœa*, orbicular, ½–1 cm. across or elongated 1–2 x ½–1 cm., thin, crustaceo-carbonaceous, black, originating beneath the epidermis, but soon bare, surface even, faintly punctulate from the minute ostiola, which are not prominent but slightly depressed, as in *Nummularia punctulata*, the opening at first filled with white farinaceous matter. Perithecia ovate-globose, small (less than ½ mm.), monostichous. Asci (p. sp.) about 25 x 3 μ, or with the short base, 45–50 μ long. Sporidia uniseriate, ends mostly slightly overlapping, subinequilaterally elliptical, pale brown, 4½–5 x 2–2½ μ.

On *Sassafras officinale*, South Carolina (Ravenel), and Ohio (Morgan and Kellerman); on *Persea*, Darien, Ga. (Ravenel).

Sec. Berkeley *N. hypophlœa* has larger ostiola and narrower sporidia. This is true as to the ostiola, but as regards the sporidia the case is exactly the opposite. The wood beneath the stroma is stained with the same olive-yellow color as in the next species, to which this is closely allied, but differs as stated.

N. hypophlœa, (B. & Rav.)

Diatrype hypophlœa, B. & Rav Grev. IV, p. 95.
Anthostoma hypophlœum, Sacc. Syll. 1137.
Nummularia hypophlœa, Cke. Grev. XII, p. 7.
Exsicc. Rav. Fungi Car. IV, 38.—Ell. & Evrht. N. A. F. 2d Ser. 1554.

Stroma thin, suborbicular, ½–1 cm. across, slate-color, originating beneath the cuticle which is soon thrown off, slightly convex, and faintly papillose from the slightly projecting ostiola. Stains the subjacent wood yellowish or yellowish-olive. Perithecia in a single layer, ovate-globose, small (½ mm.), abruptly contracted above into a slender neck piercing the superficial, carbonaceous layer of the stroma. Asci slender (100 x 4 μ), with a thread-like base, p. sp. 55–60 μ long. Sporidia uniseriate, lying mostly end to end, narrow-elliptical, pale brown, 2-nucleate, about 7 x 2½–3 μ.

N. subapiculàta, E. & E. Journ. Mycol. V, p. 23.

Subcuticular, erumpent, 1–2 cm. across, convex, 1 mm. thick, or a little more in the center, with the sterile margin thinner. Ostiola slightly papillose, prominent like those of *Nummularia Bulliardi.* Perithecia monostichous, oblong, about ¾ mm. high, closely packed, and more or less laterally compressed. Asci cylindrical, 90–100 μ (p. sp.), with a short, stipitate base, and with long, stout paraphyses, as in *N Bulliardi.* Sporidia uniseriate, oblong-navicular or inequilaterally elliptical, pale yellowish-brown, 12–16 x 5–7 μ, mostly with a single nucleus and a faint, bead-like apiculus at each end.

On bark, Kansas (Cragin).

N. rúmpens, Cke. Grev. XII, p. 8.

Diatrype rumpens, Cke. Ann. N. Y. Acad. Sci. I, p. 185.
Exsicc. Rav. F. Am. 354.

Orbicular or elliptical, ¾–1 cm diam., or by confluence 2 cm. or over, and then more or less irregular in shape, thin, black, surrounded by the ruptured epidermis, roughened by the slightly prominent ostiola. Perithecia monostichous, ovate, ¾ mm. high. Asci cylindrical, 100–115 x 10 μ. Sporidia uniseriate, hyaline, then opake, elliptical, with ends subacute or rounded, 12–15 x 7–9 μ.

On bark, Galveston Bay, Texas (Ravenel).

This description is drawn from the specimens in Rav. F. Am. This seems to differ from *N. Bulliardi* in its less prominent ostiola and rather more acutely pointed sporidia: nor are there, in the specimens we have seen, any very perceptible remains of the overlying membrane. In our collections are specimens of what appears to be the same as those in F. Am., from British Columbia and Louisiana, as well as several of the original Texas specimens from Dr. Ravenel.

N. exùtans, Cke. Grev. XII, p. 8.

Diatrype exutans, Cke. in Ann. N. Y Acad. l. c.

Broadly effused, black, subcuticular, soon erumpent, thin (about ½ mm.), papillose from the slightly prominent ostiola. Two or three inches long, with an irregular outline, thinner than *N. rumpens.* Perithecia monostichous, depressed-globose, less than ½ mm. diam. In our specimen of this species from Dr. Ravenel, from his Texas collection, the asci have disappeared. The free sporidia are acutely elliptical or almond-shaped, rather variable in size, 10–15 x 6–8 μ.

On bark, Galveston Bay, Texas (Ravenel).

Differs from *N. rumpens* in its more broadly effused, thinner stroma and depressed-globose perithecia.

N. punctulàta, (B. & Rav.)

Diatrype punctulata, B. & Rav. Grev. IV, p. 94.
Hypoxylon punctulatum, Cke. Syn. 995.
Nummularia punctulata, Sacc. Syll. 1534.
Exsicc. Rav. F. Car. III, No. 51.—Rav. F. Am. 652.—Ell. N. A. F. 84.

Originating beneath the cuticle which is soon thrown off, closely adnate, black, smooth and polished, effused and spreading for 5–20 cm. or more, but not projecting above the bark. Ostiola punctiform, depressed, appearing like minute punctures made with the point of a pin, margin sterile, thin. Perithecia monostichous, elongated-ovoid, rather more than ½ mm. high, covered above by the thin, carbonaceous stroma. Asci cylindrical, with a slender base, 100 x 7 μ, with filiform paraphyses, (p. sp. 75–80 μ long). Sporidia uniseriate, elliptical, yellowish-hyaline, 2-nucleate, 7–8 x 5 μ, ends flattened while lying in the asci. We have not seen them free, and cannot say whether they become opake. The asci and sporidia are generally poorly developed.

On bark of dead oak, common.

N. tínctor, (Berk.)

Sphæria tinctor, Berk. Hook. Lond. Journ Bot. IV, p 311.
Hypoxylon tinctor, Cke. Syn. 996.
Exsicc. Ell. & Evrht. N. A. F. 2d Ser. 1789.

Stroma effused, dull black, very hard, exhibiting all the inequalities of the matrix, 1 mm. thick, 5–20 cm. long, and 2–5 cm. wide, margin thin and sterile, surface nearly smooth, but under the lens distinctly papillose from the the slightly prominent ostiola. The subjacent wood is deeply tinged orange-red, and is rendered very hard. Perithecia monostichous, crowded, elongated (¾ mm.), covered above with the hard, brittle, shining black stromatic layer. Asci 112 (p. sp. 90–100) x 7–8 μ, with abundant, filiform paraphyses. Sporidia uniseriate, pale brown, with a single rather large nucleus, oblong-navicular, 15 x 6 μ, with the ends subobtuse.

On dead trunks and limbs of various deciduous trees, from Ohio west to Kansas, and south to Louisiana, Florida, and Texas.

The stroma originates under the cuticle which is soon thrown off. The general appearance is that of *H. punctulatum*, B. & Rav., and it has the same hard, brittle stroma as that species.

N. clýpeus, (Schw.)

Sphæria clypeus, Schw. Syn. Car. No. 42.
Diatrype clypeus, B & C. Grev. IV, p. 95.
Nummularia clypeus, Cke. Grev. XII, p. 6.

Elliptical in outline, flattened, immersed, shining-black, roughened by the conical, prominent ostiola, 2–6 mm. diam., margin surrounded by the undulate-elevated substance of the wood. Asci cylindrical. Sporidia elliptical, attenuated at each end, almond-shaped, continuous, brown, 20 x 8 μ.

On branches of *Catalpa*, &c., North America (Curtis); on oak, South Carolina (Ravenel).

Schw. in Syn. N. Am. 1219, makes this a synonym of *N. Bulliardi*, which, externally, at least, it must closely resemble.

N. mácula, (Schw.)

Sphæria macula, Schw. Syn. Car. No. 38.
Nummularia macula, Cke. Grev. XII, p. 6.

Suborbicular, erumpent, flattened-convex, black, marked with the minute, punctiform ostiola. Asci cylindrical. Sporidia broad-oval, dark brown, 12 x 9 μ.

On bark of *Platanus*, Carolina (Schw.), much smaller than *N. clypeus*.

HYPÓXYLON, Bulliard.

Champignons, I, p. 168.

Stroma of woody-corky consistence, dark brown or black within and without, free from the first or erumpent-superficial, sometimes more or less sunk in the wood, globose, semiglobose or more or less effused and crustaceous, at first covered by a conidial growth, finally bare. Perithecia peripherical, in a single layer or sometimes in several layers concentrically arranged, globose, ovate or oblong, coriaceous or corneo-coriaceous, sunk in the stroma, but generally with the upper part more or less projecting, with a papilliform or umbilicate ostiolum. Asci cylindrical, 8-spored. Sporidia uniseriate, elliptical or fusoid, inequilateral or curved, continuous, brown. (Winter in Die Pilze).

A. *Large, irregular, fibrous within.* (*Macroxylon*).

* *Perithecia monostichous.*

H. Broomeiànum, B. & C. Grev. IV, p. 94.

Irregular in shape, suborbicular, 3–4 cm. diam., or oblong, 5–8 x 3–4 cm. and ½–1 cm. thick, convex, rusty drab color or dirty purplish, surface more or less uneven, margin partially free, in some specimens distinctly so, and then abrupt, black and indistinctly zonate. Perithecia in a single superficial layer, elongated, more or less angular from compression, about 1 mm. long by ¼ mm. wide, covered above with a thin, stromatic layer which is pierced by the punctiform ostiola as in *Nummularia punctulata*, B. & Rav. Asci 100–110 x 6–7 μ. (p. sp. 70–75 μ long). Sporidia obliquely uniseriate, navicular-oblong or inequilaterally elliptical, rather pale brown, 2-nucleate, 10–14 x 4–5 μ. Substance of the stroma compact, light slate-color with a tinge of umber and a silky luster when fractured.

On rotten wood, Louisiana (Langlois).

** *Perithecia stratose.*

H. ovinum, Berk. Grev. XI, p. 129.

Hemispherical or confluent-elongated, dark purple, hard, smooth, subshining, dark within. Perithecia stratose, black, subglobose. Ostiola obsolete. Asci cylindrical. Sporidia elliptical, dark, 16–18 x 7 μ.

On wood, Orizaba, Mexico.

H. Petérsii, B. & C. Journ. Linn. Soc. X, p. 384.

Stroma pulvinate, depressed-obconical, centrally attached with a spreading margin, 3–4 x 2½–3 cm. across, covered at first by a thick, coriaceo-membranaceous veil which soon disappears except around the margin; substance corky-fibrous, hard, dull, umber-color, becoming darker outside. Perithecia crowded in several layers, subglobose or subelongated, ½–¾ mm., with slender necks ending in distinctly prominent, papilliform ostiola. Sporidia uniseriate or subbiseriate above, narrowly-elliptical, brown, 6–8 x 3½–4 μ. Asci cylindrical, p. sp. about 40 x 5 μ or, including the slender base, 60 μ long.

On rotten oak, Alabama (Peters), on dead wood, Cuba (Wright), on oak logs, Ohio and Kentucky (Morgan).

The foregoing description is from Morgan's Ohio specimens, which have been compared with specimens in Herb. Berk. In the original description, in Linn. Journ., no mention is made of the thick, membranaceous veil, which is a striking and unusual character.

B. *Stroma superficial, globose or subglobose.* (*Sphæroxylon.*)

* *Externally colored, not black.*

H. coccineum, Bull. Champ. p. 174, tab. 345, fig. 2.

Lycoperdon variolosum, Lin. Syst. Nat. Ed. XII, tom. III, Append. Veg. p 204.
Valsa fragiformis, Scop. Carniol. II, p. 399.
Sphæria lycoperdoides, Weigel Obs Bot. p. 47.
Sphæria rubra, Willd. Flora Berol. p. 415.
Sphæria radians, Tode Fungi Meckl. II, p. 29.
Sphæria tuberculosa, Sow. Eng. Fungi, III, tab. 374, fig. 8.
Sphæria fragiformis, Pers. in Usteri. N. Ann. Bot. fasc. V, p. 21.
Sphæria bicolor, DC. Flor. Franc, II, p. 286.
Sphæria lateritia, DC. l. c. VI, p. 137.
Stromatosphæria fragiformis, Grev. Scott, Crypt. Flora, III, tab. 136.

Exsicc. Fckl. F. Rh. 1056.—Rab. Herb. Mycol. 145, 146.—Rab. F. E. 950.—Thum. F. Austr. 258.—Ell N. A. F. 1178.—Cke. F. Brit. Ser. II, 466.

Stroma erumpent-superficial, subglobose, generally from $\frac{1}{4}$–$\frac{3}{4}$ cm. diam., deep brick-red when mature, often paler when young, solitary or subconfluent. Perithecia peripherical in a single layer, small, subglobose, slightly prominent. Asci cylindrical, spore-bearing part 70–80 x 6–7 μ. paraphyses abundant, simple. Sporidia uniseriate, opake, inequilaterally elliptical, 10–12 x 4–5 μ.

Generally on bark of dead beech trees, but also on oak, willow, birch, and some other trees.

Common throughout the United States and Canada, as well as in Europe. This and the next species are often accompanied by an abnormal growth (*Institale acariforme*, Fr.) consisting of a spreading fringe of somewhat flattened, ochraceous or rust-colored, more or less branched processes surrounding the base of the stroma, and about equal in length to its diameter, and bearing an abundance of very minute, obovate, subhyaline conidia. Whether this should be considered the true conidial stage of the *Hypoxylon* is doubtful, as its occurrence is exceptional. The case is in some respects analogous to that of *Sphæria flabelliformis*, Schw. and the *Xylaria* from which it springs, but with this difference: the affected *Xylaria* is always abortive, while the *Hypoxylon* surrounded with its conidial fringe matures its fruit.

H. Howeiànum, Pk. 24th Rep. N. Y. State Mus. p. 98.

Stroma depressed-globose, 5–15 mm. across, light brick-red, nearly smooth, but punctate from the minute, black òstiola, solitary or subconfluent. Perithecia peripherical, monostichous, minute, ovate, $\frac{1}{4}$–$\frac{1}{3}$ mm. high. Asci (spore-bearing part) 45–50 x 5 μ, with a slender, thread-like base, 35 μ long. Sporidia uniseriate, opake, subinequilaterally elliptical, 6–7 x 3–3$\frac{1}{2}$ μ.

On dead limbs of deciduous trees, New York (Peck), on *Ostrya Virginica*, Iowa (Holway), on dead standing shrubs and fallen limbs of oak, New Jersey (Ellis), on dead limbs, Pennsylvania (Everhart & Rau), Ohio (Morgan), Nebraska (Miss L. S. Doud).

The substance of the stroma is of a blue-black color, and a vertical section shows a radiate-fibrous structure with one or two faint concentric zones. The interior of the stroma in *H. coccineum* is homogeneous in structure, and of an even gray-black color. That species is also distinguished from this, by its smaller stroma roughened by the slightly projecting perithecia ($\frac{1}{3}$–$\frac{1}{2}$ mm. diam.), and by its larger asci and sporidia. In the Nebraska specimens the perithecia are distinctly prominent, but in other respects they do not differ from the normal form.

H. commutàtum, Nitschke, var. Holwayanum, S. & E., Mich. II, p. 570; Sacc. Syll. 5969.

Stroma erumpent-superficial, solitary or subconfluent, subglobose, hemispherical or oblong, $\frac{1}{4}$–$\frac{3}{4}$ cm. across, dull purplish-red, becoming black, grayish-black within, roughened by the distinctly prominent, ovate, monostichous, $\frac{3}{4}$ x $\frac{1}{2}$ mm. perithecia. Asci (p. sp.) 75–80 x 7–8 μ, with abundant paraphyses. Sporidia uniseriate, opake, inequilaterally-elliptical, 10–12 x 4½–5½ μ (12–14 x 6–6½ μ, Sacc.).

On bark of dead oak, Decorah, Iowa, and on bark of dead plum trees, and (maple)? Vermillion Lake, Minn. (Holway).

According to Saccardo, the perithecia are larger and more prominent than in the typical form, which is described by Nitschke as having the stroma pulvinate, depressed, rarely hemispherical or nearly globose, solitary or connate, with globose, crowded, subdistichous peripherical perithecia, and sporidia 10–12 x 6 μ. The smaller stromata resemble those of *H. fuscum*, from which it is distinguished by its smaller sporidia. From *H. multiforme*, it is distinguished by its rather larger, darker sporidia.

H. enterómelum, (Schw.)

Sphæria enteromela, Schw. Journ Acad. Phila Vol. V, p. 10.

Stromata pulvinate, often longitudinally confluent for six inches in length, rusty-red, surface not granulated, variable in shape, subcompressed, very black within, covered above with a furfuraceous, pulverulent, rust-colored bark. Immersed in the stroma are a few perithecia of larger size, the others being minute, peripherical, globose and black. The stroma stains the inner bark black. In the nature of the outer layer of the stroma, this is allied to *H. coccineum*. Sec. Cooke in Grev. XI, p. 123, the sporidia are 10 x 4 μ.

Erumpent from cracks in the bark of dead chestnut trees, Bethlehem, Pa. (Schw). Rather rare.

H. Vèra Crùcis, Berk. & Cke. Grev. XI, p. 129.

Subglobose, superficial, often confluent, 1–2 cm. diam., bright rust-color, sooty black within. Perithecia of medium size, ovate, peripherical, somewhat prominent. Asci cylindrical. Sporidia elliptical, attenuated at each end, brown, 20 x 8 μ.

On rotten wood, Vera Cruz, Mexico (Salle).

H. quadràtum, (Schw,)

Sphæria quadrata, Schw. Syn. N. Am. 1223.

Stroma round, flattened-pulvinate, reddish-black, 2–3 mm. diam., margin abrupt all round. Perithecia scarcely prominent. Ostiola papilliform. Sporidia overlapping-uniseriate, brown, continuous, 15–20 x 8–10 μ.

On the bark of various species of *Rhus*, around Bethlehem, Pa.

The above notes are from the spec. in Herb. Schw. The diagnosis given in Syn. N. Am. is as follows: Thick, elevated, abbreviated, more or less four-sided, often irregularly sublobate, base attached to the bark, with the margin rather acute, flattened-undulate and rough above. Perithecia rather large, pyriform, deeply immersed, but not sunk to the base of the whitish stroma, slightly prominent above. Ostiola very short, open, perforated; stroma at first dirty-olive, then black. The small stromata are so narrow and thick as to resemble a thick-stemmed *Peziza*. It is evident that the specc. issued in N. A. F. and elsewhere as *Diatrype quadrata*, Schw., cannot be the species here described.

H. argillàceum, (Pers.)

Sphæria argillacea, Pers. Syn. p. 10.
Hypoxylon argillaceum, Berk. Outl. p. 387.
Exsicc. Rab. F. E. 247.

Stromata erumpent-superficial, subglobose, solitary, rarely connate, clay-color, becoming black within. Perithecia in a single layer (monostichous), rarely irregularly polystichous, ovate, small, crowded, somewhat prominent, minutely mammillose, conidial layer white, becoming stag-color or clay-color; conidia small, ovate, hyaline on long, sparingly branched, septate sterigmata. Asci cylindrical, with very long, slender pedicels, spore-bearing part 140 x 16 μ. Paraphyses simple, thread-like, longer than the asci. Sporidia uniseriate, broad ovate, elliptical or subinequilateral, obtuse, opake, 18–22 x 9–10 μ, (22–24 x 10–12 μ. Sacc. in Syll).

On trunks of ash; more rarely on beech and birch, Bethlehem, Pa. (Schw.), Canada (Maclagan), on beech, New York (Peck).

This species, of which we have seen no specimens except those sent from England by Dr. Plowright, seems to be easily recognized by its clay-colored stroma and large sporidia.

H. notàtum, B. & C. Grev. IV, p. 50.

Exsicc. Rav. F. Car. IV, No. 36.

"Perithecia few, rather large, crowded into a little pulvinate mass clothed with rubiginous powder. Ostiola at length prominent, truncate, with a central perforation. The sporidia, which are shortly cymbæform, vary a little in size."

On bark of *Celtis*, Carolina (Ravenel), on *Viburnum*, Pennsylvania (Michener).

In the specimens in Rav. Exsicc. (the only ones we have seen), the little pulvinate, erumpent stroma are 1–2 mm. across, each containing 2–6 perithecia having thick, coriaceous walls, and about ½ mm. diam. The asci are surrounded by abundant paraphyses, and have the spore-bearing part 55–60 x 8 μ long. Sporidia uniseriate, short cymbiform, opake, 12–14 x 8 μ, as noted by Cke. in Grev. XI, p. 123. The interior of the stroma shows a slight yellowish tint, like that of *H. Sassafras*, Schw., but not as distinct. The substance of the stroma is quite soft, almost carnose.

H. fúscum, (Pers.)

Sphæria fusca, Pers. Syn. p. 12.
Sphæria fragiformis, Hoff. Veg. Crypt. I, p. 20.
Sphæria confluens, Willd. Flora Berol. p. 416.
Sphæria tuberculosa, Bolt. Fungi Hal. p. 123.
Sphæria castorea, Tode Fungi Meckl. II, p. 28.
Sphæria Coryli and *S. glomerata*, DC. Fl. Fr. II, p. 287.
Hypoxylon fuscum, Fr. Summa Veg. Scand. p. 384.

Exsicc. Fckl. F. Rh. 1054.—Rab. F. E. 628.—Rehm Asc. 221.—Sydow, M. March. 165.—Thum. F. Austr. 664.—id. M. U. 1861.—Ell. N. A. F. 678.—Desm. Pl. Cr. Ed. I, 476.—Cke. F. Brit. Ser. I, 246, Ser. II, 467.

Stromata erumpent-superficial, solitary or subconnate, depressed-pulvinate or hemispherical, generally 1–3 mm. diam., dark purplish-red, finally black, somewhat uneven from the slightly projecting, small, closely packed, irregularly monostichous, subglobose perithecia with minute, mammilliform ostiola. Conidia very minute, borne singly at the extremities of short, sparingly branched sterigmata. Asci cylindrical, on long pedicels, spore-bearing part 80–90 x 7–8 μ. Paraphyses filiform. Sporidia uniseriate, subinequilaterally elliptical, opake and, in the specimens examined, 11–14 x 5–6 μ, (12–16 x 5–7 μ, Sacc.).

On dead alder, birch, hazel, beech and other deciduous trees, common throughout the United States and Canada.

H. bótrys, Nitsch, Pyr. Germ. p. 34.

Sphæria botryosa, Fckl. in F. Rh. 959.

Stromata erumpent, aggregated and subconnate or oftener tuberculiform, 1–2 mm. diam., consisting of simple aggregations of perithecia with very little stromatic material interposed, golden-yellow at first, finally black, about ½ mm. diam., about ¼ of the upper part of the perithecia projecting. Asci cylindrical, 8-spored, with filiform paraphyses. Sporidia uniseriate, narrow-elliptical, brown, mostly 2-nucleate, 12–14 x 5–7 μ. The inner substance of the bark under the stroma is whitened.

On bark of a dead willow tree, Pointe à la Hache, La. (Langlois).

We have no authentic specimens of this species, but the Louisiana specimens agree so well with the description of *H. botrys*, Nits., that we have little hesitation in referring them to it.

H. bicolor, E. & E. Journ. Mycol. II, p. 88.

Stroma tubercular-hemispherical, about 2 mm. across, scattered, somewhat uneven from the slightly prominent perithecia, dull ferruginous-purple, becoming darker within, yellow, becoming darker with age. Ostiola impressed, punctiform. Perithecia subperipherical, closely packed, about ¼ mm. diam. Asci narrow-cylindrical, with a slender base, about 100 x 6 µ.. Sporidia in a single series, narrow-elliptical or subnavicular, pale yellowish at first, then opake, 1–2-nucleate, 9–12 x 3½–4½ µ, ends subacute.

On dead limbs of *Quercus virens*, Pointe à la Hache, La. (Langlois).

Allied to *H. fuscum*, but differs in its impressed ostiola and smaller stroma yellow inside. Sec. Cooke, in Grev. XI, p. 127, *Hypoxylon bicolor*, B. & C., is a *Diatrype*.

** *Stroma externally black.*

H. multiforme, **Fr.** Summa Veg. Scand. p. 384.

Sphæria multiformis, Fr. S. M. II, p. 334.
Sphæria peltata, DC. Flore Fr. II, p. 287.
Hypoxylon granulosum, Bull. Champ. p. 176, tab. 487, fig. 2.
Sphæria rubiformis, Pers. Syn. p. 9.
Exsicc. Fckl. F. Rh. 1052.—Rab. F. E. 919—Thum. M. U. 1660, 2174.—Ell. N. A. F. 575.
Desm. Pl. Cr. Ed. I, 1251.—Lin. F. Hung. 181.—Sydow, M. March. 1451, 2954.

Stroma erumpent and often margined by the ruptured bark, of various shapes, but on birch usually transversely elongated, oblong or elliptical, somewhat flattened above, 1–1½ cm. long by ½–¾ cm. wide, or by confluence 4 or more cm. long, dull rusty-red at first, finally black and smooth. Perithecia irregularly monostichous, rather large, globose, distinctly prominent, with papilliform ostiola. Conidial layer dirty yellowish, becoming darker, conidia very small, obovate. Asci cylindrical, on long pedicels, spore-bearing part 70–90 x 6 µ. Paraphyses slender, simple, longer than the asci. Sporidia uniseriate, inequilateral-oblong, pale brown, 9–10½ x 3½ µ (10–12 x 4–5 µ, Sacc.).

On dead birch, New Hampshire (Farlow), New York (O. F. Cook), Michigan (Miss Minns), Minnesota (Holway), Canada (Macoun). *Alnus*, *Sorbus*, *Quercus*, and *Castanea* are also given as habitats of this species.

Specimens on *Alnus*, sent from British Columbia by Dr. Macoun, have the stroma depressed-hemispherical, 1–½ cm. across, and the perithecia less prominent, but the asci and sporidia are the same. This is a widely-diffused species, being found throughout Europe, also in Kamtschatka and the elevated region of Nepal in Central Asia. Its range appears to be northward. It is generally found on limbs from which the bark has not yet fallen, but is also said to grow on decorticated limbs and is then more effused. The specimens we have seen of this effused form seem rather to belong to *H. rubignosum*.

H. mallèolus, B. & Rav. Grev. IV, p. 49.

Exsicc. Rav. Fungi Car. IV, 32.—Ell. N. A. F. 861.—Rav F. Am. 181.

Stroma globose, sessile, 1½ cm. diam., black, ornamented by the papillose ostiola, each sunk in a shallow, circular depression about ½ mm. across. A vertical section of the stroma shows the same radiate-fibrous, subzonate structure and shining black color seen in *H. Howeianum*. Perithecia peripherical, oval or elliptical in outline, forming a layer about 1 mm. thick, which readily separates from the inner mass of the stroma. The asci (which appear to be evanescent) have, in our specimens, disappeared, but there is an abundance of brown, fusoid, nearly straight sporidia, 18–22 x 3–3½ μ, ends subobtuse.

On oak trees, Carolina (Ravenel), Florida (Martin, Calkins, and Rau).

H. Murràyi. B. & C. Grev. IV, p. 49.

"Gregarious, subglobose, a line or more broad, black without and within, densely papillose with the minute ostiola. It resembles externally *H. bomba*, Mont., except the densely papillose surface." Sporidia sec. Cke. in Grev. XI, p. 123, 13–15 x 5–7 μ.

On dead bark, Massachusetts (Murray).

H. glomifórme. B. & C. Grev. IV, p. 49.

"Gregarious, hemispherical, nearly ¼ inch wide, at first clothed with ferruginous powder, then black and shining, even. Perithecia hidden without any external trace of ostiola. Stroma dark brown." Sporidia sec. Cke. Grev. l. c. 14–15 x 3½ μ.

On bark of *Quercus nigra*, Connecticut (Wright).

H. cohærens, (Pers.)

Sphæria cohærens, Pers. Syn. p. 11.
Hypoxylon cohærens, Fr. Summa Veg. Scand. p. 42.
Exsicc. Fckl. F. Rh. 1053.—Rab. F. E. 918.—Thum. F. Austr. 1267.—Rav. Fungi Car. III, 48.—Cke. F. Brit. Ser. I, 666.—Rav. F. Am. 651.—Plowr. Sph. Brit. 217.

Stromata erumpent-superficial, 2–4 mm. diam., gregarious or crowded, and often confluent, hemispherical or globose, mostly flattened above, at first dirty-brown, becoming nearly black. Perithecia in a single layer, 6–10 in a stroma, rather large and distinctly prominent, with papilliform ostiola. Asci cylindrical, 8-spored, p. sp. about 22 x 6 μ. Sporidia uniseriate, ovate, inequilateral, brown, 10–12 x 4–6 μ.

On bark of dead beech trees, common.

The conidial hymenium, which clothes the young stromata, is of a pale clay-color, becoming cinereous. Conidia obovate-subglobose, very small. The species is widely diffused and is found also on oak, *Nyssa* and maple. A small form, var. *minor*, is mentioned on decaying *Polyporus* in Borneo. In the old and blackened state, this species resembles outwardly some forms of *H. coccineum*, Bull., from which it differs in its smaller, connate stromata and larger perithecia, and in the different color of the young stroma.

H. turbinulàtum, (Schw.)

Sphæria turbinulata, Schw. Syn. N. Am. 1204.

Turbinate-pulvinate, applanate, subconfluent, but with the stromata (pulvinuli) always distinct. Perithecia larger than usual, not peripheric but scattered through the entire stroma even to the base; external surface granulated, pulverulent, rugose with the minute, rather prominent ostiola. Stroma scanty, dirty whitish. Stromata arranged in a seriate manner so as to bear some resemblance to Hebrew letters, and seated on a black crust which overspreads the bark.

On beech wood, Mt. Pocono, Pa. (Schweinitz), Ohio (Morgan), New York (Fairman).

H. Bagnisii, Sacc., can hardly be distinct from this. The spec. in Herb. Schw. has the stromata subturbinate, 3–4 mm. diam., flattened-convex above, mammillose from the slightly projecting perithecia. Asci cylindrical, p. sp. 40–45 x 5–6 μ. Sporidia uniseriate, navicular-elliptical, dark brown, 8–10 x 3½–4½ μ. Cooke in Grev. makes them 12 x 3½ μ. The species very much resembles outwardly *H. cohærens*, Pers., only the stromata are more prominent and mostly a little narrowed below.

H. tères, (Schw.)

Sphæria teres, Schw. Syn. N. Am. No. 1178, new Am. Sph. Journ. Acad. p. 10, tab. II, fig. 7.
Hypoxylon teres, Sacc. Syll. 1493, Cke. Syn. 896.

Pulvinate, subterete-cylindrical, apex obtuse, rounded, surface tuberculose-undulate, rust-colored. Stroma sooty-black, surrounded and roughened by the immersed, peripherical perithecia. The cylindrical, pulvinulate, scattered stromata are about 3 lines high and 1½ lines thick. In some respects allied to *H. rubiginosum*.

On bark, locality unknown.

The spec. in Herb. Schw. is too imperfect to give an idea even of the outside appearance of this species.

C. *Stroma pulvinate, more or less convex, but not effused.* (*Clitoxylon*).

* *Stroma externally colored, not black.*

H. xanthócreas, B. & C. Grev. IV, p. 51.

Hypoxylon Peckianum, Sacc. Syll. I, p. 360.

"At first distinct, pulvinate, then by confluence forming a mass half an inch broad, black, papillate from the projection of the minute perithecia. Asci linear. Sporidia uniseriate, elliptical, .0003 of an inch long." (10 x 5 μ, Cke.).

On alder, New York (Peck), New England (Sprague).

Peck, in 31st Rep. p. 49, says: "Our specimens (ou prostrate, dead alders) agree with those received from Dr. Curtis under this name, but they do not agree with the description of the species as published in Grevillea. In our specimens the young plant is covered with a compact, yellow, conidiiferous stratum bearing elliptical conidia 4–5 μ long. As the stroma increases in size, it becomes naked above and of a purple-brown or chestnut color, which contrasts beautifully with the yellow margin. When old, it becomes darker, but we have not seen it black, as described. The surface is generally irregular or uneven. The stroma is whitish or pallid within, but near the surface it is yellow. The sporidia vary from 10–15 μ long."

The specimens of this species in Rav. Car. V, No. 57, have the stroma 3–4 mm. broad, brown above with the margin and inside yellow. Asci (p. sp.) about 60 x 6 μ, with a slender, stipitate base of about the same length. Sporidia uniseriate, elliptical, pale brown, 7–9 x 4–5 μ. There does not seem to be much doubt that the specimens found by Peck in New York are the genuine *H. xanthocreas.*

H. epiphlœum, B. & C. Grev. IV, p. 52.

The early conidial stage consists of small (1–2 mm.), thin patches of brick-red tomentum consisting of erect hyphæ suboppositely branched above and minutely roughened, bearing ovate, hyaline, 3-3½ x 2½ μ conidia. In the midst of these conidial patches soon appear small clusters of 3–12 perithecia ¾–1 mm. diam., and either scattered singly, or more or less connate, the different groups or clusters more or less confluent but not continuous, covered at first with the brick-red, conidial layer, then bare and black. Asci cylindrical, 80 x 4½ μ, (spore-bearing part about 60 μ long). Sporidia uniseriate, navicular, deep brown, 7–8 x 3 μ. The perithecia have a distinct papilliform ostiolum.

On *Magnolia glauca*, Carolina (Ravęnel), New Jersey (Ellis).

H. suborbiculàre, Pk. 30th Rep. p. 63.

"Stroma thin, flattened, erumpent, surrounded by the ruptured epidermis, growing from the inner bark, purplish-brown, then black, the surface slightly uneven as if areolate-rimose. Perithecia monostichous, subglobose. Ostiola sunken, perforate, sometimes whitish. Spores unequally elliptical, colored, .0004–.0005 of an inch long." Mr. Peck considers this an ally of *H. Laschii*, Nits.

On bark of *Acer saccharinum*, New York (Peck). Different from *H. suborbiculare*, Welw. & Curr., and with that probably referable to *Nummularia*.

H. Mórsei, B. & C. Grev. IV, p. 51.

Hypoxylon Blakei, B. & C. Grev. IV, p. 52.
Hypoxylon pauperatum, Karst. Enum. Fungi Lapp. p. 211.
Sphæria mammata, Nyl. Not pro Fauna & Flora Fenn. p. 88.
Exsicc. Ell. & Evrht. N. A. F. 2d Ser. 2070, 2609.—Ell. N. A. F. 1181.—Rab. F. E. 2955.

Stroma erumpent, orbicular, 3–5 mm. diam., closely embraced by the ruptured epidermis, flattened above, brownish-black and papillose from the prominent ostiola, surrounded by a black, circumscribing line. Perithecia large (about 1 mm.), submonostichous, mostly only slightly prominent, 4–15 in a stroma. Asci linear-cylindrical, 110–120 x 12 μ. Sporidia uniseriate, oblong-elliptical, brown, 1–2-nucleate, 17–22 x 8–10 μ.

On *Alnus*, Maine (Blake), New Hampshire (Mrs. Harrison), New York (Peck); on *Pyrus Malus*, *Carpinus*, and *Betula*, Iowa (Holway).

The Iowa specimens on birch have the stroma elliptical and larger (1 x 1½ cm.) and the perithecia have a tendency to crack away from each other and separate.

H. decorticàtum, (Schw.)

Sphæria decorticata, Schw. Syn. N. Am. No. 1179.
Hypoxylon decorticatum, Berk. Grev. IV, p. 50.

Subpulvinate, flattened when on the wood, less so on the bark, surface rusty-gray and densely covered with rough, sphæriiform tubercles resembling ostiola, so as to appear roughened with black granules. Perithecia peripherical in several layers, ovate-globose, immersed in the dark rust-colored stroma. Pulvinuli subrotund or irregular, about 5 mm. across, often confluent. Sporidia (sec. Cke. in Grev. XI, p. 123) 12–14 x 4 μ.

On wood and bark, Bethlehem, Pa. (Schw.), New England (Torrey).

The spec. of this species in Herb. Schw. has sporidia 8–10 x 4½ μ, and has the same general appearance as *H. perforatum*, (Schw.).

H. pruinàtum, (Klotszch).

Sphæria pruinata, Kl. in Linnæa, 1883, p. 489.
Rosellinia pruinata, Sacc. Syll. I, p. 259.
Hypoxylon Holwayii, Ell. in Am. Nat. Feb. 1883, p. 193.
Hypoxylon pruinatum, Cke. Syn. 925.
Exsicc. Ell N. A. F. 1182.

Stroma ¼–½ cm. diam., rather thin, orbicular, black within, surface covered with a white-pruinose coat, except the projecting, acutely papillose, black ostiola. Perithecia in a single layer, 20–30 in each stroma. Asci cylindrical. Sporidia uniseriate, oblong, brown, 1–2-nucleate, 22–27 x 11 μ, resembling the sporules of a *Sphæropsis*.

On the bark of trees, North America (Dr. Richardson), on the bark of dead poplars, Iowa (Holway).

In the Iowa specc., surrounding the stroma and standing out obliquely like a coarse fringe, are short, coarse, black, bristle-like teeth, like the teeth of a *Hydnum* or *Irpex*. This curious growth also arises from the surface of the inner bark for some distance around the stroma, soon throwing off the epidermis and leaving the blackened surface of the inner bark exposed. This growth is analogous to that of *Institale acariforme*, Fr., in connection with *Hypoxylon coccineum*.

We have not seen the original spece. of *H. pruinatum*, Kl., but as the peculiarity just mentioned seems to be the only character separating *H. Holwayii* from that species, we have placed the latter as a synonym, as has been done by Cooke in his synopsis, No. 925. The conidiiferous growth around the stroma may be only accidental, as it was not found in all the specimens.

** *Stroma externally black.*

H. leucócreas, B. & Rav. Grev. IV, p. 51.

"Small, about ½ a line across, black, papillate from the projection of the perithecia. Stroma snow-white. Asci linear. Sporidia in a single row, minute, elliptical, brown." Sporidia (sec. Cke. in Grev.) 5 x 2½ μ.

On limbs of oak, South Carolina (Ravenel).

H. exíguum, Cke. Grev. XI, p. 130.

"Pulvinate, convex-applanate, black, oval or discoid (2–3 mm. broad), here and there confluent. Perithecia minute, numerous, papillate. Asci cylindrical. Sporidia very minute, elliptical, dark, 3½ x 2 μ. A most distinct species, easily recognized by the exceedingly minute sporidia, which are a little larger in the American specimens."

On rotten wood, Alabama and Carolina, also in Mauritius.

H. pállidum, E. & E. Journ. Mycol. IV, p. 68.

Perithecia globose, about 1 mm, diam., suberose-coriaceous, connate in tuberculiform clusters 2–5 mm. diam., of a coffee-brown color, smooth, but uneven from the slightly projecting, flattened apices of the perithecia, which have a small, black, papillose ostiolum surrounded by a light-colored ring. Stroma scarcely any except as formed by the connate walls of the perithecia. Asci cylindrical, 150 x 6 μ, including the substipitate base, with abundant paraphyses. Sporidia uniseriate, navicular, opake, about 12 x 6 μ.

On bark of dead oak limbs, Catahoula, La. (Langlois, No. 1273).

H. marginàtum, (Schw.)

Sphæria marginata, Schw. Syn. N. Am. No. 1176.
Sphæria durissima, Schw. Syn. Car. No. 46.
Hypoxylon durissimum, Cke. Grev. XI, p. 131.
Sphæria truncata, Schw. Syn. Car. 174 (fide Cke. Grev. XV, p. 80).
Hypoxylon marginatum, Berk. Grev. IV, p. 49.
Exsicc. Ell. N. A. F. 471 (sub nomine *H. annulati*).—Ell. & Evrht. N. A. F. 2d Ser. 2352.—Rav. Fungi Car. I, No. 47.—Rav. F. Am. 182.

Stroma pulvinate, 1–3 cm. across, or by confluence more than that, convex-hemispherical, covered at first with the olivaceous conidial layer, finally black, surface slightly roughened by the projecting perithecia with their black, papilliform ostiola, which arise from the center of a small, flat, circular depression or disk which, however, does not appear in the earlier stage of growth. Perithecia monostichous, peripherical, about 2 mm. diam., ovate. Asci cylindrical, 75–80 x 6–7 μ. Sporidia uniseriate, navicular, brown, 7–9 x 3–3½ μ (mostly 7–8 μ long). This has been issued in Ravenel's Fungi Car. Ex. Fasc. I, No. 47, and in Ellis' N. A. F. No. 362, as *Hypoxylon annulatum*, Schw., but it agrees with specimens of *Sphæria marginata*, Schw., in Herb. Schw., at Philadelphia, and also with the description of that species in Syn. N. Am. The *S. marginata*, in Fries' Elenchus, II, p. 69, is evidently a different thing—probably, as Saccardo, in Syll. I, No. 371, suggests, *Nummularia discreta*, Schw.

On dead limbs and trunks, from Maine to Florida and west to Ohio.

H. annulàtum, (Schw.)

Sphæria annulata, Schw. in Fr. Elench. II, p. 64, and in Schw. New Am. Sph. p. 11, tab. 2, fig. 8.
Hypoxylon annulatum, Mont. Syll. Crypt. p. 213.
Exsicc. Ell. N. A. F. 472.—Ell. & Evrht. N. A. F. 2d Ser, 2353.—Rav. F. Am. 183.

Stroma hemispheric-tuberculiform (about ½ cm. across) or irregularly effused and interruptedly confluent-tuberculose, purplish-black.

Perithecia subglobose, monostichous, large (1 mm.), from $\frac{1}{4}$–$\frac{1}{2}$ of the upper part free, finally annulate-truncate above, with the black, papilliform ostiolum in the center of the truncate disk. Asci narrow-cylindrical, (p. sp.) 75 x 6 μ or, including the slender base, 100–112 μ long. Sporidia oblong-navicular, uniseriate, brown, mostly 2-nucleate, 7–9 x $3\frac{1}{2}$ μ, with their extremities rather more obtuse than in the preceding species. Var. *b. depressa*, Fr. l. c., appears to be the effused form above mentioned. This species is not mentioned in Schw. Syn. N. Am. It is readily distinguished from *H. marginatum* by its larger perithecia, much more prominent and sometimes nearly free, and its smaller, purplish-black stromata. No. 182 in Rav. F. Am. (in the copies we have seen) is *H. marginatum*.

On dead limbs, and having about the same range as the preceding species.

The conidial stage is *Verticillium puniceum*, C. & E. Grev. XVIII, p. 68. Tufts scarlet, elliptical, pulvinate, often confluent in patches 1 cm. in extent. Hyphæ slender, branching, septate; branches verticillate, short, tinged with rose-color. Conidia elliptical, minute, continuous, profuse, hyaline, 4 x 2 μ. The perithecia appear in the midst of these conidial tufts, and are at first covered by them.

H. obèsum, Fr. Nova Symb. p. 129.

Hard-carbonaceous, bare, black. Stroma slightly exceeding the short, very thick stipe, of radiate structure and cinereous-black within. Perithecia immersed, peripherical, bullate-prominent. Ostiola papillate, surrounded by an elevated, orbicular margin. Fries, who described this species from a single specimen, says it is allied to *H. annulatum*, that it is very hard, an inch high and, at least when mature, quite bare, glabrous and shining black. The sterile base or stipe is $\frac{1}{2}$ an inch high, rugose outside and attenuated below, covered above with a horizontal, slightly convex layer of globose, immersed, monostichous, bullate-prominent perithecia, like an immarginate pileus an inch across. The bullate projections of the perithecia are surrounded with a prominent orbicular margin, and in the middle of this circular area emerge the papilliform ostiola. The specimen seen by Fries was old and entirely without sporidia.

On trunks, in Costa Rica (Oersted).

H. Sássafras, (Schw.)

Sphæria Sassafras, Schw. Syn. Car. No. 87
Hypoxylon Sassafras, Berk. Grev. IV, p. 54.
Exsicc. Ell. N. A. F. 473.—Rav. F. Am. 345.—Rav. F. Car. I, No. 53.—Rab. F. E. 3459.

Perithecia large ($1\frac{1}{2}$ mm.), the internal cavity nearly 1 mm. diam., occurring either singly and quite evenly scattered over the matrix or loosely aggregated in clusters or groups of 3–8 perithecia standing side by side, their bases united in a thin stroma of a dirty brownish-black outside, and rusty-yellow within, with $\frac{1}{3}$–$\frac{1}{2}$ their upper part free, subtruncate above, with a minute, papilliform ostiolum. Asci, including the slender base, 110–120 x 4 μ. Sporidia uniseriate, oblong, pale brown, 1–2-nucleate, 7–9 x 3 μ. Paraphyses filiform, abundant.

On dead limbs and trunks of Sassafras, from New York to Florida, and west to Ohio, mostly on the bark, but also on the wood.

H. smilacícolum, Howe, Bull. Torr. Bot. Club, VI, p. 31.

"Small, black, pulvinate, roundish or elliptical, irregular when confluent. Perithecia subglobose. Asci cylindrical or subclavate. Sporidia brown, subcymbiform, 15–20 x 7$\frac{1}{2}$ μ, usually with several nuclei.

On dead stems of *Smilax*. The sporidia are rarely elliptical at maturity, but sometimes pointed at both extremities."

H. culmòrum, Cke. Grev. VII, p. 51.

Exsicc. Rav. F. Am. 351.—Ell. & Evrht. N. A. F. 2d Ser. 2116.

Stroma convex, 2–4 mm. across, olive-gray, then black, at first nearly even, then tuberculose from the projecting perithecia, finally deciduous, appearing first as olive-gray, appressed, thin, rather indefinitely limited patches 2–4 mm. across, consisting of closely-packed, erect, subsimple, brownish hyphæ 15–20 x 2–2$\frac{1}{2}$ μ, bearing at their tips oblong or ovate-elliptical, hyaline conidia 4–6 x 2–2$\frac{1}{2}$ μ. These patches soon become tuberculose from the scattered, incipient perithecia (3–15 in number), soon enclosed in the dull black stroma, whose surface is tuberculose-roughened by their obtuse, projecting apices. In the specimens in Rav. F. Am. 351, the perithecia are mostly solitary but still enclosed in a stroma more or less distinct. The inner cavity of the perithecia is $\frac{1}{3}$–$\frac{1}{2}$ mm. diam. Asci subcylindrical, 75–85 x 8–10 μ (p. sp.), with a short stipitate base and with evanescent paraphyses. Sporidia oblong-navicular or fusoid-navicular, mostly obliquely uniseriate, 2–3-nucleate, brown, 15–18 x 6 μ. Resembles in some respects *H. Sassafras*, Schw.

On dead culms of *Arundinaria*, Georgia (Ravenel), Florida (Calkins), Louisiana (Langlois).

H. polyspérmum, Mont. Syll. Crypt. No. 736.

Exsicc. Rav. F. Am. Nos, 346 and 347.—Ell. & Evrht. N. A. F. 2d Ser. 1788.

Stroma effused, applanate, abruptly limited, of a purplish rust-color, becoming black, outline irregular, mostly elongated (3 x 1 cm.) and about 1 or 1½ mm. thick, surface even or subtuberculose, closely papillate from the abundant ostiola, which are surrounded by an annular depressed area as in *H. marginatum* and *H. annulatum*, smaller, however, as well as the perithecia themselves, than in either of these species. Asci narrow-cylindrical, about 40 x 4 μ (p. sp.). Sporidia oblong-elliptical, uniseriate, 4–5 x 1½–2 μ, pale brown, sometimes a little bulging on one side. Perithecia monostichous, ½ mm. or less in diam. The general appearance is that of *H. rubiginosum*, from which, as well as from the two above-named species, it is distinguished by its much smaller sporidia. The specimens in Rav. F. Am. are labeled *H. marginatum*, Schw., but they cannot be that species.

On wood and bark of various deciduous trees, *Quercus*, *Myrica*, etc., Georgia (Ravenel), Florida and Tennessee (Calkins).

H. callostròma, (Schw.)

Sphæria callostroma, Schw. Syn. N. Am. No. 1208, and New Am. Sph. tab. XI, fig. 9.
Hypoxylon callostroma, Berk. Grev. IV, p. 51.

Irregularly effused, 2–3 inches long and wide, or in subturbinate groups of smaller size and seriately arranged, but not really confluent, in this case resembling *H. turbinulatum*. The effused specimens resemble at first sight some simple *Sphæria* with large perithecia closely crowded together, but a section shows that they are joined below in a common stroma which, on the outside, is black. The surface is uneven, granulose and punctate-rugose from the slightly prominent perithecia, which have their apices truncate with an obtusely subconic ostiolum immersed below in a grumose, bright ochraceous-red stroma of varying thickness. The perithecia themselves are oval or irregular in shape, consisting of an outer bark or shell enclosing the shining black, ascigerous nucleus. The colored stroma is always present, even when reduced to the simplest form enclosing but a single perithecium. Sporidia 12 x 5 μ (sec. Cooke in Grev. XI, p. 125).

On wood and bark of *Laurus æstivalis*, Bethlehem, Pa. (Schweinitz).

H. xanthostròma, (Schw.)

Sphæria xanthostroma, Schw. Syn. N. Am. 1212.
Hypoxylon xanthostroma, Sacc. Syll. 1507.

Seated on a thin crust which is not at all effused. In a simple series emerge distinct tubercles which are sometimes confluent for an inch or more, brown-black, rugose, larger mixed with smaller ones in the same group. Ostiola indistinct. A vertical section of the tubercles shows one or more rather large, globose perithecia enclosed in the grumose yellow stroma which on the outside is black. Sporidia (sec. Cke. Grev. XI, p. 125) 12 x 6 μ.

Seriately erumpent in cracks of decorticated oak limbs, Bethlehem, Pa. (Schw.).

The spec. in Herb. Schw. is without fruit. The outside appearance is like that of *H. Sassafras*, (Schw.).

H. Catálpæ, (Schw.)

Sphæria Catalpæ, Schw. Syn. N. Am. 1214.
Hypoxylon Catalpæ, Sacc. Syll. 1509.

Seriately erumpent through cracks in the bark, of a rusty color at first, then black. Tufts or pulvinuli longitudinally confluent. Surface of the stroma granular from the underlying perithecia, finally black and rugose. Perithecia abundant in the scanty black stroma. Ostiola papilliform, deciduous. Sporidia (Cke. Grev. XI, p. 125) 13 x 6 μ.

On bark of *Catalpa*, Bethlehem, Pa. (Schw.).

The sporidia in spec. in Herb. Schw. are 6–7 x 3½–4½ μ, short-navicular, brown. Perithecia rather less than ½ mm. diam.

H. transvérsum, (Schw.)

Sphæria transversa, Schw. Syn. N. Am. 1180.
Hypoxylon transversum, Sacc. Syll. 1505.

Large, subpulvinate, subimmersed in the bark and protruding in a pulvinate manner above, sometimes angular-turbinate. Surface irregularly rugose or even, black. Perithecia peripherical, ovate, shining-black inside. Stroma dark brown, pulverulent, 1 inch long, ¼ inch thick. Ostiola distinctly prominent, plano-conical. Sporidia (Cke. l. c.), 12 x 4 μ.

Transversely erumpent through the bark on a trunk of *Betula carpinifolia*, Mauch Chunk, Pa. (Schw.).

The spec. in Herb. Schw. has some of the perithecia large and prominent, but mostly only the apex and the papilliform ostiolum projecting. Asci 75–80 μ long (p. sp. 50 x 7 μ). Sporidia navicular, pale brown, 7–8 x 3½ μ.

H. ramòsum, (Schw.) in Herb. Berk. Grev. XI, p. 132.

Convex, erumpent, pulvinate, black, 1 cm. across. Perithecia subglobose, scattered, black, not prominent, pierced above. Asci cylindrical. Sporidia sublanceolate, continuous, brown, straight or curved, 16–18 x 3½ μ. This is a different thing from *Sphæria ramulosa*, Schw., which appears referable to *Xylaria*.

On branches, Indiana.

D. *Stroma broadly effused* (*Placoxylon*).

* *Externally colored, not black.*

H. perforàtum, (Schw.)

Sphæria perforata, Schw. Syn. Car. No. 45.
Hypoxylon perforatum, Sacc. Syll. 1431.
Exsicc. Rav. F. Car. V, 54.—Rav. F. Am. 349, 350.—Thum. M. U. 368.

Stroma (on the bark or wood) superficial, effused or tubercular. convex (2–4 mm.), often interruptedly confluent for several cm. in extent, dark or purplish rust-color, dotted with the minute, white-margined, punctiform ostiola. Conidial layer cinereous-white, pulveraceous. Conidia minute, ovoid or subglobose on short, subsimple or branching hyphæ. Perithecia submonostichous, globose, small (¼–⅓ mm.), lying near the surface of the stroma, crowded, mostly not distinctly prominent. Asci cylindrical, 60–90 x 7–9 μ (p. sp.), with a long, filiform base and overtopped by the filiform paraphyses, 8-spored. Sporidia obliquely uniseriate, ovate, with the ends mostly obtuse. nearly straight or subinequilateral, dark brown, 10–14 x 5–7 μ. Bears a general resemblance to *H. rubiginosum*.

On dead oak, maple, ash and other limbs, common; also on dead petioles of *Sabal serrulata*, Florida (Calkins).

H. rubiginòsum, (Pers.)

Sphæria rubiginosa, Pers. Syn. p. 11.
Hypoxylon rubiginosum, Fr. Summa Veg. Sc. p. 384.
Exsicc. Thum. M. U. 1071.—Rav. F. Am. 654, 741 —Plowr. Sph. Brit. 21.

Stroma mostly broadly effused, but also occurring in small patches 2–4 mm. across, bright ferruginous-red, finally black, tolerably thick (1–2 mm.), surface nearly even or distinctly mammillose from the projecting perithecia. Conidial layer pulverulent, thin, at first dirty olivaceous-yellow, then bright ferruginous. Conidia obovate or oval, very small, acrogenous on short, sparingly branched sterigmata. Asci cylindrical, long pedicellate, 8-spored, with slender, filiform paraphyses, 60 x 6 μ (p. sp.). Sporidia monostichous, ovate, inequilateral, or nearly straight, dark brown, 10 x 15 μ.

On decorticated limbs of various deciduous trees, common in this country and in Europe; around Newfield, N. J., mostly on *Acer* and

Quercus; on beech and *Liriodendron*, Pennsylvania (Everhart); on various dead limbs, Florida (Calkins).

The perithecia appear first in the middle of the stroma, and spread towards its margin, which thus remains for some time sterile. The perithecia are larger than in *H. perforatum* and more evenly effused, and the stroma is of a brighter color. At first, and around the margin of the stroma, the perithecia stand quite separate, but they are finally closely packed.

H. subchlòrinum, Ell. & Calkins, Journ. Mycol. IV, p. 86.

Exsicc. Ell. & Evrht. N. A. F. 2d Ser. 2115.

Stroma suborbicular, thin (1 mm.), flat, ½–1 cm. diam., sometimes continuous or interruptedly confluent for 5 or more cm., purplish rust-color, with a thin, sterile margin at first, but this generally disappears, leaving the margin abrupt and rounded, surface papillose from the slightly prominent, rounded apices of the perithecia, which are in a single layer, subglobose, small (¼ mm.), numerous, but not crowded so as to be much compressed, covered above with a thin stromatic layer which is of a dirty greenish-yellow within, at least in the young, fresh growing state, and finely white-punctate from the minute ostiola, but both the internal yellow color and the white-punctate ostiola finally disappear. Asci (p. sp.) 60–65 x 7 μ, with a stipitate base 30–40 μ long. Sporidia uniseriate, elliptical or subnavicular, opake, 7–8 x 3½–4 μ.

On bark of dead limbs of some deciduous tree, Florida (Calkins).

The general appearance, color and mode of growth is that of *H. rubiginosum*, (Pers.), from which it differs in its yellow stroma, smaller perithecia and sporidia; nor can it be referred to any of the species already enumerated, having the internal substance of the stroma yellow. The yellow stroma and smaller sporidia will also distinguish this from *H. perforatum*, (Schw.).

H. Fendlèri, Berk. Grev. XI, p. 132.

Effused, determinate, thick, rugose, yellow, finally black-brown ("atrofuscum"). Perithecia distinct, globose, elevated, with black papilliform ostiola. Asci cylindrical. Sporidia narrow-elliptical, straight or somewhat curved, dark, 12–13 x 4 μ. Somewhat like an effused state of *H. multiforme* or a thick form of *H. rubiginosum*, at length nearly black.

On rotten wood, Venezuela. Extra limital, but will probably be found in Central America.

H. atropurpùreum, Fr. Summa Veg. Scand. p. 384.—Nitsch. Pyr. Germ. p. 48.—Sacc. F. Ital. tab. 577.

Sphæria atropurpurea, Fr. S. M. II, p 340, Fr. Obs. I, p 174.
Exsicc. Ell. N. A. F. 1180.

Stroma broadly effused, continuous or interrupted, thin, purplish-black, becoming nearly black, surface minutely papillate from the slightly prominent perithecia, which are of medium size and are closely packed in a single layer. Asci (p. sp.) cylindrical, 50–60 x 7–8 μ. Sporidia obliquely monostichous, ovate, subacute at each end and slightly inequilateral, opake, 10–14 x 5–6 μ.

On bark of *Tilia*, Iowa (Holway), on bark, British Columbia (Macoun), New York (Peck).

H. albocínctum, E. & E. Proc. Phil. Acad. July, 1890, p. 229.

Stroma thin (1 mm.), flat, carbonaceous, mostly orbicular, $\frac{1}{4}$–1 cm. diam., light cinereous at first, soon purplish-black except the margin, which remains light-colored for some time, surface uneven from the projecting vertices of the perithecia which are ovate-globose, small ($\frac{1}{3}$–$\frac{1}{2}$ mm.), monostichous, moderately crowded, sunk nearly to the base of the stroma, contracted above into short necks terminating in the minute, papilliform ostiola. Asci cylindrical, 80–100 x 5–6 μ (p. sp. about 60 μ long), with abundant paraphyses. Sporidia uniseriate, narrowly elliptical, brown, 1–2-nucleate, subacute, 7–8 x 3$\frac{1}{2}$–4 μ.

On bark of dead *Cratægus*, Hamilton Co., Ohio (Morgan).

The bark beneath the stroma is whitened and surrounded by a black, circumscribing line. The general appearance is like that of orbicular forms of *H. serpens*, from which it differs in its purplish stroma and smaller perithecia and sporidia.

H. cinèreum, E. & E. (in Herb.)

Stroma oblong, 3–4 x 1 cm., thin, cinereous, surface even and smooth as if polished. Perithecia depressed-globose, minute (200 μ or less). Ostiola punctiform, slightly depressed, orbicular, white. Asci not seen. Sporidia oblong-elliptical, 2-nucleate, nearly opake, 8–10 x 3$\frac{1}{2}$–4 μ. The wood beneath the stroma is bleached and surrounded by a black line which penetrates deeply.

On rotten wood, near St. Martinsville, La. (Langlois, 2278).

Has the general appearance of *Diatrype stigma*. The color of the stroma on the surface is about that of marbleized iron ware, but nearly black within.

H. sublùteum, E. & E. (in Herb.)

Stroma effused, mostly elongated to 5 or more centimeters in length and 2 or more cm. wide, about 1 mm. thick, surface cinereous, roughened by the projecting, obtusely conical, darker colored ostiola, black inside except the layer next the wood, which is pale yellow with a greenish tint. Perithecia monostichous, ovate, about ⅓ mm. wide and a little more than that in height, their vertices slightly raising the surface of the stroma in a pustuliform manner. Asci cylindrical, 75–80 x 6 μ (p. sp.), with a slender base about 40 μ long. Sporidia uniseriate, navicular, 2-nucleate, becoming opake, 10–12 x 4–4½ μ.

On rotten wood, near St. Martinsville, La. (Langlois, 2276).

H. Mórgani, E. & E. (in Herb.)

Effused in patches 1–2 cm. long. Perithecia in a single layer, touching each other but hardly confluent, slightly depressed-globose, about ¾ mm. diam., connected and covered, except the black, subhemispherical, broadly perforated ostiolum, by a thin, furfuraceo-tomentose, glabrescent, light tawny yellow crust. Asci subcylindrical, 150 x 10–12 μ, with evanescent, filiform paraphyses. Sporidia overlapping-uniseriate, navicular, brown, 35–38 x 10–12 μ.

On rotten wood, Ohio (Morgan).

The apices of the perithecia project, rendering the surface of the stroma mammillate.

H. Ohiénse, E. & E. (in Herb.)

Stroma effused for several centimeters and about 2 mm. thick, surface uneven, colliculose, umber color. Perithecia ovate, subdistichous, ¾–1 mm. broad and 1–1¼ mm. high. Ostiola papilliform, umbilicate-collapsing at the apex, and soon hidden by the heaps of olive-brown, discharged sporidia. Asci (p. sp.) clavate, 22 x 6 μ, with a filiform stipe 30–40 μ long, paraphysate, 8-spored. Sporidia obliquely uniseriate or biseriate, narrow-elliptical, brown, 4–5 x 2½ μ.

On rotten wood, Ohio (Morgan, 883 and 965).

This might perhaps be considered a var. of *H. Petersii*, but it differs from that species in its effused stroma without any membranaceous veil, in its larger perithecia and smaller asci and sporidia. In both, the color and texture of the stroma is the same—umber color, paler inside.

H. platýstomum, E. & E. (in Herb.)

Stroma effused, reaching as much as 6 cm. diam., thin, showing all the inequalities of the wood beneath, about 1 mm. thick, of a dull reddish color at first, becoming nearly slate-color, the margin still retaining the reddish hue. Perithecia crowded, erect, oblong, $\frac{3}{4} \times \frac{1}{2}$ mm., the broad, orbicular, discoid ostiola erumpent, as in *Diatrype platystoma*, (Schw.). Asci cylindrical, paraphysate, 50×4 μ or, including the short stipe, 55 μ long. Sporidia uniseriate, oblong-elliptical, subinequilateral, pale brown, $5–6 \times 2\frac{1}{2}–3$ μ.

On the end of a decaying log of *Melia*, St. Martinsville, La. (Langlois).

This seems easily distinct from all the allied species, on account of its minute sporidia and discoid ostiola.

H. fuscopurpùreum, (Schw.)

Sphæria fuscopurpurea, Schw. Syn. N. Am. 1209.
Hypoxylon fuscopurpureum, Berk. Cuban Fungi, 835.
Exsicc. Rav. F. Am. 653.

Variously effused, margin generally sterile. Outer crust rather hard, black and shining within, surface elegantly purple, at length dark purple, regularly granulose from the subjacent perithecia which are oblong ovate, polystichous, numerous, small, immersed in the shining-black stroma, staining the wood or bark around it black, inseparably adnate, extending for an inch or more in length and preferring depressions in the surface of the wood. Sec. Cooke Grev. XI, p. 124, the sporidia are 14×7 μ. The specimen in Rav. F. Am. 653, on bark of ash, seaboard of South Carolina, has sporidia $9–11 \times 4\frac{1}{2}–6$ μ and looks more like a smooth form of *H. rubiginosum*.

On rotten wood and bark, Carolina and Pennsylvania (Schweinitz).

H. pìceum, Ell. Am. Nat. Feb., 1883, p. 194.

Stroma effused, subelliptical or elongated, often by confluence forming patches 4–8 cm. long by half as wide, dark brown, nearly black within, surface wrinkled and covered with a dull yellow conidial growth, which also spreads over the surface of the wood adjacent, and consists of short, rudimentary, irregularly branched hyphæ covered with the minute, dust-like conidia. Perithecia in 2–3 layers, densely crowded and angular by compression, the lower layer much elongated. Ostiola minute, scarcely visible. Asci? Sporidia navicular, brown, $11–12 \times 4$ μ. The stromata resemble blotches of black pitch dusted over with yellow meal, and are of about the consistency of beeswax.

On rotten wood, Iowa (Holway).

H. jecórinum, B. & Rav. Grev. IV, p. 50.

Effused, an inch or more long and broad, at first covered with a tawny yellow powder, then liver-colored, dotted with the dark ostiola. Sporidia (sec. Cooke l. c.) 9 x 4 μ. The specimens in Rav. Fungi Car. IV, 37, have the stroma subelliptical, 1–2 x 1 cm. and sporidia 7–8 x 3–4 μ. Florida specimens, collected by Col. Calkins during the winter of 1887, have the stroma 1½–3 x 2–2½ cm.

Ravenel's specimens are on bark of *Acer rubrum*, and the Florida specimens are also on bark of some deciduous tree.

The perithecia form a single layer on the surface of the black carbonaceous, 1 mm. thick stroma, and are oval in shape and closely packed, about ½ mm. high, with their apices slightly projecting, thus making the surface of the stroma finely papillose.

H. iánthinum, Cke. Grev. XI, p. 132.

Stromata thin, elliptical, subconfluent, 1–2 x ½–1 mm., grayish or cinereous, flat, surface papillose from the slightly projecting apices of the crowded perithecia, which lie in a single layer and are very small, less than ½ mm. high, penetrating to the wood and only slightly covered by the scanty stroma. Ostiola minute, papilliform. Asci cylindrical. Sporidia elliptical, obtuse, brown, 12–16 x 4–5 μ.

On decaying wood, Canada (Macoun), Louisiana (Langlois), Carolina (Ravenel), New York (O. F. Cook).

The Canada and Louisiana specimens have been submitted to Dr. Cooke for examination, and he pronounces them to be *H. ianthinum*, Cke., of which the original was collected in Potsdam, N. Y., many years ago. The name is badly chosen and misleading, for the surface of the stroma in all the specimens (unless it be the Potsdam specimen, which is now lost or mislaid) is of a glaucous or grayish-white, about the same as in *H. atropunctatum*, (Schw.), or *H. pruinatum*, (Kl.), without any purplish shade whatever. The description above quoted applies in other respects tolerably well.

H. atropunctàtum, (Schw.)

Sphæria atropunctata, Schw. Syn. Car. No. 44.
Anthostoma atropunctatum, Sacc. Syll. 1102.
Hypoxylon atropunctatum, Cke. Syn. 977.
Exsicc. Ell. N. A. F. 576.—Rab. F. E. 3159.

Broadly effused, smooth, white, dotted with the smooth, convex, black ostiola, and surrounded with a black, sterile margin, substance very hard and rigid, black inside. Perithecia in a single layer, not crowded, ovate, about ½ mm. high. Asci cylindrical, abruptly contracted below into a short stipitate base, about 150 x 10–12 μ. Sporidia uniseriate, acutely elliptical or almond-shaped, opake, 25–30 x 10–12 μ.

On dead trunks of oak, from New York to Florida.

According to Schweinitz this species is sometimes interruptedly continuous for 20 feet along the standing trunks of oak (*Q. falcata*), which are also nearly surrounded by it.

H. crocopéplum, B. & C. Grev. IV, p. 49.

Nearly ½ inch broad, irregular, depressed, clothed with a dense coat of red ferruginous powder. Perithecia rather prominent, with a minute ostiolum. Sporidia dark, shortly cymbiform, 13–14 x 8 μ, (sec. Cooke, l. c.).

On decayed bark, South Carolina (Ravenel).

H. florídeum, B. & C. Grev. IV, p. 50.

Effused for many inches, undulate, wine-colored, pulverulent. Perithecia hidden. Sporidia cymbiform, uninucleate, 9–10 x 3½ μ. Asci linear.

On *Acer rubrum*. Carolina (Ravenel).

** *Externally black.*

H. stigmáteum, Cke. Grev. VII, p. 4.

Exsicc. Rav. F. Am. 649.

Effused, black, crustaceous, thin (1–1½ mm.), papillose from the prominent ostiola, 3–5 or more cm. broad, originating beneath the cuticle of the bark which it throws off in the same manner as *Nummularia Bulliardi*, Tul., which it much resembles. Asci linear cylindrical. Sporidia uniseriate, elliptical, with the ends subacute, sometimes navicular, dark, 28 x 8 μ (sec. Cke.); 20–23 x 10–12 μ in the Louisiana specc.; 20–25 x 10–12 μ in the F. Am. specc.

On an old log, Louisiana (Langlois), on fallen logs, South Carolina (Ravenel), on bark of dead oak, California (Harkness), on beech bark, Ohio (Morgan).

H. epirrhòdium, B. & Rav. Grev. IV, p. 51.

Effused, thin, forming small black patches about two lines across, papillose from the slightly prominent ostiola. Asci linear. Sporidia uniseriate, elliptical. Sporidia sec. Cke. l. c. 9 x 3½ μ.

On branches of rose, South Carolina (Ravenel).

H. effùsum, Nitschke, Pyr. Germ. p. 48.

Exsicc. Ell. & Evrht. N. A. F. 2d Ser. 2114.—Sacc. M. Veneta, 1470.

Stroma superficial, thin, forming black, crust-like patches of various size and shape, 3–4 mm. across or often confluently seriate, 3–4 cm. or more by ½–1 cm. wide. Perithecia in a single layer, rather large (the central cavity being about ½ mm. diam.), prominent, but mostly flattened above with a central papilla much as in *H. annulatum*, Schw., but not so distinctly annulate depressed. The specimens were old and the asci dissolved, but the sporidia were still tolerably abundant, ovate-oblong and subnavicular, pale brown, 6–8 x 3–3½ μ, rounded and obtuse at the ends. The perithecia and sporidia were rather larger than in Saccardo's specimen in M. V. 1470, and the stroma thinner, but there can hardly be any doubt that our specimens are correctly determined.

On decaying wood of *Ulmus*, Missouri (Demetrio), Kansas (Kellerman), Louisiana (Langlois).

H. concúrrens, B. & C. Grev. IV, p. 93.

Perithecia connate, forming a thin, black, uniform stratum, very minutely granulated, the upper part only exposed. Ostiola minute, papilliform. Sporidia shortly cymbiform, uninucleate. (10 x 5 μ, Cke.).

Carolina (Ravenel) without habitat; on *Acer macrophyllum*, California (Harkness).

H. crustàceum, Nitschke, Pyr. Germ. p. 49.

(Sec. Cooke, not *Sphæria crustacea*, Sow.).
Exsicc. Rab. F. E. 2433.

Stroma superficial, blackening the wood around it both on the surface and within, more or less effused, tolerably thick, sooty black or sometimes gray-pruinose, formed apparently only by the connate perithecia which are about ¾ mm. diam., globose, and either densely crowded or loosely aggregated, or even partially free, their rounded apices with distinct papilliform ostiola free, with only their bases united; rarely perithecia occur only half as large as usual. Asci cylindrical, long pedicellate, with abundant, long filiform paraphyses. Sporidia obliquely monostichous, ovate, obtuse at each end, inequilateral or nearly straight, light brown, 8–10 x 4–5 μ.

On decorticated wood, British Columbia (Macoun).

Macoun's spece. agree accurately with the above description, except the perithecia are subferruginous-pulverulent, and the sporidia oblong-navicular. Asci 150 x 5 μ (p. sp. 80 x 5 μ). Clusters of connate perithecia (stromata) 2–5 x 2–3 mm., or interruptedly confluent for 2 cm. long. The specimen in Rab. F. E. 2433, has the perithecia more sparingly connate and black, but there is no other difference.

H. sérpens, (Pers.)

Sphæria serpens, Pers. Syn. p. 20, Obs. Myc. I, p. 18.
Sphæria Macula, Tode Fungi Meckl. II. p 33, fig. 106.
Hypoxylon serpens, Fr. Summa Veg. Scand. p. 284,
Exsicc. Fckl. F. Rh. 960.—Rav. Fungi Car. IV, 34.—Sydow, M. March, 2241.—(Ell. N. A. F. 164) ?—Desm. Pl. Crypt. Ed. I, 377.

Stroma effused, thin, applanate, black, variable in form and size, often in narrow, elongated strips 2–3 mm. wide and 3–6 cm. long, but also in small subelliptical or irregular shaped patches 1–2 cm. long by ½–1 cm. wide. Perithecia subglobose, crowded, rather large, rounded and prominent above or rarely slightly depressed around the central papilla, then only slightly prominent, and the surface of the stroma not so distinctly roughened. Conidial layer pulverulent, cinereous. Conidia subglobose, minute, acrogenous, on rather long, branching septate sterigmata. Asci cylindrical, long-pedicellate, 75–100 μ long (p. sp.) by 6–8 μ wide, with abundant paraphyses. Sporidia obliquely uniseriate, subcylindrical, rounded at the ends, oblong-cylindrical, sub-inequilateral or almost curved, seldom straight, becoming dark, 12–16 x 5–6 μ.

On decaying wood (seldom on the bark) of various deciduous trees.

This is called a common and widely diffused species, but as we have some doubt as to whether we properly understand it, we have taken the above description from Nitschke's Pyr. Germ. The specimens distributed in N. A. F. under this name are certainly not the typical form, for the perithecia are small, mostly ½ mm. or less, and only slightly prominent, and the sporidia are mostly only 8–10 x 3–4 μ. The N. A. F. specimens appear to be the same as the *H. colliculosum*, Schw. in Rav. F. Am. 742, which cannot be distinguished from *Sphæria* (*Hypoxylon*) *insidens*, in Herb. Schw.

H. ínsidens, (Schw.)

Sphæria insidens, Schw. Syn. Car. No. 122, Fr. S. M. II, p. 422.
Fuckelia insidens, (Schw.) Cke. Grev. XII, p 52.
Exsicc. Ell. N. A. F. 164.

Stroma innate, effused, nearly round, brown-black, partly sterile, apparently superficial, but the base immersed in the matrix and surrounded by a faint circumscribing line. Perithecia more or less prominent, flexuous, subpapillate, half as large as a mustard seed. Asci cylindrical. Sporidia uniseriate, elliptical, pale brown, 8 x 4 μ.

On rotten wood or oftener on bark, Carolina and Pennsylvania (Schw.), on rotten wood and bark of *Magnolia glauca*, Newfield, N. J.

Whether the *Sphæria insidens*, Schw. in litt., which in Fr.

Elench. II, p. 68, is quoted as a synonym of *Sphæria atramentosa*, Fr. in Kze. Myc. Hefte, 2, p. 38, and Fr. S. M. II, p. 344, is the same as the *Sphæria insidens*, Schw. in Syn. Car. 122, and Fr. S. M. II, p. 422, is not certain, but as far as the diagnoses go they may be the same; in that case the specific name, *atramentosa*, has precedence.

H. colliculòsum, (Schw.)

Sphæria colliculosa, Schw. Syn. Car. No. 82.
Hypoxylon colliculosum, Cke. Syn. 1010.
Exsicc. (Rav. F. Am. 742)?

Effused thin, colliculose, rugose, black. Perithecia very large, covered with a thin crust which is papillate from the minute ostiola, and with flattened bases not immersed in the wood or surrounded by any circumscribing line, subdistant but connected by a stromatic crust. Margin various, shining as if oiled, surface very uneven and rimose. Sporidia 12–13 x 5 μ (Cke.).

On rotten oak wood, Carolina and Pennsylvania (Schw.).

As already stated, the specimens in Rav. F. Am. do not agree with the description of *H. colliculosum*, having both perithecia and sporidia too small, and are probably referable to *H. insidens*, Schw. We do not find any specimen of *Sphæria colliculosa* in Herb. Schw.

H. illìtum, (Schw.)

Sphæria illita, Schw. Syn. N. Am. 1205.
Hypoxylon illitum, Sacc. Syll. 1511, Cke. Syn, 1014.

Widely effused, confluent, the layers often superimposed, so as to imitate a sculptured surface, the material of the stroma appearing as if smeared on the decaying wood. Surface undulate and uneven, at first of a fine olive-green, but finally black. Perithecia rather large, slightly prominent, with ostiola indistinct or acutely conical and thick walls, surrounded with a sparing white stroma. Sporidia fusoid, navicular, very pale brown, acute, 10–12 x $2\frac{1}{2}$–$3\frac{1}{2}$ μ in spec. in Herb. Schw. (14–16 x 4 μ. Cke.).

Not infrequent on standing trunks, especially of *Platanus*, investing them almost completely with its broad, uneven, confluent stromata, Bethlehem, Pa. (Schw.).

H. invéstiens, (Schw.)

Sphæria investiens, Schw. Syn. N. Am. 1210.
Hypoxylon investiens, Berk. Cuban Fungi, No. 837.
Exsicc. Rav. Fungi Car. IV, 33.

Seated on a thick sterile crust that spreads over and blackens the wood, following all the inequalities of its surface. On this crust stand, densely crowded in a single series, the regularly oblong perithecia, forming a continuous layer about $\frac{3}{4}$ mm. thick and 4–9 cm. long and wide. The stroma is very scanty, covering the perithecia with a thin, black stratum mammillose above from the slightly projecting perithecia with their papilliform, deciduous ostiola. In the specimens in Rav. Car., as well as in the Louisiana specc., the surface of the stroma has a distinct purplish tinge. We have not seen the asci, but the sporidia are oblong, pale brown, 6–10 (mostly 6–8) x 3–4 μ. *H. effusum*, Nitschke, is closely allied to this.

On rotten wood, Carolina and Pennsylvania (Schw.), Alabama (Beaumont), on *Salix*, Louisiana (Langlois), Texas (Wright).

H. càries, (Schw.)

Sphæria caries, Schw. Syn. N. Am. 1222.
Hypoxylon caries, Sacc. Syll. 1510, Cke. Syn. 1011.

Stroma effused, black within and without, colliculose and uneven from being composed apparently of many smaller stromata 3–10 mm. diam., fused together laterally more or less perfectly into a continuous or partially interrupted crust, irregular in outline and several centimeters in extent. Perithecia subglobose, $\frac{1}{2}$–$\frac{3}{4}$ mm. diam., their apices slightly prominent with a subacute, papilliform ostiolum surrounded by an indistinct, lighter colored ring which, however, is not impressed or sunk in the stroma as in *H. annulatum*. In the specimens examined, the asci had disappeared. Sporidia navicular-fusoid, (subhyaline) pale smoky-brown, ends subacute, 10–12 by about 3 μ.

On rotten wood, Bethlehem, Pa. (Schw.), on rotten oak, Newfield, N. J., on rotten elm (*Ulmus Americana*), Missouri (Demetrio).

H. Ravenélii, Rehm, Hedwigia, 1882, p. 137.

Hypoxylon confluens, Fr. in Rav. F. Am. 348.

Perithecia single or concrescent, 2–8 together, occasionally seriate, 6–12 in a series 3–6 mm. long, nearly globose, $\frac{3}{4}$–1 mm. diam., with their bases slightly sunk in the wood (our spec. is on wood and not on bark). Ostiolum distinct, papilliform, black and shining. The perithecia are of a dead grayish-black. Asci very long, cylindrical, with abundant well-developed paraphyses. Sporidia elliptical, obtuse, pale brown, with 1–2 large nuclei, uniseriate, 10 x 5 μ.

On bark of decaying oak, Darien, Ga. (Ravenel).

This is entirely different from *H. Ravenelii*, Sacc. Syll. I, p. 389, (*H. erinaceum*, B. & Rav.) which sec. Cke. Grev. XI, p. 128, is a *Valsa* with long-necked perithecia and hyaline, allantoid sporidia. Whether the above described fungus is the *Sphæria confluens*, Tode, can not perhaps now be certainly decided. It agrees tolerably with Tode's figure, but it is not that species as understood by Nitschke, and described by him (under *Hypoxylon semiimmersum*) as having sporidia 16–20 x 8–10 μ, and by Fckl. (under the name of *H. udum*) as having sporidia 28 x 10 μ. We have therefore accepted *H. Ravenelii*, Rehm, as a distinct species.

H.? atrofúscum, B. & C.

Fuckelia atrofusca, B. & C. Grev. XII, p. 51.

Pustules erumpent, very small (hardly ½ mm. diam.), elliptical, margined by the ruptured bark. Perithecia unequally distributed in the black, depressed stroma. Asci cylindrical, stipitate. Sporidia elliptical, brown, 13 x 7 μ.

On bark of *Rhus glabra*, mountains of Virginia.

H. hydnícolum, (Schw.)

Sphæria hydnicola, Schw. Syn. N. Am. 1207.
Hypoxylon hydnicolum, Cke. Syn. 1034.

Stroma thick, short, subrepand, here and there confluent; externally very black, granulose. Perithecia large, subdistant, immersed in the light yellow substance of the stroma, monostichous, furnished with a brown veil or sack. Ostiola prominent, papilliform. The teeth of the *Hydnum* are often concrescent with the stroma, which then appears stipitate. Substance of the stroma distinctly suberose. The diameter of the stroma scarcely exceeds 4–6 mm.

Rather rare; on the teeth of decaying *Hydnums*, Bethlehem, Pa. (Schw.).

From an examination of the spece. in Herb. Schw. we add the following notes: Stroma subtubercular, 2–4 mm. across. Perithecia only slightly prominent. Ostiola papilliform. Asci 75–80 μ long, p. sp. 60 x 6 μ, cylindrical. Sporidia uniseriate, elliptical, pale brown, with a single large nucleus, 10–11 x 4 μ.

H. exaràtum, (Schw.)

Sphæria exarata, Schw. Syn. N. Am. 1206.
Hypoxylon exaratum, Cke. Syn. 1032.

Effigurately effused, surface marked with parallel, longitudinal furrows, at first covered by the epidermis which is persistent in the furrows. Perithecia very prominent on the ridges, irregular, black-brown, with a black papillate ostiolum, rather large, monostichous, surrounded with a scanty stroma, which has a sterile, subrepand margin, and rests on a crust formed of the blackened substance of the bark.

On the bark of young dead branches of *Juglans tomentosa*, Bethlehem, Pa. (Schw.).

The spec. in Herb. Schw. has the stroma ½–1 cm. long, by 1½–2 mm. wide. Ostiola slightly prominent. Sporidia (free spores, no asci seen) navicular-elliptical, 7–9 x 3½–4 μ.

H. sphæriόstomum, (Schw.)

Sphæria sphæriostoma, Schw. Syn. N. Am. 1213.
Hypoxylon sphæriostomum, Cke. Syn 1033.

Short, subpulvinate, carbonaceous, very black, 4–6 mm. long, oblong, acuminate at each end, surface longitudinally striate. On this blackened surface are sphæriiform, globose, scattered ostiola, perforated with a round opening and connected by a rather long tube with the subjacent perithecia, which are buried in the wood without any real stroma, ovate, rather large, filled with a black mass which, under the lens, appears to be composed of spores like those of a *Melanconium*.

On soft, rotten wood, Bethlehem, Pa. (Schw.).

The fructification of this species is unknown.

Hypόxylon Beaumόntii, B. & C.

Hypόxylon nudicόlle, B. & C.

Hypόxylon gemmàtum, B. & C.

Hypόxylon gregàle, (Schw.)

have already been described under *Valsaria*. Of these, it is probable that *H. Beaumontii* should be removed to *Hypoxylon*. Berkeley in Grev. makes the sporidia uniseptate, but Cke. Grev. XII, p. 134, finds the sporidia in the original specimens, in Herb. Berk., continuous. *H. gregale*, (Schw.) also (sec. spec. in Herb. Schw.) has continuous sporidia and is a true *Hypoxylon*. *Hypoxylon miniatum*, Cke. in Journ. Mycol. IV, p. 87, is (sec. Cke.) not that species, and must be rejected from the list of North American species.

Hypόxylon glòmus, B. & C. Grev. IV, p. 51, has the appearance of a *Diatrype*, but is sterile (Cke. Grev. XII, p. 126).

H. spondýlinum, (Fr.)

Sphæria spondylina, Fr. S. M II, p. 347.
Hypoxylon spondylinum, Fr. Summa Veg. Scand p. 383.
Nummularia spondylina, Sacc. Syll. 1542

Erumpent, variable, convex, subrugose, black outside and inside. Although erumpent, it becomes entirely free, not even adnate at the base, rather small, 2–4 mm. broad and thick, bullate, densely gregarious and often confluent, assuming various shapes from mutual pressure; when standing singly, regular, convex, covered with a smooth, hard stratum. Perithecia buried, ovate.

On oak branches; rare, Bethlehem, Pa. (Schw.).

BOLÍNIA, Nitschke.

Pyren. Germ. p. 26 (as a subgenus).

Stroma superficial, effused. Perithecia deeply immersed in the stroma, with elongated necks and umbilicate ostiola. Asci cylindrical, 8-spored, paraphysate. Sporidia ovoid continuous, brown.

B. Tubùlina, (Alb. & Schw.)

Sphæria Tubulina, A. & S. Consp. Nisk. p. 6, tab. IV, fig. 4.
Hypoxylon Tubulina, Fr. Summa Veg. Sc p. 383.
Bolinia Tubulina, Sacc. Syll. 1332 Cke. Syn. 807.

Stroma effused, determinate, oblong or oval, superficially adnate, when on wood still sound and hard, but with the base more or less immersed where the wood is more decayed and softer, surface uneven, with variously-shaped depressions and prominences, or in the smaller stromata, flat pulvinate and smooth, at first of a dirty-ferruginous color, finally black, fragile, margin repand. Perithecia very large, monostichous, densely crowded, ovate or angular from mutual pressure, gradually attenuated into long necks, with the perforated ostiola not prominent, and giving the surface of the stroma a porous appearance. Asci cylindrical, long-pedicellate, 8-spored, p. sp. 36–45 x 5 μ. Sporidia obliquely monostichous, small (6–7 x 3–4 μ), ovate, ends obtusely rounded, becoming nearly black.

On dead *Juglans*, Carolina and Pennsylvania (Schw.).

The stroma is 2–4 inches or more long, 1½–2 inches or more wide, and ¼–½ an inch thick, with the sides abrupt. When old and broken down, it resembles a *Tubulina*, hence the specific name.

PORÒNIA, Willdenow.

Flor. Berol. Prodr. p. 400

Stroma carnose-suberose, at first clavate, then cup-shaped or discoid, stipitate or subsessile, light-colored. Perithecia immersed in the upper, discoid surface of the stroma, carbonaceous, black. Asci cylindrical, 8-spored. Sporidia ellipsoid, brown, with a hyaline, gelatinous envelope. Stroma at first clothed with the conidial hymenium. Fimicolous.

P. punctàta, (Linn.)

Peziza punctata, Linn. Flor. Suec. Ed. II, p 458.
Sphæria nivea, Haller, Stirp. Helvet. tom. III, p. 121.
Sphæria truncata, Bolt. Fungi Halif. III, tab. 127.
Sphæria punctata, Sowerby, Eng. Fungi, tab. 54.
Sphæria Poronia, Pers. Syn. p. 15.
Poronia Gleditschii, Willd. Flor. Berol. Prodr. p 400.
Poronia fimetaria, Pers. Champ. Comest. p. 154.
Poronia punctata, Fr. Summa Veg. Scand. p. 382.
Exsicc. Rab. F. E. 2020.—Rehm, Asc. 168.—Cke. Fungi Brit. Ser I, 468, id. Ser. II, 213. Roum. F. Gall. 566 —Linht. Fungi Hung. 183.

Stroma at first clavate, soon expanded and discoid above, brown outside, the inner substance and disk white. Stipe subelongated (reaching 1–2 cm.), mostly penetrating the matrix and hidden. Disk generally 2–5 mm. diam., exceptionally 10–15 mm., grayish-white, pulverulent at first. Perithecia globose, black within and without, about $\frac{1}{3}$ mm. diam., their rather large, black, papilliform ostiola dotting the white disk. Asci cylindrical, 120–150 x 15–17 μ (p. sp. 80 90 μ long), with abundant, but imperfectly developed paraphyses. Sporidia uniseriate, elliptical, subinequilateral, soon opake, surrounded at first by a hyaline, gelatinous coat, 15–20 x 10–12 μ (10–26 x 10–14 μ, Winter).

On horse dung, Kansas (Cragin).

The measurements of asci and sporidia are from the Kansas specimens.

P. Œdipus, Mont. Syll. p. 209.

Sphæria (Poronia) punctata, var *œdipoda*, Mont. Ann Sci. Nat Ser II, tom VI, p. 333.
Hypoxylon œdipus, Mont. Cuba, p 346, tab. XIII, fig. 2
Sphæria incrassata, Jungh. Flor. Crypt. Javæ, p. 87.
Poronia macropoda, *b. cladonioides*, Ces in Klotzsch-Rab Herb. Mycol. No. 1946.
Exsicc. Rab. F. Eur. 630 —Rav Fungi Car. III, No 46

Stroma erect, simple (or sometimes branched), with a distinct, smooth, light brown stipe 2–3 cm. high, 2–3 mm. thick at the clavate-swollen base, enlarged above into a dull white, suborbicular, 2–3 mm. disk, which is at first concave, then plane, and pierced by the slightly prominent, black, papilliform ostiola. Perithecia ovate, sunk in the stroma. Asci cylindrical, short-stipitate, 8-spored, 100–120 x 18–20 μ, with long, filiform pseudoparaphyses. Sporidia mostly uniseriate, elliptical, 28–30 x 16 μ, with a thick hyaline gelatinous coat at first, finally nearly black.

On horse dung. Alabama (Peters), Texas (Wright).

Readily distinguished from *P. punctata* by its long, clavate-swollen stipe.

P. leporina, E. & E. Proc. Acad. Nat. Sci. Phil. July, 1890, p. 229.
(Plate 39)

Exsicc. Ell. & Evrht. N. A. F. 2d Ser. 2354.

Stipitate. flesh-colored, small, stipe 1–2 mm. long, $\frac{1}{2}$ mm. thick, expanding above into a discoid stroma 1–2 mm. diam. and mammillose from the slightly prominent perithecia which are ovate-globose, about $\frac{1}{3}$ mm. diam., 6–20 in a stroma. Ostiola large, black, convex, Asci clavate-cylindrical, 80–100 (p. sp. 75–80) x 10–12 μ, with obscure paraphyses. Sporidia at first greenish-hyaline, 1-2-nucleate, becoming opake, subinequilaterally elliptical, mostly uniseriate, 12–15 x 6–7 μ.

On rabbit-dung, Missouri (Demetrio).

Distinguished from *P. Œdipus* by its smaller size.

The three following species are placed under *Poronia* by Schweinitz in Syn. N. Am. p. 189.

Sphæria pocula, Schw. Syn. N. Am. 1167 (*Enslinia pocula*, Fr. Summa, p. 399). This is a *Polyporus* (*P. cupulæformis*, B. & Rav. in Rav. F. Car. I, No. 10, Grev. I, p. 38; *P. pocula*, Cke. in Grev. XII, p. 85).

Sphæria candida, Schw. Syn. N. Am. 1165, and *Sphæria intermedia*, Schw. l. c. 1166. We know nothing of these. Fries places them in his genus *Enslinia*.

DALDÍNIA, Ces. & De Not.

Schema Sferiac. Ital. in Comm. Critt. I, p. 197.

Stroma superficial, subglobose, external layer carbonaceous, becoming black, fibrous within and concentrically zoned. Asci cylindrical, 8-spored, pedicellate. Sporidia ovoid or oblong, dark-colored. Perithecia immersed in the stroma.

D. concéntrica, (Bolt.) **(Plate 38)**

Sphæria concentrica, Bolt. Fungi Hal tab. 180.
Lycoperdon atrum, Schæff. Fungi Bavar. IV, p. 131, tab. 329.
Valsa tuberosa, Scopoli, Flora Carniol, p. 399.
Sphæria tunicata, Tode, Fungi Meckl. II, p. 59, tab. XVII, fig. 130.
Sphæria fraxinea, Withering, Arrang. Brit. Plants, IV, p. 393.
Stromatosphæria concentrica, Grev. Flora Edinb. p. 355.
Hypoxylon concentricum, Grev. Scott. Crypt. Flora, VI, tab 324.
Daldinia concentrica, Ces. & De Not. Schema Sferiac. p. 24.

Exsicc. Rab Herb. Mycol. 600.—Thum. F. Austr. 1154 —Thum. M. U. 69.—Cke F Brit. Ser I, 669, id Ser II, 216.—Plowr Sph. Brit. 17.—Allesch & Schnabl, F. Bav. 76.

Stroma subspherical or hemispherical, rarely obovoid, subferruginous and softer at first, at length black and carbonaceous, 2–4 cm. diam., softer inside, of a radiate-fibrous structure and concentrically zoned. Perithecia monostichous, obovoid-oblong, 1 mm. or a little more in length and about ½ mm. broad, more or less angular from mutual pressure. Ostiola slightly prominent, punctiform, minute. Sporidia obliquely uniseriate, inequilaterally elliptical, dark brown and finally opake, 12–15 x 7–10 μ. Asci long-pedicellate, 80–100 x 8–10 μ (p. sp.), with long, filiform paraphyses.

On dead trunks of various deciduous trees, common from New England to California, and from Canada to Louisiana and Mexico.

D. vernicòsa, (Schw.)

Sphæria vernicosa, Schw. Syn. N. Am. 1175.
Daldinia vernicosa, Ces. & De Not. l. c.
Exsicc. Ell. N. A. F. 166

Stroma large (2½–3 x 1½ cm.), subturbinate, suddenly contracted below into a thick, stipe-like base which is sometimes concentrically wrinkled; surface of the stroma ferruginous at first from the conidial layer, finally black and shining. Perithecia peripherical, subglobose (sec. Schw.), but in all the specimens we have seen, ovoid-oblong, about the same in size and shape as in the preceding species. Saccardo in Sylloge says perithecia polystichous, but we have never found them so, though a vertical section through one side of the stroma shows them *apparently* in more than one layer; but this is only apparent, as may be seen in a vertical section through the center of the stroma. We find the asci and sporidia about as in the preceding species, though in the Sylloge they are said to be longer and narrower. This is distinguished from *D. concentrica* by its shining-black stroma, and the looser texture of the radiate-fibrous inner substance which is cut by 8–12 dark colored, membranaceous horizontal layers or plates. These are very noticeable in a vertical section even in the young plant, while it is still covered with the conidial layer and before the terminal, subglobose, ascigerous stroma has begun to appear. In the mature state, the fibrous inner substance and the horizontal membranes disappear to a greater or less extent, and leave the stroma more or less hollow, so that it may be easily crushed with the fingers, but in *D. concentrica* the inner substance remains firm and is also of a darker color.

On fence pickets, Salem, North Carolina, (Schw.), on trunks of dead oak trees, Newfield, N. J.; also sent from New England and New York.

D. cingulàta, (Lev.)

Sphæria cingulata, Lev. Ann. Sci. Nat. Ser. III, (1845), p. 47.
Daldinia cingulata, Sacc. Syll. 1521.

Obovate, erect, substipitate, outer layer laccate-crustaceous, of a shining brown-black color. Perithecia buried in the stroma, zonately arranged, white inside. Ostiola obsolete.

On trunks, near New York (Menaud). Stroma 1–2 dec. high, 1 dec. thick.

This seems doubtfully distinct from *D. vernicosa.*

D. loculàta, (Lev.)

Sphæria loculata, Lev. l. c.

Globose, substipitate, black, opake. Perithecia obovate, immersed in the stroma. Ostiola subprominent, shining, subhemispherical. Asci and sporidia as in the other species. Stipe short, somewhat rough.

On trunks, America.

USTÙLINA, Tul.

Sel. Carp. II, p. 23.

Stroma superficial, subeffused, rather thick, determinate, at first carnose-suberose and clothed with a pulverulent, cinereous, conidial hymenium, finally rigid, carbonaceous, black and bare and generally more or less hollow. Perithecia immersed, large, with papilliform ostiola. Asci pedicellate, 8-spored, paraphysate. Sporidia ovoid-fusiform, continuous, dark-colored.

U. vulgàris, Tul. l. c. tab. III, figs. 1–6. **(Plate 39)**

Sphæria deusta, Hoff. Veg. Crypt. I, p. 3, tab. I, fig. 2.
Sphæria versipellis, Tode. Fungi Meckl. II, p. 55.
Hypoxylon ustulatum, Bull. Champ. d. France, I, p. 176, tab. 478, fig. 1.
Hypoxylon deustum, Grev. Scot. Crypt. Flora, IV, tab. 324, fig. 2.
Exsicc. Fckl. F. Rh. 1063.—Kze F. Sel. 154 —Rab. Herb. Mycol. 145.—Thum. F. Austr. 665.—Ell. N. A. F. 860

Stroma superficial, subeffused, 3 cm. diam., repand, pulvinate, thick (3–4 mm), surface even, white and subtomentose, finally undulate-colliculose and black; substance almost gelatinous at first, then hard and tough, at length very brittle and hollow, centrally attached. Perithecia large, ovate, densely crowded, monostichous, the punctiform ostiola only projecting. Asci narrow-cylindrical, pedicellate, 8-spored, 250 x 8–10 μ (p. sp.). Paraphyses slender, evanescent. Sporidia obliquely uniseriate, fusoid, inequilateral or slightly curved, finally opake, 32–40 x 8–10 μ. Tode, l. c., gives a very minute and accurate account of this fungus.

On roots of decaying stumps; found in Europe, America and

Australia.; common throughout the eastern United States and reported by Dr. Harkness from California.

CAMÍLLEA, Fr.

Summa Veg. Scand. p. 382, Mont. Syll. Crypt. p. 207.

Stromata vertical, oblong, carbonaceous, stipitate or sessile, stromatically connected at base. Perithecia linear or bottle-shaped, membranaceous, included in the upper part of the stroma. Asci obovate, 8-spored, with capillary paraphyses. Sporidia conglobate, oblong, continuous, brown.

Some of the extra-limital species have the stroma truncate or cup-shaped above, and the sporidia appendiculate.

C. Sagræàna, (Mont.)

Hypoxylon Sagræana, Mont. Cuba, p 341, tab 12, fig. 4.
Phylacia Sagræana, Mont. Syll. Crypt. No. 921.
Camillea Sagræana. B. & C. Exot. Fungi, p. 285.

Stromata cespitose-connate, oblong-obovate, stipitate, carbonaceous, black, fragile; fertile head about 1 cm. long by 5–6 mm. wide, obtusely pointed at the apex, divided by a horizontal partition across the middle, the space above being occupied by the perithecia, and the cavity below loosely filled with pseudoparenchymatic matter. Stipe thick, about 1 cm. long. Perithecia membranaceous, subcylindrical, about 5 mm. long, with a slender neck piercing the crustaceous outer layer of the stroma and terminating in obscurely punctiform ostiola. Asci obovate-clavate, subsessile, 21–28 x 10–14 μ, 8-spored, (paraphysate)? Sporidia conglobate, oblong-elliptical, obtuse, almost truncate at the ends, continuous, brown, 10–17 x 10–12 μ.

On fallen branches, Nicaragua (Wright).

The above diagnosis is from spece. in Herb. U. S. Dept. of Agriculture, collected by Wright in Nicaragua. The young stromata are at first entirely enclosed in a common carbonaceous stroma (see Pl. 38, fig. 2) from which they finally emerge separate and distinct. The specimens differ somewhat from those collected in Cuba by Pöppig, and figured as *Camillea Sagræana* by Dr. Rehm, in Hedwigia, 1889, pp. 300 and 301, in being distinctly stipitate, with the stromata not constricted in the middle; the sporidia also, in the Cuban specimens, are smaller (9–10 x 4 μ). Currey in his Compound *Sphærias* (plate XLV, fig. 24) also figures the sporidia of *M. Sagræana* and makes them 10 μ long, but whether the Nicaragua spece. are specifically distinct, can only be determined by the examination of a more complete set of specimens.

XYLÀRIA Hill.

Hist. Plant. p. 62 and 63.

Stroma erect or ascending, cylindrical, clavate, filiform, often compressed, simple or branched, of a corky, leathery or fleshy consistence, black outside, mostly white within. Perithecia sunk in the stroma but more or less prominent, globose or ovate, with a short neck and a papilliform ostiolum. Asci cylindrical, 8-spored. Sporidia elliptical or fusoid, continuous, black, mostly inequilateral.

A. *Head fertile throughout.* (*Xyloglossa.*)

* *Head clavate; stipe slender, elongated.*

X. euglóssa, Fr. Nov. Symb. p. 124.

Stroma clavate, bare, thickened above, obtuse, smooth, argillaceous, black-punctate from the minute ostiola, light-cinereous in the center, darker towards the surface. Perithecia entirely immersed, subglobose, black. Stipe slender elongated, smooth, becoming black. Asci linear. Sporidia oblong, brown.

In Costa Rica (Oersted).

Has the form of *Geoglossum difforme*, but larger, 3 inches or more high; when dry, often arcuate-incurved or twisted, and very hard, almost like stone, longitudinally rugose when dry. In the form and color of the club (which is of a dirty pallid hue) it resembles *Clavaria ligula*, thickened above and obtuse, properly black, but apparently smeared over with alutaceous-clay color, subdistinct from the stipe. Sporidia uniseriate, subacute at the ends, occasionally curved.

X. olóbapha, Berk. (in Herb. Kew), Cke. in Grev. XI, p. 84.

Stroma erect, clavate, rufous, attenuated below into a short, slender, glabrous, equal stipe. Perithecia globose, black. Ostiola papilliform, flattened. Asci cylindrical, stipitate. Sporidia lanceolate, straight or curved, brown, 20–22 x 8½ μ.

On trunks, Brazil and Mexico.

X. rhopalòdes, (Kunze).

Sphæria rhopalodes, Kze. Exs.
Xylaria rhopalodes, Mont. Ann. Sci. Nat. 1855, III, p. 99. Sacc. Syll 1234.

This is said to be found in Mexico, Carolina and Texas, but no description is given except that the asci are cylindrical, short-stipitate, 8-spored. Sporidia 8–10 μ long.

X. pròtea, Fr. Nov. Symb. p. 125.

Stroma suberose-indurated, lanceolate, obtuse, corrugated, bare, black, white inside; stipe slender, equal, glabrous. Perithecia globose, subimmersed, peripherical. Ostiola prominent, depressed-hemispherical. Asci linear. Sporidia uniseriate, oblong, curved, opake.

On trunks, Costa Rica (Oersted).

Resembles *X. corniformis*, but has a slender, very smooth, varnished, very fragile stipe, about 2 mm. thick, and when dry, cavernose-rugose. Club or head about 1½ inches long, ¼ of an inch thick, obtuse, bare, black, apparently rimose-corrugated, but really the surface is only densely colliculose from the slightly prominent perithecia with their papillose ostiola depressed. Perithecia rather large, exactly globose, peripherical, subirregularly arranged, with the nucleus black.

X. tentaculàta, B. & Br. Grev. IV, p. 48, Cuban Fungi, 796.

Stem an inch high, not a line thick. Head cylindrical, 1–2 lines (2–4 mm.) long. Ostiola prominent, tending upwards, crowned by several tentacular processes about ½ an inch long. Allied to *Xylaria comosa*, Mont.

In shaded swamps, among mosses and rotten wood, South Carolina (Ravenel).

** *Head subclavate; stipe short, thick or obsolete.*

X. polymórpha, (Pers.)

Sphæria polymorpha, Pers. Comm. p. 149.
Valsa clavata, Scopoli, Flora Carniol. p. 398.
Xylaria clavata, Schrauck, Bayrische Flora, II, p. 566.
Clavaria digitata and *hybrida*, Bull. Champ de Franc. I, p. 192 and 194.
Sphæria digitata, Muller in Flora Danica, XV, p. 6, tab. 900.
Xylaria polymorpha, Grev. Flora Edin. p. 355.
Exsicc. Fckl. F. Rh. 1061, 2267.—Rab. Herb. Mycol. 428 —Rehm Asc. 427 —Ell N. A. F. 1300.—Plowr. Sph. Brit. 12.—Roum F. G. 666.—Sydow, M March, 183, 1655, 1736

Stromata solitary or 2–6 or more cespitose-connected at base, upright, either simple or cylindrical, subattenuated above and below, mostly obtuse; or obovote, compressed, more or loss sublobate-divided, globose or otherwise irregular in shape, thick, bare, at first dirty-brown, becoming black, not shining, very variable in size from 2–4 cm. high and ½ cm. thick, to 8–11 cm. high and 2½–3 cm. thick. Perithecia crowded, tolerably large, ovate or globose, with a papilliform ostiolum. Asci cylindrical, long-stipitate, 8-spored, 140–180 (p. sp) x 8–10 μ. Sporidia uniseriate, elliptical or fusoid, subacute at the ends, subinequilateral or curved, continuous, brown, 20–30 x 6–9 μ.

On decaying stumps and logs, common.

The head or perithecia-bearing part is much longer and larger than the stipe which is very short or almost none.

X. conocéphala, B. & C. Cuban Fungi, No. 781.

Very large, cespitose from an obtusely conical base, umber color, rimulose, here and there contracted in drying. Ostiola scattered, subprominent. Stipe short, longitudinally sulcate.

On rotten wood, Ohio (Morgan).

The Ohio spece. were determined by Cooke, and from them we add the following notes: Fertile heads 6–9 cm. long, about 1 cm. thick, subcylindrical, dirty-brown outside, quite hard when dry and lacunose-wrinkled, arising from a common, subtuberculiform base 4–5 x 2–3 cm. Perithecia sunk in the stroma, ovate, about 1 mm. high by ¾ mm. wide, necks short, terminating in the scattered, hemispheric-prominent, rather large, black ostiola (about like those of *Nummularia Bulliardi*). Asci cylindrical, p. sp. 80–85 μ long. Sporidia uniseriate, navicular-fusoid, opake, 15–18 x 4½–5 μ.

X. castòrea, Berk. Fl. Nov. Zel. p. 204, tab. 105, fig. 10.

Stipe short, at first spongy-velutinous, then bare, rugose. Head clavate, elliptical or ovate, obtusely rounded at the apex, much compressed, minutely areolate and roughened by the more or less prominent ostiola. Asci narrow. Sporidia ovoid-oblong, fuliginous, 10 μ long.

On rotten wood, Ohio (Morgan).

The above diagnosis is from Saccardo's Sylloge. The Ohio spece. (det. by Cke.) are cespitose, arising from a common spongy-velutinous, irregular-shaped base 1–2 cm. high, stipes less than 1 cm. long; heads 1½–3 cm. long, ½–1 cm. wide, flattened, brown. Perithecia ovate-globose, about ½ mm. high, monostichous, buried. Ostiola papilliform or subhemispherical, prominent in the mature spece. Asci (p. sp.) 50–60 x 5 μ, with a slender stipe about one-half as long. Sporidia uniseriate, elliptical, slightly inequilateral, 7–8 x 4–5 μ. The surface of the clubs, especially in the immature, dry spece., is cracked into minute areolæ. Resembles *X. corniformis*, Fr., only broader and compressed.

X. Tìtan, B. & C. Grev. IV, p. 47.

"Five inches long, 2 inches wide, sausage-shaped, convex on one side, hollow on the other, hard, solid, dirty-white, stained with the sporidia and dotted with the prominent ostiola."

(On wood)? Texas (Lindheimer).

X. fulvélla, B. & C. Cuban Fungi, No. 788.

Clavate, rubiginose, papillate. Perithecia subprominent. Ostiola black. Stipe cylindrical, pale yellow, lineate-rugose. Sporidia oblong, 7½ μ long."

On rotten wood, Cuba and Alabama.

*** *Head subglobose.*

X. cudònia, B. & C. l. c.

Slightly laccate, shining; stem ½ an inch long, nearly 2 lines thick above. Head semiglobose, five-twelfths of an inch across, slightly papillose from the projecting perithecia. Ostiola very small.

On a dead tree, Santee canal, South Carolina (Ravenel).

X. clávulus, B. & C. Grev. IV, p. 47.

" Gregarious, seriate. A miniature of the preceding. Stem with the head about 1 line high, not laccate, rather thick for the size of the plant, penetrating the convex, papillate head. A very curious little species."

On the dead stem of some grass, Texas (Wright).

B. *Head fertile throughout, stipe villous (Xylocoryne).*

* *Head clavate, stipe slender, elongated.*

X. Geoglóssum, (Schw.)

Sphæria Geoglossum, Schw. in Journ. Phil. Acad. V, tab. 1, fig. 4.
Xylaria Geoglossum, Sacc. Syll. 1245, Cke. Syn. 691.

Carnose-suberose, simple, very black. Head tongue-shaped, compressed, somewhat furrowed, falcate, obtuse at the apex. Perithecia oblong, black, subprominent, white inside. Ostiola minute, scarcely prominent. Stipe three times longer than the head, squamulose, subhirsute at the base, slender, suberose, black outside, white within. About an inch high. Resembles a *Geoglossum*.

Sent from New York by Dr. Torrey. No habitat given.

X. multífida, (Kunze).

Sphæria multifida, Kze. sec. Leveille in Ann. Sci. Nat. 1845, III, p. 45.
Xylaria multifida, Cke. Grev. XI, p. 85.

Conidial stroma erect, furcately and palmately divided, whitish. Ascigerous stroma simple, erect, black, clavate. Stipe as long as the head, slender, black, (glabrous)? Perithecia globose, black, subprominent. Asci cylindrical, stipitate. Sporidia fusiform, obtuse, inequilateral, brown, 10–12 x 4–5 μ.

On trunks, Java and Central America.

Greatly resembles *X. Hypoxylon*, of which it may be a variety.

X. múltiplex, (Kze. & Fr.)

Sphæria multiplex, Kze. & Fr. in Linn. 1830. p. 536.
Xylaria multiplex, B. & C. Cuban Fungi, No. 795.

Cespitose, suberose, dark brown, fertile heads terete-compressed, subdivided, smooth, white inside. Stipes elongated, leprose-villose. Perithecia entirely immersed, globose, crowded. Ostiola punctiform, then subdilated. Sporidia ovoid, 20–22 μ long.

On trunks, in Mexico (Högberg).

X. fastigiàta, Fr. Nov. Symb. p. 127.

Stipes densely cespitose-fasciculate, joined at the base, and often grown together so as to appear branched, very variable, compressed, angular, or often torulose and flexuous, an inch or more long, about a line thick, not villose, but covered at first with an appressed, scaly brown coat which finally disappears. Head not separated from the stipe, on which also scattered perithecia occur, slightly swollen, scarcely 2 lines (4 mm.) thick, unequal, bare, fastigiate, black. Perithecia small, in a thin, black, peripherical layer, globose and slightly prominent. Asci (in the spece. examined) dissolved. Sporidia oblong, somewhat curved, opake.

On trunks, in Costa Rica (Oersted).

Allied to *X. scruposa* and *X. multiplex.*

** *Head clavate, stipe short.*

X. cornifórmis, Fr. Summa Veg. Sc. p. 381.

Sphæria corniformis, Fr. Elench. II, p. 57.
Exsicc. Rav. Fungi Car. IV, 30.—Ell. N. A. F. 82, id. 83 (conidia).

Stromata scattered or subgregarious, sometimes two or three connected at base, simple, clavate, not compressed, obtuse at the apex, 3–5 cm. high and 4–5 mm. thick, white at first, becoming brownish-black, surface often minutely areolate-rimose. Head clavate, 2–3 cm. long, surface roughened by the slightly prominent, papilliform ostiola, white inside. Stipe short, black, arising from a spongy tubercular base. Perithecia monostichous, peripherical, small ($\frac{1}{3}$ mm.). Asci cylindrical, stipitate, 8-spored, p. sp. 60–70 x 5–6 μ. Sporidia obliquely uniseriate, inequilaterally elliptical, obtuse at the ends, brown, 8–10 x $4\frac{1}{2}$–5 μ.

On decaying trunks of magnolia, maple, &c., from New York to Michigan and Texas.

Sphæria flabelliformis, Schw. Syn. N. Am. 1164, is an abortive form in which the head consists of a tuft of flattened, palmately-divided, flesh-colored branches, thickly dusted over with innumerable oblong-elliptical, hyaline conidia about 3 x 1½ μ. Peck finds in New York State a variety of the ascigerous form much flattened and irregular in shape. The same is found at Newfield, N. J.

C. *Head with the apex sterile; stipe glabrous* (*Xylostyla*).

* *Head clavate, simple.*

X. graminícola. Gerard, in Peck's 26th Rep. p. 85.

Slender, simple, 3–5 cm. high. Head cylindrical, 1½ cm. long, 2 mm. thick, brown-black, colliculose-roughened by the prominent perithecia, apex sterile. Stem slender, glabrous, 1½ cm. long, arising from a brown, felt-like subiculum. Perithecia small, ⅓ mm. or less, peripherical, prominent, with black, papilliform ostiola. Asci cylindrical, stipitate, p. sp. 55–60 x 5 μ, with abundant paraphyses. Sporidia uniseriate, elliptical, brown, continuous, 7–9 x 3½–4 μ.

On decaying roots of grasses, New York (Gerard).

The diagnosis is from spece. sent by Gerard. The head or club is said to be greenish-pulverulent at first.

X. mucronàta, (Schw.)

Sphæria mucronata, Schw. Syn. Car. No. 4, Journ. Acad. Phil. V, tab. 1, fig. 1.
Xylaria mucronata, Sacc. Syll. 1279.

Carnose, simple, liver-color; head thickened, irregular, becoming yellow, apex sterile, mucronate. Stipe subsquamose, inflexed, compressed, an inch high, ⅓ of an inch thick. Asci moniliform, becoming cylindrical. Sporidia globose, black.

Rare; on a trunk of *Liriodendron*, North Carolina (Schw.).

X. grándis, Pk. 26th Rep. p. 85.

Large, blackish-brown, irregular, obtusely pointed and rusty-brown at the sterile tip, abruptly narrowed at base, central substance white. Perithecia subglobose. Sporidia subfusiform, pointed at each end, straight or slightly curved, 20–23 μ long. Stem branched, radicate, often greatly elongated. Plant 3–5 inches high; heads 1½–3 inches long, ½–1 inch thick.

On the ground, Portage, N. Y. (Clinton).

We have seen no specimens.

** *Heads connate or branched.*

X. digitàta, (Linn.)

Clavaria digitata, Linn. Syst. Veg. XV, p. 1010.
Clavaria Hypoxylon, Schæff. Icon. Fung. tab. 265 (sec. Fr.).
Sphæria clavata, Hoff. Veg. Crypt. 1, tab. 4, fig. 2.
Sphæria digitata, Ehr. Beitrag, VI, p. 7.
Xylaria digitata, Grev. Flora Edin. p. 356.
Hypoxylon digitatum, Link. Handbk. III, p. 348.
Exsicc. Fckl. F. Rh. 2547.—Rab. Herb. Mycol. 46.—Rav. Fungi Car. V, 50

Stromata erect, tufted, connate below, thick, dark-brown, leprose-velutinous, becoming glabrous, round and simple, gradually attenuated above, rarely obtuse or 2–3-dichotomously divided, rarely more or less compressed and forked, covered at first with the white conidial hymenium. Conidia obovate-subglobose, very small; fertile head occupying the middle of the stroma, attenuated below into a short stipe and ending above in a sterile apex. Perithecia numerous, densely crowded, slightly prominent, with papilliform ostiola. Asci cylindrical, long-pedicellate, 8-spored, p. sp. 100–120 x 7 μ. Sporidia overlapping-uniseriate, navicular-fusoid, subobtuse and slightly curved, dark brown, 12–16 x 5–6 μ.

On rotten wood, New York (Peck), Ohio (Morgan), Carolina (Ravenel), Texas (Lindheimer).

All the spece. we have seen have the fertile stroma distinctly compressed, and sporidia shorter than the measurements (18–20 x 5–6 μ) given for the European specc.

*** *Stroma filiform.*

X. filifórmis, (A. & S.)

Sphæria filiformis, A. & S. Conspect. p. 2, tab. III, fig. 5.
Xylaria filiformis, Fr. Summa Veg. Scand. p. 382.
Exsicc. Rab. F. E. 57, 917.—Ell. N. A. F. 163.—Ell. & Evrht. N. A. F. 2d Ser. 1948. Roum. F. G. 2091.—Sydow, M. March. 2242.—Rav. Fungi Car. II, 55.

Stromata scattered, erect, filiform, 2–3 inches high, and in the fertile part about 1 mm. thick, at first reddish (rose-colored) at the elongated, sterile apex, fertile head shorter than the stipe, roughened by the strongly prominent, globose, ½ mm. diam. perithecia which are almost superficial. Asci cylindrical, 80–100 μ long, p. sp. 70–75 μ long. Sporidia 1–2-seriate, 2-nucleate, fusoid, subinequilateral, olive-brown, 14–18 x 3–3½ μ (13–14 x 5–6 μ, Winter).

On decaying leaves of *Magnolia glauca*, Newfield, N. J., on decaying leaves, South Carolina (Ravenel), New York (Peck).

The stroma is often entirely sterile, appearing then like a black rhizomorphoid thread.

X. subterrànea, (Schw.) **(Plate 39)**

Sphæria subterranea, Schw. Syn. N. Am. 1162.
Xylaria subterranea, Sacc. Syll. 1281.
Exsicc. Ell. N. A. F. 771.

Filiform, simple or branching from the base or (the sterile part) anastomosely branched, attached to the matrix by a thin, felt-like subiculum; stems 3 inches to nearly a foot long, hardly more than 2 mm. thick, light-colored at first from the conidial layer, then black, sterile at the apex. Perithecia ovate-globose, about ½ mm. diam., unequally crowded, or scattered at intervals along the stem, singly or 2–4 together, broadly prominent, with a conic-papilliform ostiolum. Asci (p. sp.) cylindrical, 75–85 x 5 μ, with a slender stipe and abundant paraphyses. Sporidia uniseriate, oblong-elliptical, attenuated at the ends, subinequilateral, 10–12 x 4–5 μ.

On decaying wood in wells and cisterns, Bethlehem, Pa. (Schw. & Rau), New Jersey (Torrey), West Chester, Pa. (Fergus).

The fungus, as found by Mr. Rau, grew on the sides of an old wooden pump standing in a well. Where the long, rhizomorphoid stems come in contact with the wood, they are attached to it (for support, apparently) by a thin, felt-like, black subiculum, and at these points, adjacent stems are often connected by anastomosing branches which are sterile, only the free portions of the stems bearing perithecia.

D. *Head with the apex sterile, stipe villose* (*Xylodactyla*).

* *Head clavate, simple.*

X. persicària, (Schw.)

Sphæria persicaria, Schw. Syn. Car. No. 9.
Xylaria persicaria, B. & C. Grev. IV, p. 48.

Cespitose. Stem flexuous, rarely branched, rooting, 3 inches long and over, about as thick as a crow's quill, at first greenish-villose. finally black (or sec. spece. examined by Fries) ferruginous. Perithecia very prominent, situated in the middle of the club or head, which is slightly flesh-colored, becoming light yellow.

On peach pits, Càrolina (Schw.).

. The length of the stem varies according to the depth at which the pits are buried in the soil—sometimes 6 inches long. The spece. in Herb. Schw. are branched above, but are immature.

X. acùta, Pk. 25th Rep. p. 101.

Gregarious or subcespitose, 1–1½ inches high. Club cylindrical or subfusiform, generally with a sterile, acute apex, blackish-brown, central substance white, with a radiating structure. Stem involved in a dense, purplish tomentum, which causes it to appear bulbous. Perithecia globose, black. Sporidia uniseriate, elliptical, sometimes slightly curved, colored, 15–17 μ long.

On mossy, decaying logs, in woods, New York (Peck).

We have seen no specimens of this, but it is probably not distinct from *X. digitata*.

X. carpóphila, (Pers.)

Sphæria carpophila, Pers. Obs. Mycol. I, p. 19.
Xylaria carpophila, Fr. Summa Veg. Scand. p. 382.
Exsicc. Fckl. F. Rh. 1066.—Kze. F. Sel. 156.—Rehm Asc. 150.—Thum. M. U. 1266. Sydow, M. March. 1063.—Roum. F. G. 1383.

Stroma erect or ascending, slender, lanceolate, often curved, simple or spathulate-enlarged at the apex and cleft or two-parted, or sometimes with two heads on one stipe, mostly round, but sometimes subcompressed, black, more or less woolly-tomentose at base. Club thicker and mostly shorter than the stem, mostly cylindric-lanceolate, bearing only a few, sometimes only one, perithecium, the sterile apex subulate, otherwise roughened by the somewhat prominent, globose or ovate perithecia, with papilliform ostiola. Asci cylindrical, stipitate, 8-spored, p. sp. 80 x 6 μ. Sporidia uniseriate, obtuse-fusoid, dark brown, inequilateral, 12–16 x 5 μ. Conidia ovate, very small.

On nuts of *Liquidambar*, South Carolina (Ravenel), on decaying hickory nuts, Pennsylvania (Everhart), on old cones of magnolia, Newfield, N. J.

The length of the stem varies according to the depth at which the nuts from which it grows are buried.

** *Stroma furcate or divided.*

X. Hypóxylon, (Linn.)

Clavaria Hypoxylon, Linn. Flora Suec. Ed. II, p. 457.
Clavaria hirta, Batsch Elench, Cont. I, p. 229.
Clavaria cornuta, Bull. Champ. tom. I, p. 193, tab. CLXXX.
Valsa digitata, Scopoli, Flora Carniol. II, p. 398.
Sphæria cornuta, Hoff. Veg. Crypt. I, p. 11.
Sphæria digitata, Bolton Fungi Hal. III, p. 130.
Sphæria Hypoxylon, Pers. Obs. Mycol. I, p. 20.
Sphæria ramosa, Dicks. Plant. Crypt. Brit. IV, p 27.
Xylaria Hypoxylon, Grev. Flor. Edin, p. 355.
Exsicc. Thum. F. Austr. 766.—Cke. F. Brit. Ser. I, 363, Ser. II, 215.—Roum F. Gall. 172 Kze. F. Sel. 155.—Kriegr, F. Sax. 141—Linht. F. Hung. 71 —Sydow, M. March. 342.—Rehm Asc. 825.—Rav. F. Am. 648.—Ell. N. A F. 162.—Desm. Pl. Crypt. de France, Ed. I, 331.

Stroma erect, simple or variously branched, round or compressed, black, woolly-tomentose at base, 5–8 cm. high, lanceolate, with a sterile tip; stem mostly short, distinct from the fertile head, which is roughened by the more or less prominent, ovate, thickly crowded, black perithecia with papilliform ostiola. Asci cylindrical, long-stipitate, 8-spored, p. sp. 75–80 x 7–8 μ. Sporidia obliquely uniseriate, fusoid, inequilateral, obtuse at each end, black, 12–16 x 5–6 μ.

On rotten wood, common.

The conidial hymenium which covers the young stroma is white; conidia fusoid, 10 x 3 μ.

X. córnu-damæ, (Schw.)

Sphæria cornu-damæ, Schw. Syn. N. Am. 1163.
Xylaria cornu-damæ, Berk. Grev. IV, p. 48.
Exsicc. Rav. Fungi Car. Fasc. I, No. 45.

Suberose, subradicate, 1–4 inches high, simple and cylindrical or often compressed and furcately divided above, sometimes three or more clubs borne at the apex of a single stipe, covered at first with the white conidial hymenium, finally very black; heads fertile to the subobtuse apex or oftener with a narrow, short, sterile tip, $\frac{1}{4}$–$\frac{1}{2}$ cm. diam., contracted rather abruptly below into the rather slender (1–2 cm. long) stipe, which arises from a purplish-black tomentum. Perithecia globose, $\frac{1}{2}$ mm. diam., rather prominent, black, with a papilliform or short-cylindrical ostiolum. Asci narrow-cylindrical, p. sp. 100–110 x 5 μ. Sporidia uniseriate, fusoid, slightly curved, brown, becoming nearly black, continuous, 14–20 x $3\frac{1}{2}$–$4\frac{1}{2}$ μ (mostly about 15 x $3\frac{1}{2}$ μ).

On rotten wood. Carolina and Pennsylvania (Schw.).

The measurements are from the spece. in Herb. Schw. and in Rav. F. Car. Differs from *X. Hypoxylon* in its more robust growth and larger sporidia.

X. pedunculàta, (Dicks).

Sphæria pedunculata, Dicks Crypt Brit IV, p. 27, tab. XII, fig. 8.
Xylaria pedunculata, Fr. Summa Veg. Scand. p. 382

Stroma emerging from the ground, rather thick, flexuous, dark brown, simple or rarely sparingly branched, covered at first by the cinereous conidial hymenium. Fertile head subglobose, roughened by the prominent perithecia, acutely conical and sterile at the apex. Asci cylindrical, short-stipitate, 8-spored. Sporidia broadly ovate, very obtuse, straight, obliquely uniseriate, brown, becoming black and opake, surrounded by a thick, hyaline, gelatinous stratum, 40 x 20 μ.

On muddy ground mixed with manure, Missouri (Engelman).

We have seen no American spece. and cannot say whether this is the genuine *X. pedunculata*, (Dicks.) or var. *pusilla*, Tul. Sel. Carp. II, p. 18, tab. II, figs. 1–28, which is the *Xylaria Tulasnei*, Nitschke Pyr. Germ. p. 8.

SUBORDER. HYSTERIÀCEÆ.

Perithecia simple, erumpent-superficial, oblong or linear, membranaceous, coriaceous, or carbonaceous, rarely subcarnose at first, becoming black, opening by a narrow crack or cleft extending along the entire length of the perithecium. Asci paraphysate, 4–8-spored. (Sacc. Syll. II, p. 721).

The *Hysteriaceæ* form the connecting link between the *Pyrenomycetes* and *Discomycetes*, but on account of their mostly carbonaceous perithecia and the character of their sporidia, seem more closely allied to the former. It was not at first intended to include the *Hysteriaceæ* in this work, and no drawings were made to illustrate the genera; but as many members of this group are widely diffused and often met with, it seems better to give a brief account of the species thus far recorded as found in North America.

KEY TO THE GENERA.

A. *Sporidia hyaline.*

* *Sporidia continuous.*

Perithecia subcarbonaceous, flattened or convex, minute. - - - - - - - *Schizothyrium.*

** *Sporidia uniseptate.*

Perithecia membranaceous, minute, simple or branched, flattened. - - - *Aylographum.*

Perithecia carbonaceous, simple or obscurely branched. - - - - - *Glonium.*

Perithecia stellate. - - - - - - *Actidium.*

Perithecia subcarnose. - - - - - - *Angelina.*

*** *Sporidia 3-pluriseptate.*

Perithecia carbonaceous. - - - - - - *Gloniella.*

" subcoriaceous. - - - - - *Dichæna.*

**** *Sporidia muriform.*

Perithecia carbonaceous. - - - - - (*Gloniopsis*).

***** *Sporidia filiform.*

(*a*) *Sporidia much shorter than the ascus.*

Perithecia membranaceous, flattened. - - *Hypoderma.*

(*b*) *Sporidia nearly as long as the ascus.*

Perithecia elongated, flattened, membranaceous. *Lophodermium.*

Perithecia elongated, conchiform, subcarbonaceous. *Lophium.*

Perithecia elongated, coriaceo-subcarnose. - *Clithris.*

Perithecia subsphæroid. - - - - - - *Ostropa.*

B. *Sporidia brown.*

* *Sporidia uniseptate.*

Perithecia coriaceous, widely dehiscent. - - *Tryblidium.*

Perithecia elongated, coriaceous. - - - (*Lembosia*).

** *Sporidia 3-pluriseptate.*

Perithecia carbonaceous. - - - - - - *Hysterium.*

" coriaceous. - - - - - (*Tryblidiella*).

Perithecia conchiform. - - - - - *Mytilidion.*
" subcarbonaceous, striate. - - *Ostreion.*
*** *Sporidia muriform.*
Perithecia carbonaceous. - - - - *Hysterographium.*

SCHIZOTHÝRIUM, Desm.

Ann. Sci. Nat. XI, p. 360.

Perithecia simple, subsuperficial, subcarbonaceous, flattened or somewhat convex, opening with a longitudinal crack or furrow. Asci 8-spored. Sporidia ovoid or subfusoid, subhyaline.

The American representatives of this genus are not well known. We have no authentic spece. of any of the species, and take the diagnoses from Duby's "Memoire sur la Tribu Des Hysterines."

S. Verbásci, (Schw.)

Hysterium Verbasci, Schw. in Duby Hyst. p. 33, not *Hysterium Verbasci*, Schw. Syn. N. Am. 2093
Schizothyrium Verbasci, Sacc. Syll. 5558.

Erumpent, scattered, straight, black, linear, narrow, acute at each end; lips swollen, narrow, obtuse, smooth, only slightly separated, leaving a deep but narrow opening between them. Asci clavate, shorter than the filiform paraphyses. Sporidia biseriate, ellipsoid, hyaline, continuous, with granular contents.

On stems of *Verbascum*, Carolina (Schw.).

This is a very different thing from the *Hysterium Verbasci* in Herb. Schw., which is a *Hysterographium* (*Gloniopsis*). Possibly Duby's specimens were immature, the sporidia not having yet become septate.

S. cineráscens, (Schw. & Duby).

Hysterium cinerascens, Schw. in Duby Mem des Hyst. p. 32, tab. 1, fig. 15.
Henriquesia cinerascens, Sacc. Syll. 5565.

Erumpent, at length innate-superficial, gregarious or crowded, black, elongated-linear, acuminate at the ends; lips swollen, obtuse, not striate, but faintly rugulose, sometimes rimose, leaving a linear, more or less narrow and deep, straight, or subflexuous opening between them. Asci ovate-clavate, with shorter, filiform paraphyses. Sporidia biseriate, ovoid, continuous, hyaline, granular.

On denuded, decaying wood, Carolina? (Schw.).

This, too, is very different from the *Hysterium cinerascens*, in Herb. Schw., which has hyaline, muriform sporidia.

ÓSTROPA, Fr.

Summa Veg. Scand. p. 401.

Perithecia immersed, orbicular, of a corky, horn-like texture, firm, with a prominent papilla, rather large, with a longitudinal dehiscence and swollen lips. Asci cylindrical. Sporidia lying parallel, closely packed, typically filiform, multiseptáte or multiguttulate, hyaline. Paraphyses slender.

Placed by Dr. Rehm among the *Discomycetes*, (Die Pilze III, p. 185).

O. cinèrea, (Pers.)

Hysterium cinereum, Pers. Syn. p. 99.
Sphæria barbara, Fr. S. M. II, p. 468
Exsicc. Moug. & Nest. 966.—Desm. Pl. Crypt. Ed. I, 621.

Perithecia scattered, the base immersed in the wood or, more rarely, in the bark, finally emergent, gray-cinereous, finally shining-black, rather large, with a prominent papilla, depressed-sphæroid, opening with an elongated fissure extending nearly across. Asci cylindrical or filiform, 180–200 x 7-10 μ, thickened at the apex, 8-spored. Sporidia filiform, 180 x $1\frac{1}{2}$ μ, multiseptate, hyaline or yellowish-hyaline. Paraphyses very slender, branching, evanescent.

Fries, in S. M. II, p. 468, doubtfully refers to this species, specimens on wood of *Liquidambar* from Carolina.

About as large as a hemp seed. The conical or papilliform ostiolum is rarely seen, the perithecium being generally split across the top like a *Hysterium*.

O. sphærioìdes, Schw. Syn. N. Am. 1829.

Perithecia scattered or aggregated, but not confluent, rather large, orbicular-elliptical, erumpent, subcompressed, black, subrugose, opening with a short transverse cleft, almost like the ostiolum of *Tremato-sphæria pertusa*.

On a piece of dry wood, New England (Torrey).

O. rugulòsa, Schw. l. c. 1830.

Perithecia arranged in long, effused, confluent groups, the single perithecia scarcely distinct, carbonaceous, very black outside, brown inside, striate-rimose, innate in the cinereous colored wood which is raised into a tubercle, at length subdehiscent. The surface of the perithecia is generally flattened and rugulose.

On decorticated spots on a decaying log of *Juglans cinerea*, Erie Co., Pa. (Schw.).

O. hysterioìdes, Schw. l. c. No. 1831.

Perithecia often longitudinally confluent, navicular, striate, black, often deformed; dehiscence transverse, lips subobtuse, adnate-erumpent, visibly contracted towards the base, so that sometimes it becomes substipitate like a *Lophium*. Asci diffluent.

Rare; on old bleached oak wood, Bethlehem, Pa. (Schw.).

O. cinerâscens, Schw. l. c. No. 1832.

Perithecia thickly scattered, minute, brown-black, raising the wood into tubercles, marked with a longitudinal cleft, elliptical, innate; at first covered with a veil. The ostioloid cleft is slightly prominent.

On bare wood of *Liriodendron*, which is faintly cinerascent, but not covered with a crust, Bethlehem, Pa. (Schw.).

The fructification of this and the three preceding Schweinitzian species is unknown. The specimens in Herb. Schw. are all sterile, and the species must be considered as very doubtful productions.

AYLÓGRAPHUM, Lib.

Crypt. Ard. No. 272.

Perithecia minute, sublinear, simple or somewhat branched, membranaceous, opening with a narrow cleft or crack. Asci short, typically 8-spored and aparaphysate. Sporidia ovate-oblong, uniseptate, hyaline or brown.

* *Sporidia hyaline.*

A. vàgum, Desm. Ann. Sci. Nat. XIX, p. 362.

Hysterium micrographum, De Not. Micr. Ital. dec. IV, fig. III.
Exsicc. Desm. Pl. Crypt. de Franc. Ed. I, 1629.

Perithecia innate-superficial, minute, black, straight or curved, simple or branching by confluence, amphigenous, scattered; lips closed, forming a narrow ridge or crust along the vertex of the perithecium. Asci ellipsoid, 25–35 x 12–15 μ. Sporidia obovate-oblong, uniseptate and constricted, hyaline, 10–12 x 4 μ.

On fallen and decaying leaves of *Ilex opaca*, Newfield, N. J., and on decaying petioles of *Sabal serrulata*, Florida (Martin).

A. Pinòrum, Desm. Ann. Sci. Nat. II, Ser. tom. 10, p. 314.

Exsicc. Desm. Pl. Crypt. Ed. I, 994.

Perithecia adnate-superficial, scattered or gregarious, linear, straight or curved, simple or branching by confluence, black; lips closed when dry, slightly open when fresh. Asci clavate, 30–40 x

7–8 μ, obscurely paraphysate, 8-spored (6–10-spored, Desm.). Sporidia subbiseriate, obovate, hyaline, uniseptate and slightly constricted at the septum, 6–7 x 2½ μ.

On very old, dry, dead leaves of *Pinus rigida*, Newfield, N. J.

The measurements of asci and sporidia are from the spec. in Desm. l. c.

A. quércinum, Ell. & Martin, in Am. Nat. Dec. 1883, p. 1283.

Exsicc. Ell. & Evrht. N. A. F. 2d Ser. 2066.

Perithecia epiphyllous, scattered or oftener collected in orbicular patches, oblong or linear, often branched, opening by a longitudinal fissure along the center, mostly less than 1 mm. long, straight or curved, bordered with a fringe of mycelium at the base. Asci ovate, 20–25 x 15–18 μ, 8-spored. Sporidia conglobate, obovate, uniseptate, hyaline, 10–14 x 6–7 μ (including the hyaline envelope).

On leaves of *Quercus virens*, Florida (Martin & Calkins), Louisiana (Langlois).

A. gràcile, Ell. & Martin, (in Herb.)

Perithecia epiphyllous, on orbicular, dark brown spots 3–4 mm. diam., linear, straight or curved, often branched, about 1 mm. long and 60–70 μ wide, opening by a longitudinal fissure along the center. Asci ovate, 20–25 x 10–12 μ, 8-spored. Sporidia conglobate, ovate, hyaline, uniseptate, 7–8 x 3–3½ μ, including the hyaline envelope.

On leaves of *Quercus aquatica*, Houston, Texas (Ravenel).

Possibly this should be considered a var. of *A. quercinum*, E. & M., but it differs in its longer, more slender perithecia on brown spots, and its smaller sporidia.

A. Arundinàriæ, Cke. Grev. XIV, p. 14.

Perithecia erumpent, at length superficial, gregarious, linear, straight, generally longitudinally confluent, 1–2 mm. long; lips closed, black, subshining. Asci pyriform, 8-spored. Sporidia subelliptical, attenuated below, uniseptate, not constricted, hyaline, 13–15 x 5 μ.

On culms of *Arundinaria*, Darien, Ga. (Ravenel).

A. reticulàtum, Phil. & Hark. l. c. p. 23.

Exsicc. Ell. & Evrht. N. A. F 2d Ser. 2065.

Perithecia suborbicular, convex, 300–350 μ diam., adnate-superficial, scattered, thin, formed of reticulately interwoven hyphæ; lips thin, at first closely connivent, forming a narrow ridge extending only partly across the perithecium, finally rimose-dehiscent. Asci at first globose, at length obovate, 15–22 x 12 μ. Sporidia inordinate, obovate, uniseptate, hyaline, 8–12 x 3–3½ μ (14 x 4 μ, Phil. & Hark.).

On the under side of leaves of *Quercus agrifolia*, California (Harkness).

A. culmígenum, Ell. Bull. Torr. Bot. Club, VIII, p. 65.

Perithecia superficial, linear, straight or curved, slender, ½–1 mm. long, but so narrow as scarcely to be visible to the naked eye; often with a sparing fringe of mycelium around the base; lips connivent, closed or nearly so, and forming only a slight ridge along the crest of the perithecium. Asci obovate, 12–15 x 10 μ, paraphysate, sometimes nearly globose and then only 10–12 μ diam. Sporidia inordinate, 8 in an ascus, ovate, yellowish-hyaline, uniseptate and slightly constricted, 5–6 x 2–2½ μ.

On basal sheaths of old, dead *Andropogon*; common around Newfield, N. J.

Readily distinguished from *A. Pinorum* and from *A. vagum*, Desm., by its obovate asci and narrower perithecia.

A. subcónfluens, Pk. 28th Rep. N. Y. State Mus. p. 70.

Perithecia small, numerous, thin, scattered or subconfluent, orbicular, elliptical or elongated, black. Asci oblong. Sporidia oblong-clavate, hyaline, 8–10 μ long.

On dead stems of herbs, New York State (Peck).

Nothing is said of the mode of dehiscence.

** *Sporidia uniseptate, brown (Lembosia).*

A. litùræ, Cke. Grev. XII, p. 38.

Lembosia Lituræ, Sacc. Syll. IX, p. 1106.

Epiphyllous. Spots orbicular, brown. Perithecia linear, curved or straight, seated on the spots; lips connivent. Asci saccate. Sporidia biglobose, uniseptate, brown, 7½ x 3½ μ.

On leaves of *Quercus aquatica*, Texas (Ravenel).

A. cæspitòsum, E. & E. Journ. Mycol. I, p. 151.

(Not *Aylographum cæspitosum*, Cke. Grev. VIII, p. 95).
Lembosia cæspitosa, Sacc. Syll. IX, p. 1107.

Growing in small (1 mm.), suborbicular clusters, on a subcrustose, slightly prominent, black stroma, presenting the general appearance of an erumpent *Sphæria*. Perithecia minute (¼ mm. or less in length), applanate, opening with a rather broad cleft, the base mostly bordered with brown, creeping threads. Asci ovate, sessile, 30 x 15 μ. Sporidia crowded, oblong-fusiform, obtuse, hyaline and uniseptate at first, becoming brown at length and often 3-septate, 15–20 x 3–5 μ, constricted at the middle septum. Differs from the usual type of *Aylographum.*

On bare wood of old cypress pickets, Louisiana (Langlois).

A. acícolum, Hark. Bull. Cal. Acad. Feb. 1884, p. 47.

Lembosia acicola, Sacc. Syll. IX, p 1107.

Perithecia scattered or gregarious, oblong, covered by the thin epidermis, ½–1 mm. long, opening by a longitudinal fissure. Asci oval, about 25 mm. in the longer diameter, 8-spored. Sporidia crowded, inordinate, oblong-elliptical, hyaline, uniseptate, scarcely constricted, finally brownish, about 15 x 6 μ.

On living leaves of *Pinus sabiniana*, Mt. Diablo, California (Harkness).

A. lùcens, Hark. l. c.

Lembosia lucens, Sacc. Syll. IX, p. 1107.

Perithecia flattened, oblong, sparsely scattered over slightly discolored spots, ½–1 mm. long, opening by a longitudinal fissure, semi-immersed. Asci ovoid, 8-spored, 37 x 28 μ. Sporidia oblong, rounded at the ends, uniseptate, slightly constricted, with a large vacuole in each cell, hyaline, slowly becoming brown, 21 x 10 μ.

On living twigs of *Garrya elliptica*, Tamalpais, California (Harkness).

GLÒNIUM, Muhl.

Fr. Syst. Mycol. II, p. 594.

Perithecia emergent, linear, elongated, rarely oblong, sometimes radiately arranged, carbonaceous or tough-membranaceous, opening by a longitudinal cleft. Asci cylindrical or clavate, 8-spored, paraphysate. Sporidia uniseptate, hyaline or finally brownish.

G. stellàtum, Muhl, in Fr. S. M. l. c.

Solenarium byssoideum, Spreng.
Exsicc. Rav. Fungi Car. III, 43.—Rav. F. Am. 639.—Ell. N. A. F. 462.

Subiculum effused, brownish-black, indeterminate, 2 or more cm. broad, thin, composed of slender, branching, interwoven fibers. Perithecia adnate to the subiculum, anastomosely radiating and forming patches 1–2 inches across, the branches crowded so as to entirely cover the subiculum, narrowly cleft, lips mostly closed. Asci cylindrical, 75–80 x 10 μ, short-stipitate, paraphysate. Sporidia 8 in an ascus, overlapping-subbiseriate, fusoid, hyaline, uniseptate and constricted at the septum, 20–22 x 5–6 μ.

On rotten wood, northern United States and Canada, and south to Carolina; probably to be met with in other parts of the country.

The large patches of radiate perithecia are generally made up of several smaller, orbicular patches, 3–4 mm. diam., confluent at their adjacent margins and presenting a very neat appearance.

G. accumulàtum, Schw. Syn. N. Am. 2016.

Subiculum scanty around the margin, coarsely fibrillose, about one inch across, rising in an irregular, truncate-pyramidal shape, very uneven and rough, on account of the ramose-radiate perithecia being crowded and lying one above the other; half an inch thick. Color the same as in the preceding species.

On rotten wood, Bethlehem, Pa. (Schw.).

Apparently a mere var. of *G. stellatum*, but regarded by Schw. as distinct.

G. tryblidioìdes, E. & E. Bull. Torr. Bot. Club, X, p. 76.

Exsicc. Ell. N. A. F. 1283.

Emergent, oblong, 1–2 mm. long, ends obtuse, lips incurved, distant, smooth, leaving the sooty disk more or less permanently exposed. Asci clavate-cylindrical, sessile, 80–90 x 9–12 μ, overtopped by the densely crowded paraphyses whose closely matted, dark-colored tips give the sooty color to the disk. Sporidia uniseriate or partly biseriate above, ovate, uniseptate, hyaline, 12–16 x 5–7 μ. In the fresh state the swollen disk entirely hides the margin.

On old fence rails, Washington (Suksdorf).

G. lineàre, (Fr.)

Hysterium lineare, Fr. S. M. II. p. 583.
Glonium lineare, Sacc. Syll. 5588.
Exsicc. Ell. N. A F. 463.—Rehm Asc. 365.—Sydow, M. March. 2951.

Perithecia more or less crowded, sometimes longitudinally confluent, mostly lying parallel, subimmersed in the wood which is often blackened, linear, rather flat, straight or flexuous, ends obtuse, black, smooth, lips slightly swollen 1–1½ mm. long, about ½ mm. wide. Asci clavate-cylindrical, 75–90 x 12–14 μ. Paraphyses conglutinate, their tips united above and forming a dark-colored, coarsely granular stratum. Sporidia uniseriate, ovate, uniseptate and slightly constricted at the septum, hyaline, 12–15 x 6–8 μ.

On old decorticated wood of various deciduous trees, common.

G. párvulum, (Ger.)

Hysterium parvulum, Ger. Bull. Torr. Bot. Club, V, p. 40 (1874).
Glonium parvulum, Sacc. Syll. 5597.
Glonium microsporum, Sacc. M. Ven. Ser. IV, p. 25 (1875), F. Ital. tab. 121.
Hysterium aggregatum and *Hysterium abbreviatum*, in Herb. Schw. (not *H. aggregatum*, Duby).
Exsicc. Ell. N. A. F. 153.—Rav. F. Am. 765.—Sacc. Myc. Ven. 1281.

Perithecia densely gregarious, or sometimes more or less scattered, seated on a thin, black crust, short (½–1 mm. long), black, subrotund or oblong, sometimes slightly curved, ends obtuse, flattened above and marked with a longitudinal groove, on each side of which, in well matured perithecia, are one or two faint striæ. Asci cylindrical, subsessile, 55–60 x 5–6 μ, paraphysate. Sporidia uniseriate, oblique, rounded at the ends, hyaline, strongly constricted, about 7 x 3 μ.

On decorticated wood of *Alnus*, New York (Gerard), on oak, Newfield, N. J.

A careful comparison of the American spece. of *Glonium parvulum*, (Ger.), with the spec. of *G. microsporum*, Sacc., in M. Veneta, leaves no doubt that the two are identical. In Sacc. Syll. the sporidia of *G. parvulum* are given as 15–18 μ long. Gerard (l. c.) makes them 5–7½ μ long (.0002–.0003 in.), which is about the average length, though some sporidia may reach a length of 8–9 μ. The young perithecia are generally quite flat on top, without any longitudinal groove, but this appears later and is more distinct in the elongated forms.

G. caryígenum, E. & E. (in Herb.)

Perithecia gregarious, lying in different directions on the matrix, small (½–1 mm. long), oblong, straight or slightly curved, ends obtuse, lips open so as to leave a tolerably wide and deep groove or furrow. Asci cylindrical, about 70 x 10–12 μ, paraphysate, 8-spored. Sporidia either uniseriate or quite as often biseriate, ovate, uniseptate, hyaline, slightly constricted at the septum, 12–15 x 5–6 μ.

On an old decaying hickory nut lying on the ground, Newfield, N. J.

This seems different from *Hysterium nucicola*, Schw., in the absence of any black crust and the oblong, not ovate or hemispherical, perithecia. The lips also leave a very distinct furrow between them.

G. gràphicum, (Fr.)

Hysterium graphicum, Fr. S. M. II, p. 581, Obs. Myc. I, p. 194.
Hysterium contortum, Dittm, in Sturm Deutschl. Fl. 3. I, p. 65.
Hysterographium conjungens, Karst. Symb. Mycol. Fenn. 259.
Glonium graphicum, Duby, Hyst. p. 35.

Perithecia scattered or gregarious, superficial, elongated, straight or curved, rarely branched, smooth, black, lips at first connivent, at length open, 1 mm. long, ⅓ mm. wide. Asci oblong-clavate, rounded above, very short-pedicellate, 60–70 x 21–25 μ, with thick, branching, septate paraphyses with brownish tips. Sporidia conglobate, ovoid or fusoid-oblong, straight or slightly curved, uniseptate, slightly or not at all constricted in the middle, at length brownish or yellowish-brown, 28–38 x 9–12 μ (21–27 x 5–7 μ, Rehm).

On bark of *Sassafras*, Bethlehem, Pa. (Schw.).

G. vàrium, (Fr.)

Hysterium varium, Fr. S. M. II, p. 582.
Glonium varium, Sacc. Syll. 5598.

Innate-superficial, subelongated, variable, black, brown inside, the subobsolete lips and disk black, lips obtuse, at first closed, finally open. Asci elongated-clavate, 8-spored, paraphysate. Sporidia subuniseriate, elliptic-oblong, uniseptate, 2-nucleate, 25 x 12–13 μ, ends subacute, hyaline, finally brownish.

Very rare; on oak, Bethlehem, Pa. (Schw.), Carolina (Berk. in Grev.) Diagnosis from Sacc. Sylloge.

G. símulans, Gerard, Bull. Torr. Bot. Club, VI, p. 78.

Gregarious, superficial, linear, oblong or subglobose, obtuse at the ends, whole surface marked with close, fine, longitudinal striæ, mostly about 1 mm. long, straight or slightly curved, lips closed. Asci subcylindrical, with an abruptly narrowed, short, stipe-like base, 75–80 x 10–12 μ. Paraphyses linear with an abundance of yellowish granular matter intermixed. Sporidia biseriate, ovate-oblong, uniseptate, hyaline, slightly constricted at the septum, 11–15 x 4–5 μ.

On decaying, but still hard wood, New York, Canada, Pennsylvania and Ohio; probably common in other sections.

G. nítidum, Ell. Grev. VIII, p. 13.

Exsicc. Ell. N. A. F. 570.

Perithecia densely gregarious, superficial, minute (¼–½ mm. long), lying parallel in the direction of the fibers of the bark, subconchiform, faintly striate; lips closed so as to form a narrow ridge or crest along the apex of the perithecium. Asci cylindrical, sessile, 35–45 x 4 μ, (paraphysate)? Sporidia uniseriate, clavate-oblong, uniseptate, hyaline, 6–7 x 2–2½ μ.

On the inner surface of the loosened bark, on cedar (*Cupressus thyoides*) stumps not much decayed; not uncommon around Newfield, N. J.

G. chlòrinum, (B. & C.)

Hysterium chlorinum, B. & C. Grev. IV, p. 12.
Glonium chlorinum, Sacc. Syll. 5595.

"Soon liberated from the cuticle, elevated from the bark, often narrowed at the base, elliptical, at first greenish from a fine, powdery coat which soon wears off; lips sulcate; disk greenish. Sporidia in two rows, oblong, uniseptate, constricted in the middle, 7½ μ long; the endochrome has frequently a little emargination."

On twigs of *Quercus aquatica*, Alabama (Beaumont).

G. mèdium, (Cke.)

Hysterium medium, Cke. Texas Fungi, p. 183.
Glonium medium, Sacc. Syll. 5604.

Perithecia oblong, about 1 mm. long, emergent, but only slightly prominent, gregarious, lying parallel; lips closed, forming a narrow, subacute ridge along the top of the perithecium, finally more or less open. Asci cylindrical, 50–60 x 4–5 μ. Sporidia ovoid, uniseptate, hyaline, 6–8 x 3 μ, (8–10 x 4 μ, Cke.).

On decorticated branches of *Berchemia*, Houston, Texas (Ravenel). Diagnosis from a specimen from Ravenel.

G. Ravenélii, Cke. & Phil. in Rav. F. Am. 763.

Perithecia narrow-cylindrical, 1–2 mm. long, subflexuous, gregarious on a very thin, dark-colored subiculum which, however, is sometimes wanting; lips closed or leaving only a very narrow cleft between them, and with 2–3 distinct, longitudinal striæ on each side. Asci clavate-oblong, 75–80 x 12 μ. Sporidia crowded-biseriate, clavate-oblong or clavate-fusoid, hyaline, uniseptate, about 20 x 5 μ.

On bark of *Platanus*, seaboard of South Carolina (Ravenel). The diagnosis is from the spec. in Rav. F. Am. and from other specc. sent by Ravenel.

G. velàtum, E. & E. (in Herb.)

Perithecia gregarious, oblong, black, rounded off above towards each end, rough and more or less distinctly longitudinally striate, hardly more than 1 mm. long; lips only slightly swollen, closed at first, finally opening partially. The perithecia are imbedded in and covered, except the apex, by a thin, black, felt-like layer (*Dendryphium*) consisting of brown, branching hyphæ, with abundant, cylindrical, multiseptate, brown, catenulate conidia 25–75 x 5–6 μ, borne on short, upright branches. The septa in the conidia are about 4 μ apart and there is a more or less distinct constriction at each septum. Asci clavate-cylindrical, paraphysate, 8-spored, subsessile, 70–75 x 8–10 μ. Sporidia biseriate, fusoid, slightly curved, hyaline, 4-nucleate, finally uniseptate and slightly constricted at the septa, 20–25 x $3\frac{1}{2}$–$4\frac{1}{2}$ μ.

On rotten wood, St. Martinsville, Louisiana (Langlois).

It is not improbable that the sporidia may become 3- or more-septate, but none were seen with more than one septum.

G. Cyríllæ, (B. & C.)

Hysterium Cyrillæ, B. & C. Grev. IV, p. 11.
Glonium Cyrillæ, Sacc. Syll. 5593.

"Scattered, elevated, opake, elliptical, even. Sporidia biseriate, oblong, 10 μ long; endochrome divided into two portions, one of which is less than the other, epispore thick."

On twigs of *Cyrilla*, South Carolina.

G. hyalósporum, Ger. in Peck's 31st Rep. p. 49.

On dead wood, Willowemoc. The name only is given, without any diagnosis.

ACTÍDIUM, Fr.

Obs. Mycol. I, p. 190, S. M. II, p. 595.

Perithecia sessile, rotund-lobate, subcarbonaceous, nearly closed: dehiscence radiate, from the center towards the margin. Asci elongated, mostly 8-spored and aparaphysate. Sporidia oblong, bilocular, hyaline. Allied to *Glonium*, but less perfectly developed. There is never any open disk as in *Hysterium* and *Phacidium*.

A. caríciuum, Schw. Syn. N. Am. 2019.

Perithecia flattened, black, radiate-stellate, lobes oblong, obtuse, rugose, at first covered by the epidermis, finally denuded.

On culms and leaves of the larger *Carices*, Bethlehem, Pa. (Schw). *Actinothyrium* occurs on the same leaves.

A. diatrypoìdes, Cke. Grev. VII, p. 49, (without any diagnosis).

On trunks of *Carpinus* and *Ostrya*, Darien, Ga. (Ravenel).

ANGELÌNA, Fr.

Summa Veg. Scand. p. 358.

Perithecia rufous or rufescent; when fresh, carnose and open (pezizoid), when dry, subcorneous and darker, with the margin paler; lips involute and disk pallid. Asci paraphysate, 8-spored. Sporidia oblong, uniseptate and hyaline.

This genus is closely allied to *Ascobolus*.

A. rufèscens, (Schw.)

Hysterium rufescens, Schw. Syn. N. Am. 2081.
Ascobolus conglomeratus, Schw. Syn. N. Am. 960.
Angelina rufescens, Duby, Hyst. p. 39.
Exsicc. Ell. N. A. F. 466.

Perithecia (ascomata) much crowded, subelliptical, orbicular or elongated and variously lobed, 1 mm. and over in diameter, disk almost convex when fresh, with a narrow white margin, involute and nearly closed when dry, in the fresh, growing state, greenish, becoming mouse-color or slate-color. Asci clavate, stipitate, 8-spored, 100–120 x 7–8 μ (including the slender base), with filiform paraphyses. Sporidia collected in the upper part of the ascus, fusoid-oblong, hyaline, uniseptate, 8–12 x 2–2½ μ.

On the decayed surface and in the cavities of decaying oak stumps, common around Bethlehem, Pa. (Schw.), and also at Newfield, N. J.

The part of the wood occupied by the fungus is quite rotten and soft, caused apparently by the mycelium of the fungus, but the decay does not at first penetrate deeply, the subjacent wood remaining for some time hard and sound.

GLONIÉLLA, Sacc.

Sylloge Fung. II, p. 765.

Perithecia emergent, oblong or linear, carbonaceous, black, dehiscing by a longitudinal cleft. Asci 4–8-spored, paraphysate. Sporidia oblong or fusoid, 2-pluriseptate, generally 3-septate, hyaline or subhyaline.

G. Curtisii, (Duby).

Hysterium Curtisii, Duby, Hyst. p. 30, tab. 1, fig. 10.
Gloniella Curtisii, Sacc. Syll. 5721.

Perithecia erumpent-superficial, variable in shape, subglobose at first and contracted at base, soon laterally compressed and conchiform, finally flattened horizontally, narrow-elliptical, 1–1¼ mm. long, with the ends subacute; lips swollen, distinctly open, sides of the perithecia rough and obscurely striate. Asci oblong-clavate, 100–150 x 25–35 μ, subsessile, paraphysate. Sporidia irregularly crowded, oblong, mostly a little curved, ends obtusely rounded, yellowish-hyaline, with a gelatinous envelope at first, 1–2-septate and constricted at the septa, 60–70 x 12–15 μ.

On dead twigs of *Vitis*, Carolina (Curtis), Louisiana (Langlois), on *Sabal serrulata*, Florida (Martin).

The above description of the species was made from the Louisiana specc., which agree well with Duby's diagnosis in all but the presence of paraphyses. The sporidia are mostly uniseptate.

G. ovàta, (Cke.)

Hysterium ovatum, Cke. Grev. XI, p. 107.
Gloniella ovata, Sacc. Syll 5717.
Exsicc. Rav. F. Am. 321.

Perithecia gregarious, superficial, ovate, ends obtuse, black, longitudinally striate, lips closed. Asci subcylindrical. Sporidia sublanceolate, rounded at the ends, hyaline, 15–18 x 8 μ, nucleate, finally pseudo-triseptate.

On old oak wood, Carolina (Ravenel).

The spec. in Rav. F. Am. (the only one we have seen) is entirely without fruit, so that we can only copy the published diagnosis. The outward appearance is about the same as that of *Glonium parvulum*, Ger.

G. sycnóphila, (Cke.)

Hysterium (*Gloniella*) *sycnophilum*, Cke. Grev. XIV, p. 14.
Gloniella sycnophila, Sacc. Syll. 7385 (Add. I, p. 269), id. II, p. 1114.

Perithecia gregarious, lanceolate, straight or flexuous, black, emergent, with a narrow, longitudinal cleft. Asci subclavate, 8-spored, Sporidia sublanceolate, multinucleate, finally 5-septate, hyaline, 36–40 x 8–10 μ.

On bark of living fig trees, South Carolina (Ravenel).

Specimens in our Herb. from Ravenel are apparently some species of *Opegrapha*. Asci oblong-clavate, 45–50 x 12 μ, subsessile. Sporidia inordinate, fusoid, hyaline, 3-septate (pseudoseptate), 20 x 3–3½ μ.

MYTILÍDION, Duby.

Mem. sur la Trib. des Hyst. p. 22.

Perithecia emergent-superficial, laterally compressed, conchiform.

dehiscing by a narrow cleft along the acute apex, thin, carbonaceous, fragile; lips acute, closely connivent. Asci paraphysate, 8-spored. Sporidia oblong, elongated, 3-multiseptate, hyaline, becoming yellowish.

M. aggregàtum, (DC.)

Hysterium aggregatum, DC. Fl. Fr. VI, p. 168.
Mytilinidion aggregatum, Duby, l. c.

Perithecia collected in groups or patches, conchiform, scarcely ½ mm. long, carbonaceous, shining-black, opening with a narrow cleft. Asci cylindrical, 120–130 x 10 μ, obtuse at the apex, 8-spored. Sporidia subuniseriate, at first uniseptate, at length fusoid, 3-septate, 20–24 x 6–7 μ, slightly constricted at the middle septum, olive-brown, the terminal cells subhyaline.

Found (sec. Gerard) on bark of red cedar, Staten Island, N. Y., but the specimens sent by him are entirely without fruit.

M. Califórnicum, Ell. & Hark. Bull. Torr. Bot. Club, VIII, p. 51.

Perithecia conchiform, about 350 μ long, faintly striate, black; lips at first closely compressed, finally slightly open. Asci clavate-cylindrical, 35–40 x 7–8 μ, 8-spored. Sporidia biseriate, oblong-fusiform, yellowish, 3-septate, sometimes slightly constricted at the septa, 12–15 μ long.

On dead foliage of *Sequoia gigantea*, California (Harkness).

Closely allied to *M. acicolum*, Winter, and possibly not distinct.

M. tórtile, (Schw.)

Hysterium tortile, Schw. Syn. Car. No. 250, id. Syn. N. Am. 2065.
Mytilidion tortile, Sacc. Syll. 5709.
Mytilinidion Juniperi, E. & E. Journ. Mycol. IV, p. 57.
Exsicc. Ell. & Evrht. N. A. F. 2d Ser. 2152.

Perithecia gregarious, superficial, lying in various directions on the matrix, membranaceo-carbonaceous, black, brittle, shaped like clam shells with the sharp edges pointing up, 1–1½ mm. long; lips closed, acute, sides of the perithecia more or less distinctly longitudinally striate. Asci cylindrical, p. sp. 75–80 x 6 μ, with a stipitate base 12–15 μ long. Paraphyses obscure. Sporidia uniseriate, oblong, 3-septate, pale brown, 12–15 x 4–5 μ, ends obtuse, only slightly or not at all constricted at the septa.

On bark of *Juniperus Virginiana*, Carolina and Pennsylvania (Schw.), also around Newfield, N. J.

This agrees with specc. in Herb. Schw. Saccardo makes the sporidia 3-septate and 28–30 x 7–8 μ while in the asci, and when free,

5-septate, 38–40 x 8–10 μ. In the Newfield specc. occasionally a free sporidium may be found with 4 septa, but this is exceptional.

M. fusísporum, (Cke.)

Lophium fusisporum, Cke. Grev. IV. p. 114.
Mytilidion fusisporum, Sacc. Syll. 5712.
Exsicc. Ell. N. A. F. 858.

Perithecia erect, expanded above from a narrow base, laterally compressed, thin and fragile, shaped like the blade of a hatchet with the corners rounded off, black and subshining, about 400 μ broad and high, finely striate both longitudinally and transversely. Asci cylindrical, short-stipitate, paraphysate, 100–110 x 8–10 μ. Sporidia irregularly biseriate, fusoid, yellowish, nearly straight, 40–50 x 4–5 μ, about 7-septate.

Not uncommon around Newfield, N. J., on old pitchy pine wood.

TRYBLÍDIUM, Dufour.

Ann. Sci. Nat. XIII, t. 10, fig. 3.

Perithecia erumpent-superficial, oblong or lanceolate, corneo-coriaceous; lips swollen and open. Asci paraphysate, elongated, 8-spored. Sporidia oblong, 1–3-septate, dark-colored.

The genus belongs really in the *Discomycetes*, but for the present we include it here.

* *Sporidia uniseptate.*

Tr. hystérinum, Dufour, l. c.

Tryblidium rufulum, Spreng. var. simplex, E. & E. Journ. Mycol. V, p. 30.
Hysterium elevatum, Pers. Myc. Eur. I, tab. I, fig. 4.
Hysterographium elevatum, Desm. XXII, Not. p. 18.
Exsicc. Desm. Pl. Crypt. Ed. I, No. 996, id. Ed. II, 296.—Moug. & Nestl. Stirps. Vog. 1070

Perithecia erumpent-superficial, the central part attached, the ends free, 2–3 mm. long, 1 mm. or more wide, slaty-black, at first subglobose, closed and covered, then erumpent and opening with a roundish aperture which soon becomes elongated, exposing the elliptical, brick-red disk; lips swollen, incurved but not closed, when dry, transversely wrinkled. Asci cylindrical, 120–160 (p. sp. 110) x 15–16 μ, overtopped by the abundant paraphyses. Sporidia uniseriate, elliptical or ovate, uniseptate and constricted, dark brown, 18 22 x 8–10 μ, (25–30 x 12–14 μ, Rehm; 20–28 x 12–15 μ, Sacc.).

On a dead limb, Ocean Springs, Miss. (Earle).

The Mississippi specc. agree with those in Desm. Exsc., but we can not make the sporidia as large as stated by Rehm and Saccardo.

Tr. inscúlptum, Cke. Disc. of the U. S. p. 32.

Exsicc. Ell N. A. F. 150.

Erumpent, slate-colored, gregarious, thin, ½–1½ mm. long, oblong, elliptical or irregular in shape, closely embraced by the epidermis and scarcely rising above it, margin thin, erect, scarcely incurved, often undulate from the pressure of the ruptured epidermis against it. Asci ventricose, sessile, overtopped by the abundant paraphyses which are olivaceous and conglutinated above. Sporidia inordinate, oblong, uniseptate and constricted, pale at first, soon dark brown, 25–35 x 12 μ.

On dead limbs of *Carya*, Newfield, N. J.

** *Sporidia 3- or more-septate (Tryblidiella).*

Tr. rùfulum, (Spreng.)

Hysterium rufulum, Spreng. in Vet. Ac. Holm. 1820, p. 20, Fr. S. M. II, p. 584.
Hysterium confluens, Kze. in Weig. Exs.
Tryblidium confluens, De Not. Pir. Ist. p. 16.
Tryblidiella rufula, Sacc. Syll. II, p. 757.
Exsicc. Rav. F. Car. II, No. 47 —Rav. F. E. 3369.—Rab. F. Am. 637.

Erumpent, oblong, flexuous or subtriangular, smooth, brownish-black; lips swollen, transversely striate, 2–3 mm. long, and 1 mm. wide, disk brick-red. Asci 150–200 x 13–15 μ, without paraphyses, clavate-thickened above. Sporidia uniseriate, oblong, 3-septate, 24–30 x 10–12 μ (30–35 x 10 μ, Sacc. in Syll.), scarcely constricted at the septa, reddish-brown, becoming nearly opake.

On bark of dead limbs, Southern States and California.

In the fresh state this resembles a *Peziza*, the disk being nearly orbicular and fully exposed, but when dry, the opposite margins roll together and the plant assumes the aspect of *Hysterium* with the lips loosely closed.

Var. *microsporum*, E. & E., *Triblidiella Ellisii*, Rehm, in Rab. Krypt. Flora, Discom. p. 235, Exsicc. Ell. N. A. F. 1285, differs from the usual form only in its smaller (18–22 x 6–7 μ) sporidia.

Tr. clavæsporum, Pk. 35th Rep. N. Y. State Mus. p. 143.

Receptacles, when moist, suborbicular, plane or slightly convex, margined, ¾–1 mm. diam., black, when dry, more or less contracted, hysteriiform, with thick lips. Asci clavate or cylindrical, 90–112 μ long. Sporidia oblong-clavate, crowded or biseriate, colored, 4-septate, 20–27 x 7½–10 μ.

On decorticated wood of willows (*Salix nigra*), Albany, N. Y. (Peck).

Tr. fúscum, E. & E.

Tryblidium rufulum, Spreng var. *fuscum*, E. & E. Journ. Mycol. V, p. 30.
Exsicc. Ell. & Evrht. N. A. F. 2d Ser. 2331.

Perithecia cespitose, erumpent, suborbicular, elliptical, triangular or otherwise irregular from crowding, 2–3 x 1½–2 mm., centrally attached, with the margin free and, when dry, with the opposite sides rolled in so as partly to hide the slate-colored disk, and strongly marked with transverse wrinkles or striæ. Asci cylindrical, 170 x 12–15 μ, with abundant, clavate-tipped paraphyses. Sporidia 8 in an ascus, uniseriate, oblong-elliptical, 3-septate and more or less constricted, 25–30 x 12–14 μ.

On dead limbs, near Jacksonville, Fla. (W. W. Calkins).

We have given this specific rank on account of the clustered perithecia, slate-colored disk and clavate-tipped paraphyses.

Tr. turgídulum, Phil. & Hark. Bull. Cal. Acad., Feb. 1883, p. 25.

Scattered, sessile, oblong-elliptical, turgid, nearly smooth, black. Asci clavate. Sporidia oblong-elliptical, uniseptate, strongly constricted in the middle, each half 5–6-pseudoseptate, reddish-brown, 60–90 x 13–20 μ.

On dead stems of *Pentstemon breviflorus*, California (Harkness). Specc. in our Herb. from Harkness are entirely sterile.

Tr. mìnor, Cke. Grev. IV, p. 182, tab. 67, fig. 9.

This can not be separated from *Opegrapha varia*, (Pers.).

LÓPHIUM, Fr.

Systema Mycol. II, p. 533.

Perithecia vertical, conchiform, compressed, submembranaceous, fragile; lips acute, closed, at length opening with a very narrow fissure.

L. mytilìnum, (Pers.)

Hysterium mytilinum, Pers. Syn. p. 97.
Lophium mytilinum, Fr. S. M. II, p. 533.
Exsicc. Rab. Herb. Mycol. 714.—Fckl. F. Rh. 762.

Subpedicellate, dilated above, transversely striate, shining-black. Asci cylindrical, 140–160 x 9–10 μ, with a short, thick stipe, 8-spored, with slender, septate, hyaline, branching paraphyses. Sporidia filiform, 120–140 x 1½–2 μ, 18–20-guttulate, hyaline.

On bark and wood of pine and spruce. Generally arising from a black, effused, indeterminate crust in which the stipe is concealed. Perithecia fragile, of medium size.

We have given the diagnosis of this species, though it is uncertain whether it has yet been found in this country. The spece. distributed in Ell. N. A. F. 858, are *Mytilidion fusisporum*, (Cke.).

L. naviculàre, Schw. Syn. N. Am. 2017.

Not pedicellate, but contracted at base, elongated-ovate, subflexuous, navicular, much like a *Hysterium*, especially in the young state, before it becomes fully emergent, being then acuminate at each end with a central fissure; but at a more advanced stage it becomes compressed and longitudinally striate, conchiform, black and obtuse at the ends.

On wood, Bethlehem, Pa. (Schw.). A doubtful species.

L. Sássafras, Schw. l. c. 2018.

Rather large, scattered, allied to *L. mytilinum*, but much larger and more obtuse; subpedicellate, dilated above, rugose-striate, black, not shining; aperture closed, nucleus white; sometimes opening in two directions.

On bark of *Sassafras*, Bethlehem, Pa. (Schw.).

The fructification of this and the preceding species is unknown, and the species are doubtful.

HYSTÈRIUM, Tode.

Fungi Meckl. II, p. 4.

Perithecia superficial or erumpent, oblong or ellipsoid, corneo-carbonaceous, opening with a longitudinal cleft. Asci clavate or cylindrical, mostly 8-spored, paraphysate. Sporidia oblong or elongated, 2-or more septate, brown.

H. pulicàre, Pers. Syn. p. 98.

Hysterographium pulicare, Cda. Icones V, p. 77, tab. V, fig. 61.
Hysterium betulignum, Schw. Syn. N. Am. 2075.
Exsicc. Rehm Asc. 215.—Kze. F. Sel. 375. Rab. F. E. 2644.—Rav. F. Am. 762.—Krieger, F. Sax 437.—Ell. N. A. F. 457.—Desm. Pl. Crypt. Ed. I, 779.

Perithecia scattered or gregarious, superficial, variable in shape, oblong, ellipsoid, longitudinally striate, black, lips obtuse, slightly open exposing the linear disk, about 1 mm. long and ½ mm. broad. Asci clavate. Sporidia subbiseriate, oblong, straight or slightly curved, 3-septate, hardly constricted, the two end cells paler, straight or nearly so, each cell with a single nucleus, 18–25 x 7–9 μ (27–33 x 8–10 μ, Sacc. and Karst.)

Common on bark of various deciduous trees.

Very variable in the shape of the perithecia, which in the typical form are short and broad, subelliptical, but elongated-oblong forms are not uncommon, though in this respect the European specc. show less variation than the American. We do not find the sporidia as large as stated by Sacc.—mostly about 22 x 7 μ. *H. betulignum*, Schw, is given as a synonym from an examination of specc. in Herb. Schw.

H. truncàtulum, C. & P. Disc. of the U. S. p. 33.

Gregarious, superficial, elliptical, abruptly rounded at the ends, straight, longitudinally striate, flattened along the apex, 1 x ½ mm., lips closely connivent. Asci clavate, shortly stipitate. Sporidia biseriate, fusiform, rounded at the ends, triseptate, ultimate cells shorter, hyaline, central cells brown, 35–40 x 10 μ.

On wood. N. York (Peck).

This can hardly be more than a large-spored var. of *H. pulicare.*

H. angustàtum, A. & S. Consp. p. 55.

H. pulicare, b. angustatum, Fr. S. M. II, p. 580.
Hysterium vulgare, De Not. Pir. Istr. p. 18.
Hysterium Eucalypti, Phil. & Hark. Grev. XIII, p. 23.
Exsicc. Rehm Asc. 214.—Sydow, M. March. 2426.

Perithecia gregarious or crowded, more or less immersed in the bark or often superficial, mostly elongated or linear, nearly smooth, black, ¾–1½ mm. long, lips more or less open. Asci 75–80 x 12–15 μ. Sporidia subbiseriate, oblong, obtuse, 3-septate, all the cells brown and usually with a single nucleus, 15–22 x 6–7 μ (18–27 x 6–7 μ Sacc.).

On wood, Vermilion Lake, Minn. (Holway), on bark and wood of dead limbs, Carolina and Pennsylvania (Schw.)

In Schw. Syn. N. Am. the following varr. are mentioned as found in Pennsylvania:—var. *lenticulare*, Fr., round or elliptical, minute, smooth, only half as large as the normal form; var. *læve*, Pes., elliptical, elongated, swollen, smooth; var. *Juglandis*, Schw., ovate, black, lips swollen, subdistant, bistriate; frequent on bark of *Juglans*, Pennsylvania. This differs from *H. pulicare*, Pers. in having the sporidia swollen and all the cells uniformly brown, with the perithecia less prominent and mostly smaller and more elongated. Fries & Duby make it a mere var. of *H. pulicare.* Specimens of *Hysterium Eucalypti*, Phil. & Hark. (from Harkness), do not differ in any way from this.

H. tères, Schw. Syn. N. Am. 2077.

Perithecia subcylindrical, 2 mm. or more long, elevated, subflexuous, narrowed at the ends and subacute, black, glabrous, lusterless, lips subinflexed, open, seated on a greenish-fuliginous crust. Asci cylindrical, 75–80 x 12 μ with abundant paraphyses. Sporidia, subbiseriate, cylindric-oblong, 3-septate, straight, not constricted, end cells hyaline, 15–22 x 6–7 μ.

On decaying wood of *Rhododendron*, Carolina and Pennsylvania, (Schw.).

Differs from *H. pulicare* only in the shape of the perithecia.

H. macrósporum, Pk. 26th Rep. p. 83.

Perithecia superficial or nearly so, oblong or elliptical, sometimes slightly flexuous, 200–400 μ long, black, opening by a narrow chink, the lips slightly striate. Asci subcylindrical, 125–150 x 25–30 μ, 4–8-spored. Sporidia crowded in the ascus, oblong or subfusiform, sometimes slightly curved, at first colorless and uniseptate, then colored and triseptate, 40–57 x 12–15 μ.

On decaying wood of pine, New York (Peck).

H. sphæriàceum, Ell. Am. Nat. 1883, p. 193.

Erumpent, minute, hemispherical, ¼–⅓ mm. diam., gregarious, black, and nearly smooth, but not polished or shining, lips closed, slightly prominent. Asci subcylindrical, subsessile, 55 x 7 μ; paraphyses obscure. Sporidia biseriate, fusiform, hyaline and nucleate at first, becoming yellowish and 3-septate, often slightly constricted at the septa, 12–20 x 3–3½ μ.

On decaying wood, Decorah, Iowa (Holway).

This approaches *Lophiostoma*. Has much the same general appearance as some of the abbreviated forms of *Glonium parvulum*, Ger., but the fruit is different. The glandular hairs mentioned in the original diagnosis are evidently some hyphomycetous growth and only accidental.

H. magnósporum, Ger. in Proc. Poughkeepsie Acad. of Nat. Sci. Feb., 1875, p. 5, pl. 1, fig. 9.

Perithecia erumpent-superficial, about 1 mm. long, straight or slightly curved, ends subobtuse, black, subshining, longitudinally striate; lips slightly open so as to leave but a narrow fissure; ususlly one or two of the striæ on each side, near the apex, are deeper and more distinct. Asci clavate, 150 x 35–40 μ, with a short stipe and with abundant paraphyses. Sporidia irregularly crowded, broad, fusoid-oblong, slightly curved or a little bulging on one side, nearly hyaline and uniseptate at first, becoming deep, clear brown and 7-septate, three of the septa close to each end, 50–60 x 15–20 μ.

On a dead hickory limb, Poughkeepsie, N. Y. (Gerard), on dry, decaying oak wood, Newfield, N. J.

The general aspect is the same as that *H. vulvatum*, Schw., or *H. lineolatum*, Cke., but the sporidia are different. Gerard (l. c.) figures the sporidia as uniseriate with the septa at equal distances, but in the specc. sent by him they are as stated above.

H. eumórphum, Sacc. Mich. II, p. 40.

Perithecia loosely gregarious, decidedly superficial, so that the base is not entirely adnate, regularly navicular, straight, acute at each end, 1½ x ½ mm., becoming black, dehiscing with a lanceolate opening and exposing the reddish disk; lips smooth, obtuse. Asci cylindrical, very short-stipitate, 150–160 x 15 μ, rounded at the apex, 8-spored, surrounded with conglutinated, filiform paraphyses with dark-colored tips. Sporidia obliquely uniseriate, fusoid-oblong, subobtuse at the ends, mostly curved, 30–38 x 11–14 μ, with three broad septa, not constricted, guttulate, dark brown, becoming nearly black.

On bark, South Carolina (Ravenel).

Agrees (sec. Cooke) with specc. of *Hysterium biforme*, Fr. Scler. Suec. (not Duby), but not with the diagnosis of *H. biforme* as given by Fries. We have not seen this, but the description tallies well with that of *Tryblidium rufulum*, (Sprengel.)

H. thujàrum, C. & P. Cke. Disc. of the U. S. p. 33.

Subgregarious, superficial. Perithecia elliptical, acuminate, elevated, almost naviculoid, longitudinally striate, black; lips prominent, closely connivent. Asci cylindrical. Sporidia broadly lanceolate, obtuse, 3–5-septate, brown, 38–40 x 10–12 μ.

On *Thuja*, New York (Peck).

H. deprèssum, B. & C. Cke. Disc. of U. S. p. 34, Grev. IV, p. 10.

Elongated, rough with little granules, depressed, disk extremely narrow. Asci clavate. Sporidia cymbiform, with about 5 septa, sometimes bulging in the center on the convex side, 30 μ long.

On dry wood, Carolina (Curtis).

H. versísporum, Ger. Bull. Torr. Bot. Club, VI, p. 78.

Perithecia scattered, minute, black, smooth and shining, elliptical;

lips well rounded, closely connivent. Sporidia variable, elongated-clavate, elliptical and ovoid, 1–3-septate, pale brown, (hyaline at first), 14–25 x 5 μ.

On decorticated oak, New York (Gerard).

H. prælóngum, Schw. in Duby, Hyst. p. 27 (not Schw. Herb.).

Perithecia lying parallel, linear, narrow, more or less elongated, faintly transversely rugulose, black, shining; lips swollen and closely connivent. Asci obovate, shorter than the filiform paraphyses which are thickened and united above. Sporidia biseriate, ovate, 1½–2 times as long as broad, rounded at the ends, hyaline and homogeneous at first, at length 3-septate, dark red-brown and finally opake.

On dry, dead wood, New Mexico (Curtis).

This is very different from the *H. prælongum* in Herb. Schw., which is *Hysterographium lineolatum*, (Cke.).

H. insidens, Schw. Syn. N. Am. 2078.

(*Hysterium Berengerii*, Sacc. F. Ven. Ser. IV, No. 50)?
Hysterium complanatum, Duby, Hyst. p. 26.
Exsicc. Ell. N. A. F. 460.

Seated on a widely effused, black crust. Perithecia short, scattered, thick, oblong-ovate, subtruncate, generally contracted into a pseudo-stipitate base, obtuse at the ends. black; lips gaping, inflexed. This is the diagnosis given by Schweinitz. The specimen in Herb. Schw. does not show any black crust, nor can the perithecia be called even pseudo-stipitate; they are only erumpent-superficial, ¾–1¼ mm. long, obtuse, mostly lying parallel. Asci about 75 x 15 μ. Sporidia overlapping, subbiseriate, fusoid, slightly curved, 6–8-septate, with one joint (about the third from the top) slightly swollen, reddish-brown, ends narrowed but subobtuse, 25–30 x 6–8 μ.

Found by Schweinitz in Carolina, on hard, decaying wood. Specimens collected at Poughkeepsie, N. Y., on old chestnut wood, have the sporidia 30–40 x 7–9 μ, which is about the size of those in a spec. on an old pine shingle found at Newfield, N. J. Sec. Cooke, Grev. XVII, p. 88, the sporidia in an authentic spec. from Schweinitz, are 45–50 x 15 μ, with one cell swollen (as above).

The specimens in N. A. F., 460, were determined by Gerard from authentic specc. of *H. complanatum*, sent him by Duby. The species seems to be a variable one, both in the form of the perithecia and in the number of septa in the sporidia. Duby says 3–5-septate. In the N. A. F. specc. the sporidia are 3–7 (mostly 5-) septate and 25–30 x 7–8 μ. Some of the specc. distributed in N. A. F. have the perithecia at first tuberculiform and slightly contracted at base ("versus basin in pseudostipitem contractis"). Duby (l. c.) speaks of a spec. from Herb. Hooker, labeled *H. insidens*, Klotzsch, and which he doubtfully refers to his *H. complanatum*, as having sporidia 3–6-celled, frequently with one cell (the 2d or 3d) swollen, just as in the specc. of *H. insidens* in Herb. Schw.

H. lineariförme, Sacc. Syll. II, p. 651.

Hysterium lineare, Berk. N. Am. Fungi in Grev. IV, p. 11.

With the habit of *Glonium lineare*, (Fr.), but sporidia fusiform, strongly constricted in the middle, 37–50 μ long, multiseptate, (colored.)?

On wood of *Quercus*, *Gleditschia*, and *Taxodium*, Carolina (Ravenel), on dead wood, New England (Russell).

H. füsiger, B. & C. Grev. IV, p. II.

Elongated, flexuous, lying in various directions. Sporidia fusiform, with about 8 septa, sometimes strongly curved, 25 μ long. Resembling somewhat *Mytilidion tortile*, and *Glonium graphicum*, but with very different sporidia.

Nothing is said of the color of the sporidia in this and the preceding species.

H. Próstii, Duby, Hyst. p. 26.

Opegrapha Prostii, Nyl. Prodr. Lich. p. 154.
Hysterium lineare, var. *corticola*, Fr. Elench. II, p. 140.
Hysterium Wallrothii, Duby, Hyst. p. 25, tab. I, fig. 5.
Hysteropatella Prostii, Rhem in Rab. Krypt. Flor. III, p. 367.
Exsicc. Rehm Asc. 75.—Kze. F. Sel. 278.—Rab. F. E. 748, 1921.—Sydow, M. March. 2425
Ell. N. A. F. 461.—Rehm Asc. 75.

Innate, at length emergent, scattered or aggregated (hardly crowded), narrow-elliptical, acute, thin, membranaceo-carbonaceous, black, brittle; lips at first incurved, then open, exposing the slaty-black, narrow-elliptical disk, so as to resemble in the fresh state a sessile *Peziza*, small (about 1 mm. long or less). Asci clavate, 70–75 x 10–12 μ. Paraphyses not abundant. Sporidia biseriate, oblong, straight or curved, pale brown, 3-septate, 12–16 x 4–5 μ.

On maple bark, Newfield, N. J., on bark of *Pyrus coronaria*, Iowa (Holway), Missouri (Demetrio), on bark, Illinois (Seymour,) on inner surface of willow bark, Ohio (Morgan), on bark of elm? New York (Fairman.) Specc. from Washington (Suksdorf) on bare wood, seem to be this species, but have the sporidia larger, 15–22 x 6–7 μ, and may be different.

H. proteifórme, Duby, Hyst. p. 27.

Erumpent from the fibers of the wood, very variable in shape, linear, oblong, oblong-globose or globose. arranged in parallel series, black, shining, very faintly transversely striate; lips narrow, their edges more or less remote, exposing the linear-lanceolate, oblong or sphæroid, black disk. Asci (sec. Rehm) ovate, thickened at the apex, 36 x 10 μ. Sporidia 8 in an ascus, biseriate, clavate, obtuse, straight or somewhat curved, 3-septate, brownish, 9–11 x 3–3½ μ. Paraphyses branched and united above in a greenish-brown layer above the asci.

On dry, decaying oak wood, Carolina (Curtis).

The appearance is peculiar. When closed, it often resembles a *Sphæria*, when open, a *Lecidea*. We have seen no specimens and take the diagnosis from Duby's Memoire. This (sec. Dr. Rehm, in Hedw., 1886, p. 187), is closely allied to *H. Prostii*, and with that species belongs in the *Patellarieæ*.

H. ellipticum, Fr. Obs. Mycol. I, p. 195.

Hysteropatella elliptica, Rehm, in Die Pilze (Discomycetes) p. 368.

Perithecia gregarious or collected in small groups, lying parallel, at first buried among the fibers of the wood, finally emergent and sessile, rounded or elongated, straight or somewhat curved, simple or stellate, ½–3 mm. long, ¼–½ mm. wide, with an elongated, finally elliptical opening, showing the thin black disk, of a wax-like consistency. Asci clavate, thick-walled, 60–70 x 12–16 μ, 8-spored, with dichotomously branched paraphyses united in a brown epithecium above. Sporidia biseriate, oblong, obtuse, straight or slightly curved, 3-septate or slightly constricted at the septa, hyaline, becoming brown. 15–17 x 5–6 μ, the two middle cells mostly larger and each with a large oil globule.

On bark, North Carolina and Pennsylvania (Schw.).

The specc. of this species in Herb. Schw. are without fruit, and as there are no reliable specc. in our collection we have taken the diagnosis from Rehm, in Die Pilze. The sporidia (sec. Fckl.) are 24–26 x 8 μ The species is closely allied to *H. Prŏstii*, and with that species and some others here included for convenience, in the *Hysteriaceæ*, really belongs in the *Discomycetes*.

The specimens of the following species in Herb. Schw. are without fruit.

H. rugulòsum, Schw. Syn. N. Am. 2079.

Perithecia small, round or of various shapes, black, generally convex, much wrinkled, aggregated or scattered on a black spot, longitudinally dehiscent, opening central, sometimes abbreviated and distinctly labiate.

On fragments of willow wood, Bethlehem, Pa. (Schw.).

H. Rhòis, Schw. Syn. N. Am. 2092.

Scattered, short, ovate or subrotund, convex-globose, black, not shining, glabrous. Lips distant, leaving a wide opening. Subimmersed among the fibers of the wood, minute, appearing to the naked eye like black specks.

Rather rare, on rotten wood of *Rhus typhina*, Bethlehem, Pa. (Schw.).

H. nucícola, Schw. Syn. N. Am. 2080.

Seated on a black crust. Perithecia ovate-hemispherical, strongly convex, the elongated orifice subimpressed, acute ,closed, otherwise glabrous, of a brownish-black color, crowded together in various positions, transverse, parallel and mixed.

On old hickory nuts, Bethlehem, Pa. (Schw.).

The specc. in Herb. Schw. have the outward appearance of *Glonium parvulum*, Ger., but are entirely without fruit.

H. Kálmiae, Schw. Syn. N. Am. 2091.

Perithecia immersed among the fibers of the wood, at length longitudinally erumpent, very long, linear, straight, acuminate at the ends, confluent, very black. Lips thin, subturgid, gaping, with a rather broad opening.

On partly rotten wood of *Kalmia*, erumpent among the fibers of the wood and covered by them, Bethlehem, Pa., (Schw.).

H. fibritéctum, Schw. Syn. N. Am. 2095.

Erumpent, gregarious, covering patches often an inch across among the loosened fibers of the wood. Perithecia semiimmersed, black, broad-ovate, abruptly acuminate at each end, generally somewhat flattened above, sides convex, opening broad, margined, at first closed.

On old willow wood, Bethlehem, Pa. (Schw.).

The specimens of the following species in Herb. Schw. do not belong to Hysterium.

H. libríncola, Schw. Syn. N. Am. 2101.

This pertains to the *Sphæriaceæ*, but can not be determined from the specimen. The spec. labeled *H. gramineum*, Fr., (No. 2108), is also some indeterminable, sterile thing.

H. Polygonàti, Schw. Syn. N. Am. 2115.

This is *Vermicularia Polygonati*, Schw. Syn. N. Am. 1847.

H. Osmúndæ, Schw. Syn. N. Am. 2113, is *Leptostroma litigiosum*, Desm.

H. Ptéridis, Schw. Syn. N. Am. 2114, is *Leptostromella filicina*, Sacc. Syll. III, p. 660, (*Cryptosporium filicinum*, B. & C. Grev. II, p. 84.)

H. sámaræ, Fr. (2112).

This spec. is apparently *Phoma Samararum*, Desm., but entirely sterile.

H. rimíncola, Schw. Syn. N. Am. 2086.

The specimen has the appearance of some undeveloped *Dothidea*, and evidently is not a *Hysterium*.

H. Sámbuci, Fr. (2098).

The specimen can not be a *Hysterium;* it is apparently some undeveloped sphæriaceous fungus, and is without fruit of any sort.

H. stictoìdeum, C. & E. Grev. VII, p. 7.

Exsicc. Ell. N. A. F. 571.

This is *Stictis hysterina*, Fr.

OSTREÌON,* Duby.

Mem. sur la Tribu des Hyst. p. 21.

Perithecia conchiform, superficial, subpedicellate, longitudinally striate. Asci large, 4-spored, paraphysate. Sporidia very large, fusoid multiseptate, colored.

O. Americànum, Duby, l. c. p. 22, tab. I, fig. 1. Cooke in Grev. IV, pl. 67.

Scattered, black and shining, obsoletely transversely striate, shaped like an oyster-shell, margin incurved. Asci cylindrical, 350–400 x 30 μ, 4-spored. Sporidia uniseriate, broad-fusoid, constricted in the middle, opake, reddish-brown, 12–20–septate, 90–100 x 25–27 μ.

*Name changed from *Ostreichnion*.

The two extreme septa at each end of the sporidium are close together and near the end of the sporidium.

On bark of a trunk of *Liquidambar*, South Carolina, (Curtis.)

HYSTEROGRÁPHIUM, Corda.

Icones Fungorum, V, p. 34.

Perithecia erumpent, sessile, elongated or elliptical, obtuse, prominent, mostly simple, opening by a narrow, elongated, longitudinal cleft, black and carbonaceous. Asci thick-walled, clavate, 8-spored, with paraphyses branching above and forming a colored epithecium. Sporidia 1–2-seriate, elliptical or ovate, obtuse, becoming muriform, brown or, in the subgenus *Gloniopsis*, hyaline. (Rehm in Die Pilze).

The genus differs from *Hysterium* in its muriform sporidia.

* *Sporidia colored.*

H. Fráxini, (Pers.)

Hysterium Fraxini, Pers. Syn. p. 98.
Hysterographium Fraxini, De Not. Pir. Ist. p. 22.
Exsicc. Rehm Asc. 26.—Kunze, F. Sel 376 —Desm. Pl. Cr. Ed. I, 83 —Roum. F. G. 2569, 5450.—Sydow, M. March. 951.—Ell. N. A. F. 997.—Rab. F. E. 58.

Perithecia scattered or gregarious, erumpent, elliptical, black, obtuse above, 1–1½ mm. long, ½–¾ mm. wide; lips swollen, smooth, partially open so as to expose the narrow disk. Asci clavate, rounded above, 150–200 x 30–40 μ, 8-spored, with filiform paraphyses. Sporidia biseriate, oblong-elliptical, scarcely constricted in the middle, 7–9-septate and muriform, dark yellow-brown, 30–40 x 15–18 μ.

On dead limbs of *Fraxinus*, New York, Pennsylvania, Iowa and Canada; probably common throughout the United States.

H. Syríngæ, (Schw.)

Hysterium Syringæ, Schw. Syn. N. Am. 2073.
Tryblidium Syringæ, Cke. Disc. of the U. S. p. 32, Grev. IV, Pl. 67, fig. 10.
Tryblidium dealbatum, Ger. Bull. Torr. Bot. Club, V, p. 40.
Hysterographium Syringæ, Sacc. Syll. 5763.

At first subimmersed, at length denuded, mostly scattered over pallid spots and often surrounded by a black crust, ovate-elliptical, acuminate or subobtuse black, not striate; lips at length widely gaping and exposing the subrugose, black disk. Asci oblong-clavate, rather abruptly contracted into a short stipe below, obtuse and rounded above, paraphysate, 8-spored, 100–112 x 20–25 μ. Sporidia subbiseriate, ovate-elliptical, 6–8-septate and muriform, yellowish-hyaline at first, becoming dark brown and almost opake, 25–34 x 12–15 μ.

On bark of *Syringa vulgaris*, Pennsylvania (Schw.), New York (Gerard), New Jersey (Ellis).

The perithecia are small, only a little larger than those of *Hysterium Prostii*, Duby, which they much resemble. The specc. in Herb. Schw. are without fruit. The above diagnosis is from specc. sent by Gerard. Cooke's figure in Grevillea represents the asci and sporidia very accurately.

H. formòsum, (Cke.)

Hysterium formosum, Cke. Grev. VII, p. 3.
Hysterographium formosum, Sacc. Syll. 5780.
Exsicc. Ell. & Evrht. N. A. F. 2d Ser. 2092.

Perithecia scattered, prominent, 1–1½ mm. long and proportionally rather broad, not depressed above, black and shining, striate, lips closed. Asci cylindrical, 80–90 x 10 μ with abundant paraphyses, 8-spored. Sporidia uniseriate, oblong-elliptical, 3-septate and faintly muriform, brown, slightly constricted at the middle septum, 18–22 x 7–8 μ.

On dead limbs of *Pinus contorta* and *Juniperus occidentalis*, Sierra Nevada Mts., California (Harkness).

H. subrugòsum, (C. & E.)

Hysterium subrugosum, C. & E. Grev. V, p. 54, Pl. 81, fig. 1.
Hysterium acuminatum, Fr. in Herb. Schw.
Hysterographium subrugosum, Sacc. Syll. 5770.
Exsicc. Ell. N. A. F. 459.

Perithecia scattered, oblong-elliptical, faintly transversely subrugose, black, not striate, about 1 mm. long, often slightly curved; lips incurved, open so as to expose a narrow strip of the black disk. Asci oblong-cylindrical, sessile, paraphysate, rounded at the apex, 75–80 x 20 μ. Sporidia inordinate, ovate-oblong, obtuse, slightly curved, 5–7-septate, with one or two partial longitudinal septa, 20–25 x 10 μ (40–45 μ long, sec. Cooke).

On dry hard wood of decaying oak stumps, Newfield, N. J.

The substance of the perithecia, when crushed under the microscope, has a reddish tinge. The spec. in Herb. Schw. labeled *Hysterium acuminatum*, Fr., is certainly the same as *H. subrugosum*, C. & E. and very different from *H. acuminatum*, Fr. The sporidia in all the specc. examined are much smaller than stated in Grevillea.

H. Nòva-Cæsariénse, (Ell.)

Hysterium Nova-Cæsariense, Ell. Bull. Torr. Bot. Club, VI, p. 133.
Mytilidion Nova-Cæsariense, Sacc. Syll. II, p. 764.
Exsicc. Rehm Asc. 313.—Roum. F. Gall. 4854.—Ell. N. A. F. 152.

Gregarious, lying in various directions on the matrix. Perithecia smooth, black, tubercular at first, becoming oblong-elliptical and about 1 mm. long, obtuse above; lips at first closed, then distinctly separated. Asci oblong-cylindrical, subsessile, paraphysate, 8-spored, p. sp. 80–100 x 30–35 μ. Sporidia inordinate, crowded, fusoid-oblong, multiseptate, (7–15-septate,) often with a longitudinal septum running through one or more of the cells, brown, 35–50 x 10–13 μ.

Common around Newfield, N. J., on outer bark of living *Pinus rigida.*

The obtuse perithecia and submuriform sporidia will remove this from *Mytilidion.*

H. cineráscens, Schw. Syn. N. Am. 2076.

Gregarious, elongated, flexuous, very black, subshining, but punctulate on the surface. The surrounding wood is of a cinereous color (hence the specific name). Perithecia densely crowded, elongated: lips thin, often breaking away in frustules so that the perithecium becomes widely dehiscent. Asci clavate-oblong, sessile, 75–90 x 15–20 μ, overtopped by abundant paraphyses, which are blackened and conglutinated at their tips. Sporidia inordinate, oblong-elliptical, 6–8-septate, with one or more longitudinal septa running through the medial cells, brown, about 20 x 8 μ.

On rotten wood of *Juglans cinerea*, Bethlehem, Pa. (Schw.).

The measurements are from spec. in Herb. Schw., but the spec. not being in good condition, they may not be exact, though they are very nearly so.

H. Mòri, (Schw.)

Hysterium Mori, Schw. Syn. N. Am. 2087.
Hysterographium Mori, Rehm Asc. 363, Sacc. Syll. 5779.
Hysterium Rousselii, De Not. Pir. Ist. p. 19.
Hysterographium Rousselii, Sacc. Syll. 5768.
Hysterium viticolum, C. & P. Disc. U. S. p. 33, Grev. IV, tab. 68, fig. 9.
Hysterium Gerardi, C. & P. l. c.
Exsicc. Fckl. F. Rh. 751.—Rehm Asc. 316.—Rab. F. E. 2958.—Ell. N. A. F. 75, 77, 78, 458, 1286.

Perithecia erumpent-superficial, elliptical, oblong, linear or cylindrical, 1–3 mm. long, and ½–1 mm. wide, mostly straight and lying parallel, gregarious and often crowded so as to cover the matrix more or less completely for some extent, more or less distinctly longitudinally striate; lips mostly closed at first, finally more or less open, exposing a narrow, linear or lanceolate disk. Asci cylindrical, about 100 x 12 μ (including the short-stipitate base), paraphysate, 8-spored. Sporidia uniseriate or subbiseriate above, ovate, varying to oblong or ovate-elliptical, 3–5-septate, constricted at the middle septum, one or two of the cells divided by a longitudinal septum, brown, 15–25 x 7–8 μ.

On decorticated, exposed wood, also (but less frequently) on dead limbs still covered with the bark; common.

Var. *Gerardi*, C. & P., has the perithecia mostly shorter and broader, with the lips more open; var. *viticolum*, C. & P., has the asci mostly shorter and broader, 75–90 x 15–18 μ, and the sporidia more or less perfectly biseriate. Cooke makes the sporidia 36 x 12 μ. We find none over 25 x 8 μ. *H. Rousselii*, De Not. (sec. specc. from Gerard, compared by him with authentic specc. from Duby), can not be distinguished in any way from *H. Mori*, Schw., as represented by specc. in Herb. Schw., nor can we, after a careful examination of many specc. from different localities, during the past ten years, find any reliable characters by which any of the so-called species above quoted can be safely separated. Specimens of *H. Rousselii*, with elongated perithecia, appear quite distinct from specc. of *H. Gerardi*, C. & P., with shorter, oblong or elliptical perithecia, but these extremes are connected by a graduated series of forms passing imperceptibly into each other so as to completely fill up the gap; nor do the sporidia furnish any distinctive characters. At first they are 3-septate, becoming almost always 5-septate, and not seldom 6–7-septate.

H. variábile, C. & P. Disc. of the U. S. p. 33.

"Erumpent, then superficial, following the interstices of the woody fibers of the matrix, narrowly elliptical or linear and elongated, straight or flexuous, often parallel, faintly striate-rugose, flattened along the apex, slightly narrowed at each end, lips closely connivent. Asci cylindrical, stipitate. Sporidia uniseriate, very variable in size and form, ovate or elliptical, or broadly clavate, constricted in the middle, 5–7-septate, at length with longitudinal divisions, dark brown."

On old chestnut posts, &c., New York (Peck).

The foregoing is the diagnosis given by Cke. (l. c.). Specc. from Peck have the perithecia crowded, 1–2 mm. long; lips incurved so as to leave a distinct furrow along the apex, often but partially closed, exposing the linear disk. Often the perithecia are marked with a single deep furrow close to and parallel with the lips, but quite as often the furrow is wanting and the flattened apex of the perithecium is faintly transversely rugulose. Asci paraphysate, short-stipitate, p. sp. about 70–75 x 10–12 μ. Sporidia subbiseriate, ovoid, 3–7-septate, somewhat constricted in the middle, about 20 x 8 μ, dark brown, with a partial longitudinal septum running through the central cells. What appears to be the same has been found at Newfield, N. J., on bare, decaying wood of *Quercus* and *Pyrus Malus*.

H. vulvàtum, (Schw.)

Hysterium vulvatum, Schw. Syn. N. Am. 2072.
Hysterographium vulvatum, Sacc. Syll. 5774.
Exsicc. Rav. F. Car. II, 48.—Ell. N. A. F. 76.—Thum. M. U. 181.—Rehm Asc. 315. (In all these under the name of *H. flexuosum*, Schw.).

Perithecia erumpent-superficial, scattered or gregarious, 1–3 mm. long, straight or flexuous; lips at first closed, then open so as to leave a wide furrow between them, mostly with 1–2 deep striæ on each side, often apparently double, i. e., one set within another. Asci oblong-clavate, obtuse, paraphysate, 8-spored, p. sp. 100–112 x 20–25 μ, with a stipe 30-35 μ long. Sporidia irregularly biseriate, broad-fusoid, slightly curved, multi- (10–15-) septate, and muriform, strongly constricted in the middle, the upper part broader, olive-brown, 50–62 x 15–20 μ.

On dead, dry, mostly decorticated limbs of oak and other deciduous trees, common.

The species here described as *H. vulvatum*, Schw., is certainly the *Hysterium vulvatum* of the Schweinitzian Herbarium, but whether the specimen so labeled in that collection is the *Hysterium vulvatum* of Schw. Synopsis N. Am. 2072, may be open to some doubt. The specc. in the Exsiccati above referred to, certainly agree well with the diagnosis of *H. flexuosum*, Schw., in his Syn. Car. No. 249.

H. Lesquereùxii, (Duby).

Hysterium Lesquereuxii, Duby, Hyst. p. 29.
(*Hysterium fibrisedum*, Ger. Bull. Torr. Bot. Club, V, p. 26)?

Perithecia scattered or aggregated, black, faintly rugulose under the lens, ovate or ovate-oblong, or sublinear, often flexuous, narrowed towards each end, subacute; lips swollen, partially open so as to leave a distinct cleft between them. Asci cylindric-clavate, 80–100 x 12–15 μ, paraphysate, 8-spored, subsessile. Sporidia subbiseriate, ovate-oblong, slightly constricted in the middle, obtuse, 5–7-septate and muriform, pale brown, 20–25 x 8–10 μ.

On corticated branches of *Gleditschia triacanthos*, Ohio (Lesquereux), Louisiana (Langlois).

The diagnosis is from the Louisiana specc. There is sometimes a distinct furrow on each side of the lips and parallel with them, but often this is entirely wanting. *H. fibrisedum*, Gerard, probably belongs here, but we have seen no specimens.

H. pròminens, (Phil. & Hark.)

Hysterium prominens, Phil. & Hark. Bull. Cal. Acad. Feb. 1884, p. 25.
Hysterium Ceanothi, Phil. & Hark. l. c.
Hysterographium prominens, Berl. & Vogl. Add. Sacc. Syll. I, p. 270.
Exsicc. Ell. & Evrht. N. A. F. 2d Ser. 2064.

Perithecia erumpent-superficial, gregarious or scattered, elliptical or oblong, 1–2½ x ½–¾ mm. black, not striate; lips incurved, nearly closed at first, becoming finally more or less open so as often to expose the narrow-lanceolate, black or slate-colored disk, generally more or less distinctly transversely wrinkled. Asci clavate-oblong, short-stipitate, 100–115 x 30–35 μ (p. sp.), paraphysate, 8-spored. Sporidia biseriate, broad-fusoid, uniseptate and constricted in the middle, then multi- (12–15-) septate, and muriform, olive-brown, slightly curved, 50–60 x 10–12 μ.

On dead limbs of *Salix lasiolepis* and *Ceanothus*, California (Harkness).

The asci and sporidia do not differ from those of *H. vulvatum*, Schw., unless the sporidia may be a little narrower, but the perithecia are more open and not striate. *H. Ceanothi*, Phil. & Hark. (sec. specc. from Harkness) does not differ in any way from *H. prominens*. Both are erumpent-superficial, with the disk finally more or less exposed, and the asci and sporidia are the same. Probably both should be considered mere varr. of *H. vulvatum*, Schw.

H. stýgium, (Cke.)

Hysterium stygium, Cke. Grev. XI, p. 107.
Hysterographium stygium, Sacc. Syll, 5775,

Superficial, scattered. Perithecia lanceolate, black, shining, smooth, ½–1 mm. long. Asci clavate. Sporidia elliptical or ovate, multiseptate-muriform, dark brown, 30 x 13 μ.

On bark of oak, North America.

We know nothing of this beyond the above brief diagnosis from Grevillea.

H. elongàtum, (Wahl.)

Hysterium elongatum, Wahl. Flor. Lapp. p. 528.
Hysterographium elongatum, Cda. Icones, V, p. 77, tab. IX, fig. 62.
Exsicc. Fckl. F. Rh. 1754.—Thum. M. U. 1862.

Perithecia gregarious, seated on the dark-colored, decorticated surface of the wood, oblong or oblong elliptical, straight, seldom somewhat curved, obtuse, prominent, mostly smooth, deeply cleft; lips nearly closed, black, carbonaceous, 2–4 x 1 mm. Asci clavate, thick-walled, 120–150 x 25 μ, 8-spored, with paraphyses branched above and forming a thick, brown layer (epithecium) above the asci. Sporidia biseriate, oblong-elliptical, obtuse, seldom somewhat constricted in the middle, hyaline, becoming brown, 9–11-septate, with a longitudinal septum, 36–40 x 12–15 μ.

On wood, Carolina and Pennsylvania (Schw.).

The specc. of this species in Herb. Schw. are without fruit. The diagnosis is from Rehm (in Die Pilze).

H. hiáscens, Rehm Asc. No. 314.

Exsicc. Rehm Asc. 314 —Rab. F. E. 3564.—Ell. N. A. F. 151

Scattered, superficial, narrow-elliptical, hardly over 1 mm. long, black; lips incurved, open so as to expose the narrow disk. Asci clavate-oblong, obtuse, subsessile, paraphysate, 8-spored, 100 x 20–25 μ. Sporidia irregularly crowded, oblong-elliptical, mostly a little curved, 6–8-septate, with a longitudinal septum generally running through all the cells, 22–27 x 8–10 μ.

On bark of *Quercus bicolor* and *Celtis occidentalis*, in Carolina, on bark of *Quercus coccinea*, Newfield, N. J.

Generally accompanied by *Hysterium pulicare*, Pers. The asci and sporidia are much like those of *H. Syringæ*, Schw., but the perithecia are rather larger and more obtuse. *Hysterium hiascens*, B. & C. Grev. IV, p. 11, is said to be different and referable to the *Discomycetes*. We have never seen it.

** *Perithecia discoid; sporidia yellowish* (*Pseudographis*).

H. elátina, (Ach.)

Lichen elatinus, Acharius Prodr Lich p. 22.
Lecanora elatina, Ach. Lich. Univ. p. 387.
Hysterium elatinum, Pers. Syn. p. 28.
Pseudographis elatina, Nyl. Herb. Mus. Fenn. p. 96.
Krempelhuberia Cadubriæ, Massal. Geneac. p. 15, No 21.
Exsicc. Rab. Herb. Mycol. 709 —Rehm Asc. 25.

Perithecia gregarious, round or irregular, rounded-angular or elongated-elliptical, obtuse, at first sunk in the bark which is raised into a pustule and then laciniately torn, suberumpent, closed at first, then with an elongated or laciniate opening above and widely exposing the reddish-yellow disk, rough, black, carbonaceous, 1–2 mm. across. Asci cylindrical, 180–250 x 15–18 μ, 8-spored. Sporidia uniseriate, oblong-ovate, straight or slightly curved, 1–7-septate, and finally muriform, hyaline or yellowish, 21–30 x 10–14 μ. Var. *crispum*, (Pers.), reported by Schweinitz as found in Carolina and Pennsylvania, on bark of pine trees, differs in having much longer and narrower, more curved and flexuous perithecia with thin, crisped lips and black disk.

This genus and species is placed by Dr. Rehm (Die Pilze) in the *Discomycetes*, Fam. *Pseudophacidieæ*, where it evidently belongs; but as the species has hitherto been classed among the *Heysterineæ*, and will quite likely be sought for in that Family, we leave it there for the present.

*** *Sporidia hyaline* (*Gloniopsis*).

H. Cookeiànum, Ger. Bull. Torr. Bot. Club, V, p. 77.

Perithecia erumpent-superficial, black, not striate, linear-elongated, often curved, lying in various directions on the matrix, ends acute; lips narrow, slightly swollen, edges somewhat remote. Asci oblong, 8-spored, subsessile, 60–75 x 20–22 μ. Paraphyses branched above. Sporidia inordinate, subelliptical, hyaline, muriform, 20–30 x 12–15 μ.

On bark of *Carya alba*, New Paltz Landing, N. Y. (Gerard), on dry, exposed wood, Ohio (Morgan), on cast-off bark of *Acer* and of *Pyrus Malus*, and on decorticated oak limbs, Newfield, N. J; common, but not abundant.

H. gloniópsis, (Gerard).

Hysterium hyalosporum, Ger. in Peck's 31st Rep. p. 49.
Hysterium gloniopsis, Ger. Bull. Torr. Bot. Club, VI, p. 78.
Hysterium hyalinum, C. & P. in Cooke's Disc. of the U. S. p. 33.
Gloniella hyalina, Sacc. Syll. 5719.
Gloniopsis Gerardiana, Sacc. Syll. 5747.

Perithecia erumpent-superficial, lying in the direction of the fibers of the wood, frequently parallel, linear, narrow, rounded at each end, 1–2 mm. long, straight, faintly striate; lips acute, nearly closed. Asci cylindric-clavate, paraphysate, 8-spored, subsessile, 57–60 x 12 μ. Sporidia biseriate, clavate-oblong, 3-septate, with one or two of the broader cells often divided by a longitudinal septum, 12–15 (exceptionally 20) x 5–6 μ, hyaline.

On dry, hard wood of deciduous trees, New York State (Peck & Fairman), Louisiana (Langlois), on old, dry oak wood, New Jersey (Ellis).

The above diagnosis is from the original specimens of *Hysterium gloniopsis*, determined by Gerard. The specc. of *H. hyalinum*, C. & P., determined by Cooke, do not differ appreciably. In the diagnosis of *H. hyalinum*, in Disc. U. S. the sporidia are said to be 26–28 x 6–8 μ, but we can only make them as stated above. In a spec. of *Hysterium hyalosporum*, Ger., from Gerard, the sporidia, when mature, become brown, though they remain for a long time hyaline, but in all other respects the spec. agrees so well with the characters given above that it cannot well be separated. In the original description of *Hysterium hyalinum*, C. & P., nothing is said of any longitudinal septa in the sporidia, but in the mature sporidia they are always present.

H. prælóngum. (Schw.)

Hysterium prælongum, Schw. in Herb. Schw. & Syn. N. Am. 2074 (not Duby).
Hysterium lineolatum, Cke. Grev. XI, p. 107.
Hysterium Verbasci, Schw. in Herb. Schw. & Syn. N. Am. 2093.
Gloniopsis Verbasci, Rehm Rev. Duby, Hyst. p. 11.
Exsicc. Rav. F. Am. 69.—Ell. N. A. F. 1321.

Perithecia linear, mostly lying parallel, straight or subflexuous, flattened above, immersed in the wood with the flattened apex exposed and slightly prominent; lips incurved, leaving a narrow crevice or furrow between them, 1½–2½ mm. long. Asci clavate-oblong, 65–70 x 15 μ, 8-spored, subsessile, overtopped by the abundant, filiform paraphyses, which are blackened and glued together at their tips. Sporidia subbiseriate, ovate-elliptical or subclavate, often a little curved, 5–7-septate and muriform, yellowish-hyaline, 20–22 x 8–10 μ. *Hysterium Verbasci*, Schw. Syn. N. Am. 2093, has the sporidia a little more attenuated below and the lips of the perithecia rather more distinctly closed, perhaps because the specc. are not as old.

On weather-beaten wood, Carolina and Pennsylvania (Schw.), Ohio (Morgan), on wood of *Quercus* and *Persea*, Texas and Florida (Ravenel), on old stems of *Verbascum Thapsus*, Bethlehem, Pa. (Schw.).

Without a microscopical examination, the specc. of *H. Verbasci* and *H. prælongum*, in Herb. Schw. can not readily be distinguished from *Glonium lineare*, Fr. It is doubtful whether *H. Verbasci* should be separated from *H. prælongum*, even as a variety; it certainly can not be specifically distinct. The specc. issued in N. A. F. 1321, as *Hysterium lineolatum*, Cke., are mostly *Gloniella Curtisii*, (Duby), but *H. lineolatum* also occurs on several of them.

H. Lonicèræ, (Phil. & Hark.)

Hysterium Loniceræ, Phil. & Hark. Grev. XIII, p. 23.
Gloniopsis Loniceræ, Berl. & Vogl. Add. Sacc. Syll. I, p. 270.

Scattered or gregarious, elliptical or oblong-lanceolate, 500–1200 μ long, parallel, subprominent, striate or smooth; lips subprominent, open at maturity. Asci cylindric-clavate, 8-spored. Sporidia elliptical or subpyriform, muriform, hyaline, mostly with a gelatinous coat, 20–30 x 7–11 μ, with indistinct paraphyses.

On decorticated branches of *Lonicera*, California (Harkness).

H. insígnis, Cke. & Hark. Grev. XIV, p. 10.

Perithecia gregarious, erumpent, oval or elliptical, ½–1 mm. long, black, subshining; lips closely approximated. Asci clavate, 8-spored. Sporidia lanceolate, acuminate at the ends, constricted in the middle, 5–7-septate and muriform, hyaline, 20 x 10 μ.

On wood of *Acacia*, California (Harkness).

H. Smìlacis, (Schw.)

Hysterium Smilacis, Schw. Syn. Car. 241. Syn. N. Am. 2096.
Hypoderma Smilacis, Sacc. Syll. 5801.
Exsicc. Ell. & Evrht. N. A. F. 2d Ser. 2375.—Rehm Asc. 318.—Thum. M. U. 662.

Perithecia erumpent, surrounded at base by the ruptured epidermis, about 1 mm. long (exceptionally 1½–2 mm.), gregarious, black, but not shining; lips nearly closed so as to leave but a narrow cleft between them, marked on each side by two distinct, longitudinal striæ. Asci oblong, subsessile, 60 x 15 μ, paraphysate, 8-spored. Sporidia biseriate or inordinate, clavate-oblong or clavate-fusoid, yellowish-hyaline, 3–5-pseudoseptate, one or two of the inner cells divided by a longitudinal septum, 12–20 (mostly 15) x 4–5 μ, or, including the gelatinous envelope, 7 μ wide.

Common, on dead stems of various species of *Smilax*.

The perithecia (sec. Schw.), are 2–3 lines (3–6 mm.) long. None of the specc. we have seen, including those in Herb. Schw., are as long as that—mostly about 1 mm. long.

FAMILY. HYPODERMÌEÆ.

Perithecia elongated-linear or elliptical, mostly simple, covered by the epidermis and mostly adnate to it, thin, membranaceous, opening by a longitudinal cleft along the summit.

HYPODÉRMA, DC.

Flore de Franc. II. p. 304.

Perithecia innate, oblong or ellipsoid, at first covered by the epidermis, membranaceous, thin, subdimidiate, opening by a longitudinal cleft along the apex. Asci clavate, with filiform paraphyses. Sporidia fusoid or oblong, guttulate, becoming mostly 2-celled, much shorter than the asci. Spermogonia (*Leptostroma* sp.), orbicular, finally deciduous. Spermatia simple, minute.

H. commùne, (Fries).

Hysterium commune, Fr. S. M. II, p. 589.
Hypoderma commune, Duby, Hyst. p. 41.
Exsicc. Rab. Herb. Mycol. II, 576.—Rav. F. Am. 323.—Sydow, M. March. 2151, 3271.—Ell. N. A. F. 464.

Perithecia superficial-innate, ellipsoid, obtuse at each end, black; lips longitudinally rugose or smooth, disk subfuliginòus, 1–1½ x ½ mm. Asci ovate-clavate, very delicate, slender-stipitate, 8-spored, 60–75 x 10–12 μ, with slender, filiform paraphyses, straight or curved above. Sporidia elongated or narrow-fusoid, straight, subobtuse, 2-celled, hyaline, 18–20 x 4 μ, lying mostly parallel in the upper part of the asci.

On *Sambucus Canadensis*, *Sedum Telephium*, and various dead herbaceous stems, in Carolina, on *Eupatorium purpureum*, Pennsylvania and New Jersey; probably common throughout the United States.

Leptostroma vulgare, Fr., is considered to be the spermogonial form of this species; sporules elongated, subobtuse, slightly curved, continuous, 7 x 1½–2 μ. The diagnoses of this and of *H. virgultorum* are almost identical, and whether the two are really distinct may well be doubted. *H. commune* seems to differ only in its more obtuse, rather smaller perithecia with less open lips. If the two are really distinct, the difference is to be sought in the spermogonia. *Hysterium confluens*, Schw. Syn. N. Am. 2101, and *H. expallens*, Schw. l. c. 2106, are apparently only forms of *Hypoderma commune*, but the specc. in Herb. Schw. are entirely sterile.

H. virgultòrum, DC. Flore Franc. VI, p. 165.

Hysterium Rubi, Pers. Obs Myc I, p. 84.

Exsicc. Fckl. F. Rh. 757.—Kunze F. Sel. 162.—M. & N. Stirp. Vog. 564.—Rehm Asc. 362, 919.—Sydow, M. March. 329 —Thum. M. U. 180.—Rav. F. Am. 324.—Roum. F. G. 270.—Ell. & Evrht. N. A. F. 2d Ser. 2378.

Perithecia scattered, subinnate, lying parallel in the direction of the axis of the stem, elongated, acute, smooth, shining, black, at length gaping, gray inside. Asci clavate-oblong, long-stipitate, 8-spored, 90 x 9 μ, with filiform paraphyses hooked or spirally coiled above. Sporidia fusoid-elongated, subobtuse, straight or slightly curved, 2-celled, 2-nucleate, hyaline, 21–24 x 3–4 μ; polystichous. The spermogonial stage is *Leptostroma virgultorum*, Sacc., with 1-celled, subcylindrical sporules 4–5 x 1 μ.

On dead stems of *Rubus*, New Jersey and Carolina, on bramble. California (Harkness).

H. ilícinum, De Not. Giorn. Bot. Ital. II, p. 35, Duby, Hyst. p. 40.

Exsicc. Ell. N. A. F. 1000.

Innate, epiphyllous, scattered, elliptical, obtuse at the ends, at length deciduous, dark gray as seen through the epidermis, 700 x 350 μ; lips slightly swollen, white-margined, at first closed, at length partially open, exposing the pallid disk. Asci clavate-stipitate, 80 x 12 μ, p. sp. about 50 x 12 μ, paraphysate, 8-spored. Sporidia oblong or clavate-cylindrical, slightly curved, hyaline, about 15–20 x 3–4 μ, continuous, indistinctly nucleate, at length faintly uniseptate.

On leaves of *Quercus virens*, Hibernia, Florida (Dr. Martin), on leaves of some species of oak, in Mexico (Dr. Eckfeldt).

The measurements of asci and sporidia are from the Florida specimens.

H. variegàtum, (B. & C.)

Hysterium variegatum, B. & C. Grev. IV, p. 12.
Hypoderma variegatum, Duby, Hyst. p. 43.

Innate-superficial, scattered or gregarious, sometimes confluent, linear-elongated, subobtuse, black; lips flat, closely connivent, their edges sometimes slightly raised, so as to leave a very narrow, straight or subflexuous cleft. Asci very minute, clavate, sessile. Sporidia hyaline, linear, obtuse, inordinate, 48–50 x $1\frac{1}{4}$ μ. Paraphyses none.

On dead branches of *Viburnum Opulus* and various species of *Andromeda*, also on *Aralia spinosa*, North and South Carolina (Ravenel).

H. rufilábrum, (B. & C.)

Hysterium rufilabrum, B. & C. Grev. IV, p. 12.
Hypoderma rufilabrum, Sacc. Syll. 5796.

"Obtuse, elliptical, growing on a pallid spot; lips rufous. Asci clavate. Sporidia shortly fusiform, hyaline, 15 μ long. Sporidia very different from those of neighboring species."

On twigs of *Acer striatum*, South Carolina (Ravenel).

H. Eucalýpti, Cke. & Hark. Grev. XIII, p. 21.

Epiphyllous, gregarious. Perithecia lanceolate or linear, erumpent-superficial, black; lips rounded, loosely connivent. Asci clavate. Sporidia inordinate, elongated-fusoid, continuous, hyaline, 30 x 3 μ.

On leaves of *Eucalyptus*, California (Harkness).

H. scírpinum, DC. Fl. Fr. p. 166.

Hysterium scirpinum, Fr. S. M. II, p. 590.
Sphæria Leptostroma, Ehr. Sylv. Berol. p. 29.
Exsicc. Kze F. Sel. 277.—Rab. Herb. Mycol. II, p. 262.—Thum. M. U. 178.

Perithecia innate, oblong, straight, depressed, very black, about 2 x $\frac{3}{4}$ mm.; lips forming an elevated ridge, parallel, finally open. Asci fusoid-clavate, 120–140 x 15–18 μ, 8-spored, with abundant filiform paraphyses. Sporidia conglobate, fusoid-bacillate, straight or curved, yellowish-hyaline, 40–56 x 5–6 μ, Sacc; 36–40 x $4\frac{1}{2}$–5 μ, Rehm. Spermogonial perithecia (*Leptostroma scirpinum*, Fr.) intermixed. Perithecia orbicular, umbonate in the center, finally deciduous, 150 μ diam.

On *Scirpus validus*, Montezuma marshes, N. Y. (Peck).

Specc. from Peck have asci stipitate, 110–125 x 10–13 μ, and sporidia 30–35 x 4–$4\frac{1}{2}$ μ.

H. Desmazièrii, Duby, Hyst. p. 42, tab. II, fig. 22.

Amphigenous, scattered, innate, ovate, convex, longitudinally striate, very black and shining, covered by the thin epidermis to which it is closely adnate; lips closed so as to leave a scarcely discernible crevice between them. Asci broad-clavate, sessile, paraphysate, 8-spored. Sporidia inordinate, oblong or clavate-oblong, hyaline, continuous, 15–22 x 2½–3 μ, rather narrower and more acute below.

On leaves of *Pinus Strobus*, London, Canada (Dearness), on pine leaves, New York (Peck & Gerard).

H. conígenum, (Pers.)

Hysterium conigenum, Pers. Syn. p. 102.
Hypoderma conigenum, Cke. Hndbk. II, p 762.

Perithecia erumpent, small, shining, elliptical or elongated, thin; lips whitening out, broadly open. Asci clavate, rather long-stipitate, 90–100 x 12–14 μ, with filiform paraphyses curved above, 8-spored. Sporidia bacillary-fusoid, often curved, 20–22 x 2½–3 μ, pluriguttulate, hyaline.

Rather rare, on cones of pine trees, Bethlehem, Pa. (Schw.).

Species imperfectly known.

H. plantàrum, (Schw.)

Hysterium plantarum, Schw. Syn. Car. 255.
Hypoderma plantarum, Sacc. Syll. 5805.

Innate, linear, smooth, becoming black; lips thin, subflexuous, forming a narrow ridge along the apex of the perithecium.

On dead stems of *Monotropa*, North Carolina (Schw.).

Differs from *H. commune* in its elongated form; the long, narrow lips slanting upwards like the roof of a house. The specc. in Herb. Schw. are without fruit.

LOPHODÉRMIUM, Chev.

Flore de Paris, I, p. 436.

Perithecia mostly on pale spots on the stems or leaves, innate-erumpent, elongated, mostly simple, sometimes forked, opening by a longitudinal cleft along the middle. Asci clavate, mostly subacute at the apex, 8-spored, with the paraphyses mostly hooked or spirally coiled at the tips. Sporidia filiform, continuous, nucleate, hyaline, lying parallel, nearly as long as the asci.

L. hysterioides, (Pers.)

Xyloma hysterioides, Pers. Syn. p. 106.
Lophodermium xylomoides, Chev. Flor. Par. I, p. 437.
Hysterium foliicolum, Fr. S. M. II, p. 592.
Hypoderma xylomoides, DC. Fl. Franc. VI, p. 164.
Hysterium Berberidis, Schleich. Exsicc. No. 182.
Aporia microtheca, Duby, Hyst. p. 52.
Schizothyrium microthecum, Sacc. Syll. 5554 (fide Rehm).
Lophodermium hysterioides, Sacc. Syll. 5808.

Exsicc. Fckl. F. Rh. 742, 759.—M. & N. Stirp. Vog. 761.—Rab. Herb. Mycol. II, 156, id. F. E. 1151.—Rehm Asc. 867.—Sydow, M March. 856, 1838.

Perithecia scattered on roughish, pale spots, innate, convex-erumpent, elliptical, sometimes suborbicular, straight or sometimes a little bent, mostly simple, subshining-black, the sharp lips nearly closed, leaving only a slight, narrow crevice, about 1 x ½ mm. Asci clavate, obtusely pointed above, stipitate, 8-spored, 80–100 x 9–10 μ, with filiform paraphyses bent into a spiral above. Sporidia filiform, curved, continuous, hyaline, 60–70 x 1½ μ, lying parallel, but somewhat curved or twisted around the longer axis of the ascus.

On leaves of *Andromeda axillaris*, North Carolina (Schw.); found in Europe mostly on leaves of the *Rosaceæ*.

L. maculàre, (Fr.)

Hysterium maculare, Fr. S. M. II, p. 592.
Lophodermium maculare, De Not. Piren. Ister. p. 40.

Exsicc. Fckl. F. Rh. 1752.—Moug. & Nestl. Stirp. Vog. 1072.—Thum. M. U. 75.

Perithecia scattered on pale spots which are sometimes limited by a narrow, black line, adnate, mostly less than 1 mm. long by ⅓–½ mm. broad, elliptical, obtuse, simple, at first depressed above, then swollen, the opposite sides slanting up towards each other, and their upper, brownish margins separated by a straight, narrow longitudinal cleft extending nearly across. Asci clavate, obtusely pointed at the apex, stipitate, 40–50 x 5 μ, with filiform paraphyses more or less bent or hooked at the tips. Sporidia 8 in an ascus, filiform, somewhat twisted together, 30–40 x 1 μ.

On leaves of various species of *Vaccinium*, around Newfield, N. J., and probably common throughout the United States and Canada.

Var. *albolabrum*, E. & E., Exsicc. Ell. N. A. F. 859 (partly). Perithecia with white-margined lips, as in *L. melaleucum*, but with asci and sporidia as in the usual form. On leaves of some unknown shrub or tree, Utah (S. J. Harkness).

In some copies of N. A. F. two kinds of leaves were distributed under No. 859. The var. is on the longer, narrower leaves, on large, irregular shaped, bleached spots bounded by a narrow, black line.

The perithecia are rather smaller than in *L. melaleucum*, and though they have the white-margined lips of that species, the fructification is exactly that of *L. maculare*.

L. melaleùcum, (Fr.)

Hysterium melaleucum, Fr. Obs Mycol. I, p. 192, tab II, fig. 1.
Lophodermium melaleucum, De Not. Pir. Istr. p. 40.
Exsicc. Fckl. F. Rh. 736, Sacc. M. V. 1478.

Perithecia on pale or whitish, subdeterminate spots, innate, convex-erumpent, elliptical or suborbicular, obtuse, mostly simple, straight, black, 1 x ½–¾ mm., at first depressed, then carinate, at length opening by a linear crevice and exposing the white (sometimes yellow or greenish) margin of the lips. Asci cylindric-clavate, obtusely pointed at the apex, stipitate, 70–75 x 9 μ, with filiform paraphyses, 2 μ thick and subundulate-bent. Sporidia filiform, straight, continuous, hyaline, 50–55 x 2 μ, lying parallel, 8 in an ascus.

On living leaves of *Rhododendron Catawbiense*, Roan Mountain, North Carolina (Scribner).

This agrees so well with specc. and diagnosis of *L. melaleucum*, that we refer it here without hesitation. The lips when first opened have a yellow margin, and it cannot therefore be the typical form.

L. sphærioìdes, (A. & S.)

Hysterium sphærioides, A. & S. Conspect. Fung. Nisk. p. 57, tab. X, fig 3.
Lophodermium sphærioides, Duby, Hyst. p. 44.

Perithecia scattered on paler spots, innate, hemispherical, about ½ mm. diam., brownish-black, pierced above with a nearly round opening or with a narrow, slightly white-margined cleft extending nearly across the apex. Asci clavate, stipitate, 60–65 x 7–8 μ, with filiform paraphyses longer than the asci and mostly recurved or hooked at the tips. Sporidia 8 in an ascus, filiform, lying parallel, hyaline, about 35 x 1½ μ.

On dead leaves of *Ledum palustre*, Vermilion Lake, Minnesota (Holway), New York State (Peck).

The diagnosis is from the Minnesota specimens.

L. tùmidum, (Fr.)

Hysterium tumidum, Fr. S. M. II, p. 591
Coccomyces tumida, De Not. Giorn. Bot. Ital. II, p. 38
Lophodermium tumidum, Rehm, Die Pilze, III, p. 40.
Exsicc. Fckl. F. Rh. 746 —Rehm Asc. 519 —Romell, Fungi Scand. 84

Perithecia scattered on pale spots, innate, swollen-erumpent, elliptical, subacute at the ends, shining-black, the sharp, slightly open lips leaving only a narrow cleft between them, and partially exposing the brownish disk, ¾–1½ mm. long by 1 mm. wide (or a little less). Asci clavate, 70–75 x 9–10 μ, obtusely pointed at the apex, short-stipitate, with filiform paraphyses hooked at the tips. Sporidia filiform. straight, often a little swollen at the tips, continuous, hyaline, 45–50 x 1½–2 μ, lying parallel.

On dead leaves and petioles of *Amelanchier alnifolia*, Klikitat Co., Washington (Suksdorf, No. 201), on oak leaves, Bethlehem, Pa. (Schw.), on fallen beech leaves, New York (Peck).

The species is specifically characterized by its swollen perithecia with walls rising like an arch over the disk. It is found in Sweden and Germany on leaves and petioles of *Sorbus Aucuparia*. The sporidia (sec. Fries) are oval, which would place the species in *Hypoderma*. The Washington specc. have filiform sporidia agreeing with the spece. issued by Rehm and Romell.

L. exáridum, C. & P. in Cke. Disc. of the U. S. p. 36.

Hysterium exaridum, C. & P. in Pk. 29th Rep. p. 63.

Perithecia scattered on pale, irregular shaped spots, suborbicular or oblong, ½–1¼ mm. long, adnate, superficial, black, obtuse, minutely rugose; lips slanting upwards, nearly closed, not swollen, hypophyllous. Asci slender-clavate, 70–85 x 6–7 μ, 8-spored; subsessile, with filiform paraphyses more or less bent and thickened at the tips. Sporidia filiform, multinucleate, hyaline, slightly thickened above, 60–70 x 1¼–1½ μ.

On fallen leaves of *Kalmia angustifolia*, New York (Peck), on dead, dried up leaves of *Kalmia latifolia*, still hanging on limbs cut some months previously, at Newfield, N. J.

Diagnosis from the Newfield specimens.

L. petiolícolum, Fckl. Symb. p. 255.

Exsicc. Fckl. F. Rh. 745.—Kunze F. Sel. 276.—Rab. F. E. 462. 2642.—Thum. M. U. 1757.

Perithecia on pale spots, innate-erumpent, subprominent, finally collapsing, elliptical, subacute, straight, simple, shining-black, cleft narrow, finally partially opening, so as partly to expose the pale disk, 1–1½ x ½ mm.; lips not swollen. Asci clavate, obtusely pointed at the apex, 8-spored, 45–70 x 6–7 μ, with filiform paraphyses hooked at the tip. Sporidia filiform, mostly straight, continuous, hyaline, 45–50 x 1½ μ, lying parallel.

On petioles and midribs of fallen oak leaves, Newfield, N. J., on petioles of *Acer saccharinum*, Caroga, N. Y. (Peck).

L. Rhododéndri, (Schw.)

Hysterium Rhododendri, Schw. Syn. N. Am. 2116, not *Lophodermium Rhododendri*, Ces. in Erb. Critt. Ital. No. 537, Rehm in Die Pilze, Disc. p. 40.
Exsicc. Ell. N. A. F. 1287.

Epiphyllous, on round, pale spots 1–2 cm. diam., with a reddish, swollen margin both above and below, punctiform at first, then orbicular, subdiscoid, depressed, $\frac{1}{2}$–$\frac{3}{4}$ mm. diam., becoming elliptical, 1–1$\frac{1}{2}$ x $\frac{3}{4}$–1 mm., with the opposite sides slanting up towards each other and separated by a very narrow cleft. Asci clavate, 110–130 x 12–15 μ, 8-spored, with filiform paraphyses broadly recurved at the tips. Sporidia linear-cylindrical, nucleate, continuous, hyaline, 60–75 x 2–2$\frac{1}{2}$ μ.

On leaves of *Rhododendron*, Bethlehem, Pa. (Schw.), Mauch Chunk, Pa. (Martindale), on leaves of *Rhododendron Californicum*, Oregon (Carpenter), New York (Peck).

On the Oregon specimen is also another fungus agreeing accurately with *Phacidium dentatum*, Kze. & Sch., but which can not be conjured into anything like the *Phacidium Rhododendri*, Schw. The diagnosis of this last mentioned species does not differ essentially from that of *Hysterium Rhododendri*, Schw., both being said to be 2 lines long and sometimes, at least, trifariously dehiscent; and as Schweinitz himself says he fears the two species may not be distinct, there is some reason to suppose that he has given two hybrid diagnoses, each combining the characters of *Phacidium* and *Hysterium*. The diagnosis given above is from the specc. collected by Mr. Martindale. None of these now accessible show any perithecia 2 lines (4 mm.) long. There can be no doubt that these specc. are a *Lophodermium* and not a *Phacidium* (Coccomyces).

L. orbiculàre, (Ehrenb.)

Hysterium orbiculare, Ehrenb F. Champ. No. 30, t. xx, fig. 15.

Spots amphigenous, dull white above, with a narrow, reddish border, 2–4 mm. diam., becoming ferruginous below. Perithecia hypophyllous, innate-erumpent, small, $\frac{1}{4}$–$\frac{1}{2}$ mm., sphæroid, more or less flattened and depressed above, becoming hemispherical or sub-elongated with a distinct cleft across the summit, at length more or less open. Asci clavate, obtusely pointed above, 70–75 x 7–8 μ, 8-spored, with abundant paraphyses curved and thickened above. Sporidia clavate-cylindrical, yellowish-hyaline, multinucleate, 60–70 x 1$\frac{1}{2}$–2 μ.

On leaves of *Andromeda calyculata*, London, Canada (Dearness).

In the original diagnosis the fungus is said to be epiphyllous, but this may be in contradistinction to epicauline. The specc. from Dearness are also paraphysate. Without authentic specc for comparison, it can not be certainly decided that this is the veritable *Hysterium orbiculare*, Ehrenb., but the locality, habitat and general characters indicate that species. The perithecia are at first covered and almost hidden by the whitish, orbicular scales of the leaf.

L. Heterómelis, (Phil. & Hark.)

Hypoderma Heteromelis, Phil. & Hark Grev. XIII, p. 23.

Gregarious, innate-erumpent, oblong, straight or curved, flattened or slightly prominent, 400–800 μ long. Asci clavate-cylindrical, 75–80 x 5–6 μ, with filiform paraphyses. Sporidia filiform, 60–65 x $1\frac{1}{2}$ μ.

On the under side of leaves of *Heteromeles arbutifolia*, California (Harkness).

The perithecia in specc. from Harkness are covered by the thin epidermis, which gives them a dull appearance. The sporidia being nearly as long as the asci, will place this in *Lophodermium*.

L. Dracǽnæ, Phil. & Hark. Grev. XII, p. 84.

Scattered, elliptical or oblong-elliptical, 300–800 μ long, superficial, deciduous, glabrous; lips adherent, then open and gaping. Asci clavate-cylindrical, 100 x 5–6 μ, 8-spored, with filiform paraphyses bent or hooked at the tips. Sporidia filiform, 70–85 x 1–2 μ.

On *Dracæna*, California (Harkness).

We have supplemented the original diagnosis by the examination of specc. from Dr. Harkness, but can not make the paraphyses dichotomous at the tips.

L. junipérinum, (Fries).

Hysterium pinastri, b. juniperinum, Fr. S. M. II, p. 588.
Lophodermium juniperinum, De Not. Piren. Istr. p. 40.
Exsicc. Fckl. F. Rh. 735, 1753.—Rab. Herb. Mycol. 445.—Rehm Asc. 128.—Thum. F. Austr. 1268.—Thum. M. U. 76.—Kriegr. F. Sax. 382.—Linht. Fung. Hung. 153. Sydow, M. March. 2424—Roum. F. G. 1270.—Ell. N. A. F., 999.—Cke. F. Brit. Ser. I, 395.

Perithecia scattered, innate-erumpent, depressed above when young, finally more prominent, elliptical, obtuse, nearly black, $\frac{1}{2}$–1 mm. long by $\frac{1}{3}$–$\frac{1}{2}$ mm. wide, with a narrow cleft and lips scarcely swollen. Asci clavate, short-stipitate, 70–85 x 10–12 μ, 8-spored, with filiform paraphyses straight or a little curved above. Sporidia filiform, continuous, multinucleate, hyaline, nearly straight, 65–75 x $1\frac{1}{2}$–2 μ.

On dead leaves of *Juniperus communis*, Iowa (Holway), on dead leaves of *Cupressus thyoides*, Newfield, N. J.

L. arundinàceum, (Schrad.)

Hysterium arundinaceum, Schrad. Journ. Bot. II, p. 68, tab. 3, fig. 3.
Lophodermium arundinaceum, Chev. Flore Paris. I, p. 435.

Perithecia scattered or gregarious, on pale spots, innate-erumpent and prominent, lying in the direction of the axis of the stem, elongated or elliptical, straight, simple, obtuse or acute, brown or black, with sharp, sometimes white-margined lips, finally a little open so as to leave a narrow crevice through which the pale brown disk is partly visible, ½–2 mm. long, ¼–½ mm. wide. Asci clavate, subsessile, 70–80 x 8–10 μ, (75–100 x 9–12 μ, Rehm), with filiform paraphyses longer than the asci and mostly undulate above. Sporidia 8 in an ascus, filiform, straight or slightly bent, continuous, multinucleate, hyaline, 40–70 x 1½–2 μ. The spermogonia (*Leptostromella hysterioides*, Sacc.), have curved, subulate, continuous sporules 16–18 μ long.

Var. *vulgare*, Fckl. Symb. Mycol. p. 256. Exsicc. Fckl. F. Rh. 737.—Rab. F. E. 1613.—Sydow. M. March. 538.—Thum. M. U. 77. Perithecia scarcely prominent, brownish elliptical, with a very narrow cleft, 1½–2 mm. long, ½–¾ mm. wide. Sporidia 70–75 μ long.
On dead leaves, sheaths and culms of *Phragmites communis*. We have seen no American specimens.

Var. *culmigenum*, (Fr.) Fckl. Symb. p. 257. *Hysterium culmigenum*, Fr. S. M. II, p. 591, and *b. gramineum*, Fr. ibid.; *Lophodermium arundinaceum*, *b. culmigenum*, Fckl. Symb. p. 257. Exsicc. Fckl. F. Rh. 738, 740, 2557.—Kze. F. Sel. 161.—Rab. Herb. Mycol. 34.—Rab. F. E. 1226.—Rehm Asc. 271.—Sydow, M. March. 25, 855.—Ell. N. A. F. 465.—Desm. Pl. Crypt. Ed. I, 85. Perithecia strongly prominent, mostly elliptical, obtuse, black, opening with a decided cleft. ½–1½ mm. long, ¼–½ mm. broad. Sporidia 45–75 x 1½ μ, subacute.

On dead culms and sheaths of *Poa compressa*, *Phleum pratense* and *Andropogon Virginicus*, common around Newfield, N. J., on wheat straw, Ohio (Kellerman); probably not uncommon everywhere.

Var. *caricinum*, (Rob.); *Lophodermium caricinum*, Duby, Hyst. p. 47; *Aporia neglecta*, Duby, l. c. p. 51 (sec. Rehm). Exsicc. Fckl. F. Rh. 1751.—Rab. Herb. Mycol. II, 723.—Sacc. M. Ven. 1280. Perithecia elliptical, acute, ½–¾ mm. long, ¼–½ mm. wide, margin of lips finally yellowish, cleft rather narrow. Asci 60 x 6–8 μ (60–80 x 6–9 μ, Rehm). Sporidia 45–55 x 1¼ μ (50–60 x 1–1½ μ, Rehm).

On dead leaves of various species of *Carex*, Newfield, N. J.

L. týphinum, (Fr.)

Hysterium typhinum, Fr. S. M. II, p. 590.
Lophodermium typhinum, Lambotte, Flor. Belg. II, p. 452.

Perithecia innate, oblong, covered by the bullate-swollen epidermis, finally bare, black, 2 lines (4 mm.) and over, long; lips swollen, whitish within.

On dead leaves of *Typha latifolia*, Guilderland, N. Y. (Peck).

The above diagnosis is from Fries. We have not seen the New York specimens, but specc. from Plowright (England) have asci 60–75 x 6–8 μ. Sporidia filiform, 40–50 x 1–1½ μ. Perithecia elliptical, rather obtuse, 1–1½ mm. long by ½ mm. broad, depressed above when young, covered by the epidermis. The specc. are evidently not mature.

L. cladóphilum, (Lev.)

Hysterium cladophilum, Lev. in Moug. and Nest. Exsicc. 1243.
Hysterium Vaccinii, Carmich. Eng. Fl. V, p. 295.
Sporomega cladophila, Duby, Hyst. p. 48.
Lophodermium cladophilum, Rehm, Die Pilze, III, p. 42.
Exsicc. Desm. Pl. Crypt. Ed. II, 564.—Ell. N. A. F. 154.—Rab. F. E. 157.—Thum. F. Austr. 507.

Perithecia scattered and lying in various directions, elliptical, oblong or short-linear, black, prominent, small (½–1½ mm.); lips swollen, convex, leaving a narrow fissure between them, covered by the blackened epidermis, but when this falls away, becoming superficial. Asci cylindrical. Sporidia filiform, hyaline, not thickened at the apex, about as long as the asci.

On dead branches of *Vaccinium Pennsylvanicum*, Newfield, N. J.

L. nervisèquium, (DC.)

Hypoderma nervisequium, DC. Fl. Franc. VI, p. 167.
Lophodermium nervisequium, Rehm, Die Pilze, III, p. 44.
Exsicc. Fckl. F. Rh. 2559.—Rab. Herb. Mycol. II, 722.—id. F. E. 2144.—Thum. F. Austr. 262.—id. M. U. 463, 1073.—Linht. Fungi Hung. 65.

Hypophyllous, at first standing singly along the midrib, at length confluent, forming a continuous black line, erumpent, convex, black, marked above with a distinct, longitudinal cleft, finally open so as to expose the pale yellow disk, 1–1½ mm. long, ⅓–½ mm. broad. Asci clavate-oblong, 60–90 x 12–15 μ, sessile, obtusely pointed above, 8-spored, with slender, filiform, undulate paraphyses. Sporidia filiform, clavate-thickened and bent at the tips, which are about 2 μ thick.

On leaves of *Abies balsamea*, Caroga, N. Y. (Peck).

The measurements are from Linhart's specc. and are about the same as those given by Dr. Rehm.

L. lineàre, (Pk.)

Rhytisma lineare, Pk. 25th Rep. p. 100, pl. 1, figs. 24–26.
Hypoderma lineare, Thum. in Diag. M. U. Cent. X–XII, p. 12, and Sacc. Syll. 5788.
Exsicc. Thum. M. U. 1073.

Hypophyllous, linear, here and there interrupted or constricted, forming a black strip along the midrib, often extending the entire length of the leaf; lips thin, closed at first, then open, exposing the pallid disk. Asci ventricose-clavate, 100–110 x 35–40 μ, aparaphysate, 8-spored. Sporidia lying in irregular order, 50–70 μ long, didymous, consisting of two oblong, granular-hyaline parts 25–30 x 4–5 μ joined end to end by a slender neck.

On leaves of *Pinus Strobus*, Guilderland, Greenbush and Sand-lake, N. Y. (Peck).

The measurements of asci and sporidia are from the spec. in Thüm. Mycotheca.

L. pinástri, (Schrad.)

Hysterium pinastri, Schrad. Journ. Bot. II, p. 69, tab. 3, fig. 4.
Aporia obscura, Duby, Hyst. p. 51.
Lophodermium pinastri, Chev. Flore Paris. I, p. 430.
Exsicc. Fckl. F. Rh. 734 —Kze. F. Sel. 371.—Rab. F. E. 1443, 1922, 2022.—Rehm Asc. 127. Sydow, M. March. 93 —Thum. F. Austr. 505, 871, 1059.—Thum. M. U. 292.

Perithecia scattered on pale spots mostly limited by a black line, innate-erumpent, oblong-elliptical, subobtuse, simple, finally shining-black, $\frac{3}{4}$–$2\frac{1}{2}$ x $\frac{1}{2}$–1 mm., with a narrow cleft, the lips (sometimes yellowish) slightly separated, revealing the pale disk. Asci clavate, obtusely pointed above, 8-spored, 90–150 x 10–14 μ, with filiform, nearly straight paraphyses 2–$2\frac{1}{2}$ μ thick and hyaline. Sporidia straight or a little bent, continuous, multinucleate, hyaline, 75–120 x $1\frac{1}{2}$–2 μ. Spermogonium (*Leptostroma pinastri*, Desm.) with hyaline, cylindrical sporules 6–8 x $\frac{1}{2}$–1 μ.

On the upper side of leaves of various species of pine trees, common everywhere.

CLÍTHRIS, Fr.

Syst. Mycol. II, p. 186.

Perithecia soft-coriaceous, elliptical, flexuous or sublinear, becoming black, subcutaneous, then erumpent, dehiscing with a longitudinal fissure, soon broadly open, but for a long time covered by the epidermis. Asci elongated, paraphysate, 8-spored. Sporidia fasciculate, filiform, hyaline, about as long as the asci, continuous at first, finally multiseptate. Differs from *Lophodermium* in its more widely dehiscent perithecia which are also of a softer texture.

As there seem to be no decided characters separating *Colpoma*, Wallr., and *Sporomega*, Corda, we have followed Dr. Rehm (Die Pilze, III, p. 101) in merging these two genera in *Clithris*, Fr., which has precedence, and with which *Colpoma*, at least, is synonymous, but have retained the genus in the *Hypodermieæ*, between which and the *Discomycetes*, its affinities are about equally divided.

* *Perithecia gray-pruinose outside* (*Colpoma*).

Cl. quércina, (Pers.)

Hysterium quercinum, Pers. Syn. p. 100.
Cenangium quercinum, Fr. S. M. II, p. 189
Triblidium quercinum, Pers. Mycol. Eur. I, p. 333.
Colpoma quercinum, Wallr. Fl. Crypt. Germ. II, p. 423.
Hysterium nigrum, Tode, Fungi Meckl. II, p. 5. tab. III, fig. 64.
Sphæria collapsa, Sow. Eng. Fungi, tab. 373. fig. 3.
Variolaria corrugata, Bull. Champ. p. 117, tab. 432, fig. 4.
Exsicc. Desm. Pl. Crypt. Ed. I, 383.—Thum. M. U. 369.—Rehm Asc. 27.—Kriegr. F. Sax. 184.—Sydow, M. March. 344.—Vize, Micr. Fungi, 269.

Perithecia ovate-oblong, transverse, convex or semicylindrical from a flattened base, dark brown, opake, variable in size, at first subcuticular and closed, then rupturing the epidermis and disclosing the elongated, pale, boat-shaped disk, soon becoming friable and falling out. Asci clavate, long-stipitate, apex acute, 135 x 8–10 μ, 8-spored, with filiform paraphyses. Sporidia filiform, fasciculate, equal, 90 x $1\frac{1}{2}$ μ, at length multicellular, hyaline. Spermogonia orbicular, covered, 1-celled. Spermatia cylindrical, curved, 8 x $1\frac{1}{2}$ μ.

On branches of oak, South Carolina (Curtis).

Cl. láctea, (C. & P.)

Colpoma lacteum, C. & P. 28th Rep. N. Y. State Mus. p. 69.

Perithecia scattered, erumpent, thin, black, the longitudinally ruptured epidermis closely appressed, disk plane, milk-white. Asci cylindrical or clavate. Sporidia filiform, 20–30 μ long.

On dead stems of *Ledum latifolium*, Sandlake, N. Y. (Peck).

When moist, the perithecium gaps widely, revealing the conspicuous white disk. This and the different habit distinguish the species from *Cl. Ledi*, (A. & S.).

Cl. Azàleæ, (Schw.)

Hysterium Azaleæ, Schw. Syn. N. Am. 2089.
Colpoma Azaleæ, Cke. Disc. U. S. p. 36.

Perithecia at first (nearly) covered by the epidermis which is raised into oblong swellings over them, and soon cleft with a narrow crack which is never more than partially open, mostly lying parallel and often seriately confluent. Perithecia 2–4 mm. long, sometimes (sec. Schw. ½ an inch, 12 mm.). From the swollen bark are formed pseudo-labia, covering the true lips which are of a reddish-brown color and open, so as to expose the rather broad disk, but this is hardly ever denuded till the fungus decays. When the epidermis is finally thrown off the flexuous perithecia, with swollen margin, are seen to be erumpent from the wood itself. Asci and sporidia as in *S. Andromedæ*, from which this is very doubtfully distinct.

Cl. Juníperi, (Karst.)

Coccomyces Juniperi, Karst. Mycol. Fenn. I, p. 254 (1871).
Colpoma juniperinum, C. & P. Bull Buff. Soc. Sept. 1875, p. 36.
Clithris Juniperi, Rehm, Die Pilze, III, p. 102.
Hysterium Petersii, B. & C. Grev. IV, p. 13, sec. Cke. Grev. XVII, p. 58
Exsicc. Rehm Asc. 272.

Gregarious, oblong, elliptical, or slightly elongated, covered by the cuticle which is ultimately fissured in an irregular manner, blackish, disk pallid, at length exposed, at first white-pulverulent outside, membranaceous, 1–3 mm. long. Asci clavate, obtusely pointed at the apex, 90–100 x 9–10 μ, with filiform paraphyses about 2 μ thick and spirally bent above. Sporidia filiform-fusoid, straight, continuous, guttulate, hyaline, 45–50 x 1½–2 μ, lying parallel.

On juniper branches, New York (Peck).

** *Perithecia dark brown, not pruinose; sporidia joined at base, (Sporomega).*

Cl. degénerans, (Fries).

Hysterium degenerans, Fr. S. M. II, p. 585.
Clithris degenerans, Rehm, Die Pilze, III, p. 104.
Sporomega degenerans, Duby, Hyst. p. 48.
Exsicc. Fr. Scl. Suec. 40.—Moug. & Nestl. 762.—Desm. Pl. Crypt. Ed II, 182.

Erumpent, gregarious, round or elongated, or variously shaped, black outside, disk open, dilated, soft, livid when fresh, flesh-color when dry, margin thin, at first adnate to the epidermis, finally separating, and erect. Asci elongated-clavate, subacute above, paraphysate, 150 x 12 μ. Sporidia filiform, acute at each end, 55–80 x 2½ μ, lying parallel in the asci.

On dry, decaying branches of *Vaccinium* and *Andromeda*, Newfield, N. J.

Cl. mórbida, (Pk.)

Tryblidium morbidum, Pk. 31st Rep. p. 48.

Perithecia seated on a thin, black crust, irregular in shape, elliptical, oblong or orbicular, rugose, black, closed at first, at length gaping widely and exposing the dingy-white or yellowish disk. Asci elongated-clavate, 100–112 x 10–12 μ, with filiform paraphyses. Sporidia filiform, nearly as long as the asci, multinucleate, becoming multiseptate, about 1½ μ thick, hyaline.

On dead wood of *Abies nigra*, Sandlake, N. Y. (Peck), on decaying wood of white cedar (*Cupressus thyoides*), Newfield, N. J.

Differs from *Cl. degenerans* in the black, crustose subiculum and sporidia thickened at the apex.

Cl. Émpetri, (Rostr.)

Sporomega Empetri, Rostr. Fungi Groenl. p. 543.

Perithecia epiphyllous, brown-black, elongated, straight or flexuous, rather thick, opening with a longitudinal cleft; lips finally remote. Asci clavate, attenuated towards the apex, 80–90 x 18 μ, 8-spored. Sporidia filiform, hyaline, simple, joined at the base, 60–64 x 2 μ.

On dry leaves of *Empetrum nigrum*, Egedesminde, Greenland.

Cl. grìsea, (Schw.)

Hysterium griseum, Schw. Syn. N. Am. 2097.
Sporomega grisea, Cke. Disc. U. S. p. 36.
Exsicc. Ell. & Evrht. N. A. F. 2d Ser. 2377.

Perithecia at first entirely covered, but visible through the transparent epidermis, grayish-black, elliptical, flat, obtuse, about 1 mm. long, at length opening with a longitudinal fissure. Asci clavate-cylindrical, 65–75 x 6–7 μ, with filiform paraphyses longer than the asci and branched above. Sporidia 8 in an ascus, filiform, 30–35 x 1 μ, multinucleate.

Common on dead stems of *Smilax*, Pennsylvania and New Jersey.

Cl. Andrómedæ, (Schw.)

Hysterium Andromedæ, Schw. Syn. N. Am. 2090.
Sporomega Andromedæ, Duby, Hyst. p. 48, tab. II, fig. 24.
Exsicc. Ell. N. A. F. 155.

Perithecia lanceolate, ovate or suborbicular, often elongated to ½ cm. in length, lying in various directions on the matrix, at first covered by the epidermis which is raised into elongated swellings, as if some larva had burrowed beneath, soon splitting in a narrow crack along the apex of the perithecium, and finally partially opening so as to expose the waxy-white disk which at length becomes black. Asci narrow-clavate, 75–85 x 8–10 μ, 8-spored, with paraphyses slightly thickened and often branched at the tips. Sporidia filiform, nearly as long as the asci.

On dead stems and branches of *Azalea viscosa* and *Andromeda racemosa*, Newfield, N. J., Bethlehem, Pa. (Schw.), New York State (Peck).

The perithecia are generally circumscribed by a black line penetrating the wood. According to Schweinitz, the perithecia arise from the surface of the inner bark and never from the wood.

Cl. Vaccínii, (Schw.)

Hysterium Vaccinii, Schw. Syn. N. Am. 2088.

Erumpent, large, elongated, ovate, brownish-black, much larger than *Sporomega Ledi*, to which it is allied. Lips thin; disk rufescent, erumpent and surrounded by the bark.

On old branches of *Vaccinium frondosum*, Bethlehem, Pa. (Schw).

The specimen in Herb. Schw. apparently belongs here, but affords no fruit.

FAMILY. DICHÆNÀCEÆ.

Perithecia round or elongated, simple, covered, raising the epidermis into pustules, finally erumpent, membranaceous or coriaceo-membranaceous, black, opening by a cleft across the apex.

DICHÆ̀NA, Fr.

Summa Veg. Scand. p. 403.

Perithecia orbicular or elongated, simple, innate-erumpent, coriaceo-membranaceous, opening by a cleft across the apex, mostly in densely crowded patches. Asci saccate or subelongated, 8-spored. Sporidia oblong or oblong-elliptical, 2–3-septate, subhyaline, mostly biogenous. The place of this genus in the mycological system is uncertain. It was formerly classed with the lichens. It differs from the *Hysteriaceæ* in its membranaceous and at first buried perithecia. The ascigerous state is seldom met with, but the pycnidial stage (*Psilospora*, Rab.) is common.

D. quércina, (Pers.)

Opegrapha quercina, Pers. in Ann. Bot. VII, p. 31, tab. 3, fig. 4.
Schizoderma quercinum, Chev. Flor. Paris, p. 438, tab. 11, fig. 21.
Opegrapha macularis, Ach. Lichen. Univ. p. 247.
Dichæna quercina, Fr. Elench. II, p. 142.
Exsicc. Fckl. F. Rh. 1966.—M. & N. Stirp. Vog. 265.—Sydow, M. March. 384.—Rav. F. Am. 71, 640.—Roum. F. G. 995.—Ell. N. A. F. 793.

Perithecia erumpent, on round or transversely elongated, black, crustaceous spots ½–1 cm. or more across, subglobose or conic-globose, rough, brownish-black, ¾–1 mm. diam., or subelongated, membranaceous, opening at the apex with a nearly round or subelongated, rather large opening, finally deciduous. Asci clavate-cylindrical, short-stipitate, paraphysate, 8-spored, 80–90 x 20 μ. Sporidia biseriate, clavate-oblong, yellowish-hyaline, uniseptate, becoming 3-septate (pseudo-septate), slightly constricted at the middle septum, 20–24 x 7–8 μ.

On living branches of *Quercus alba*, *Q. coccinea*, &c., common.

The diagnosis is from the specimens distributed in N. A. F. The asci and sporidia differ considerably from Dr. Rehm's figure in Die Pilze; the former, especially, being longer and narrower. The sporidia remain a long time with only one septum.

D. faginea, (Pers.)

Opegrapha faginea, Pers. in Annal. Bot. VII, p. 32.
Hysterium fagineum, Rab. Pilze, p. 155.
Schizoderma fagineum, Chev. Flor. Par. p. 438.
Opegrapha epiphega, Ach. Meth. Lich. p. 24.
Hysterium rugosum, Fr. Summa Veg. Scand. p. 402.
Dichæna rugosa, Rab. Pilze, p. 472.
Dichæna faginea, Fr. Elench. II, p. 141.

Exsicc. Fckl. F. Rh. 1569.—Rab. Herb. Mycol, 450.—Rav. Car. II, 66.—Rav. F. Am. 335. Sydow, M. March. 486.—Ell. & Evrht. N. A. F. 2d ser. 2067.

Perithecia as in *D. quercina*, but more elongated and hysterii-form, opening above by a more distinctly elongated cleft, erumpent, as in that species, on sharply defined, black, crustaceous spots which are usually transversely elongated, 5–6 x 1 cm., or often longer. The inner surface of the perithecia is lined with stout basidia 15–25 μ long, bearing at their tips elliptical, hyaline sporules 12–15 x 7–8 μ, with granular contents.

Common on trunks of living beech trees.

D. strumòsa, Fr. Nov. Symb. p. 132.

Exsicc. Rav. Fungi Car. II, 67.

Perithecia about as in *D. faginea*, crowded on orbicular, raised spots about 1 cm. diam. The hymenial cavity is lined with stout sporophores 15–25 μ long, bearing, as in the preceding species, terminal, elliptical, hyaline sporules 20–25 x 12–15 μ, with granular contents.

Common on limbs and trunks of living *Quercus coccinea* and *Q. nigra*, Carolina (Ravenel), Mexico (Liebman), New Jersey (Ellis).

This species is very common around Newfield, N. J., on *Quercus coccinea*, and very injurious, finally killing the trees on which it grows. The round, black spots on which the perithecia are seated are at first only slightly raised above the bark, but each succeeding year they increase in circumference and rise higher, soon forming globose, knob-like swellings, at first bulging out on one side of the limb, but finally surrounding it like a broad, convex, thick ring blackened and roughened by the broken stromatic crust and the abundant perithecia. Often a limb 6–10 ft. long will have half a dozen or more of these swellings scattered along at intervals and varying in size from 2–8 inches in diameter. Sometimes they appear on the trunk of a tree, forming swellings 6–12 inches thick, or even larger, when, as often happens, they surround the trunk. We have never found ascigerous specimens of this or of *D. faginea.*

D. cæspitòsa, Schw. Syn. N. Am. 1828.

Perithecia cespitose-erumpent through an innate veil, generally four together, rounded-subcompressed, sooty-black, at length brownish-pulverulent, sometimes regularly elongated at the apex, or dehiscing by a short cleft. Perithecia surrounded by the subcinereous, ruptured epidermis, black, elevated and collected in groups of considerable extent.

On beech bark, New England (Torrey).

The diagnosis is from Schw. Synopsis, and is all we know of this species.

ADDITIONS AND CORRECTIONS.

Page 11.—Add to habitats of **Erysìphe commùnis**—

Lupinus argenteus, Pœonia, Draba hirta, Vicia Americana, Astragalus adsurgens, and *A. hypoglottis.*

Page 31.—Add to synonyms of **Dimerospòrium Collínsii**—

Plowrightia phyllogena, Hark. Fungi of Pac. Coast, p. 106.

Page 35.—After **Dimerospòrium anómalum,** insert—

D. balsamícolum, (Pk.)

Meliola balsamicola, Pk. 34th Rep. p. 52, plate 1, figs. 22–27.

Perithecia few, gregarious, minute, ovate or subconical, free, black, seated on a small, blackish-brown, spot-like subiculum. Asci generally oblong, rarely subcylindrical and elongated. Sporidia mostly crowded or biseriate, rarely uniseriate, uniseptate, colorless, 9–11 μ long, generally 2–3-nucleate, and one cell a little narrower than the other.

On living or languishing leaves of balsam fir, Catskill Mts., N. Y. (Peck). Associated with *Peziza balsamicola*, Pk.

Page 68.—After **Acrospérmum Ravenélii,** insert—

A. álbum, Pk. Bull. N. Y. State Mus. No. 2 (1887), p. 24.

Perithecia elongated, subfusiform, somewhat compressed, pointed at the apex, narrowed below into a short, terete, stem-like base, white. Sporidia very long, filiform.

On dead stems of *Aralia racemosa*, Catskill Mts., N. Y. (Peck). Resembles *A. compressum* in size, but persistently white.

Page 83.—Add to **Hypócrea melaleùca** the habitat—

On a decaying oak limb, Newfield, N. J.

Page 86.—After **Hypócrea consimilis,** add the habitat—

On *Eucalyptus*, California (Harkness).

Page 122.—After **Melanóspora chrysomálla,** insert—

CLEISTOSÒMA, Hark.

Bull. Cal. Acad. Feb. 1884, p. 41.

Perithecia orbicular, membranaceous. Asci borne on branching threads, globose, evanescent. Sporidia hemispherical, echinulate.

This genus (sec. Sacc. Syll. IX, p. 943) is probably not distinct from *Inzengæa*, Borzi.

Cl. purpùreum, Hark. l. c.

Perithecia purple-black, very delicate, soon dehiscent, developed within the heaps of *Thecospora bifida* (its conidial stage), which it stains purple. Asci globular, hyaline, 8-spored, 9–12 μ diam. Sporidia hemispherical, purple, long-echinulate around the margin, 3–4 μ.

On rotting leaves of *Eucalyptus*, San Francisco, Cal. (Harkness).

Page 186.—After **Melanómma sporádicum**, insert—

M. Verrucària, (Fr.)

Sphæria Verrucaria, Fr S. M. II, p 496.
Melanomma Verrucaria, Sacc Syll. 3255.

Minute, scattered or emerging in dense groups from the epidermis or cracks in the bark, or on the truncate ends of limbs. Perithecia without any manifest stroma, sphæroid, obtuse, astomous, or marked at the apex with a minute, impressed point, dark brown, rugulose and subpulverulent, rather thick-walled, carbonaceous, fragile, opake. Asci terete, elongated, 8-spored. Sporidia oblong, 4-celled, somewhat constricted at the septa, pale fuliginous, diaphanous, 15 μ long or a little less. On bark of *Betula*, Bethlehem, Pa. (Schw.).

Page 189.—After **Zígnoélla subvestìta**, insert—

Z. pállida, (Ell.)

Lophiostoma pallidum, Ell. Bull. Torr. Bot. Club, X, p. 52.

Perithecia scattered or gregarious, often two or three standing close together, but not confluent, erumpent-superficial, subglobose, black, rough, about ⅓ mm. diam., collapsing above. Ostiolum papilliform, minute. Asci subcylindrical, sessile, 75–80 x 12–15 μ, with rather scanty, filiform paraphyses. Sporidia biseriate, clavate-oblong, yellowish-hyaline, about 7-septate, slightly curved, obtuse, 18–22 x 5–7 μ.

On weather-beaten wood of "Service bush," Utah (S. J. Harkness).

On account of the minute, papilliform ostiolum, this can hardly be a *Lophiostoma*.

Page 234.—After **Lophióstoma præmórsum**, insert—

L. nùcula, (Fr.)

Sphæria nucula, Fr. S. M. II, p. 466.
Lophiotrema nucula, Sacc. Mich. I, p. 338.
Lophiostoma nucula, Ces. & De Not. Schema, p. 46.

Perithecia scattered or subgregarious, innate-superficial, ovoid, smooth, at first with a short, cylindrical or compressed ostiolum, black, 300–500 μ diam. Asci cylindric-clavate, 90–125 x 10–12 μ, 8-spored. Sporidia biseriate, elongated or oblong, 3-septate, constricted in the middle, greenish-hyaline, 20–26 x 5–8 μ (sometimes 35 μ long).

On branches, Bethlehem, Pa. (Schw.).

The ostiolum is finally deciduous; the perithecium is then rather broadly perforated.

Page 239.—After **Cuccurbitària Fráxini**, add—

Var. *effusa*, on decorticated ash limbs, London, Canada (Dearness), has the perithecia densely gregarious and effused in patches more or less crowded for 1 cm. or more in extent. Asci and sporidia as in the typical form.

Page 243.—After **Cucurbitària Labúrni**, insert—

C. Ravenèlii, Ck. & Massee, Grev. XVI, p. 25.

Perithecia subcutaneous, erumpent, cespitose, black, subglobose, papillate, seated on a pulvinate stroma. Asci cylindrical, 8-spored. Sporidia lanceolate, 3–5-septate, cells divided by longitudinal septa, olivaceous, 50 x 15–18 μ.

On *Ailanthus glandulosa*, Aiken, South Carolina (Ravenel).

Evidently different from *C. Ailanthi*, Rabh. Sacc. Syll. 3958.

Page 276.—After **Sphærélla Gaulthèriæ**, insert—

S. stemmàtea. (Fr.)

Sphæria stemmatea, Fr. S. M. II, p. 528.
Depazea stemmatea, Fr. Summa Veg. Sc. p. 422.
Septoria stemmatea, Sacc. Syll. III, p. 493.
(*Sphærella brachytheca*, Cke. Grev. VII, p. 88)?

The spec. in Herb. Schw. labeled *Sphæria stemmatea*, Fr., "from Bartrams," is apparently a *Phyllosticta* or a *Septoria*, but without fruit; it is, however, different from a spec. of *S. stemmatea*, Fr., in the same collection from Fries, in having larger perithecia. The Friesian spec. shows the asci very distinctly, but the sporidia are immature and not well defined. The general appearance of this spec. is exactly that of *Sphærella Gaultheriæ*, C. & E. Grev. VII, p. 42. Probably *S. Gaultheriæ* and *S. brachytheca* are both synonyms of *S. stemmatea*, Fr., but this can not be definitely stated without further observations.

Page 280.—After **Sphærélla Pinsàpo**, insert—

S. Andersòni, E. & E.

Sphærella congena, E. & E. Proc. Acad. Nat. Sci. Phil. July, 1890, p. 230.

Perithecia gregarious on the back of the exposed tip of the scale, minute (74 μ), buried, except the black, smooth, conic-papilliform apex. Asci narrow clavate-cylindrical, gradually attenuated below, 75–80 x 5 μ, paraphyses none. Sporidia uniseriate, ovate, uniseptate and constricted at the septum, hyaline, 6–7 x 3–3½ μ.

On scales of dead cones of *Abies Douglasii*, Belt Mts., Montana, Sept., 1889 (F. W. Anderson, 612).

This is near *S. Pinsapo*, Thüm., but differs in its habitat, its longer, narrower asci and smaller sporidia not constricted at the septum. *S. conigena*, Pk., has broader asci and crowded, longer (10–12 μ) sporidia. We have changed the specific name, *Sphærella conigena* having precedence.

Page 285.—Add to habitats of **Sphærélla Stellarineàrum**—

Stellaria longipes, S. longifolia, S. crassifolia, Arenaria pungens, Cerastium arvense and *C. nutans*, Montana (Anderson).

Page 292.—After **Sphærélla Lactùcæ**, E. & K., insert—

S. melæ̀na, (Fr.)

Sphæria melæna, Fr. S. M. II, p. 431.
Sphærella melæna, Sacc. Syll. 1986, Cke. Syn. 5569.

Perithecia black, densely crowded, and connate, pierced above, 66–80 μ diam. Asci obovate, sessile, 8-spored, 27 x 14 μ. Sporidia 2–3-seriate, crowded, obovate-oblong, rounded at the ends, uniseptate below the middle, not constricted, pale yellow or subhyaline.

Common on herbaceous stems, Pennsylvania (Schw.).

Page 292.—After **Sphærélla sabalígena**, insert—

S. allícina, (Fr.)

Sphæria allicina, Fr. S. M. II, p 437.
Sphærella allicina, Awd. Mycol. Eur Pyren p. 19, fig 69.

Perithecia amphigenous, covered by the gray epidermis, densely gregarious, sometimes confluent, globose, perforated, black, 80 μ diam. Asci slightly narrowed above from a broad base, sessile, 8-spored, 55–58 x 14 μ. Sporidia biseriate, oblong, rounded at the ends, nearly straight, uniseptate, not constricted, hyaline, 16 x 4–5 μ.

On leaves of *Allium schœnoprasum*, Nazareth, Pa. (Schw.).

Page 302.—After **Lizònia Sphàgni**, insert—

L. Thalíctri, Rostr. Fungi Grönl. p. 556.

Perithecia ovoid, coriaceo-membranaceous, with a conoid papilla, collapsing when dry. Asci thick-clavate, very thick-walled at the apex, 110 x 45 μ, with a very short pedicel, 8-spored. Sporidia 2-3-seriate elongated-fusoid, uniseptate, 4-guttulate, 45–50 x 10–13 μ, with a hyaline, gelatinous envelope.

On dry stems of *Thalictrum alpinum*, Umanak-Fiord, Greenland.

Page 305.—Add to habitats of **Physalóspora megástoma**—

Astragalus adsurgens and *A. hypoglottis*, Montana (Anderson).

Page 317.—After **Didymélla Raùii,** insert—

D. Smìlacis, E. & E. (in Herb.)

Perithecia gregarious, globose, about ⅓ mm. diam., closely covered by the cuticle which is blackened over them and raised into little pustules barely pierced by the papilliform ostiolum. Asci clavate-cylindrical, 35–40 x 5–6 μ, paraphysate. Sporidia biseriate above, uniseriate below, ovate-oblong, hyaline, uniseptate and slightly constricted at the septum, 8–10 x 3 μ.

On dead stems of *Smilax*, Newfield, N. J.

Differs from *Didymosphæria polysticta*, (B. & C.), in its hyaline sporidia, and from *Physalospora disrupta*, (B. & C.), in its uniseptate sporidia and different asci. The habit is exactly that of *Anthostomella sepelibilis*, (B. & C.). *Sphærella smilacina*, E. & E., grows on stems not as much decayed, and has smaller perithecia and asci without paraphyses.

Page 334.—After **Didymosphæria adélphica,** insert the three following new species—

D. Manitobiénsis, E. & E. (in Herb.)

Spots orbicular, brown, 2–3 mm. diam., surrounded by a light-colored border, fainter and purplish below. Perithecia clustered in the center of the spots, mostly 3–8 together, erumpent-superficial, subhemispherical, black, rough, 150–200 μ diam., the apex and papilliform ostiolum smoother and shining. Asci clavate-cylindrical, subsessile, 8-spored, paraphysate, 65–75 x 8–10 μ. Sporidia subbiseriate, obovate-oblong, uniseptate, the septum nearer the narrower end, 12–15 x 4–5 μ, brown.

On raspberry leaves (*Rubus*), banks of the little Saskatchawan river, Manitoba, Oct. 3, 1891 (Dearness).

D. Arundinàriæ, E. & E. (in Herb.)

Perithecia gregarious, depressed-globose or subhemispherical, about $\frac{1}{3}$ mm. diam., white inside, covered by the epidermis which is blackened directly over them and raised into little pustules which are perforated at the apex by the papilliform ostiolum. Asci clavate, p. sp. 50–70 x 12–15 μ, short-stipitate, 8-spored, with abundant paraphyses. Sporidia biseriate, fusoid-oblong, ends slightly curved while lying in the asci, 4-nucleate, becoming obtuse and uniseptate near the middle, yellowish-brown, 16–25 x 5–6 μ.

On dead canes of *Arundinaria*, Louisiana (Langlois, 2338).

Comes near *Metasphæria sabalensis*, (Cke.), but sporidia smaller and constantly only 1-septate. *D. eumorpha*, (B. & C.), has the perithecia seriate ("linear").

D. euryásca, Ell. & Galw. Journ. Mycol. V, p. 67.

Perithecia scattered, suberumpent, minute (80–100 μ), perforated above. Asci inequilaterally ovate, sessile, 35–40 x 12–15 μ, paraphysate? Sporidia 2–3-seriate, ovate-oblong, uniseptate, constricted at the septum, rounded at the ends, brown, 12–15 x $3\frac{1}{2}$–5 μ. The perithecia remain partly covered by the epidermis.

On dead leaves of *Pinus Murrayana*, Mt. Helena, Montana (Anderson).

Page 337.—**Pleóspora láxa,** Ell. & Galw., is on *Carex straminea* and *C. stipata*, and not on grasses.

Page 350.—After **Pyrenóphora chrysóspora**, insert—

Pyr. pellìta, (Fr.)

Sphæria pellita, Fr. S. M. p. 503
Pleospora pellita, Rab. Herb. Mycol. Ed. II, 749.
Pyrenophora pellita, Sacc. Syll. 3846.
Exsicc. Fckl. F. Rh. 2315.—Rab. F. E. 1447.

Perithecia gregarious, rounded-conical, black, clothed with slender, dark brown hairs, or bare above, 300 μ diam., with a fringe of hyphæ around the base. Asci long-clavate, gradually narrowed into the stipe, 8-spored, 100–120 x 10–12 μ. Sporidia obliquely uniseriate, oblong, attenuated at each end so as to become broad-fusoid, 3-septate, constricted at the middle septum, 17–21 x 9 μ, the two middle cells, or sometimes only one of them, divided by a longitudinal septum. *Brachycladium penicillatum*, Cda., is the conidial stage.

On thistle stems, Bethlehem, Pa. (Schw.).

Pyr. trichóstoma, (Fr.)

Sphæria trichostoma, Fr. S. M. II, p. 504.
Pyrenophora trichostoma, Fckl. Symb. p. 215.
Exsicc. Fckl. F. Rh. 904.—Rab. Herb. Mycol. 535.—Rab. F. E. 1868.—Rehm Asc. 180, 592.
Kze. F. Sel. 265.—Sydow, M. March. 98.—Plowr. Sph. Brit. 287.

Perithecia scattered or loosely gregarious, at first sunk in the matrix, at length emergent and even superficial, globose with a flat base, clothed with stiff, black bristles standing out on all sides, but especially around the conical ostiolum, black and tolerably large. Asci oblong-clavate, substipitate, 8-spored, 200–230 x 44–52 μ. Sporidia subbiseriate, oblong, attenuated at the ends and rounded, 3-septate, the second cell often a little broader, one or both the middle cells generally divided by a longitudinal septum, yellow, slightly constricted at the septa, 44–50 x 17–20 μ.

On rye straw, Bethlehem, Pa. (Schw.).

Pyr. Penicíllus, (Schmidt).

Sphæria Penicillus, Schmidt, in Fr. S. M. II, p. 508.
Pleospora Penicillus, Fckl. Symb. Nachtr. II, p 23.
Pyrenophora Penicillus, Sacc. Syll. 3852.
Exsicc. Fckl. F. Rh. 2522.

Perithecia scattered or subgregarious, depressed-globose, covered by the epidermis, surrounded at base by brown, undulate, creeping hyphæ, clothed above with stiff, black bristles which, on the ostiolum, are collected in an acute or spreading, brush-like tuft, membranaceous, dark brown, about 200 μ diam. Asci cylindrical, short-stipitate, 8-spored, 60–80 x 12–14 μ. Sporidia obliquely uniseriate, oblong, ends rounded, constricted in the middle, 5-septate, with one (mostly partial) longitudinal septum, deep yellow-brown, 15–17 x 8 μ.

On dead stems of *Humulus*, Pennsylvania (Schw.).

The diagnosis of this and the two preceding species are from Winter's Pilze.

Page 354.—

Leptosphæria taxícola, Pk.

In Pk. 39th Rep. p. 58, the sporidia are said to be hyaline, 20–22 x 4–5 μ, and the species is referred to *Metasphæria*.

Page 355.—To habitat of **Leptosphæria Thalictri,** add—

On *Thalictrum polygamum*, London, Canada (Dearness).

Page 361.—After **Leptosphæria Utahénsis,** insert—

L. nigrélla, (Rabh.)

Cucurbitaria nigrella, Rabh. Hedw. 1873, p. 140, and 1887, p. 59.

Perithecia innate-superficial, subglobose, constantly seated on broad, black spots. Asci numerous, 80 μ long, sublinear-clavate from a narrow base, 6–8-spored. Sporidia overlapping-uniseriate, honey-yellow, obovate-oblong, constantly 3-septate, constricted at the septa, the second cell thicker, 20 x 5 μ.

On dead herbaceous stems, Bethlehem, Pa. (Schw.).

Page 405.—After **Massària Gerárdi**, insert the two following—

M. fœdans, (Fr.)

Sphæria fœdans, Fr S M. II, p 480.
Sphæria amblyospora, B & Br Ann. Nat Hist. No. 627, tab. 10, fig. 10.
Massaria fœdans, Fr. Summa Veg. Sc p. 396
Exsicc. Rab. F. E 41. 257.—Rehm Asc 437.—Thum. M. U 2061.—Sydow, M March. 350.

"Scattered, scarcely visible externally. Asci large, clavate. Paraphyses flexuous. Sporidia large, at first hyaline, consisting of two subconical articulations placed base to base; one of these gradually increases in diameter and becomes very obtuse; a septum is then formed at the base of the smaller articulation, and sometimes, though rarely, there is a septum in the other cell. In every stage except in extreme age, they have a gelatinous coat. Distinguished from *M. inquinans* by the peculiar form of the sporidia, and especially in their mode of formation." (Sporidia 48–54 x 19–23 μ, Winter).

On dead branches, Carolina and Pennsylvania (Schw.).

Diagnosis from Cke. Hndbk.

M. pùpula, (Fr.)

Sphæria pupula, Fr S M. II, p. 484.
Massaria pupula, Tul. Sel Carp. II, p 225.
Exsicc. Kze. F. Sel. 93 —Rab. F. E. 543, 1928.—Rehm Asc. 187 —Sydow, M. March. 2162.

Scattered. Perithecia covered, depressed, concentrically striate, black, mouth denuded, whitish, with a yellow papilla. Asci oblong, 8-spored, 180 x 96 μ, with branching paraphyses. Sporidia subbiseriate, oblong-clavate, 3-septate, brown, with a didymous, hyaline coat, 54–58 x 16–18 μ. Conidial stage, *Steganosporium pyriforme*, Cda.

Under the epidermis of *Platanus*, Pennsylvania (Schw.).

Page 406.—After **Massària scoriàdea**, insert—

ÉNCHNOA, Fr.

Summa Veg. Scand. p. 410.

Perithecia scattered or gregarious, covered by the bark, thin, fragile, seated on or enveloped in a dense, brown tomentum, connate with the epidermis which is pierced by the punctiform vertex. Asci cylindric-clavate, 8-spored. Sporidia cylindrical, guttulate, hyaline or olivaceous. Ramicolous.

E. infernàlis, (Kunze).

Sphæria infernalis, Kze. in Fr. S. M. II, p. 371.
Sphæria Glis, B. & Br. Not. Brit. Fungi, No. 884.
Enchnoa infernalis, Sacc. M. Ven. Spec. p. 210.

Covered, subemergent, the effused, dark brown, floccose-strigose stroma enveloping the globose, thin, collapsing perithecia with obsolete ostiola. Asci clavate, paraphysate, long-stipitate, 8-spored, 100–120 x 12–16 μ (p. sp. 60 μ long). Sporidia biseriate or conglomerate, cylindrical, curved, subacute at the ends, continuous, guttulate, pale olivaceous, 20–24 x 5 μ.

On (oak branches)? Bethlehem, Pa. (Schw.).

We have seen no specimens, and insert this on the authority of Schweinitz.

E. lanàta, (Fr.)

Sphæria lanata, Fr. S. M. II, p. 482.
Enchnoa lanata, Sacc. Syll. 374, Cke. Syn. 4064.

Perithecia covered, free, globose, woolly, rusty-brown, with erumpent, black ostiola. Asci ample, broad-elliptical, 8-spored. Sporidia cylindrical, obtuse at the ends, curved, hyaline, with a central nucleus resembling a septum, 9 x 3 μ.

On *Betula nigra*, Bethlehem, Pa. (Schw.).

E. floccòsa, (Fr.)

Sphæria floccosa, Fr. S. M. II, p. 375.
Enchnoa floccosa, Karst. Symb. Mycol. Fenn. IV, p. 187.

Perithecia scattered or crowded, sometimes seriately aggregated, at first covered by the epidermis, then denuded, sphæroid, collapsing below, covered with a brown or umber-colored tomentum fine as a spider's web, 200–300 μ diam. Asci aparaphysate, clavate, 30–36 x 8–9 μ. Sporidia tristichous, oblong or cylindrical, 2-nucleate, curved, greenish-hyaline, 10–12 x 2–2½ μ.

On dead stems of *Sambucus*, Bethlehem, Pa. (Schw.).

Page 411.—After **Clypeosphæria Hendersònia**, insert—

C. Notarísii, Fckl. Symb. p. 117.

Sphæria clypeata, Nees. Syst. fig. 355.
Clypeosphæria Notarisii, Sacc. Syll. 3189.

Perithecia depressed so as to appear like convex, flattened disks but slightly prominent, covered by the adnate, blackened epidermis, with the ostiolum emergent, conic-truncate. Asci stipitate, narrow-cylindrical, 8-spored, 152 x 8 μ. Sporidia obliquely uniseriate, lanceolate-oblong, obtuse at the ends, finally distinctly triseptate, brown, 22–24 x 4–5 μ.

Common on *Rosa* and *Rubus* stems, Carolina and Pennsylvania (Schw.).

Page 413.—After **Hypóspila pústula,** insert—

H. bìfrons, (DC.)

Xyloma bifrons, DC. Flor. France, VI, p. 156.
Sphæria bifrons, Kze. and Schm. Deutschl. Schwamme, No. 204.
Hypospila bifrons, Sacc. Syll. 3535.

Perithecia amphigenous, innate, arranged in circular groups, flattened, black, at length circumscissile, convex-prominent, finally umbilicate. Asci cylindric-clavate, 50–55 x 10–11 μ. Sporidia biseriate, oblong, inequilateral, 10–12 x 4 μ, guttulate, subobtuse, slightly curved, hyaline, with a single septum near the lower end, 10–12 x 3½–4 μ.

On oak leaves, Bethlehem, Pa. (Schw.).

Page 426.—After **Diapórthe subcóngrua,** add—

More perfectly developed spece. on dead limbs of *Acer saccharinum*, London, Canada (Dearness), have the sporidia broader (5–7 μ).

Page 429.—To synonyms of **Diapórthe leiphæmia,** add—

Sphæria Micheliana, Fr. S. M. II, p. 414 (sec. Cke. Grev. XV, p. 80).

Page 434.—After **Diapórthe tuberculòsa,** insert—

D. staphýlina, E. & E. (in Herb.)

Stroma formed from the slightly altered substance of the bark, light-colored inside, orbicular, elliptical or elongated, 2–4 mm. in the longer diameter, flattish-pulvinate, with a very narrow, faint, circumscribing line which does not penetrate the wood, covered above by the epidermis, which is not ruptured, but simply perforated by the scattered ostiola and only slightly elevated. Perithecia buried in the stroma, 3–10, globose, about ½ mm. diam., their papilliform, finally umbilicate ostiola erumpent, sometimes singly, but mostly joined in a small, black, irregular-shaped disk which barely pierces the epidermis without rising above it. Asci clavate, stipitate, p. sp. 50–55 x 7–8 μ, paraphysate, 8-spored. Sporidia subbiseriate, oblong-elliptical, slightly curved, scarcely constricted, 3–4-nucleate, becoming uniseptate, hyaline, 12–15 x 4–5 μ.

On dead limbs of *Staphylea trifolia*, London, Canada (Dearness).

Closely allied to *D. tuberculosa*, but has the sporidia rather narrower, and the circumscribing line very faint and not penetrating the wood as in that species.

Page 438.—After **Diapórthe Cratægi**, insert—

D. acervàta, (E. & E.)

Diatrype acervata, E. & E. Journ. Mycol. IV, p. 75.
Exsicc. Ell. & Evrht. N. A. F. 2d Ser. 2124.

Stromata small (½ mm.), tobacco brown, becoming black, soft, either single or oftener in compact groups, erumpent in the center of elliptical (½–4 cm. long), dirty-white, dead spots with a definite, dark red-brown border. Perithecia subcircinately arranged, 5–10 in a stroma, white inside, 75–100 μ diam., subglobose, with a short, subcylindrical ostiolum which is hardly discernible on the surface of the stroma. Asci oblong, 35–40 x 7–8 μ, without any distinct paraphyses. Sporidia biseriate, oblong-cylindrical, slightly curved, hyaline, obtuse, slightly constricted in the middle and uniseptate, 12–18 x 3 μ, exactly resembling the sporidia of a *Sphærella*. The clusters of stromata resemble the sori of a *Puccinia*.

On dead spots in living leaves of *Yucca filamentosa*, Newfield, N. J.

D. furfuràcea, (Fr.)

Sphæria furfuracea, Fr. S. M. II, p. 409.

Irregularly circinate. Perithecia globose, surrounded by a bran-like substance. Ostiola very short, joined together, obsoletely prominent. Asci cylindrical. Sporidia (sec. Cke.) uniseptate, narrow-elliptical, hyaline, 25–30 x 11 μ.

On branches of *Tilia*, Pennsylvania (Schw.).

Page 458.—After **Diapórthe Murràyi**, insert—

D. cláviceps, E. & E. (in Herb.)

Perithecia buried in the wood, globose, ¾–1 mm. diam., scattered or subseriately arranged, the separate groups surrounded by a black line penetrating the wood. Ostiola erumpent, cylindrical, rough, brittle, black, subflexuous, mostly tuberculose-enlarged at the base and somewhat swollen at the tips, ½–1 mm. long. Asci clavate, 35–45 x 4½–5 μ. Sporidia biseriate, oblong, hyaline, uniseptate, constricted at the septum, 4-nucleate, 11–13 x 3½–4 μ.

On decorticated, partly decayed wood of *Ostrya Virginica*, London, Canada (Dearness).

The stromata are irregular in outline, mostly elongated and variously confluent, and the wood inside the circumscribing line is whiter than the surrounding parts.

Page 460.—After **Diapórthe racèmula,** (C. & P.), insert—

D. umbellatàrum, (Schw.)

Sphæria umbellatarum, Schw. Syn. N. Am. 1467.

Covered; spots effused far and wide, forming figures of various shapes, reminding one of a geographical map, surface of the spots crustose, black, but not shining. Perithecia scattered, deeply buried under the crust, depressed-globose. Ostiola emergent, short cylindrical, rugose. Asci 35–40 x 5–6 μ. Sporidia biseriate, oblong, uniseptate, slightly constricted, 10–12 x 3 μ.

On dead stems of *Umbelliferæ*, Bethlehem, Pa. (Schw.).

Measurements of asci and sporidia from spec. in Herb. Schw.

Page 469.—After **Válsa æquilineàris,** insert—

V. haustellàta, Fr. in Cooke's Valsei of the U. S. p. 115.

Sphæria haustellata, Fr. S. M. II, p. 383, Schw. Syn. N. Am. 1320.
Exsicc. Rav. F. Car. III, 53.

Pustules with their base deeply sunk in the bark and loosely circumscribed by a narrow, black line penetrating to the wood. Perithecia deeply buried, their long necks converging and joined in a narrow, convex, prominent disk. Ostiola short, distinct, smooth, sometimes dilated at the apex. Asci 8-spored. Sporidia allantoid, strongly curved, slightly umber-colored, 6–8 x 2 μ.

On *Alnus serrulata* and *Ostrya Virginica*, Carolina (Curtis & Ravenel), New York (Peck), on oak limbs, Pennsylvania (Schw.).

Page 470.—**Válsa Línderæ, Pk.** should be transferred to subgenus *Leucostoma.*

Page 502.—After **Eùtypa élevans,** insert—

Eu. crustàta, (Fr.)

Sphæria crustata, Fr. S. M. II, p. 376.
Massaria crustata, Fr. Summa Veg. Sc. p. 596.
Valsa crustata, Nitsch. Pyr. Germ. p. 135.
Eutypa crustata, Sacc. F. Ven. Ser. IV, p. 16.

Stromata widely effused, often surrounding the branches, sunk in the bark and forming with it a crust which becomes blackened on the surface and very rough or even spiculose from the strongly projecting ostiola. Perithecia sunk in the bark, monostichous, tolerably large, numerous, often crowded or, especially around the margin of the stroma, scattered and standing singly, more or less prominent, with large, obtusely conical or depressed-hemispherical, entire, or 3–4-radiate-cleft ostiola. Asci narrow-clavate, long-stipitate, 8-spored, p. sp. 32 x 4–5 μ. Sporidia biseriate, cylindrical, straight, brownish, 6–12 x 2 μ.

On branches, Carolina (Schw.).

Page 507.—After **Eùtypa sepúlta,** insert—

Eu. scabròsa, (Bull.)

Hypoxylon scabrosum, Bull. Champ. p. 179, tab. 468, fig. 5.
Sphæria scabrosa, DC. Flore France, II, p. 288.
Diatrype scabrosa, Fr. Summa Veg. Sc. p. 385.
Valsa scabrosa, Nits. Pyr. Germ. p. 131.
Eutypa scabrosa, Fckl. Symb. p. 171.
Exsicc. Fckl. F. Rh. 1039, 1045.—Kze. F. Sel. 151.—Rab. F. E. 1139.

Stromata forming suborbicular tubercles, oval or convex, separate or confluent, rarely widely effused, rimose or undulate, seated on the wood or on the surface of the bark under the epidermis, and then soon erumpent, black throughout, roughened by the numerous ostiola. Perithecia lying in the stroma in irregular order, densely crowded, globose, small, with necks of variable length furnished with hemispherical or subconical, entire, very minute ostiola. Asci cylindric-clavate, long-pedicellate, 8-spored, p. sp. 40–50 x 4–5 μ. Sporidia subbiseriate, cylindrical, slightly curved, pale brown, 6–12 (mostly 8–10) x 1½–2 μ.

On wood, around Bethlehem, Pa. (Schw.).

Page 513.—After **Calosphæria éxpers,** insert—

C. vibrátilis, (Fr.)

Sphæria vibratilis, Fr. S. M. II, p. 482. Nitsch. Pyr. Germ. p. 97.
Calosphæria vibratilis, Sacc. Syll. 411.

Perithecia solitary, scattered, sphæroid or ovoid, mostly collapsing when dry, smooth or at first sparingly pilose, black, shining, 300–800 μ diam., with very short necks and rounded ostiola very minute and scarcely rising above the epidermis. Asci clavate, 24 x 4 μ (p. sp.), with long paraphyses. Sporidia subbiseriate, fusoid-elongated, slightly curved, hyaline, 4–6 x 1 μ.

Under the bark of *Prunus Virginiana*, Bethlehem, Pa. (Schw.).

Page 553.—After **Melográmma Bulliárdi,** insert—

M. spiníferum, (Wallr.)

Sphæria spinifera, Wallr. Crypt. Flora, No. 4073.
(*Sphæria podoides*, Pers. Syn. p. 22)'
Diatrype podoides, Fr. Summa Veg. Sc. p. 385.
Melogramma asperum, Ces. & De Not, Schema, p. 30.
Melogramma spiniferum, De Not, Sfer. Ital. p. 53.
Exsicc. Fckl. F. Rh. 1000.—Kze. F. Sel. 153.—Thum. M. U. 1860.

Stromata densely gregarious or subconfluent, erumpent-superficial, swollen-pulvinate, black. Perithecia immersed, subglobose, with their cylindrical, rough, subtortuous ostiola more or less projecting. Asci cylindric-clavate, 8-spored, 160 x 16 μ. Sporidia biseriate, cylindric-fusoid, slightly curved, 55–70 x 7–8 μ, 6-septate, cells fuliginous, guttulate, the terminal ones shorter and subhyaline.

On beech bark, Carolina and Pennsylvania (Schw.).

Page 563.—After **Valsària Phoradéndri,** insert—

V. cornícola, E. & E. (in Herb.)

Perithecia globose, minute (¼ mm,), 4–8 loosely grouped in a cortical stroma or gregarious without any distinct stroma, slightly raising the epidermis, and when this falls away, slightly blackening the bark above and around them, but without any circumscribing line. Ostiola papilliform, minute. Asci cylindrical, short-stipitate, 65–75 x 7–8 μ. paraphysate, 8-spored. Sporidia uniseriate, elliptical, uniseptate and constricted, brown, 9–12 x 4½–5½ μ.

On dead limbs of *Cornus*, London, Canada (Dearness).

Page 564.—After **Valsària Beaumóntii,** insert—

V. anserina, (Pers.)

Sphæria anserina, Pers. Icon. p. 5, tab. I, figs. 8–10; Cke. Hndbk. No. 2637.
Valsaria anserina, Sacc. Syll 2842.

Stroma effused, blackening the surface of the wood or bark in which it is immersed. Perithecia gregarious but separate, sphæroid, scarcely ½ mm. diam., deeply or entirely immersed, black, with their ostiola rising to the surface and raising it into pustules. Asci cylindrical, short-stipitate, 8-spored. Sporidia uniseriate, ovoid or suboblong, didymous, slightly constricted at the septum, 16–20 x 6–7 μ, sometimes one cell a little narrower than the other.

Bethlehem, Pa. (Schw.). No habitat given.

Page 570.—Add to synonyms of **Diatrỳpe albopruinòsa**—

Sphæria euphorea, Fr. S. M. II, p. 354 (sec. specc. in Herb. Schw.).

Page 582.—After **Anthóstoma tuberculòsum**, insert—

A. hiáscens, (Fr.)

Sphæria hiascens, Fr. S. M. II, p. 477.
Anthostoma hiascens, Nitsch. Pyr. Germ. p. 113.

Stroma effused, immersed in the wood, which at length whitens out on the surface. Perithecia immersed, depressed-globose, monostichous, scattered or crowded, with very short necks. Ostiola large, mostly deeply 4-sulcate, sometimes almost cup-shaped. Sporidia obtusely fusoid, dull black, straight, inequilateral, 32–36 x 8 μ.

On (beech)? wood, Bethlehem, Pa. (Schw.).

Page 609.—Insert after the genus **Dothidélla**—

SCÍRRHIA, Nitschke.

in Fckl. Symb. Mycol. p. 220.

Stromata linear, often crowded or subconfluent, forming elongated swellings erumpent through parallel cracks in the epidermis. Cells immersed, polystichous. Asci elongated, 8-spored. Sporidia oblong, uniseptate, hyaline. The conidial stage is *Hadotrichum.*

S. ostiolàta, Ell. & Galw. (in Herb.)

Stromata narrow-elliptical, raising the epidermis into elongated swellings 3–6 x 1–1½ mm. Cells numerous, white inside, the conic-tuberculiform ostiola seriate-erumpent, through parallel, longitudinal cracks in the overlying epidermis. Asci clavate, paraphysate, 55–65 x 6–7 μ. Sporidia biseriate, fusoid-oblong, uniseptate and constricted at the septum, 12–15 x 3–4 μ, yellowish-hyaline, cells guttulate.

On dead culms of *Cyperus articulatus*, College Station, Texas (Jennings).

Has the habit and general appearance of *S. rimosa*, (A. & S.), but differs in its smaller asci and sporidia and its larger, tuberculiform ostiola.

Page 636.—After **Hypóxylon turbinulàtum**, insert—

H. atroriùfum, E. & E. (in Herb.)

Stromata oblong or suborbicular, erumpent-superficial, 2–3 mm. diam., or subseriately confluent for ½–1 cm., pulvinate, rounded above, reddish-brown and mamillose from the prominent apices of the perithecia, contracted at the base and black. Perithecia peripherical, ovate, small, (½ mm.). Stroma brownish-black inside. Ostiola acutely papilliform and perforated, black. Asci cylindrical, 100 x 5 μ (p. sp. 55–60 x 5 μ), paraphysate, 8-spored. Sporidia uniseriate, subinequilaterally elliptical, brown, 6–7 x 3 μ.

On bark of dead (oak)? limbs, Michigan (Hicks).

Has the general appearance of *H. cohærens* and *H. turbinulatum*, but the perithecia and sporidia are smaller than in either of these species. The stromata are so much contracted below as to appear only centrally attached.

Page 657.—After **Hypóxylon sphærióstomum**, insert—

H. griseum, (Schw.)

Sphæria grisea, Schw. Syn. N. Am. 1252.

Oblong, effused, subconcave, innate-immersed in the wood, acuminate at each end, surrounded by a deeply penetrating, black line, tolerably thick in the middle, and whitish-gray, thinner towards the margin. Perithecia numerous, flattened, very small, immersed in the stroma. Ostiola subumbonate, obtuse, gray, prominent. The surface of the stroma cracks into frustules.

On decorticated wood, Bethlehem, Pa. (Schw.); rather rare. The stroma is often an inch in diameter. The spec. in Herb. Schw. is without fruit.

Page 673.—After Xylária **pedunculàta**, insert—

X. bulbòsa, (Pers.)

Sphæria bulbosa, Pers. Obs. Mycol. II, p. 63, tab. 1, fig. 1.
Xylaria bulbosa, B. & Br. in Berk. Outl. of Brit. Fungol. p. 385, tab. 24.

Stroma erect, thick, simple or forked above, round or rarely a little dilated and compressed towards the summit, glabrous, becoming black, base tuberous. Fertile head sterile at the apex, sometimes 3-parted, mammillose from the somewhat prominent, densely crowded perithecia. Asci cylindrical, pedicellate, 8-spored, pseudoparaphysate, 80–84 x 6–7 μ (p. sp.). Sporidia obliquely uniseriate, fusiform, obtuse, inequilateral, brown, 12 x 4 μ.

On the ground, Carolina and Pennsylvania (Schw.).

The following notes are from specimens in the Schweinitzian Herbarium at Philadelphia:

1203.* *Sphæria afflata*, Schw.—The spec. in Herb. Schw. is a mere sterile crust.

1255. *Sphæria enteroxantha*, Schw.—Looks like *Hypoxylon Sassafras*, but immature or sterile—no fruit.

*The numbers refer to Schw. Syn. of N. Am. Fungi.

1273. *Sphæria subconfluens*, Schw.—This is a *Haplosporella.* Stromata pulvinate-tuberculiform, subseriate, closely embraced by the ruptured epidermis or, when on decorticated limbs, by the fibers of the wood. Sporules oblong, continuous, brown, 15–20 x 5–6 μ.

1274. *Sphæria obscura*, Schw.—This is the spermogonial stage of some *Valsa* or *Diatrype.* Spermatia oblong, hyaline, minute, 3–4 x 1 μ.

1309. *Sphæria sacculus*, Schw.—Spermogonia. Spermatia allantoid, hyaline, curved, 4–5 x 1 μ.

1422. *Sphæria junipericola*, Schw.—Sec. specc. in Herb. Schw. and Cke. Grev. XV, p. 80, this is a *Sphæropsis.*

1424. *Nectria dematiosa*, Schw.—The spec. shows only the conidial stage (*Tubercularia*). Conidia oblong, 5–6 x 1½ μ, borne terminally and laterally on the long, slender, semicircularly curved sporophores. (See page 96).

1425. *Sphæria Sumachi*, Schw.—*Haplosporella Sumachi*, (C. & E.), *Sphæropsis Sumachi*, C. & E. Grev. V, p. 31.—Sporules oblong, brown, uninucleate, 12–15 x 5–6 μ—*Botryosphæria Sumachi*, sec. Cke. Grev. XV, p. 80.

1426. *Sphæria pubens*, Schw.—This is *Camarosporium Robiniæ*, (West). Sporules oblong-elliptical, 4–5-septate and muriform, brown, 15–20 x 6–7 μ.

1428. *Sphæria Hyperici*, Schw.—Allied to *Botryosphæria fuliginosa*, (M. & N.), if not identical with it.

1430. *Sphæria parasitans*, Schw.—The spec. is too imperfect to give one an accurate idea of this species.

1438. *Sphæria fissa*, Pers.—The specimen in Herb. Schw. is a *Phoma* with sporules 3–4 x 2½ μ.

1439. *Sphæria mutila*, Fr.—*Haplosporella.* Sporules 12–16 x 5–7 μ, 1-nucleate, brown, (See p. 546).

1442. *Sphæria fuliginosa*, Pers.—*Haplosporella.* Sporules brown, 15–20 x 8–10 μ. The general appearance is that of *Botryosphæria fuliginosa*, of which it is doubtless the pycnidial stage. (See p. 546).

1444. *Sphæria Hibisci*, Schw.—*Dothiorella.* Sporules hyaline, 20–22 x 12 μ. (See p. 547).

1446. *Sphæria gallæ*, Schw.—*Dothiorella.* Sporules globose-elliptical, 15–18 x 12–14 μ. Stroma black, orbicular, 1–1½ mm., nearly superficial. *Sphærella gallæ*, E. & E., is different, having simple, scattered perithecia.

1451. *Sphæria Zeæ*, Schw.—The spec. in Herb. Schw. is the same as *Diplodia Zeæ*, Lev. in Ell. N. A. F. 31. (See p. 453).

1452. *Sphæria linearis*, Nees.—*Rhabdospora.* Sporules linear, curved above, about 20 μ long.

1458. *Sphæria chloromela*, Fr.—Sterile.

1462. *Sphæria Cimicifugæ*, Schw., and

1463. *Sphæria euphorbiicola*, Schw., are both apparently young *Sphærellas*, but no asci or sporidia can be made out.

1464. *Sphæria iridicola*, Schw.—Stem blackened by some mycelium, with minute, sterile perithecia.

1465. *Sphæria fumosa*, Schw.—Mere discoloration.

1466. *Sphæria Peponis*, Schw.—Minute, sterile perithecia. *Phoma?* No sporules.

1470. *Sphæria Silphii*, Schw.—*Phoma* or young *Sphærella.* Cooke finds elliptical spores 12 x 2 μ.

1472. *Sphæria nervisequia*, Schw.—Something like a *Leptostroma*, but entirely sterile.

1473. *Sphæria fuscata*, Schw.—This is *Leptosphæria doliolum*, Pers.

1474. *Sphæria epiphylla*, Schw.—The specimen in Herb. Schw. is *Ravenelia glanduliformis*, Berk. On leaves of *Tephrosia.*

1475. *Sphæria Scirporum*, Schw.—This is *Hypocrella Hypoxylon*, (Pk.). Sporidia filiform, septate. On leaves of *Carex?*

1477. *Sphæria conferta*, Schw.—Looks very much like *Sphærella maculiformis*, but sterile. This is a different thing from the *Sphæria conferta*, Schw. Syn. Car. 187, which is *Amphisphæria.* (See p. 206).

1483. *Sphæria punctum*, Schw.—Apparently a sterile form of *Phyllachora graminis*, with a minute, punctiform stroma. On dry leaves of *Panicum nitidum.*

1484. *Sphæria andropogicola*, Schw.—The stromata are larger and not as narrow as in *S. Andropogi*, and the specc. resemble *Phyllachora graminis* more closely than *S. Andropogi* does. There are also asci, but the sporidia are still immature. This is common around Newfield, but so far, always sterile. It is no doubt a small form of *Phyllachora graminis*, Pers.

1485. *Sphæria Andropogi*, Schw.—Small (1 x ½ mm.), elongated strips of sterile crust with some *Vermicularia* or *Colletotrichum?*

1487. *Sphæria canaliculata*, Schw.—This (sec. Cooke, Grev. XIII, p. 43) is *Puccinia cellulosa*, Berk.

1490. *Sphæria Panici*, Schw.—On withered leaves of species of *Panicum*, and

1491. *Sphæria Agrostidis*, Schw.—On *Agrostis filiformis*, are apparently referable to *Phyllachora*. Schweinitz himself remarks that the latter may be only a Var. of *P. graminis*.

1492.* *Sphæria Iridis*, Schw.—Frequent (sec. Schw.) on leaves of Iris, at Kaighn's Point, Philadelphia, Pa. Affords elliptical, hyaline stylospores 10 x 5 μ, and is different from 1464, *Sphæria iridicola*, Schw.

1568. *Sphæria Lecythea*, Schw. Syn. Car. 155, Fr. S. M. II, p. 460.—This (sec. Cke. Grev. XVI, p. 98) is a *Sphæropsis*.

1820. *Sphæria æsculicola*, Fr., is entirely sterile.

1811. *Sphæria Pyrolæ*, Fr.—Black, sterile spots on leaves of *Chimaphila umbellata*. *Sphærella Pyrolæ*, Rostr. (p. 282) may not be distinct from this.

1813. *Sphæria frondicola*, Fr., on a leaf of *Sassafras*, is entirely sterile; small, subcuticular stromata (or perithecia)? on spots ½ cm. across.

1815. *Sphæria carpinicola*, Fr.—The spec. in Herb. Schw. is a *Glœosporium*. Spores linear, 8–10 x 1½ μ, curved. Acervuli flesh-colored, minute, thickly scattered over the under side of the leaf, but not on any spots.

1817. *Sphæria Dianthi*, A. & S., var. *Saponariæ*, Kze.—The spec. in Herb. Schw. is a *Septoria*, on round, white spots with a reddish margin, epiphyllous, thickly scattered. Sporules cylindrical, nearly straight, some of them becoming obscurely 3-septate. *Septoria Dianthi*, Desm.?

*In the Schweinitzian Herbarium, at the Philadelphia Acad., the numbers 1600–1800 are mostly missing.

1822. *Sphæria Tulipiferæ*, Schw., on fallen leaves of *Liriodendron*, is entirely sterile—only spots.

1823. *Sphæria dryophila*, Schw.—Sterile perithecia on large, dirty-white spots. On oak leaves. *Sphærulina dryophila*, Cke. & Hark., is probably not distinct from this. (See p. 312).

1824. *Sphæria catalpicola*, Schw.—On the upper side of fallen leaves of *Catalpa*. Perithecia on large brown spots. Sporules 6–7 x 3–4 μ. *Phyllosticta catalpicola*, (Schw.). (*P. Catalpæ*, E. & M., Journ. Mycol. II, p. 14).

1825. *Sphæria smilacicola*, Schw.—On leaves of *Smilax rotundifolia*. Sterile, dark-colored, subbullate spots, 1½–2 mm. diam., but no perithecia.

From an examination of authentic specc. in Herb. Berk. Cooke, in Journal Bot., 1883, excludes the four following Schweinitzian species from the Pyrenomycetes:

1784. *Sphæria collapsa*, Schw.—Probably an imperfect *Pocillum*.

1786. *Sphæria Mori-Albæ*, Schw.—Perithecia carbonaceous, no fruit.

1803. *Sphæria excipulans*, Schw.—Specimen without fruit.

1769. *Sphæria Fragariæ*, Schw., is also without fruit; nor did the spec. show any cylindrical ostiola.

APPENDIX.

In the following Schweinitzian species the fructification is unknown. The diagnoses are from Schweitz' Synopsis of North American Fungi.

Sphæriæ Villosæ.

1513. *Sphæria penicillata*, Schw.—Perithecia scattered, but forming extensive groups and arranged in subflexuous series, superficial, the base firmly fixed to the epidermis, rarely fasciculate and subconfluent. Perithecia obovate-globose, flattened, densely villose, the hairs towards the base shorter and dark green, from the middle up, dense and longer, penicillate-divergent and very white, hiding the minute, obtuse, black ostiola. Walls of the perithecia black and thin, sometimes collapsed, resembling a white-villose *Peziza*.

On fallen twigs of *Ribes aureum*, Bethlehem, Pa.

1531. *Sphæria intonsa*, Schw.—Gregarious, black, erumpent, subhemispherical, apex obtuse, ostiolum indistinct. Perithecia small, clothed with short, rigid hairs like a recently shaven beard.

On decorticated pickets of *Robinia*, Bethlehem, Pa.

1532. *Sphæria involuta*, Schw.—Densely aggregated, loosely attached. Perithecia smooth, subglobose, subpapillate, sometimes subcorrugated, dark brown, shining, entirely enveloped in a dense coat of white wool which can be rubbed off; apex only slightly denuded.

On old, decaying trunks, Bethlehem, Pa. Reminds one of *Myriococcum*, (M. Everhartii)?

1536. *Sphæria cæspitulans*, Schw.—Cespitulose-concrescent or fasciculate in elongated-linear strips. Perithecia very small, ovate, becoming irregular, acutely narrowed into the ostiolum, very black, rugose, covered all over with minute, short, thick (almost tuberculiform) hairs, often irregularly collapsing, not cespitose, substance soft. The smallest of the tribe.

On branches of *Rubus Idæus*, Bethlehem, Pa.

1537. *Sphæria viridiatra*, Schw.—Scattered or subcespitose. Perithecia minute, globose or ovate, carbonaceous, rugose, obtuse, scarcely ostiolate, bare below, densely covered above with a short, yellowish-green villose coat. Parasitic on various old compound *Sphærias*, Bethlehem, Pa.

Reminds one of *Calonectria chlorinella*, Cke.

1539. *Sphæria monstrosa*, Schw.— Gregarious, seated on a black, woody crust, rather large, sometimes very large and deformed, swollen at the base, narrowed above into a thick, sulcate, pyramidal ostiolum. Substance carbonaceous. Covered entirely, even the ostiolum, with a dense coat of brown, subrigid hairs. Has some resemblance to some forms of *Eutypa spinosa;* occasionally denuded.

On chestnut wood, Bethlehem, Pa.

Pertusæ.

1587. *Sphæria inclinata*, Schw.—Scattered or aggregated, erumpent through the fibers of the lower stratum of the bark where the epidermis has fallen away, at first immersed, then exposed. Perithecia ovate-globose, minute, obliquely inclined, with the ostiolum comparatively large and finally deciduous; then the perithecia are simply perforated, black and rugose.

On twigs of *Viburnum*, Bethlehem, Pa. (Schw.).

1589. *Sphæria glandicola*, Schw.—This (sec. Cke. Grev. XVI. p. 91) is a *Phoma* with sporules 5 x 3 μ, basidia 20 x 3 μ.

1590. *Sphæria pericarpii*, Schw., and

1583. *Sphæria Surculi*, Fr., are also (sec. Cke. l. c.) only species of *Phoma.*

1591. *Sphæria tingens*, Schw. (not *Lophidium tingens*, Ell.)—Scattered, conically beaked at first immersed, then adnate with a flattened base. Perithecia compressed-conical, scarcely rugose, rostrate, with a thick, deformed, subshining ostiolum. The bark is blackened around the perithecia.

In cracks of the bark on young branches of *Sassafras*, Bethlehem, Pa. (Schw.).

1595. *Sphæria elliptica*, Schw.—Rather large, elliptical, elongated, scattered, only slightly elevated, flattened, base subimmersed, black, at length perforated with a central pore, at first crowned with a papilliform ostiolum which is finally deciduous.

On *Viburnum*, Mauch Chunk, Pa. (Schw.).

1596. *Sphæria deformata*, Schw.—Widely scattered, minute, protruding and denuded, of an irregular-cylindrical shape, very black, surface subrugose and uneven; ostiolum indistinct. At length perforated. With the preceding species.

Obtectæ.

**Lignatiles.*

1632. *Sphæria lævigata*, Schw.—Forms a continuous, smooth, black crust widely effused (6 inches), determinate, surrounding the limb. Perithecia scattered, large, globose-depressed, deeply immersed, their ostiola appearing as mere points on the surface of the crust.

On soft, rotten wood, Bethlehem, Pa. (Schw.).

1633. *Sphæria inundatorum*, Schw.—Gregarious, covered by a widely effused, rimose crust, so that the wood appears charred. Perithecia scattered, buried in the wood beneath the crust, globose-depressed, without any distinct neck, but with flattened-globose ostiola erumpent through a crustaceous tubercle, and finally pezizoid-umbilicate.

On wood lying in the Delaware, at Kaighn's Point, N. J. (Schw.).

1634. *Sphæria excussa*, Schw.—Perithecia large, thickly scattered under the partially loosened epidermis, immersed in the inner bark. Ostiolum punctiform, barely piercing the epidermis, but finally elongated into a distinct beak perforated at the apex. The large, flask-shaped rugose perithecia surrounded with a distinct furrow, also emerge, and when mature, easily fall out, leaving pits in the bark two lines across.

Frequent on young branches of *Pyrus Malus*, Bethlehem, Pa. (Schw.). Allied to *Massaria fœdans*, (Fr.).

1625. *Sphæria denudans*, Schw.—Scattered and aggregated, oblong-globose, depressed. Perithecia immersed, at length subprominent, inner cavity elliptical, enclosed in a kind of cinerascent membrane finally black. Ostiola very small, papilliform, situated in a slight depression in the top of the perithecium, sometimes hysteriiform.

On limbs, Bethlehem, Pa. (Schw.).

1636. *Sphæria Rosæ*, Schw.—Scattered on a dark brown crust widely effused under the epidermis, finally denuded. Perithecia subimmersed in the crust, hemispheric-prominent, oblong-globose, rugose, black. Ostiola at first scarcely piercing the epidermis, at length short-conical.

On tender branches of rose bushes, New Jersey (Schw.).

1637. *Sphæria ampelos*, Schw.—Minute, distantly seriately scattered, circumscissile, brownish, flattened-globose. Ostiola prominent through cracks in the bark.

On much decayed shoots of grape vines, Bethlehem, Pa. (Schw.).

** *Corticolæ.*

1657. *Sphæria albofarcta,* Schw.—Scattered or gregarious. Perithecia covered, very small, entirely immersed in the bark, white inside, globose, horizontally striate, indistinctly ostiolate, the ostiola visible through cracks in the bark which is blackened (by the discharged sporidia)?

On young branches of *Sassafras,* which are extensively stained and blackened, Bethlehem, Pa.

1658. *Sphæria tenella,* Schw.—Scattered, buried in the bark under the epidermis. Perithecia minute, elliptic-oblong or globose, not glabrous. Ostiola very minute, visible through cracks in the epidermis, apparently rising from a disk. Sometimes several ostiola appear to rise from one perithecium.

In the fibrous bark of *Hibiscus roseus,* Bethlehem, Pa.

1659. *Sphæria Daphnidis,* Schw.—At first covered, finally bare above, but immersed in the bark. Perithecia globose-depressed, dark brown, tomentose, immersed in a blackish, floccose crust. Ostiola at first papillate, finally larger and deformed, especially in the denuded perithecia which finally become rugose. Stains the bark black.

On *Daphnis Mezerei,* Bethlehem, Pa.

1660. *Sphæria fuscescens,* Schw.—Aggregated or scattered, at first entirely covered by the epidermis, but when this is thrown off, seated free on the surface of the inner bark. Perithecia minute, ovate-conical, much wrinkled, black or dark brown, sometimes collapsing, the papilliform ostiola visible through cracks in the epidermis. The black, spermatic contents exuding stain the surface of the bark around the ostiolum.

On chestnut limbs.

1661. *Sphæria palliolata,* Schw.—Subseriate in cracks of the epidermis, collected in small, elliptical groups, sometimes confluent. At first covered by the epidermis, at length free, but always covered by the thin, white inner membrane of the epidermis. Perithecia comparatively large, only a few in a cluster, globose-flattened, very black, punctate-rugose. Ostiolum papilliform.

On smooth branches of *Rosa corymbosa.*

1662. *Sphæria rhuina,* Schw.—Gregarious; extensively erumpent through the epidermis and closely surrounded by it. Perithecia very black, flattened, with a central papilla through which the copious spores exude. When the epidermis falls away, there is seen a contiguous, black, cortical crust raised into numerous pustules by the hemispherical, black, smooth perithecia.

Under the epidermis of *Rhus glabra.*

1663. *Sphæria conspersa*, Schw.—Gregarious or scattered, at first covered by the epidermis which finally falls off. Ostiola erumpent in the form of a minute, concave disk. Perithecia numerous, very black, of medium size, flattened at the base and when the ostiolum falls off, perforated. Perfect specimens are crowned with a sphæriiform ostiolum half as large as the perithecium. The base of the perithecium is immersed in the bark and the whole is stained and blackened by the exuding spores.

On *Robinia viscosa*, Nazareth, Pa. (Schw.).

Obturatæ.

1664. *Sphæria Sclerotium*, Schw. Syn. Car. 163.—Scattered, erumpent. Perithecia subovate, dark brown. Ostiolum impressed, opening at first narrow, then round. At first sight resembles a *Sclerotium*, but it is a true *Sphæria*, regular ovate-globose, a line high, nearly free, seated in a kind of receptacle under the epidermis. Color when dry, dirty white, with a small, yellowish sack inside.

On young branches, Carolina (Schw.).

1666. *Sphæria erumpens*, Schw. Syn. Car. 209.—Subsimple, scattered, erumpent. Perithecia ovate-depressed, ashy-brown, obsoletely papillate, squamose below, smooth above.

On dead branches of *Smilax*, Carolina.

1677. *Sphæria Ruborum*, Schw.—Scattered, gregarious or seriate, soon throwing off the epidermis. Perithecia ovate-globose or hemispherical, subimmersed in the bark, black, rugose, comparatively large, crowned with a punctate-rugose tubercle half as large as the perithecium or sometimes much smaller. The upper part of the perithecium finally breaks away, leaving the cup-shaped base.

Common on dead stems of *Rubus.*

1678. *Sphæria olivascens*, Schw.—Almost always covered by the epidermis, causing an olive-black spot in the bark which is raised by the minute, hemispheric-globose perithecia with their perforated ostiola alone visible.

On unknown twigs, Salem, N. C.

1680. *Sphæria amorphula*, Schw.—Densely aggregated or even cespitose, seated in the substance of the bark under the epidermis, amorphous or polymorphous, subconfluent, slightly raised, carbonaceous, black, astomous or indistinctly ostiolate.

In the bark of young branches of *Juglans.**

1681. *Sphæria capsularum*, Schw.—Simple, scattered, flattened, hemispherical, minute, with a deciduous papilla, surrounded at base by the thin, cinerascent epidermis of the capsule in which at first it is immersed.

On capsules of various plants; *e. g. Convolvulus purpureus*, &c.

1782. *Sphæria druparum*, Schw.—Closely aggregated, black and crustaceo-confluent. Perithecia numerous, ovate, rugose, indistinctly ostiolate, at length partially denuded and then covered by adherent fragments of the matrix.

On decaying black walnuts.

1683. *Sphæria pomorum*, Schw.—Scattered or aggregated, seated on a black crust under the thin epidermis, which is finally ruptured and falls off in fragments. Perithecia ovate, rugose, black, minutely papillate.

On dried up apples and quinces hanging on the trees through the winter, Carolina and Pennsylvania.

1685. *Sphæria Azaleæ*, Schw.—Perithecia seriately erumpent through flexuous cracks in the epidermis, immersed in the inner bark, scarcely confluent, subdistant, rather small, black, rugose, with a pezizoid-umbilicate ostiolum.

On trunks and branches of *Azalea nudiflora*.

1686. *Sphæria concomitans*, Schw.—Scattered, erumpent, minute, punctiform, subglobose, astomous, black, finally pezizoid-collapsing.

On the petioles of large leaves (of trees) where they are enlarged from the stings of gall-producing insects.

1687. *Sphæria lineolans*, Schw.—Rather large, gregarious, erumpent, closely surrounded by the epidermis. Perithecia confluent in irregular lines, black and rough outside, globose-flattened, indistinctly ostiolate, discharging the dark brown spores with which the perithecia are filled. When young, covered by the raised epidermis.

On willow branches.

1688. *Sphæria obtusa*, Schw.—At first covered by the raised

*This and the remaining numbers not otherwise noted, were collected by Schweinitz, at Bethlehem, Pennsylvania.

epidermis, at length denuded, scattered, but thickly covering the stems. Ostiola obtuse, subprominent. Perithecia black, obovate, rugose, minute, bare or surrounded at base by the epidermis. When young, brown, subpellucid, surrounded by a whitish-pulverulent mass which finally disappears.

On dead stems of *Rubus villosus.*

Subtectæ.

1689. *Sphæria sphærocephala*, Schw. Syn. Car. 166.—Perithecia innate and rising with the yellowish epidermis to which they are closely attached, suberumpent, of medium size, black; when empty, cinereous and obsoletely cellulose. Ostiola formed of many crowded tubercles, one of which exudes pellucid globules as in *Sphæronema.*

On branches of *Hydrangea*, Carolina.

1700. *Sphæria vacciniicola*, Schw.—Scattered, covered. Perithecia depressed-globose, subrugose, empty or filled with dark-colored spermatic gelatine, visible through minute cracks in the epidermis, scarcely papillate, black, very minute.

On small branches of *Vaccinium.*

1701. *Sphæria Kalmiarum*, Schw.—Scattered, very black; when immature, swollen and brown, rarely exuding short, simple, white cirrhi. Perithecia hemispherical, innate.

On fallen leaves of *Kalmia* and *Rhododendron.*

1703. *Sphæria samaræ*, Schw.—Scattered; at first covered by the epidermis which is soon stellately ruptured. Perithecia obtuse, subastomous, punctate-rugose, dark brown, nucleus white; sometimes extruding a small, straight, dark brown cirrhus.

On samaræ of *Fraxinus.* Apparently different from *Pleospora samaræ*, Fckl.

1704. *Sphæria Jasmini*, Schw.—Scattered, minute, brown, white at the apex, covered by the raised epidermis. Perithecia elliptical or subrotund, at length collapsing with the epidermis still adherent, the collapsed disk subrugose.

On shoots of *Jasminus* (cult.).

Caulincolæ.

1731. *Sphæria tecta*, Schw.—Permanently covered, scattered extensively over the slightly blackened stems. Perithecia depressed-elliptical, crowned with a globose, deciduous papilla.

On large, herbaceous stems.

1732. *Sphæria malvicola*, Schw.—Minute, subseriate, erumpent from the bark and finally free, variable in shape; when perfect, conical or globose, obtuse and deformed. Ostiolum confluent with the perithecium which is clothed with grayish-black, divergent hairs.

On stems of *Malva Alcea.*

1734. *Sphæria tenuissima*, Schw.—Covered, scattered, shining-black. Perithecia astomous, dark greenish, very delicate; when fresh, hemispherical, soon collapsing, visible through the epidermis, finally denuded.

On stems of *Polygonatum latifolium.*

1735. *Sphæria navicularis*, Schw.—Perithecia sometimes scattered, but mostly on a black, elliptical or boat-shaped spot abruptly contracted at each end, not glabrous, indistinctly papillate, finally irregularly collapsing, at length partly loosened or only slightly attached; substance carbonaceo-pulveraceous.

On herbaceous stems, Salem, N. C. (Schw.).

1737. *Sphæria obtusata*, Schw.—At first covered, but finally free, staining the substance of the stem yellow. Perithecia variable in shape but always obtuse at the apex, cylindrical, globose or round, moderately raised, glabrous, finally subcollapsing. Ostiolum hysteriiform, transverse.

On various herbaceous stems.

1738. *Sphæria platypus*, Schw.—Scattered, very black, minute, covered when young, but generally found denuded. The perithecia are seated on a round, flattened base with the margin subinflexed, and often sublobate when dry, and easily separated from the stem. Perithecia conic-globose, apparently punctate from the discharged spores scattered over them; otherwise glabrous and almost shining.

On stems of *Anemone Virginiana.*

1739. *Sphæria Polygoni sagittati*, Schw.—Scattered thickly over the stems, black, at length entirely free, at first covered, flattened-globose, much wrinkled. Ostiolum cylindric-papillate, brown. Base of the perithecia sometimes effused.

On dead stems of *Polygonum sagittatum.*

1740. *Sphæria Brassicæ*, Schw.—Scattered, erumpent between the fibers of the stem, rather large, globose or hemispherical, black or brownish-black, astomous, at length ruptured, leaving the lower part of the perithecium irregularly torn and empty still attached to the stem. Perithecia glabrous or only apparently punctate from the discharged spores.

On cabbage stems in cellars.

1741. *Sphæria Cannabis*, Schw.—Scattered or seriate, at length bursting through the thin epidermis, very minute, flattened-globose, rugose. Ostiolum indistinct. Perithecia easily deciduous.

Under the epidermis of stems of *Cannabis*.

1742. *Sphæria lactescentium*, Schw.—Densely aggregated, rather large, at first covered by the epidermis, lying among the fibers of the stem and easily deciduous, at length free. Perithecia black, rugose, subconfluent, subconical and irregular, finally subcollapsing, but always crowned with a white, spermatic globule. The stems are often covered with the perithecia for a foot in length.

On the lower part of the stems of *Asclepias Syriacus*.

1743. *Sphæria Asclepiadis*, Schw.—Scattered, covered, shining-black and visible through the epidermis, oblong-applanate, only a little raised, subconfluent, rough, astomous, nucleus black. Although there are no bristles, Schweinitz thinks this may belong to the genus *Exosporium*.

On the upper part of stems of *Asclepias Syriacus*.

Different from *Diaporthe Asclepiadis*. (See p. 459).

1744. *Sphæria Daturæ*, Schw.—At first covered, minute, lying scattered and hidden under the epidermis, hardly ever entirely denuded. Perithecia flattened, rugulose, subcollapsed, papillate-ostiolate, dark brown.

On stems, capsules and spines of *Datura*, Salem, N. C., and Bethlehem, Pa. (Schw.), New York (Peck).

1745. *Sphæria scapincola*, Schw.—Very minute, orbiculate, scarcely covered, appearing as minute, black specks thickly scattered over the thin epidermis, flattened, subcollapsed, rugulose, with a very minute central point or papilla scarcely visible. The minute perithecia are scattered over the entire length of the scape.

On scapes of *Yucca filamentosa*.

1746. *Sphæria ampliata*, Schw.—Covered, broadly effused on a subcuticular spot. Perithecia compressed, nestling among the fibers of the stem, in which little cavities are left after the perithecia have disappeared. Ostiolum thick, cylindrical, perforated at the enlarged apex.

On stems of *Umbelliferæ*.

1747. *Sphæria rubicunda*, Schw.—Spots determinate, of a uniform red color, covered. Perithecia minute, black, papillate (the papillæ seriately prominent), seated on the inner bark and covered by the red-tinted epidermis.

On stems of *Solanum* and *Chenopodium*, Carolina and Pennsylvania (Schw.). Compare *Leptosphæria rubicunda*, on p. 360.

1748. *Sphæria lilacina*, Schw.—Covered, spots indeterminate. always lilac-colored. Perithecia loosely scattered under the spots. very minute, black, here and there subprominent.

On stems of *Asclepias Syriacus*.

1749. *Sphæria tageticola*, Schw.—Covered, at length nearly free, scattered on a cinerascent spot in the epidermis. Perithecia minute, often 2–3 confluent-subseriate, distinctly raising the epidermis. Ostiola cylindrical, becoming globose, very distinctly prominent.

On stems of *Tagetes* (cult.).

1752. *Sphæria evulsa*, Schw.—Scattered, half a line (in diam.). at first entirely covered by the fibers with which it is easily torn away. being only slightly attached. Perithecia depressed, subconical from a rather broad base, often collapsed below. Ostiolum minute, perforated.

On stems of climbing herbaceous plants.

1753. *Sphæria meloplaca*, Schw.—Gregarious, covered at first by the thin epidermis, at length denuded, immersed in a black or dark brown crust which it finally penetrates, the numerous irregularly globose, nearly free, black, rugose, astomous, empty perithecia coming in sight. Sometimes they are subostiolate-impressed at the vertex.

On stems of the larger herbs.

1754. *Sphæria Myrrhis*, Schw.—Perithecia scattered, minute, globose-conical, dark brown, very thin, glabrous, subcollapsing, confluent with the ostiola, erumpent-superficial, punctiform, seated on a dark cinereous, thin, subdeterminate, effused spot an inch in diameter. not surrounding the stem. Somewhat resembles *Phoma nebulosa*.

On branches of *Myrrhis Canadensis*.

1756. *Sphæria sulcigena*, Schw.—Seriately scattered in the furrowed surface of the culm, very minute, at first covered by the thin epidermis, tinged with black, subimmersed, subrotund, flattened, crowned with a deciduous papilla.

On old culms of *Zizania*, Philadelphia, Pa.

1757. *Sphæria Pastinacæ*, Schw.—Perithecia scattered but

effigurate-approximate, yellowish-brown, at first covered by the epidermis, at length denuded, subrotund-obovate, sometimes navicular, minute, rugose-punctate, papillate, finally irregularly collapsing.

On stems of *Pastinaca.*

Foliicolæ.

1767. *Sphæria pyramidalis*, Schw.—At first covered by the epidermis, then emergent, spinuliform, smooth, glabrous. Ostiola rather long, black when dry (*Gnomonia*).

On the lower surface of leaves, Carolina.

The following species mentioned by Schweinitz, in his Synopsis of North American Fungi, are imperfectly known and doubtful, and only their names are given here:

Sphæria Anethi, Pers. Syn. p. 30.
" *araneosa*, Pers. Syn. p. 67.
" *calva*, Tode, Fungi Meckl. II, p. 16.
" *circumscissa*, Pers. Disp. Meth. p. 4.
" *clandestina*, Fr. S. M. II, p. 484.
" *crypta*, Fr. S. M. II, p. 479.
" *foveolaris*, Fr. S. M. II, p. 499.
" *galbana*, Fr. l. c. p. 512.
" *inversa*, Fr. l. c. p. 414.
" *Lingam*, Tode, Fungi Meckl. II, p. 77.
" *mucosa*, Fr. S. M. II, p. 425.
" *oppilata*, Fr. S. M. II, p. 493.
" *pætula*, Fr. l. c. p. 483.
" *palina*, Fr. l. c. p. 411.
Sphæria polita, Fr. l. c. p. 426.
" *pleurostoma*, Kze. in. Fr. S. M. II, p. 456.
" *plinthis*, Fr. S. M. II, p. 511.
" *pyrina*, Fr. S. M. II, p. 494.
" *rhytostoma*, Fr. V. A. H. 1816, p. 148.
" *sarmentorum*, Fr. S. M. II, p. 498.
" *Solani*. Pers. Syn. p. 62.
" *subradians*, Fr. S. M. II, p. 525.
" *sulcata*, Fr. S. M. II. p. 498.
" *tenacella*, Fr. S. M. II, p. 492.
" *tephrotricha*, Fr. S. M. II, p. 448.
" *varia*, Pers. Syn. p. 52.

GLOSSARY.

Abnormal. Differing from the usual form.
Aculeate. Prickly; beset with prickles.
Aculeolate. Beset with diminutive prickles
Adnate. Grown to; firmly united.
Amorphous. Shapeless; without form.
Amphigenous. Growing on both sides.
Anastomosing. Connected by transverse branches so as to form a more or less perfect network.
Annular. In the form of a ring.
Aparaphysate. Without paraphyses.
Areolate. Marked out into small spaces.
Ascospores. Spores produced in asci.
Ascus. The membranaceous sac containing the sporidia.
Astomous. Mouthless.
Basal. At or pertaining to the base.
Beaked. Ending in a prolonged tip or beak.
Bifid Two-cleft.
Bilocular. Two-celled.
Biogenous. Growing on living organisms
Biseriate. In two rows or series
Canescent. Gray or hoary.
Capitate. Collected in a head or furnished with a head.
Capsule. A dry, dehiscent fruit composed of more than one carpel
Carinate. Furnished with a keel or projecting, longitudinal line.
Cartilaginous. Firm and tough; like a cartilage.
Caudate. Having a slender, tail-like appendage.
Cespitose. Growing in tufts.
Chartaceous Like paper.
Cinereous. Ash-colored.
Circinate. Lying in a circle.
Circumscissile. Splitting around horizontally.
Clavate. Club-shaped
Cirrhus, pl. *Cirrhi.* Tendrils, elongated, tendril-like collections of exuded spores.
Colliculose. Covered with small, rounded prominences.
Concrescent. Growing together.
Conglobate. Collected in a subspherical mass.
Continuous. Not divided by septa.
Crustaceous. Of hard and brittle texture.
Cuneate. Wedge-shaped.
Cymbiform. Boat-shaped.
Deciduous. Not persistent: falling off.
Dendroid. Shaped like a tree.
Dichotomous. Dividing regularly by pairs.
Didymous. Double, of two equal parts.
Digitate. Furnished with fingers, dividing like the fingers of a hand.
Dimidiate. Halved. In a dimidiate perithecium the lower half is wanting.
Distichous. In two rows or layers.
Echinate. Beset with prickles.
Epiphyllous. Growing on the upper side of a leaf.
Epithecium. The layer sometimes formed above the asci by the concrescent tips of the paraphyses.
Erumpent. Bursting out.
Exserted. Projecting: standing out.
Falcate. Scythe-shaped.
Farinose. Covered with a meal-like powder
Fastigiate. with branches erect and close together.
Fascicle. A small bundle.
Ferruginous. Rust-colored.
Filiform. Thread-like.
Fimbriate. Fringed
Flexuous. Bending alternately in opposite directions.
Forked. Divided into two equal parts.
Friable. Easily crumbling.
Fugacious. Disappearing very early.
Fuligmous. The color of soot.
Glabrous. Smooth, not rough, pubescent, or hairy
Gregarious. Growing in groups.
Guttulate Containing nuclei resembling small drops of water or oil.
Hirsute. Clothed with rather coarse, stiff hairs.
Hispid Clothed with rigid hairs or bristles.
Hyaline. Transparent like glass.
Hypophyllous. Growing on the under side of a leaf.
Innate. Grown into.
Inordinate. Not in regular order.
Lacerate. Irregularly torn.
Laciniate. Cut into narrow, pointed lobes.
Lenticular. Lens-shaped.
Linear. Long and narrow with sides parallel.
Membranaceous. Thin and soft and more or less transparent.
Moniliform. Like a string of beads.
Muricate. Rough with short, hard points.
Muriculate. Finely muricate.
Obconical. Inversely conical.
Oblique. Unequal-sided or slanting.
Obovate. Inverted ovate.
Obsolete. Not evident; rudimentary.
Opake. Dull; not smooth or shining.
Operculate. Furnished with a lid.
Ostiolum. A little mouth; the opening in the apex of the perithecium.
Ovate. Egg-shaped.
Papillose. Having minute, rounded projections.

Paraphyses. Slender, thread-like bodies growing with the asci.
Parasitic. Growing on and deriving nourishment from another plant
Pedicellate Furnished with a stem.
Penicillate. Like a brush
Peripherical. Lying around the circumference.
Perithecium. The case or hollow shell which contains the spores.
Polymorphous. Having many forms.
Polystichous In several rows or layers.
Pubescent. Covered with hairs, mostly short, soft hairs.
Pulverulent As if covered with dust.
Punctulate Dotted with minute depressions, as if punctured with the point of a pin.
Pycnidium A stylosporous perithecium forming an early stage in the development of some Pyrenomycetes.
Radiate Spreading from a common center.
Ramicolous Growing on branches.
Reticulate. Forming a network.
Rostrate. Having a beak.
Rudimentary But slightly developed.
Rufous. Reddish-brown.
Rugose. Wrinkled.
Saprogenous. Growing on decaying substances.
Septum. Any kind of partition.
Sessile. Without any stem or foot-stalk.
Sinuate. With a wavy outline.
Spermogonium. Differs from the pycnidium in its smaller spores.
Sporidium. A spore produced in an ascus.
Spores. A general term applied to the fruit among the fungi.
Sporules. Spores produced in perithecia but not in asci.
Squamose. Scaly.
Squamulose. With minute scales.
Sterile. Unproductive; without fruit.
Stipe. A stem or foot-stalk.
Stroma. A bed; that which supports or includes the perithecia.
Stylospores. Spores produced in a perithecium and borne on pedicels (basidia), but not contained in asci.
Suberose. Like cork; corky.
Sub-erose. Slightly erose.
Terete. Having a circular, transverse section.
Tomentose Densely pubescent with matted wool.
Trimorphous. Having three forms.
Turbinate Top-shaped; inverted conical.
Undulate Wavy
Verrucose. Warty.
Verticillate Arranged in a whorl.
Vesicular. Composed of vesicles or small bladder-like cavities.

ABBREVIATIONS.

det. Determined by.
diam. Diameter.
fide. On the authority of.
l c. In the place (book or work) cited.
mm. Millimeter or Millimeters.
sec. According to.
spec. Specimen.
specc. Specimens.

CORRECTIONS.

Page 300, last line, for "shiny," read slimy.

Page 609, fifth line from top, for "*gramius*," read *graminis*.

Page 45, bottom line, for "3-septate" read 4-septate.

" 46, 7th line from bottom, for "3-septate" read 4-septate.

" 166, 5th line from bottom, for "1—1 ½ micr." read 1—1 ½ mm.

" 219, 4th line from top strike out "not."

" 300, bottom line, for "shiny" read slimy.

" 429, 19th line from top for "3 micr." read 3—4 and second line from bottom, 1890 for "1891."

" 452, 16th line from bottom, for "85×7" micr. read 45×7.

" 493, 4th line from bottom for "30×35" read 30—35.

" 538, add measurements of sporidia of *Ps profusa*, 45—50×15 micr.

" 581, 11th line from bottom, for "¼—½ mm." read ¼—½ cm.

" 624, add measurements of sporidia of *N. glycyrrhiza*, 9—10×5 micr.

" 640, 18th line from bottom for "2 mm." read ½ mm.

" 645, 3d line from bottom for "10×15" read 10×4—5.

to the bottom of the stroma or distributed through it. Ostiola not convergent. Sporidia allantoid, hyaline or brownish.

contains the sn

To the "Corrections" on page 760 add the following:

Page 111. For "Attractium" read Atractium.
" 143. " "A." read V. before "Alchemillæ."
" 171. " "Shæropsis" read Sphæropsis.
" 173. " "Helmithosphæria" read Helminthosphæria.
" 189. " "Carlise" read Carlisle.
" 192. " "Wallwrothiella" read Wallrothiella.
" 194. " "hystricina" read hystricinum.
" 285. " "Stellarinearun" read Stellarinearum.
" 384. " M. "sqmata" read M. squamata.
" 482. Put V. before "Cypri."
" 554. " "Silia" read Sillia.
" 577. " "grandinea" read grandineum.
" 584. " "E." read D.
" 603. " "Cheonpodii" read Chenopodii.
" 605. " 15th line, for "1855" read 1955, and in the next line for "1041" read 1941, and for "Rosæa" read Rosæ.
" 730. " For "Cuccurbitaria" read Cucurbitaria.
" 763. " "Ticothecium" read Tichothecium.
" 763. " "Trichosphæriaceæ" read Trichosphærieæ.

In "Keys to the genera," pp. 4 and 58, *Uncinula and Hyponectria* should have been included.

The genera *Botryosphæria to Valsaria* inclusive, constitute the Family Melogrammeæ, and *Diatrype* to *Diatrypella* inclusive, the *Fam. Diatrypeæ.* The diagnoses of these Families were accidentally omitted and are given below.

Page 546, before Botryosphæria, insert:

Family. Melogrammeæ.

Stroma valsoid or diatrypoid, subpulvinate, globose or irregular, mostly erumpent, and becoming superficial. Perithecia sunk in the stroma or often with the upper part free. Sporidia various.

Page 565, before Diatrype, insert:

Family. Diatrypeæ.

Stroma effused or valsoid, subglobose or pulvinate, erumpent. Perithecia sunk to the bottom of the stroma or distributed through it. Ostiola not convergent. Sporidia allantoid, hyaline or brownish.

INDEX OF GENERA.

INDEX OF SPECIES.

PLATE 1.

Erysipheae.

Sphaerotheca Castagnei, Lev.

Fig. 1. Natural size on the leaf.

" 2. A group slightly magnified.

" 3. A conidiophore bearing three conidia, with one at the base germinating, the germ tube entering one of the stomata of the leaf.

" 4. A perithecium having one of the appendages bifid at the tip.

" 5. Perithecium with the single 8-spored ascus escaping.

" 6. Four sporidia.

NOTE.

Figures on Plates 1, 2, 3, 4, 6, 9, 11, 13, 14, 15, 18, 21, not drawn to any scale.

PLATE 1.

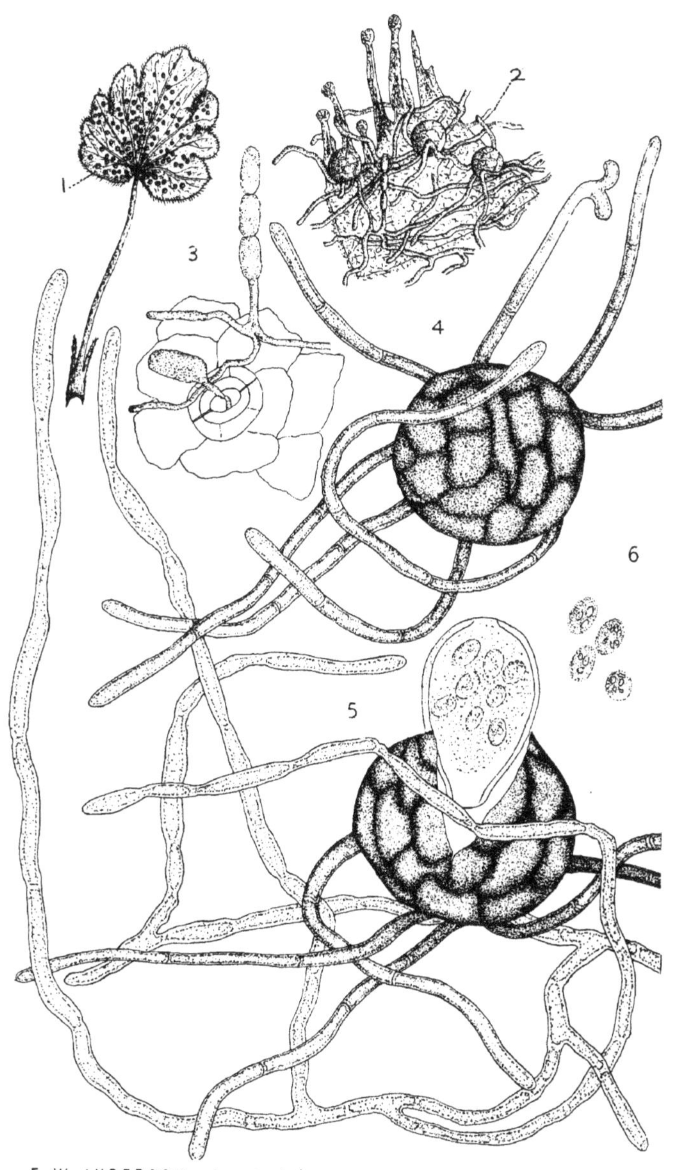

F. W. ANDERSON, ad. nat. del

PLATE 2.

Erysipheae.

Uncinula Salicis, (D C.).

Fig. 1. Natural size on leaf of Salix flavescens, var. Scouleriana, from Montana.

" 2. Part of a leaf with perithecia magnified.

" 3. A perithecium highly magnified.

" 4. An ascus.

" 5. Three sporidia.

" 6. A conidia-bearing hypha.

" 7. Five conidia.

" 8. Initial hyphae forming a perithecium.

PLATE 2.

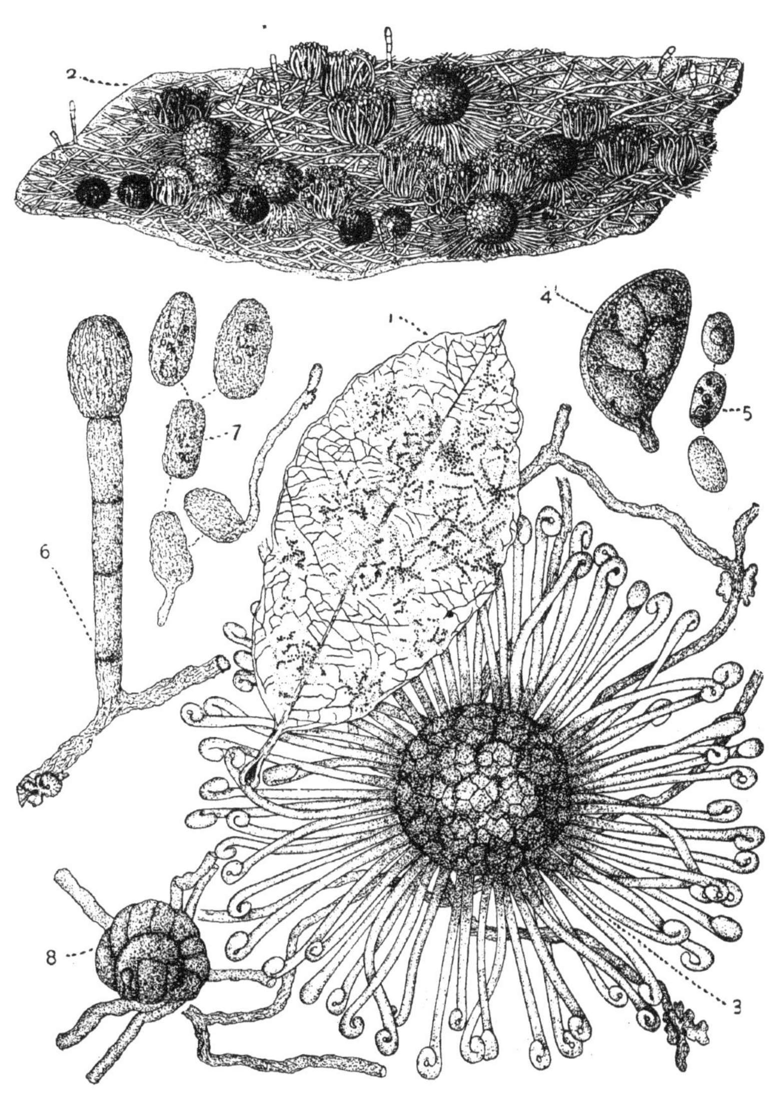

F. W. ANDERSON, ad. nat. del.

PLATE 3.

Erysipheae.

Phyllactinia suffulta, (Reb.).

Fig. 1. Natural size, on chestnut leaf.

" 2. Perithecium enlarged.

" 3. Two asci.

" 4. Three sporidia.

" 5. Conidia-bearing hyphae.

" 6. Conidium germinating.

PLATE 3.

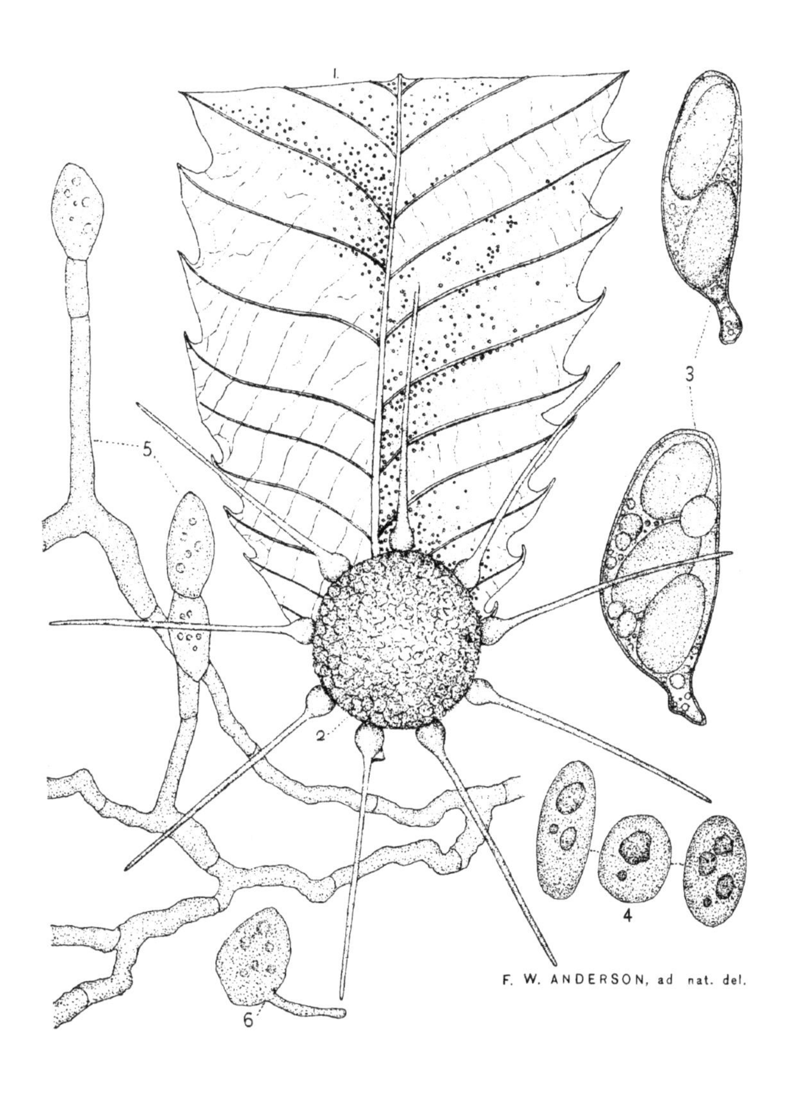

PLATE 4.

Erysipheae.

Podosphaera tridactyla, (Wallr.).

Fig. 1. Natural size on leaf of Morello cherry.

" 2. An enlarged perithecium.

" 3. Two asci.

" 4. Four sporidia.

" 5. A chain of conidia.

" 6. Two conidia germinating.

PLATE 4.

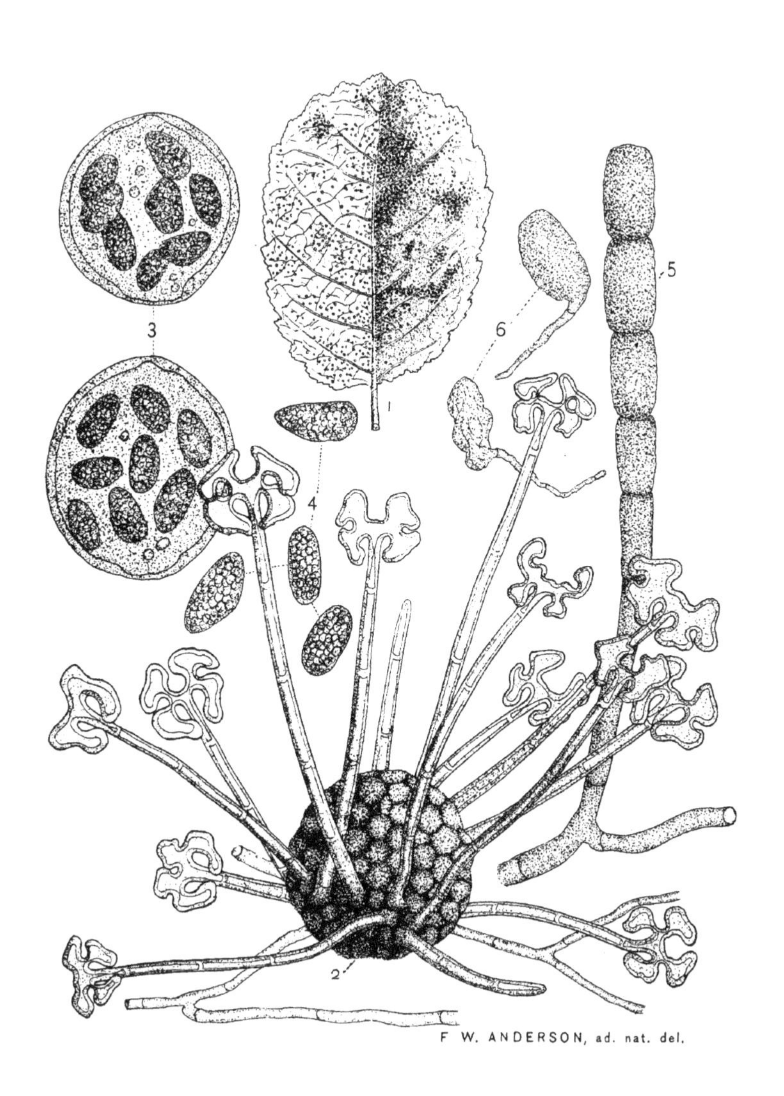

PLATE 5.

Perisporieae.

Fig. 1. Enlarged perithecium of Meliola bidentata, Cke.

" 2. An ascus.

" 3. A sporidium.

" 4. A branch of mycelium with hyphopodia.

" 5. Tips of three appendages magnified.

" 6. **Meliola palmicola,** Winter, four magnified appendage-tips.

" 7. **Meliola furcata,** Lev. three appendage-tips, from specimens collected in Nicaragua by Wright.

" 8. **Meliola bicornis,** three appendage-tips, copied from Rabh-Winter's Fungi Eur. No. 3547.

N. B.—In this plate and all succeeding plates, where not otherwise noted, all asci are drawn to a scale of 28 μ to the inch and all enlarged sporidia to a scale of 14 μ to the inch.

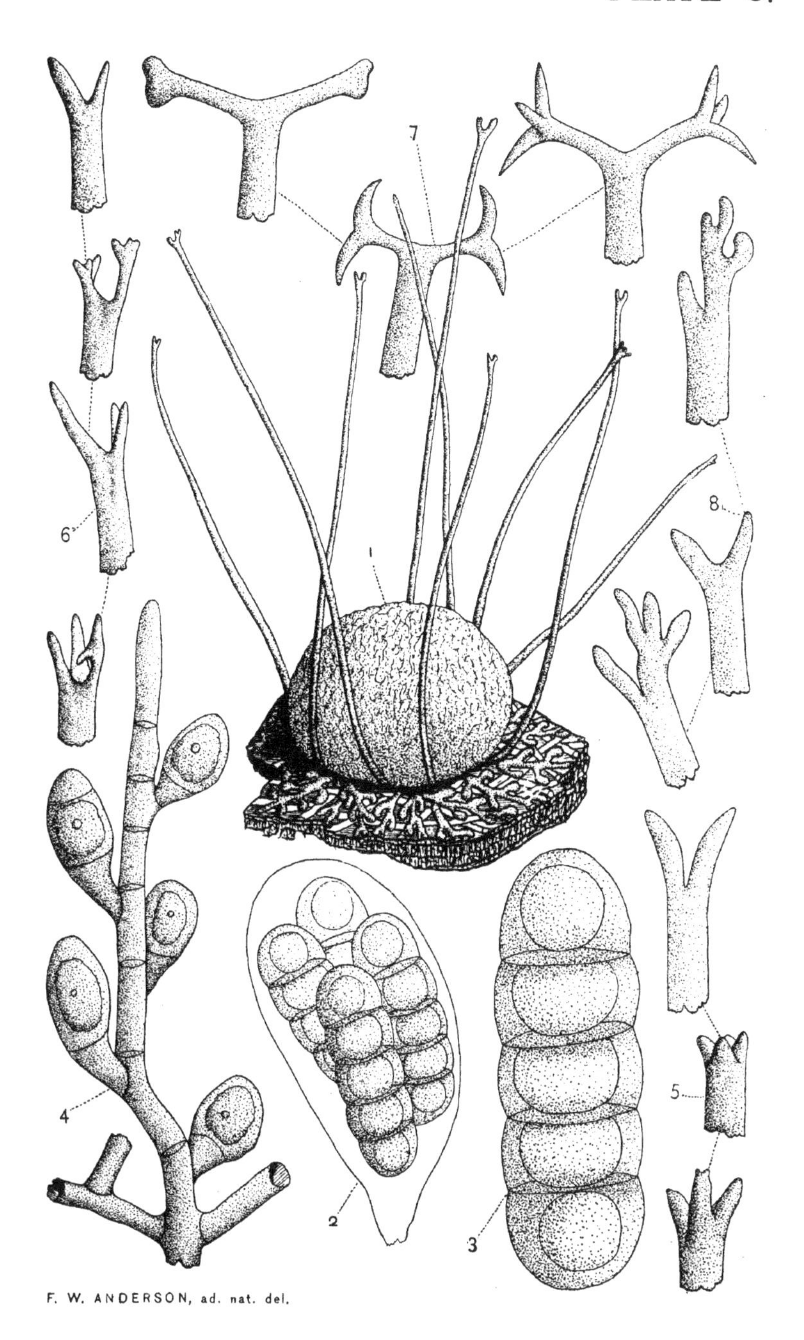

F. W. ANDERSON, ad. nat. del.

PLATE 6.

Perisporieae.

Asterina delitescens, E. & M.

Fig. 1. Natural size on leaf.

" 2. Portion of a leaf with the fungus somewhat magnified.

" 3. Four perithecia, one entire and three showing the stellate mode of dehiscence.

" 4. Two asci.

" 5. Two sporidia.

" 6. Perithecium in its early stage of growth.

" 7. Hypha from which arises a conidiophore and conidia?

" 8. Conidium?

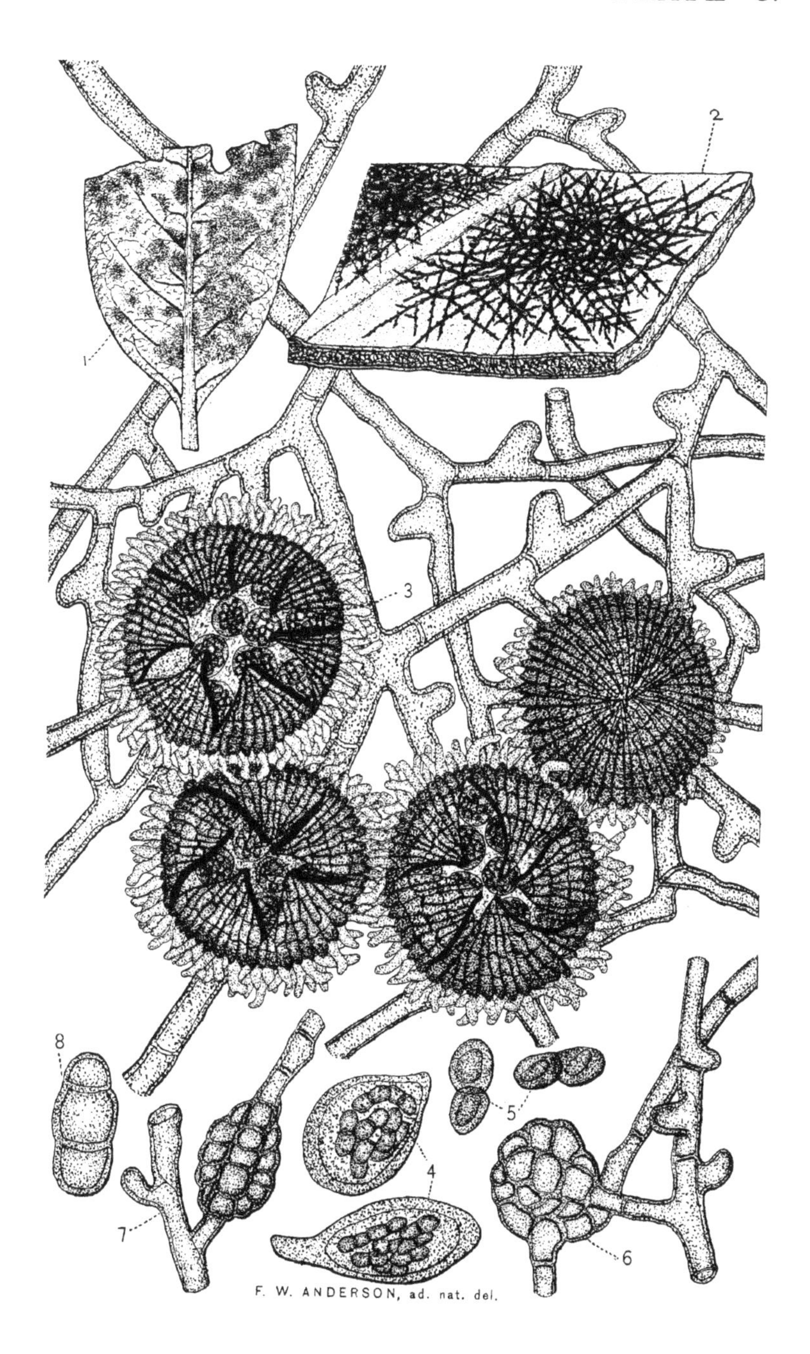

1
2
3
4
5
6
7
8
F. W. ANDERSON, ad. nat. del.

PLATE 7.

Perisporieae. (figs. 1-13.)

Fig. 1. **Perisporium funiculatum,** Preuss, natural size. The specimens are from Kriegers Saxon Fungi 626, on straw of an old bee hive and were figured because no American specc. were at hand.

" 2. Enlarged section of a perithecium.

" 3. An ascus.

" 4. A sporidium.

" 5. **Capnodium grandisporum,** E. & M. natural size, on leaf of *Gelsemium sempervirens*, Florida.

" 6. Perithecium enlarged.

" 7. Ascus and sporidia.

" 8. A sporidium.

" 9. Pycnidial perithecium.

" 10. Pycnidial spores.

" 11. **Microthyrium Smilacis,** De Not, natural size, on stem of *Smilax*, Newfield, N. J.

" 12. Two asei with paraphyses, the ascus on the left mature, the other immature.

" 13. Two sporidia.

" 14. **Polystigma rubrum,** (Pers.)

" 15. An ascus.

" 16. Two sporidia.

" 17. Three spermatia.

NOTE.—Figures 15, 16 and 17 enlarged from the drawing in Briosi & Cavara's Fungi Parasiti, Fasc. 1, No. 12.

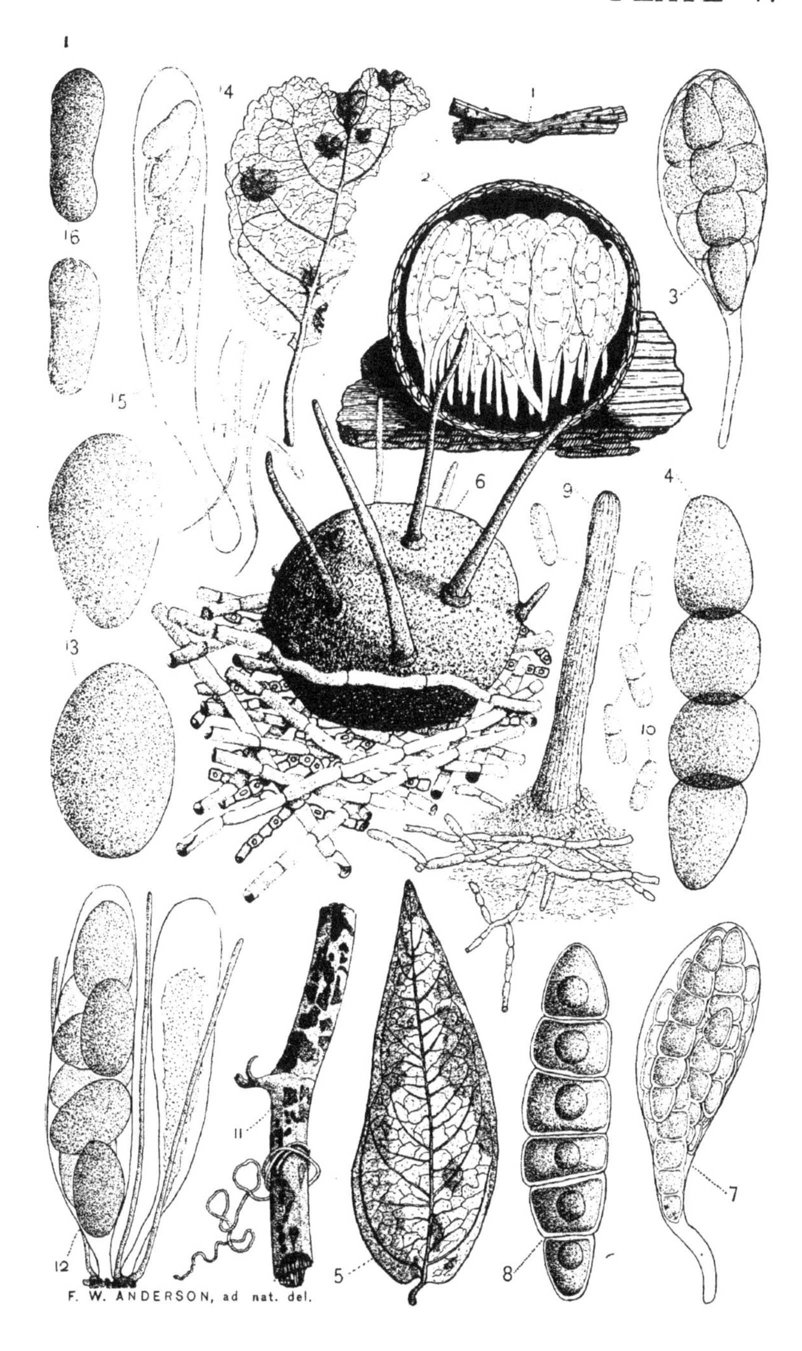
1
14
1
2
16
3
15
6
9
4
13
10
11
12
5
8
7
F. W. ANDERSON, ad nat. del.

PLATE 8.

Perisporieae.

Fig. 1. **Eurotium herbariorum,** Lk. natural size on a barrel hoop, in a cellar.

" 2. A piece of the hoop with the fungus enlarged.

" 3 An enlarged perithecium.

" 4. A cluster of sporidia as they lie in the ascus.

" 5. Three sporidia seen from different angles.

" 6. **Apiosporium erysipheoides,** S. & E., an enlarged perithecium.

" 7. **Lasiobotrys Lonicerae,** Kze. on leaf of Lonicera—an enlarged stroma showing perithecia around its margin.

8 & 9. Enlarged from Winters Pilze.

" 10. **Dimerosporium erysipheoides,** E. and E., natural size, on Cynodon dactylon.

" 11. A group enlarged.

" 12. Sectional view of an enlarged perithecium.

" 13. An ascus with paraphyses.

" 14. Three sporidia.

PLATE 8.

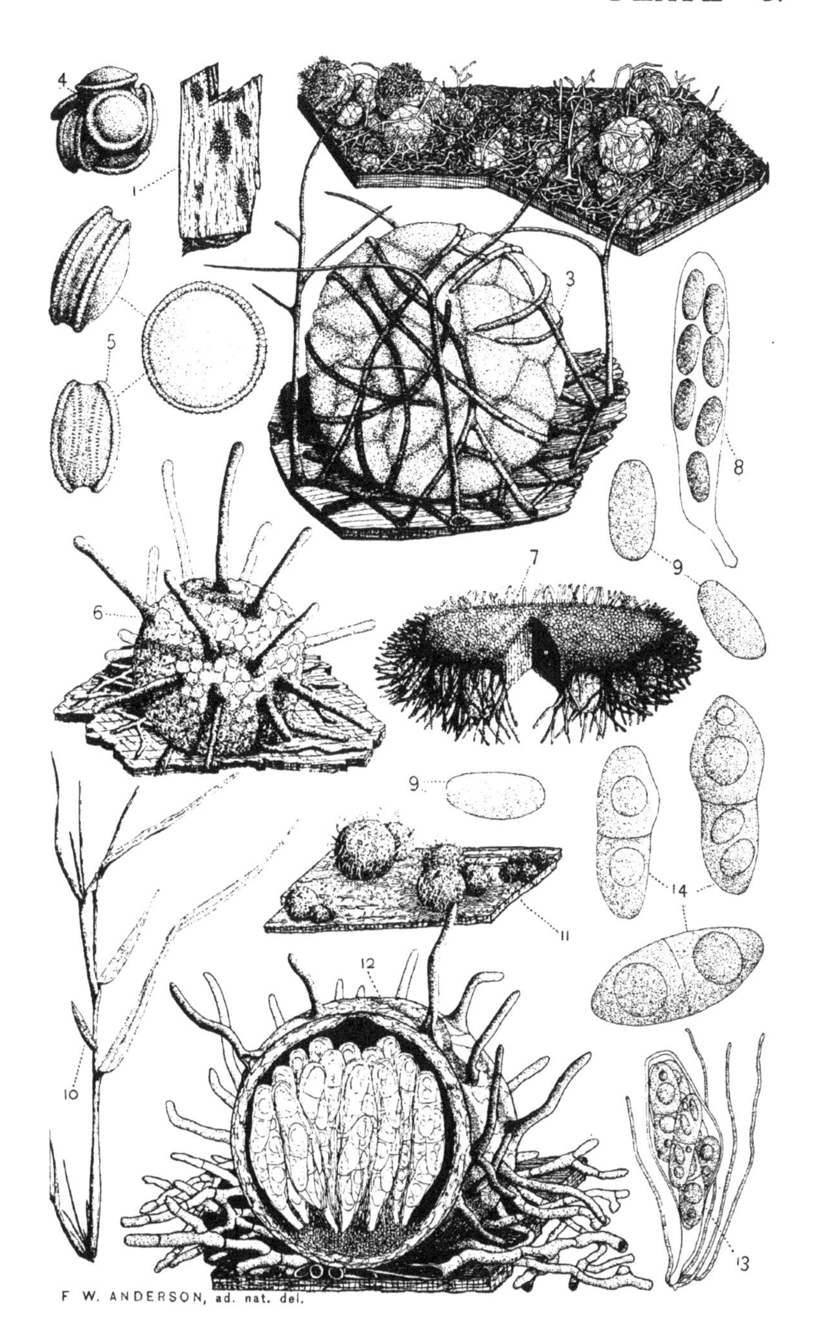

PLATE 9.

Perisporieae.

Fig. 1. **Saccardia Martini,** Ell. & Sacc. natural size, on leaf of *Quercus laurifolia.*

" 2. Part of a leaf with a perithecium somewhat enlarged.

" 3. Perithecium enlarged with appendages and mycelium.

" 4. Two asci.

" 5. Six sporidia.

" 6. **Myriococcum Everhartii**, S. & E., on rotten wood.

" 7. A group of perithecia enlarged.

" 8. A perithecium highly magnified.

" 9. A perithecium with the outer wall partly removed, showing the arrangement of the interior cells.

" 10. Eight variously shaped cells.

" 11. A few of the superficial hairs that cover the perithecia.

NOTE.—The figures on this plate not drawn to any scale.

PLATE 9.

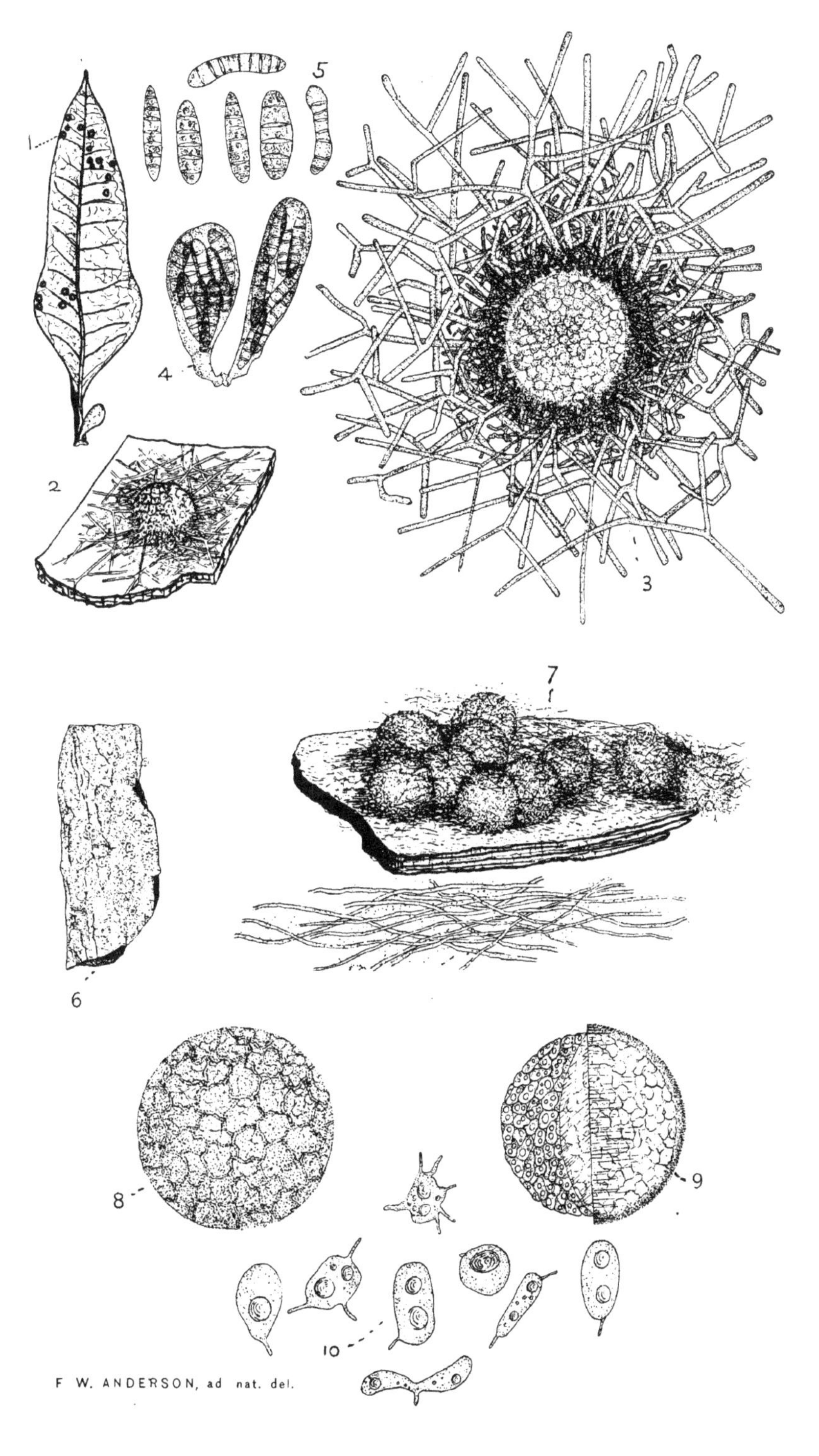

F. W. ANDERSON, ad nat. del.

PLATE 10.

Perisporieae.

Fig. 1. **Scorias spongiosa,** Schw. natural size on limb of *Alnus serrulata.*

" 2. Part of a fertile branch enlarged, showing spermogonial and ascigerous perithecia.

" 3. An ascus.

" 4. Three sporidia.

" 5. **Capnodium salicinum,** (Pers.). On bark of *Negundo aceroides*; a group of spermogonial and ascigerious perithecia considerably magnified.

" 6. An ascus.

" 7. Three sporidia.

" 8. **Capnodium axillatum.** Cke. natural size on fragment of leaf of *Catalpa.*

" 9. A group of spermogonial perithecia, considerably magnified.

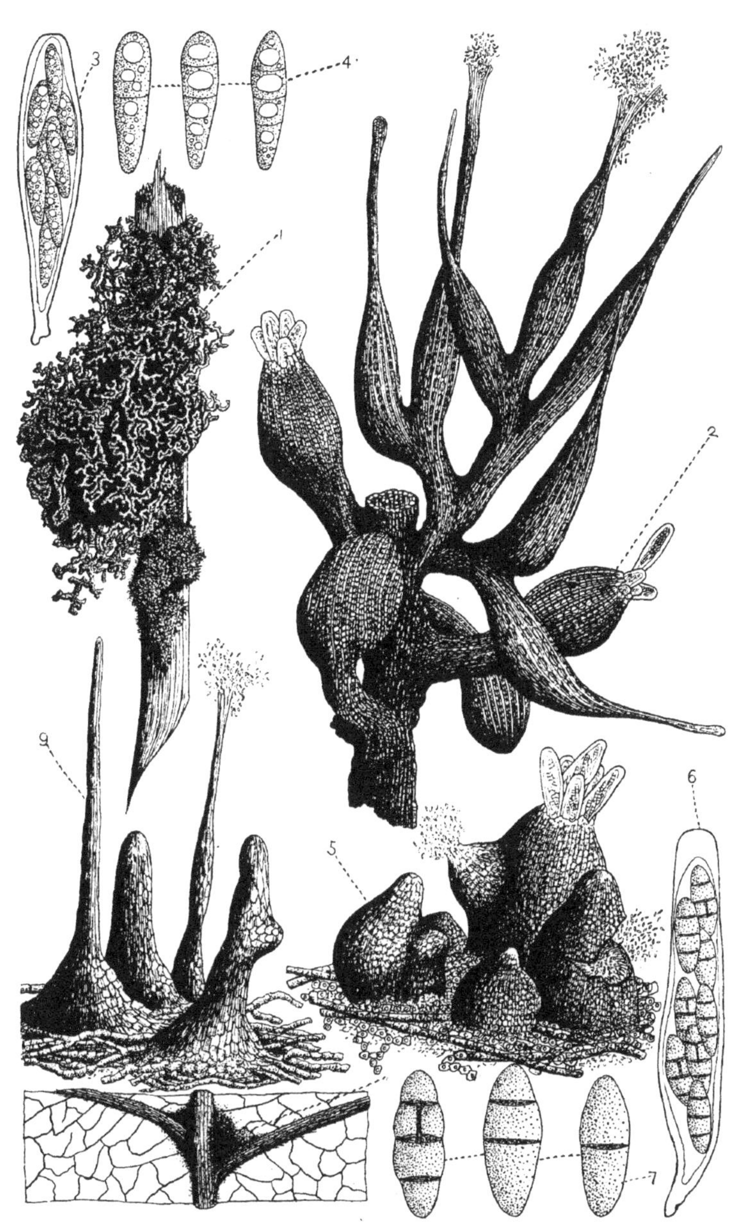

F. W. ANDERSON, ad nat. del

PLATE 11.

Hypocreaceae.

Fig. 1. **Hypocrea lichenoides**, (Tode) natural size on a decaying limb.

" 2. An ascus.

" 3. Two sporidia.

" 4. **Hypocrea citrinella,** Ell. on a dead twig of *Vaccinium* somewhat enlarged.

" 5. A section through the edge of the stroma, showing the enlarged perithecia and prominent ostiolum.

" 6 An ascus.

" 7. Two sporidia.

" 8. **Ascus of Hypocrea consimilis**, Ell.

" 9. Three sporidia.

" 10. **Hypocrea corticiicola**, E. & E., an ascus.

" 11. Two sporidia.

" 12. **Hypomyces Lactifluorum,** (Schw.), section of host enlarged showing the imbedded perithecia.

" 13. Three asci, the middle one containing mature sporidia.

" 14. Two sporidia much enlarged.

NOTE.—Figures on this plate not drawn to any definite scale.

PLATE 11.

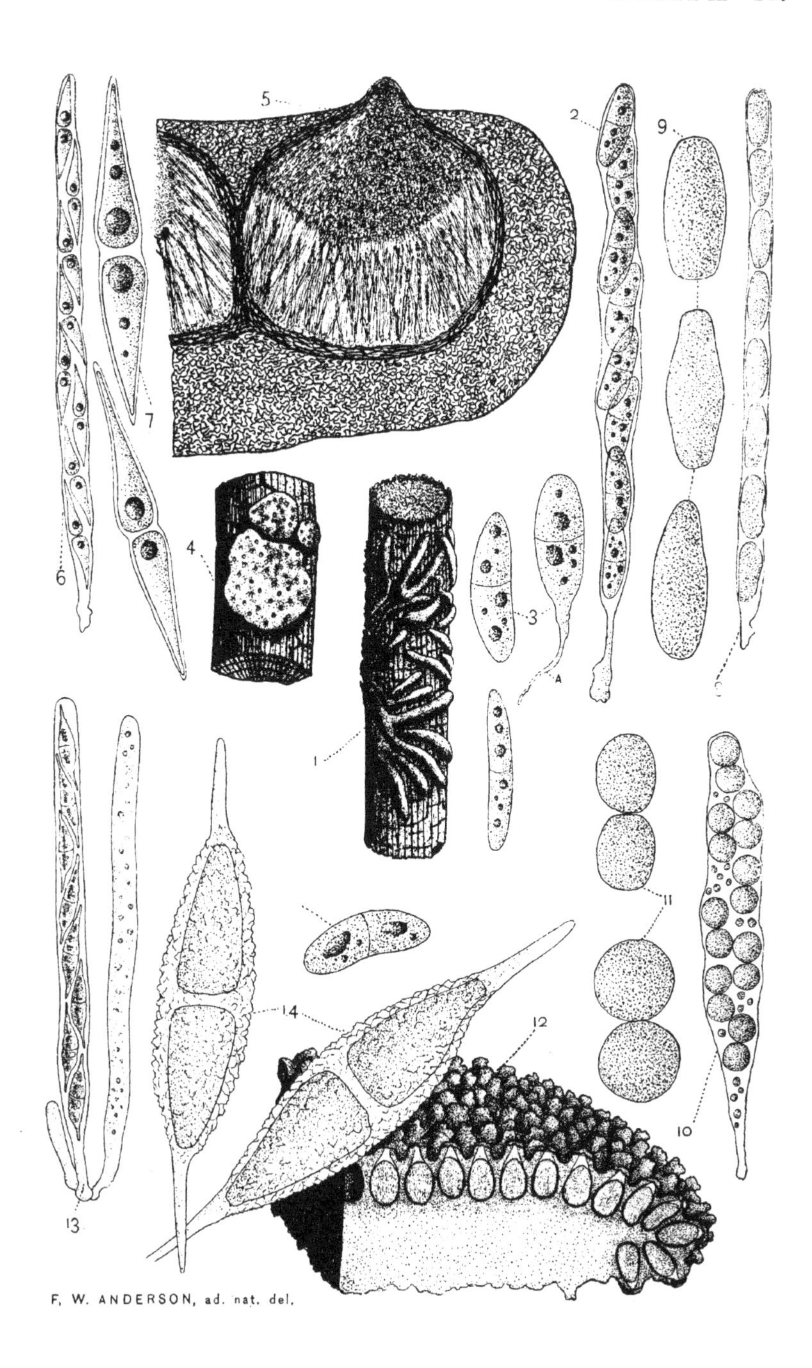

PLATE 12.

Hypocreaceae.

Fig. 1. **Sphaerostilbe gracilipes,** Tul.

" 2. Conidia (Stilbum), one of which rises from the center of the cluster of perithecia.

" 3. An ascus.

" 4. Two sporidia.

" 5. Conidia on their branching basidia, enlarged.

" 6. Three conidia highly magnified.

" 7. **Pleonectria Berolinensis,** Sacc., an ascus.

" 8. Three sporidia.

" 9. **Chilonectria cucurbitula,** (Curr.), an ascus.

" 10. One of the cylindrical bodies contained in the asci with sporidia escaping from a rupture in one side.

" 11. Cluster of basidia (Tubercularia) with terminal conidia.

" 12. Conidia more highly magnified.

" 13. **Nectria verrucosa,** (Schw.), natural size on piece of limb of *Morus alba.*

" 14. A cluster of magnified perithecia.

" 15. The depressed stroma (Tubercularia).

" 16. Group of basidia and conidia taken from the stroma and enlarged.

" 17. Conidia enlarged.

" 18. An ascus.

" 19. Two sporidia.

NOTE.--Asci drawn to a scale of 28 μ to the inch and the enlarged sporidia and conidia to a scale of about 14 μ to the inch.

PLATE 12.

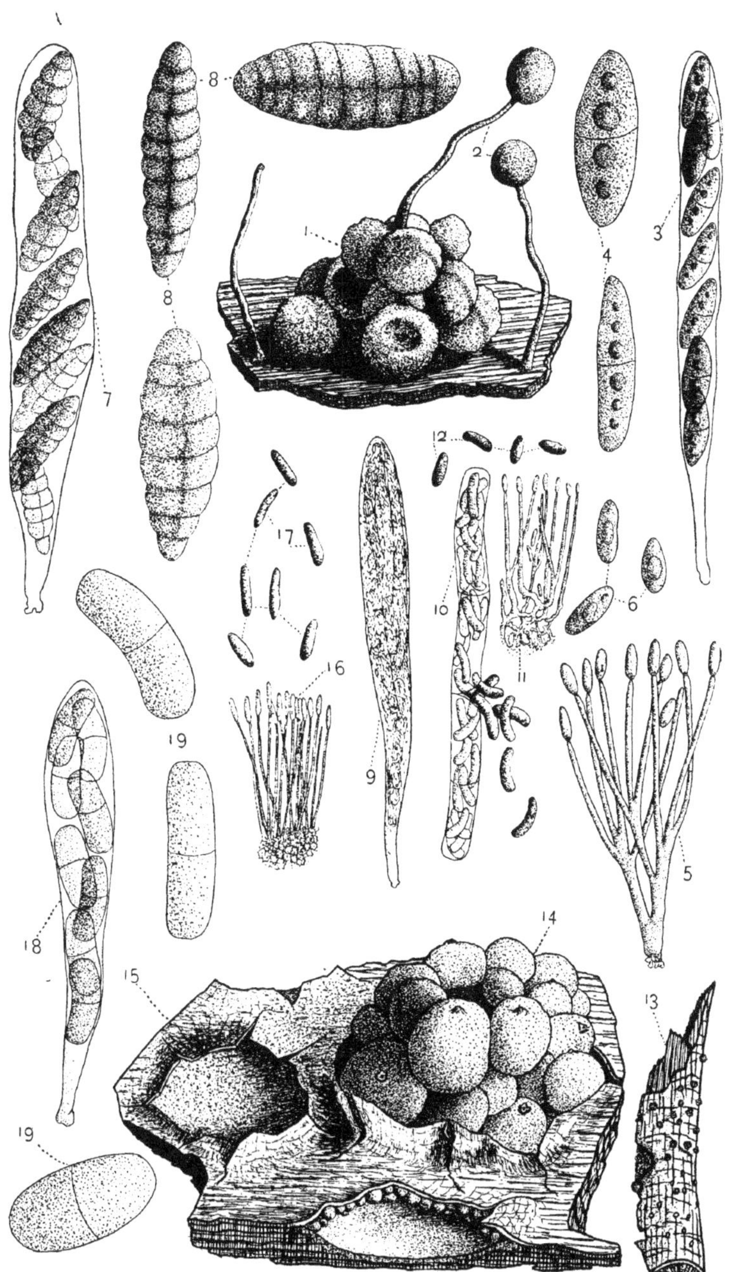

F. W. ANDERSON, ad nat. del.

PLATE 13.

Hypocreaceae.

Fig. 1. **Gibberella pulicaris**, Fr., natural size, on dead corn stalk.

" 2. A group of perithecia somewhat enlarged.

" 3. Three perithecia highly magnified. (A.) an entire perithecium. (B.) a broken down and empty one. (C.) a vertical section of a perithecium showing asci.

" 4. A group of asci in various stages of development.

" 5. Four mature sporidia.

" 6. Three sporidia germinating.

" 7. **Calonectria Canadensis**, E. & E., natural size on dead elm.

" 8. Cluster of perithecia around the base of the conidial stroma.

" 9. Vertical section of perithecium.

" 10. Two mature asci.

" 11. Three mature sporidia.

" 12. Young perithecia around the base of the conidia-bearing stroma, the central head cut vertically showing the superficial conidial layer.

" 13. Basidia and conidia enlarged.

" 14. Two conidia enlarged.

NOTE.—Figures on this plate not drawn to a definite scale.

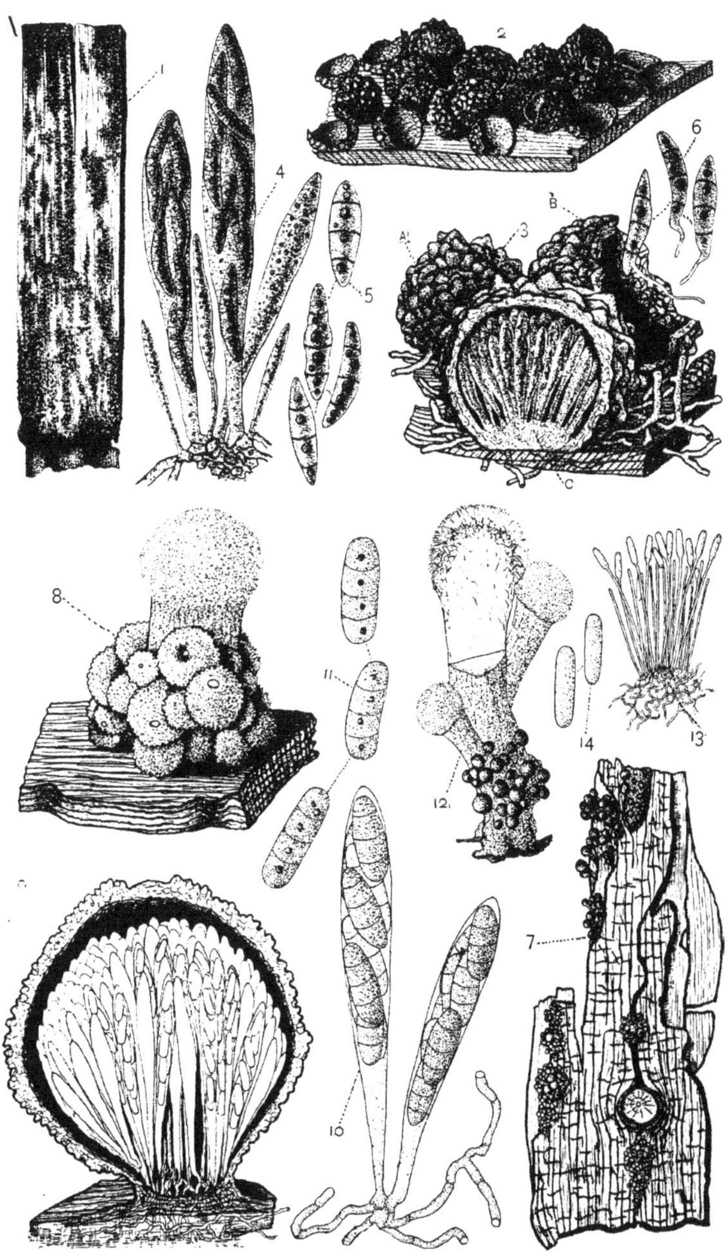

F. W. ANDERSON, ad nat. del.

PLATE 14.

Hypocreaceae.

Fig. 1. **Melanospora chionea,** (Fr.) on decaying leaf.

" 2. Portion of the same enlarged.

" 3. Sectional view of a perithecium highly magnified.

" 4. Three asei, one of them mature.

" 5. Mature sporidium.

" 6. **Eleutheromyces subulatus,** (Tode) natural size on decaying Agarie.

" 7. A fragment somewhat magnified.

" 8. Section of a perithecium highly magnified.

" 9. Several asci, one mature, with the peculiarly jointed paraphyses.

" 10. Three sporidia.

" 11. A stout jointed bristle from the outside of the ostiolum.

" 12. A portion of one of the hair-like filaments composing the inner lining of the ostiolum.

NOTE.—Figures on this plate not drawn to any scale.

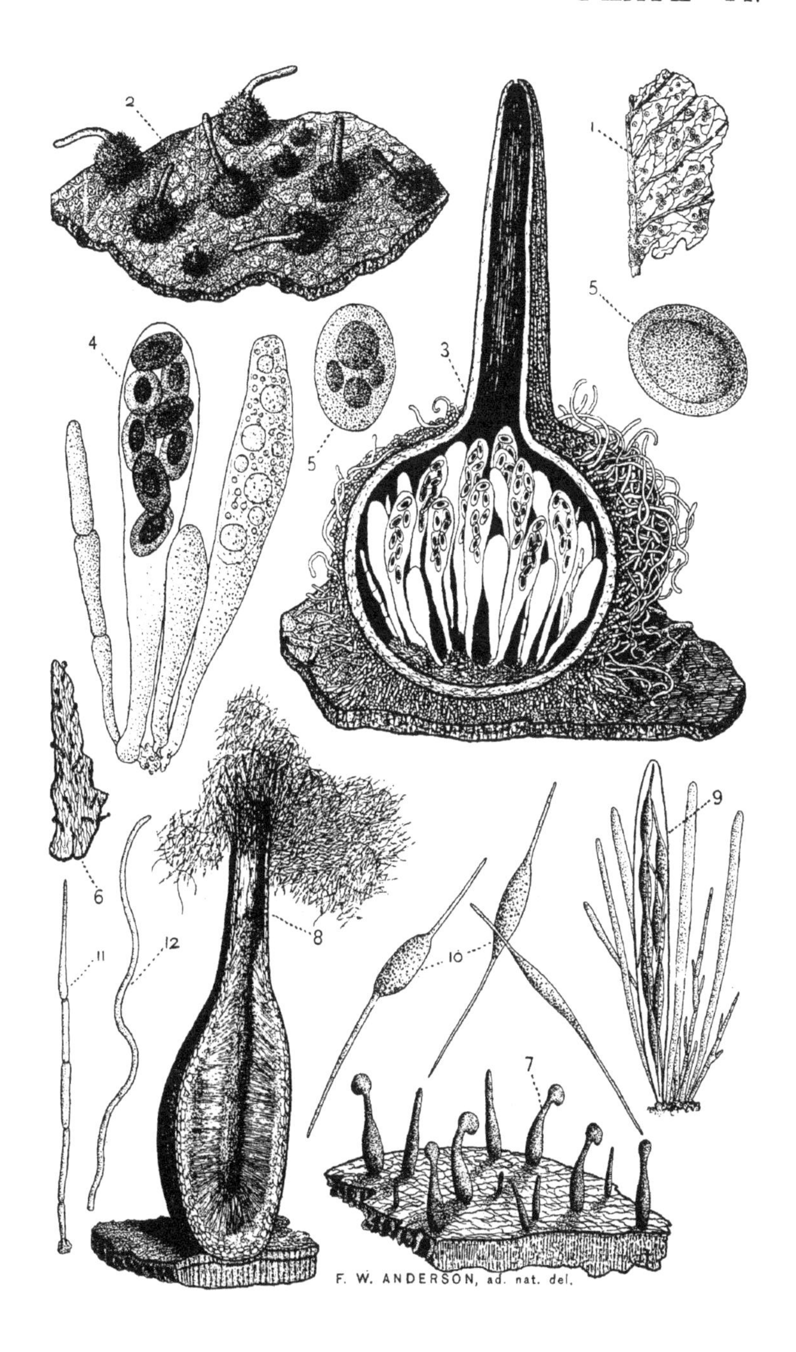
1
2
3
4
5
5
6
7
8
9
10
11
12
F. W. ANDERSON, ad. nat. del.

PLATE 15.

Hypocreaceae.

Fig. 1. **Ophionectria Everhartii**, E. & G., enlarged perithecium.

" 2. An ascus.

" 3. A sporidium.

" 4. **Cordyceps Sphingum**, (Tul.) natural size growing from a dead larva in a cocoon.

" 5. An enlarged piece of a stroma bearing perithecia.

" 6. A perithecium highly magnified.

" 7. An ascus.

" 8. **Epichloe typhina**, (Pers.), natural size, on culm of grass.

" 9. Enlarged section through the stroma and perithecia.

" 10. An ascus with the lower part broken away showing the protruding sporidia.

" 11. **Cordyceps clavulata**, (Schw.) on dead scale insects on a living twig of *Fraxinus*, somewhat enlarged.

" 12. An enlarged head of one of the stromata showing the perithecia.

" 13. A sporidium.

" 14. **Claviceps microcephala**, Tul. somewhat enlarged.

" 15. An ascus.

" 16. A sporidium.

NOTE.—Figures on this plate not drawn to any definite scale.

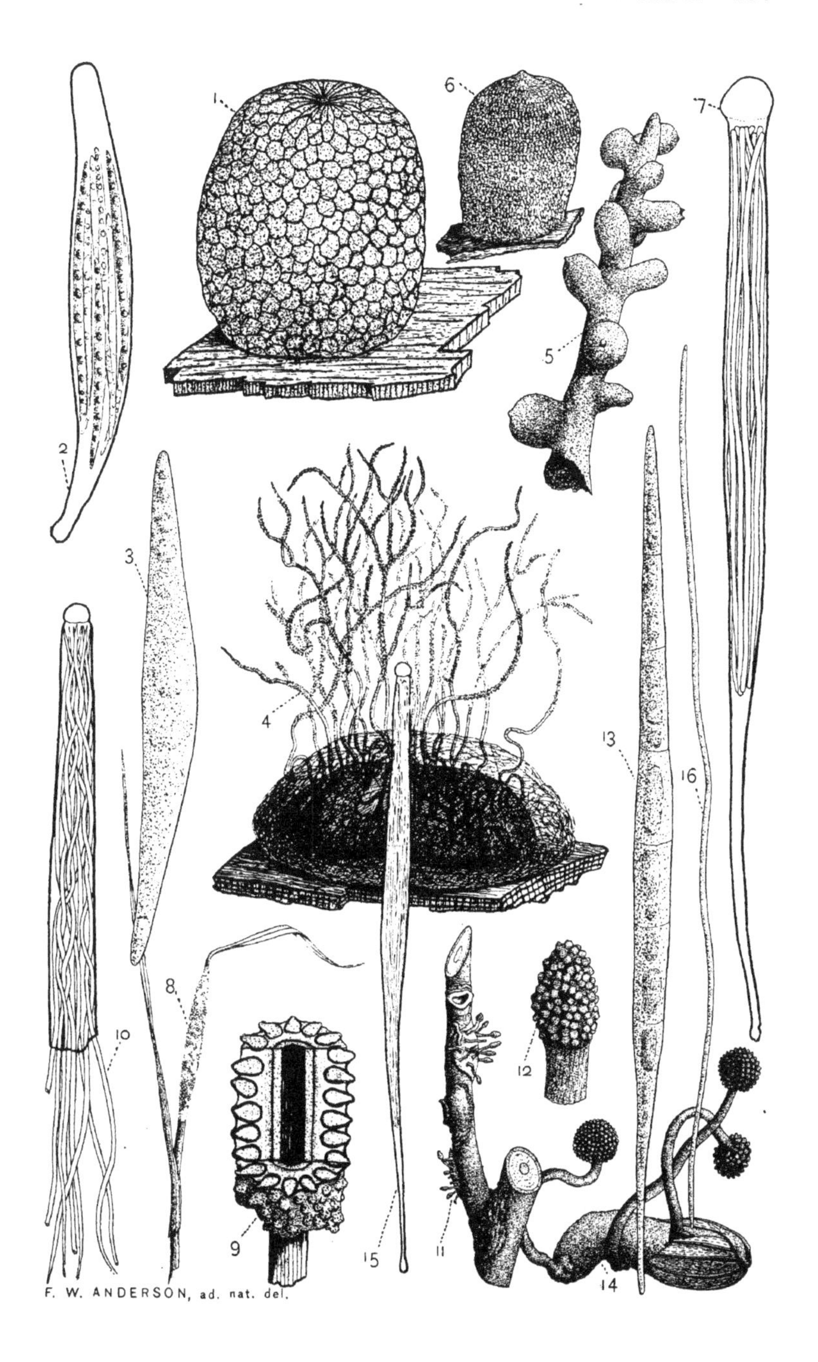
1
2
3
4
5
6
7
8
9
10
11
12
13
14
15
16
F. W. ANDERSON, ad. nat. del.

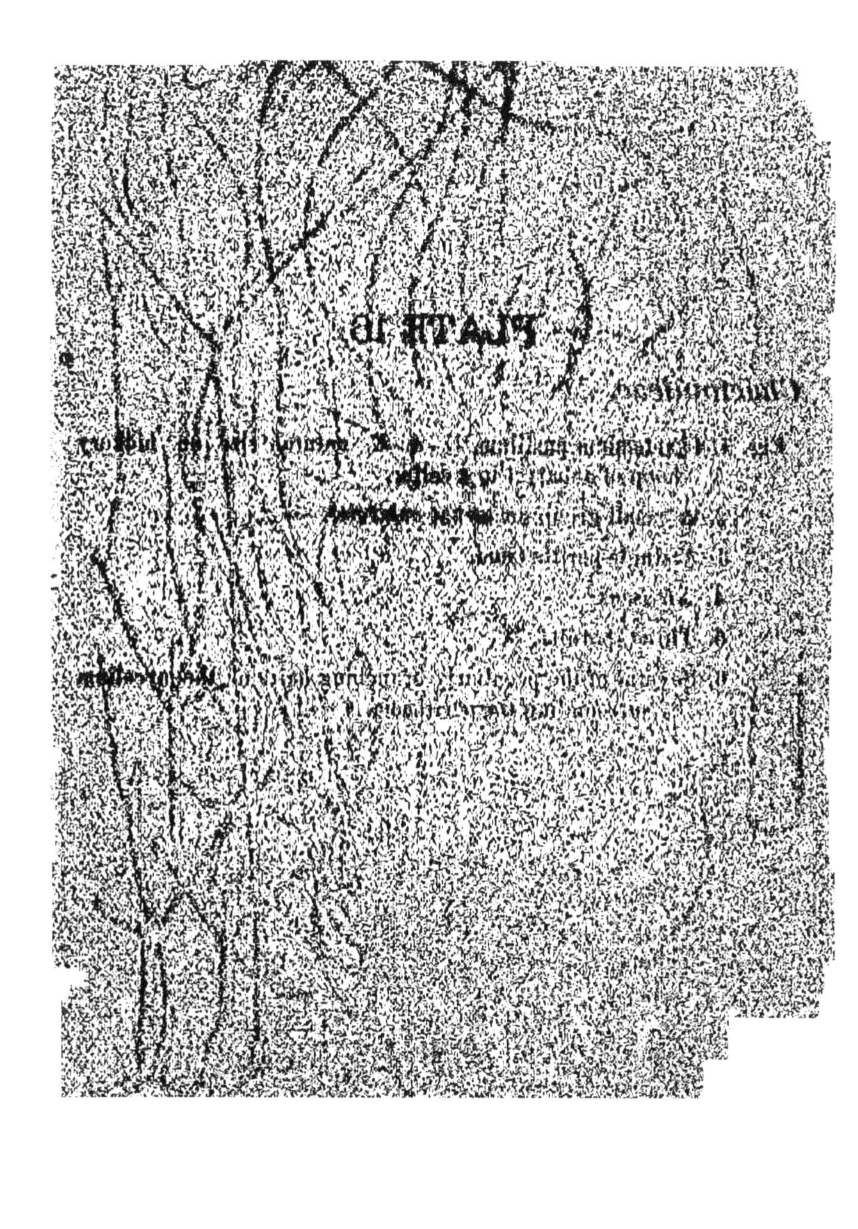

PLATE 16.

Chaetomieae.

Fig. 1. **Chaetomium pusillum,** E. & E. natural size on hickory hoop of a barrel in a cellar.

" 2. A small group somewhat enlarged.

" 3. A single perithecium.

" 4. An ascus.

" 5. Three sporidia.

" 6. Several of the peculiarly branching hairs of the mycelium surrounding the perithecia.

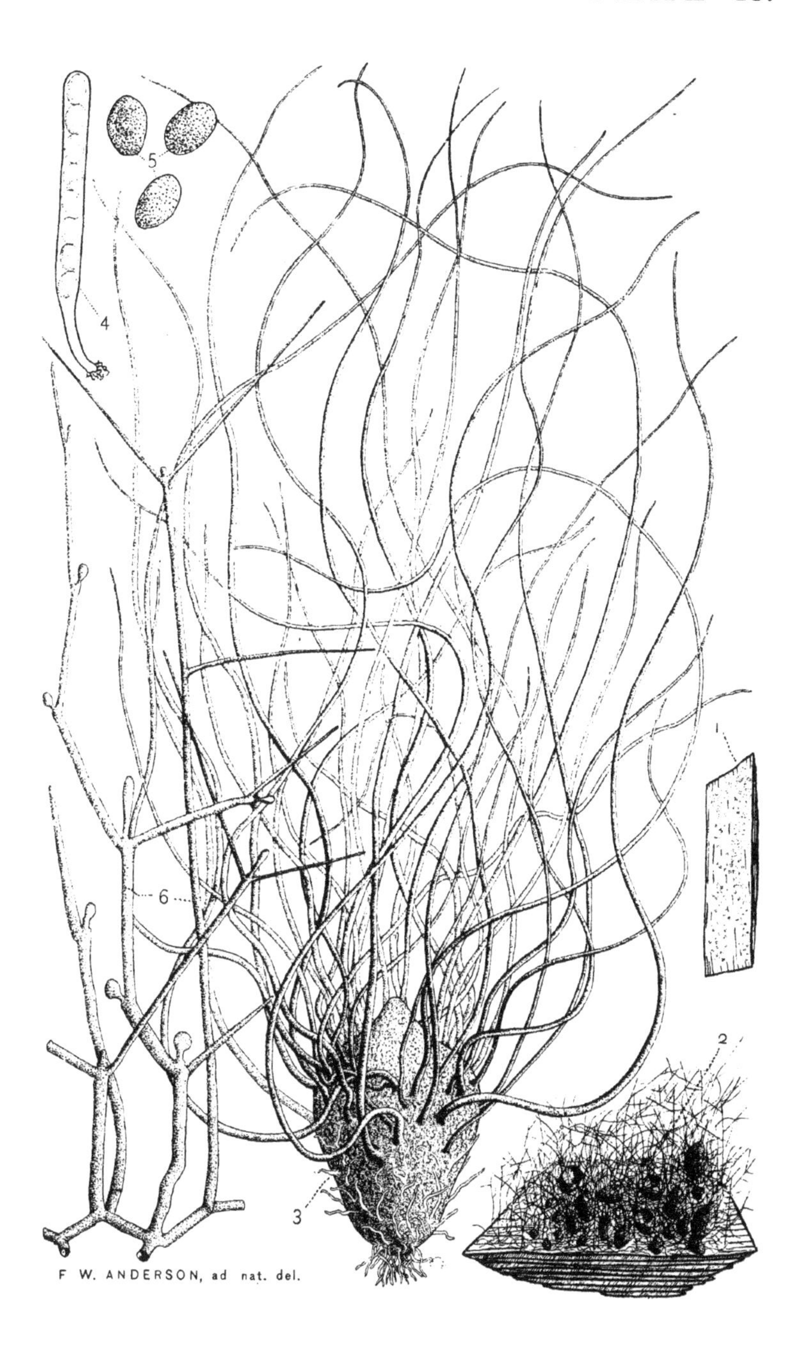
1
2
3
4
5
6
F. W. ANDERSON, ad nat. del.

PLATE 17.

Sordarieae.

Fig. 1. **Sordaria humana,** sectional view, somewhat enlarged.

" 2. An ascus.

" 3. A sporidium.

" 4. **Hypocopra equorum,** (Fckl.) somewhat enlarged sectional view showing the asci in the perithecia and the thin black superficial stroma above.

" 5. **Oomyces Langloisii,** E. & E. natural size on stem of *Vigna luteola.*

" 6. An enlarged stroma.

" 7. Vertical section through the stroma and enclosed perithecia.

" 8. An ascus with two branching paraphyses.

" 9. A sporidium cut in two.

" 10. **Delitschia bisporula,** (Cronan) perithecium enlarged.

" 11. An ascus and paraphysis.

" 12. A sporidium.

NOTE.—Figures 10, 11 and 12 from Hansen's Fungi fimicoli danici.

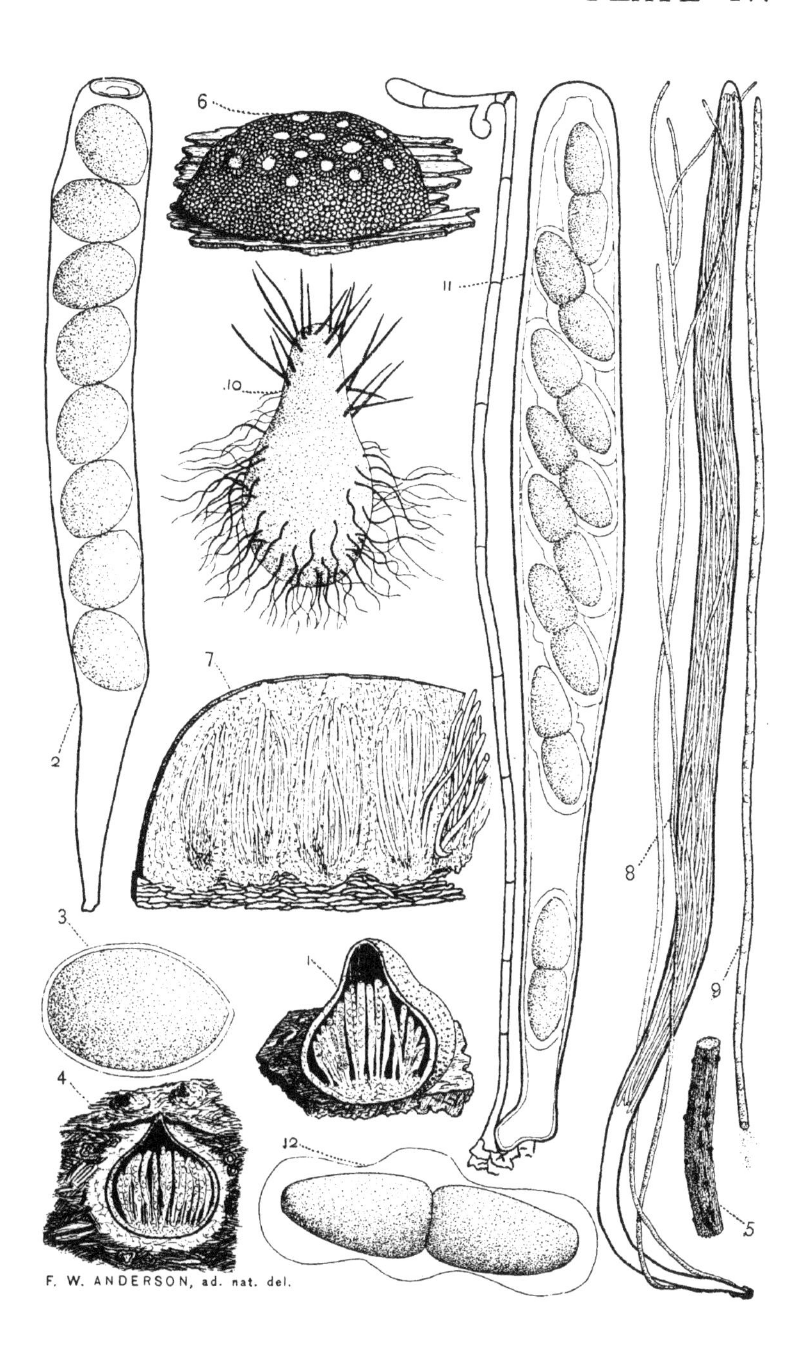
6
11
.10
7
2
8
3
1
9
4
12
5
F. W. ANDERSON, ad. nat. del.

PLATE 18.

Sordarieae.

Fig. 1. **Podospora lutea,** E. & E. natural size on decaying bark.

" 2. The same somewhat magnified.

" 3. Sectional view of a perithecium highly magnified.

" 4. An ascus.

" 5. Four sporidia, A, B, C and D, in different stages of development, C and D mature.

" 6. **Sporormia minima,** Awd. somewhat magnified, on goat's dung.

" 7. A perithecium highly magnified and broken open, showing the asci within.

" 8. A cluster of asei, middle one mature.

" 9. Two sporidia.

Note.---Figures on this plate not drawn to any scale.

PLATE 18.

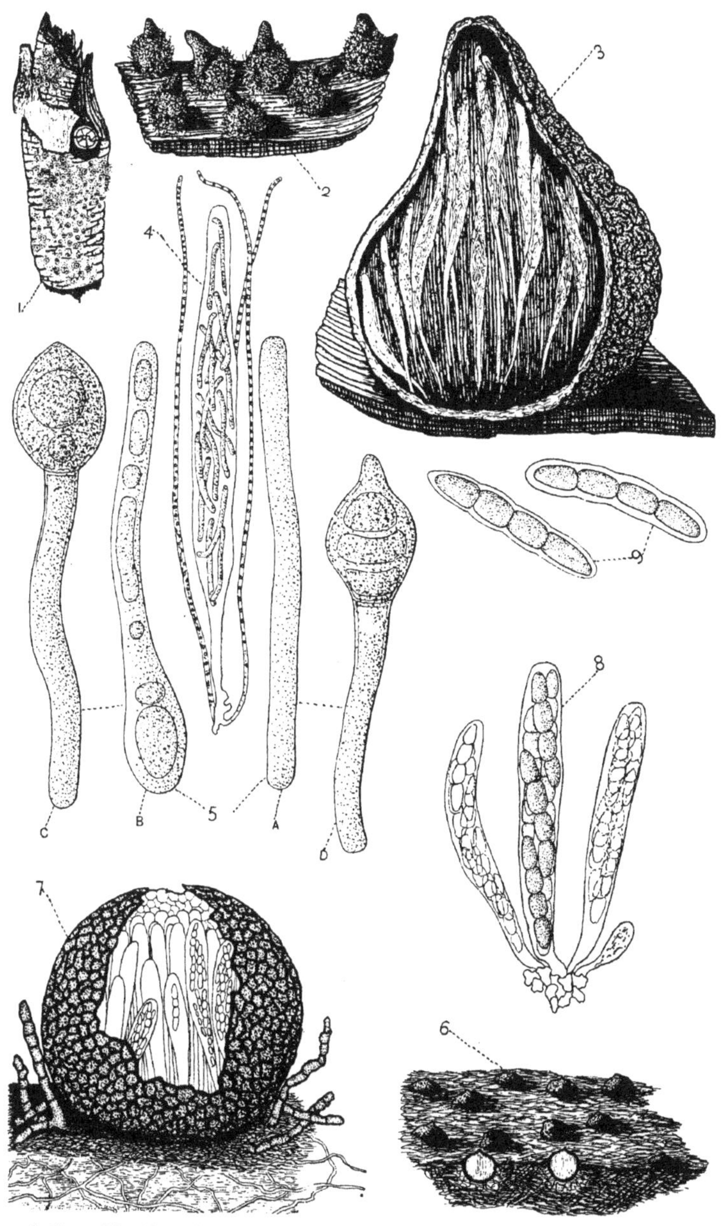

[illegible]

PLATE 19.

Lasiosphaerieae.

Fig. 1. **Acanthostigma decastylum,** (Ckc.) slightly magnified, on bark of rotten magnolia.

" 2. Section of a perithecium more highly magnified.

" 3. An ascus.

" 4. A sporidium.

" 5. **Lasiosphaeria stupea,** E. & E. natural size on wood of Tsuga Pattoniana, Mt. Paddo, Wash.

" 6. Three perithecia moderately enlarged.

" 7. Section of a perithecium highly magnified.

" 8. An ascus.

" 9. A sporidium.

" 10. A sporidium of L. hispida, (Tode.).

" 11. **Chaetosphaeria pannicola,** B. & C. on decaying bark, natural size.

" 12 A perithecium somewhat enlarged.

" 13. An ascus.

" 14. A sporidium.

" 15. **Herpotrichia Rhenana,** Fckl. on culm of some grass.

" 16. A perithecium somewhat enlarged.

" 17. A sporidium.

" 18. Sporidium of H. pinetorum, (Fckl. 15, 16, 17 and 18 from Winter's Pilze).

" 19. **Trichosphaeria pilosa,** (Pers.) two asci.

" 20. Two sporidia, (19 and 20 from Winter's Pilze.)

PLATE 19.

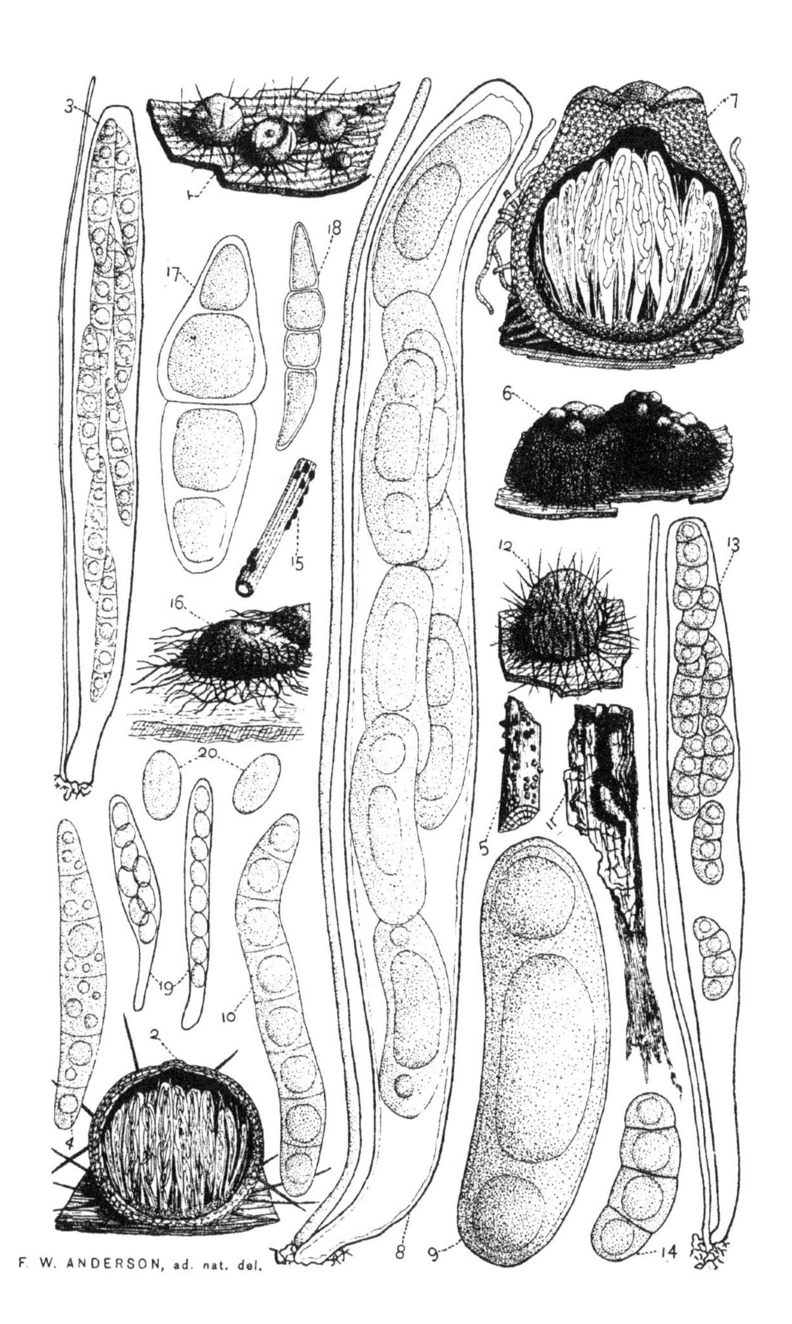

F. W. ANDERSON, ad. nat. del.

PLATE 20.

Melanommeae.

Fig. 1. **Rosellinia Clavariae,** (Tul.) natural size on living Clavaria.
" 2. Branch of same somewhat magnified.
" 3. Section of a perithecium highly magnified.
" 4. A cluster of asci.
" 5. Three sporidia.
" 6. Conidiophores with conidia.
" 7. Three conidia highly magnified.
" 8. **Rosellinia ovalis,** Ell. on piece of "sage brush," natural size.
" 9. Piece of the same enlarged.
" 10. Sectional view of a perithecium.
" 11. Two asei.
" 12. Three sporidia.

NOTE.—Figures not drawn to a definite scale.

PLATE 20.

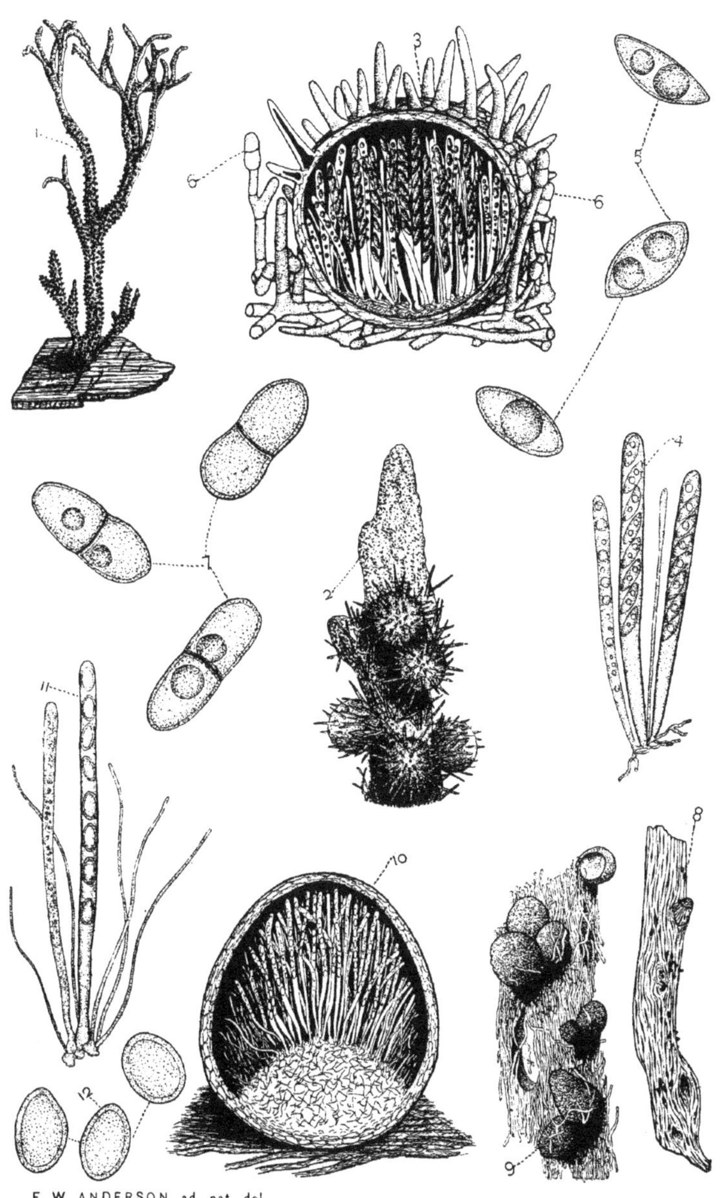

PLATE 21.

Melanommeae.

Fig. 1. **Melanomma occidentale,** Ell. natural size on "sage brush."

" 2. Fragment of same enlarged.

" 3. Sectional view of a perithecium.

" 4. An ascus.

" 5. Two sporidia.

" 6. **Bombardia fasciculata,** Fr. natural size on rotten wood.

" 7. Same enlarged.

" 8. Section of a perithecium more highly magnified.

" 9. Three asci.

" 10. Four sporidia, (A) mature, the others in various stages of development.

NOTE.—Figures on this plate not drawn to any definite scale.

PLATE 21.

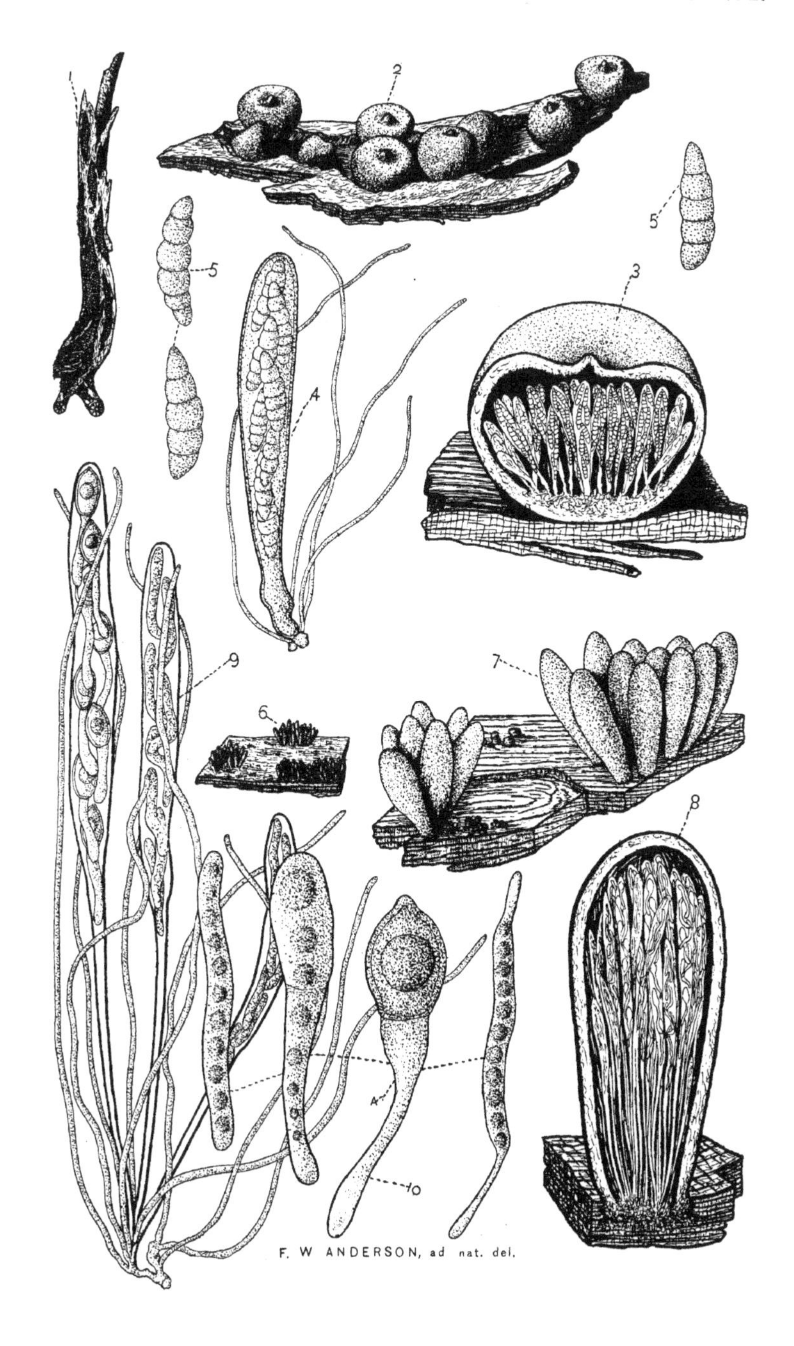

PLATE 22

PLATE 22.

Ceratostomeae. (Figs. 1 to 13)

Fig. 1. **Hypsotheca subcorticalis,** (C. & E.) natural size on oak bark.

" 2. Same enlarged.

" 3. A perithecium broken open showing the asci attached to the surface of the inner wall.

" 4. An ascus.

" 5. Two sporidia.

" 6. **Ceratosphaeria microdoma** E. & E. natural size, on bark of elder.

" 7. Same enlarged.

" 8. An ascus.

" 9. A sporidium.

" 10. **Ceratostoma subrufum,** E. &. E on wood of dead oak limb.

" 11. Portion of same enlarged.

" 12. An ascus.

" 13. Two sporidia.

" 14. **Trematosphaeria pertusa,** (Pers) natural size, on dead wood.

" 15. Portion of the same enlarged.

" 16. An ascus.

" 17. A sporidium 1–septate.

" 18. " " 2–septate.

" 19. " " 3–septate.

PLATE 22.

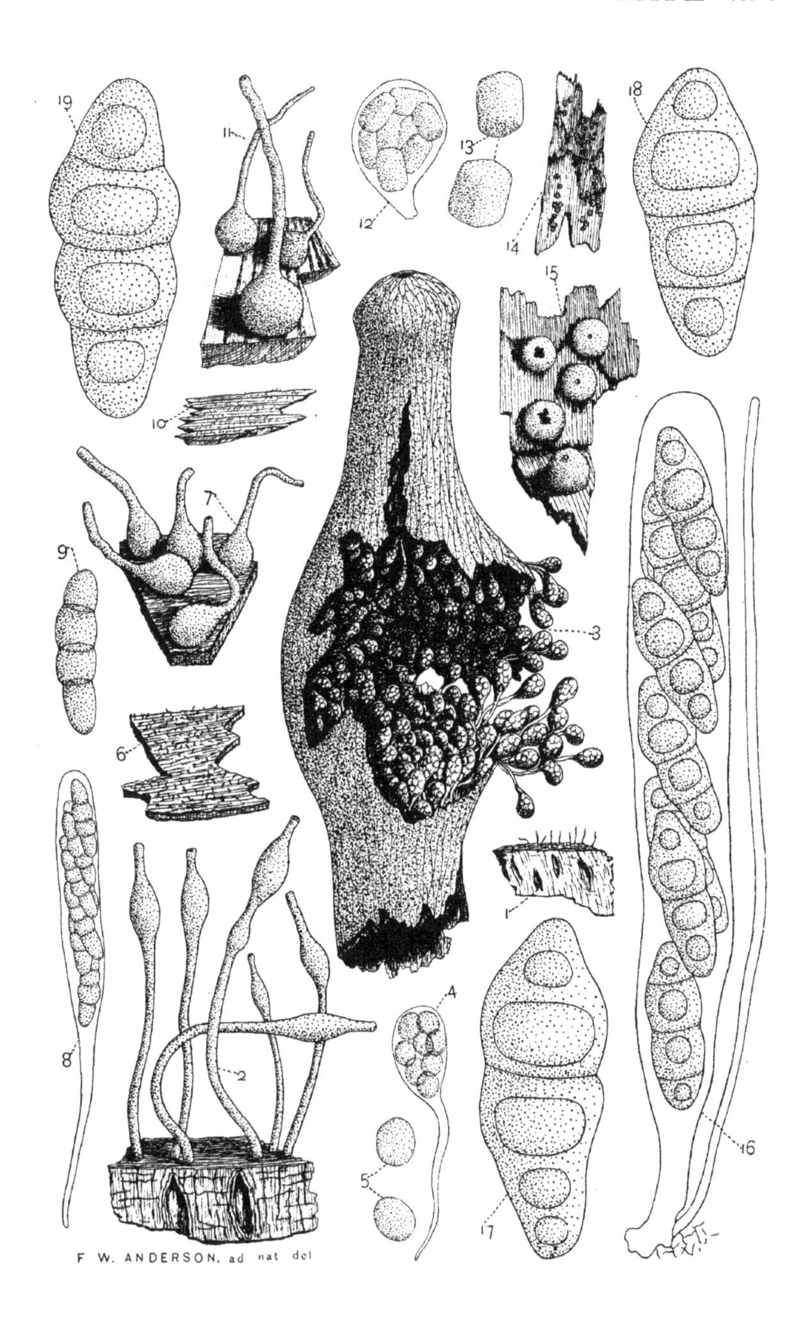

PLATE 23.

Amphisphaerieae. (Figs, 6-18).

Fig. 1. **Herpotrichia diffusa,** (Schw.) natural size.
" 2. Portion of the same somewhat magnified.
" 3. Enlarged section of a perithecium.
" 4. An ascus.
" 5. A sporidium.
" 6. **Ohleria rugulosa,** Fckl. natural size.
" 7. Same somewhat magnified.
" 8. An ascus.
" 9. A sporidium.
" 10. **Winteria rhoina,** E. & E. natural size.
" 11. Portion of the same enlarged.
" 12. An ascus.
" 13. A sporidium.
" 14. **Teichospora Helenae,** E. & E. natural size, on dead limb.
" 15. Portion of same enlarged.
." 16. Enlarged section of perithecium.
" 17. An ascus.
" 18. A sporidium.

PLATE 23.

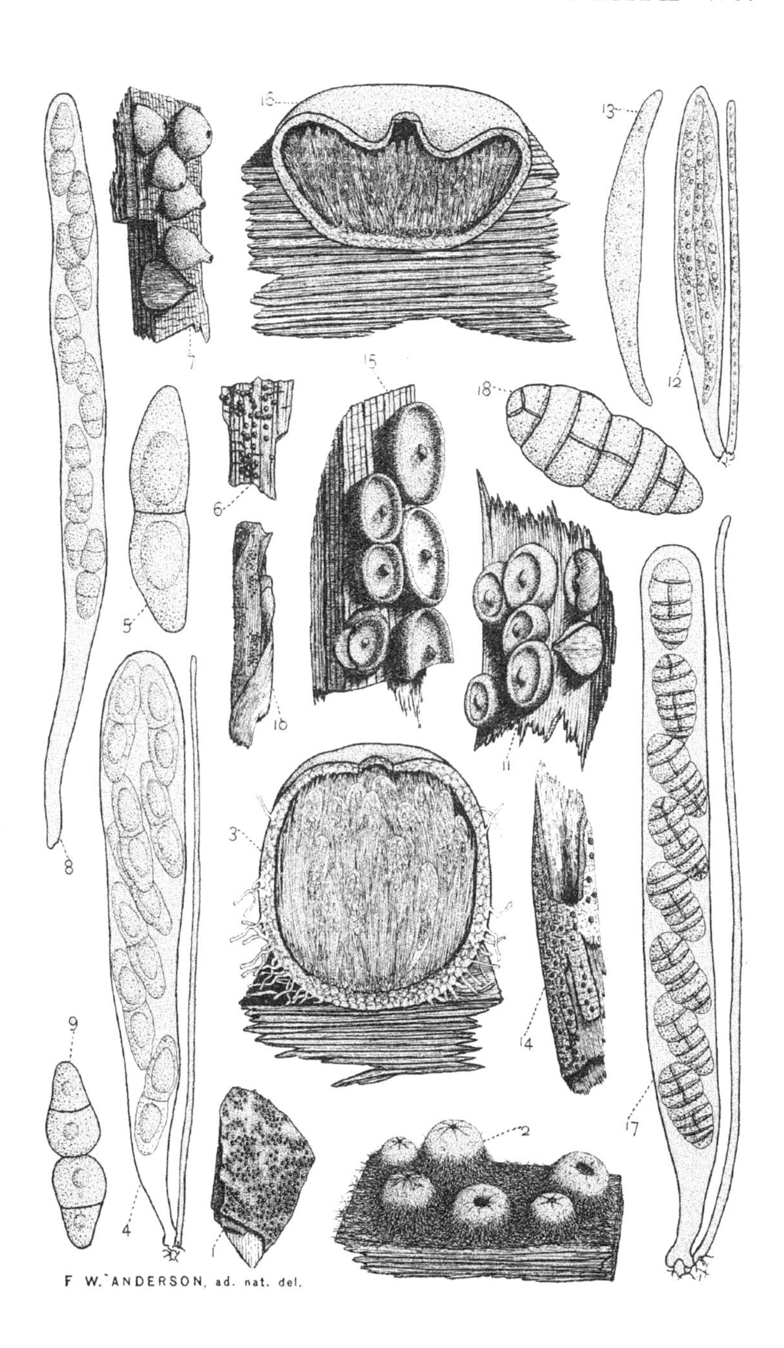

F. W. ANDERSON, ad. nat. del.

PLATE 24.

Amphisphaerieae.

Fig. 1. **Caryospora putaminum,** (Schw.) natural size, on peach pits.

" 2. Same enlarged.

" 3. Section of a perithecium more highly magnified.

" 4. A part of an ascus containing two sporidia.

NOTE.—Ascus and sporidia drawn to a scale 28 μ to the inch.

PLATE 24.

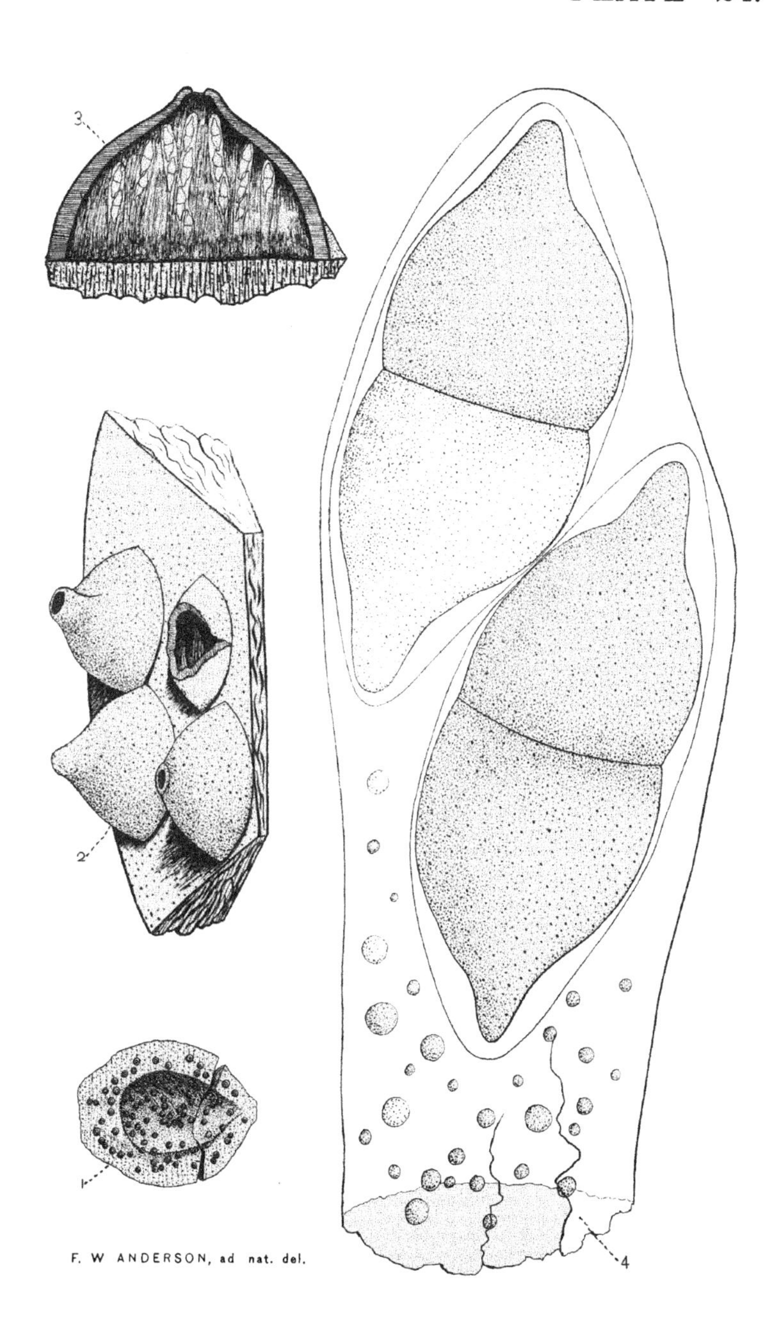

F. W ANDERSON, ad nat. del.

PLATE 25.

Lophiostomeae.

Fig. 1. **Lophionema vermisporum,** (Ell.) somewhat enlarged, on dead stem of *Oenothera.*

" 2. Enlarged section of a perithecium.

" 3. An ascus.

" 4. A sporidium.

" 5. **Lophiostoma Pruni,** E. & E. natural size, on dead limb of cherry.

" 6. Part of same enlarged.

" 7. An ascus.

" 8. A sporidium.

" 9. A chain ot spermogonia.

" 10. **Lophidium tingens,** E. & E. on dead limb of maple—a section through a perithecium somewhat enlarged.

" 11. An ascus.

" 12. A sporidium.

PLATE 25.

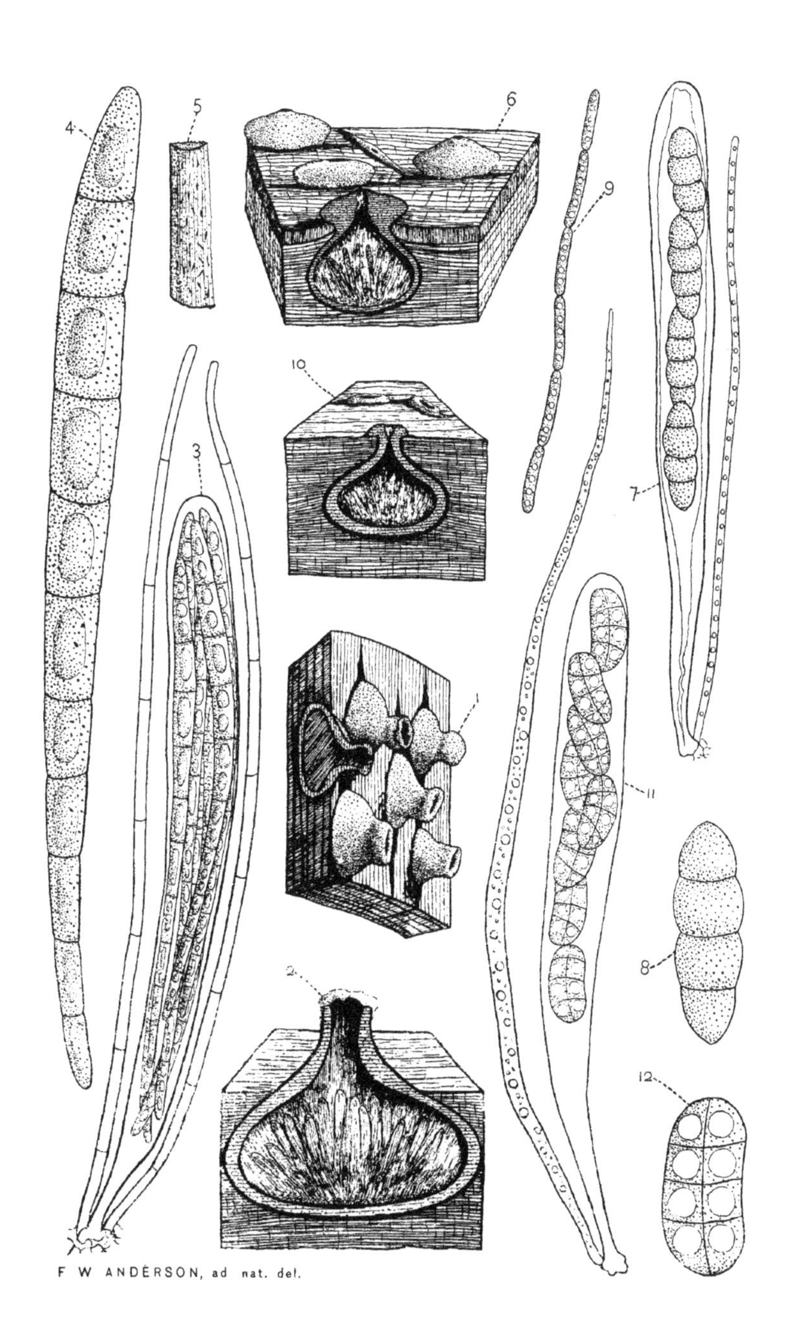

PLATE 26.

Cucurbitarieae.

Fig. 1. **Cucurbitaria Fraxini,** E. & E. natural size, on dead ash limb.

" 2. Portion of same somewhat enlarged.

" 3. Sectional view through the center of a cluster of perithecia.

" 4. An ascus.

" 5. A sporidium.

" 6. **Gibbera Vaccinii,** (Sow.) natural size, on Vaccinium twig.

" 7. A cluster of perithecia somewhat enlarged.

" 8. An ascus.

" 9. A sporidium.

" 10. **Otthia hypoxyloides,** E. & E. natural size on dead wood.

" 11. A cluster of perithecia enlarged.

" 12. An ascus.

" 13. A sporidium.

" 14. **Nitschkia cupularis,** (Pers.) natural size on dead limb of horse chestnut.

" 15. A cluster of perithecia somewhat enlarged.

" 16. An ascus.

" 17. Two sporidia.

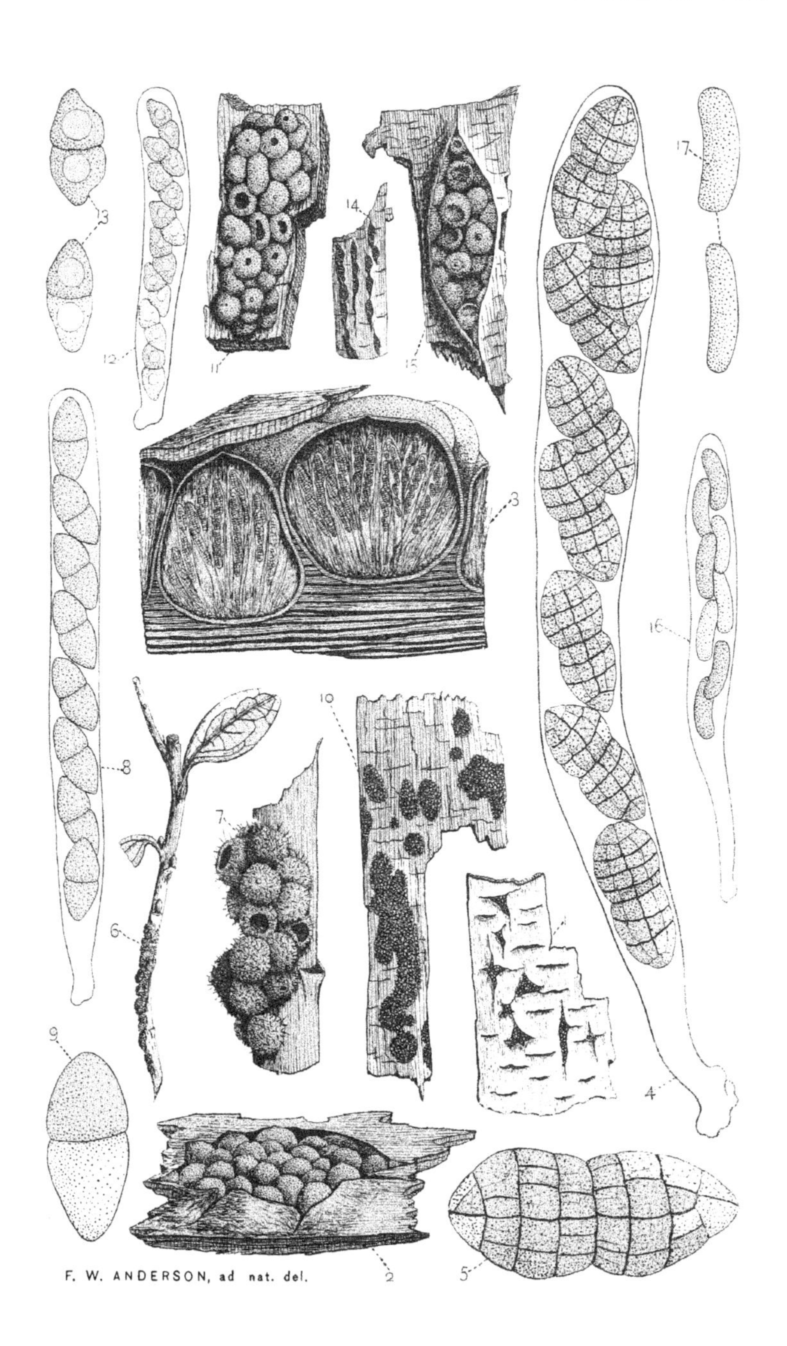

F. W. ANDERSON, ad nat. del.

PLATE 27.

Sphaerelloideae.

Fig. 1. **Physalospora aurantia,** E. & E. on dead leaves of *Astragalus*, natural size.

" 2. Part of a leaf somewhat enlarged.

" 3. Enlarged section of a perithecium.

" 4. An ascus with uniseriate sporidia.

" 5. An ascus with partly biseriate sporidia.

" 6. A sporidium.

" 7. **Sphaerella Oenotherae,** E. & E. natural size, on old capsule of *Oenothera biennis*.

" 8. A small piece enlarged.

" 9. Enlarged section of a perithecium.

" 10. An ascus.

" 11. A sporidium.

" 12. **Stigmatea Robertiana,** Fr. natural size, on leaf of *Geranium Robertianum*.

" 13. Piece of same enlarged.

" 14. Sectional view of perithecium.

" 15. An ascus.

" 16. A sporidium.

" 17. **Sphaerulina myriadea,** (Pers.) natural size on dead oak leaf.

" 18. Portion of same enlarged.

" 19. Sectional view of a perithecium more highly magnified.

" 20. An ascus.

" 21. A sporidium.

PLATE 27.

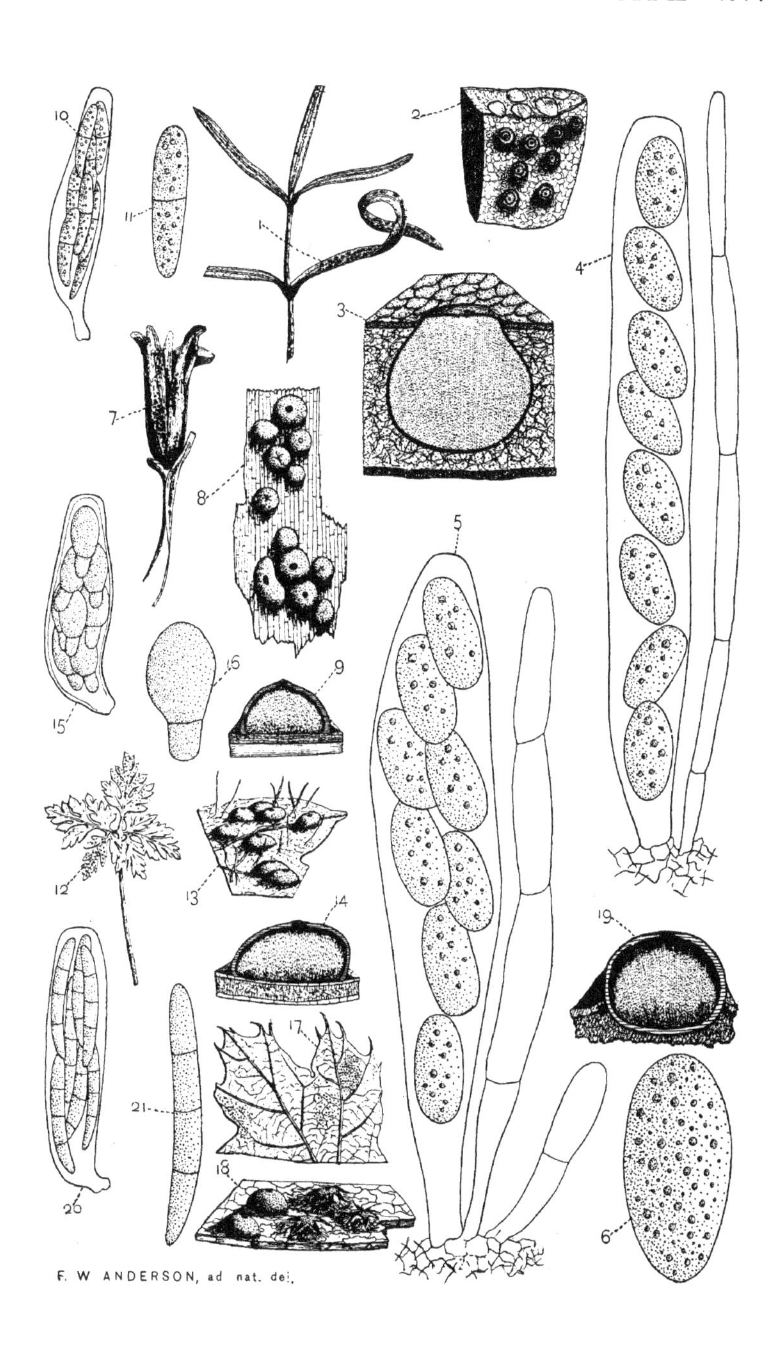

F. W ANDERSON, ad nat. del.

PLATE 28.

Pleosporeae. (except 6 & 7.).

Fig. 1. **Leptosphaeria Harknessiana,** E. & E. natural size, on dead stem of *Frasera speciosa.*

" 2. Piece of the same moderately enlarged.

" 3. Sectional view of a perithecium more highly magnified.

" 4. An ascus.

" 5. A sporidium.

" 6. **Rhopographus clavisporus,** (C. & P.),—an ascus.

" 7. A sporidium.

" 8. **Didymosphaeria cupula,** Ell. slightly enlarged, on petiole and ribs of dead oak leaf.

" 9. An ascus.

" 10. A sporidium.

" 11. **Ophiobolus olivaceus,** Ell. somewhat magnified, on dead herbaceous stem.

" 12. An ascus.

" 13. An enlarged sporidium.

" 14. **Pleospora aurea,** Ell. somewhat magnified, on dead herbaceous stem.

" 15. An ascus.

" 16. A sporidium—front view.

" 17. A sporidium—side view.

PLATE 28.

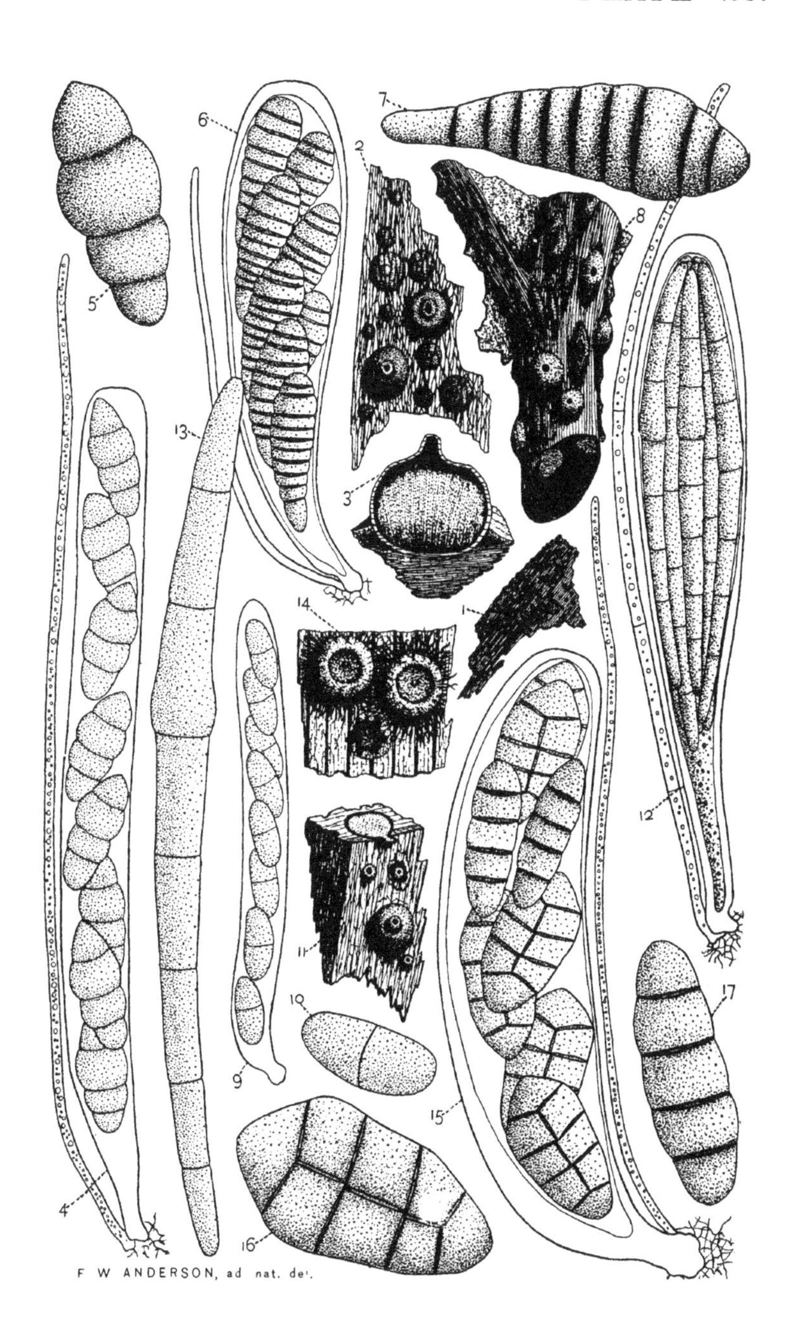

PLATE 29.

Massarieae.

Fig. 1. **Massaria vomitoria,** B. & C. somewhat enlarged, on maple limbs.

" 2. A perithecium more highly magnified.

" 3. An ascus.

" 4. An enlarged sporidium showing the hyaline envelop.

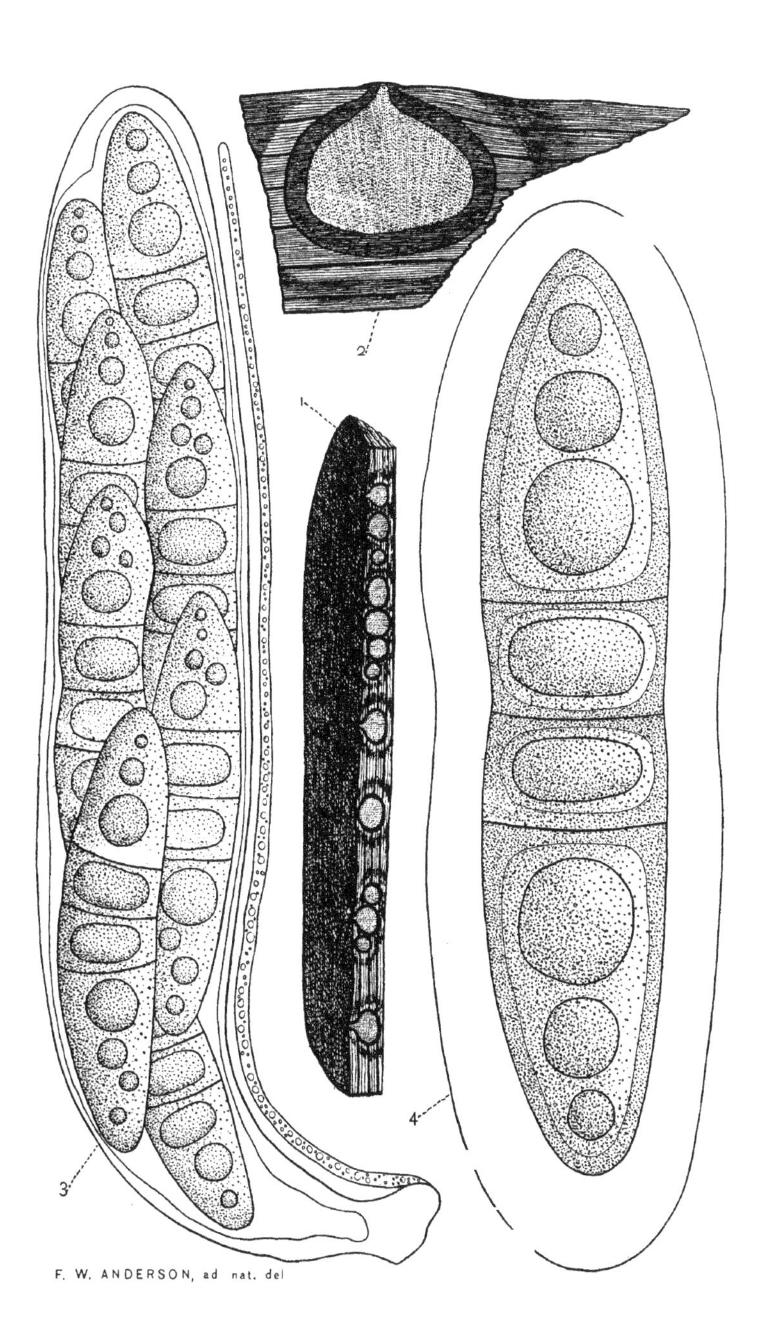

F. W. ANDERSON, ad nat. del

PLATE 30.

Massarieae.

Fig. 1. **Massariovalsa sudans,** (B. & C.) showing the perithecia in vertical section somewhat enlarged.

" 2. Horizontal section of perithecia and stroma.

" 3. An ascus.

" 4. A sporidium.

" 5. **Pleomassaria rhodostoma,** (A. & S.), a sporidium.

" 6. **Massariella bufonia,** (B. & Br.) on white oak bark, showing vertical section of perithecia somewhat enlarged.

" 7. An ascus.

" 8. A sporidium.

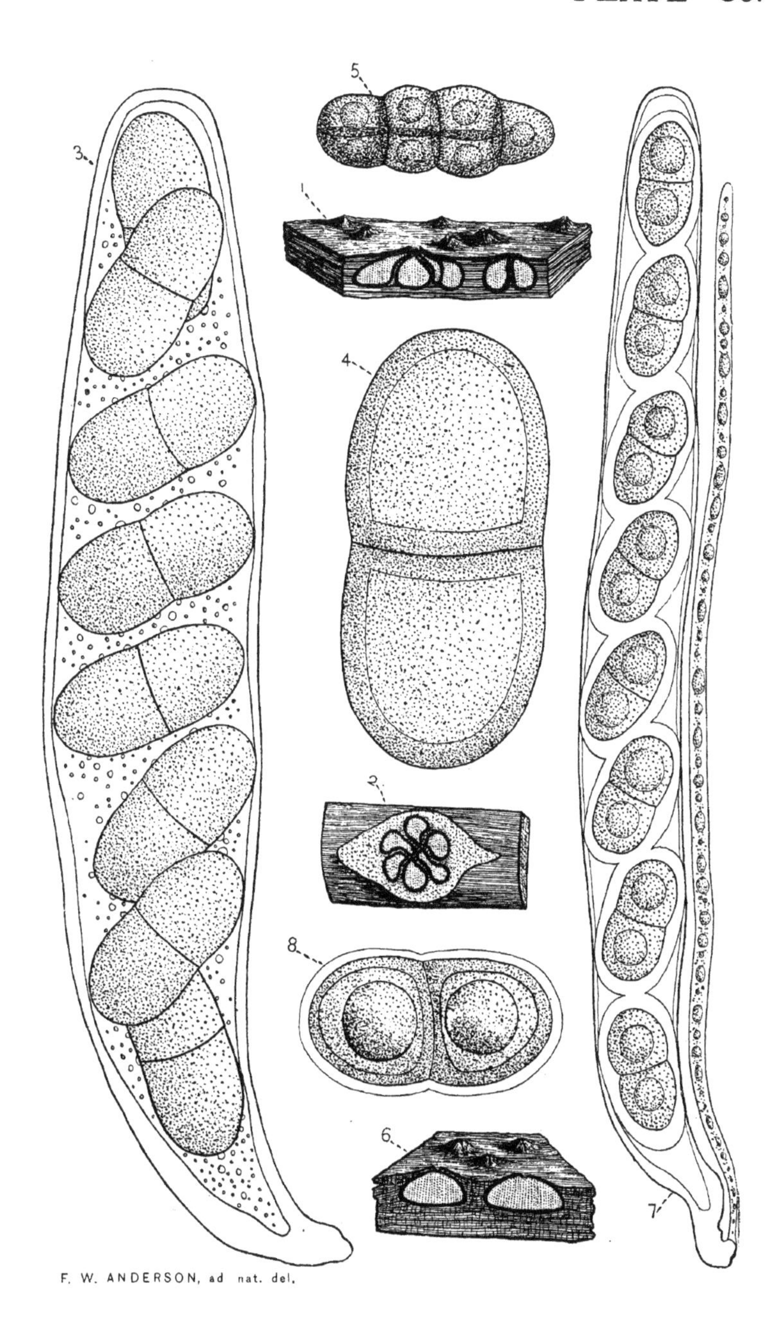

F. W. ANDERSON, ad nat. del.

PLATE 31.

Clypeosphaerieae.

Fig. 1. **Anthostomella Magnoliae,** E. & E. slighly enlarged, on leaf of magnolia.

" 2. An ascus.

" 3. Two sporidia.

" 4. **Linospora Palmetto,** E. & E. natural size on leaf of Palmetto.

" 5. A piece of the same enlarged.

" 6. An ascus.

" 7. A sporidium.

" 8. **Trabutia quercina,** (Fr.) natural size, on oak leaf.

" 9. Portion of same enlarged.

" 10. An ascus.

" 11. A sporidium.

" 12. **Hypospila pustula,** (Pers.) natural size on oak leaf.

" 13. Portion of same enlarged.

" 14. An ascus.

" 15. A sporidium.

" 16. **Clypeosphaeria Hendersonia,** Ell. natural size on dead *Rubus* stems.

" 17. Same enlarged.

" 18. An ascus.

" 19. Two sporidia.

NOTE.—Figures 12—15 drawn from a specimen in Linharts Fungi Hungarici 467.

PLATE 31.

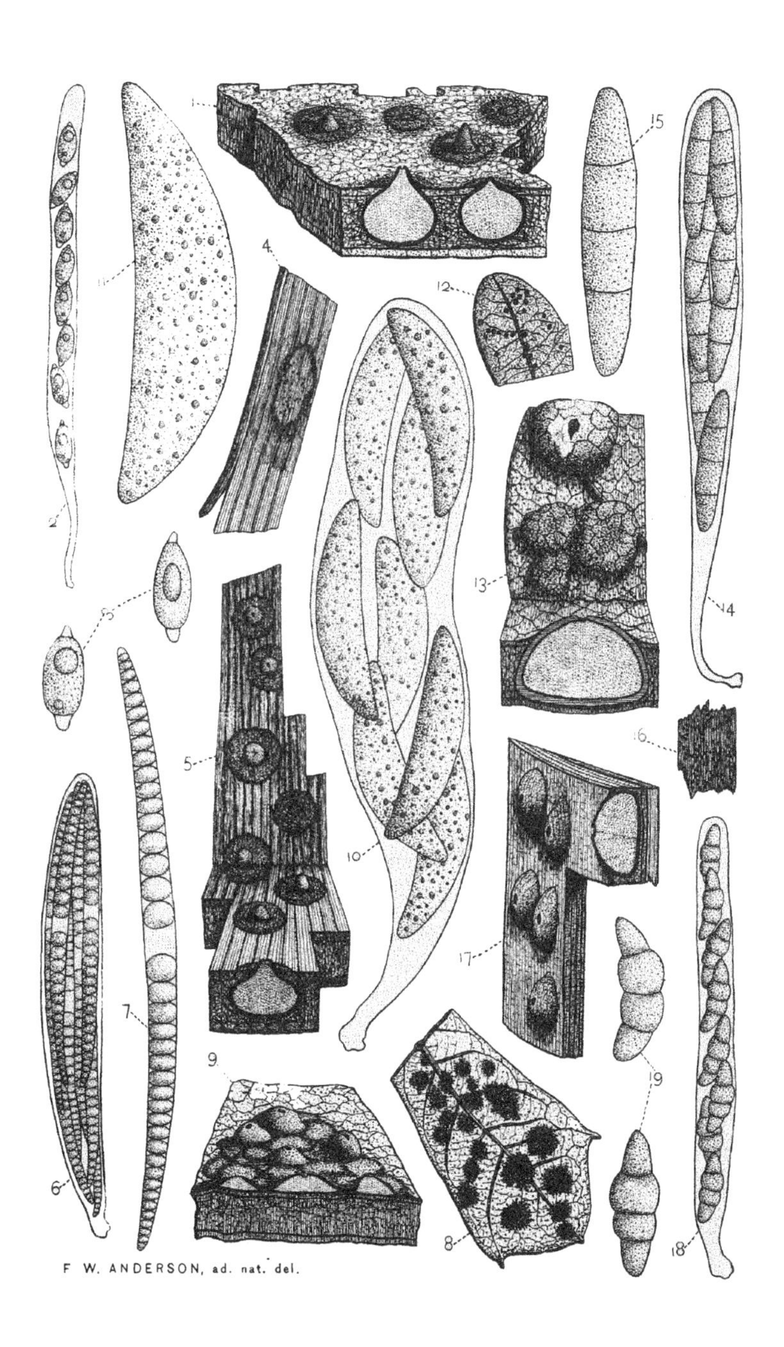

PLATE 32.

Gnomonieae.

Fig. 1. **Gnomonia Magnoliae,** Ell. slightly enlarged, on leaf of *Magnolia glauca.*

2. Section of perithecium enlarged.

" 3. An ascus.

" 4. Two sporidia.

" 5. **Gnomonia tenella,** E. & E. somewhat enlarged on petiole of *Acer rubrum.*

" 6. Section of perithecium enlarged.

" 7. An ascus.

" 8. Two sporidia, one with the appendages almost straight, the other having them more or less bent.

" 9. **Gnomonia setacea,** (Pers.), an ascus.

" 10. Two sporidia.

" 11. **Ditopella fusispora,** DeNot. somewhat enlarged. Drawn from No. 286 in Kriegers Saxon Fungi, on ***Alnus*** *glutinosa.*

" 12. Section of perithecium enlarged.

" 13. An ascus.

" 14. Three sporidia.

" 15. **Ceriospora xantha,** Sacc. Drawn from French specimens —section of perithecia enlarged.

" 16. An ascus.

" 17. Two spores.

PLATE 32.

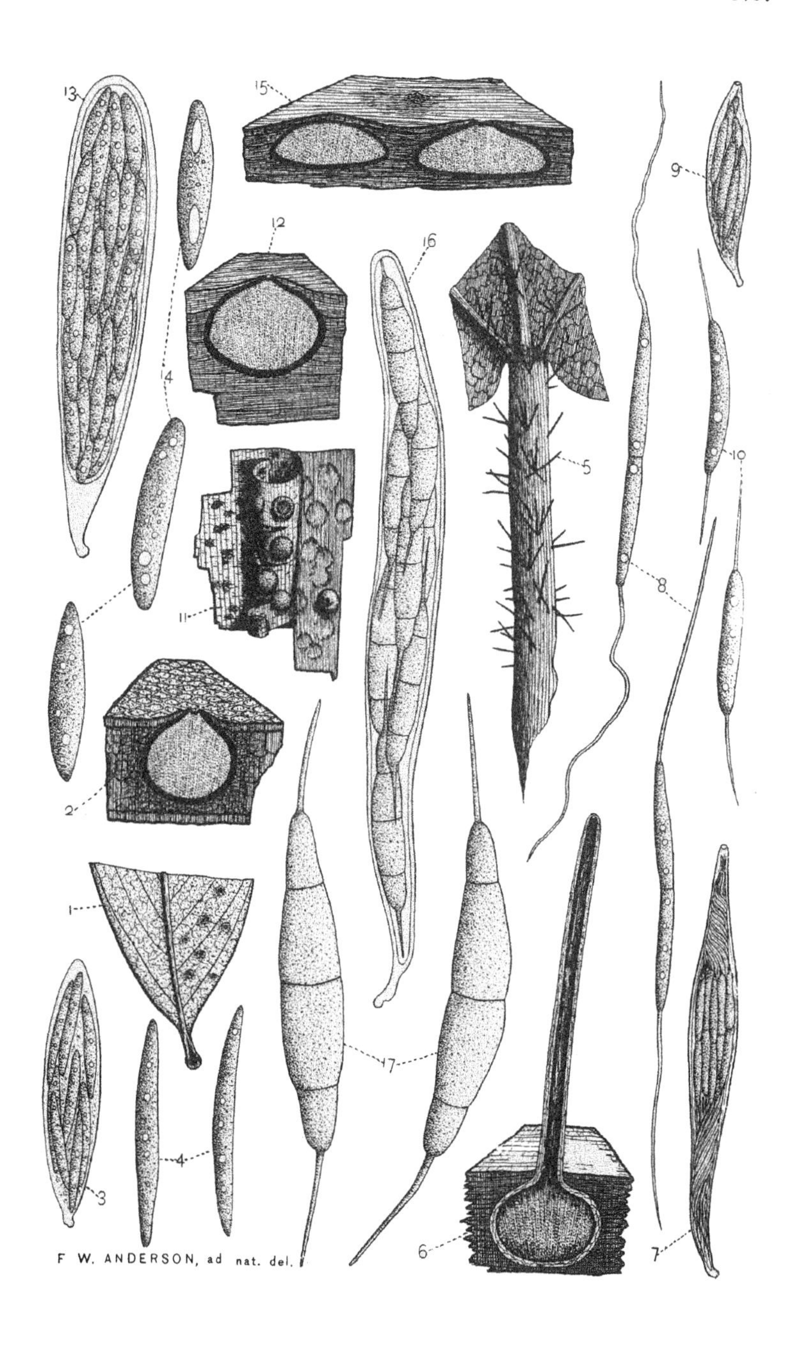

PLATE 33.

Valseae.

Fig. 1. **Valsa floriformis,** E. & E. somewhat enlarged, on bark.

" 2. Horizontal section through a stroma.

" 3. Vertical section, much enlarged, showing the perithecia lying around the labyrinthiform, central spermogonial cavities.

" 4. An ascus.

" 5. Two sporidia.

" 6. Two spermogonial spores.

" 7. **Calosphaeria microsperma,** E. & E. somewhat enlarged, on bark of *Carpinus.*

" 8. A cluster of perithecia somewhat enlarged and exposed by the removal of the bark.

" 9. Three clusters of asei arising from interwoven, branching filaments in the bottom of the perithecia and accompanied by very long stout paraphyses.

" 10. Four sporidia.

" 11. **Eutypa echinata,** E. & E. somewhat enlarged, on *Fraxinus.*

" 12. An ascus.

" 13. Three sporidia.

" 14. **Diaporthe tuberculosa,** Ell., an ascus.

" 15. A sporidium.

" 16. **Diaporthe densissima,** Ell., two sporidia.

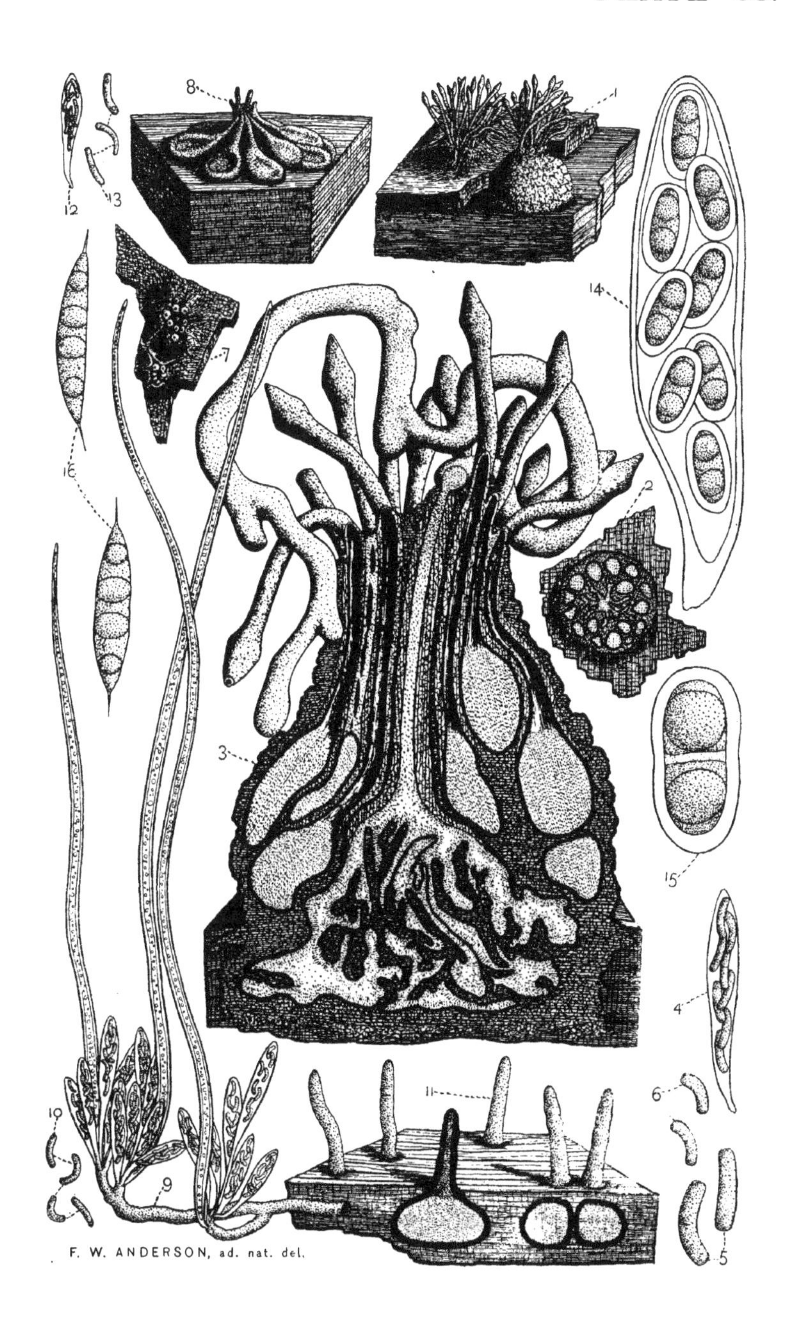
1
2
3
4
5
6
7
8
9
10
11
12
13
14
15
16
F. W. ANDERSON, ad. nat. del.

PLATE 34.

Diatrypeae.

Fig. 1. **Diatrype Hochelagae,** E. & E. moderately enlarged, on dead elm.

" 2. Vertical section through a stroma, more highly magnified.

" 3. An ascus.

" 4. Three sporidia.

" 5. **Diatrype platystoma,** (Schw.) somewhat enlarged, on dead maple.

" 6. Vertical section through a portion of a stroma, more highly magnified.

" 7. **Diatrype virescens,** (Schw.), on dead wood of *Carpinus*, a single stroma somewhat enlarged.

" 8. Vertical section through a rather small stroma of the same species.

" 9. **Diatrypella hysterioides,** E. & E. on dead poplar wood, somewhat enlarged.

" 10. Vertical section through a stroma more highly magnified.

" 11. An ascus.

" 12. Three sporidia.

" 13. **Anthostoma Ontariense,** E. & E. on bark of dead willow, slightly enlarged.

" 14. Vertical section through a piece of bark, showing the imbedded perithecia.

" 15. An ascus.

" 16. Two sporidia.

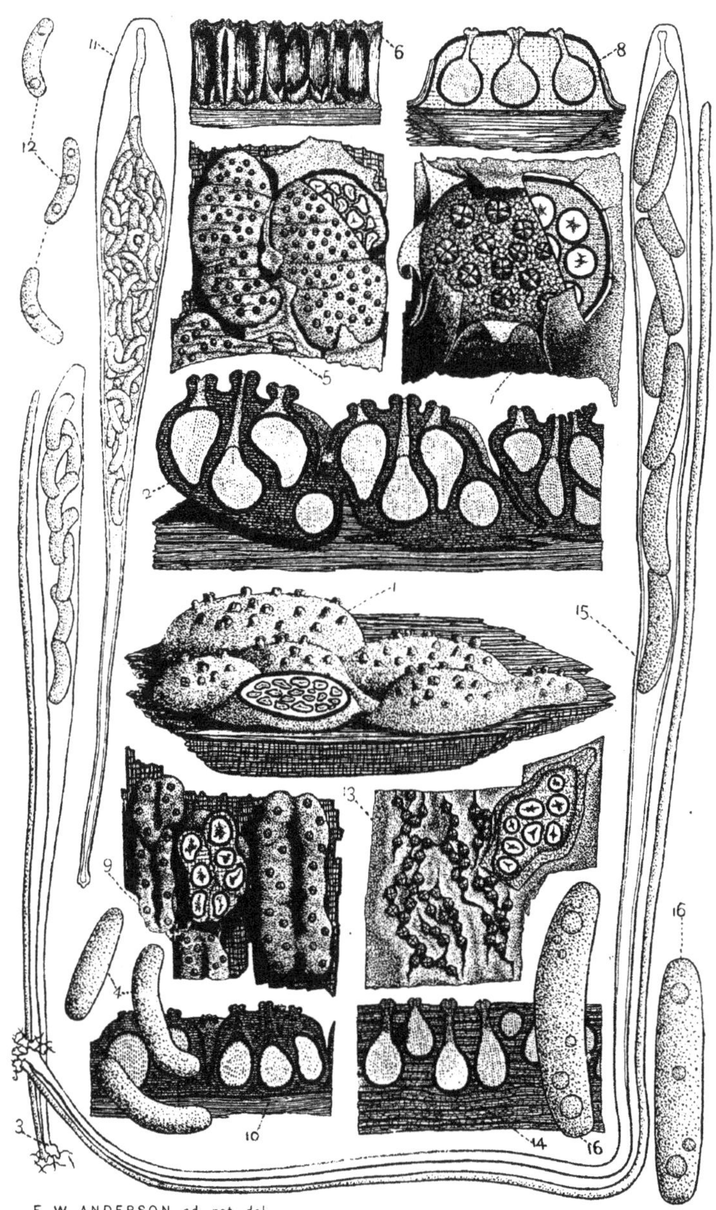

F. W ANDERSON, ad nat. del

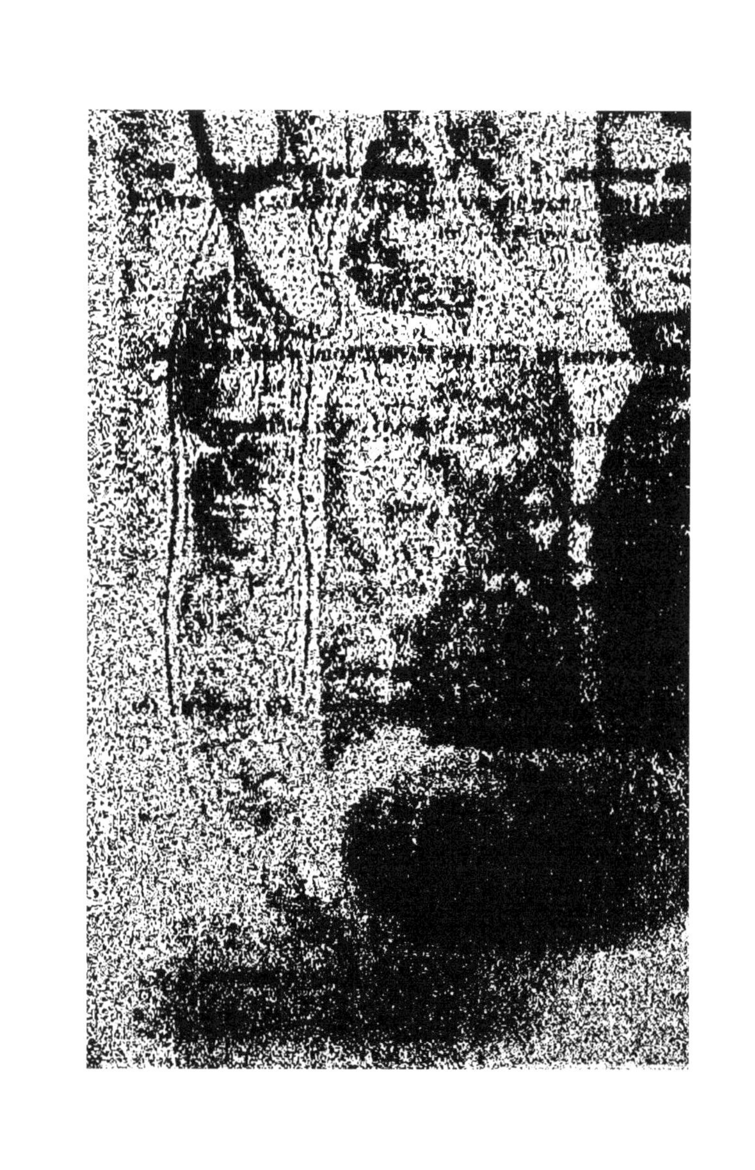

PLATE 35.

Melanconideae.

Fig. 1. **Fenestella amorpha,** E. & E. moderately enlarged, on hickory limb, showing three entire stromata, also a vertical and transverse section.

" 2. An ascus.

" 3. A sporidium.

" 4. **Melanconis Everhartii,** Ell. the stroma somewhat enlarged, on maple.

" 5. The same showing vertical and transverse sections.

" 6. A sporidium.

" 7. **Melanconis apocrypta,** Ell., an ascus.

" 8. A sporidium.

" 9. **Pseudovalsa stylospora,** E. & E., an ascus.

" 10. A sporidium.

" 11. A stylospore, on maple. (*Acer spicatum*).

NOTE.—The dotted lines with fig. 6 should run up towards the left to the large appendiculate sporidium.

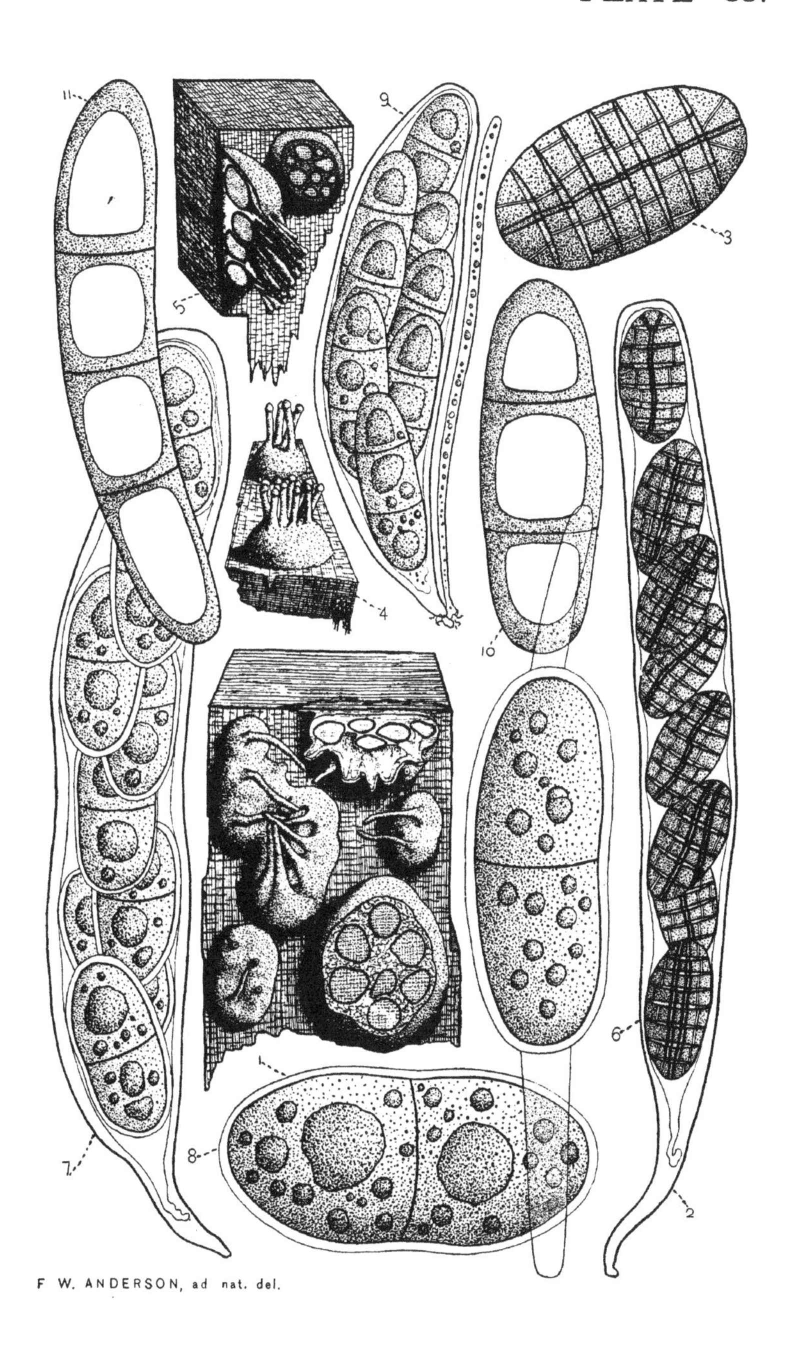

F. W. ANDERSON, ad nat. del.

PLATE 36.

Melogrammeae.

Fig. 1. **Botryosphaeria fuliginosa,** (M. & N.) somewhat enlarged, on an oak gall.

" 2. Same species somewhat enlarged, on decaying limb of *Quercus coccinia.*

" 3. An ascus.

" 4. A sporidium.

" 5. Two mature stylospores?

" 6. **Endothia gyrosa,** (Schw.) somewhat enlarged on bark.

" 7. An ascus.

" 8. Three sporidia.

" 9. **Valsaria Farlowiana,** Sacc. enlarged on dead limb of *Berberis.* sp.

" 10. Sectional view of the same.

" 11. An ascus.

" 12. A sporidium.

" 13. **Melogramma vagans,** De Not. on dead *Carpinus*, enlarged.

" 14. An ascus.

" 15. A sporidium.

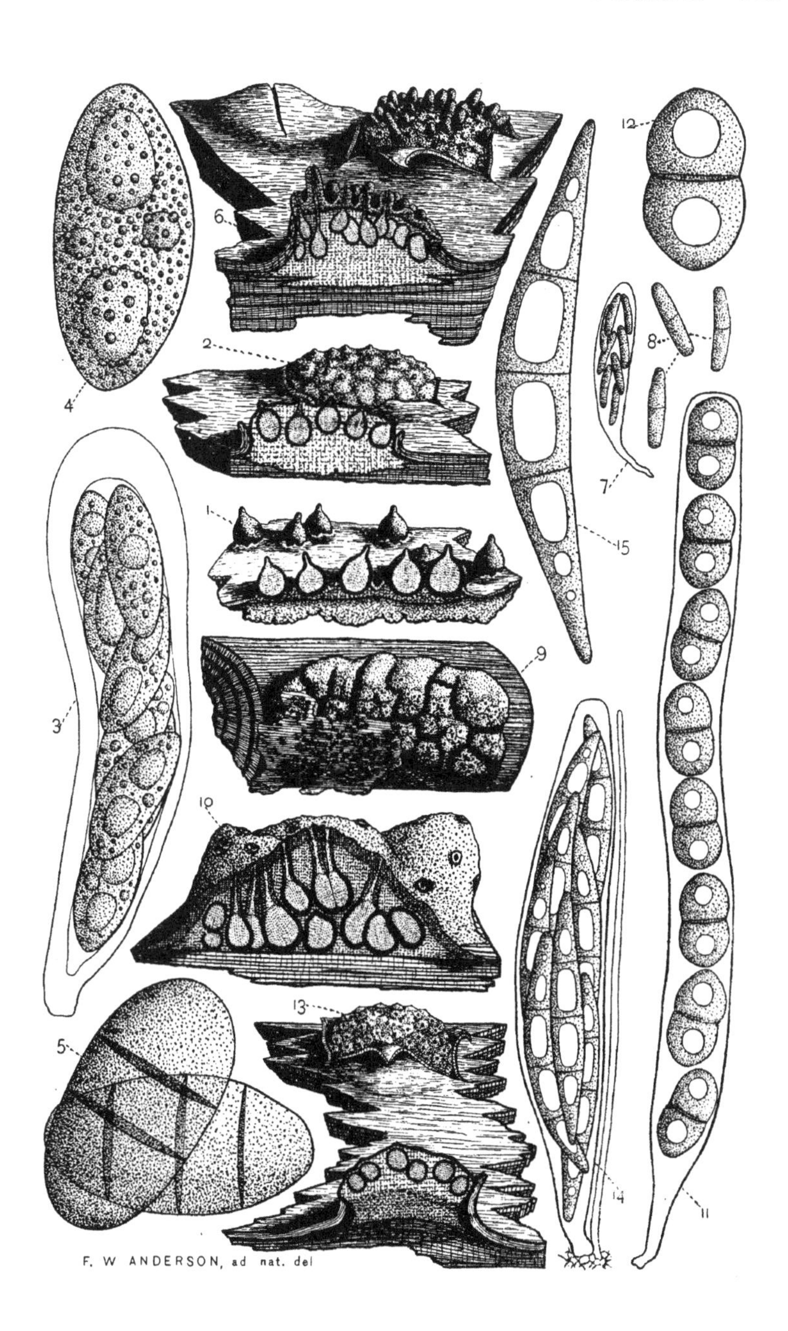
1
2
3
4
5
6
7
8
9
10
11
12
13
14
15
F. W ANDERSON, ad nat. del

PLATE 37.

Xylarieae.

Fig. 1. **Hypoxylon perforatum,** (Schw.), natural size, on dead oak limb.

" 2. Same magnified.

" 3. A smaller portion containing two perithecia highly magnified.

" 4. A cluster of asei in different stages of growth.

" 5. A single mature ascus.

" 6. Five sporidia.

" 7. **Hypoxylon marginatum,** (Schw.) natural size, on dead oak bark.

" 8. Vertical section of part of a stroma considerably magnified.

" 9. A cluster of asci, one of which is mature.

" 10. Three sporidia.

NOTE.—Figures on this plate not drawn to any definite scale.

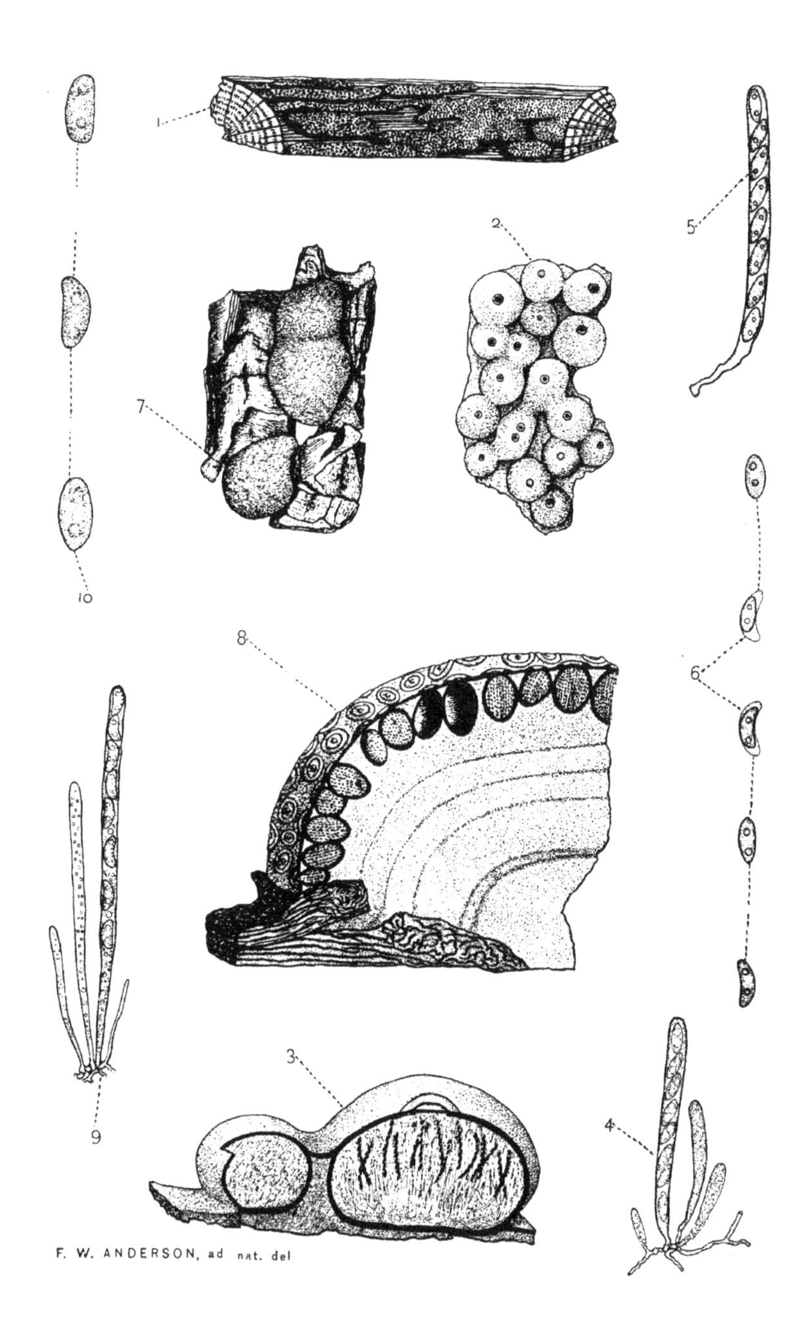
1
2
3
4
5
6
7
8
9
10
F. W. ANDERSON, ad nat. del

PLATE 38.

Xylarieae.

Fig. 1. **Camillea Sagraeana,** (Mont.), magnified two diameters. Drawn from a specimen collected in Nicaragua, by Wright.

" 2. Sectional view of a young stroma.

" 3. Section of a mature stroma.

" 4. Two asci.

" 5. Two sporidia.

" 6. **Daldinia concentrica,** (Bolt.), natural size, from Mexico.

" 7. Same, natural size, from Missouri.

" 8. Vertical section, natural size.

" 9. Portion of section magnified.

" 10. An ascus.

" 11. A sporidium.

PLATE 38.

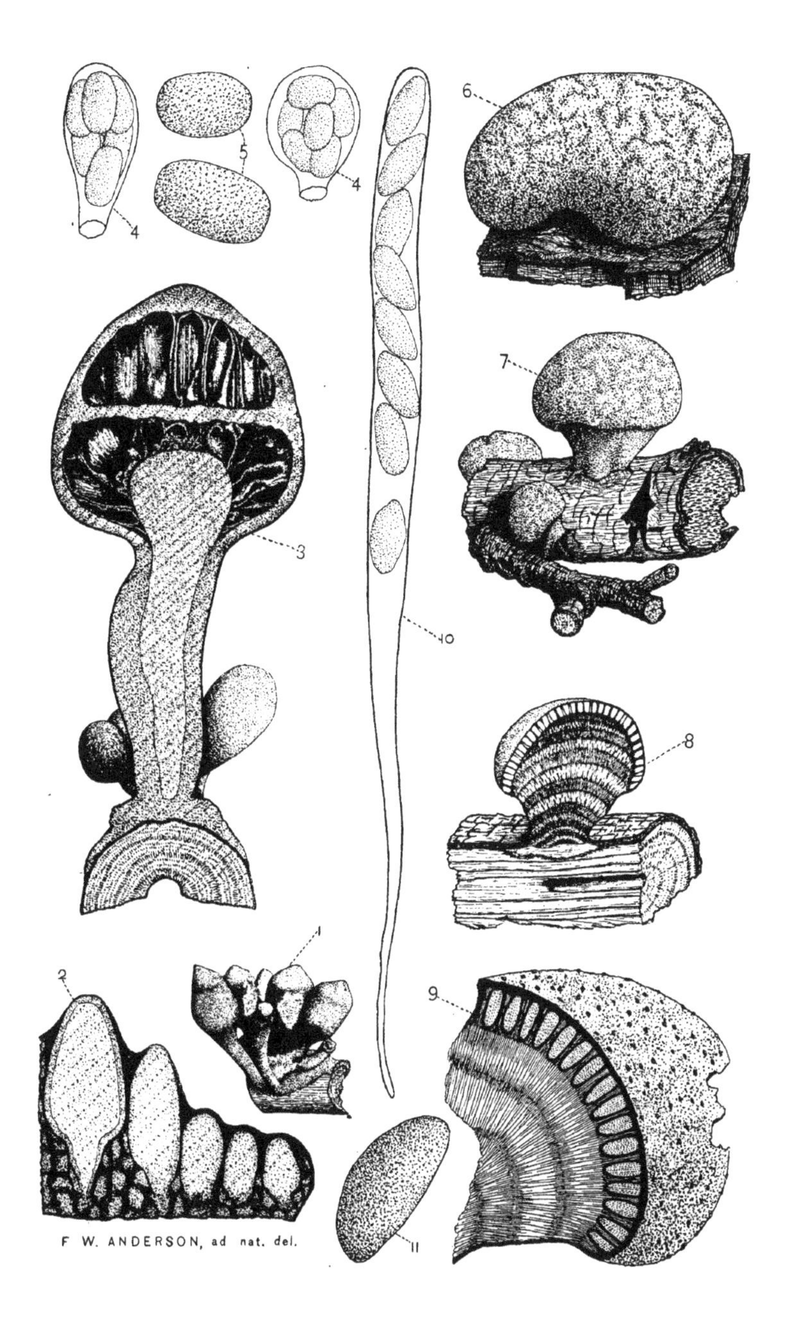

Geo May Powell, Phila.

PLATE 39.

Xylarieae.

Fig. 1. **Xylaria subterranea,** (Schw.) natural size on wood of an old pump standing in a well at Bethlehem, Pa.

" 2. Sectional view of a piece of the stroma somewhat enlarged.

" 3. An ascus.

" 4. Two sporidia.

" 5. **Ustulina vulgaris,** Tul. natural size.

" 6. Portion of stroma enlarged.

" 7. Part of an ascus containing six sporidia.

" 8. A sporidium.

" 9. **Nummularia discreta,** (Schw.) natural size on wood of "Mountain ash."

" 10. Sectional view of enlarged stroma.

" 11. An ascus with two paraphyses.

" 12. A sporidium.

" 13. **Poronia leporina,** E. & E. natural size on rabbit's dung (Missouri).

" 14. Sectional view of a stipitate stroma.

" 15. An ascus with three jointed paraphyses.

" 16. Two sporidia.

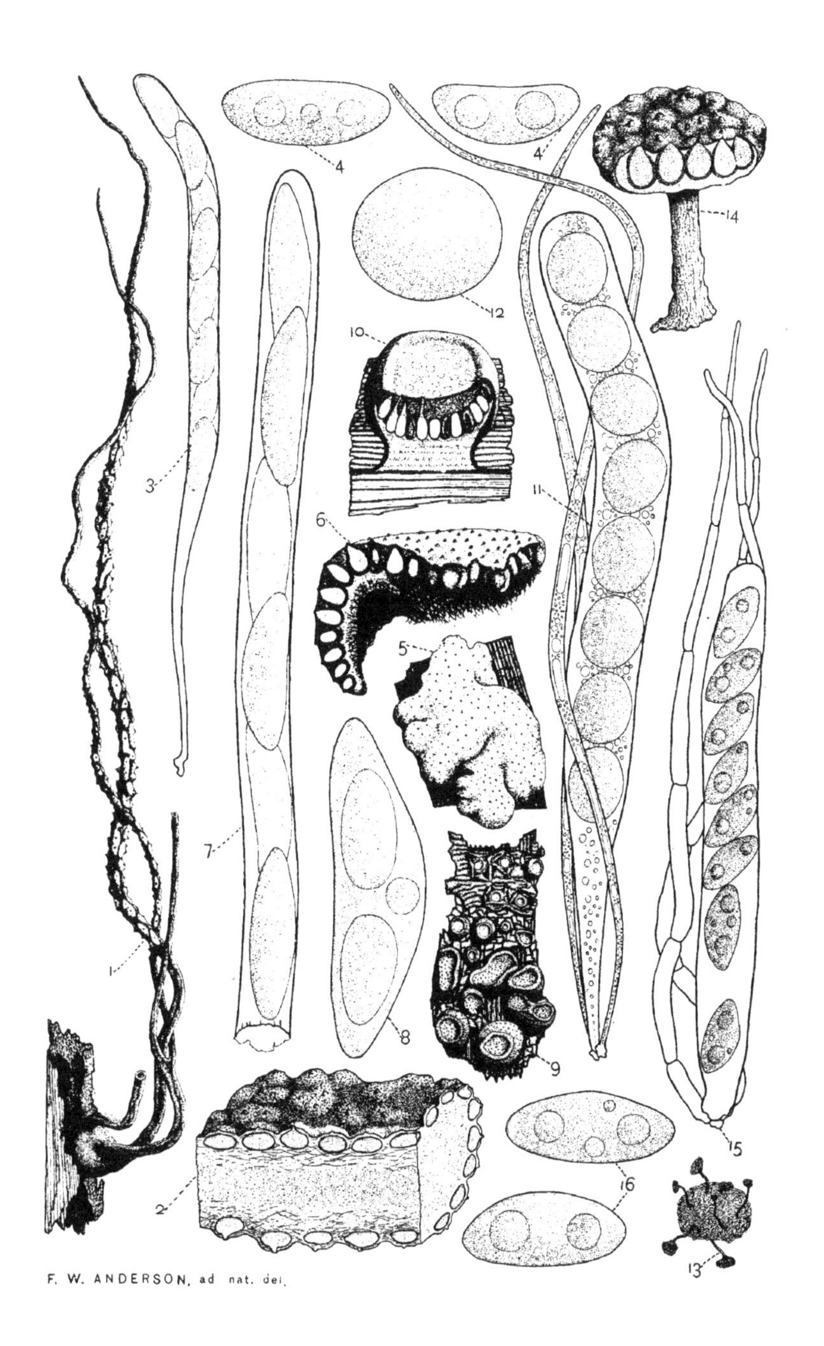

F. W. ANDERSON, ad nat. del.

PLATE 40.

Dothideaceae.

Fig. 1. **Dothidea Montaniensis,** E. & E. several stromata somewhat enlarged, on bark of *Bigelovia*?

" 2. Vertical section of a stroma.

" 3. An ascus.

" 4. A sporidium.

" 5. **Phyllachora graminis,** (Pers.) somewhat enlarged, on leaf of grass.

" 6. Vertical section of the same.

" 7. An ascus.

" 8. A sporidium.

" 9. **Homostegia Kelseyi,** E. & E. somewhat enlarged on dead stem of *Ribes rotundifolium.*

" 10. Vertical section through a stroma.

" 11. An ascus.

" 12. Two sporidia.

" 13. **Rhopographus filicinus,** (Fr.), somewhat enlarged, on *Pteris aquilina.*

" 14. Vertical section of the same.

" 15. An ascus.

" 16. Two sporidia.

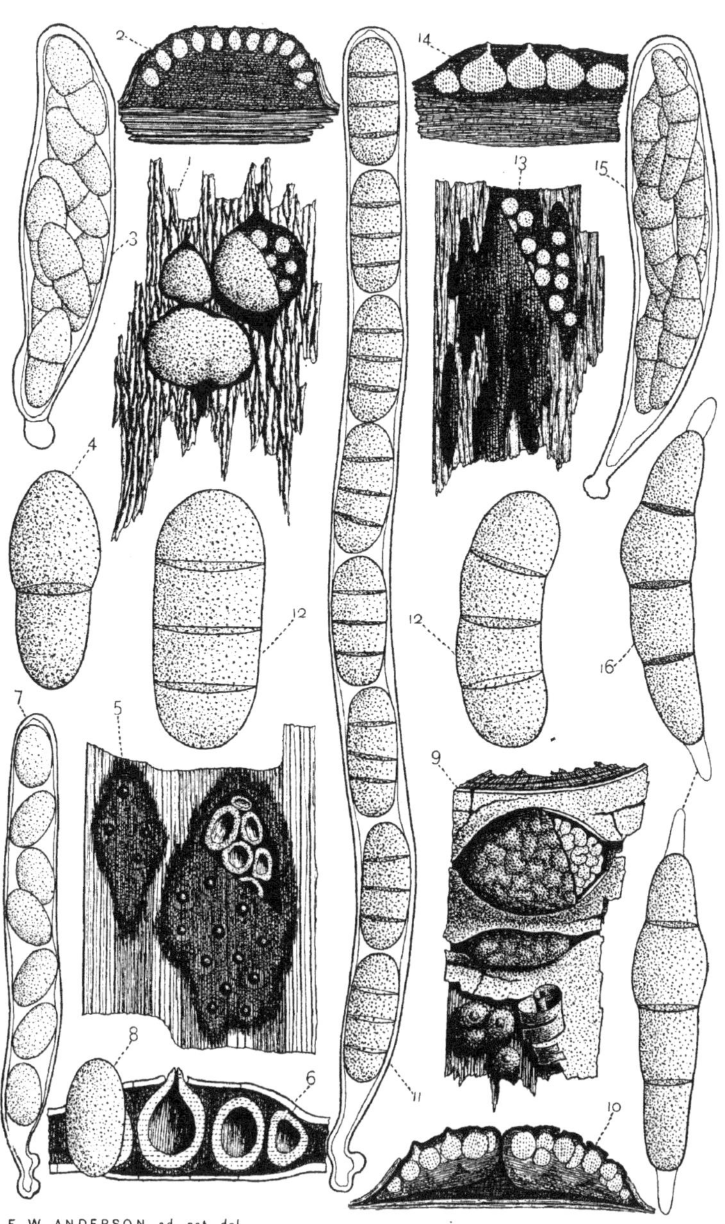

F W ANDERSON, ad nat. del.

PLATE 41.

Cucurbitarieae.

Fig. 1. **Montagnella Heliopsidis,** (Schw.) natural size, on stem and leaf of ***Helianthus divaricatus***, somewhat enlarged.

" 2. Same on a piece of the stem, still more enlarged.

" 3. Vertical section through several perithecia.

" 4. An ascus.

" 5. A sporidium, front view.

" 6. A sporidium, side view.

" 7. **Parodiella grammodes,** (Kze.) about natural size, on *Desmodium paniculatum.*

" 8. Fragment of leaf with enlarged perithecia.

" 9. Vertical section of a perithecium more highly magnified.

" 10. An ascus.

" 11. Front view of a sporidium.

" 12. **Otthia staphylina,** E. & E. about natural size, on bark of *Staphylea.*

" 13. A cluster of perithecia somewhat magnified.

" 14. Vertical section through a cluster more highly magnified.

" 15. An ascus.

" 16. Four sporidia.

N. B.—In this and the preceding plates, where not otherwise noted, all asci drawn to a scale of 28 μ to the inch and all enlarged sporidia to a scale of 14 μ to the inch.

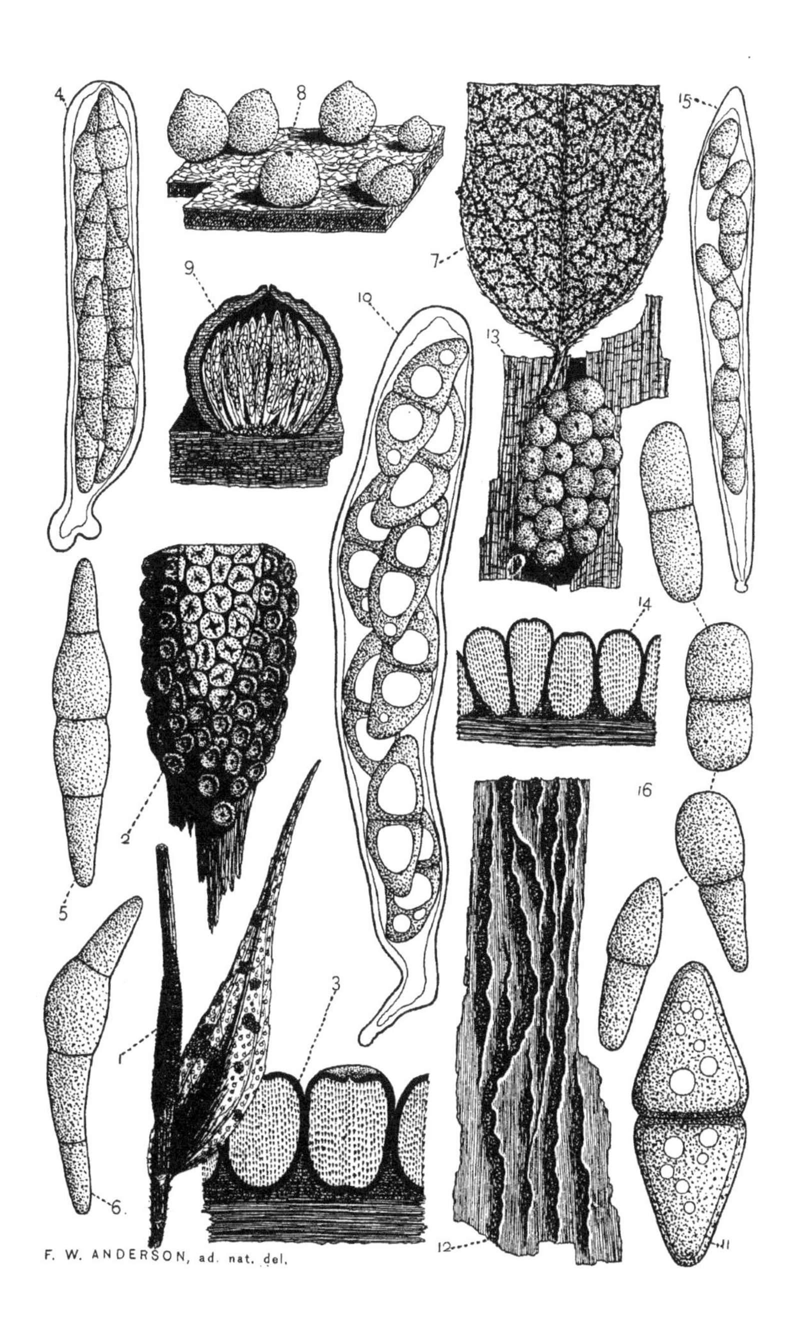
4
8
15
7
9
10
13
14
16
2
5
1
3
6.
12
11
F. W. ANDERSON, ad. nat. del.

Lightning Source UK Ltd.
Milton Keynes UK
UKHW041817250219
337978UK00011B/747/P